# Sustainable Innovations in the Textile Industry

# The Textile Institute Book Series

Incorporated by Royal Charter in 1925, The Textile Institute was established as the professional body for the textile industry to provide support to businesses, practitioners and academics involved with textiles and to provide routes to professional qualifications through which Institute members can demonstrate their professional competence. The Institute's aim is to encourage learning, recognise achievement, reward excellence and disseminate information about the textiles, clothing and footwear industries and the associated science, design and technology. It has a global reach, with individual and corporate members in over 80 countries.

*The Textile Institute Book Series* supersedes the former *Woodhead Publishing Series in Textiles* and represents a collaboration between The Textile Institute and Elsevier aimed at ensuring that Institute members and the textile industry continue to have access to high-calibre titles on textile science and technology.

Books published in *The Textile Institute Book Series* are offered on the Elsevier website at store.elsevier.com and are available to Textile Institute members at a discount. Textile Institute books still in print are also available directly from the Institute's website at www.textileinstitute.org.

To place an order, or if you are interested in writing a book for this series, please contact Sophie Harrison, Acquisitions Editor: s.harrison2@elsevier.com.

## Recently Published and Upcoming Titles in *The Textile Institute Book Series*:

*Smart Textiles from Natural Resources,* 1st Edition, Md. Ibrahim H. Mondal, 978-0-44-315471-3

*Advances in Plasma Treatment of Textile Surfaces,* 1st Edition, Shahid Ul Islam, Aminoddin Haji, 978-0-44-319079-7

*The Wool Handbook: Morphology, Structure, Property and Application,* 1st Edition, Seiko Jose, Sabu Thomas, Gautam Basu, 978-0-32-399598-6

*Natural Dyes for Sustainable Textiles,* 1st Edition, Padma Shree Vankar, Dhara Shukla, 978-0-32-385257-9

*Digital Textile Printing: Science, Technology and Markets*, 1st Edition, Hua Wang, Hafeezullah Memon, 978-0-44-315414-0

*Textile Calculation: Fibre to Finished Garment*, 1st Edition, R. Chattopadhyay, Sujit Kumar Sinha, Madan Lal Regar 978-0-32-399041-7

*Advances in Healthcare and Protective Textiles*, 1st Edition, Shahid Ul Islam, Abhijit Majumdar, Bhupendra Singh Butola, 978-0-32-391188-7

*Fabrication and Functionalization of Advanced Tubular Nanofibers and their Applications*, 1st Edition, Baoliang Zhang, Mudasir Ahmad, 978-0-32-399039-4

*Functional and Technical Textiles*, 1st Edition, Subhankar Maity, Kunal Singha, Pintu Pandit, 978-0-32-391593-9

The Textile Institute Book Series

# Sustainable Innovations in the Textile Industry

*Edited by*

*Roshan Paul*
RWTH Aachen University, Germany

*Thomas Gries*
RWTH Aachen University, Germany

The Textile Institute

Woodhead Publishing is an imprint of Elsevier
50 Hampshire Street, 5th Floor, Cambridge, MA 02139, United States
125 London Wall, London EC2Y 5AS, United Kingdom



ISBN: 978-0-323-90392-9

For information on all Woodhead Publishing publications visit our website at https://www.elsevier.com/books-and-journals

*Publisher:* Matthew Deans
*Acquisitions Editor:* Sophie Harrison
*Editorial Project Manager:* Rafael Guilherme Trombaco
*Production Project Manager:* Fizza Fathima
*Cover Designer:* Greg Harris

Typeset by TNQ Technologies

# Contents

# List of contributors

**Thomas Bechtold** Research Institute for Textile Chemistry and Textile Physics, University Innsbruck, Dornbirn, Austria

**Nuno Belino** Universidade da Beira Interior, Covilhã, Portugal

**Mirela Blaga** 'Gheorghe Asachi' Technical University of Iasi, Faculty of Industrial Design and Business Management, Iasi, Romania

**Babita U. Chaudhary** Department of Fibres and Textile Processing Technology, Institute of Chemical Technology, Mumbai, Maharashtra, India

**Edith Classen** Hohenstein Institute, Boennigheim, Germany

**Frederik Cloppenburg** Institut für Textiltechnik der RWTH Aachen University, Aachen, Germany

**Caroliny Minely da Silva Santos** Micro and Nanotechnologies Innovation Research Group, Federal University of Rio Grande do Norte, Natal, Rio Grande do Norte, Brazil

**Lauren M. Degenstein** Department of Human Ecology, University of Alberta, Edmonton, AB, Canada

**Patricia I. Dolez** Department of Human Ecology, University of Alberta, Edmonton, AB, Canada

**Jose Heriberto Oliveira do Nascimento** Micro and Nanotechnologies Innovation Research Group, Federal University of Rio Grande do Norte, Natal, Rio Grande do Norte, Brazil

**Thiago Felix dos Santos** Micro and Nanotechnologies Innovation Research Group, Federal University of Rio Grande do Norte, Natal, Rio Grande do Norte, Brazil

**Mohamed Elshafie** Analysis and Evaluation Department, Egyptian Petroleum Research Institute (EPRI), Nasr City, Cairo, Egypt

**Thomas Gries** Institut für Textiltechnik der RWTH Aachen University, Aachen, Germany

**Md. Saiful Hoque** Department of Human Ecology, University of Alberta, Edmonton, AB, Canada

**Saniyat Islam** School of Fashion and Textiles, Royal Melbourne Institute of Technology University, Melbourne, VIC, Australia

**Laxmikant Jawale** Apparel Impact Institute, Kalyan West, Maharashtra, India

**Jan Jordan** Institut für Textiltechnik der RWTH Aachen University, Aachen, Germany

**Zafar Juraev** Department of Metrology, Standardization and Product Quality Management, Andijan Machine-Building Institute, Andijan, Uzbekistan

**Ravindra D. Kale** Department of Fibres and Textile Processing Technology, Institute of Chemical Technology, Mumbai, Maharashtra, India

**Mirza Mohammad Omar Khyum** Department of Environmental Toxicology, The Institute of Environmental and Human Health, Texas Tech University, Lubbock, TX, United States

**Takeshi Kikutani** School of Materials and Chemical Technology, Tokyo Institute of Technology, Yokohama, Kanagawa, Japan

**Sea-Hyun Lee** Institut für Textiltechnik der RWTH Aachen University, Aachen, Germany

**Henning Löcken** Institut für Textiltechnik of RWTH Aachen University, Aachen, Germany

**Jan W.G. Mahy** Saxion University of Applied Sciences, School of Creative Technology, Enschede, The Netherlands

**Jayesh V. Malanker** Malanker Enterprises, Bilimora, Gujarat, India

**Avinash Pradip Manian** Research Institute for Textile Chemistry and Textile Physics, University Innsbruck, Dornbirn, Austria

**Yasser M. Moustafa** Analysis and Evaluation Department, Egyptian Petroleum Research Institute (EPRI), Nasr City, Cairo, Egypt

**Jens Oelerich** Saxion University of Applied Sciences, School of Creative Technology, Enschede, The Netherlands

**Rudrajeet Pal** The Swedish School of Textiles, University of Borås, Borås, Sweden; Department of Industrial Engineering and Management, University of Gävle, Gävle, Sweden

**Manoj Kumar Paras** National Institute of Fashion Technology, Kangra, Himachal Pradesh, India

**Prakash Pardeshi** Associated Chemical Corporation, Mumbai, Maharashtra, India

**Siva Rama Kumar Pariti** BluWin Ltd, Huddersfield, United Kingdom

**Sujata Pariti** Independent Consultant, Navi Mumbai, Maharashtra, India

**Roshan Paul** Institut für Textiltechnik der RWTH Aachen University, Aachen, Germany

**Tung Pham** Research Institute for Textile Chemistry and Textile Physics, University Innsbruck, Dornbirn, Austria

**Seshadri Ramkumar** Department of Environmental Toxicology, The Institute of Environmental and Human Health, Texas Tech University, Lubbock, TX, United States

**Jianhua Ran** State Key Laboratory of New Textile Materials and Advanced Processing Technologies, Wuhan Textile University, Wuhan, China

**Narendra Reddy** Center for Incubation Innovation Research and Consultancy, Jyothy Institute of Technology, Bengaluru, Karnataka, India

**Leon Reinsch** Institut für Textiltechnik der RWTH Aachen University, Aachen, Germany

**Arash Rezaey** Institut für Textiltechnik der RWTH Aachen University, Aachen, Germany

**Sanjay Kumar Sahu** Clearity Specialities LLP, Thane West, Maharashtra, India

**Mukhitdin Sattarov** Department of Metrology, Standardization and Product Quality Management, Andijan Machine-Building Institute, Andijan, Uzbekistan

**Sascha Schriever** Institut für Textiltechnik der RWTH Aachen University, Aachen, Germany

**Nagaiyan Sekar** Department of Speciality Chemicals Technology, Institute of Chemical Technology, Mumbai, Maharashtra, India

**Rakesh Seth** Om Tex Chem Pvt. Ltd., Thane West, Maharashtra, India

**Umesh Sharma** Sweet Merchandising Services Ltd, Kent, United Kingdom

**Felix Y. Telegin** G.A. Krestov Institute of Solution Chemistry, Russian Academy of Sciences, Ivanovo, Russia

**Srishti Tewari** Department of Fibres and Textile Processing Technology, Institute of Chemical Technology, Mumbai, Maharashtra, India

**Jenny Underwood** School of Fashion and Textiles, Royal Melbourne Institute of Technology University, Melbourne, VIC, Australia

**Sherif A. Younis** Analysis and Evaluation Department, Egyptian Petroleum Research Institute (EPRI), Nasr City, Cairo, Egypt

**Yi Zhao** College of Textiles, Donghua University, Shanghai, China

# Preface

The textile sector provides one of the three basic needs for the survival of human beings and is also the third-largest employer worldwide after the food and housing sectors. Actually, fibres and textiles are much more than just clothing, and offer themselves as versatile technical materials for a wide range of applications in almost every field of life. The textile supply chain is also complex, as it spans from harvesting or fibre forming to the end consumers of both domestic and industrial sectors.

The textile industry also has been long labelled as one of the biggest polluters of the environment, and as we strive for a more sustainable world, there lies a huge responsibility ahead, opening up enormous opportunities for textile technologists. Currently, the industry is at crossroads, and it is important to take all the possible remedial measures to rectify this tarnished polluter image, and to move forward further in the long road towards sustainability. This destination is distant, but definitely not unreachable.

Several governments the world over, including the EU, are convinced that textiles is one of the most important industries economically, and there should be a clear roadmap to achieve the goal of textile sustainability. This requires technological, social, and business innovations, supported by policies and the adaptation of consumer behaviour towards responsible consumption. The circular economy, safe and sustainable by design (SSbD) approach, alternative raw materials, slow fashion, regenerative manufacturing and biotransformation hold the keys to textile sustainability. Circular business models and the eco-designs can reduce the negative impacts of textile production and consumption, and it is high time that all textile goods produced, be they clothing or technical textiles, should be reusable or recyclable. Thus, it is possible to retain the value of textiles, extend their life cycles and enhance the usage of recycled materials by enlightened consumers. Moreover, a biotransformation from petroleum-based to bio-based products is inevitable.

This book on *Sustainable Innovations in the Textile Industry* is an attempt to accelerate this sustainable transformation. It covers the complete textile chain in a single book, the chapters of which are written by eminent technologists from around the world, so that it should live up to the expectations of present and future readers, and help to continually inspire them in the decades to come. It was not an easy task to bring all of these authors together on this single platform, but it was the need of the hour for there to be a single book covering all the sustainability aspects of the complete textile chain. We are glad to be able to present this book to the professionals, industrialists, researchers, innovators and more over to students, who are the future of the textile industry.

We are confident that this book will serve as an authentic resource to support the ongoing innovation activities and also will remain as a reference for the future, accelerating the transformation and evolution of the industry. It will definitely help the industry to shed its grey aspects, opening up an optimistic and futuristic course. The book is poised to motivate the younger generations to remain in the textile field, and to contribute to making it a sustainable one for future generations, thus reserving for them their share of raw materials and a pollution-free environment. It will in turn ignite the passion for the sustainability drive in textiles and we can foresee how this will further trigger multiple innovations. We are convinced that a completely green and sustainable textile industry is within reach.

**Roshan Paul and Thomas Gries**

# Sustainable innovations in textiles

1

*Roshan Paul and Thomas Gries*
Institut für Textiltechnik der RWTH Aachen University, Aachen, Germany

## 1.1 Introduction

The textile industry has long been considered one of the biggest polluters. Over the years there have been initiatives to make the industry more sustainable and environmentally friendly. However, most such innovations have taken place in discrete areas of the textile value chain. To enable a lasting global impact, it is important to bring the whole textile chain under the umbrella of sustainability and move forward, shedding the pollutant image.

Currently, petroleum based fibres, colourants and textile products dominate the market. In addition, the textile and apparel industry produces huge volumes of polymer, colourant and chemical waste. Based on global concerns such as climate change, the greenhouse effect, water scarcity and the limited availability of virgin raw materials, several governments and policymakers are promoting sustainable transformation to green the industry. Such innovations should lead to the sustainability, cost-effectiveness, energy savings, resource efficiency, biotransformation and circularity of the industry. Regenerative manufacturing, Safe and Sustainable by Design (SSbD) and digitalisation are important aspects to be considered for the complete transformation of the textile value chain. It is important to join forces across the world in an interdisciplinary manner to promote the use of bio-based, secondary and recyclable raw materials to create a global, resilient, circular and sustainable textile industry.

## 1.2 Innovations in textile manufacture

Natural cellulosic and protein fibres have been widely used since ancient times and still hold a good share of the fibre market. The production of natural fibres faces various challenges involving sustainability and processes, and involves the huge use of land and natural resources. It is impossible to increase land use for fibre crops owing to conflicts with food production, climate change and socioeconomic factors. Hence, several approaches are being pursued to make natural fibres more sustainable and reduce their environmental impact. Furthermore, alternative sources such as agricultural residues, and byproducts of agricultural production containing cellulose, hemicellulose and lignin, as well as different biomasses, are being considered to produce natural fibres.

**Sustainable Innovations in the Textile Industry. https://doi.org/10.1016/B978-0-323-90392-9.00013-6**

Novel concepts such as fibres from $CO_2$, algal fibres and a policy of zero agricultural waste could synergistically contribute to the sustainability of the natural fibre industry.

In the case of man-made and synthetic fibres, raw materials such as biomass and agricultural residue streams, as well as microbes, hold the key to the sustainable future of environmentally friendly man-made fibres and bio-based synthetic fibres such as polylactic acid, polyester, and polyamide. Increased consumer awareness about the environmental and social impacts of purchasing decisions accelerates this biotransformation in developing functional fibres from renewable, recycled, bio-based or biodegradable materials, with low carbon and water footprints. Moreover, it is fuelled by regulations on chemical use and safe disposal, emissions and waste control, and consumer protection.

The development of novel functional fibres meeting the expectations of customers and stakeholders enables the industry to achieve long-term viability and enhance its sustainability and environmental accountability. Among fibre spinning processes, melt spinning is considered to be more sustainable than solution spinning for man-made fibres. Cellulose, microfibrillated cellulose and nanocellulose are reinvented as a renewable and sustainable alternative to petroleum-based synthetic polymers. The implementation of a circular economy of fibres with renewable carbon sources can reduce the environmental impacts and create value from waste. Biomass, $CO_2$ and recycling are considered the pillars for developing future fibres, which can create an eco-efficient circular economy. Fig. 1.1 shows examples of sustainably developed functional automotive and home textiles.

In the spinning process as well as in the machinery, there are several sustainable innovations. Basically, ring and rotor spinning can be considered the most common methods of converting fibres into yarns. To make spinning more sustainable, the most important parameters to be considered are a reduction in cost, energy efficiency, increased process efficiency and waste reduction. Moreover, the complete recovery and reuse of spinning waste as a valuable raw material should be considered. Several innovations are taking place on spinning machinery, including spinning techniques such as magnetic ring spinning to increase spinning speeds. Other techniques such as air-jet spinning or a hybrid system can provide better performance and lower costs.

**Figure 1.1** Sustainably developed functional automotive and home textiles.

Increasing environmental restrictions and the limited availability of resources also promote sustainable innovations in spinning.

Weaving, knitting and nonwoven machinery are undergoing a steady transformation in addition to sustainable innovations in these processes. The automated inspection of machines using cameras with machine learning power, advanced systems for light analysis as well as the development of artificial intelligence (AI)-enhanced machines can ensure quality, increase productivity and boost sustainability. Energy efficient machines along with improved production processes and eco-designs can contribute immensely to enhance sustainability in textile production. Modern machinery offers great possibilities for process configurations and material and product portfolios. Digitalisation is an effective tool to adapt machinery to ever-changing legislative requirements and market needs and to stay competitive. Upgradation and retrofitting of used textile machinery are other options to enhance productivity and sustainability.

Sustainable innovations in woven textiles are focusing on reducing energy, chemical and water use, avoiding waste generation, and promoting safe working conditions. Considering the multitude of environmental, social and economic impacts of weaving activities, the evolution of the woven sector is critical for global efforts towards sustainable development. Weaving innovation should be built on the pillars of sustainability, environmental, social and economic opportunities. Other important areas are the development of bio-based textiles and the use of environmentally friendly sizing agents. It is also important to adapt new technologies quickly and to forge collaboration within and along the value chain and among different disciplines. Fig. 1.2 shows examples of sustainably finished and coated technical textiles.

Innovations in knitted textiles and knitting processes cover a vast area from weft knitting, warp knitting, three-dimensional and four-dimensional knitting to whole-garment knitting. Major areas of focus are reduced resource consumption, energy efficiency, the reduction of process times, use of bio-based and recycled yarns and digitalisation of the process chain. Digitalisation is considered the main driving force in optimising the industry towards green goals. Consumer awareness of sustainable products and consideration of the environmental impact are also contributing to purchase decisions regarding knitwear.

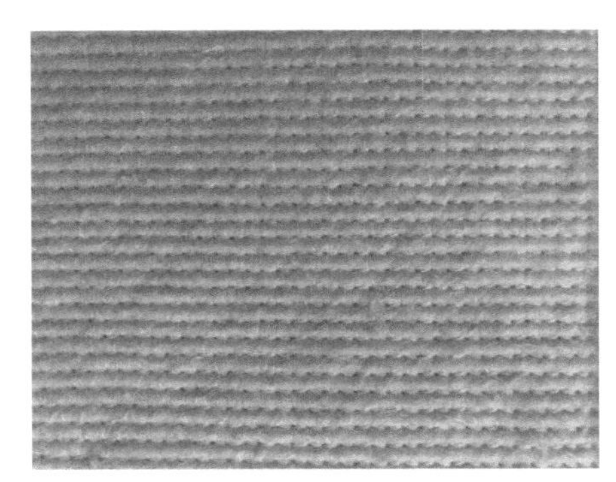

**Figure 1.2** Sustainably finished and coated technical textiles.

There are several innovations in the production and application of nonwoven textiles. Innovative nonwovens are developed by needle punching, thermal bonding, eco-friendly chemical bonding and hydro entanglement. The use of bio-based and recycled raw materials and overall product circularity are important parameters for achieving sustainability. Thus, sustainability can be enhanced by using renewable fibres such as lyocell, cotton, bamboo, algae, microbial fibres, bast fibres or other sustainable sources such as recycled paper, recycled polyester or polylactic acid. Biodegradability, reusability, recyclability and safe disposal are major areas for achieving sustainability in various application areas of nonwovens, such as civil engineering, filtration, insulation, oil spill remediation, agrotextiles, battery materials, medical textiles, wound-healing sutures and sanitary and hygiene products.

## 1.3 Sustainability of textile processing

An important and highly unsustainable area of the textile chain is the colouration industry, which has a huge environmental impact from effluent loaded with dyes and chemicals. Thus, the use of natural and sustainable synthetic dyes, as well as the chemoinformatics approach that can lead to sustainable innovations in dyes are highly relevant. Algal dyes, fungal and microbial pigments, natural mordants, eco-friendly dyeing techniques, as well as completely absorbing reactive dyes, high tinctorial strength dyes and functional dyes lead the way to a sustainable future. Digital colouration using high tinctorial power dyes can result in zero effluent discharge. Multifunctional dyes with incorporated functional properties can lead to sustainable one-step dyeing and multifunctional finishing. The structural colouration of textiles is expected to grow further as the sustainable colouration method of the future. The chemoinformatics approach can lead to sustainable dye innovations. Therefore, future dye molecules could be made sustainable and environmentally friendly by green chemistry and the chemoinformatics approach.

Novel molecules of speciality chemicals can impart unique properties to textiles, even when used in small quantities. Enzymes have already had a sustainable impact on textile processing by replacing several harmful and toxic chemicals. Enzymatic treatments are in place for several industrial processes such as desizing, degumming, scouring, denim wash and the treatment of synthetics. In the field of textile finishes, several novel molecules are being evaluated for developing functional properties such as flame retardance, and hydrophobic and oleophobic properties by replacing halogenated flame retardants, and perfluoroalkyl and polyfluoroalkyl substances (PFAS). Similarly, formaldehyde-based formulations are being replaced by carboxylic acids in easy-care textile finishes. Sustainable nontoxic, biodegradable and biocompatible antimicrobial finishes such as chitosan are also being developed.

The textile processing machinery is another important area for sustainable innovations. Textile wet processing depends on several complex variables. Each production batch needs to be handled carefully and depends on a well-trained workforce, apart from the machinery. Several innovations are taking place in machinery manufacture that can provide sustainable solutions to the huge consumption of water, energy, chemicals, and emissions into water. Several resource-saving, environmentally friendly and sustainable machine technologies exist, such as plasma, ozone bleaching, supercritical $CO_2$ dyeing, electrochemical dyeing, digital printing, foam finishing, enzymatic processing and ultrasonic processing. The integration of eco-design, process technology and production based on intelligent machines with electronic programme control could lead to sustainable wet processing by reducing production costs and material waste.

Sustainable innovations in textile pretreatment processes are mainly related to desizing, mercerising, bleaching and cleaning. The basic problems in the textile industry are the low efficiency of textile pretreatments and a poor understanding of the life cycle of the chemicals involved. Achieving sustainability in textile pretreatment processes is essential to obtain value-added textile products that can be branded as environmentally friendly. Manufacturing steps in wet processing should use green technologies and chemicals to meet the requirements of Industry 4.0. It is important to follow an ecologically correct approach based on the principles of green chemistry and clean technology to produce environmentally compatible textile goods. This can lead to the economic viability of the industry and thus help to preserve natural resources for future generations.

Sustainable innovations in dyeing, printing and digital colouration are needed for the reinvention of these well-established processes. The possibility of the reuse, recycling and biodegradability of dyes and speciality chemicals could open new avenues of sustainability. To achieve biotransformation, an intelligent design of sustainable processes is needed for the optimal consumption of resources and release of the minimum quantity of biodegradable waste. The environmental impact could be reduced thorough a reconsideration of current processes, chemicals and colourants, leading to circularity. Printing techniques with a reduced consumption of pigments and chemicals, as well as digital printing, exhibit a high potential to take an important role in the colouration of products with a low consumption of chemicals and dyes. Considering the textile sustainability as a whole, the colouration of biodegradable fibres should be achieved with biodegradable dyes and finishes. Textile fabrics digitally coloured in a sustainable manner are shown in Fig. 1.3.

Functional finishing and smart coating can add high value to textiles. Innovative functional finishes offer possibilities for developing technical textiles by finishing of nontechnical textiles towards the end of the manufacturing process. A wide variety of sustainable materials and bio-based finishing chemicals are being developed that meet or exceed the expectations of consumers. The future direction in functional finishing is to develop multifunctional textiles that are highly efficient, durable and cost-effective and are manufactured in an environmentally sustainable manner.

Smart functional coatings can be created on textiles by using anchor peptides as a tool. Anchor peptides are a highly diverse class of small amphipathic peptides able to bind substrates such as textiles with high selectivity and binding strength. Bifunctional anchor peptides consist of two peptides separated by a linker to prevent intermolecular interactions between the anchor peptides. One anchor peptide binds specifically to the textile surface and the other interacts selectively with the desired functional molecule, leading to a smart functional coating.

**Figure 1.3** Digital colouration of textiles.

Both the garment machinery and manufacturing technologies have a key role in the transition of the textile fashion industry to a circular one with zero emissions. Regenerative approaches in garment production, dyeing and finishing demonstrate how garment production needs to be considered as part of a holistic system driven by processes rather than products. To achieve the goal of sustainability, the fashion industry has to adopt innovative garment processes, including ecological colouration and finishing, leading to a circular approach, the reduction of waste and minimised use of natural resources. The application of AI can lead to smart garment manufacturing by considering a range of factors such as sustainability, resource efficiency and cost-effectiveness.

Textile care and maintenance is a water-intensive process. Effluent from a cleaning or washing process loaded with detergents has a negative impact on the environment. To enhance sustainability, biodegradable detergents developed from sustainable or bio-based raw materials are necessary, but without a reduction in cleanness. Moreover, ecological detergents and energy and water-efficient washing machines are leading the way to more sustainable textile care and maintenance. Regarding the sustainability of washing machines and dryers, longevity, repairability and the recycling are important aspects to reduce the use of raw materials. The development of high-performance filters can reduce fibre contamination in the wastewater and effectively prevent microplastic pollution. Whereas water is used as the solvent in home laundering, organic solvents are used in the industrial laundry, particularly for dry cleaning. The wet cleaning process for industrial laundry involving water as the solvent adds sustainability. $CO_2$ is considered another environmentally friendly solvent for dry cleaning.

## 1.4 Postproduction sustainability and circularity

Sustainability is highly desirable in the supply chain and logistics of the fashion business. Reusable, repairable, recyclable and durable textile and fashion products can be achieved through the SSbD of products, an increased reuse and recycle initiative, waste collection, green procurement, efficient production and a sustainable lifestyle. Moreover, a digital system can enable access to product-specific characteristics for the manufacturer, consumer and other stakeholders of the supply chain. Because the fashion business involves raw material sourcing, distribution and retailing, it involves a forward supply chain from manufacturing to the consumer. A closed-loop supply chain is necessary to enhance the sustainability of the fashion business because it involves the high consumption and disposal of fashion products. A reverse supply chain involving collection, sorting, resale or redesign can close this loop. Such a closed-loop supply chain can increase consciousness among consumers, who may purchase good-quality fashion products from the downstream channel.

On the other hand, closed-loop postconsumer textile recycling can significantly decrease the environmental impact of textiles by shifting to a circular system. The current situation of fast fashion, use and discard, overconsumption and following unsustainable production needs to be replaced. It is important to reverse this trend by slow fashion, with a transparent and unconstrained use of sustainable, repairable and recycled materials. The use of SSbD to develop long-lasting textile products with high quality and an efficient recycling process is inevitable. Designers should use recycled fibres in fashion products. Fig. 1.4 shows high-end fashion products from recycled textiles and functional sportswear containing sustainable components.

Current mechanical, thermomechanical and chemical recycling technologies should be further innovated to reduce their environmental impact. Rather than removing colour, it is highly sustainable to recycle fibres with dyestuffs so that there is no need for a second colouration. This also avoids dye disposal issues. Another sustainable innovation is to blend recycled fibres with varying colours to develop new products with different and vivid colours. Such innovations in postconsumer recycling can lead to significant reductions in the use of virgin fibres, and enhances the circularity of textiles.

**Figure 1.4** High-end fashion from recycled textiles and functional sportswear with sustainable components.

It is also possible to converge recycling and reuse initiatives synergistically with eco-designs and social innovations to develop value-added artistic products for niche markets, as shown in Fig. 1.5. There is a global trend in this direction. Skill development and recycling programmes of social development organisations such as Social Vision, India involve the complete reuse of discarded textiles for new product development and the empowerment of women. Another social innovation initiative is the Social Fashion Factory, Greece, which runs an eco-sustainable circular fashion manufacturing studio.

The treatment, reuse and recycling of textile effluents containing high concentrations of dyes, speciality chemicals, detergents and heavy metals is the final step to closing the sustainability chain of the textile industry. It is important to reduce effluent loads by making the textile processes more sustainable, as well as to develop high-performance effluent treatment technologies that can treat hazardous textile pollutants economically and efficiently for safe disposal. The characteristics of textile effluent cannot be generalised. They vary from country to country and depend on the type of fibre and textile, the quality of the water, the process technology, and the effectiveness of wastewater treatment plants.

Innovative technologies such as advanced chemical oxidation processes, membrane bioreactors, photocatalytic membrane reactors and direct contact membrane distillation are highly efficient in reducing the negative impacts of pollutants in textile effluent. An alternative approach is to achieve zero-liquid discharge by substantially reducing effluents. Considering global challenges such as climate change, water shortages, environmental degradation and public health protection, effluent recycling strategies should be promoted. The sustainable use of recycled chemical-free textile effluents can also occur in other sectors such as agriculture, horticulture and fisheries.

To ensure, monitor and maintain the sustainability of the textile industry, clarity regarding sustainability aspects and regulations as well as the implementation of innovative tools such as SSbD, Life Cycle Analysis (LCA) and eco-labels is important.

**Figure 1.5** Artistic home textile products from discarded garments.
Design by Sally George Panthiruvelil.

Sustainable textile innovations should address current demands efficiently, but without exploiting resources that should be kept for future generations. In the scenario of climate change, global warming and environmental deterioration, sustainability has attracted much attention. An LCA is highly effective because it includes all stages of the life of a product - from cradle-to-grave - covering raw material extraction through materials processing, manufacture, distribution, use, repair and maintenance, and disposal or recycling. Several regulations can affect textile production. Registration, Evaluation, Authorisation and Restriction of Chemicals (REACH) is one that is applied to improve the protection of human health and the environment through the efficient and earlier identification of the intrinsic properties of chemical substances. Moreover, several eco-labels have been developed to meet the needs of consumers and protect their right for eco-friendly and sustainable products.

## 1.5 Conclusion

The textile industry is constantly striving for innovative production technologies to improve the quality of products. It is also important for functional and technical textile products as well as fashion articles to be developed in an environmentally friendly way. Regulations impose increased restrictions concerning chemicals and chemical processes, which in turn promote sustainable innovations. Sustainability is a multifaceted concept involving environmental, social, health and economic factors. Sustainable practices must be implemented in all stages of the industry, including production, consumption, energy, transportation, recycling and disposal. A more equitable, resilient and sustainable future can be achieved only by continuously innovating and adopting technologies that can continuously reduce emissions and the use of virgin resources. Bioeconomy, biotransformation, regenerative manufacturing, recycled fashion, digitalisation and circular economy are some of the futuristic business models that can ensure the sustainable future of textiles.

## Sources of further information

Gries, T., Veit, D., Wulfhorst, B., 2015. In: Textile Technology: An Introduction, second ed. Carl Hanser Verlag GmbH & Co. KG, Munich, Germany. ISBN: 9783446229631.

https://environment.ec.europa.eu/strategy/textiles-strategy_en.

https://www.socialvisionindia.org.

https://www.soffa.gr.

Paul, R. (Ed.), 2014. Functional Finishes for Textiles: Improving Comfort, Performance and Protection. Woodhead Publishing Ltd (Elsevier), Cambridge, UK. ISBN: 9780857098399.

Paul, R. (Ed.), 2015. Denim: Manufacture, Finishing and Applications. Woodhead Publishing Ltd (Elsevier), Cambridge, UK. ISBN: 9780857098436.

Paul, R. (Ed.), 2019. High Performance Technical Textiles. John Wiley & Sons Ltd, Chichester, UK. ISBN: 9781119325017.

# Production and processing of natural fibres

**2**

*Narendra Reddy*[1] *and Sanjay Kumar Sahu*[2]
[1]Center for Incubation Innovation Research and Consultancy, Jyothy Institute of Technology, Bengaluru, Karnataka, India; [2]Clearity Specialities LLP, Thane West, Maharashtra, India

## 2.1 Introduction

Natural fibres have been a major part of human life since civilisation. From cradle to grave, natural fibres have been used for protection, preservation, aesthetics and decoration. Cotton and silk have been the predominantly used natural cellulosic and protein fibres, respectively, for millennia, and continue to hold precedence over other natural fibres despite the ease of production and processing and lower cost of synthetic fibres which makes them preferable over natural fibres. However, the source for synthetic fibres and the environmental constraints on their production, use and disposal, coupled with the increasing awareness on using biodegradable and sustainable sources are renewing interests in natural fibres. Unfortunately, production of conventional natural cellulosic fibres such as cotton, jute, flax, etc. requires extensive and dedicated use of natural resources such as land, water and energy, which are becoming increasingly scarce. In addition, the input costs required to grow natural fibres are becoming prohibitive, hence leading to a decline in the extent of cultivation and production of most natural fibers including cotton, the most common natural cellulosic fibre. The global production of cotton was between 118 and 123 million bales (480 lb/bale) between 2016–2021 (https://www.fibre2fashion.com/industry-article/8632/outlook-for-global-cotton-production). Cotton cultivation is also notorious for the extensive use of fertilisers and there is a substantial push to move towards organic and pesticide-free cotton. A similar scenario exists for the other natural cellulosic fibres also, but bast fibres generally require lower energy inputs for growth compared to cotton (La Rosa and Grammatikos, 2019). The constraints on growing cotton or other natural fibres are expected to increase in the coming decades. Hence, it is imperative that sustainable and low-energy alternatives for both cotton and bast fibres are developed and adopted.

Similar to cellulosic fibres, the production of wool and silk, the two most common protein fibres, has also been constrained by changes in the environment, climate, input costs and profitability. Wool and silk production have been stagnating between 1.0 and 1.1 metric tonnes and 0.7–0.9 metric tons per year, respectively. Silk production, which is dominated by China, has seen good demand but not much value addition. Although both wool and silk are preferred fibres and are considerably expensive, there are several restraints that inhibit increasing the production of natural fibres. Wool of

**Sustainable Innovations in the Textile Industry. https://doi.org/10.1016/B978-0-323-90392-9.00003-3**

only specific lengths is suitable for processing into textiles and garments, and much of the short wool is disposed of as waste. The production and processing of silk also creates large amounts of byproducts and coproducts starting from the undigested leaves to the litter from the worms. The byproducts and coproducts generated during the processing of silk are estimated to have a selling price of up to three to five times that of the actual silk fibres produced (Reddy et al., 2021). Hence, impetus on the use of the agricultural residues and coproducts as sources for fibres, materials and bioproducts is inevitable.

## 2.2 Historical aspects and current scenario

Natural fibres have been the primary source of protection and for making textiles since time immemorial. Fibres of both plant and animal origin were extensively grown and processed for eventual conversion into textiles. As civilisation progressed, natural fibres such as cotton, flax and jute were cultivated as major crops. Cotton and linen grown in Asia and Africa were traded across the globe and led to the establishment of industries in Europe and the Americas. In fact, the entire colonisation and subsequent foreign rule of several countries could be linked to the trade of natural fibres.

Similar to cotton and other natural cellulose fibres, exclusive availability and trading of silk from China across the globe created what is even today regarded as a trade route (the silk route). The cultivation, processing, trading and use of textiles have led to the establishment and even destruction of kingdoms. During and after the world wars, there was considerable constraint on the cultivation of fibre crops. Simultaneously, there was the advent of synthetic fibres, particularly polyester, rayon and nylon, which completely transformed the natural fibre industry. Despite the price, availability and other advantages of synthetic fibres, natural fibres have persisted and even today are preferable for several textile applications. Sustainable innovations in farming, processing and developments in technologies to reduce and reuse raw materials are ensuring that natural fibres remain relevant in the near future and for decades to come.

## 2.3 Sustainable production of natural cellulosic fibres

### *2.3.1 Sustainable production of cotton fibres*

#### *2.3.1.1 Production of cotton through farm management*

Cotton is the most predominant natural cellulosic fibre and is also equally infamous for the extensive use of natural resources and pesticides required for its growth. It has been reported that 10%–16% of the world's pesticides and 25% of insecticides are used to grow cotton, which accounts for only 2%–4% of global agricultural land (Hansen and Schaltegger, 2016). Hence, considerable efforts have been made to reduce the use of natural resources and develop sustainable approaches to grow and even process cotton.

For example, the 'Better Cotton Initiative' (BCI) which was started in 2009 has gained considerable interest across the globe and today is practiced by 2.4 million licenced farmers in 23 countries and accounts for nearly 23% of global cotton production (Figs. 2.1 and 2.2) with the main aim being to make cotton growth more sustainable. This initiative operates by following seven key principles (https://bettercotton.org/wp-content/uploads/2019/06/Better-Cotton-Principles-Criteria-V2.1.pdf) as listed below

- Minimise harmful impact of crop protection practices
- Promote water stewardship
- Care for soil health
- Enhance biodiversity and use land responsibly
- Care for and preserve fibre quality
- Promote fair work
- Operate an effective management system

Adopting the best cotton practices is reported to assist in reducing farming costs, increase returns and also contributes towards resource conservation and promoting social benefits and hence sustainable development (Zulfiqar and Thapa, 2018). Quantitatively, adopting the better cotton principles was reported to increase gross farming margins by 37%, cotton yield by 9%, and reductions in the use of pesticides by 7% and water by 14%, but with approximately 3% increase in labour costs (Zulfiqar et al., 2019). Newer concepts such as using information and communication technology (ICT)-based intelligent pest and disease warning systems, accurate prediction of weather patterns and price and market information are also expected to supplement efforts towards sustainable cotton production (Madasamy et al., 2020).

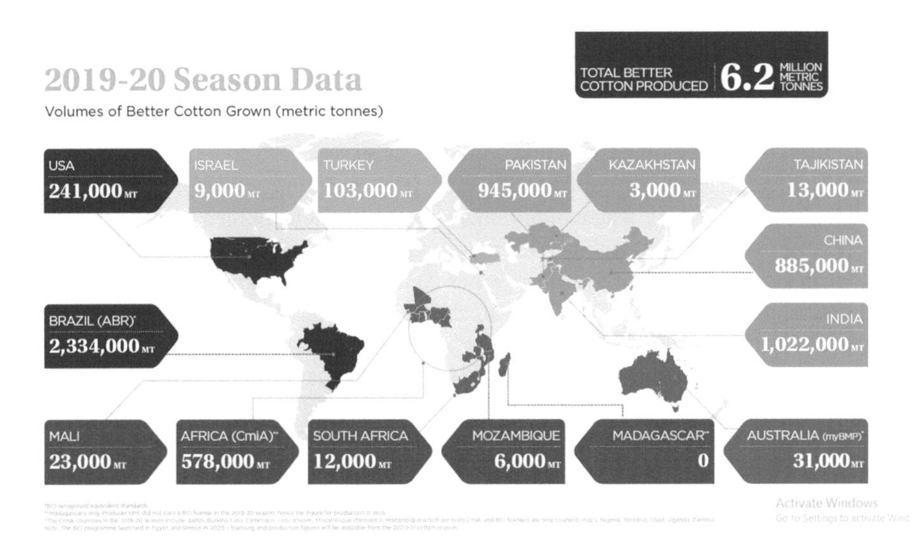


**Figure 2.1** Production of cotton through the better cotton initiative in different countries https://bettercotton.org/.

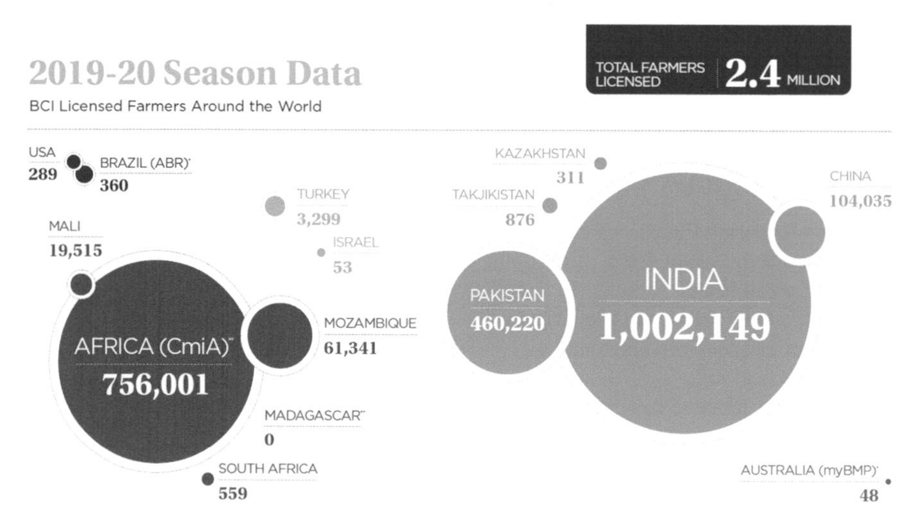


**Figure 2.2** Number of farmers enrolled in the better cotton initiative in different countries https://bettercotton.org/.

Changes in the conventional approaches to cultivation of cotton are also suggested to be useful to improve the sustainability and profitability of cotton. Intensive farming technologies that include seedling, transplanting, plastic mulching, double-cropping and using high-yield varieties have resulted in more than 30%−50% increases in productivity. Adopting narrow row spacing and ultra narrow row cotton where the distance between plants is between 7 and 30 inches, using plant growth regulators, plastic films and cotton sowing on beds are also reported to make cotton cultivation more sustainable. Furthermore, rational use of resources, simplifying management and increase in mechanisation would be necessary to further improve the production and profitability of cotton cultivation (Dai and Dong, 2014). For instance, managing and improving water requirements would benefit immensely in reducing the overall costs and sustainability of cotton. Based on a 2 year field study, it was found that using 50% of the available water content did not significantly reduce the yield of cotton (Ahmad et al., 2021). A substantial reduction in water was also possible by using drip irrigation compared to regular irrigation without affecting the boll number shedding percentage or cotton yield (Ertek and Kanber, 2003). Under Australian conditions, it has been estimated that producing one tonne of cotton lint from cradle to port has a $CO_2$ emission load of 1601 kg, which is higher than many other crops. Use of nitrogen-based fertilisers (46%), electricity and diesel used for irrigation (10%) and fuel used for farm machinery (9%) were the major contributors to the $CO_2$ emissions. Use of controlled-release nitrogen-containing fertilisers, adopting solar power instead of diesel, intercropping with legumes and nitrogen fertigation were suggested to be helpful in decreasing the greenhouse gas emissions from cotton cultivation (Fig. 2.3) (Hedayati et al., 2019). Several other approaches have also been

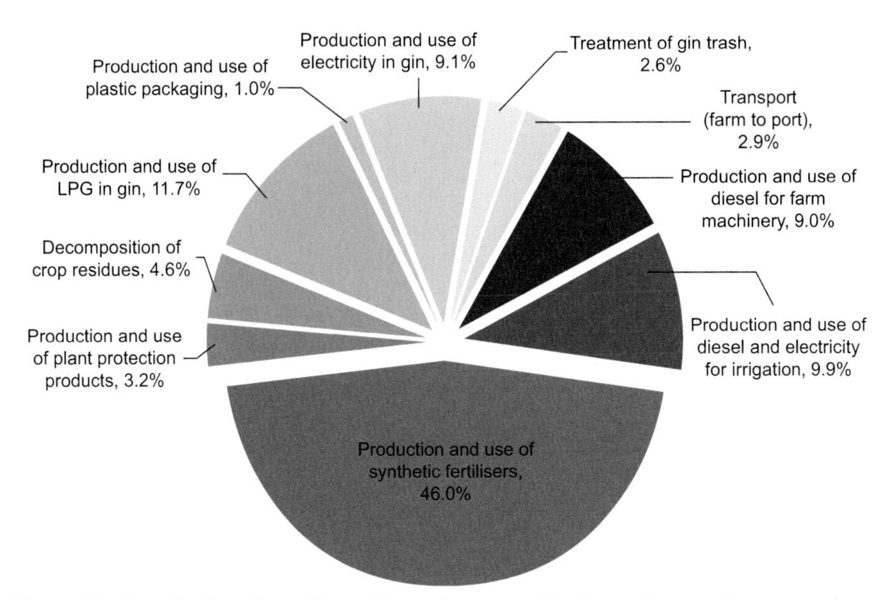


**Figure 2.3** Contribution of specific activities of cotton cultivation and processing to greenhouse gas emissions (Hedayati et al., 2019).
Reproduced with permission from Elsevier.

suggested to improve the sustainability of cotton from the production, use and disposal perspectives (Table 2.1) (Radhakrishnan, 2017). Similarly, the various stages at which sustainable cotton production can be adopted are provided in Table 2.2 (Zulfiqar et al., 2021).

### 2.3.1.2 *Production of organic cotton*

Another approach to produce sustainable cotton is through the organic farming approach. Here, crop rotations, and the use of natural fertilisers and natural pesticides are followed. However, the yield of the crop reduces and it is argued that the offset in yield makes the life cycle assessment of organic cotton similar to that of conventional cotton production. Hence, the amount of organic cotton cultivated is considerably low and accounts for 0.93% of the total cotton produced. Organic cotton is grown in about 19 countries, with a major share from India (51%) and China (17%) (Fig. 2.4, Table 2.3). Nevertheless, consumption and demand for organic cotton have been increasing rapidly and organic cotton trade was worth about $2 billion in the United States during 2018−19. Organic cotton can substantially decrease the cultivation costs and reduce the environmental burden of cotton. For instance, the production of conventional cotton requires about 2446 kg $CO_2$ equivalent compared to 978 for organic cotton (La Rosa and Grammatikos, 2019). However, the price for organic cotton (US$16−3.43/kg) was generally higher than that for conventional cotton (US$1.61−2.19/kg). In addition to growing organic cotton, substantial efforts have been made to develop standards for certifying organic cotton.

**Table 2.1** Inputs to improve the sustainability of cotton production (Radhakrishnan, 2017).

| Choice of cultivar/selection of variety | Growth and environment | Pre-harvest technologies | Post-harvest technologies | Social intervention |
|---|---|---|---|---|
| **Technological, agricultural and production inputs for sustainability** | | | | |
| • Agricultural biotechnology<br>• Plant architecture<br>• Adaptive cotton hybrids<br>• Improved fibre traits | • Water management<br>• Soil enhancement<br>• Crop protection<br>• Habitat and ecosystem safety | • Integrated nutrition management<br>• Integrated pest management<br>• Integrated irrigation management<br>• Mechanisation: sowing to harvesting<br>• Reduction in hybrid seed costs | • Modernisation of gins<br>• Technology-based bale classification<br>• Electronic trading<br>• Better pricing<br>• Pigmented cotton<br>• Ecofriendly processing and natural dyeing<br>• Gossypol free seeds<br>• Cotton biorefinery<br>• Cotton seed oil biorefinery | • Employment conditions<br>• Occupational health and safety<br>• Prevention of forced labour<br>• No child labour<br>• Basic treatment and disciplinary practices<br>• Non-discrimination<br>• Freedom of association and collective bargaining |

**Sustainable cotton production**

| Farmer | Manufacturer | Retailer | Consumer | Government: domestic and global |
|---|---|---|---|---|
| • Choice of cotton variety<br>• Choice of standards and certification<br>• Local assistance and healthy practices<br>• Education and training | • Use of certified organic cotton<br>• Clean technologies: spinning, weaving and knitting<br>• Ecofriendly processing<br>• Transparency in production techniques<br>• Product development with sustainability certification | • Sustainable supply chain<br>• Green marketing<br>• Life cycle assessment and reporting<br>• Focus on reduction in carbon footprint, GHG emissions<br>• Fair pricing to manufacturer<br>• Promotion of reuse and recycling<br>• Transparency in supply chain<br>• Benefits for consumers for sustainable choices | • Choice of product<br>• Preference for eco-friendly certified products<br>• Participation in recycling programs<br>• Willingness to pay more for sustainable products<br>• Influencing others to purchase sustainable products | • Facilitate better conditions for farmers through financial assistance and subsidies<br>• Rendering help during natural calamities and rehabilitation<br>• Promoting training programmes on latest development<br>• Facilitating global and domestic sustainable marketing<br>• Member of global organisations for sustainability<br>• Undertaking global events and fairs for promoting sustainable cotton trade<br>• Agricultural tie-ups with leading research institutions and global facilitators |

**Individual and collective inputs for sustainability**

**Table 2.2** Stages at which sustainable cotton production can be adopted (Zulfiqar et al., 2021).

| Stage of cotton production | Designed sustainable practices | |
|---|---|---|
| Sustainable land preparation and sowing practices | - Bed and furrow<br>- Furrow<br>- Laser levelling in last 3 years<br>- Deep ploughing in last 3 years<br>- Rotavator use<br>- Manual sowing<br>- Drill use<br>- If drill use, drill convert into ridges? | - Planter sowing<br>- If less germination, gap filling<br>- Germination test<br>- Grading of home-used seed<br>- Registered variety<br>- Seed treatment<br>- Already treated |
| Sustainable production practices | - Furrow irrigation<br>- Fill furrows<br>- Water scouting<br>- Soil nutrient testing<br>- Split doses of fertiliser<br>- Use of fertilisers by ridges<br>- Use of fertiliser by drum, fertigation<br>- Use of organic manure<br>- Use of compost<br>- Identification of beneficial insects<br>- Pest scouting | - No use of endosulfan<br>- Pest-specific spray<br>- Use of border crops<br>- Pest and insect control through biological methods<br>- Registered and labelled pesticide use<br>- Pesticides applied by skilled adults<br>- Protective and safety equipment for pesticide spray<br>- Pesticide equipment stores appropriately<br>- Pesticides applied in proper weather<br>- Buying of used bottles |
| Sustainable picking and post-harvest practices | - Covered head<br>- Picking stars after dew subsides<br>- Picking from bottom up<br>- Picking after 50% bolls open<br>- Cotton storage at dry place | - Cotton storage on cloth or sheet<br>- Variety-wise storage<br>- Cotton storage in small heaps<br>- Cotton heaps are covered |

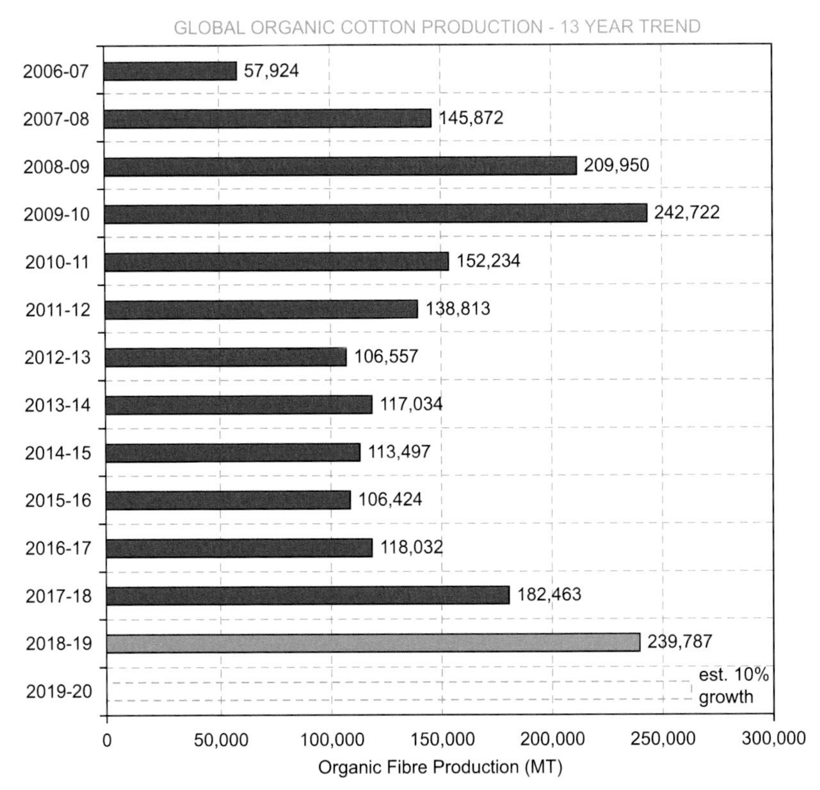


**Figure 2.4** Organic cotton grown in different countries.
Reproduced from Textile Exchange (https://textileexchange.org).

The Textile Exchange Organic Content Standard (OCS) and Global Organic Textile Standard (GOTS) which operates in about 70 countries are some of the agencies that certify organic cotton production and use (https://ota.com/advocacy/organic-standards/fiber-and-textiles/get-facts-about-organic-cotton). Adopting sustainable agricultural practices/better management practices (SAPs/BMPs) and effective dissemination of such practices have also been reported to decrease input costs and increase the profitability from organic cotton cultivation (Awan et al., 2015). Some of the practices used for organic cotton production in the United States are provided in Table 2.4 (Delate et al., 2021). However, challenges associated with the cultivation of organic cotton such as seed availability, chemical drift, climate change and marketing are some of the points that have to be considered for increasing the organic cotton cultivation. Many initiatives to certify the production of authentic organic cotton have been established. Among them, organic content standard (OCS) is an international voluntary standard now certified at about 6181 facilities. A more comprehensive global organic textile standard (GOTS) has also been developed which covers post-harvest processing including weaving and dyeing. About 17 independent accredited

**Table 2.3** Production and share of organic cotton in different countries (https://textileexchange.org/knowledge-center/reports/2020-organic-cotton-market-report/).

| | Organic Cotton Fiber (MT) | Fiber Year-on-Year | Share of global organic cotton production |
|---|---|---|---|
| Global | 239,787 | 31% | 100% |
| India | 122,668 | 43% | 51.15% |
| China | 41,247 | 7% | 17.20% |
| Kyrgyzstan | 23,637 | 6% | 9.86% |
| Turkey | 22,839 | 77% | 9.52% |
| Tajikistan | 12,178 | 35% | 5.08% |
| Tanzania | 5,281 | 8% | 2.20% |
| USA | 5,175 | 2% | 2.16% |
| Uganda | 2,581 | 238% | 1.08% |
| Greece | 1,168 | 12% | 0.49% |
| Benin | 998 | 40% | 0.42% |
| Peru | 558 | 11% | 0.23% |
| Burkina Faso | 453 | -16% | 0.19% |
| Pakistan | 398 | (new) | 0.17% |
| Egypt | 287 | -34% | 0.12% |
| Ethiopia | 130 | 115% | 0.05% |
| Brazil | 97 | 335% | 0.04% |
| Mali | 84 | 9% | 0.03% |
| Argentina | 11 | 575% | 0.005% |
| Thailand | 6 | -9% | 0.003% |

certification bodies are available and more than three million workers are reported to be working in GOTS-certified facilities.

Although organic cotton provides lower yields, the properties of the cotton obtained are similar to those of regular cotton (Table 2.5). Similarly, yarns made from organic cotton have properties similar to those of conventional cotton (Sinclair, 2014) (Table 2.6).

**Table 2.4** Practices adopted for organic cultivation of cotton (Delate et al., 2021).

| **Nutrient and pest management practice in organic cotton production** | **Environmental effects of practice** | **Nutrient and pest management practice in conventional cotton production** | **Environmental effects of practice** |
|---|---|---|---|
| Cover crops (e.g., cereal rye, crimson clover) | Carbon and nitrogen fertility added to soil | Synthetic nitrogen fertiliser | Acidification of soils; detrimental effect |
| Rotational crops (e.g., chickpea, lentil, sunflower, soybean) | Soil fertility enhanced; insect and disease pests mitigated by varying host crops; weed management assistance | No rotational crops; herbicides and herbicide-tolerant cotton | Monoculture system supporting build-up of insect and disease pests; resistance development in GM crops |
| Trap crops (e.g., okra, sunflowers) | Trap insect pests to isolate cotton crop and/or trap for organic-compliant treatments | Synthetic insecticides; *Bt* cotton | Some insecticides with toxicity to bees; potential harmful effect on beneficial insects who help keep pest populations in check; resistance development in GM crops |
| Entomopathogens: *Bacillus thuringiensis*; *Steinernema* spp. | Natural treatments of beneficial bacteria and nematodes that can manage bollworms and armyworms | Synthetic insecticides; *Bt* cotton | Some insecticides with toxicity to bees; potential harmful effect on beneficial insects who help keep pest populations in check; resistance development in GM crops |

*Continued*

**Table 2.4** Practices adopted for organic cultivation of cotton (Delate et al., 2021).—cont'd

| Nutrient and pest management practice in organic cotton production | Environmental effects of practice | Nutrient and pest management practice in conventional cotton production | Environmental effects of practice |
|---|---|---|---|
| Insect- and disease-resistant or tolerant varieties | Enhancement of beneficial organisms unharmed by pesticides | Synthetic insecticides; *Bt* cotton | Some insecticides with toxicity to bees; potential harmful effect on beneficial insects who help keep pest populations in check; resistance development in GM crops |
| Planting later and in warm soils | Avoiding soil-borne fungal attack of seedlings; enhancement of beneficial organisms unharmed by pesticides | Synthetic seed treatments | Detrimental effect on beneficial soil biota |



**Table 2.5** Comparison of the properties of organic cotton and regular cotton (Sinclair, 2014).

| Properties | Organic cotton | Regular cotton |
|---|---|---|
| Length (mm) | 27.9 | 29.2 |
| SFC (<16 mm, W%) | 10.5 | 9.1 |
| Uniformity ratio (%) | 81.9 | 82.9 |
| Fineness (dtex) | 18.4 | 17.4 |
| Micronaire (mic) | 4.9 | 4.5 |
| Tenacity (cN/dtex) | 2.7 | 2.8 |
| Maturity ratio | 1.65 | 1.8 |
| Immature content (%) | 6.8 | 6.3 |
| Impurity (%) | 1.7 | 1.1 |

**Table 2.6** Comparison of properties of the yarns (18.2 tex) made from organic and regular cotton (Sinclair, 2014).

| | Tenacity (cN/tex) | Weight variation (%) | CV of weight (%) | CV (%) | Thin place (Cnt/ 1000 m) | Thick place (Cnt/ 1000 m) | Neps (Cnt/ 1000 m) |
|---|---|---|---|---|---|---|---|
| Organic yarn | 13.3 | 1.39 | 2.28 | 14.44 | 3 | 21 | 19 |
| Regular yarn | 14.6 | 1.13 | 2.05 | 13.93 | 1 | 15 | 15 |

#### 2.3.1.3 *Genetically modified or transgenic cotton*

Although several controversies and considerable resistance have been seen across the globe, transgenic cotton has slowly gained acceptance as sustainable and profitable cotton for farmers, particularly in developing countries (Joseph and Paul, 2007). In addition to an increase in production and the use of fewer chemicals, cultivation of transgenic cotton has been found to improve the physical and chemical properties of soil and increase the accumulation of nitrogen without affecting the microbial population in the soil (Tian et al., 2020; Paul and Joseph, 2003). However, some countries, after several years (8–10) of adoption, have restricted or banned the use of transgenic cotton citing poor lint quality and other socio-political issues (Luna and Dowd-Uribe, 2020). Some studies have shown that GM crops have modified the structure and functions of the indigenous soil microbial community and soil heterogeneity while most studies have reported that the impacts of GM crops on soil or ecological risks were considerably low (Mandal et al., 2020). However, several limitations and risks of using genetically modified, particularly *Bt* cotton have been reported. For instance, it has been documented that although *Bt* cotton provides substantial resistance to conventional bollworm infections a considerable increase in secondary worm infections has been noticed (Zhao et al., 2011; Catarino et al., 2015). Although the use of pesticides to control bollworms has decreased, *Bt* resistance and increases in other non-target pests are considered to negate the benefits of using transgenic cotton. Based on 20-year data, it has been suggested that farmers spend higher on controlling pests compared to that used before the introduction of *Bt* cotton (Kranthi and Stone, 2020) (Figs. 2.5 and 2.6). Hence, in addition to transgenic cotton, integrated pest management systems are suggested to be necessary to reduce pesticide use and improve the sustainability of cotton.

### 2.3.2 *Non-cotton natural cellulosic fibres*

Despite the predominance of cotton in the natural fibre market, the role and importance of the other cellulosic fibres cannot be ignored. Conventionally, jute, flax, ramie, hemp

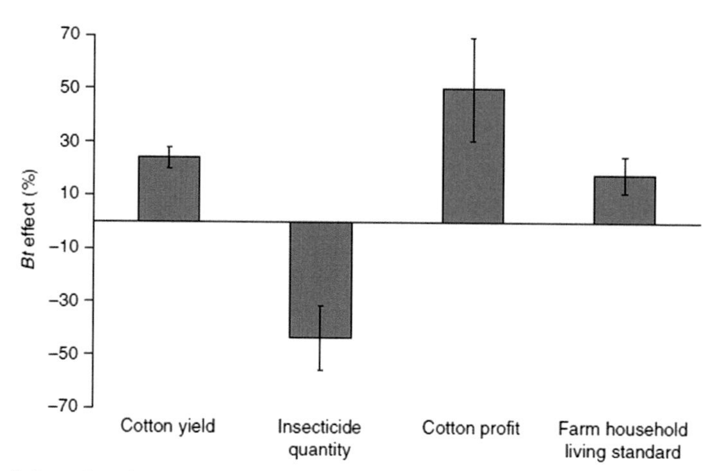


**Figure 2.5** Benefits of using *Bt* cotton in India based on a survey between 2002−08 (Kranthi and Stone, 2020).
Reproduced with permission from Springer Nature.

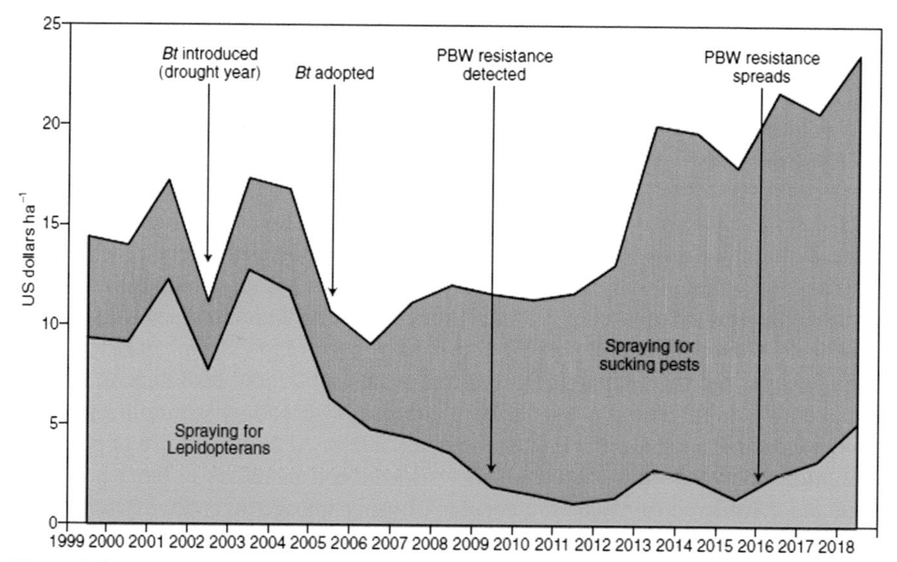


**Figure 2.6** Changes in pesticides applied onto cotton plants before and after the introduction of *Bt* cotton (Kranthi and Stone, 2020).
Reproduced with permission from Springer Nature.

and kenaf are allied fibres that have supported the textile industry to develop textiles with unique characteristics and functionalities. For example, jute fibres are the major commodity crop and a main source of income in Bangladesh, and jute-based products are preferred for food grain packaging. Linen fibres extracted from flax stems have regained their prominence and are now preferred as a source for premium textiles and apparel. However, both jute and linen require land and other natural resources, and their processing also needs considerable energy and often leads to environmental pollution. Retting of jute fibres is not only tedious and time consuming, but the retted water is reported to pollute freshwater ecosystems. Hence, attempts are being made to make jute and other non-cotton cellulosic fibres to be more sustainable and environmentally friendly (Biswas et al., 2019). A proposal was suggested to make production and processing of jute more environmentally friendly in Bangladesh. Fig. 2.7 shows the steps to be followed for the eco-friendly processing of jute. Quality control during retting and utilising the jute stems remaining after extraction of the fibres were

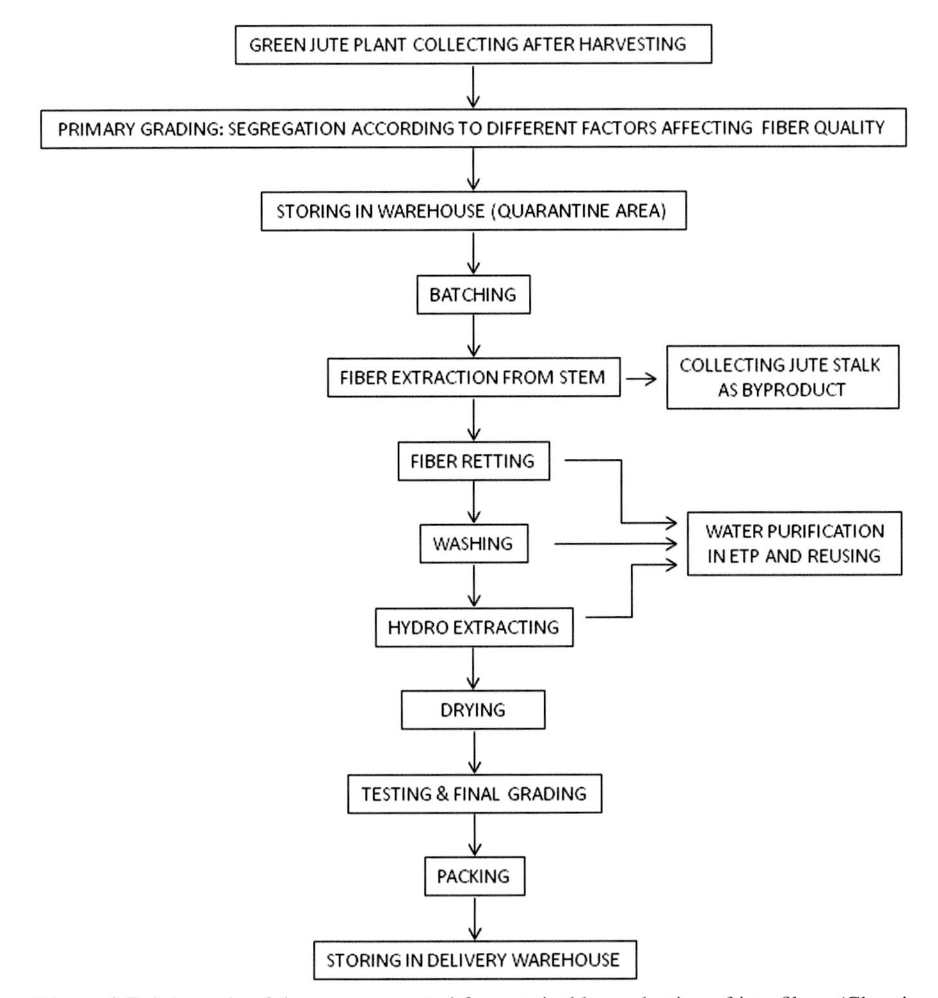


**Figure 2.7** Schematic of the steps suggested for sustainable production of jute fibres (Ghorai and Chakraborty, 2020).
Reproduced with permission through open access publication.

necessary to make jute processing more efficient. Increasing scarcity of water and other natural resources required for separation of jute fibres necessitate the development of alternative approaches. To overcome the lack of water and resulting poor fibre quality, a comprehensive retting technology using micro-ponds was developed. In such ponds, retting could be completed using about 43 L water/kg of fibre compared to 693 L water/kg in traditional retting. Using organic manure, diversified production and integrated farming systems were suggested to provide up to 28 times higher income to farmers compared to conventional jute farming (Ghorai and Chakraborty, 2020). The addition of eco-friendly chemicals during retting is also reported to reduce the retting time from 21 to 12 days and the quality and appearance of jute had also improved (Kumar et al., 2014). Fibre yields of up to 30–35 quintals per hectare were obtained when sustainable practices and crop rotation of 120 days were followed (Kumar et al., 2014).

Similar to jute, several approaches have been used for sustainable production of flax, hemp and other fibres. The selection of appropriate cultivars with biological resistance, environmental conditions such as soil, climate and agronomic treatments are suggested to be necessary for sustainable flax fibre production. Some of the advantages of flax fibres and the steps necessary to improve the quality and quantity of the fibres produced are provided in Table 2.7 (Heller et al., 2012). Adopting transgenic flax lines that require fewer chemical herbicides, the ability to grow on soils contaminated with sulphonylurea herbicide residue and increased yields were also possible (McHughen and Holm, 1995). Instead of using chemicals and microorganisms, attempts have

**Table 2.7** Advantages and diversified applications of flax fibres to ensure sustainability (Heller et al., 2012).

| |
|---|
| Suitable soil conditions for flax |
| Appropriate breeding, ensuring cultivation of cultivars with high economical value |
| Farming technology, skill and experience |
| High yield value |
| Flax fibre is a renewable, biodegradable raw material |
| Flax fibre has excellent hygienic and use value |
| Flax fibre is free from pesticide residues (2–3 treatments are carried out, on average 0.25 kg of active ingredients per hectare, while in cotton there are over 10 treatments) |
| High value and versatility of flax products. These include raw and processed materials. Fibre, seeds and shives are the raw materials, while woven fabrics, knitted fabrics, oil, fodder, food products, semi-pharmaceuticals (with anticancer and antisclerosis actions), paints, lignocellulosic boards, biocomposites, biofuels and biolubricants are included amongst the processed products. Flax is a good plant for diversifying crop rotation, because it improves soil structure |
| A favourable employment market for developments in agriculture and industry |
| Possibility of reclaiming land contaminated by industrial activity – flax is a non-food plant, which can be grown on polluted soils with technical use of the yield |
| Flax is a good crop for sustainable agriculture |

Table 2.8 Yield and properties of enzymatically retted flax with different enzyme formulations.

| Treatment | Fine fibre yield (%) | Strength (g/tex) | Fineness (mic) |
|---|---|---|---|
| Control | 13.5 ± 2.3 c | 43.6 ± 4.4 a | 8.0 ± 0.0 a |
| Viscozyme (0.05%)[a] | 19.5 ± 3.4 ab | 33.3 ± 2.5 c | 7.1 ± 0.6 bc |
| Lyvelin (0.05%)[a] | 10.4 ± 0.7 c | 29.4 ± 3.0 d | 6.6 ± 0.2 c |
| Lyvelin (0.1%)[a] | 12.7 ± 1.1 c | 24.5 ± 1.0 e | 5.7 ± 0.2 d |
| BioPrep (0.05%)[a] | 17.1 ± 0.7 b | 41.0 ± 0.8 ab | 8.0 ± 0.0 a |
| BioPrep (0.01%)[a] | 13.4 ± 1.4 c | 39.9 ± 1.3 ab | 8.0 ± 0.0 a |
| BioPrep (0.05%) + STPP[b] | 18.1 ± 1.6 b | 39.7 ± 0.1 b | 7.7 ± 0.5 ab |
| PGase I (*A. niger*)[a,c] | 22.7 ± 2.5 a | 40.6 ± 1.9 ab | 6.8 ± 0.7 c |
| PGase II (*Rhizopus* spp.)[a,c] | 13.4 ± 2.8 c | 23.4 ± 1.5 e | 7.2 ± 0.3 bc |

a,b,c,d,e: values within columns with different letters differ at $P < .05$.
[a]Mayoquest 200 was used to provide 20 mM EDTA.
[b]Sodium tripolyphosphate (100 mM).
[c]Data adapted from Evans et al.
Reprinted by permission from Elsevier through open access publishing (De Prez et al., 2018).

also been made to obtain natural cellulose fibres from plant stems using enzymes as the retting agents. Enzymes for flax retting have been developed and are available on a commercial scale. Fibres obtained through enzyme retting have shown similar or better properties than chemical retting (Table 2.8) (De Prez et al., 2018). Recently, it has been reported that using radio frequency (RF) treatment during enzyme retting will improve fibre properties by faster degradation of non-cellulosic substances. Fibres obtained were whiter and also marginally weaker compared to non-RF-treated fibres depending on the extent of RF pretreatment (Ruan et al., 2020).

Hemp is one of the most important fibre crops, particularly in Europe. However, the cultivation of hemp has faced considerable challenges and the production of hemp decreased substantially mainly due to the poor fibre quality and poor fibre homogeneity. A major initiative (HEMP-SYS: Design, development and up-scaling of a sustainable production system for hemp textiles: an integrated quality system approach) was started between 2002–06 to develop comprehensive strategies throughout the entire hemp production chain and make production of hemp sustainable and to enable the development of value-added hemp products (https://cordis.europa.eu/project/id/QLK5-CT-2002-01363). Another initiative (FIBNATEX: Production and technical development of natural fibres for the textile industry in South-West Europe) is an INTERREG SUDOE IV B-funded project during 2008–12, aiming at creating innovative and environmentally friendly technical textiles from hemp fibres grown in Europe. In this project, efficient chemical and enzymatic processes for the extraction and separation of hemp fibres were developed for subsequent production of hemp yarns and textile fabrics (http://4.interreg-sudoe.eu/contenido-dinamico/libreria-ficheros/10E78093-D41F-DEC5-257F-37052E2A988F.pdf) (Paul et al., 2009).

Green decortication, green retting and adopting new washing techniques have been suggested to be necessary to make hemp production and processing sustainable

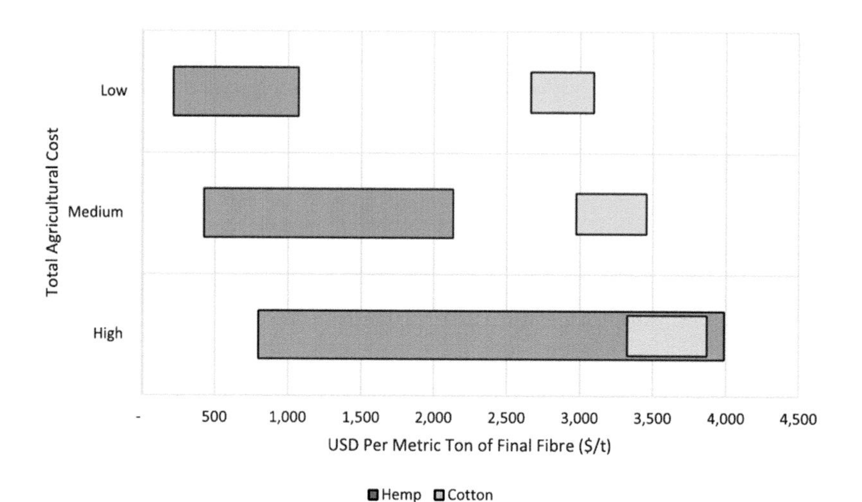


**Figure 2.8** Comparison of agricultural costs for producing hemp and cotton (Schumacher et al., 2020).
Reproduced with permission from Elsevier.

(Riddlestone et al., 2006). When sustainable technologies are adopted, the production of hemp increased to 15 tons per hectare, which was equivalent to reducing about 26 tons of $CO_2$. Hemp-based building materials were also found to improve thermal resistance and environmental performance, and reduced greenhouse gas emissions (Zampori et al., 2013). In addition, major changes in legislation, particularly in the United States, have also been passed in the last few years which allow growing hemp for commercial applications. Vermont and several other states in the United States have realised the potential of hemp and are now promoting the cultivation and processing of hemp for fibres and other applications. Overall hemp production costs are also considerably lower compared to cotton, even considering the vastly different production rates (Fig. 2.8). It has been found that hemp requires only one-third of the land compared to that required to produce the same amount of cotton, and more importantly cotton needs nearly 2.5 times the water required to grow hemp. Overall, it has been suggested that agricultural costs for growing hemp were about 78% lower compared to cotton and because of the higher yield (three times) and lower input costs for hemp. Hence, hemp was suggested as an eco-friendly alternative to cotton and other bast fibres (Schumacher et al., 2020). Not only cotton, life cycle assessment (LCA) has shown that the production of hemp is more beneficial than six other major crops in France. Hemp cultivation showed eutrophication potential of 20.5 kg $PO_4$ equivalent, global warming potential of 2330 kg $O_2$ equivalent and acidification potential of 9.8 kg $SO_2$ equivalent, and an energy use of 11.4 GJ, which was lower compared to sugar beets and sweet potato (Van der Werf, 2004; Van der Werf and Lea, 2008). The productivity and yield of hemp fibres were also dependent on the type of retting used (Table 2.9). Similarly, the properties of fibres are affected by the strain of bacteria used during retting (Zheng et al., 2001). Hemp fibres produced

**Table 2.9** Comparison of the yields (kg) per hectare of the intermediate and final products based on different retting conditions (Van der Werf and Lea, 2008).

| | Hemp | | | |
|---|---|---|---|---|
| **Products** | **Water retting** | **Bio-retting** | **Baby hemp** | **Flax dew retting** |
| Green stem | 8000 | 8000 | – | 6000 |
| Retted stem | 6480 | – | 3250 | 5400 |
| Green scutched long fibre | – | 1000 | – | – |
| Green scutched short fibre | – | 1000 | – | – |
| Grain yield (9% humidity) | 0 | 0 | 0 | 600 |
| Green scutched long fibre after retting | – | 658 | – | – |
| Scutched long fibre | 583 | – | 293 | 972 |
| Scutched short fibre | 1490 | – | 748 | 594 |
| Shives | 2592 | 3600 | 1300 | 2970 |
| Yarn | 236 | 213 | 119 | 512 |



through enzymatic retting have shown better properties and higher yields but are considered more expensive than natural or chemical retting. In one study, pectinase was able to degum hemp fibres and produce fibres with better tensile properties than those obtained by dew retting (Zimniewska, 2022). Enzymes (BioPrep3000L/BAYLASE EVO) are available commercially which can degum and provide hemp fibre with properties comparable to alkaline-separated fibres (Dreyer et al., 2002).

### 2.3.3 Sustainable fibres from renewable resources

Although substantial efforts have been made to improve the sustainability and ecological impacts during natural cellulosic fibre production, processing and use, the extent of use of eco-friendly fibres is very low. Another approach to improve the sustainability of fibre production is to use renewable resources instead of the conventional fibres which require dedicated land and considerable natural resources. Fibres such as linen and jute are lignocellulosic and contain cellulose, hemicellulose and lignin as the major components. Similar to these lignocellulosic fibres, agricultural residues such as stems, stalks, leaves and husks are also lignocellulosic in nature. Most of these residues are considered as waste and are disposed of either by burning or burying and such agricultural residues are available in large quantities at low cost and almost everywhere around the world.

Lignocellulosic, agricultural residues contain cellulosic fibres that are made of single cells that are about 80–150 μm in length and a few micrometres in diameter. Such smaller dimensions single cells are not suitable for textile applications and are hence generally considered for paper and pulp production. However, it has been shown that

**Table 2.10** Comparison of the properties of natural cellulosic fibres obtained from various agricultural residues in comparison to cotton and linen (Reddy et al., 2008, 2009).

| Fibre | Mechanical properties | | |
|---|---|---|---|
| | **Strength (g/den)** | **Elongation (%)** | **Modulus (g/den)** |
| Cornhusk | 2.0 ± 0.3 | 11.9 ± 1.1 | 49 ± 3.7 |
| Cornstalk | 2.2 ± 1.0 | 2.2 ± 0.7 | 127 ± 56 |
| Rice | 3.5 ± 1.2 | 2.2 ± 0.4 | 200 ± 33 |
| Switchgrass | 5.5 ± 1.2 | 2.2 ± 0.7 | 240 ± 74 |
| Velvet leaf | 2.9 ± 0.7 | 2.5 ± 0.7 | 194 ± 69 |
| Pineapple leaf | 0.7–3.8 | 2–6 | – |
| Sugarcane | 1.8–3.3 | 5.5–11.8 | – |
| Oil palm fibres | 1.9 | 14 | – |
| Bamboo | 5.1 ± 2.7 | 5.3 ± 1.3 | – |
| Cotton | 3.0–3.5 | 8.0–9.0 | 50–55 |
| Linen | 4.0–8.0 | 2.0–3.0 | 190–220 |

**Table 2.11** Comparison of the properties of natural cellulosic fibres obtained from the leaves and stems of switchgrass with cotton, linen and kenaf (Reddy and Yang, 2007a,b).

| | **Switchgrass** | | | | |
|---|---|---|---|---|---|
| | **Leaf** | **Stem** | **Linen** | **Cotton** | **Kenaf** |
| Fineness (denier) | 30 ± 12 | 60 ± 20 | 1.7–17.8 | 1–3.3 | 50 |
| Length (cm) | 6.5 ± 4.3 | 5.8 ± 3.3 | 20–140 | 1.5–5.6 | 150–180 |
| Strength (g/den) | 5.5 ± 1.2 | 2.7 ± 0.8 | 4.6–6.1 | 2.7–3.5 | 1.0–2.3 |
| Elongation (%) | 2.2 ± 0.7 | 6.8 ± 2.1 | 1.6–3.3 | 6.0–9.0 | 1.3–5.5 |
| Modulus (g/den) | 240 ± 74 | 70 ± 23 | 203 | 55 | 92–230 |
| Work of rupture (g/den) | 0.16 | 0.23 | 0.09 | 0.19 | 0.03–0.3 |
| Moisture regain (%) | 10.0 | 9.3 | 12.0 | 7.5 | 9.5–10.5 |



the single cells or ultimates in the residues can be extracted in the form of fibre bundles suitable for textile production. Fibre bundles have been extracted from cornhusks, cornstalks, wheat straw, rice straw, soybean straw and from the byproducts of most major agricultural crops. As can be seen from Table 2.10, the properties of the fibres obtained from the residues are similar to those of cotton and linen. In addition to the residues, biomass crops such as switchgrass, sabai grass, etc. can be used to obtain natural cellulosic fibres with properties suitable for textiles. In fact, the leaves and stems of switchgrass provide considerably distinct and unique fibres, as can be seen from Table 2.11. The fibres from the leaves have strength and elongation similar to linen and fibres from the stems are similar to cotton (Reddy and Yang, 2007a,b).

Fibres extracted from the stems of *Cajanus cajan* (pigeon pea) showed high strength (984 MPa) and elongation (3.4%) similar to the properties of flax fibres. These fibres had good thermal stability and showed potential to be useful for textile, composite, and other applications. Such applications will provide a considerably high value addition to the stems of *C. cajan* and also be a renewable and sustainable source for fibres (Navada et al., 2022). Several other unique fibres with inherent antimicrobial properties have also been obtained from agricultural wastes such as turmeric leaves, castor plant stems and outer shell of *Kigelia africana*. In an interesting study, it has also been shown that natural cellulosic fibres with properties similar to linen and better than those of jute can be obtained from the cotton stalks left as residues after harvesting cotton fibres. Using these agricultural residues or biomasses as a source for fibres not only provides a renewable or sustainable source for fibres but also helps to add value and improve the income from the crops.

Sabai grass, *Eulaliopsis binata*, being a perennial grass of the *Poaceae* family, is one of the best fibre grass plants for its high fibre quality and production (Zou et al., 2013). Easy planting, good perennial growth, wide adaptability, stress resistance, well-developed root system and dense populating propensity are the noteworthy attributes of sabai grass to become a suitable species for soil and water conservation and wasteland construction (Huang et al., 2004; Duan and Zou, 2006). In addition, its good flexibility and strength, high leaf fibre content (>55%), low lignin content (<14%), excellent average fibre length (20 mm), enable it to be one of the best raw materials among fibre grass plants for the paper industry and as rayon and woven materials (Han et al., 2008). Its large-scale cultivation in barren hills and slopes has already been practiced and proved to have fast ecological benefits (Huang et al., 2003; Duan et al., 2003). These grass family characteristics resemble a relative of cereal crops and are wildly seen in China, India and countries in southeast Asia with different ecological habitats. Sabai grass is also popularly used as a construction material for thatches, walls, roofs and ropes, and as a filler material in plastics and in mud matrix. Some of the properties of sabai grass are provided in Table 2.12.

**Table 2.12** Selected properties of sabai grass fibres.

| Particulars | Sabai grass |
|---|---|
| Colour | Brownish yellow |
| Fibre length (L) in mm | 2.4 |
| Fibre width (D) in μm | 9.90 |
| Lumen width (d) in μm | 5.75 |
| Cell wall thickness (w) in μm | 2.11 |
| Flexibility coefficient (d/D × 100) | 57.67 |
| Slenderness ratio L/D | 266 |
| Rigidity coefficient 2w/D | 0./42 |
| Wall fraction (2w/D) × 100 | 42 |
| Runkel ratio 2w/d | 0.73 |

### 2.3.4 Environmentally friendly fibre extraction processes

The conventional approach of extracting natural cellulosic fibres from plant stems is through natural retting, bacterial degumming, etc. Such practices require considerable time, water, chemicals and other resources. Hence, attempts have been made to study alternative methods to make fibre extraction more sustainable and environmentally friendly. It has been shown that using controlled steam explosion can separate the components in the biomass and provide natural cellulosic fibres with good properties. Cotton stalks exposed to a combination of steam explosion, potassium hydroxide and peroxide treatments provided fibres with fineness of 27 dtex and with cellulosic content as high as 82% and tensile properties similar to those of cotton (Dong et al., 2014). Presoaking raw kenaf with hydrogen peroxide, sulphuric acid and sodium hydroxide and later subjecting the material to steam explosion provided higher removal of lignin and hemicellulose. Hydrogen peroxide retained fibre strength but treating with sulphuric acid caused a decrease in fibre strength (Song et al., 2017). Treating lotus petioles with 5% sodium hydroxide at a pressure of 1.5 MPa for 160 s resulted in slender soft and fragrant natural cellulosic fibres with good moisture absorption and mechanical properties (Yuan et al., 2013). In another study, a deep eutectic solvent system (DES) combined with microwave and alkaline-ultrasonication was followed to effectively remove the non-cellulosic components in kenaf bast and obtain fibres with fineness of about 4 dtex, breaking tenacity of 13.7 cN/tex and with projected decreases in chemical, water and time by 48.9%, 66.7% and 66.8% compared to a conventional two-step alkaline extraction process (Nie et al., 2020). Three DES solvents, namely choline chloride-urea (CU), choline chloride-imidazole (CI) and ethylamine hydrochloride-ethylene glycol (EE), were used to obtain natural cellulosic fibres from ramie stems. The stems were treated for 2 h in the DES system at the boil and compared with the traditional alkaline (TAL) method. The fibres obtained had a breaking tenacity of 6.5 cN/dtex and lower residual gum content of 3.8%. It was also suggested that the DES could be recycled and reused several times without affecting the fibre quality, making the process environmentally friendly and low-cost (Huang et al., 2021). Similar to DES, an organic solvent system made up to glycol and acetic acid was used to degum ramie stems and obtain fibres. Various proportions of the two solvents were used to obtain fibres with a 44% decrease in hemicellulose and 54% decrease in lignin content. Treating the stems only with glycol at 200°C for 80 min resulted in fibres with a strength of 6.5 cN/dtex and elongation of 5.8% lower than the properties of fibres (Fig. 2.9) obtained through conventional degumming. However, the organosolv system provided a higher yield of fibres (Qu et al., 2020).

Microwave-assisted ultrasonic degumming was also followed to obtain natural cellulosic fibres from *Apocynum venetum* (Fig. 2.10). The fibres obtained had 3% lignin, fineness of 4 dtex and breaking tenacity of 7.7 cN/tex, which are considerably better than those obtained through conventional degumming (Li et al., 2020). In another study, a combination of microwave and lactic acid treatment was able to remove 88% of lignin and the cellulose obtained had a higher molecular weight of

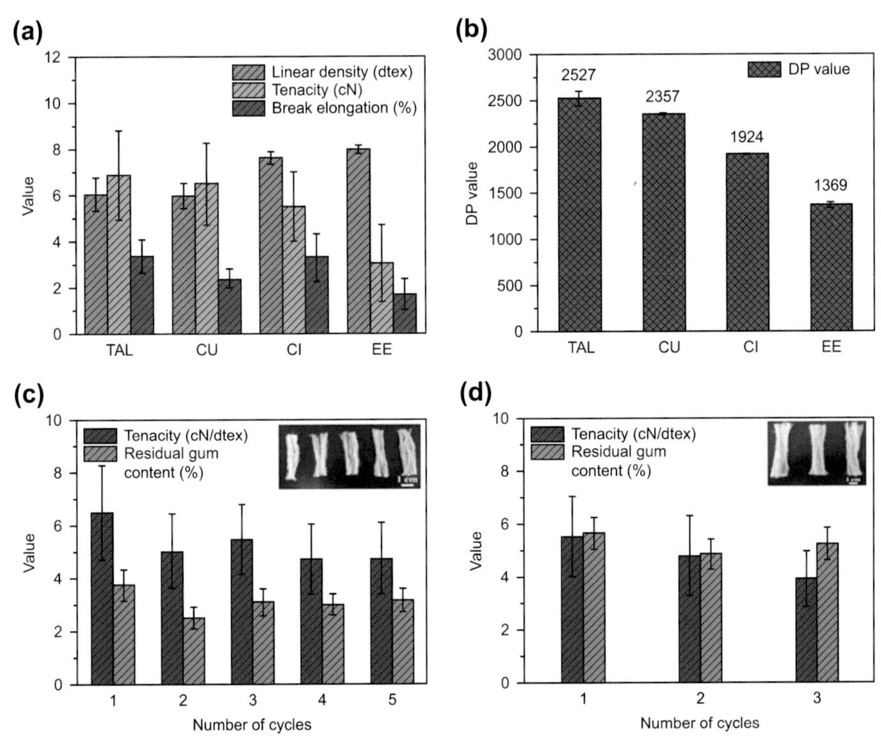


**Figure 2.9** Comparison of the properties of fibres obtained using DES and TAL processes: (a) mechanical properties, (b) DP values, (c) residual gum content against number of cycles and (d) gum content and tenacity of fibres using recycled DES (Huang et al., 2021).
Reproduced with permission from Elsevier.

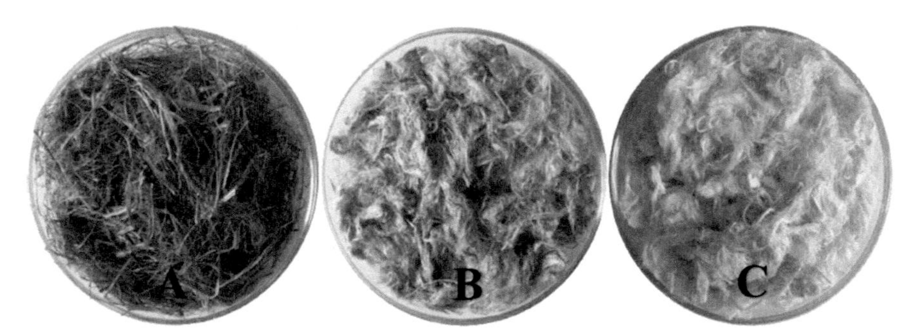


**Figure 2.10** Images of the untreated *Apocynum venetum* basts, fibres obtained from conventional alkaline degumming, and fibres obtained through microwave-assisted ultrasonic degumming (Li et al., 2020).
Reproduced with permission from Elsevier.

**Table 2.13** Effect of various strains of *Bacillus* sp. on the ability to degum ramie basts and corresponding tenacity of fibres obtained (Zheng et al., 2001).

| Strain | Pectin (%) | Hemicellulose (%) | Residual gum (%) | Brightness, (ISO) | Breaking tenacity (g/denier) |
|---|---|---|---|---|---|
| – | 3.5 ± 0.1 | 14.7 ± 0.2 | 17.6 ± 0.4 | 32.4 ± 0.2 | 6.3 ± 0.3 |
| NT-39 | 0.8 ± 0.2 | 10.5 ± 0.3 | 11.2 ± 0.3 | 35.2 ± 0.2 | 6.1 ± 0.2 |
| NT-53 | 1.1 ± 0.3 | 12.3 ± 0.4 | 12.8 ± 0.2 | 34.7 ± 0.1 | 6.2 ± 0.2 |
| Nt-39+NT53 (1:1) | 0.9 ± 0.2 | 8.8 ± 0.3 | 9.4 ± 0.4 | 35.0 ± 0.1 | 6.1 ± 0.3 |
| NT-76 | 1.2 ± 0.1 | 11.0 ± 0.2 | 12.5 ± 0.3 | 37.8 ± 0.2 | 6.0 ± 0.2 |
| NT-80 | 3.0 ± 0.2 | 14.2 ± 0.4 | 16.6 ± 0.4 | 33.0 ± 0.2 | 6.1 ± 0.2 |

Reproduced with permission from Elsevier.

1174 kg/mol compared to 607 kg/mol for those obtained through conventional alkaline fibre extraction (Lv et al., 2021).

In addition to the organic chemicals, microwave and ultrasonication approaches, extensive studies have also been considered to obtain fibres through environmentally friendly and green approaches using microorganisms. For instance, flax fibres were treated with *Bacillus subtilis* HR5 (0.16%) at 55°C for 1 h as a replacement for the traditional sulphuric acid and sodium chlorite treatment used to extract fibres. Up to a 27% decrease in the use of chemicals and fibres with inherent antimicrobial activity could be obtained using the new approach (Xiang et al., 2020). The effect of various bacterial strains to degum ramie and decrease the non-cellulosic contents and provide fibres with good strength was studied by Zheng et al. (Zheng et al., 2001). As can be seen from Table 2.13, a considerable decrease in hemicellulose and pectin was possible with a corresponding decrease in gum content as well as no major decrease in the brightness or bundle strength of the fibres. Recently, a bacterial strain, *Pectobacterium wasabiae* (PW), with broad degumming capabilities was developed and used to degum ramie, hemp, flax and kenaf. This bacterial strain was able to release pectinase, mannase and xylanase with an enzyme activity of up to 157 U/ml. A degumming time of 12 h at 33°C was required and using the bacteria for degumming also had lower pollution levels (60%) in terms of chemical oxygen demand (Duan et al., 2021).

### 2.3.5 Yarns and fabrics from unconventional fibres

Fibres obtained from agricultural residues have been processed into yarns and fabrics. As can be seen from Tables 2.14 and 2.15, the properties of cornhusk fibre blended yarns are similar to those of 100% cotton fibres. Yarns made from cornhusk fibres

**Table 2.14** Properties of cotton–cornhusk fibre blended yarns in comparison to 100% cotton open end and ring spun yarns (Reddy et al., 2006).

| Count (tex) | Blend proportion | Strength | | | Elongation | |
|---|---|---|---|---|---|---|
| | | g/tex | CV% | % Retention[a] | Percent | % Retention[a] |
| Ring spun yarn | Cotton/ cornhusk | 10.7 | 20 | 97 | 4.6 | 150 |
| 30 | 70:30 | 12.2 | 16 | 90 | 4.9 | 72 |
| 42 | 70:30 | 12.6 | 26 | 87 | 6.6 | 92 |
| 50 | 70:30 | | | | | |
| 30 | 50:50 | 8.9 | 19 | 81 | 4.2 | 136 |
| 30 | 70:30 | 10.7 | 20 | 97 | 4.6 | 150 |
| 30 | 80:20 | 9.7 | 30 | 88 | 4.3 | 140 |
| Open end yarn | | | | | | |
| 84 | 65:35 | 8.7 | 22 | 64 | 6.9 | 83 |
| Ring spun yarn | Polyester/ cornhusk | | | | | |
| 27 | 65:35 | 17.6 | 24 | 117 | 15.7 | 104 |

[a]Compared to 100% cotton yarn of the corresponding count for all cotton/cornhusk blends and to 65/35 polyester/cotton yarn for the polyester/cornhusk blends.

**Table 2.15** Comparison of the properties of cotton stalk fibre–cotton blend yarns with those of jute–cotton blend yarns (Dong et al., 2016).

| Yarns | Blending ratio | Hairiness index (number/m) | Breaking strength (cN/tex) | Breaking elongation (%) |
|---|---|---|---|---|
| 100% cotton | – | 10.8 ± 2.2 | 17.8 ± 1.6 | 6.5 ± 0.5 |
| Cotton stalk–cotton blend | 30/70 | 22.5 ± 2.2 | 14.1 ± 1.5 | 5.8 ± 0.3 |
| | 10/90 | 12.6 ± 1.5 | 16.6 ± 1.5 | 6.2 ± 0.5 |
| Jute–cotton blend | 10/90 | 16.5 ± 3.3 | 11.9 ± 1.3 | 4.4 ± 0.4 |



were made into a sweater and dyed using reactive dyes (Fig. 2.11). Similarly, fibres extracted from rice straw could also be converted into cotton-blended fabrics. Similarly, fibres extracted from cotton stalks were blended with cotton and made into yarns and fabrics (Dong et al., 2016).

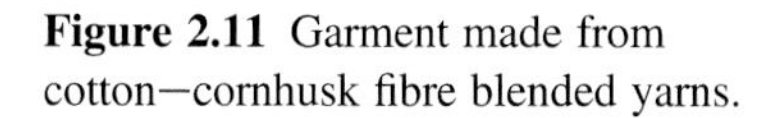
**Figure 2.11** Garment made from cotton—cornhusk fibre blended yarns.

## 2.4 Sustainable production of protein fibres

Cellulosic and lignocellulosic fibres have been dominating the natural fibre market. However, protein fibres have their unique features and niche market which neither the cellulosic nor synthetic fibres can match. Hence, protein fibres have retained their prominence in the fibre market. Silk and wool are the two most common natural fibres. Although silk and wool are available and are converted into value-added textiles, these protein fibres are facing considerable challenges in terms of availability and cost. For example, growing silk and rearing sheep for wool requires considerable and dedicated resources including land, energy and water. The availability of these required resources is becoming a challenge and hence there is a need to find approaches to make wool and silk production more sustainable. Also, alternatives to wool and silk such as the azlons that were produced in a commercial scale during the 1930s should be reconsidered.

### *2.4.1 Sustainable wool production*

There has been a continued decrease in the quantity and quality of the range lands and graze lands available for sheep due to various environmental, socio-economic and political reasons. To overcome this challenge, it has been proposed that a holistic management system comprising rotational grazing, careful planning, goal setting, etc. should be followed. Such management is being adopted in the Falkland Islands where more than half a million sheep graze (Tourangeau and Sherren, 2020). In many cases, wool fibres become the byproduct and sheep are mainly raised for their meat. Even in such instances, the quality of wool matters and provides substantial income to farmers. In addition, changing environmental conditions and natural disasters such as cyclones, floods and droughts are burdening farmers and make it necessary to find increased sources of revenue. Hence, the production of wool and sheep milk is helpful even in sheep which are mostly reared for meat. The production of wool from farm to mill is reported to require about 48 MJ/kg of energy, which is about 50% lower than polyester and 75% less than nylon. Also, LCA assessments have shown that the production of wool requires considerably less fossil energy and water compared to producing meat or processing of the fibres (Wiedemann et al., 2019) (Fig. 2.12). Several strategies (Fig. 2.13) have been proposed to make wool production and processing more sustainable (Gowane et al., 2017). In India for instance, cross-breeding among native breeds, cross-breeding with exotic breeds for wool production, developing value-added and new applications particularly for the low-quality and discarded wool are being attempted to make wool production more sustainable (Kumar and Lesile, 2017). Such attempts have produced encouraging results with substantial improvements observed in the quality of wool fibres obtained (Table 2.16) (Kumar and Lesile, 2017).

### 2.4.2 Sustainable silk production

Similar to wool, the production and processing of silk are also facing considerable challenges. Silk production requires considerably higher inputs to grow compared to other common fibres (Table 2.17). Several measures have been proposed to improve the sustainability of mulberry plants and hence production of silk fibres. In Brazil, for example, making silk production more circular and increasing the use of renewable energy, building community digesters, establishing agro-industrial co-operatives, etc. have been proposed to meet the sustainable development goals (SDGs) proposed by the United Nations (Barcelos et al., 2021). Although efforts have been made to promote and increase the production of silk in India, it has been found that there could be a potential scarcity of water and decrease in food security due to increased mulberry cultivation. Mulberry cultivation in India requires about 16,000 $m^3$ of water per hectare per year and often is done through irrigation. Such high water availability is said to be doubtful in many areas, which necessitates adopting various measures. Proponents of silk suggest that cultivation of mulberry helps in substantial carbon mitigation. Based on a study in Brazil, it has been found that mulberry plants can mitigate $CO_2$ of about 735 times that which is required to produce silk in that specific area (Giacomin et al., 2017). Other studies based on silk produced in India have shown that the production of silk requires higher inputs and is more taxing on the environment than other textile fibres (Table 2.18). Several approaches have been proposed to improve the yield of mulberry leaves and enhance the quality and economic returns. Using vermicompost and vermiwash and integrated nutrient management increased

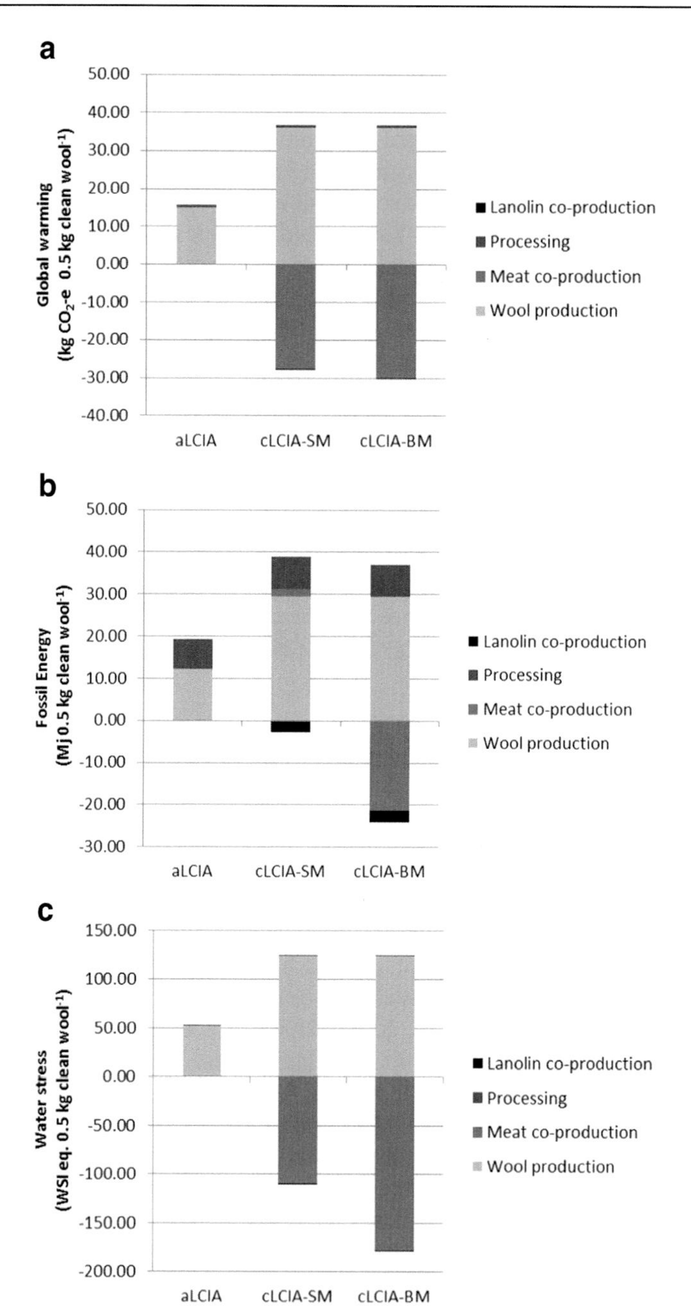


**Figure 2.12** Contribution to global warming, energy and water requirements for the production of wool compared to processing wool and sheep for meat production (Wiedemann et al., 2019). Reproduced with permission from Springer Nature.

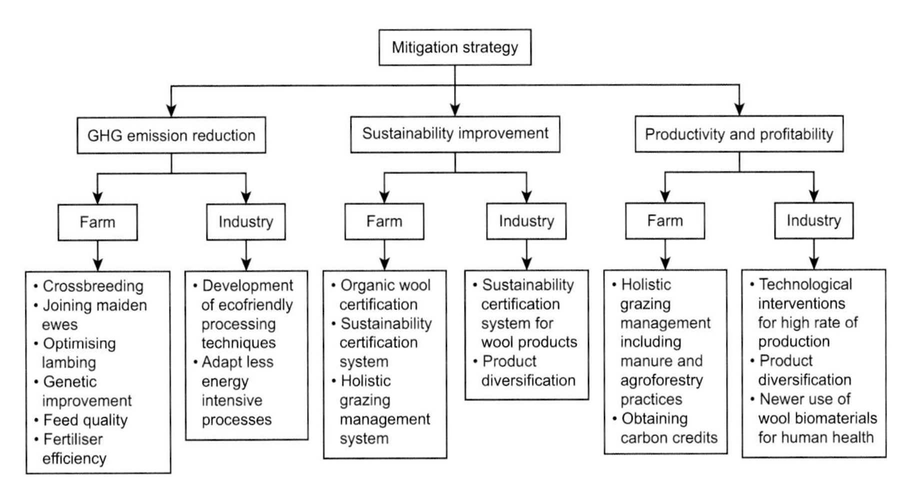


**Figure 2.13** Strategies proposed for the sustainable production of wool fibres (Gowane et al., 2017).
Reproduced with permission from Springer Nature.

**Table 2.16** Fleece characteristics of wool fibres obtained from new sheep strains developed for sustainable wool production (Kumar and Lesile, 2017).

| Breed | Greasy fleece weight (kg) | Average staple length (cm) | Average fibre diameter (μm) | Medullation (%) |
|---|---|---|---|---|
| Hissardale | 1.46–2.72 | 5.52–6.15 | 21.53–24.45 | 0.0–0.55 |
| Kashmir Merino | 2.45–2.80 | 4.77–5.60 | 20.40–20.94 | 0.0 |
| Nilgiri | 0.20–1.40 | 7.01–8.70 | 21.57–27.40 | 8.40–18.0 |
| Deccani Merino | 1.50 | 5.75 | 22.1 | 1.6 |
| Nilgiri synthetic | 1.0–1.5 | 4.6–5.0 | 20.6–21.6 | 1.00–3.20 |
| Patanwadi synthetic | 1.2–1.7 | 4.6–5.2 | 19.3–23.2 | 9.0–20.8 |
| Bharat Merino | 2.61–2.8 | 2.88–9.40 | 17.61–20.24 | 1.00–1.08 |
| Avivastra | 1.90 | 3.50 | 21.6 | 7.60 |
| Avikalin | 1.72 | 4.26 | 25.2 | 24.9 |

Reproduced with permission from Elsevier.

**Table 2.17** Environmental impact of silk production in comparison with other fibres (Astudillo et al 2015).

| Category | GWP (kg $CO_{2eq}$/kg) | Renewable CED (MJ/kg) | Non-renewable CED (MJ/kg) | Eco-toxicity ($CTU_e$/kg) | ALO ($m^2a$/kg) | BWF ($m^3$/kg) | FE ($gP_{eq}$/kg) |
|---|---|---|---|---|---|---|---|
| Raw silk (India) | 51.5 | 1349.9 | 110.1 | 522.8 | 19.7 | 24.6 | 4.8 |
| Cotton (China) | 3.4 | 19.7 | 0.1 | 71.2 | 7.8 | 7 | 0.8 |
| Nylon 66 | 8 | 1.3 | $7*10^{-4}$ | $6.0*10^{-4}$ | $2.0*10^{-4}$ | 0.2 | 0.3 |
| Wool (US) | 18.5 | 81.7 | 0.1 | 3.4 | 53.5 | 0.2 | 0.5 |

*ALO*, agricultural land occupation; *BWF*, blue water footprint; *CED*, cumulative energy demand; *FE*, freshwater eutrophication; *GWP*, global warming potential.


leaf yield by up to 40% and hence higher net profits from the fields (Chowdhury et al., 2013). Similarly, adopting low-cost drip fertigation (LCDF) systems increased leaf yield by 26% and leaf quality by 56%. Up to 24% reduction in water requirement was also possible by adopting the LCDF system (Mahesh et al., 2021). In addition to using as feed for silkworms, mulberry leaves have been found to be useful for medical, energy, environmental remediation and other applications and hence are considered to be an ideal plant for sustainable development. Conditions during plant growth such as soil properties, temperature and humidity during rearing also contribute to the quality and quantity of the silk produced. Amount of organic carbon and nitrogen available in the soil were found to significantly affect the tasar silkworm yields.

### 2.4.3 Alternative protein fibres

Wool and silk have been the two most common protein fibres since ancient times. However, various other plant and animal sources have been used as sources for silk fibres. Regenerated protein fibres were developed from plant sources such as soyproteins, corn zein, etc. and commercialised in the 1930s (Table 2.18). However, the price, complicated production systems and quality of the fibres obtained could not sustain the competition from synthetic fibres and they were hence discontinued. However, recent interest in the production of biofuels from oil seeds has led to the availability of byproducts and coproducts containing proteins. The proteins (up to 40%) in these byproducts, generally called oil meals or oil cakes, are low-cost, sustainable and renewable sources of fibres. Studies have been carried out to utilise the proteins in

**Table 2.18** Environmental impact of silk compared to other fibres (Karthik and Rathinamoorthy, 2017).

| Category | GWP (kg $CO_{2eq}$/kg) | Renewable CED (MJ/kg) | Non-renewable CED (MJ/kg) | Eco-toxicity ($CTU_e$/kg) | ALO ($m^2a$/kg) | ULO ($m^2a$/kg) | BWF ($m^3$/kg) | FE ($gP_{eq}$/kg) |
|---|---|---|---|---|---|---|---|---|
| Farm practices | 80.9 | 1613.6 | 244.4 | 1043.1 | 35.6 | 1.37 | 54.0 | 7.0 |
| Recommended practices | 52.5 | 1350.6 | 116.7 | 522.9 | 19.8 | 1.13 | 26.7 | 4.8 |
| Cotton | 3.4 | 19.7 | 0.1 | 71.2 | 7.8 | 0.02 | 7.0 | 0.8 |
| Nylon 66 | 8.0 | 1.3 | $7 \times 10^{-4}$ | $6 \times 10^{-4}$ | $2 \times 10^{-4}$ | $4 \times 10^{-4}$ | 0.2 | 0.3 |
| Wool | 18.8 | 81.7 | 0.1 | 3.4 | 53.5 | 0.36 | 0.2 | 0.5 |

*ALO*, agricultural land occupation; *BWF*, blue water footprint; *CED*, cumulative energy demand; *FE*, freshwater eutrophication; *GWP*, global warming potential; *ULO*, urban land occupation.
Reproduced with permission from Elsevier.

the oil meals for textile, medical and other applications. These oil meals can also be used for the production of regenerated protein fibres. Cereal proteins such as soyproteins, zein in corn and wheat gluten (gliadin and glutenin) have also been used to produce regenerated protein fibres. Since the regenerated protein fibres from these sources have poor stability in aqueous conditions and low mechanical properties, the fibres have been cross-linked with biocompatible and efficient carboxylic acids such as citric acid and butanetetracarboxylic acids (BTCA) (Reddy and Yang, 2007a,b, 2008; Reddy et al., 2009) (Table 2.19). The cereal proteins such as soyproteins and wheat gluten do not melt or dissolve in common solvents and hence it is difficult to process them into fibres. Reducing the proteins using sodium sulphite in the presence of urea and precise control of the ageing time, protein concentration, viscosity and spinning conditions enable extrusion of the fibres into coagulation baths and subsequent drawing, annealing and drying for complete fibre formation. Alternatively, proteins such as corn zein and wheat gliadin are prolamins that dissolve in aqueous alcohol solutions. These proteins can be made into fibres relatively easily using the dry spinning approach. However, the fibres obtained have relatively poor strength and need to be cross-linked. In addition to textiles, the regenerated protein fibres have shown potential to be useful for tissue engineering, controlled release and other medical applications (Li et al., 2008).

Another unique source of protein fibres are poultry feathers which are made of more than 90% keratin and have unique structure and properties (Table 2.20). Several studies have reported the possibilities of converting the keratin in chicken feathers into high-quality regenerated protein fibres. For instance, keratin extracted from feathers was dissolved using a sodium carbonate–sodium bicarbonate buffer system and made into fibres with a strength of 0.5 cN/dtex and elongation of 28% (Ma et al., 2016), which was considered low for practical applications. To further improve the properties of regenerated protein fibres from feather keratin, a new approach using dithiol reducing agents and preserving the long intermolecular cross-linkings was developed (Fig. 2.14). Fibres obtained through this system had a strength of 161 MPa and elongation of 14%, similar to that of natural feathers depending on the stretching ratio, extent of cross-linking, etc. Furthermore, a controlled cleavage and assembly of disulfide cross-linkages in keratin enabled continuous production of fibres on a pilot scale with properties similar to or better than most fibres (Table 2.21). The production of regenerated keratin fibres using this process was estimated to cost about $0.83 per kilogram which is considerably lower compared to selling price of wool at $7–8 per kilogram and silk at $45–80 per kilogram (Mu et al., 2020). Similar to keratin from feathers, keratin in wool, particularly discarded wool, has been used to develop regenerated fibres (Fig. 2.15). The fibres obtained were blended with other polymers and cross-linked to improve the strength and stability (Mu et al., 2020).

**Table 2.19** Properties of regenerated protein fibres from different protein sources compared to common textile fibres (Lebedytė and Sun, 2021).

| | Dry state | | | Wet state | | |
|---|---|---|---|---|---|---|
| **Fibre source** | **Tenacity (g/denier)** | **Initial modulus (g/denier)** | **Breaking extension (%)** | **Tenacity (g/denier)** | **Initial modulus (g/denier)** | **Breaking extension (%)** |
| Fibrolane (casein) | 1.1 | 40 | 63 | 0.35 | 2 | 60 |
| Ardil (peanut) | 0.8–1.0 | 30 | 10–110 | 0.3 | 0.5 | 90 |
| Vicara (zein) | 1.0 | 50 | 28 | 0.6 | 15 | 028 |
| Soybean (Drackett co) | 0.6 | 40 | 40 | 0.12 | 4 | 40 |
| Merino wool | 1.6 | 25 | 43 | 1.1 | 10 | 57 |
| Cotton | 3.6 | 30 | 9 | 4.0 | 10 | 10 |
| Silk | 3.7 | 120 | 16 | 3.4 | 30 | 26 |
| Polyester (Terylene 45/24) | 5.3 | 120 | 15 | 5.3 | 15 | 120 |
| Nylon 6 (Grilon 30/7) | 5.4 | 19 | 31 | 4.7 | 19 | 26 |
| Polypropylene (Ulstron) | 7.1 | 80 | 17 | 7.4 | 80 | 17 |

**Table 2.20** Comparison of properties of protein fibres from cereals and animal sources (Reddy and Yang, 2007b).

| | Breaking tenacity (g/den) | | Breaking elongation (%) | | |
|---|---|---|---|---|---|
| **Protein source** | **Dry** | **Wet** | **Dry** | **Wet** | **Moisture regain (%)** |
| **Cereal proteins** | | | | | |
| Soyproteins | 0.5–0 0.6 | 0.2–0.3 | 30–40 | 60–70[c] | – |
| Wheat gluten | 0.6–1.1 | – | 25 | – | 18 |
| Zein | 0.8 | 0.6 | 30–40 | 60–70 | 13 |
| Peanut | 0.7 | 0.5 | 50 | 100 | 15 |
| **Animal proteins** | | | | | |
| Casein | 0.5–0.6 | 0.2–0.3 | 30–50 | 85–120 | 14 |
| Collagen[a] | 5.3 | 4.1 | 18 | 21 | 15 |
| Feather barbs | 0.4–1.9 | – | 7–16 | – | 10 |
| Wool | 1.2–1.7 | 0.7–1.5 | 25–35 | 25–50 | 16 |
| Silk | 3.1–3.6 | 1.9–2.5 | 15–25 | 27–33 | 11 |
| Pupal protein[b] | 1.5–1.6 | 0.7–0.8 | 18–22 | 25–29 | 12 |
| **Genetically engineered proteins** | | | | | |
| Spider silk | 6.4–9.6 | – | 46 | – | – |

[a]protein content not reported.
[b]about 20% protein content.
[c]data unavailable.

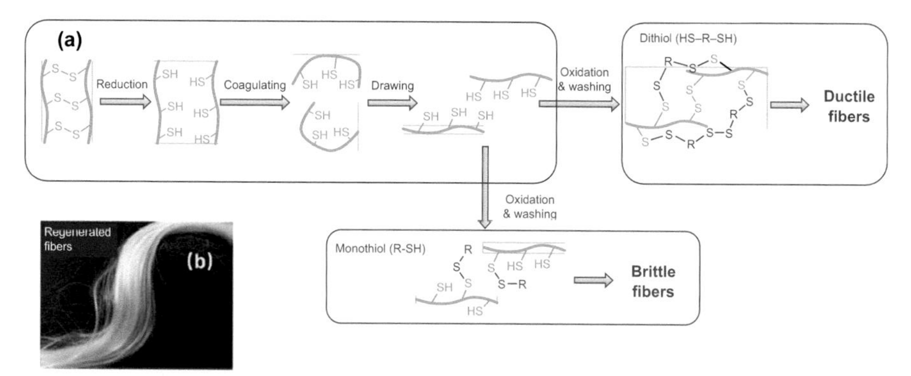


**Figure 2.14** Comparison of (a) cross-linking reactions between keratin filaments, DTT and cysteine; (b) picture of the fibres obtained (Mi et al., 2020).
Reproduced with permission from Elsevier.

**Table 2.21** Comparison of the properties of regenerated keratin fibres with other common fibres (Mu et al., 2020).

| Fibre source | Dry state | | | Wet state | | |
|---|---|---|---|---|---|---|
| | Strength (MPa) | Breaking strain (%) | Toughness (J/cm) | Strength (MPa) | Breaking strain (%) | Toughness (J/cm) |
| Feather barbs | 161 ± 27.5 | 9.7 ± 3.3 | 21 ± 2 | 1.27 ± 24.5 | 18 ± 3 | 31.5 ± 3 |
| Keratin fibres | 138 ± 30.5 | 11 ± 2.8 | 18.5 ± 2 | 81 ± 23 | 25 ± 3 | 28.7 ± 4 |
| Wool | 173 ± 23 | 36 ± 5 | 32 ± 5 | 140 ± 22 | 46 ± 5 | 36 ± 4 |
| Cotton | 420 ± 46 | 6.2 ± 1.5 | 10.5 ± 5 | 472 ± 34 | 9 ± 1 | 24.6 ± 6 |
| Linen | 700 ± 45 | 3.1 ± 0.4 | 5 ± 1 | 800 ± 40 | 5 ± 1 | 7 ± 1 |
| Viscose | 276 ± 20 | 21 ± 5.2 | 24 ± 6 | 120 ± 20 | 25 ± 4 | 21.1 ± 4.3 |

Reproduced with permission from Royal Society of Chemistry.

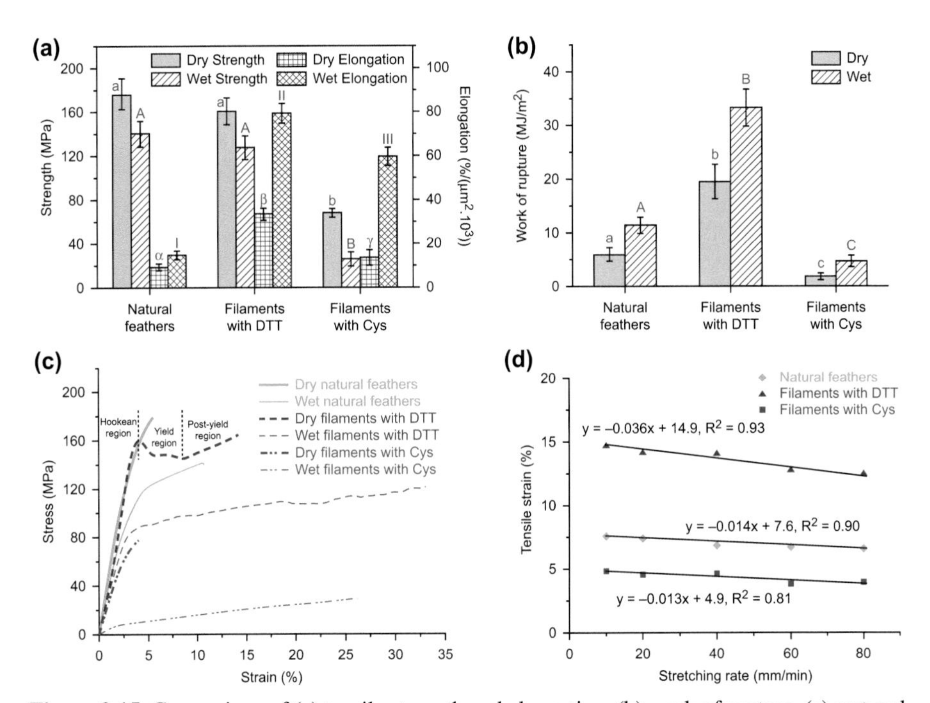


**Figure 2.15** Comparison of (a) tensile strength and elongation, (b) work of rupture, (c) wet and dry strength, (d) relationship between strain and stretching rate (draw ratio) during fibre production (Mi et al., 2020).
Reproduced with permission from Elsevier.

## 2.5 Future perspectives

Increasing population, rapidly changing consumer perceptions, affordability and spending potential will inevitably lead to higher demand for fibres and textiles. The last 2 decades have seen nearly exponential growth in the market for synthetic fibres, particularly polyester, whereas the share of natural fibres has, on an average, remained stagnant. Increasing costs for cultivation, decreasing availability of natural resources and the impacts of climate change have put the natural fibre industry in distress. It is quite impossible to allocate or increase the natural resources exclusively for fibre crops, across the globe.

In this scenario, it is inevitable to look into alternatives for natural fibres. Agricultural residues are generated on a large scale at low cost and contain cellulose, hemicellulose and lignin as major constituents. Hence, these agricultural residues can be considered as an inexpensive resource for the production of natural cellulosic and protein fibres. In tandem, it is also necessary to ensure that the production and processing of conventional natural fibres increases and the cost of production decreases. In this regard, the use of green and sustainable processing and increasing efficiency of processing and decreasing the use of natural resources will be critical. Concepts such as circular economy and zero agricultural wastes that are gaining rapid attention should be made part of the textile industry to ensure sustainability in the long term.

## 2.6 Conclusion

Production and processing of natural fibres and textiles across the globe will face considerable challenges due to the limited natural resources, changing climatic conditions and socio-economic factors. Data from the past 2 decades show that several promising technologies such as adopting genetically modified cotton, reducing water and pesticide requirements to grow cotton, increasing the production of lignocellulosic fibres such as hemp and flax have met with limited success. Not only cellulosic fibres but also the production of protein fibres such as wool and silk are facing considerable constraints. In this scenario, it is imperative to look for alternatives for the long-term sustainability of the fibre and textile industry. Adopting sustainable agricultural practices, green and clean processing and complete utilisation of the byproducts and co-products obtained during fibre and textile processing offer hope towards developing a sustainable textile industry. Agricultural residues and biomass have to be exploited to develop new cellulosic and protein fibres. Researchers have demonstrated the feasibility of using the agricultural residues for fibres and textiles. Such efforts have to be increased and technologies developed and adopted on a larger scale and eventually commercialised.

## Acknowledgements

Narendra Reddy thanks the Center for Incubation Innovation Research and Consultancy for their support to complete this work.

# References


Ahmad, H.S., Imran, M., Ahmad, F., Shah, R., Muhammad Ikram, R., Muhammad Rafique, H., Iqbal, Z., Alsahli, A.A., Alyemeni, M.N., Ali, S., 2021. Improving water use efficiency through reduced irrigation for sustainable cotton production. Sustainability 1 (7), 4044.

Astudillo, M.F., Thalwitz, G., Vollrath, F., 2015. Life cycle assessment of silk production—a case study from India. In: Handbook of Life Cycle Assessment (LCA) of Textiles and Clothing. Woodhead Publishing, pp. 255–274.

Awan, S.A., Ashfaq, M., Naqvi, S.A.A., Hassan, S., Kamran, M.A., Imran, A., Makhdum, A.H., 2015. Profitability analysis of sustainable cotton production: a case study of cottonwheat farming system in Bahawalpur District of Punjab. Bulgarian Journal of Agricultural Science 21 (2), 251–256.

Barcelos, S.M.B.D., Salvador, R., Barros, M.V., de Francisco, A.C., Guedes, G., 2021. Circularity of Brazilian silk: promoting a circular bioeconomy in the production of silk cocoons. Journal of Environmental Management 296, 113373.

Biswas, M.K., Ridwan-Ul-Risty, K.M., Datta, A., 2019. A proposal of sustainable and integrated plant for jute fiber extraction in an eco-friendly manner. International Journal of Scientific Engineering and Research 10 (1), 801–809.

Catarino, R., Ceddia, G., Areal, F.J., Park, J., 2015. The impact of secondary pests on Bacillus thuringiensis (Bt) crops. Plant Biotechnology Journal 13 (5), 601–612.

Chowdhury, P.K., Setua, G.C., Ghosh, A., Kar, R., Maity, S.K., 2013. Sustainable quality leaf production in S 1635 mulberry (morus alba) under irrigated condition through organic nutrient management. Indian Journal of Agricultural Sciences 83 (5), 529–534.

Dai, J., Dong, H., 2014. Intensive cotton farming technologies in China: achievements, challenges and countermeasures. Field Crops Research 155, 99–110.

De Prez, J., Van Vuure, A.W., Jan, I., Guido, A., Van de Voorde, I., 2018. Enzymatic treatment of flax for use in composites. Biotechnology Reports 20, e00294.

Delate, K., Heller, B., Shade, J., 2021. Organic cotton production may alleviate the environmental impacts of intensive conventional cotton production. Renewable Agriculture and Food Systems 36 (4), 405–412.

Dong, Z., Hou, X., Sun, F., Zhang, L., Yang, Y., 2014. Textile grade long natural cellulose fibers from bark of cotton stalks using steam explosion as a pretreatment. Cellulose 21 (5), 3851–3860.

Dong, Z., Hou, X., Haigler, I., Yang, Y., 2016. Preparation and properties of cotton stalk bark fibers and their cotton blended yarns and fabrics. Journal of Cleaner Production 139, 267–276.

Dreyer, J., Müssig, J., Koschke, N., Ibenthal, W.-D., Harig, H., 2002. Comparison of enzymatically separated hemp and nettle fibre to chemically separated and steam exploded hemp fibre. Journal of Industrial Hemp 7 (1), 43–59.

Duan, W.J., Zou, D.S., Luo, J.X., 2003. The soil and water conservation efficiency of eulaliopsis binata in the deserted sloping field of purple soil in South China. Journal of Hunan Agricultural University 29, 204–206.

Duan, S., Xu, B., Cheng, L., Feng, X., Qi, Y., Zheng, K., Gao, M., Liu, Z., Liu, C., Peng, Y., 2021. Bacterial strain for bast fiber crops degumming and its bio-degumming technique. Bioprocess and Biosystems Engineering 1–10.

Duan, W.J., Zou, D.S., 2006. The amelioration of microenvironment of purple soil in south China by planting eulaliopsis binata. Ecology and Environment 15, 124–128.

Ertek, A., Kanber, R., 2003. Effects of different drip irrigation programs on the boll number and shedding percentage and yield of cotton. Agricultural Water Management 60 (1), 1–11.

Ghorai, A.K., Chakraborty, A.K., 2020. Sustainable in-situ jute retting technology in low volume water using native microbial culture to improve fibre quality and retting waste management. International Journal of Current Microbiology and Applied Sciences 9 (11), 1080–1099.

Giacomin, A.M., Garcia, J.B., Zonatti, W.F., Silva-Santos, M.C., Laktim, M.C., Baruque-Ramos, J., 2017. Silk industry and carbon footprint mitigation. IOP Conference Series: Materials Science and Engineering 254 (19), 192008. IOP Publishing.

Gowane, G.R., Gadekar, Y.P., Prakash, V., Kadam, V., Chopra, A., Prince, L.L.L., 2017. Climate change impact on sheep production: growth, milk, wool, and meat. In: Sheep Production Adapting to Climate Change. Springer, Singapore, pp. 31–69.

Han, P., Song, G.J., Xu, S.T., Sun, G.B., Yuan, F., Han, Y.H., 2008. Properties of new natural fibers: Eulaliopsis binata fibers. Journal of Qingdao University (Natural Science Edition) 23, 44–47.

Hansen, E.G., Schaltegger, S., 2016. Mainstreaming of sustainable cotton in the German clothing industry. Sustainable Fibres for Fashion Industry 1, 39–58.

Hedayati, M., Brock, P.M., Nachimuthu, G., Schwenke, G., 2019. Farm-level strategies to reduce the life cycle greenhouse gas emissions of cotton production: an Australian perspective. Journal of Cleaner Production 212, 974–985.

Heller, K., Baraniecki, P., Praczyk, M., 2012. Fibre flax cultivation in sustainable agriculture. In: Handbook of Natural Fibres. Woodhead Publishing, pp. 508–531.

Huang, Y., Zou, D.S., Wang, H., Yu, Y.L., Luo, J.X., 2003. Ecological benefit of eulaliopsis binata grown in slope wasteland. Journal of Agro-Environmental Science 22, 217–220.

Huang, Y., Zou, D.S., Wang, H., 2004. The benefit of soil and water conservation of eulaliopsis binata. Chinese Journal of Eco-Agriculture 12, 152–154.

Huang, H., Qi, T., Lin, G., Yu, C., Wang, H., Li, Z., 2021. High-efficiency and recyclable ramie cellulose fiber degumming enabled by deep eutectic solvent. Industrial Crops and Products 171, 113879.

Joseph, M., Paul, R., 2007. Genetic engineering of novel qualities in fibers. Melliand International 13 (1), 20.

Karthik, T., Rathinamoorthy, R., 2017. Sustainable silk production. In: Sustainable Fibres and Textiles. Woodhead Publishing, pp. 135–170.

Kumar, M., Bera, A., Gotyal, B.S., Naik, M.R., Kumar, S., 2014. Technologies for sustainable jute fibre production. Popular kheti 2.

Kumar, A., Lesile, L., 2017. Prince, and seiko Jose. Sustainable wool production in India. In: Sustainable Fibres and Textiles. Woodhead Publishing, pp. 87–115.

Kranthi, K.R., Stone, G.D., 2020. Long-term impacts of Bt cotton in India. Nature Plants 6 (3), 188–196.

La Rosa, A.D., Grammatikos, S.A., 2019. Comparative life cycle assessment of cotton and other natural fibers for textile applications. Fibers 7 (12), 101.

Lebedytė, M., Sun, D., 2021. A review: can waste wool keratin be regenerated as a novel textile fibre via the reduction method? The Journal of the Textile Institute 1–17.

Li, Y., Reddy, N., Yang, Y., 2008. A new crosslinked protein fiber from gliadin and the effect of crosslinking parameters on its mechanical properties and water stability. Polymer International 57 (10), 1174–1181.

Li, C., Liu, S., Song, Y., Nie, K., Ben, H., Zhang, Y., Han, G., Jiang, W., 2020. A facile and eco-friendly method to extract Apocynum venetum fibers using microwave-assisted ultrasonic degumming. Industrial Crops and Products 151, 112443.

Luna, J.K., Dowd-Uribe, B., 2020. Knowledge politics and the Bt cotton success narrative in burkina faso. World Development 136, 105127.

Lv, W., Xia, Z., Song, Y., Wang, P., Liu, S., Zhang, Y., Ben, H., Han, G., Jiang, W., 2021. Using microwave assisted organic acid treatment to separate cellulose fiber and lignin from kenaf bast. Industrial Crops and Products 171, 113934.

Ma, B., Qiao, X., Hou, X., Yang, Y., 2016. Pure keratin membrane and fibers from chicken feather. International Journal of Biological Macromolecules 89, 614–621.

Madasamy, B., Balasubramaniam, P., Dutta, R., 2020. Microclimate-based pest and disease management through a forewarning system for sustainable cotton production. Agriculture 10 (12), 641.

Mahesh, R., Anil, P., Debashish, C., Sivaprasad, V., 2021. Improved mulberry productivity and resource efficiency through low-cost drip fertigation. Archives of Agronomy and Soil Science 1–15.

Mandal, A., Sarkar, B., Owens, G., Thakur, J.K., Manna, M.C., Khan Niazi, N., Jayaraman, S., Patra, A.K., 2020. Impact of genetically modified crops on rhizosphere microorganisms and processes: a review focusing on Bt cotton. Applied Soil Ecology 148, 103492.

McHughen, A., Holm, F.A., 1995. Transgenic flax with environmentally and agronomically sustainable attributes. Transgenic Research 4 (1), 3–11.

Mi, X., Li, W., Xu, H., Mu, B., Chang, Y., Yang, Y., 2020. Transferring feather wastes to ductile keratin filaments towards a sustainable poultry industry. Waste Management 115, 65–73.

Mu, B., Hassan, F., Yang, Y., 2020. Controlled assembly of secondary keratin structures for continuous and scalable production of tough fibers from chicken feathers. Green Chemistry 22 (5), 1726–1734.

Navada, A.P., Guna, V., Paul, R., Belino, N., Tavares, M., Reddy, N., 2022. Residues from cajanus cajan plant provide natural cellulose fibers similar to flax. Journal of Natural Fibers 19 (16), 14539.

Nie, K., Liu, B., Zhao, T., Wang, H., Song, Y., Ben, H., Arthur, J.R., Han, G., Jiang, W., 2020. A facile degumming method of kenaf fibers using deep eutectic solution. Journal of Natural Fibers 1–11.

Paul, R., Joseph, M., 2003. Genetic engineering for cotton development. Asian Textile Journal 12 (5), 47.

Paul, R., Surribas, A., Brouta, M., Alaman, M., Esteve, H., 2009. Hemp: an ecological textile alternative. Revista de Quimica Textil 195, 30.

Qu, Y., Yin, W., Zhang, R.Y., Zhao, S., Liu, L., Yu, J., 2020. Isolation and characterization of cellulosic fibers from ramie using organosolv degumming process. Cellulose 27 (3), 1225–1237.

Radhakrishnan, S., 2017. Sustainable cotton production. In: Sustainable Fibres and Textiles. Woodhead Publishing, pp. 21–67.

Reddy, N., Yang, Y., Mc Alister III, D.D., 2006. Processability and properties of yarns produced from cornhusk fibres and their blends with other fibres. IJFTR 31 (4), 537–542.

Reddy, N., Yang, Y., 2007a. Natural cellulose fibers from switchgrass with tensile properties similar to cotton and linen. Biotechnology and Bioengineering 97 (5), 1021–1027.

Reddy, N., Yang, Y., 2007b. Novel protein fibers from wheat gluten. Biomacromolecules 8 (2), 638–643.

Reddy, N., Tan, Y., Li, Y., Yang, Y., 2008. Effect of glutaraldehyde crosslinking conditions on the strength and water stability of wheat gluten fibers. Macromolecular Materials and Engineering 293 (7), 614–620.

Reddy, N., Yang, Y., 2008. Self-crosslinked gliadin fibers with high strength and water stability for potential medical applications. Journal of Materials Science: Materials in Medicine 19 (5), 2055–2061.

Reddy, N., Li, Y., Yang, Y., 2009. Alkali-catalyzed low temperature wet crosslinking of plant proteins using carboxylic acids. Biotechnology Progress 25 (1), 139–146.

Reddy, R., Jiang, Q., Aramwit, P., Reddy, N., 2021. Litter to leaf: the unexplored potential of silk byproducts. Trends in Biotechnology 39 (7), 706–718.

Riddlestone, S., Stott, E., Blackburn, K., Brighton, J., 2006. A technical and economic feasibility study of green decortication of hemp fibre for textile uses. Journal of Industrial Hemp 11 (2), 25–55.

Ruan, P., Raghavan, V., Du, J., Gariepy, Y., Lyew, D., Yang, H., 2020. Effect of radio frequency pretreatment on enzymatic retting of flax stems and resulting fibers properties. Industrial Crops and Products 146, 112204.

Schumacher, A.G.D., Pequito, S., Pazour, J., 2020. Industrial hemp fiber: a sustainable and economical alternative to cotton. Journal of Cleaner Production 268, 122180.

Sinclair, R. (Ed.), 2014. Textiles and Fashion: Materials, Design and Technology. Elsevier.

Song, Y., Han, G., Jiang, W., 2017. Comparison of the performance of kenaf fiber using different reagents presoak combined with steam explosion treatment. The Journal of the Textile Institute 108 (10), 1762–1767.

Tourangeau, W., Sherren, K., 2020. Leverage points for sustainable wool production in the falkland Islands. Journal of Rural Studies 74, 22–33.

Tian, W.-hui, Yi, X.-long, Liu, S.-shan, Zhou, C., Wang, A.-ying, 2020. Effect of transgenic cotton continuous cropping on soil bacterial community. Annals of Microbiology 70 (1), 1–10.

Van der Werf, H.M.G., 2004. Life cycle analysis of field production of fibre hemp, the effect of production practices on environmental impacts. Euphytica 140 (1), 13–23.

Van der Werf, H.M.G., Lea, T., 2008. The environmental impacts of the production of hemp and flax textile yarn. Industrial Crops and Products 27 (1), 1–10.

Wiedemann, S.G., Simmons, A., Watson, K.J.L., Biggs, L., 2019. Effect of methodological choice on the estimated impacts of wool production and the significance for LCA-based rating systems. International Journal of Life Cycle Assessment 24 (5), 848–855.

Xiang, M., Bai, Y., Li, Y., Wei, S., Tong, S., Wang, H., Li, P., Yu, T., Yu, L., 2020. An eco-friendly degumming process of flax roving without acid pickling and NaClO2-bleaching. Process Biochemistry 93, 77–84.

Yuan, B.Z., Han, G.T., Pan, Y., Zhang, Y.M., 2013. The effect of steam explosion treatment on the separation of lotus fiber. Advanced Materials Research 750, 2307–2312. Trans Tech Publications Ltd.

Zampori, L., Dotelli, G., Vernelli, V., 2013. Life cycle assessment of hemp cultivation and use of hemp-based thermal insulator materials in buildings. Environmental science and technology 47 (13), 7413–7420.

Zhao, J.H., Ho, P., Azadi, H., 2011. Benefits of Bt cotton counterbalanced by secondary pests? perceptions of ecological change in China. Environmental Monitoring and Assessment 173 (1), 985–994.

Zheng, L., Du, Y., Zhang, J., 2001. Degumming of ramie fibers by alkalophilic bacteria and their polysaccharide-degrading enzymes. Bioresource Technology 78 (1), 89–94.

Zimniewska, M., 2022. Hemp fibre properties and processing target textile: a review. Materials 15 (5), 1901.

Zou, D., Chen, X., Zou, D., 2013. Sequencing, de novo assembly, annotation and SSR and SNP detection of sabaigrass (eulaliopsis binata) transcriptome. Genomics 102, 57–62.

Zulfiqar, F., Thapa, G.B., 2018. Determinants and intensity of adoption of "better cotton" as an innovative cleaner production alternative. Journal of Cleaner Production 172, 3468–3478.

Zulfiquar, S., Yasin, M.A., Bakhsh, K., Ali, R., Munir, S., 2019. Environmental and economic impacts of better cotton: a panel data analysis. Environmental Science and Pollution Research 26 (18), 18113–18123.
Zulfiqar, F., Datta, A., Tsusaka, T.W., Yaseen, M., 2021. Micro-level quantification of determinants of eco-innovation adoption: an assessment of sustainable practices for cotton production in Pakistan. Sustainable Production and Consumption 28.

## Sources of further information

Andrade, S., Rebecca, Torres, D., Ribeiro, F.R., Chiari-Andréo, B.G., Oshiro Junior, J.A., Iglesias, M., 2017. Sustainable cotton dyeing in nonaqueous medium applying protic ionic liquids. ACS Sustainable Chemistry and Engineering 5 (10), 8756–8765.
Bianchini, R., Cevasco, G., Chiappe, C., Pomelli, C.S., Douton, M.J.R., 2015. Ionic liquids can significantly improve textile dyeing: an innovative application assuring economic and environmental benefits. ACS Sustainable Chemistry and Engineering 3 (9), 2303–2308.
Reddy, N., Yang, Y., 2005. Properties and potential applications of natural cellulose fibers from cornhusks. Green Chemistry 7 (4), 190–195.
Reddy, N., Yang, Y., 2015. Innovative Biofibers from Renewable Resources. Springer, Berlin.
Ricciardi, L., Chiarelli, D.D., Karatas, S., Rulli, M.C., 2021. Water resources constraints in achieving silk production self-sufficiency in India. Advances in Water Resources 154, 103962.

# Innovations in man-made and synthetic fibres

*Takeshi Kikutani*[1], *Sascha Schriever*[2] *and Sea-Hyun Lee*[2]
[1]School of Materials and Chemical Technology, Tokyo Institute of Technology, Yokohama, Kanagawa, Japan; [2]Institut für Textiltechnik der RWTH Aachen University, Aachen, Germany

## 3.1 Introduction

With the advancement of processing and manufacturing technologies, the time span for breakthroughs and the choice of materials in the textile industry has become shorter. Modern textiles are considered to be barely 70 years old and they are strongly connected to the start of commercial polymer production using petrochemicals. The development of textiles from earlier polymers depended mainly on polymer synthesis, which was important to the properties of fibre and new fibre spinning methods.

With the rise of commercial polymers, the shift from natural fibres to man-made and synthetic fibres took less than 100 years. Whereas the world produces more fibres than before from petroleum origin, the old ways of using bio-based materials are making a comeback. Spinning processes that are distinguished in the production of man-made fibres are solution spinning and melt spinning. Each process works according to the following principle. A spinning pump conveys the liquid/viscous mass through the spinneret and forces the material out of a nozzle. Through reactions such as coagulation or solidification, the final fibres are formed and wound to further processing (Gries et al., 2015).

Fibres created not for aesthetics but for a certain functionality are defined as technical fibres. There can be various functionalities depending on the end uses, such as agriculture, construction, medical, transportation and protection. High mechanical properties and high thermal resistance are typical functionalities for various applications. For synthetic fibres, functionalities can be realised through the selection of polymers and by applying proper fibre formation conditions.

Fibres can be allocated into two categories, textile and technical, depending on their fineness. Textile fibres such as smooth yarn and texture yarn can vary between 15 and 150 dtex or up to 300 dtex. Technical fibres used in tyre cord, industrial yarns or high-strength fibres typically vary between 400 and 2000 dtex. Depending on the linear mass density of fibres, a differentiation can be made between the categories. Translating this information to a standard polymer such as polyamide or polyester fibre, both could be categorised as textile and technical fibres if the titres are within the respective ranges.

**Sustainable Innovations in the Textile Industry. https://doi.org/10.1016/B978-0-323-90392-9.00021-5**

Technical fibres normally exhibit a much higher tensile modulus, tensile strength and thermal resistance compared with textile fibres even though the same type of polymer is used. From the viewpoint of environmental friendliness, high mechanical properties lead to a reduction in the amount of fibre for certain applications. For example, with the use of high-strength fibres, the weight of tyres can be reduced by decreasing the amount of fibre to strengthen the tyre (i.e. tyre cords). Eventually, along with the reduction of material consumption, energy consumption can be reduced through an increase in fuel efficiency.

Achieving the sustainability of man-made and synthetic fibres is important, so there is need for a circular economy based on sustainable raw materials and energy. Because the textile industry depends on carbon sources, the route is to increase the supply of renewable carbon sources and stop the supply of fossil carbon in the future.

## 3.2 Historical aspects and current scenario

The history of man-made fibres begins as early as 1665, when Robert Hooke had the idea of producing artificial threads from a viscous mass. From then until the actual realisation of the idea, more than two centuries passed with many failures in between (Shor, 1944). Known as the father of artificial silk, in 1857 Mathias Eduard Schweizer discovered that cellulose is soluble in ammoniacal copper salt solution.

The actual breakthrough came with Count Hilaire de Chardonnet, between 1878 and 1884, through the production of the first natural man-made fibre from dinitrocellulose, through the derivatisation of cellulose. This type of fibre was called artificial silk or nitro artificial silk, and started commercial production in 1890 (Luft, 1925). Because nitrocellulose is explosive and highly flammable, research sought alternatives. In 1857 Matthias Eduard Schweizer discovered that cellulose could be dissolved in copper oxide and ammonia. Schweizer was responsible for the principle of dissolving cellulose in cuoxam, but the regeneration of a fibre through coagulation had yet to be discovered. From the initial ideas of Schweizer, Max Fremerey and Johann Urban joined to produce the first cupro fibre in 1892. The discovery that cellulose could be dissolved in acetic anhydride, creating cellulose acetate, was made in 1865 by Paul Schützenberger. However, the actual use of cellulose acetate for continuous fibres did not begin before 1913.

With an understanding of derivatisation and regeneration, production of the well-established viscose fibres began within the same time frame as artificial silk. The formation of viscose fibres from cellulose xanthate, which is soluble in a sodium hydroxide solution, was developed by Charles F. Cross, Edward J. Bevan and Clayton Beadle in 1892 (Cross et al., 1893). Realising that viscose needs an ageing process to spin and wind it for use as a textile, around 1900 C.H. Stearn and C.F. Topham found the missing piece to start commercial production. Commercialisation started with the formation of the Viscose Syndicate Ltd in 1894. Within 41 years, viscose achieved an 88% market share of all man-made fibres produced in 1935.

By contrast, synthetic fibres such as polyamide 6,6 or polyamide 6, polyacrylonitrile (PAN) and polyester are completely independent of natural raw material cellulose because they originate from petroleum. Polyamides were created in the 1930s by Carothers. In 1935 Carothers successfully synthesised polyamide 6,6 with a combination of hexamethylene diamine and adipic acid, which gave the polymer the mechanical properties to be cold drawn and thus viable for producing fibres. The first implementation came from DuPont in 1938, with small-scale production. In a similar time frame, Paul Schlack worked on polyamides and successfully created polyamide 6 via caprolactam in 1938 (Viswanathan, 2010).

In 1941, Whinfield and Dickson produced useable polyester filaments at the Calico Printers Association, Ltd (CPA) for the first time. Three years later, Imperial Chemical Industries, Ltd (ICI) started laboratory-scale production of fibres and reached an agreement with CPA on the patent rights. The patent rights for the United States were acquired from CPA in 1946 by E.I. DuPont de Nemours & Co. In 1953 DuPont began producing poly(ethylene terephthalate) fibre under the trade name Dacron. In Germany polyester production started in the 1950s under the names Trevira and Diolen. After the patent expired in 1966, large-scale production of polyester began throughout the world (Jaffe et al., 2020).

Acrylonitrile was considered technically insignificant owing to its chemical and mechanical properties, which made working with the polymer difficult. In 1930, H. Fikentscher suggested using acrylonitrile for fibre production. In 1939, IG Farben produced the first filaments from PAN under the brand name PAN-Fasern. The commercialisation of PAN fibres began in 1950 through DuPont. The term man-made fibres can be considered a general description of fibres made by humans and thus differs from natural fibres such as wool and cotton. Man-made fibres can broadly be split into three categories, as depicted in Fig. 3.1.

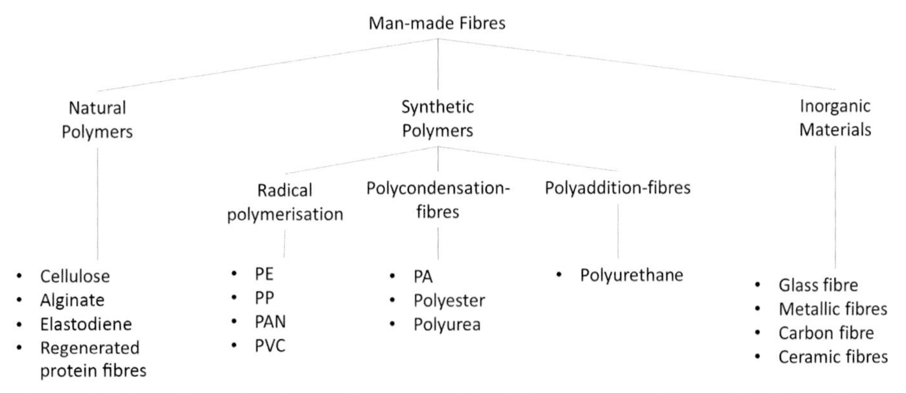


**Figure 3.1** Overview of various polymers considered for man-made fibres. *PA, Polyamide; PAN*, polyacrylonitrile; *PE*, polyethylene; *PP*, Polypropylene; *PVC*, Polyvinyl chloride. From the Institut für Textiltechnik der RWTH Aachen.

With the rise in natural man-made fibres, a new competitor for natural fibres came into place. To compete, natural man-made fibres had to achieve results comparable to or better than the technical performance and wear comfort set as a benchmark by cotton and wool. From a commercial standpoint, cellulose has seen the most advances with man-made cellulose fibres (MMCF), holding the largest market share in the category of natural man-made fibres. Cellulose is a versatile polymer that fulfils the needs of the textile, pulp and paper and chemical sectors. Fig. 3.2 shows various methods in which cellulose can be used.

The market share of fibres including those that are natural and man-made can be separated into MMCF, which has a total market share of ~6.5%, whereas the share of man-made synthetic fibres is ~62%. The share of synthetic fibres is divided into 52% polyethylene terephthalate (PET), 5% PA, 5.2% polyethylene (PE), 2.7% PP, and 1.7% PAN. The total share of man-made fibres is continuously growing whereas natural fibres are mostly stagnating owing to the limitation in their areas of cultivation (Textile Exchange, 2021). With the current production of raw material base of man-made fibres, 91% of all man-made fibres are crude oil-based. Combined with a linear economy, waste disposal through landfilling and incineration is composed of end-of-life scenarios leading to tremendous sustainability issues. In the case of the clothing sector, less than 1% of closed-loop recycling is performed (Morlet et al., 2017).

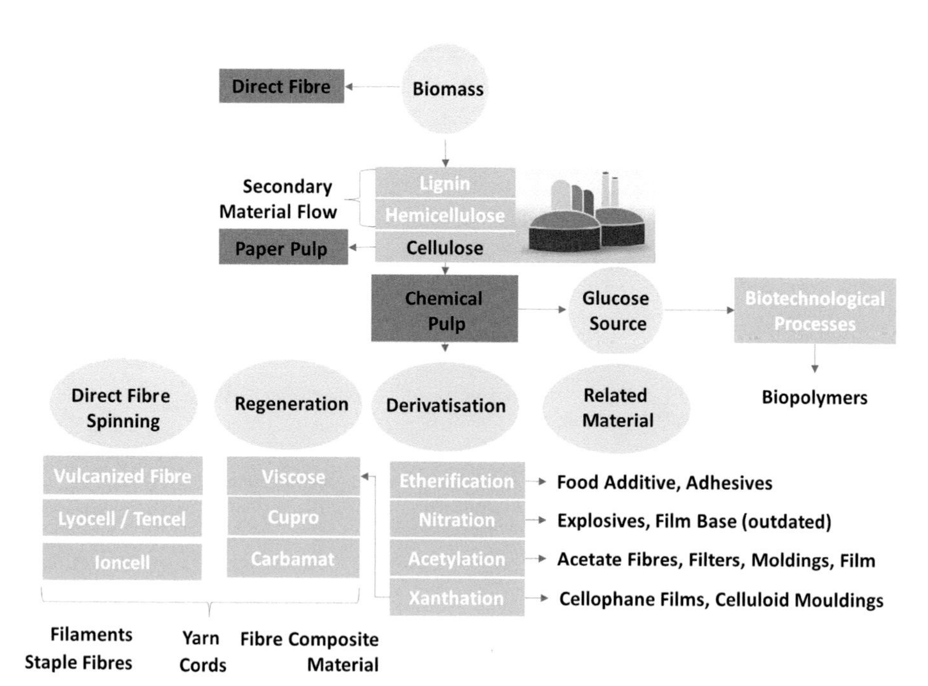


**Figure 3.2** Technologies and applications using cellulose.
From the Institut für Textiltechnik der RWTH Aachen.

## 3.3 Impacts of fibres on health and environment

Impacts of man-made and synthetic fibres on health and the environment can occur during the production and processing phase of fibres and in the use phase and their end of life. The impact of fibres on the health and the environment is structured along the phases of the product's life cycle of fibres and products derived from them.

- Production of polymers (production phase)
- Production of fibres (production phase)
- Processing of fibres (processing phase)
- Use of textiles containing fibres (use phase)
- End of life of textiles containing fibres (end-of-life phase)

Fig. 3.3 shows the stages of man-made fibre production and processing allocated to phases of the product's life cycle. In general, the impact on health and the environment is determined by the raw material base used to produce polymers and auxiliary materials, and substances used to produce and process fibres, the energy sources, the amount of freshwater used, the amount of wastewater generated and its type of contamination, and the release of microfibres. Especially for raw materials and the sources of energy used in the different phases of the product's life cycle, the release of fossil carbon has a high impact on health and the environment. Possible impactful influences on health and the environment during the product's life cycle of fibres are:

- Raw material (fossil-based)
- Energy consumption/source (fossil-based)

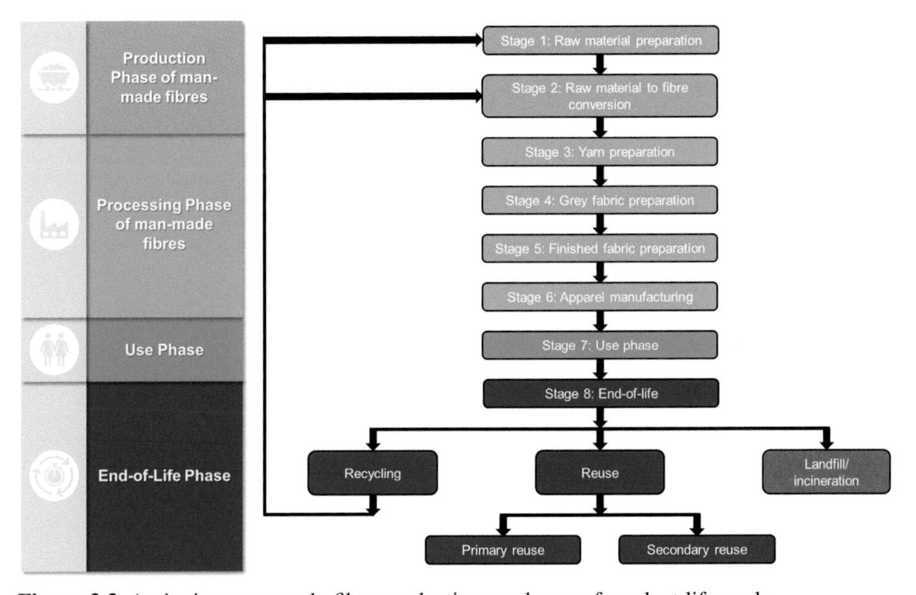


**Figure 3.3** Assigning man-made fibre production to phases of product life cycle. Adapted from Muthu (2020).

- Hazardous/toxic substances
- Freshwater use
- Contaminated wastewater
- Microfibre release

These impactful influences on health and the environment can be assigned to all phases in the product's life cycle of fibres. This does not indicate that the impact of a certain influence is equal for all phases of the product's life cycle. The impact of a particular influence can vary widely with regard to the processes and technologies used in the different phases as well as the type of product and the materials used. The impact can be quantified with different methods such as the life cycle assessment. One of the most discussed topics in public and politics is the microfibre release of textiles. An overview of microfibre release during the product's life cycle of textiles is shown in Fig. 3.4.

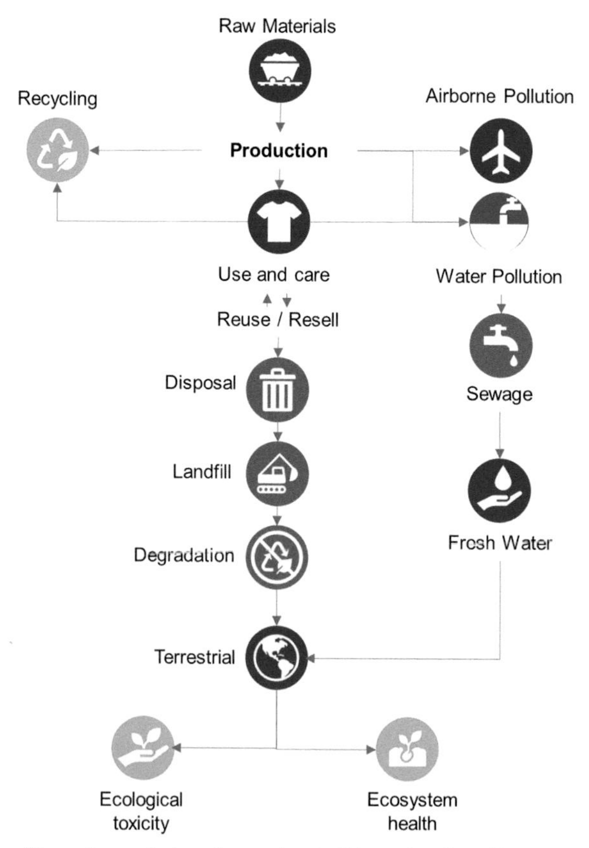


**Figure 3.4** Microfibre release during the product's life cycle of textiles. Adapted from Henry et al. (2019).

### 3.3.1 Production phase

The production of fibres can be divided into two phases:

- Production of the polymer
- Production of fibres

Polymers for fibres can be divided into two groups: synthetic and natural. The synthetic polymer is a polymer obtained by the polymerisation of chemical molecules (monomers) such as PET (Sandin et al., 2019), in which monomers are also chemically synthesised. On the other hand, natural polymers are produced by cells of living organisms, such as cellulose and other polysaccharides. Natural polymers are based on biomass, which can be defined as a renewable carbon source. These polymers are mostly extracted from bio-based resources in biorefineries.

Most synthetic polymers are based on crude oil, a fossil carbon source that is not renewable. Examples of crude oil-based synthetic bulk polymers are PET, PA, PP, PE, and PAN. Because of the steadily increasing demand for sustainable synthetic polymers, they can also be produced from biomass (renewable carbon), such as polylactide (PLA) and poly(3-hydroxybutyrate-co-3-hydroxyhexanoate) (PHA). Here, monomers are mostly synthesised via biotechnological fermentation. For example, PLA is mostly synthesised via ring opening polymerisation of lactide, in which the lactide is synthesised from starch, for instance. On the other hand, PET, PA, PE, PP and so can be synthesised from bio-based monomers (drop-in solutions).

Regarding the impact on health and the environment, a renewable carbon source is most likely preferred. However, the use of fertilisers to generate biomass also needs to be considered, as does as the use of fresh water. In addition, more or less toxic or hazardous substances are used to extract natural polymers or synthesise polymers. To extract cellulose from wood and other bio-based feedstocks with a high cellulose content, the Kraft process is used. During the Kraft process mainly sodium hydroxide (NaOH) and sodium sulphide ($Na_2S$) are used to break down binding forces among cellulose, lignin and hemicellulose. Both chemicals are strongly alkaline, and $Na_2S$ can react with acids to form hydrogen sulphide ($H_2S$). $Na_2S$ and $H_2S$ can have toxic and hazardous impacts on health and the environment. For synthetic polymers, monomers as well as catalysts used for the monomer and/or polymer synthesis can be hazardous and toxic. An example of a hazardous and toxic monomer is acrylonitrile, which is used to synthesise PAN. The synthetic polymer PAN is used to produce acrylic fibres and carbon fibre precursors. Toxic and hazardous catalysts include antimony-based (Sb) ones and are conventionally used as condensation catalysts, such as to polymerise PET. These hazardous and toxic catalysts and monomers can have a negative impact on health and the environment if they are in contact with humans or animals and the environment at certain concentrations. Therefore, it is important for these substances to be removed after polymerisation. Different systems in place regulate the use of hazardous and toxic substances affecting health and the environment. One of these regulating systems is Registration, Evaluation, Authorisation and Restriction of Chemicals in the European Union (EU).

After the extraction of natural polymers and the synthesis of synthetic polymers, products need to undergo purification steps to get rid of nonreacted educts, byproducts and side products as well as auxiliaries such as solvents. For this, the use of fresh water is necessary in bigger amounts, and so contaminated wastewater is generated. This wastewater needs to be treated to recover auxiliaries and valuable substances, to feed them back into production. On the one hand, the wastewater could affect health and the environment if it is released owing to leaks in production plants or accidents. The release of wastewater should be prevented in all cases. On the other hand, energy for wastewater treatment should be obtained from renewable energy sources to reduce the impact on $CO_2$ emissions compared with energy from fossil resources. However, renewable energy should be used more intensively in all industrial processes to reduce $CO_2$ emissions to the atmosphere.

To produce fibres derived from natural and synthetic polymers (including bio-based and crude oil-based polymers) there are two general processes: melt spinning and solution spinning (Fig. 3.5). Melt spinning is used for thermoplastic polymers whereas solution spinning is used for nonthermoplastic or temperature-sensitive polymers. Table 3.1 lists commercial bulk and upcoming polymers according to the two spinning processes.

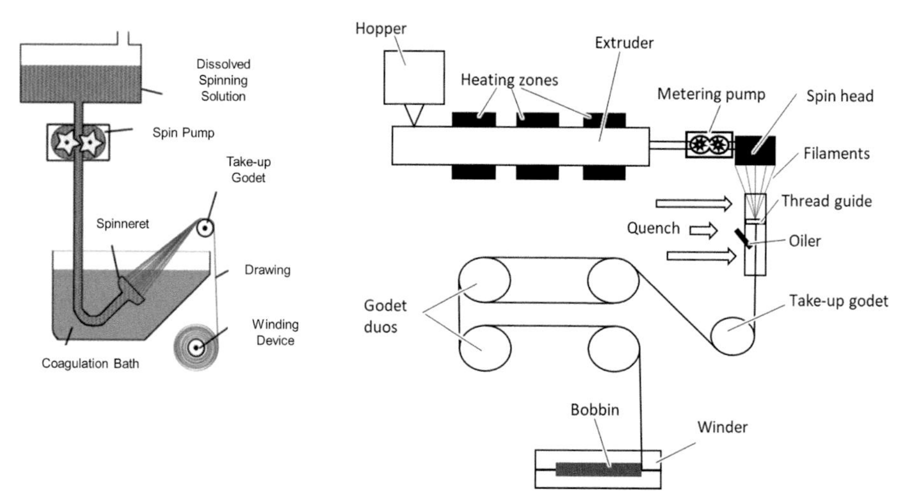


**Figure 3.5** (*Left*) Solution spinning (wet spinning); (*right*) melt spinning. From Gries et al. (2014).

**Table 3.1** Assignment of different polymers to spinning processes for fibre. *PUR*, polyurethane; *TPU*, thermoplastic polyurethane.

| Process | Commercial bulk polymers | Upcoming polymers |
|---|---|---|
| Melt spinning | Polyethylene terephthalate, PP, polyethylene, PA, PUR | Polylactide, TPU, poly(3-hydroxybutyrate-co-3-hydroxyhexanoate) |
| Solution spinning | Polyacrylonitrile, cellulose/ cellulose derivate, aramid (*p*-/*m*-), | Various polysaccharides (e.g. chitin/ chitosan/glucan/lignin blends), proteins (e.g. silk, collagen) |

Regarding the impact of melt and solution spinning on health and the environment, melt process has a lower impact than solution spinning. In melt spinning, the amount of auxiliaries needed is limited to additives used to adjust polymer properties and spin finishes to process textiles more easily. In solution spinning different types of auxiliaries are used in larger amounts, starting from preparation of the spinning solution, in which the polymer is dissolved in a suitable solvent. Moreover, in some processes additives are used to make the polymer dissolvable, such as in the viscose process, in which carbon disulphide ($CS_2$) is added to a mixture of cellulose and a strong base to form xanthate, which is later generated to cellulose fibres during coagulation. The coagulation bath, where formation of the fibre takes place, consists of a solvent/antisolvent mixture and additives, spinning auxiliaries, to adjust diffusion processes during the coagulation of the fibres. The antisolvent is mostly water.

Afterwards, the as-spun fibres need to be washed and should be solvent- and additive-free. These processes require huge amounts of fresh water and generate wastewater that needs to be treated to recover fresh water and additives and auxiliaries to be fed back into fibre production. Owing to the recovery of solvents via distillation, energy consumption is also high. The release of microfibres into environment during fibre production is limited compared with the processing and use phases. Microfibres may be released into the environment through ventilation systems of the production sides by passing air filters or through wastewater by passing through filters in wastewater treatment plants.

The amount of hazardous and toxic substances is mostly higher in solution spinning. Moreover, the toxicity of substances used is higher. For instance, $CS_2$ used in the viscose process is toxic and hazardous to the environment and humans. Organic solvents such as Dimethylacetamide (DMAC) and Dimethylformamide (DMF), used to produce PAN fibres, are toxic and hazardous to the environment and humans.

Melt spinning is less impactful to health and the environment compared with solution spinning. Melt spinning needs fewer additives and auxiliaries and almost no fresh water or toxic or hazardous substances. Energy consumption is also lower than for solution spinning. In both processes, microfibres may be released into the environment.

### 3.3.2 Processing phase

The processing of fibres includes:

- Yarn spinning
- Fabric production
- Finishing of fabric

For the yarn spinning and fabric production, mostly dry processes are used. Thus no fresh water or solvents are used and no wastewater is generated. However, several auxiliaries are used for the easier processing of fibres. These auxiliaries are natural or synthetic formulations such as waxes or silicone oils. The interaction between fibres and machine parts is positively influenced by these auxiliaries to prevent fibre breakage and reduce abrasion. In particular, the abrasion of fibres leads to the release of microfibres during yarn spinning and fabric production, in which microfibres may be released into the environment by passing through the filters of ventilation systems at the place of production.

The finishing of fabrics made from fibres includes dyeing and functionalisation. Both processes can also be performed at the staple fibre or yarn level. However, these processes are mostly wet processes in which huge amounts of fresh water, additives, and auxiliaries are used. Therefore the generated wastewater needs to be treated, and there is always the risk of a high impact on health and the environment owing to poor wastewater treatment and release into rivers, groundwater, oceans and so on. The release of microfibres through wastewater may also occur. As in most production processes energy is needed, and it is preferably generated from renewable energy sources.

### 3.3.3 Use phase

During the use phase of fibre-containing textiles, certain processes have a main influence on health and the environment:

- Use of textiles
- Care of textiles

Using and caring for textiles are successive processes in which the continuous shedding of microfibres occurs. Shedding can be categorised as dry or wet. Wet shedding occurs mainly during the washing of textiles, and dry shedding mostly while drying (e.g. in a tumbler) and while using textiles. The release of microfibres occurs when they are mechanically stressed, such as owing to friction. The amount and size of the released microfibres depend strongly on the polymer type and on the yarn and fabric construction. When textiles are washed, microfibres are released into wastewater treatment plants, bypassing the filter system of the washing machine. From there, they can be released into the environment. When textiles are dried, such as with a tumbler, microfibres may be released into the environment by passing through filters of the exhaust system. When textiles are used, microfibres are released directly into the environment.

Next to microfibre release during the washing of textiles, the use of huge amounts of fresh water and the generation of wastewater from them affect health and the environment. Furthermore, various chemical formulations are added during washing, such as detergents, softeners, oxy-products, and bleach. In addition, the release of dyes occurs, and the resulting wastewater needs to be treated.

### 3.3.4 End-of-life phase

Depending on the scenario, the end-of-life phase of textiles can also have a huge negative impact on health and the environment. A linear economy and a circular economy may be distinguished; the international textile industry is mostly a linear economy. However, the overall future goal is to come to a fully circular textile economy on the local and international levels. Therefore different regulations and action plans are in place in the EU and worldwide. The EU Waste Framework Directive sets basic waste management principles and the waste hierarchy, as in Fig. 3.6. Preventing waste is the preferred option and sending waste to landfill should be the last resort. With regard to a linear economy, when textile products and materials are not employed for material reuse, there are two scenarios:

- Recovery (e.g. energy recovery)
- Disposal (e.g. landfill)

In the case of energy recovery, energy is gained from end-of-life textiles, but $CO_2$ of fossil fuel origins is released into the atmosphere, with negative impacts. In the case of disposal or a landfill, the impact on health and the environment is tremendous. On the one hand, the release of microfibres or plastic into the soil and water systems occurs as a result of fragmentation by photomechanical processes. Biodegradable fibres might prevent environmental pollution caused by microfibres and plastic. On the other hand, additives, auxiliaries and other chemical substances can leach into soil and water systems. In this case, nontoxic and nonhazardous auxiliaries, additives, and other chemical formulations should be used to prevent ecotoxicity. With regard to a circular economy, in which textiles are fed back into a material cycle, there are three scenarios:

- Prevention (e.g. nonwaste products)
- Preparation for reuse (e.g. second-hand market)
- Recycling (e.g. upcycling)

For these three scenarios the 6R rules can be applied as a guideline for sustainable textile product life cycles: rethink, reduce, refuse, repair, reuse and recycle. When talking about recycling textiles, it is important that recycling technologies be economic, ecological and socially relevant. Otherwise, the industrial application of these technologies will be impossible in a sustainable way.

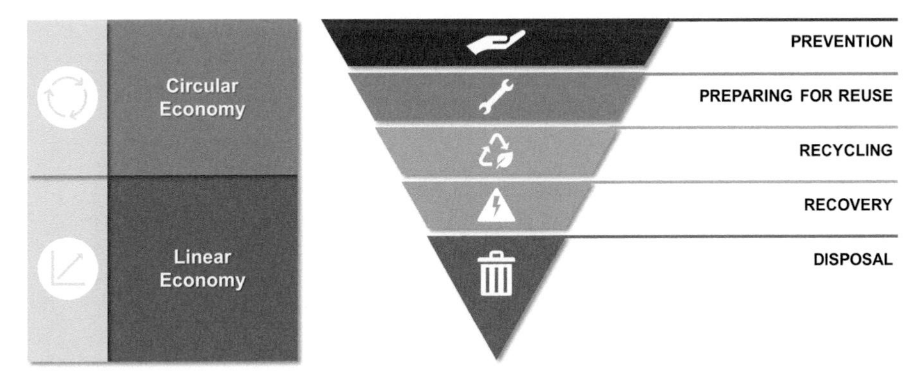


**Figure 3.6** Waste hierarchy proposed by European Union.
Adapted from European Union (2008).

## 3.4 Renewable carbon approach in fibre industry

To create a sustainable future, dynamics are changing towards approaches using renewable carbon and the coupling of various sectors, as depicted in a simplified overview in Fig. 3.7. The approach is categorised into $CO_2$, biomass and recycling, creating the pillars for a diverse platform on which future products can be built. The diversity of $CO_2$ is exhibited through the current landscape of industry players and joint ventures. Startups such as Sunfire GmbH can turn $CO_2$ into a gas mixture of CO and $H_2$ called syngas. This gas mixture can be used as building blocks to create a base for future products such as ethanol, methanol, ammonia, waxes, and olefins and polymers such as PE, PP, PVC, and even TPU (dos Santos and Alencar, 2019). With biomass, the possibilities are even greater because of the complex composition that is characteristic of each plant.

Wood has been used commonly throughout the decades as cellulose feedstock, but the expected rise in the need for cellulose poses a problem for the ability to fill the need with sustainable sourcing, so the current focus is slowly shifting towards alternatives such as agricultural residue, exotic plants and recycling processes. A major challenge to alternative resources is availability for industry. The differentiation between theoretical and technical potentials should always be weighed in to prevent a false understanding of biomass availability as most waste streams are already utilised by various industry sectors (Brosowski et al., 2015). In addition, to bridge seasonality, key technologies are needed to process multiple streams, including further downstream side products. With more complex compositions and the need for smart positioning in the market, the branding of a product becomes increasingly important. There are many ideas for using renewable materials, but the main problem is the uneven distribution of costs and benefits, from the need for extensive R&D to scaling a process to commercial standards and an underdeveloped supply chains.

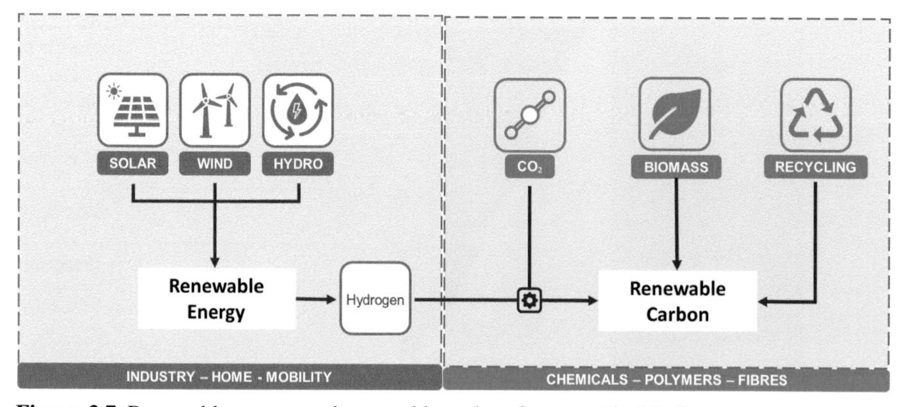


**Figure 3.7** Renewable energy and renewable carbon for a sustainable future. Adapted from nova-Institute (2020).

### 3.4.1 Sector coupling and cascade use of products and materials

To fix the current supply chain in a still undeveloped bioeconomy, the key to success lies in the efficient and transparent coupling of sectors. A simple overview of the process is depicted in Fig. 3.8. In this process chain, overlapping sectors are agriculture, logistics, pulping, chemical, textiles, and energy. Especially when working with bio-based materials, the cascade use of output streams for a diversified portfolio of products is essential to be economically viable. Implications involve the high-value use of cellulose, hemicellulose and lignin with further downstream applications for the latter two side streams. Cellulose can be implemented directly in the textile industry but hemicelluloses need further modification to synthesise platform chemicals, which can be used as building blocks for monomers and consequently polymers offering new and drop-in solutions for the textile industry. Exemplary polymers from hemicellulose may be PLA and polyethylene furanoate (PEF) (Jong et al., 2022). Lignin has been mainly used as a fuel source, but research is working on methods to fractionate lignin into its main components to lay the foundation for higher-value products.

For the effective combination of all sectors, the current trends work towards a circular economy with a partially open innovation approach connected through a framework exhibited through joint commitments, public programs, alliances and nonprofit initiatives. First, the coupling of different stakeholders is essential because knowledge about an entity is broadly limited to a field of expertise. Farmers do not know the specific properties of cellulose, and pulp producers lack expertise in textiles to figure out the necessary precise parameters for spinning. To overcome the knowledge gap, a transparent exchange of information is needed between partners. Second, the design of the product needs to be made for a closed-loop environment with recycling or possible biodegradability in mind. This process is also known as Design4-Recycling. Depending on the market and product specifications, degradability could be greatly useful. Especially in the agriculture and food sectors, where most polymers and coatings are synthetic in origin, nonbiodegradable and prone to create microplastic pollution, biodegradable textiles, agrotextiles, geotextiles, irrigation systems, packaging, and foils from bio-based origin offer a valid solution to a more sustainable future.

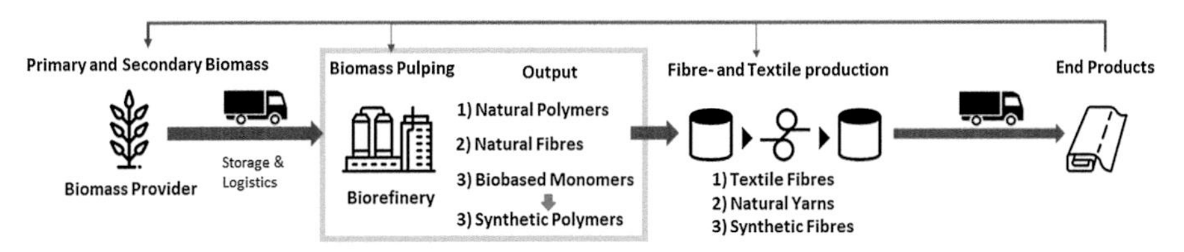


**Figure 3.8** Simplified overview of bio-based circular economy connecting agriculture, food, and textile industries.
From Institut für Textiltechnik der RWTH Aachen University.

### 3.4.2 Availability of cultivated land for plant growth

With the rise in alternative feedstocks for the textile industry, a major point that is highly influential to the cause is the availability of land for plants. Different from energy crops for bio-fuel production, using only residual material solves the dilemma of whether to use arable land for food or technical purposes and offers all stakeholders within the system a win—win result. With the rising population, one would automatically assume that the need for food would rise, leading to efforts to increase the crop yield or allocate more arable land for the food sector. However, the truth shows a decrease of total worldwide arable land of 1% per year owing to urbanisation. In addition, countries such as Germany have made it mandatory for individual farms to set aside 4% of their entire arable land, beginning in 2023. Similar actions have been taken by other countries in the EU (Jong et al., 2022; Michalczyk, 2023).

An increase in production requires much greater use of inorganic fertilisers and pesticides, as well as significant improvements in irrigation systems. However, all of these efforts lead to a long-term decline in crop yields as agricultural soils are steadily losing fertility. The use of synthetic fertilisers and pesticides can be reduced or avoided, which will have a significant impact on yield. The only viable method for reducing the use of pesticides while increasing yields is to use genetic modification (Haemmerle, 2011; Searchinger et al., 2019).

### 3.4.3 Mass-scale production to on-demand small-scale production

In the 1980s, multiple economists outlined the significant advantages of moving production facilities offshore, especially for the textile industry (Dunning, 1981). With the rise of automation and data science, new tools have emerged that companies can use to achieve an agile production cycle specifically matched to on-demand production, which will lead to smaller production scales, a reduction in overstock and a rise in disruptive innovations, most likely in the formation of new startups and large initiatives from established players. A key focus for on-demand production is the speed of information and processing, which leads to the topic of digitisation. Although some sectors are transitioning well, the textile sector has still a way to go to catch up. To plan efficiently, process a strategy for a future product and maintain agility, various areas can be identified:

- End-to-end process management
- Virtual sampling
- Virtual libraries
- Supply chain transparency and traceability
- Capacity planning

Technical textiles are well-established in the EU. Germany is the lead technical textile supplier, and apparel- and home textiles are generally produced offshore in developing countries (Hedrich et al., 2021). There are several reasons for on-demand production from the consumer side, creating a pull for companies:

- Consumers decide what they want to buy and wear
- Consumers will use social media
- Consumers are the creators of new trends, not the companies

These factors are beneficial to small on-demand production because large design and production processes might not respond well to sudden and detrimental changes to a product. This is the problem as conflicting priorities among cost and consumer demands, high-quality materials and functional performance stand between new product lines. Furthermore, the use of small decentralised production sites offers the inclusion of consumers through their creative input, enabling cocreation (Dana et al., 2007; Harper and Pal, 2022). Small-scale production also offers the opportunity to create a new supply chain switching from offshore to onshore or near shore. Whereas the offshore concept has been implemented by well-known apparel companies such as H&M, Zara, Mango, and Primark, onshore and near-shore concepts have been gaining traction (Amed et al., 2022). A survey from McKinsey in 2021 stated that 24% of sourcing and procurement executives plan to increase reshoring in their strategy whereas 71% of chief product officers plan to increase their near-shoring share, with some sources claiming a near-shoring boom in the EU (Hedric et al., 2021).

## 3.5 Innovations leading to sustainability, energy and resource efficiency

As shown in Fig. 3.9, there is an ever increasing demand for man-made and synthetic fibres because of the limitation for improving the productivity of natural fibres such as cotton. The production of cotton in the world has been nearly constant for several decades because of limited land for cultivation and limited improvement in the breed. Animal fibre such as wool is not eco-friendly. This is because the concept that vegetarians are more eco-friendly than carnivorous (meat-eating) people can be applied.

Around 82% of synthetic fibres in the world is PET, normally called polyester. This is mainly because of its excellent properties and low cost. Thus the impact of innovations leading to sustainability, energy and resource efficiency is high for polyester. Current and future technologies for improving the sustainability of the fibre and textile industry is analysed in a concrete manner, adopting PET fibre as an example.

### 3.5.1 *Production of bio-based polyesters*

From the viewpoint of the eco-friendliness of materials, PET is polymerised from fossil-based monomers such as ethylene glycol (EG) and terephthalic acid (TPA) (or dimethyl terephthalate [DMT]). The production of partially bio-based PET from bio-based EG and fossil-based TPA is expanding quickly. Although only 30% of the weight is bio-based in this case, the use of the total amount of bio-based material is huge. Among various bio-based and partially bio-based polymers, the largest amount of bio-based material is used to produce partially bio-based PET.

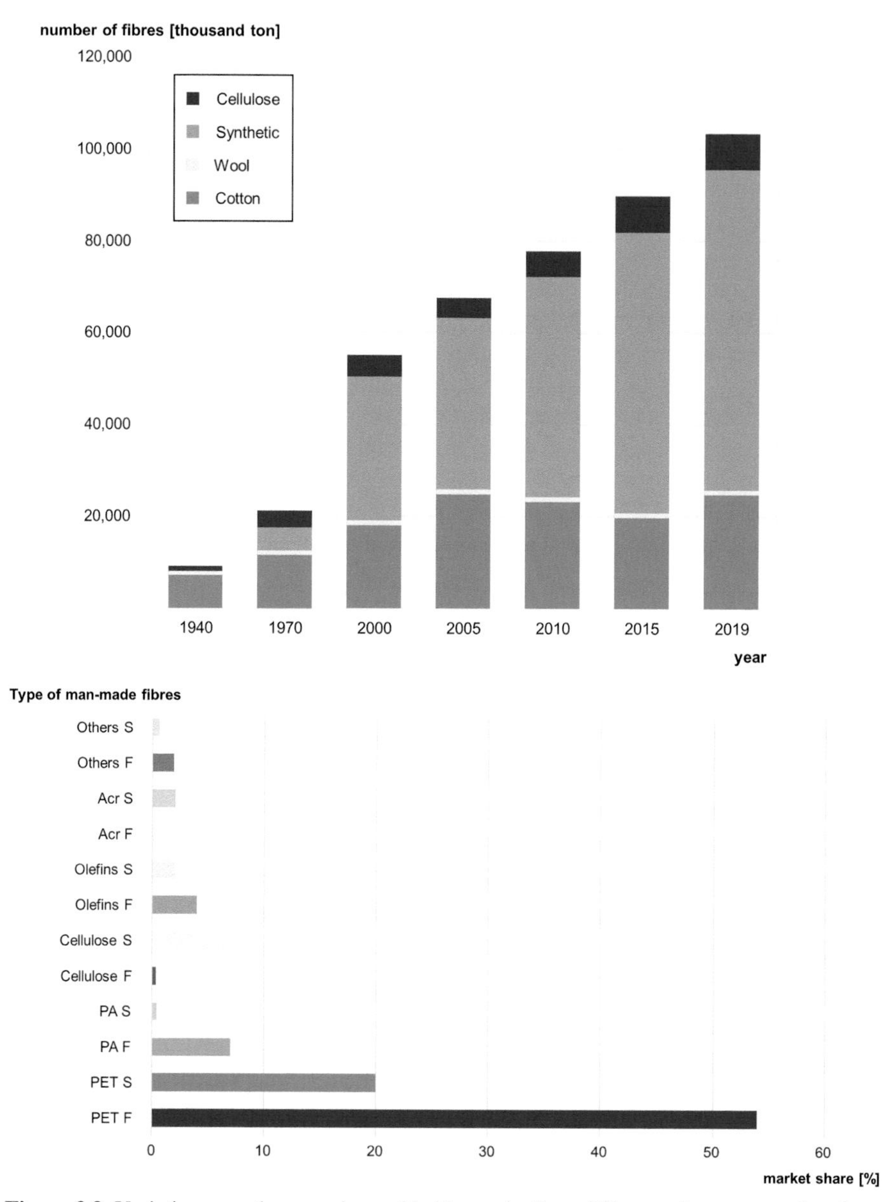


**Figure 3.9** Variation over the years in worldwide production of fibres and current market share of man-made fibres. *PET*, polyethylene terephthalate.
From Textile Handbook (2022); Japan Chemical Fibres Association, p.24.

There have been technological developments for the production of fully bio-based PET through the preparation of bio-based TPA. For example, Toray Inc announced the development of a technology for producing fully bio-based PET fibres with the conversion of bio-based para-xylene developed by Gevo to TPA in 2015 (Aoyama and Tanaka, 2016). More recently, technologies for producing TPA from bio-cellulose (wood chips) were announced by Suntory and Anellotech Inc in 2022 (Sudolsky and Mendes-Jorge, 2019).

Because of the difficulty of using bio-based TPA in terms of the technology and cost, the application of a monomer unit with a similarity in chemical structure to TPA was proposed. PEF is a promising polymer for this concept. PEF can be synthesised using furan dicarboxylic acid (FDCA) instead of TPA. FDCA can be prepared through a bio-process from abundant C6 sugars such as fructose and glucose. Therefore, PEF synthesised from EG and FDCA can be regarded as a 100% bio-based polyester alternative to PET. Like PET, PEF can be applied to produce bottles, films and fibres. Compared with PET, PEF has a similar glass transition temperature ($T_g$) of around 75°C and a slightly lower melting temperature of around 225°C (PET: 255–260°C). High–gas barrier properties are an important characteristic of PEF (Codou et al., 2016). In 2020, approximately 600 million tons of PET fibres was produced in the world, whereas the total amount of bio-based polymer, excluding partially bio-based PET, was only 6 million tons. We may need to consider the limitations in terms of the available amount of natural resources for producing bio-based monomer units to produce fibres of bio-based polymers.

There are various bio-based and/or biodegradable aliphatic polyesters. Characteristics of these polymers are different from those of the semiaromatic polyester PET. Aliphatic polyesters are normally melt-processable but have much lower $T_g$ and melting temperatures compared with PET. Among those, poly(glycolic acid) (PGA) (fossil-based and highly biodegradable) and PLA (bio-based and degrades mainly through hydrolysis) have a $T_g$ higher than room temperature. This leads to a significant advantage for the processability of these polymers because normally aliphatic polyesters have a low crystallisation rate. The combination of a low crystallisation rate and low $T_g$ leads to the sticking of fibres on the takeup bobbin after melt-spinning because fibres are still in an amorphous state. PGA and PLA fibres are reported to have good mechanical properties (Saigusa et al., 2020). The melting temperature of PLA can be significantly improved through the formation of the stereocomplex crystals of poly-L-lactide (PLLA) and poly-D-lactide (PDLA) (Roungpaisa et al., 2022). On the other hand, melt processing a PHA [poly(3-hydroxybutyrate-co-3-hydroxyhexanoate) (PHBH)], is difficult because of the low crystallisation rate and low $T_g$. However, the production of PHBH fibres through melt spinning is widely attempted (Qin et al., 2017) because this polymer has the advantage of being biodegradable in seawater. There are some commercialised biodegradable aliphatic-aromatic copolyesters. For example, elastomeric fibres can be produced by melt-spinning poly(butylene adipate-co-terephthalate) (Shi et al., 2005).

### 3.5.2 Energy reduction in dyeing of polyester

From the viewpoint of sustainability, the reduction of energy consumption to produce PET fibres is another important issue. It is well-known that only a disperse dye can be used to dye PET fibres, in which a temperature of around 120°C and high pressure are required. The consumption of a huge amount of energy and water is a serious negative aspect of PET fibre dyeing. Various technologies have been proposed for colouring PET fibres. Dyeing PET fibres in supercritical carbon dioxide has been proposed. In this process, dyeing can be performed without using water. Thus the treatment of wastewater contaminated with dyestuff is unnecessary. Carbon dioxide used in the process can be fully recycled. More than a 40% reduction in energy consumption was reported compared with ordinary dyeing.

Colouring of PET fibres using the concept of interference colour has been proposed. In this case, a multilayered structure of PET and polyamide with a thickness of each layer of around 70–100 nm are embedded in the cross-section of the fibre. Colour can be controlled by adjusting the thickness of layers. The use of dyestuff can be eliminated and there is no colour fading in principle (Kamiyama et al., 2012). The application of polyester swelling agents and enzymatic treatments in the dyeing of polyester/cotton blended textiles has resulted in a variety of shades (Teli and Paul, 2006).

Another proposal is the concept of infusion to produce coloured fibres. Bye drawing undrawn yarn (UDY) in ethanol, fibre structure development (i.e. the development of molecular orientation and crystallisation) can be accomplished at room temperature. In the course of drawing, the uptake of ethanol through the mechanism of infusion occurs, and crystallisation proceeds because of the solvent-induced crystallisation. Ethanol cannot be absorbed into fibres simply by immersing UDY or fully drawn yarn into ethanol. If dyestuff is dissolved in ethanol, the dyeing of fibre proceeds simultaneously under the mechanism of infusion. In other words, simultaneous fibre structure formation and dyeing can be accomplished using this process (Go, 2019).

### 3.5.3 Improvement in mechanical properties

High-strength PET fibres are widely used as reinforcing fibres for tyres (tyre cord) and belts, and high-strength textiles such as seat belts. From the viewpoint of improving the eco-friendliness of PET fibres, making mechanical properties stronger is another important research subject to save energy and material consumption. Among the various high-strength fibres, the strength of PET fibre is limited. Commercial production of high-strength fibres with a tensile strength higher than 2 GPa has been accomplished for various polymers. Typical examples are high–molecular weight (Mw) PE, polyvinyl alcohol (PVA), thermoplastic liquid crystalline polymers (TLCP) such as polyarylate, poly(p-phenylene terephthalamide) (PPTA) and polybenzobisoxazole (PBO). The molecular chains of PE and PVA are flexible whereas those of PPTA, TLCP and PBO are rigid. The development of high-strength fibres from polymers with semiflexible molecular chains, such as PET and PA, is difficult (Fig. 3.10).

The general concept for improving the mechanical properties of PET fibres by homogenising the state of molecular entanglement in the UDY has been proposed. In this case, melt spinning PET fibres under the condition of a low Deborah number is the key to preparing UDY, which possesses a high potential for becoming high-strength and high-toughness fibres after the application of drawing and annealing processes (Kikutani, 2016).

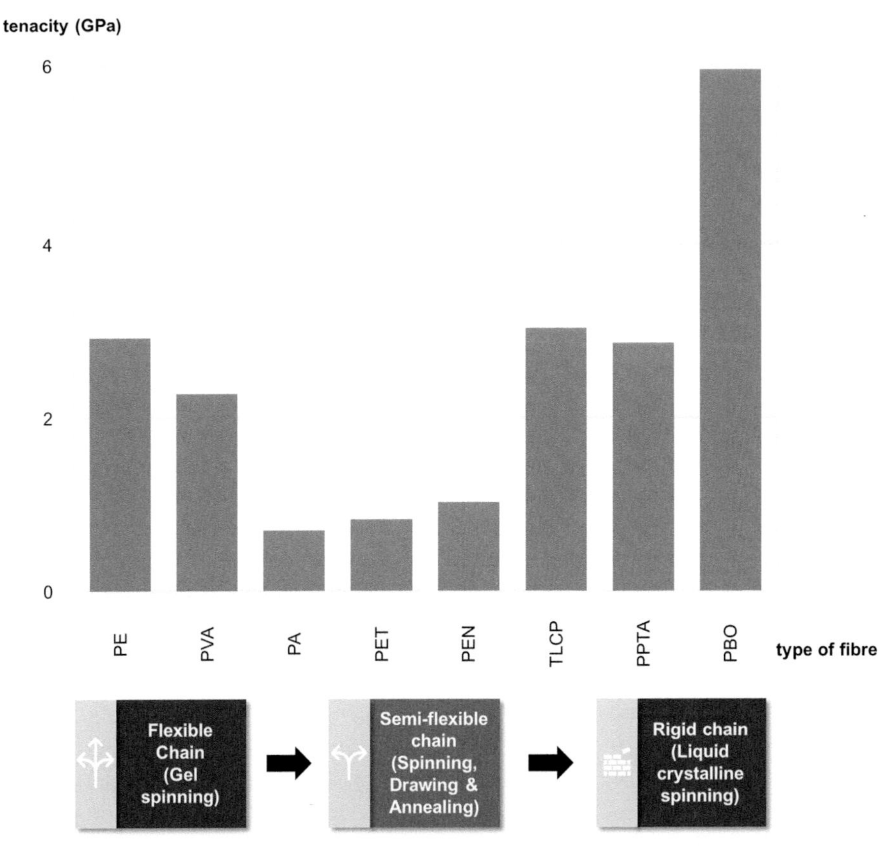


**Figure 3.10** Tensile strength of various high-strength fibres on the market. *PBO*, polybenzobisoxazole; *PE*, polyethylene; *PEN*, polyethylene naphthalate; *PET*, polyethylene terephthalate; *PPTA*, poly-p-phenylene terephthalamide; *PVA*, polyvinyl alcohol; *TLCP*, thermoplastic liquid crystalline polymers.
From Kikutani (2016).

To produce high-strength PET fibres, the use of high-Mw PET is necessary. High-Mw PET is produced through the solid-state polymerisation (SSP) of PET chips under vacuum conditions at high temperatures. A long reaction time is required for a sufficient increase in Mw. Therefore this is considered a process with high energy consumption. If the increase in Mw in a molten state through melt-state polymerisation (MSP) is accomplished, energy consumption can be significantly reduced. Another advantage of this process is that the continuous process of polymerisation, MSP and fibre production can be achieved. The MSP of PET is possible using the gravitational flow of a thin layer of the polymer melt in a vacuum. Energy consumption for producing high-strength polyester fibres can be reduced by 30% compared with the conventional process with SSP (Xuming, 2010; Chen et al., 2013). A plant with a 0.5 million ton capacity was reported to be operating in China since 2015.

### 3.5.4 Recycling of polyester

In terms of the carbon neutralisation of polyester fibres, if the availability of bio-based polymer is limited, recycling is the most important approach. Recycling of PET has been investigated using various approaches. Compared with the material recycling of polyolefins such as PE and PP, PET has the advantage of the high controllability of Mw after recycling. In polyolefins, processability can be affected by the Mw as well as by the Mw distribution. For example, the processability of polymers can be affected significantly by a small amount of high-Mw components. Therefore difficulty exists in recycling polyolefins. On the other hand, there is no need to consider Mw distribution for recycling PET in principle. Theoretically, Mw distribution expressed by the ratio of weight-average to number-average Mws becomes two even though PETs of different Mws are blended because transesterification occurring in the polymer melt leads to the automatic control of Mw distribution.

On the other hand, in the recycling of PET, Mw can be controlled simply by adjusting the pressure at a vent-port during twin-screw extrusion for pelletising the polymer. Even an increase in Mw can be achieved by introducing a chain extender, which connects the ends of two molecular chains, into an extruder for pelletising. Modifying polymer characteristics by incorporating long-chain branching is also possible using a chain extender with three functional units such as trimesic acid and trimellitic acid.

Material (mechanical) recycling of PET bottles has been widely used. It was reported that more than 95% of PET bottles are recycled in Japan. Compared with PET bottles, material recycling of PET fibres and textiles is not simple, mainly because of the difficulty of collecting pure PET fabrics. Thus the production of monomaterial fabric is a solution to this issue. In the textile market, however, a large amount of blend fibres, such as polyester/cotton 65%/35%, are widely used. There is a proposal to upcycle PET/cotton blend fibres using ion-liquid to dissolve only cotton. The dissolved dope can be used to produce cellulosic fibres applying the concept of upcycling, whereas the remainder can be used as a PET material (Haslinger et al., 2019).

For the chemical recycling of PET, technologies using catalysed degradation have been investigated. There are various drawbacks to the current technologies for PET decomposition. For example, after alkali decomposition, neutralisation using a large amount of acid is required. A type of catalysed degradation of polyesters to monomers via transesterification was reported to be a technology under development. Using this technology, DMT and EG can be obtained through the transesterification reaction (Abe et al., 2022).

### 3.5.5 Eco-friendliness

The catalyst design is an important elemental technology for producing PET. PET for bottles has been produced mainly using the catalyst of germanium compound, whereas antimony compounds have been used for PET for textiles. Catalyst added in polymer by several tens to hundreds of parts per million affects the properties of fibre products. For example, titanium compounds have higher activity. There have been attempts to use aluminium compounds (Abe et al., 2022). Alkali treatment is applied to a large amount of PET fabric to introduce softness and flexibility to textiles. In this process, up to 20% of the fabric is dissolved and the waste contains the residue of the catalyst. Therefore, the development of a technology using eco-friendly catalysts is another important issue for the PET textile industry.

Microplastic pollution is an important issue from the viewpoint of the eco-friendliness of polymer waste. Microfibres from fabrics during laundry are considered one of the major sources of microplastics in seawater. There may be two ways to overcome this issue. One is to minimise the release of microfibres using filament fibres instead of staple fibres. The technology for preparing spun-like yarns from filament yarn is an important research subject in the textile industry, and many patents have been applied in this field (Oguchi et al., 1979; Nakatsuka, 2011).

Another approach is to modify PET to introduce degradability in the marine environment. The polymer is supposed to be stable during ordinary use and starts to decompose by a specific trigger. To achieve this, the concept of multilock type decomposition was introduced. Technologies for designing copolymers from nonedible bio-based monomers, introducing the concept of a decomposition accelerator, and controlling the release of decomposition accelerators are key issues. In this type of polymer, hydrolysis (primary decomposition) and the biodegradation of oligomers (secondary decomposition) need to be controlled (Teijin Limited, 2022).

## 3.6 Future perspectives

The future of man-made and synthetic fibres is influenced by a range of factors such as consumer demand, technological innovation, environmental regulations, social responsibility and market competition. As the fibre industry strives to meet the evolving needs and expectations of customers and stakeholders, it faces both opportunities and challenges. Consumers are becoming increasingly conscious of the environmental and social impacts of their purchasing decisions. As a consequence, companies and consumers alike are seeking textile products made from renewable, recycled, bio-based or biodegradable materials with low carbon and water footprints, durability and proper end-of-life functionalities.

Through new technologies, the development of fibres with enhanced properties and functionalities can be achieved. However, the man-made and synthetic fibre industry is subject to numerous regulations and standards aimed at ensuring the quality and safety of its products. These include regulations on chemical use and management, emissions and waste control, product labelling and certification and consumer protection. The fibre industry must comply with these regulations and standards by adopting best practices and improving transparency and traceability, thus inevitably shifting towards natural sources.

To seize opportunities and overcome challenges, the fibre and textile industry must focus on developing sustainable innovations that address environmental and social issues throughout its value chain. This includes creating new types of fibres with enhanced properties and functionalities that meet the changing needs and expectations of customers and stakeholders. This approach enables the industry to achieve long-term viability and enhance its sustainability performance.

## 3.7 Conclusion

This chapter delves into the impact of natural and synthetic polymers on the production and applications of man-made and synthetic fibres. The sustainability of these fibres is influenced by the choice of carbon source, whether renewable or fossil. Water and energy consumption as well as hazardous and toxic substances involved in polymer and fibre production hinder sustainability. Melt and solution spinning were compared and their differences in terms of efficiency and environmental performance were highlighted. It was suggested that man-made and synthetic fibre production should aim to minimise the use of harmful chemicals and maximise the use of renewable resources. Melt spinning was also identified as a preferable process compared with solution spinning for fibre production.

New technologies and synthetic fibres have transformed the textile industry but have also caused environmental and social problems. Thus cellulose has been reintroduced as a renewable and versatile alternative to synthetic polymers. Natural, regenerated man-made and synthetic fibres were compared and their advantages and disadvantages were discussed. The health and environmental impacts of fibres were assessed from production to disposal, and a circular economy with renewable carbon sources was suggested as a way to reduce these impacts and create value from waste.

New innovations and increased sustainability are needed to succeed in the future, in which man-made and synthetic fibres could have an important role in this transition if they are produced and used responsibly. Using $CO_2$, biomass and recycling as the pillars for a diverse platform of future products can reduce dependence on fossil fuels and create a more circular economy for the chemical industry. $CO_2$ exhibits a diverse platform and thus is an interesting feedstock for numerous chemicals and polymers. Biomass as a feedstock shows high potential but alternative sources and technologies have yet to be developed. Highlighting the importance of branding and positioning for renewable products, barriers and opportunities for their market entry, it is proposed that $CO_2$, biomass use, and recycling are promising ways to create a more sustainable and circular economy.

The production and characteristics of bio-based polyester and its alternatives were compared with synthetic and natural sources of carbon, and their health and environmental impacts, while differentiating between linear and circular economies of the textile industry and the scenarios for end-of-life textiles. Biodegradable fibres and their properties, such as nontoxic, renewable sourcing, and efficient technologies, are essential for a more sustainable textile industry. Innovative technologies for colouring polyester fibres without using water or dyestuff are also becoming important.

The production and characteristics of high-strength polyester fibres were compared with other high-strength fibres, and it was proposed that their mechanical properties might be improved by homogenising molecular entanglement. The need for high-Mw polyester for producing high-strength polyester fibres and the conventional and alternative processes for increasing the Mw of polyester were highlighted. High-strength polyester fibres can save materials and energy, and liquid-state polymerisation can reduce energy consumption to produce high-strength polyester fibres.

Man-made and synthetic fibres have a significant impact on the environment and human health but also offer opportunities for innovation and sustainability. Recycling strategies are crucial for reducing the carbon footprint of polyester fibres, especially when bio-based sources are limited. Innovative solutions such as monomaterial fabrics, ionic liquids, catalysed degradation and degradable polyester can overcome challenges associated with recycling polyester fabrics and prevent microplastic pollution. The careful design of catalysts and solvents is essential to ensure the eco-friendliness and efficiency of these solutions. By implementing these solutions, the textile industry can achieve carbon neutrality and sustainability.

## References


Abe, et al., 2022. La(iii)-Catalysed degradation of polyesters to monomers via trans-esterifications. Chemical Communications 58, 8141–8144. Available at: https://doi.org/10.1039/D2CC02448A.

Amed, et al., 2022. The State of Fashion 2022. Available at: https://www.mckinsey.com/~/media/mckinsey/industries/retail/ourinsights/stateoffashion/2022/the-state-of-fashion-2022.pdf.

Aoyama, Tanaka, 2016. History of polyester resin development for synthetic fibres and its forefront. Japan the Society of Fibre Science and Techno. In: High-Performance and Specialty Fibres. Springer Japan, Tokyo, pp. 67–80. https://doi.org/10.1007/978-4-431-55203-1_4.

Brosowski, et al., 2015. Biomassepotenziale von Rest- und Abfallstoffen Status Quo in Deutschland. In: F.N.Rohstoffe e.V. (FNR) (Ed.). (FNR), Fachagentur Nachwachsende Rohstoffe e.V.

Chen, et al., 2013. Ein Fertigungsverfahren von Pet-Industriegarn Durch Direktes Spinnen aus Schmelzen. China. Available at: https://patentimages.storage.googleapis.com/68/12/03/3044c706db29fb/DE112013006789B4.pdf.

Codou, et al., 2016. Glass transition dynamics and cooperativity length of poly(ethylene 2,5-furandicarboxylate) compared to poly(ethylene terephthalate). Physical Chemistry Chemical Physics 18 (25), 16647–16658. Available at: https://doi.org/10.1039/C6CP01227B.

Cross, Bevan, Baedle, 1893. Herstellung Eines in Wasser löslichen Derivats der Zellulose, Genannt "Viskoid". Reichspatentamt Berlin, Germany.

Dana, Hamilton, Pauwels, 2007. Evaluating offshore and domestic production in the apparel industry: the small firm's perspective. Journal of International Entrepreneurship 5 (3), 47–63. Available at: https://doi.org/10.1007/s10843-007-0015-1.

Dunning, 1981. International Production and the Multinational Enterprise. George Allen & Unwin Limited.

European Union, 2008. Waste Framework Directive. Available at: https://environment.ec.europa.eu/topics/waste-and-recycling/waste-framework-directive_en#contact.

Go, et al., 2019. Continuous drawing of poly(ethylene terephthalate) in ethanol and agile functionalization through infusion. Materials Today Communications 19, 98–105. Available at: https://doi.org/10.1016/j.mtcomm.2019.01.001.

Gries, Veit, Wulfhorst, 2014. Textile Fertigungsverfahren. Carl Hanser Verlag GmbH & Co. KG.

Gries, Veit, Wulfhorst, 2015. Textile Fertigungsverfahren. Carl Hanser Verlag GmbH & Co. KG.

Haemmerle, 2011. The Cellulose Gap (the Future of Cellulose Fibres).

Harper, Pal, 2022. Small-series supply network configuration priorities and challenges in the EU textile and apparel industry. Journal of Fashion Marketing and Management. Available at: https://doi.org/10.1108/FJMM-07-2021-0173.

Haslinger, et al., 2019. Upcycling of cotton polyester blended textile waste to new man-made cellulose fibres. Waste Management 97, 88–96. Available at: https://doi.org/10.1016/j.wasman.2019.07.040.

Hedric, et al., 2021. Revamping Fashion Sourcing: Speed and Flexibility to the Fore. Available at: https://www.mckinsey.com/industries/retail/our-insights/revamping-fashion-sourcing-speed-and-flexibility-to-the-fore.

Hedrich, et al., 2021. Revamping Fashion Sourcing: Speed and Flexibility to the Fore. McKinsey&Company. Available at: https://www.mckinsey.com/~/media/mckinsey/industries/retail/our-insights/revamping-fashion-sourcing-speed-and-flexibility-to-the-fore/revamping-fashion-sourcing-speed-and-flexibility-to-the-fore.pdf.

Henry, Laitala, Klepp, 2019. Microfibres from apparel and home textiles: prospects for including microplastics in environmental sustainability assessment. The Science of the Total Environment 652, 483–494. Available at: https://doi.org/10.1016/j.scitotenv.2018.10.166.

Jaffe, Easts, Feng, 2020. Polyester fibres. In: Jaff, M., Menczel, J.D. (Eds.), Thermal Analysis of Textiles and Fibres. Woodhead Publishing Limited, pp. 133–149. Available at: https://doi.org/10.1016/B978-0-08-100572-9.00008-2.

Jong, et al., 2022. The road to bring FDCA and PEF to the market. Polymers 14 (5), 943. Available at: https://doi.org/10.3390/polym14050943.

Kamiyama, et al., 2012. Development and application of high-strength polyester nanofibres. Polymer Journal 44, 987–994. Available at: https://doi.org/10.1038/pj.2012.63.

Kikutani, 2016. Development of high-strength poly(ethylene terephthalate) fibres: an attempt from semiflexible chain polymers. The Society of Fibre Science and Technology. In: High-Performance and Specialty Fibres. Springer Japan, Tokyo, pp. 133–147. https://doi.org/10.1007/978-4-431-55203-1.

Luft, 1925. Artificial silk. - with special reference to the viscose process. Industrial and Engineering Chemistry 17 (10), 1037–1042. Available at: https://doi.org/10.1021/ie50190a016.

Michalczyk, 2023. Agrarreform 2023 - ein Überblick, Landwirtschaftskammer Nordrhein-Westfalen. Available at: https://www.landwirtschaftskammer.de/foerderung/hinweise/agrarreform-2023.htm#:~:text=Kultur-angebaut-wird.-,Flächenstilllegung-ist-Pflicht,an-einer-solchen-Brache-liegen.

Morlet, et al., 2017. A New Textiles Economy - Redesigning Fashion's Future. Ellen MacArthur Foundation. Ellen MacArthur Foundation.

Muthu, 2020. Assessing the Environmental Impact of Textiles and the Clothing Supply Chain, second ed. Woodhead Publishing Limited.

Nakatsuka, 2011. Structure Processed Yarn (Japan).

nova-Institute, 2020. Renewable Energy and Renewable Carbon for a Sustainable Future. Available at: https://renewable-carbon.eu/publications/product/renewable-energy-and-renewable-carbon-for-a-sustainable-future—graphic/.

Oguchi, et al., 1979. Mold for Continuous Casting (Japan).

Qin, Takarada, Kikutani, 2017. Fibre structure development of PHBH through stress-induced crystallization in high-speed melt spinning process. Journal of the Fibre Science and

Technology 73 (2), 49–60. Available at: https://www.jstage.jst.go.jp/article/fibrest/73/2/73_2017-0007/_article/-char/ja/.

Roungpaisan, Takasaki, Takarada, Kikutani, 2022. Mechanism of fibre structure development in melt spinning of PLA. In: Poly(lactic acid): Synthesis, Structures, Properties, Processing, and Application, second ed.". John Wiley & Sons, pp. 425–438. Available at: https://onlinelibrary.wiley.com/doi/abs/10.1002/9781119767480.ch18.

Saigusa, Takarada, Kikutani, 2020. Improvement of the mechanical properties of poly(glycolic acid) fibres through control of molecular entanglements in the melt spinning process". Journal of Macromolecular Science Part B 59 (6), 399–414. Available at: https://www.tandfonline.com/doi/full/10.1080/00222348.2020.1730600.

Sandin, Roos, Johansson, 2019. Environmental Impact of Textile Fibres - What We Know and What We Don't Know. Fibre bible part 2. https://doi.org/10.13140/RG.2.2.23295.05280.

dos Santos, R.G., Alencar, A.C., 2019. Biomass-derived syngas production via gasification process and its catalytic conversion into fuels by Fischer Tropsch synthesis: a review. International Journal of Hydrogen Energy 45 (36). https://doi.org/10.1016/j.ijhydene.2019.07.133.

Searchinger, et al., 2019. In: Matthews, E. (Ed.), Creating a Sustainable Food Future. World Resources Institute. Available at: https://research.wri.org/sites/default/files/2019-07/WRR_Food_Full_Report_0.pdf.

Shi, X.Q., Ito, H., Kikutani, T., 2005. Characterization on mixed-crystal structure and properties of poly(butylene adipate-co-terephthalate) biodegradable fibres. Polymer 46, 11442–11450. Available at: https://www.sciencedirect.com/science/article/pii/S0032386105015338.

Shor, 1944. Nylon. Journal of Chemical Education 21 (2), 88. Available at: https://doi.org/10.1021/ed021p88.

Sudolsky, Mendes-Jorge, 2019. 100% Renewable Plastic Bottle Significantly Closer to Reality after Successful Production of Bio-Based Paraxylene From Non-Food Biomass Bio-Based Benzene is Next. Anellotech. Available at: https://anellotech.com/press/100-renewable-plastic-bottle-significantly-closer-reality-after-successful-production-bio.

Teijin Limited, 2022. Development of highly degradable polyester-based multi-lock type bio-tough polymer and its fibres. In: Moonshot Goal 4 Annual Report 2022. Available at: https://www.nedo.go.jp/content/100958162.pdf.

Teli, M.D., Paul, R., 2006. The role of swelling and enzymatic finishing in developing new shades on polyester/cotton blends. International Dyer 191 (8), 31.

Textile Exchange, 2021. Preferred Fibre & Materials - Market Report 2021.

Viswanathan, 2010. Wallace Carothers: more than the inventor of Nylon and Neoprene. World Patent Information 32 (4), 300–305. Available at: https://doi.org/10.1016/j.wpi.2009.09.004.

Xuming, 2010. Tubular Film Evaporator (China).

# Innovations in spinning processes and machinery

*Yi Zhao*[1] *and Narendra Reddy*[2]
[1]College of Textiles, Donghua University, Shanghai, China; [2]Center for Incubation Innovation Research and Consultancy, Jyothy Institute of Technology, Bengaluru, Karnataka, India

## 4.1 Introduction

Spinning is an essential and important part of the textile production process to convert fibres into fabrics and subsequent applications. Not only is it a complex engineering process, efficient spinning is also critical for economic and technical reasons. It is estimated that the spinning process accounts for up to 50%–60% of energy consumption in a fibre to fabric conversion process. In the spinning process, the spinning machine itself accounts for 49%–55% of the total energy consumed, with about 1.32 kWh consumed per kilogram of yarn spun (30s Ne count) (Dhayaneswaran and Ashok Kumar, 2013). Furthermore, the amount of machinery involved and the time taken to pass through the entire spinning process is considerably higher compared to weaving, dyeing, finishing and other textile processes. Hence, it is critical that the spinning process be understood and optimised, and efforts made to conserve energy, manpower and other resources.

Substantial efforts have been made not only to optimise the spinning process but also to make spinning equipment energy efficient, generate minimum waste and increase the productivity and life of machinery. The extent of decrease in energy consumption varies depending on the processes and machinery used and also on the machinery manufacturer. For example, Reiter, one of the largest textile machinery manufacturers, reports that their processing lines require considerably fewer machines and are highly energy efficient compared to machinery and processes used by other manufacturers (https://www.textileexcellence.com/news/using-rieter-systems-for-sustainable-yarn-production/). Despite the progress in machinery and processes, spinning is a relatively energy- and resource-intensive process. Hence, it is necessary to develop technologies and machinery to make spinning more environmentally friendly and sustainable. This chapter provides an overview of the latest developments in spinning and allied machinery and attempts that are focused towards energy conservation, waste reduction, increase in efficiency and lowering of costs.

**Sustainable Innovations in the Textile Industry. https://doi.org/10.1016/B978-0-323-90392-9.00004-5**

## 4.2 Historical aspects and current scenario

Conversion of fibres into yarns and further into various types of textiles is a skilled and ancient art. Yarn spinning where fibres are twisted and made into threads to be woven into textiles was done using hands by skilled workers. Eventually, mechanical spindles and wheels were invented where the process became semi-automatic. Once the larger looms and dyeing equipment were invented, it was necessary to produce yarns also on a large scale. During the latter part of the 18th century and early 19th century, machines that could process fibres into yarns were invented. These machines were able to produce yarns in a limited number of counts and as coarse yarns.

The advent of the Industrial Revolution saw substantial progress in textile machinery development. Machines to process the fibres into yarns were developed and most of these machines were processing cotton through the ring spinning system. Later, the rotor spinning and air-jet technologies were developed which not only increased productivity but also produced unique and high-quality yarns.

By the mid-19th century, ultra and super fibre count yarns and yarns from blends of various fibres became common. In the 1990s, concepts such as compact spinning were invented and resulted in considerably higher efficiencies in yarn production. Currently available spinning machinery from bale opening, blow room, carding and spinning are able to operate at very high production levels with minimum waste leading to high efficiencies. However, no major breakthrough technologies in yarn processing have been invented in the last decade or so. Significantly higher investment in research and development and the adoption of newer technologies may be necessary to ensure that yarn processing remains sustainable in the future.

## 4.3 Sustainable innovations in spinning processes

### 4.3.1 *Blowroom*

Innovations in spinning machinery are mainly focused on a reduction of the amount of energy consumed, waste generated and increasing the process efficiency. In this regard, major spinning companies have developed unique technologies. For example, Reiter had introduced a new blowroom line (VARIOline) which can process 2000 kg per hour due to the variable opening and cleaning and modular arrangement of the machinery. This blowroom line is reported to provide a 1% material saving and 40% lower power consumption, as shown in Table 4.1. The material saving is achieved by ensuring that small (micro) tufts are efficiently removed and the extent of waste removal can also be controlled. A vision shield optical magic eye system, detects all types of contaminants and foreign materials, which improves yarn quality. Furthermore, the UNIfloc A 12 automatic bale opener from Rieter has increased the bale opening capacity to 2400 kg/h and the opening is done in microtufts using a patented take-off roller system. Similarly, the BLENDOMAT BOA from Trutzchler has considerably wider width and can process 2500–3000 kg per hour. This machine occupies

**Table 4.1** Possible process benefits and energy conservation by Reiter machines (https://www.textileexcellence.com/news/using-rieter-systems-for-sustainable-yarn-production/).

| Yarn production (400 kg/h) | Other machines | Rieter | Savings/ improvements |
|---|---|---|---|
| **Main deviation in spinning** | | | |
| Number of compact spinning machines | 23 | 22 | 1 less machine |
| Production speed, rpm | 22,000 | 23,000 | +4.5% |
| Spindles per machine | 1824 | 1824 | |
| **Main deviation in preparation** | | | |
| Number of cards | 12 | 10 | 2 cards less |
| Number of drawframes | 8 | 6 | 2 drawframes less |
| Number of combers | 10 | 9 | 1 comber less |
| Comber noil, % | 18 | 17 | 1% less |
| Waste in blowroom and carding | 6.1 | 5.8 | 0.3% less |

up to 45% less space and provides 25%−40% better blending. Up to five to seven bales can be laid side by side and the machine can operate unattended for up to 48 h. Although with lower production, the Marzoli super blender B12 can handle four different fibre mixes and with a working width of 2250 mm and an output capacity of 1600 kg/h.

### 4.3.2 Carding

Carding is the heart of the spinning process and hence considerable efforts have been made to improve the productivity and lower costs during carding, as shown in Table 4.2. One of the major parameters critical for the performance of carding is the

**Table 4.2** Specifications and performance of the latest carding machines in comparison to conventional cards (Shi et al., 2020).

| Parameters | Conventional card | Latest card 1 | Latest card II |
|---|---|---|---|
| Can diameter, mm | 1000 | 1200 | 1200 |
| Can height, mm | 1200 | 1200 | 1200 |
| Filling quantity, kg | 52 | 76 | 80 |
| Space required for 5 cans, $m^2$ | 120.5 | 110.9 | 110.9 |
| Card production, kg/h | 90 | 90 | 90 |
| Production at can change, kg/h | 24 | 90 | 90 |
| Delivery speed at can change, m/min | 80 | 300 | 300 |
| Can change 1/h | 1.9 | 1.3 | 1.2 |
| Card efficiency, % | 97.5 | 99.6 | 99.8 |

distance between the flats and the cylinder. Trutzchler has introduced the new Gap Optimiser T/GO which ensures a constant, ideal carding gap even under changing production conditions. Extremely narrow flat settings of 3/1000″ have been possible. This results in 40% less yarn imperfections and up to 40% higher production. Trutzchler also has a new TC 19i card which is specifically made to process waste from yarns or garments generally called hard waste. This card has wired licker-in with stationary carding segments to ensure removal of disruptive particles and minimise loss of good fibres. The cards also have MULTI WEBCLEAN technology where fast, reliable and customised adjustments can be made. Similarly, Reiter's card C70 and C80 claim up to 30% higher output (card sliver) due to the larger carding area (40 active flats) and precise control over the carding gap (https://www.rieter.com/products/systems/fibre-preparation/card-c-80).

### 4.3.3 Drawframes

All new drawframes are preferably integrated to the carding machine. The new drawframes contain autolevellers and latest scanning technology for autolevelling and monitoring the card sliver. The number of drawframe passages is also significantly reduced and, in fact, the drawframe sliver can be directly fed into the rotor spinning machines. Furthermore, the systems come with a four over three draughting system with two draught zones which enables setting up to five-fold draughts. A CLEANtube is attached to the coiler drive which removes the trash particles and short fibres and improves the quality of the sliver.

### 4.3.4 Combing

Combing is still a preferred step to produce yarns with higher counts and better performance. The major improvement in combing is with respect to the number of nips per minute, which has increased to up to 600 per min. The latest combing machines also have disc monitor and count monitoring systems which regulate the main draught and balance count variations. Additionally, the machines are equipped with a piecing optimiser which regulates the piecing time and reduces the combing cycle leading to higher productivity. Some manufacturers provide automatic lap transport and a contactless system leading to increased quality of laps and lower manpower requirements. With these developments, current combing capacity has reached up to 90 kg/h.

### 4.3.5 Roving

Technological advancements have led to highly efficient roving machines where the entire doffing can be done automatically in 1.5 min. Each roving unit can be monitored individually, and the electronic draughting system can be used to adjust the roving count through a touchscreen system. Such innovations are reported to provide up to 20% energy savings. One of the major spinning machinery manufacturers, Saurer, has adopted the $E^3$ concept where energy optimisation, economics and ergonomics are combined. The $E^3$ concept is implemented through the following steps (https://saurer.com/en/products/machines/spinning/roving/autospeed):

***Energy***
Up to 20% less energy
- Powerful drive concept
- Energy-optimised bobbin rail drive
- Maximum cost transparency through energy monitoring

***Economics***
Maximum productivity
- Long machines for all raw materials
- Doffing time less than 2 min
- Short lot change times
- Automatic roving lift and transport system

***Ergonomics***
Reduced handling input
- User-friendly settings with EasySpin
- Ergonomically adjustable screen
- Made-to-measure automation

# 4.4 Sustainable innovations in spinning systems

## 4.4.1 Ring spinning

Ring spinning has remained the most critical and technologically challenging process in spinning. Nevertheless, the latest developments have ensured high spinning speeds and process efficiencies leading to better yarn quality and increased range of counts. In addition to spinning speed, draughting parameters, roving quality, etc. the geometry and conditions in the spinning triangle are critical for the quality of yarns produced. It has been recently shown that providing a pre-twist and specific level of pre-tension contributed to the higher final yarn strength obtained. Having the pre-twist at the edge of the spinning triangle was found to provide the maximum yarn strength and increasing inclination angle or height of spinning triangle leading to improved yarn strength, as depicted in Fig. 4.1. Introducing a contact surface during spinning, as in Fig. 4.2, has also provided substantial improvements to the yarn properties.

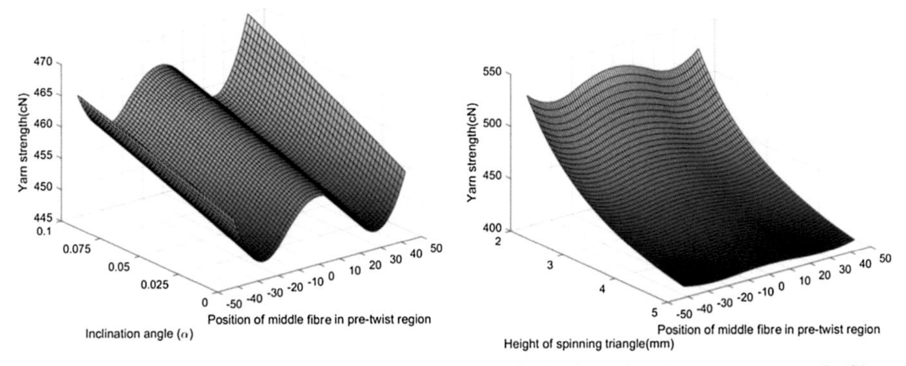


**Figure 4.1** Influence of inclination angle and height of spinning triangle on yarn strength (Shao et al., 2022).
Reproduced with permission from Taylor and Francis.

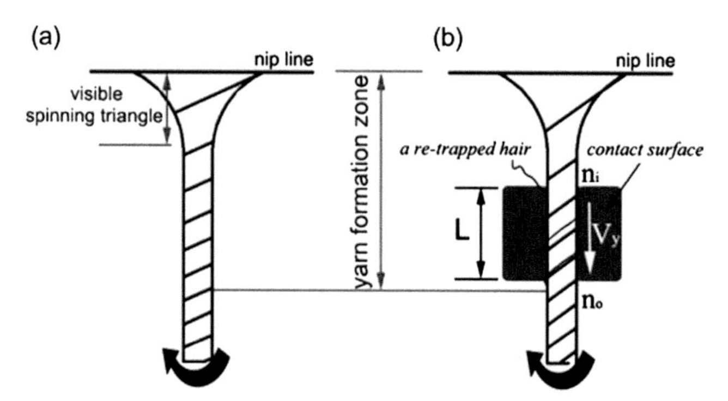


**Figure 4.2** Illustration of contact spinning principle: (a) conventional yarn formation zone during spinning; (b) a contact surface added in the yarn formation zone to re-wrap hairs onto or into yarn body (Xia et al., 2012).
Reproduced with permission from Springer Nature.

With the contact surface, loose fibres are re-wrapped onto the yarns and hence less wastage, loose fibres, neps, etc. (Xia et al., 2012).

Variations to the spinning frame, particularly the draughting zone, have improved both the productivity and economics. For instance, a novel device installed in the drawing zone of ring spinning machines has provided better yarn evenness and higher yarn strength (Quan et al., 2022). Similarly, the developments at Saurer provide twin-suction and opti-suction features on the ring spinning frame leading to up to a 66% reduction in energy consumption. Additionally, an electronically controlled draughting system with adjustable break draught, intelligent speed optimisation and individual spindle monitoring has also been made as an integral part of the system.

Although the advent of compact spinning has revolutionised yarn spinning and the properties of the yarns obtained, unlike compact spinning where air suction is used to reduce the spinning triangle, a new technology called ProSPIN has been developed, where a unit is installed in the ring spinning machine, which bisects the fibre mass in a controlled manner followed by separate compaction. The system can be installed on existing compact spinning machines and is suitable for both carded and combed yarns (https://www.prospin.com.tr/en/prospin-technology). Another development in ring spinning is the automatic piecing robot which along with the individual spindle monitoring (ISM) system tends to broken ends on the ring frame. One robot is attached to each side of the machine and the units can be retrofitted to existing ring frames. Substantial reductions in machine down time, improvements in quality and reduced manpower are reported to be the benefits of the new system.

#### 4.4.1.1 Advancements in ring and traveller systems

Additional changes to the spinning frame include an innovative ring and traveller system where the orbit system has replaced the conventional T-flange system which has enabled reaching traveller speeds of up to 41 m/s (https://www.bracker.ch/fileadmin/bracker/products/EN/Spinning_Rings/orbit1.pdf). Similarly, the use of steel travellers and wires has enabled achieving higher spinning speeds and also a 50% longer life. Some of the other developments in the ring and traveller system are provided in Table 4.3 (Yin et al., 2021).

**Table 4.3** Innovative technologies and principles used for a ring and traveller system (Yin et al., 2021).

| Technology | Principle | Ring/traveller |
|---|---|---|
| ORBIT | Enlarged contact surface | Both |
| New falcon | Special heat and surface treatment | Traveller |
| Blackspeed | Lubrication film | Traveller |
| Champion | Fine grained ball bearing alloy steel | Ring |
| Cruzer | High-tech finish | Traveller |
| Sapphire plus | Special diffusion treatment | Traveller |
| Steelhawk | Non-metallic dry lubrication coating | Ring |

In another approach, the complete elimination of the ring and traveller system is also being considered. A friction-free superconducting magnetic bearing (SMB) system is the new alternative to replace the ring/traveller system. In this system, a repulsive magnetic force is applied to keep a permanent passive magnetic bearing floating ring which enables the fibres to be twisted and form a yarn, as shown in Fig. 4.3. Since friction between the metals (ring and traveller) is avoided, a considerably high spinning efficiency and increase in productivity by 200%–300% are achieved. Angular spinning speeds of up to 50,000 rpm have been reported as possible with this system (Hossain et al., 2018a, 2018b). The properties of polyester yarns obtained on the magnetic spinning system were found to be similar to those of conventional ring spun yarns. In another study, the SMB system was found to provide better yarn properties for 100% cotton and polyester when spun into 15, 20 and 30 tex with angular speeds of 5000, 10,000 and 15,000 rpm (Hossain et al., 2018a,b).

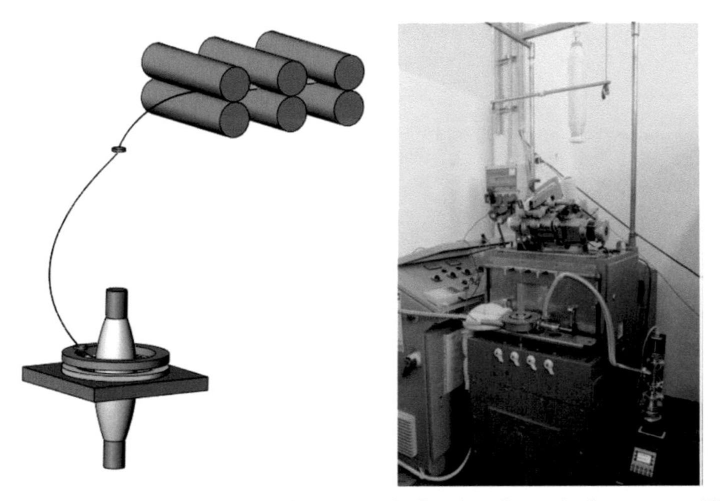

**Figure 4.3** Principle of a superconducting magnetic floating ring spinning system (Hossain et al., 2018a,b).
Reproduced with permission through open access publication.

### 4.4.2 Rotor spinning

The distinct advantages and economics of producing rotor spun yarns have made it the preferred method of yarn production, particularly for coarser counts (5–40 Ne). Rotor spinning is 5–7 times faster than ring spinning and coupled with output in the form of wound bobbins instead of spindles make it very convenient for further processing. Recent developments in rotor spinning provide 7% higher productivity mainly due to the simultaneous piecing of individual spinning positions. Similarly, newer cleaning technologies enable processing fibres with higher trash content and machines are equipped with auto-doff mechanisms leading to increased production. Current machines are equipped with up to 200 spinning positions and are capable of producing yarns at a delivery speed of 200 m/min. Ensuring readiness for Industry 4.0, the latest rotor spinning machines are integrated with turbostart, increased doffing capacity, increased piecing efficiency and higher waste tolerance. Rieter and Saurer are the two major manufacturers of rotor spinning, dominating the market with similar technology and production capabilities. However, Savio and Lakshmi Machine Works (LMW) India also offer rotor spinning machines with competitive features.

### 4.4.3 Torque spinning systems

No major progress in ring spinning has been reported since the development of compact spinning. A team from Hong Kong Polytechnic University have introduced a new method of physical yarn manufacturing where yarns are produced through torque balancing. The new technology called Nu-Torque, as shown in Fig. 4.4, is suitable for the production of yarns in the range of 7–100 Ne and can impart up to 25%–40% lower twists than traditional yarns. An increase of 25%–40% in productivity without

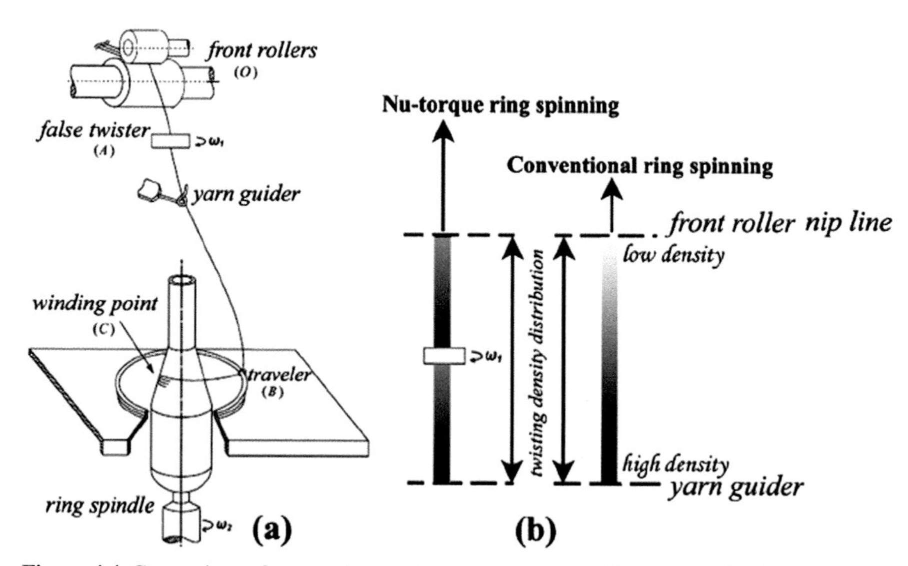


**Figure 4.4** Comparison of conventional spinning with the new NU-torque spinning system (Xia and Xu, 2013).
Reproduced with permission from Taylor and Francis.

compromising yarn strength and further benefits during knitting and chemical processing have been proposed. About 337 kW of energy savings per ton of yarn produced has been reported (Xia and Xu, 2013).

### 4.4.4 Friction spinning

Friction spinning or DREF spinning was considered to be the best available method to process fibres into yarns at low cost and with desirable properties. However, DREF spinning (DREF-1, 2 or 3) was not as popular as ring or rotor spinning. Studies have been conducted to further improve the friction spinning process and DREF 2000 and DREF 3000 systems are now in existence. The DREF 3000 system is capable of producing yarns from 0.3 Ne to 14.5 Ne in both Z and S configurations. Yarns produced on the DREF 3000 are reported to be ideal for high-tenacity and flame resistance protective clothing, fibres for the composite industry and for the aviation and construction sectors. However, friction spun yarns have only 60% of the strength of ring spun yarns and about 90% of rotor spun yarns depending on the type of fibres processed. Friction spun yarns also have higher hairiness and also more imperfections, etc. (Bhowmick et al., 2018).

### 4.4.5 Vortex and air-jet spinning

The concept of air-jet or vortex spinning introduced in the 1990s has gained gradual acceptance and also seen major developments in the machinery and also quality of yarns obtained. Some of the properties of vortex spinning machinery are provided in Table 4.4. Air-jet spun yarns have relatively lower hairiness, and better resistance

**Table 4.4** Specifications of the vortex spinning system (Begum et al., 2018).

| Specifications/parameters | Types/values |
|---|---|
| Raw materials | 100% cotton, synthetics and blends |
| Yarn count range | 15–60 Ne |
| Fibre length | 38 mm (1.5″ max) |
| Number of units | 16–96 |
| Spinning speed | 500–550 m/min |
| Total draught ratio | 65–400 |
| Brake draught ratio | 1.5–5.0 |
| Intermediate draught ratio | 1.1–5.0 |
| Main draught ratio | 15–60 |
| Feed ratio | 0.9–1.1 |
| Take-up ratio | 0.9–1.1 |
| Spindle inner diameter | 1–1.4 mm |
| Nozzle pressure ($kg/cm^3$) | 4–6 |
| Distance between front roller and spindle, mm | 20 |

to pilling and abrasion. Air-jet yarns are specifically suited as core yarns for apparel and other applications. Yarns from 10 to 80 s (Ne) counts can be spun, and both natural and man-made fibres with fibre lengths of up to 51 mm can be spun. Murata and Reiter have been the two major manufacturers of air-jet spinning machinery. Reiter's J26 is an automatic air-jet spinning machine with 200 rotors and six robots and a delivery speed of up to 500 m/min. Comparatively, the latest model VORTEX 870EX from Murata has been shown to have high speeds of up to 550 m/min for similar yarn counts (http://www.muratec-vortex.com/history/). Combined with compact yarns with good strength, air-jet spinning seems to be a viable approach for large-scale yarn spinning. The Reiter J26 system is capable of producing 100% combed yarns in both the S and Z twist configurations. Properties of yarns obtained on the air-jet spinning machine were found to be dependent on various parameters such as delivery speed, spinning air pressure and number of fibres in the yarn cross-section, etc., as depicted in Fig. 4.5. The wrapper-fibre angle and wrapper fibre ribbon width were reported to be the most significant factors affecting yarn tenacity during air-jet and vortex spinning (Mertová and Moučková, 2020).

Studies have shown that the tenacity of air-jet spun acrylic/cotton blended yarns increased with an increase in spinning speeds and ribbon width, but with 20%–30% lower tenacity compared to ring spun yarns (Eldessouki et al., 2015). Although vortex and air-jet spinning are similar, it has been found that vortex spun yarns provide greater strength. However, air-jet yarns showed better evenness compared to vortex yarns for polyester-cotton blends, as observed from Table 4.5 (Basal and Oxenham, 2003). Vortex spun yarns are also reported to have lower yarn and fabric wicking values compared to ring spun yarns due to their crimped yarn axis and tighter wrapping along the yarn length. Knitted fabrics made from vortex spun yarns also had a lower water absorbency rate (Erdumlu and Saricam, 2013).

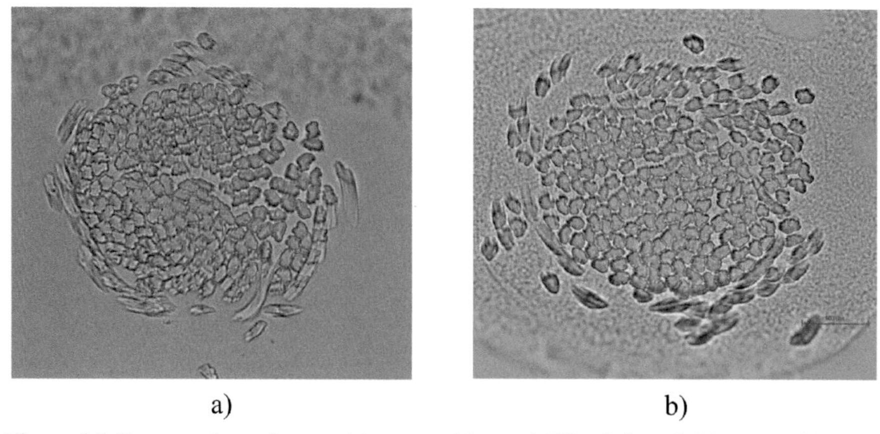


**Figure 4.5** Cross-sections of yarns: (a) spun at 4 bar and 400 m/min and (b) spun at 6 bar and 500 m/min (Mertová and Moučková, 2020).

**Table 4.5** Properties of vortex spun (MVS) compared to air-jet spun (MJS) polyester/cotton (P/C) yarns (Basal and Oxenham, 2003).

| | Tenacity (cN/tex) | | | | Elongation (%) | | | | Work (cN.cm) | | | |
|---|---|---|---|---|---|---|---|---|---|---|---|---|
| | Tensorapid | | Tensojet | | Tensorapid | | Tensojet | | Tensorapid | | Tensojet | |
| Blend ratio (P/C) | MVS | MJS | MVS | MJS | MVS | MJS | MVS | MJS | MVS | MJS | MVS | MJS |
| 100/0 | 23.43 | 23.4 | 24.9 | 24.0 | 9.6 | 12.9 | 9.3 | 10.5 | 1080 | 1281 | 1032 | 1124 |
| 83/17 | 22.07 | 18.9 | 22.5 | 20.5 | 9.6 | 11.4 | 8.9 | 9.7 | 1031 | 913 | 889 | 904 |
| 67/33 | 18.2 | 14.6 | 18.7 | 16.4 | 8.4 | 10.5 | 7.8 | 8.8 | 746 | 668 | 659 | 666 |
| 50/50 | 14.36 | 12.2 | 15.7 | 13.5 | 6.1 | 9.7 | 6.4 | 7.5 | 432 | 511 | 436 | 469 |
| 33/67 | 12.47 | 8.9 | 13.7 | 9.9 | 4.9 | 8.2 | 5.5 | 6.5 | 293 | 297 | 308 | 277 |
| 17/83 | 10.67 | | 12.2 | | 3.9 | | 4.9 | | 196 | | 240 | |

### 4.4.6 Comparison of spinning systems

The three major spinning systems (ring, rotor and air-jet) have several advantages and limitations, as shown in Table 4.6. Each type of spinning not only affects productivity and costs, but also the downstream processing and eventual end use. Ring spinning and rotor spinning are different, but have their own strengths and weaknesses and are well established spinning systems, particularly for cotton and denim (McLoughlin et al., 2015). Yarns produced by each system have structural and performance differences, as shown in Tables 4.6 and 4.7. An overview of the major developments in spinning machinery is provided in Table 4.8.

## 4.5 Reduction in waste generation

The spinning process from the blowroom to the spinning unit generates about 20%–25% of fibre as waste. Most of this waste is discarded in landfills. Hence, efforts have been made not only to optimise the process but also to choose appropriate machinery. For instance, vortex spinning is reported to generate less waste and also consume lower energy per kilogram of yarn produced (Goyal and Nayak, 2020). In addition to reducing waste generation, attempts have also been made to recycle various wastes generated during the spinning process. For instance, used clothes and garments wastes are collected and converted into fibres (reclaimed/regenerated fibres) and blended with virgin cotton and made into melange yarns. The blending of the fibres is done in the blowroom, drawframe or even on the speed or roving frame. Production of melange yarns has been increasing not only due to the impetus to reduce wastage but also to conserve energy and other resources required for processing yarns. It has been shown that benefits of melange yarn production are also dependent on the type and colour of yarns and fibres being combined. Cradle-to-grave life cycle assessment (LCA) of the production of melange yarns has shown that spinning and dyeing provide substantial energy savings (Liu et al., 2020a,b). Also, melange yarns are reported to have similar properties compared to normal yarns of the same twist and the same yarn count and hence are acceptable for practical applications, as shown in Fig. 4.6 (Wang et al., 2020).

## 4.6 Process automation and optimisation

Improving the production and sustainability of yarn production also relies on the use of innovative automation and optimisation techniques. For instance, adopting lean manufacturing techniques as part of the system of systems (SOS) approach has provided substantial benefits in terms of cost and reduction in wastage of materials. The results of adopting the lean manufacturing approach at a textile mill in South Asia are shown in Fig. 4.7. New software and mill management systems such as ARENA have been proposed to simulate actual spinning mill conditions before commencement of actual production (Halife and Alshukur, 2022).

**Table 4.6** Comparison of the advantages and limitations of the three major types of spinning systems (Goyal and Nayak, 2020).

| Spinning machine | Positive points | Quantitative data | Sustainability issues |
|---|---|---|---|
| Ring spinning | • Versatility in producing a wide range of yarn counts<br>• Uniform helical yarn structure<br>• The strongest yarns are produced using this system | • The highest energy cost/kg of yarn irrespective of yarn count<br>• The ring frame itself consumes 58% of the total energy of ring yarn production system Winding and roving making processes consume significant energy, which is absent in rotor and air-jet spinning<br>• Very low delivery speed in the range of 25–30 m/min | • The highest amount of electricity is consumed<br>• Significant waste generated during yarn production |
| Rotor spinning | • Suitable for producing coarse to medium yarn count range<br>• Bipartite structure comprising a helical core of fibres with outer zone of wrapper fibres, which occur irregularly along the core length<br>• Lower yarn strength than ring and air-jet yarns but regular, bulkier, and fewer imperfections than ring yarn | • Lowest energy cost/kg of yarn for coarser count like 30s but higher than air-jet for finer counts (40–50s) and always lower than ring yarns irrespective of count<br>• Only rotor spinning machine consumes 80% of the total energy, which is lower than for ring spinning<br>• Delivery speed is in the range of 200–250 m/min<br>• Production rate is 5–8 times higher than ring-spun yarn | • Low energy requirement and hence, more sustainable, particularly for coarser yarn count<br>• Low waste generation and also consumes waste from ring spinning process |

*Continued*

**Table 4.6 Continued**

| Spinning machine | Positive points | Quantitative data | Sustainability issues |
|---|---|---|---|
| Air-jet spinning | • Suitable for medium to finer count range<br>• Fasciated yarn structure comprised of parallel core fibres wrapped by helical sheath fibres<br>• The yarn strength is higher than rotor yarn but lower than ring yarn. Yarn stiffness is the highest | • The lowest energy cost/kg of yarn for finer counts (40–50s) but higher than rotor yarn for coarser count (30s) and always lower than ring yarns irrespective of yarn count<br>• Only air-jet spinning machine consumes 68% of the total energy<br>• Very high delivery speed in the range of 400–500 m/min<br>• Production rate is 10–15 times higher than ring-spun yarn | • Low energy requirement and hence, more sustainable, particularly for medium and fine yarn count<br>• Low waste generation |

**Table 4.7** Characteristics of the yarns produced by the four common spinning systems.

| Com4Ring | Com4Compact | Com4Rotor | Com4jet |
|---|---|---|---|
| • Most flexible in raw material, counts<br>• High tenacity<br>• High hairiness<br>• Pleasant soft touch and drape<br>• Good opacity | • Highest tenacity<br>• Even yarn structure<br>• Low hairiness<br>• High yarn density<br>• High fabric strength<br>• Clearly defined structures | • High optical evenness<br>• Low strength variation<br>• Designable hairiness<br>• High abrasion resistance<br>• Highest volume<br>• Uniform fabric appearance<br>• High abrasion resistance<br>• Good appearance after raising | • Unique low hairiness<br>• High volume<br>• Low fluffing<br>• High abrasion resistance<br>• Low pilling tendency<br>• High water absorption<br>• High wash resistance |

**Table 4.8** Overview of improvements in spinning machinery.

| Stage of spinning | Major developments | Implications |
|---|---|---|
| Blowroom | - Automatic bale openers with production up to 1600 kg/h; double plucking rollers; mote knife with suction instead of grid bar<br>- Automatic cotton sorter and removal of contaminants in raw cotton<br>- Reciprocating bale pluckers<br>- Multimixers with width up to 2000 mm<br>- Integral dedusters | Decrease tuft size up to 0.1 mg; lower neps and fibre loss<br>Ability to detect foreign particles, metals and incandescent particles<br>- Lower fibre loss, better evenness and dispersion<br>- Openers with capacity of up to 3000 kg/h<br>- Ability to operate with cotton having >2.5% impurities |
| Carding | - Gap optimisers for a constant, ideal carding gap even under changing production conditions<br>- Minimum carding gap of 3/1000″<br>- Low suction pressure of 740 Pa and low air requirement of only 4200 $m^3$/h<br>- Automatic sliver filling and can movement system<br>- Carding angle increased from 212 degrees to 247 degrees and arc length increased from 2.4 to 2.8 mm | Higher (57%) nep removal efficiency<br>- 40% lower yarn imperfections and 40% higher production<br>- 2% less waste<br>- Production rates of up 300 kg/h<br>- Web doffing speeds of up to 400 m/h<br>- Card coiling systems can save up to 30% space |

*Continued*

**Table 4.8 Continued**

| Stage of spinning | Major developments | Implications |
|---|---|---|
| Drawframes | - Automatic pin draughting<br>- Autoleveller drawframe tongue<br>- Autobreak draught setting<br>- Clean coil<br>- Tension measuring system<br>- Short term autolevellers<br>- Sliver monitoring cameras<br>- Production speed of up to 2 × 1000 m/min | - Clean coil reduces cleaning frequency<br>- Better control on sliver dimensions<br>- Reduced fibre imperfections, irregularities<br>- Automatic reduction in machine speed<br>- Output diameter of up to 1000 mm<br>- Draughting ratios of up to 0.001<br>- 10%–15% higher running efficiency |
| Simplex/ roving frame | - Automatic winding, draughting, twisting and transportation<br>- Spindle speeds up to 2000 rpm<br>- Roving bobbin transport system<br>- Electronic driving system | - Lower labour costs<br>- Lower power consumption |
| Ring spinning | - Spindle speed up to 20,000–25,000 rpm<br>- Shorter spinning triangle<br>- Superconducting magnetic bearings | - Lower end breaks<br>- Increase productivity<br>- Lower energy consumption<br>- Stable rotation at high speeds |

Adapted from Shi et al., (2020).

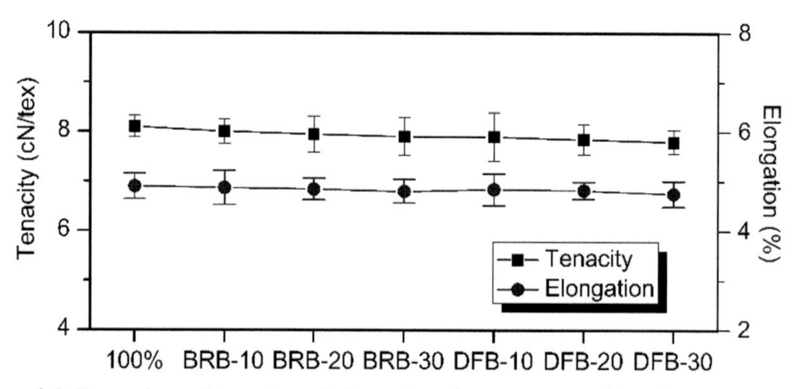


**Figure 4.6** Comparison of tenacity and elongation of cotton yarns with melange yarns containing various proportions of recycled fibres (Wang et al., 2020).
Reproduced with permission through open access publication.

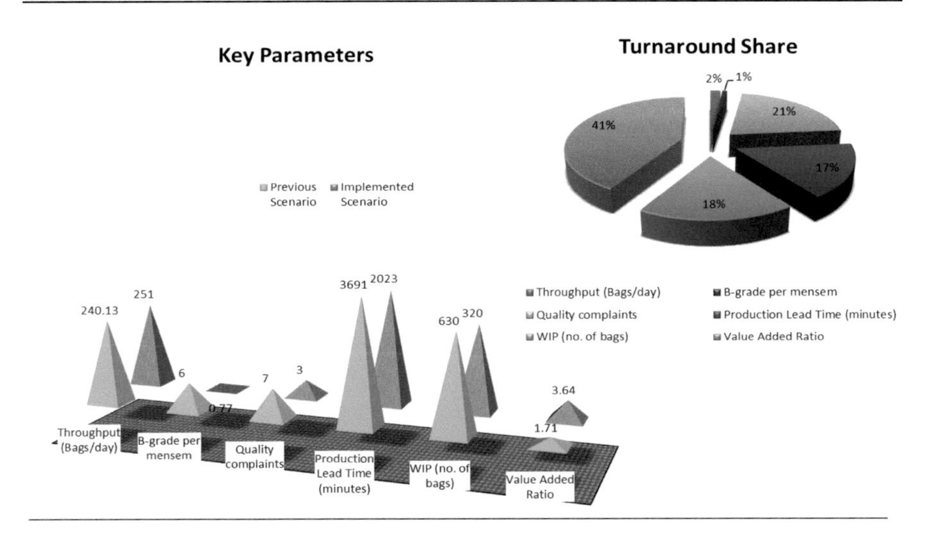


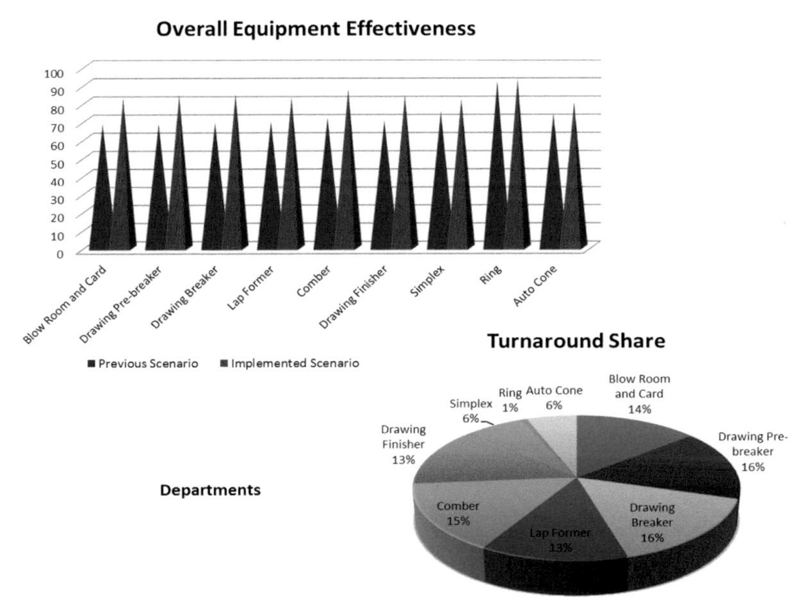


**Figure 4.7** Steps in adopting lean manufacturing principles and the benefits at different stages of spinning (Mahmood, 2020).

Not only entire plants, but even individual machines have been monitored and automated to ensure efficiency and decrease machine downtime and yarn quality issues. For instance, the total productive maintenance (TPM) approach has increased overall equipment efficiency (OEE). Using a fishbone diagram, the efficiency of ring frame was increased from 75% to 86% (Hossen, 2016).

## 4.7 Future perspectives

Ring and rotor spinning continue to be predominant technologies for converting fibres into yarns. Despite technical and economic challenges, the two systems have prevailed and proven their suitability for a wide range of yarn production. However, the need to further increase efficiency, lower costs and improve quality are posing challenges to both machinery manufacturers and fibre processors.

There have been no breakthrough developments in spinning machinery in the last decade, although process efficiencies and cost savings are being achieved, thus contributing to sustainability. Techniques such as air-jet spinning or a hybrid system that can provide better performance and lower costs should be considered for sustaining good margins during spinning. Likewise, efforts should be focused towards process optimisation and cost savings from the blowroom to the spinning equipment.

## 4.8 Conclusion

Progress in spinning processes and machinery has led to considerable improvements in yarn quality, productivity and reductions in cost. Nevertheless, challenges in terms of making the spinning industry more sustainable persist. The high cost of the machinery, variations in the quality of raw materials and quick adaptation to changing yarn quality and counts are constraints that need to be addressed. Increasing environmental restrictions and limited availability of resources will increase the pressure on the textile industry, and particularly spinning, to further innovate and ensure optimum use. Efficient recovery and reuse of the spinning process wastes should also be considered. Overall, the developments in spinning machinery and processes provide a positive outlook to improve the sustainability of the textile industry.

## Acknowledgements

Narendra Reddy thanks the Center for Incubation Innovation Research and Consultancy for their support to complete this work.

## References

Basal, G., Oxenham, W., 2003. Vortex spun yarn vs. air-jet spun yarn. Autex Research Journal 3 (3), 96–101.

Begum, H.A., Khan, Md K.R., Rahman, Md M., 2018. An overview on spinning mechanism, yarn structure and advantageous characteristics of vortex spun yarn and fabric. Advances in Applied Sciences 3 (5), 58–64.

Bhowmick, M., Rakshit, A.K., Chattopadhyay, S.K., 2018. Dref-3 yarn structure with plied staple fibrous core. Research Journal of Textile and Apparel 22 (3), 235–246. https://doi.org/10.1108/RJTA-09-2017-0044.

Dhayaneswaran, Y., Ashok Kumar, L., 2013. A study on energy conservation in textile industry. Journal of the Institution of Engineers (India): Series B 94 (1), 53–60.

Eldessouki, M., Ibrahim, S., Farag, R., 2015. Dynamic properties of air-jet yarns compared to rotor spinning. Textile Research Journal 85 (17), 1827–1837.

Erdumlu, N., Saricam, C., 2013. Wicking and drying properties of conventional ring-and vortex-spun cotton yarns and fabrics. The Journal of The Textile Institute 104 (12), 1284–1291.

Goyal, A., Nayak, R., 2020. Sustainability in yarn manufacturing. In: Sustainable Technologies for Fashion and Textiles. Woodhead Publishing, pp. 33–55.

Halife, H., Alshukur, M., 2022. Simulation for the optimization of technical study and scheduling required to design a new spinning mill. Arabian Journal for Science and Engineering 47 (2), 1375–1386.

Hossain, M., Abdkader, A., Cherif, C., Berger, A., Sparing, M., Hühne, R., Schultz, L., Nielsch, K., 2018a. Potential of high performance ring spinning based on superconducting magnetic bearing. International Journal of Industrial and Manufacturing Engineering 12 (5), 512–516.

Hossain, M., Abdkader, A., Cherif, C., 2018b. Analysis of yarn properties in the superconducting magnetic bearing-based ring spinning process. Textile Research Journal 88 (22), 2624–2638.

Hossen, J., 2016. Improvement of Overall Equipment Efficiency (OEE) of Ring Frame Section of a Spinning Mill a Case Study.

Liu, Y., Zhu, L., Zhang, C., Ren, F., Huang, H., Liu, Z., 2020a. Life cycle assessment of melange yarns from the manufacturer perspective. International Journal of Life Cycle Assessment 25 (3), 588–599.

Liu, Y., Huang, H., Ren, F., Wang, Y., Liu, Z., Ke, Q., Li, X., Zhang, L., 2020b. Cradle-to-gate water and carbon footprint assessment of melange yarns manufacturing. Procedia CIRP 90, 198–202.

Mahmood, A., 2020. Smart lean in ring spinning—a case study to improve performance of yarn manufacturing process. The Journal of The Textile Institute 111 (11), 1681–1696.

McLoughlin, J., Hayes, S., Paul, R., 2015. Chapter 2 - Cotton fibre for denim manufacture. In: Paul, R. (Ed.), Denim. Woodhead Publishing, pp. 15–36.

Mertová, I., Moučková, E., 2020. Effect of the selected machine variables on the structural characteristics of air-jet spun yarn. The Journal of The Textile Institute 111 (11), 1556–1566.

Quan, J., He, Q., Cheng, L., Yu, J., Xue, W., 2022. Investigation into novel drafting systems on ring spinning frame for improving yarn properties. Textile Research Journal, 00405175211073824.

Shao, R., Cheng, L., Xu, Z., He, S., 2022. Effect of pre-twist in the spinning triangle on yarn strength. The Journal of The Textile Institute 1–7.

Shi, J., Liang, W., Wang, H., Memon, H., 2020. Recent advancements in cotton spinning machineries. Cotton Science and Processing Technology: Gene, Ginning, Garment and Green Recycling 165–190.

Wang, H., Memon, H., Abro, R., Shah, A., 2020. Sustainable approach for mélange yarn manufacturers by recycling dyed fibre waste. Fibres and Textiles in Eastern Europe 3, 141–149.

Xia, Z., Xu, W., 2013. A review of ring staple yarn spinning method development and its trend prediction. Journal of Natural Fibers 10 (1), 62–81.

Xia, Z., Xu, W., Zhang, M., Qiu, W., Feng, S., 2012. Reducing ring spun yarn hairiness via spinning with a contact surface. Fibers and Polymers 13 (5), 670–674.

Yin, R., Ling, Y.L., Fisher, R., Chen, Y., Li, M.J., Mu, W.L., Huang, X.X., 2021. Viable approaches to increase the throughput of ring spinning: a critical review. Journal of Cleaner Production 323, 129116.

# Innovations in weaving, knitting and nonwoven machinery

*Jan Jordan, Arash Rezaey, Leon Reinsch and Frederik Cloppenburg*
Institut für Textiltechnik der RWTH Aachen University, Aachen, Germany

## 5.1 Introduction

Textiles are manufactured using various well-established technologies such as weaving, knitting and nonwovens. Weaving is the oldest manufacturing technique and the most popular types of woven textiles are with plain, twill or satin weaves. In plain weave, the warp yarns are held parallel under tension while a crosswise weft yarn is shot over and under alternate warps across the width of the web. The twill weave is distinguished by diagonal lines and is created by the weft crossing over two warp yarns, then under one, the sequence being repeated in each succeeding pick. Twills with more warps than wefts floating on the fabric face are called warp faced and those with wefts predominating are called weft faced. In general, satin weave resembles twill, but there are no clear diagonal lines. Satin fabrics are very smooth and are generally considered as luxury fabrics.

In knitting, the fabric is produced by employing a continuous yarn or set of yarns to form a series of interlocking loops. Knitted fabrics can generally be stretched to a greater degree than woven fabrics. The two main types of knits are weft knits, which include plain, rib, purl, pattern and double knits, and warp knits, which include tricot, raschel and milanese. Knitted fabrics can be produced in both flat and tubular forms. Weft knits are most often tubular and warp knits are usually flat. Flat filling knits can be shaped by a process called fashioning, in which stitches are added to some rows to increase width, and two or more stitches are knitted as one to decrease width. Tubular knits are shaped by tightening or stretching stitches (Britannica, 2023).

Nonwovens are innovative, high-tech, engineered fabrics made from fibres, and have the most complex production processes in the textile sector. Although nonwovens are considered as high-tech textiles, their production is often quite conventional. They are used in a wide range of consumer and industrial products, either in combination with other materials or alone. Nonwovens are designed for their specific application, ranging from thin, light-weight nonwovens to strong and durable nonwovens, either for consumer or industrial applications. The combination of their specific characteristics through the raw materials selection, formation and bonding methods used or the applied finishing treatments, such as printing, embossing, laminating, etc. allow the delivery of high-performance products (EDANA, 2023).

Regarding the weaving and knitting machinery, whereas 94% of all worldwide deliveries of shuttle-less weaving machines and 81% of flat and circular knitting machines are sold to Asia and Oceania (ITMF, 2021), the drivers, initiatives and

**Sustainable Innovations in the Textile Industry. https://doi.org/10.1016/B978-0-323-90392-9.00006-9**

concepts related to sustainability in textile machinery can be investigated by exploring the changes in sustainability-related legislation in the EU. The nonwovens industry is also driven by sustainable technology developments in machinery, process control and materials.

## 5.2 Historical aspects and current scenario

Over the course of the industrial revolutions, a main target of innovation in textile machinery has been an increase in productivity. Nevertheless, it can be assumed that during the early beginnings of textile manufacturing, the speeds were rather moderate. Woven fabrics, for example, were manufactured as early as 27,000 years ago according to archaeological findings (Whitehouse, 2000). For the manufacturing of woven fabrics in the Neolithic period among others a warp-weighted loom can be assumed to have been used (Barber, 1991).

The further development of weaving mechanisms resulted in automated, faster selection and movement of the heddle frames or harness yarns during shedding (e.g., with a treadle loom, dobby head or Jacquard head). The weft insertion, which was at first carried out manually, was accelerated with the invention of the flying shuttle. With the availability of steam power, transmitted power from water wheels, as well as nowadays electrical power, the automation and acceleration of machinery parts evolved. The development of shuttle-less weft insertion enabled the usage of weft-carrying devices of reduced weight, as in projectile or rapier weaving, or even relying only on streaming media as a weft carrier as in air jet and water jet weaving.

Whereas knitting is considered to have emerged later in history than weaving, a technique called Nålebinding (needle-binding), which makes use of a single needle for the formation of yarn loops, resulted in textile structures that have been found at archaeological sites dating back to as early as 6500 BC (Andersen, 1985). The invention of the first knitting machine in the fashion of a flat-knitting machine dates back to 1589, when William Lee filed a patent for his 'Stocking Frame' (Amos, 2010). The first circular arrangement of knitting needles was introduced by Marc Brunel in 1816. The next evolution in 1878, by Henry Josiah Griswold, was the addition of a second set of needles which were placed in a disc horizontally on the top of the circular knitting machine. This addition enabled rib knitting and the cut-off or welt for socks (Hawkins, n.d.). Further evolutions resulted, e.g., in individual motions of the knitting needles as well as computerised patterning.

In periods and regions in which resources were scarce, the aspects of resource efficiency and product lifetime extension have already played an important role before the current increased awareness of the need for sustainable consumption and production (FTR, 2022). Besides the renewed formation of yarns, the manufacturing of nonwovens is an important process for end-of-life textile materials. The term nonwoven fabrics was first coined in 1942 and they were produced in the United States by adhesively bonding fibre webs. The first definition of nonwoven fabrics came from the

American Society for Testing and Materials in 1962 which defined them as textile fabrics made of carded web or fibre web held together by adhesives.

As with the growing significance of End Producer Responsibility, the standards and regulations in Europe have made an impact on the global value chains (and hence to a certain extent on the machinery used). The European Textile Strategy (European Commission, 2022a,b) and the 'scenarios towards co-creation of a transition pathway for a more resilient, sustainable and digital textiles ecosystem' (European Commission, 2022a,b) lay out a number of competitive strengths and challenges of the European textile industry which may provide orientation to textile machinery manufacturers globally on the demands and requirements that their customers might require. Among the sustainability-related competitiveness, strengths and challenges identified (European Commission, 2022a,b) as relevant to textile machinery manufacturers are:

- Rapid integration of new and innovative materials
- Increasing experience with business models based on reuse, recycling and circularity
- Integration of sustainable and circular practices in the value chain to reduce the environmental footprint of the sector
- Skills gaps and shortages
- Ageing workforce

Another set of sustainability-related aspects is listed by the United Nations Environment Programme (UNEP) (United Nations Environment Programme, 2015) on 'Sustainable Consumption and Production':

1. Improving the quality of life without increasing environmental degradation and without compromising the resource needs of future generations.
2. Decoupling economic growth from environmental degradation by:
   - Reducing material/energy intensity of current economic activities and reducing emissions and waste from extraction, production, consumption and disposal
   - Promoting a shift of consumption patterns towards groups of goods and services with lower energy and material intensity without compromising quality of life.
3. Applying life-cycle thinking which considers the impacts from all life-cycle stages of the production and consumption process.
4. Guarding against the re-bound effect, where efficiency gains are cancelled out by resulting increases in consumption.

The sustainability-related aspects (European Commission, 2022a,b; United Nations Environment Programme, 2015) can be translated to following Key Objectives (KO) addressing the sustainability-related requirements for textile machinery manufacturers:

A. Reduce resource and energy consumption during production
B. Avoid the emission of hazardous noise, particles or substances
C. Enable the processing of 'sustainable' yarn materials
D. Enable the manufacturing of products which contribute to sustainability
E. Support skills-building for machine operators during use
F. Enable repair, maintenance and retrofitting of machinery

In the following sections, textile machinery design and production characteristics for weaving, knitting and nonwoven manufacturing are discussed with regard to their relations to the above listed sustainability-related key objectives.

## 5.3 Innovations leading to sustainability, energy and resource efficiency

### *5.3.1 Weaving machinery and textile manufacture*

According to ISO 3572, a woven fabric is produced by interlacing (by weaving on a loom or a weaving machine) a set of warp threads and a set of weft threads normally at right angles to each other. Weaving processes are distinguished in particular according to their weft insertion system and shedding mechanism. The weaving process has a decisive influence on the fabric properties, so that the required fabric properties must also be taken into account when selecting the weaving process (Adanur, 2001). Furthermore, the number of repeats differs depending on the weaving machine. A repeat is the repetition of the thread course of warp and weft in the fabric (Cherif, 2011).

The shedding is usually done by means of a cam, dobby or jacquard. In addition, various shedding mechanisms are available for the production of leno fabrics. Furthermore, there are special procedures for shedding in order to separately interlace individual warp threads with the inserted weft yarn. For narrow woven fabrics, a high-speed shedding mechanism based on the use of a linear motor is available.

The influence of different types of shedding mechanisms on sustainability depends on the capability of processing sustainable yarn materials (e.g., recycled or biobased yarns →KO C), avoiding the release of hazardous particles (e.g., sizing or microfibre release/defibrillation →KO B), limiting the emission of excessive noise and reducing energy consumption. A very significant impact on the sustainability-related characteristics is determined by the selection of weft insertion methods. Hence, selected methods of weft insertion are discussed below with regard to their sustainability-related characteristics.

#### *5.3.1.1 Shuttle weaving*

The weft insertion method with the longest history might at the same time (not necessarily from an economic perspective) be the most sustainable method: shuttle-weaving on handlooms. Due to the formation of a true selvedge with high quality, there is no necessity to cut off the edges and the energy consumption is limited to the energy consumption of human metabolism (→KO A). Sustainable yarn materials are easily processable due to the rather low insertion velocities and accelerations causing a low load on yarns, also making the processing of sensitive yarns possible (→KO C).

The processing of paper machine clothing (PMC), which has been used already for decades in the manufacturing of paper from recycled cellulose pulp (→KO D), require the weft insertion to be carried out over wide widths (up to 31 m). This weft insertion cannot be carried out in hand weaving and is instead realised via a powerful hydraulic propulsion system. In order to limit energy consumption, machinery manufacturers of such broadlooms apply their machinery with a capacitive energy recovery system (→KO A).

### 5.3.1.2 Projectile weaving

Projectile weaving is the most energy-efficient weft insertion method of automated looms, as the energy required to accelerate the light projectile (40 g) to the desired speed is almost negligible compared to the energy required for shed formation and reed beat-up (→KO A). Due to the high peaks of acceleration in the initiation of weft insertion, sensitive yarns cannot be processed with this insertion method. The guiding dents required to keep the projectile on a stable trajectory can have, furthermore, a negative impact on the fabric's optical appearance due to friction with the warp yarns during reed beat-up and shedding. For these reasons projectile weaving has a limited range of processable yarn materials.

Regardless of the method of weft insertion, the support of skills-building for machine operators during use (→KO E) and enabling repair, maintenance and retrofitting of machinery (→KO F) can be supported via digital technologies in weaving. In the following, a methodology on how the most suitable digital technologies can be selected for achieving the key objectives is described.

### 5.3.1.3 Digital technologies in weaving

Digitalisation and the digital integration of processes in textile production is a topic that has gained importance in the context of sustainability from the production point of view. The conversion of data, which describe the real world in a number-based system, is called digitalisation. The converted data can be processed, saved and transferred easily in a digital system. Increasing effectivity and efficiency using digital technologies in the textile production chain reduces waste production and addresses directly the first key objective: Reduce resource and energy consumption during production (→KO A). Building skills used during machine operation (→KO E) could also be supported by digital technologies.

The introduction and implementation of digital technologies in a production chain needs transformation of business processes. Surveys have shown that decision-making processes for digital transformation needs to be supported by methodical approaches (Jeglinsky and Winkler, 2020). The technological complexity of the textile production chain due to the variety of materials and products makes it difficult to find suitable concepts with best effects on the reduction of waste production. The 'Guideline for the Implementation of Smart Factory Concepts' is a tool supporting this decision-making (Rezaey et al., 2022). The aim of this tool is to increase effectiveness and efficiency by avoiding quality-related errors. The approach is an analytical complexity management one to determine the causes and effects of problems. The requirements of the digital solution are determined during the application of the guideline.

This procedure is easy to understand and is suitable for being independently transferred by the respective company into a continuous improvement process (CIP) in terms of digitalisation. The company in question is initially supported in carrying out the steps by an external team of experts (referred to below in simplified terms as the 'expert team'). In a participative approach, the employees are involved by the management in determining the requirements of the digital solution. The expert team (e.g., Mittelstand digital competence centre) acts as a neutral supporter. The individual steps are discussed below.

Description of requirements: At the beginning, a discussion is held between the expert team and the company at the executive level. During this meeting, the already-known problems that cause additional production costs due to production errors are discussed. The aim is to work out the company's requirements for a continuous improvement process (CIP). In the first step, a general overview of the existing production processes and the already-known bottlenecks as well as the improvement and solution possibilities proposed by the company and their scope of functions is obtained. In this step, a sample product is also defined as the focus for the improvement process. In addition, the system boundary of consideration is also drawn and the project team is determined. Subsequently, the already-known system properties of the digital solution required for the CIP are determined.

Description of the system: The goal of this phase is to create a detailed description of the production system. With the participation of employees, the team of experts conducts detailed analyses of the company to identify the causal chain between the errors and their causes. To do this, a process flow chart is first drawn up for the example product according to the phase model of production (Schulze and Ebert, 2002). The phase model of production provides a generalised overview of the factors influencing the production processes and the quality-relevant properties of the products. This model is used in the following steps to carry out the root cause analysis for the production errors and to create a process data model.

Error analysis: In the next stage, a Failure Mode and Effects Analysis (FMEA) for products and processes is carried out. The errors, their causes and consequences are recognised in this stage. For each error a risk priority number based on its importance is then calculated. In this way the most important errors can be identified.

Determination of critical error paths: Subsequently, a process data model which includes all input and output variables of the processes and the quality-relevant properties of the products is generated. Here, the critical error paths between the errors and the parameters that cause them (input variables) and detection possibilities (output variables, quality-relevant properties) are determined.

Concept overview, concept selection, implementation and validation: In the final step, concepts for digitised solutions for reducing the severity of the most important errors are developed. The concepts are analysed regarding their implementation time and costs, and the most relevant ones are prioritised for implementation. Here, it is important to strongly involve the stakeholders on the company side in order to ensure that the requirements from the user's point of view are considered. The digital solution can then be implemented and validated.

### 5.3.2 *Knitting machinery*

Knitting machines not only have to produce innovative, high-quality and sustainable products in demanding production environments, but the production process itself must be increasingly sustainable too. In general, knitting machines have a great impact on energy and material consumption, product quality and wastage, subsequent processes as well as major implications for social sustainability, e.g., for machine operators. Overall, additional aspects of sustainability and individualisation need to be considered (Elmogahzy, 2019) and knitting machinery needs to be equipped to address those aspects (Elmogahzy, 2019).

Furthermore, the design and selection of machine types and equipment are decisive factors for the overall properties of products and their environmental impact throughout the production process and the product life cycle. As knitting machines are multifaceted in characteristics, underlying working principles and individual setups, new machine developments introduce additional potential for sustainable usage in a circular economy. The German manufacturer Mayer & Cie. GmbH & Co. KG, Albstadt, Germany, recently presented their strategy to foster digital continuity in the value chain of machine manufacturing. The diversity of machine types and parts is increasingly organised with the help of product lifecycle management.

#### 5.3.2.1 *Weft and warp knitting machines*

The fundamental distinction between knitting machines and the resulting products can be made between weft and warp knitting. Each of these types is further classified into additional subdivisions, manly determined by characterisation of the machine elements and resulting products. Important characteristics are the needle beds, the needles themselves and the movement and control of the needles since those are the loop-forming elements and the degrees of motion (Ray, 2012). In weft knitting machines, needles are moved one after the other, while needles on warp knitting machines are moved simultaneously. These working principles lead to dissimilar loop structures, as shown in Fig. 5.1.

#### 5.3.2.2 *Productivity*

Production efficiency and fabric quality have long been the focus of topics, e.g., in circular knitting, that also have a high impact on sustainability aspects currently (Au, 2011) The redesign of machine parts in order to achieve higher productivity has reached its limit to a large extent (Ray, 2012; Spencer, 2001). On the one hand, limitations of materials and components have been exploited in the last decade already. On the other hand, a redirection of efforts from a technology-based industry towards innovation has a major impact on current developments concerning machinery and its usage. Productivity certainly remains an important factor, but constraints from design choices, product requirements and quality specifications limit the choice of options. The determination of realistic machine speeds is important both within existing machine stocks and, in particular, for new product developments and machine purchases.

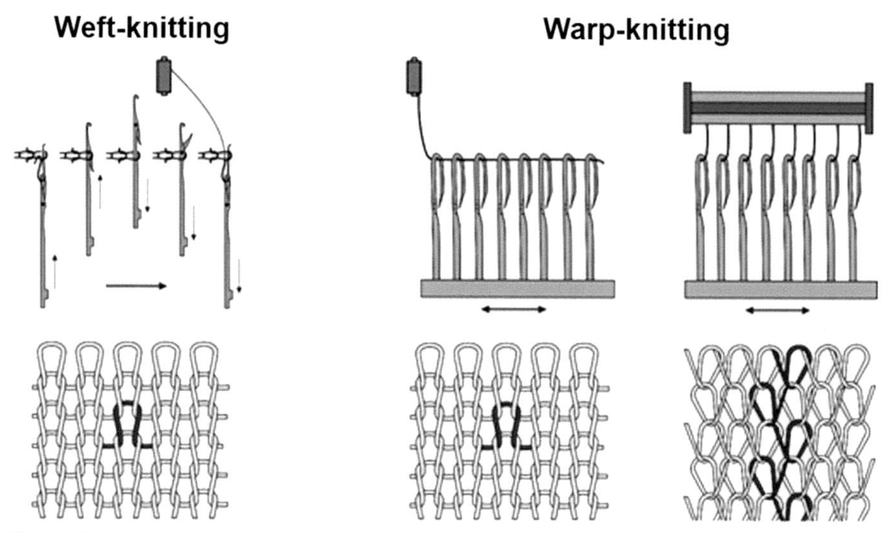


**Figure 5.1** Working principles and textile structures from weft and warp knitting machines. Adapted from Saggiomo et al. (2017).

Due to the huge variety of machine types, product types, potential process conditions and controls, the predictive estimation of machine efficiency can be a challenging task.

Circular knitting machines are generally characterised by very high productivity. This high productivity follows from the high number of knitting systems placed on the circumference of the formed needle carriers. One row of stitches can be produced on each knitting system per machine revolution. Therefore, the productivity parameters can also be related to the cylinder circumferential speed and the number of knitting systems on the cylinder circumference in relation to the nominal diameter (Cherif, 2011).

However, generalised or optimal efficiency values are not sufficient if sustainability metrics are to be determined or optimised for individual products and processes. Innovative developments of machinery and equipment constantly expand the range of what is possible. A comparison of common knitting technologies is shown in Fig. 5.2.

Maximum output on a particular machine can be achieved when all available knitting systems are in usage, machine speed is set to a maximum and with the highest fineness that fits into the needle heads and in the needle gaps. For example, design choices or quality aspects with important implications for sustainability can limit productivity. For technical yarns it may be advisable to reduce the machine speed to avoid broken needle heads, downtime and fabric wastages.

In turn, machine and process conditions as well as resulting product quality have major implications for the sustainable production of knitted textiles. Firstly, those aspects have to be taken into account for innovation in knitting machinery and products, and secondly those values need to be accurately tracked for transparent production.

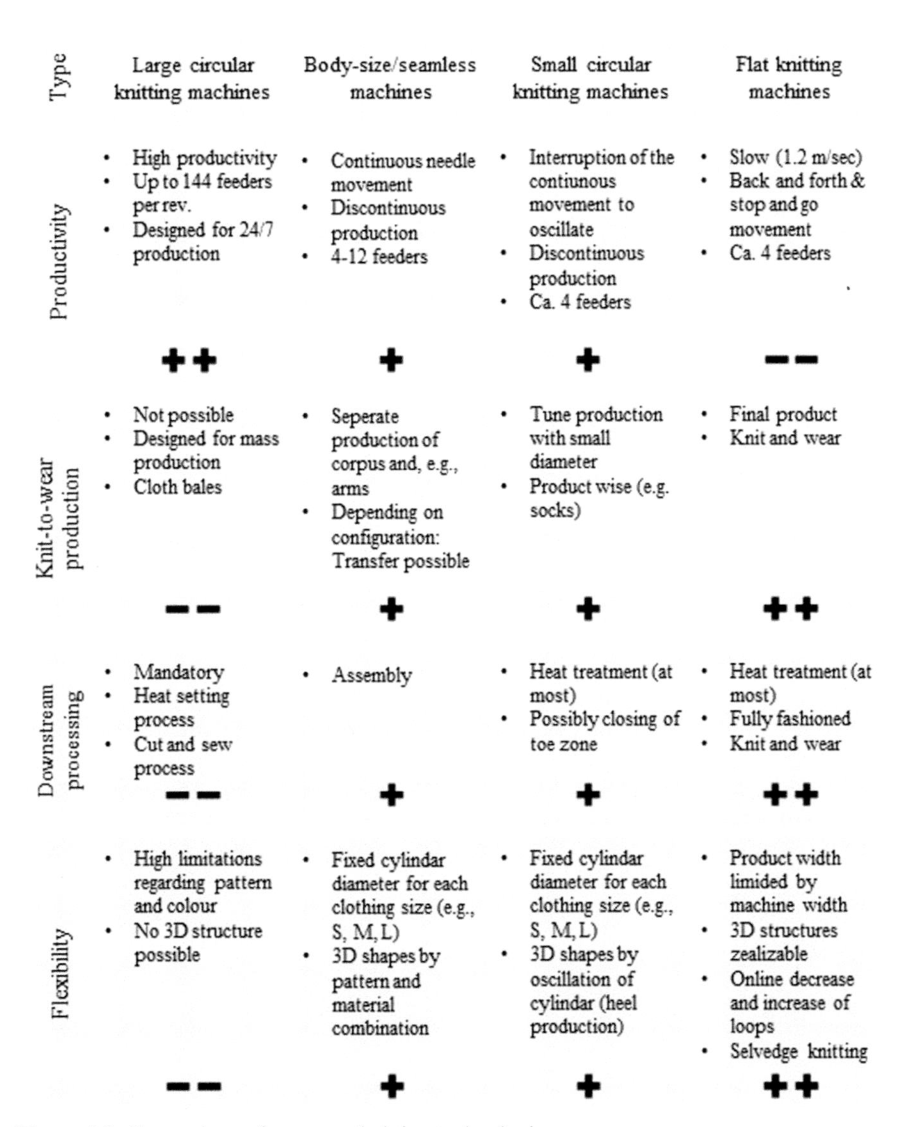


**Figure 5.2** Comparison of common knitting technologies.
Based on Saggiomo et al. (2017).

With the introduction of computerised knitting machines some decades ago, a gain in productivity was already unlocked, e.g., by enabling traversing of the cam carriages only as required (Spencer, 2001). Karl Mayer Stoll Textilmaschinenfabrik GmbH, Obertshausen, Germany, a manufacturer of flat knitting machines, introduced the so-called 'ADF generation'. ADF stands for 'Autarc Direct Feeding' and allows for a decoupled movement of cam and yarn carriages. The actual exploitation for productivity gains and product versatility of similar innovations depends on the application

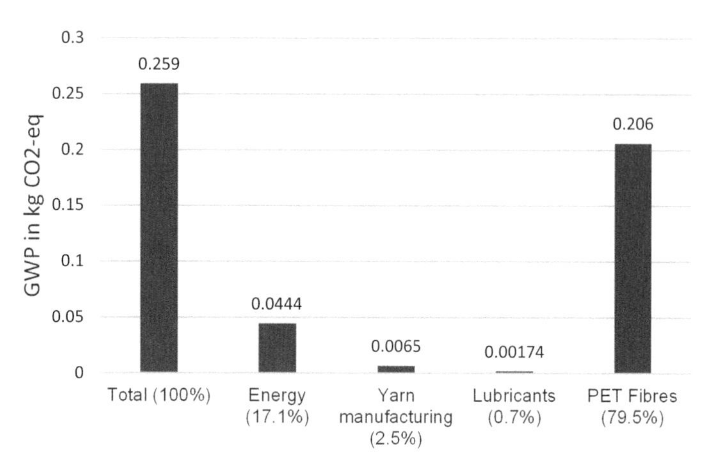


**Figure 5.3** Global warming potential of an example knitting process.

and machine control. Current research is directed towards planning and optimisation of needle transfer (Lin et al., 2018) and yarn connection paths (Huh and Kim, 2022). These are only two examples of the need for simultaneous development of hardware and software.

The knitting process itself is not considered to contribute very much to the overall footprint of knitted textiles. A study at RWTH-ITA found that with 80% of the $CO_2$ footprint stems from the yarn material itself, while only about 17% can be attributed to the production of raw materials. In Fig. 5.3, the global warming potential (GWP) of an example knitting process is given.

This calculation only illustrates a small excerpt from the full process chain, namely the production of a grey fabric, and also considers only a single process state. For sustainable innovations in knitting machinery and production, additional factors need to be considered, such as:

- Current and upcoming innovations in yarn materials might reduce the relative impact of yarn materials
- The versatility of knitting machinery that must be examined individually
- The implications for the knitting process on subsequent processes like transportation or finishing
- The general availability of process data for life cycle assessments (LCAs) and digital process shadows.

For a transparent record or a realistic prediction of sustainability metrics, technical hardware of knitting machinery as well as data collection and processing are decisive elements. For every LCA the area under consideration and the actual data that need to be acquired are largely determined by the knitting machinery and need to be facilitated by its sensory equipment and software. The actual productivity and energy consumption alone can vary widely between different knitting technologies and even between different process conditions on the same machine.

The specific decision on required and unitised data needs to be made for each LCA individually. However, it becomes evident that the exact impact of knitting processes may very well be impacted by the type of the knitting technology, its usage and process condition compared to an established manufacturing process; starting a new production series while the machine is in a 'cold state' the energy consumption will be increased. Knitting technology not only must contribute to reducing negative environmental impacts in the future but also needs to support data acquisition for sustainability assessments and digital product passports.

### *5.3.2.3 Weft knitting: Resource saving with zero waste*

Knitting in principle is suitable for producing any conceivable shape with targeted placement of desired materials and stitch architecture. In this science, an impressive community has formed to investigate the capabilities of knitted textiles. For instance, the knittability of complex mathematical shapes generally is feasible. The unlimited design freedom specifically can be exploited to sustainably produce any desired product with minimal waste as well as additional and elaborate process steps. In industrial settings, the realistic capabilities are limited by the available knitting machines and the surrounding knitting technology compared to the unbound freedom of movement in hand knitting (McCann et al., 2016; Albaugh, 2016; Matsumoto, 2021).

Special types of knitting machines and technologies are designed and evolving to further unlock the capabilities of knitting for specific products, applications and sustainable production of the same. Established techniques are near-net-shape production on flat knitting machines by means of needle transfer, decreasing and increasing, different kinds of spacer fabrics, tubular structures and 3D shapes by, e.g., held loops or partial knitting (Underwood, 2009; Liu and Hu, 2015; de Araujo et al., 2011). Most these principles have been well described in literature and implemented by all leading manufacturers of flat knitting machines such as Karl Mayer Stoll Textilmaschinenfabrik GmbH, Obertshausen, Germany, Shima Seiki Mfg., Ltd., Wakayama, Japan and Steiger Participations Sa, Vionnaz, Switzerland.

Current tasks regarding the potential of zero waste production for sustainability and a circular economy include the facilitation and scalable utilisation of resource-saving production. A variety of examples exist, including applications such as well-established sweater production and innovative techniques in architecture (Popescu et al., 2021). However, knitting technology in general and 3D knitting in particular act as versatile implementation tools for sustainable textiles and act as enablers on different levels. The specific adaptation of machine components for the integration of new materials is just as important as the capability to integrate knitting technology into local production environments.

With respect to new materials, Memminger IRO GmbH, Dornstetten, Germany, recently added a positive feeding device for flat knitting machines to their portfolio (MTD Feeder). Shima Seiki Mfg. Ltd. announced the prototype of a 'Multi-Axial Computerised Flat Knitting Machine' which allows for knitted reinforcement materials in 3D shape (exhibit at JEC World, 2022). Stoll by Karl Mayer flat knitting machines are also adapted for local reinforcements with glass, thermoplastics and high-strength fibres. Consequently, the trend is directed towards new materials and functionalisation in an integrated production technology.

On the other hand, versatile knitting machines can be further optimised to support local production and recycling in micro-factories and small studios, and therefore has a tremendous impact on sustainability aspects beyond material savings.

The design, capability and accompanying business models of knitting machines are influencing fast fashion, over-production, transportation and recycling among others. Hennes & Mauritz AB (H&M), Stockholm, Sweden, presented an in-store recycling line described as loop. Adidas AG, Herzogenaurach, Germany, sold individualised sweaters in a so-called storefactory and the startup, Kniterate, is offering a reduced version of an industrial knitting machine for semi-professionals and even hobbyists.

#### *5.3.2.4 Onloom fabric inspection systems*

Another contribution to resource-saving production is provided by technologies of onloom fabric inspection systems. Such systems support detecting failures in fabrics instantly after their occurrence. Hence, correction measures can be applied to the process in order to avoid the manufacturing of defective fabrics. Usually, only during the step of fabric inspection after the production of a complete roll of fabric is the detection of failures implemented. Having a system in place which makes it possible to correct the process almost in real time can help in preventing the discarding of complete fabric rolls. One example of such an onloom fabric inspection system is shown in Fig. 5.4. Another potential of such onloom fabric inspection systems is exploited if the localisation and types of failures can be tracked and recorded with sufficient accuracy. In Fig. 5.5, detected fabric defects as they are shown to the operator are illustrated.

**Figure 5.4** An onloom fabric inspection system installed on a circular knitting machine. Smartex Europe Unipessoal Lda.

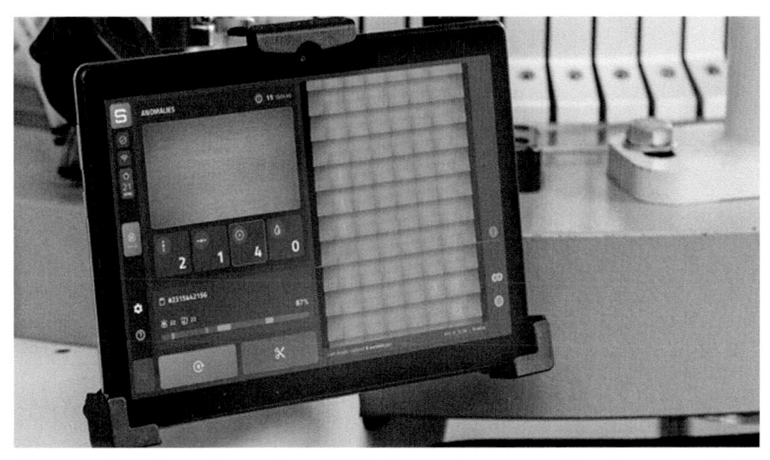

**Figure 5.5** User interface of an onloom fabric inspection system showing potential fabric defects.
Courtesy of Smartex Europe Unipessoal Lda.

If this information on recorded failure types and localisation on the fabric roll is stored on an information carrier such as a QR code, a digital failure map can be obtained and transported to the following process steps such as dyeing, finishing and cutting. In the cutting process, the information on the localisation of defective areas can be used for automatic adoption of the cutting process. This adoption allows the use of areas of fabric without defects which would otherwise have been cut out together with the defective areas of fabric. In this way, the discarding of complete fabric rolls can be prevented, as in certain cases the manufacturer decides to discard a complete roll even if only a few areas of the fabric are defective. An example of a QR code including the introduced concept of a fabric defect map is illustrated in Fig. 5.6.

## 5.3.3 Nonwoven machinery

A nonwoven line consists of many machines in line. The actual set of machines is highly dependent on the product to be produced. Fig. 5.7 shows a standard nonwovens line for carded and needlepunched nonwovens. Depending on the product to be produced, the technology to build the fibre web and to consolidate the nonwoven differ. Fig. 5.8 shows the manufacturing processes for nonwovens.

Within the nonwoven machinery industry, sustainable innovations often enable resource or energy efficiency (sustainability key objective A – reduce resource and energy consumption during production). Therefore, academia as well as manufacturers of nonwoven machines have stressed developments to decrease the energy needed to run the production and the amount of waste resulting from the production and the variation of the basis weight.

**Figure 5.6** QR code as an information carrier for fabric defect mapping. Courtesy of Smartex Europe Unipessoal Lda.

### 5.3.3.1 *Material efficiency*

While the reduction of waste or energy demand is clearly sustainable, the sustainability impact of a reduced variation of the basis weight needs to be explained. Customers of nonwoven producers typically order a minimum base weight of the nonwoven. Variations of the basis weight are unavoidable due to the nature of the process. The lower the process-inducted variation of the basis weight is, the lower the produced mean basis weight can be set to guarantee the minimum base weight. The effect of basis weight variation on raw material savings is illustrated in Fig. 5.9.

Even before the fibre web is built, innovative processes were set up to increase the uniformity of the later product. The more uniform the feeding of a nonwoven card is, the less is the variation of the basis weight in the product. To reduce the variation of the basis weight perpendicular to the direction of production (CD: cross-machine direction) Trützschler Nonwovens patented a system to react to naturally occurring variations of the web weight, which is fed into the nonwoven card.

The Scanfeed system uses the fact that air automatically flows into the direction with lower pressure. The air transports the previously opened fibre flakes. The air constantly flows through the fibre chute and escapes through vents at the base of the chute, where the fibres accumulate. If more fibres accumulate in a particular area, the differential pressure rises, due to the fact that the packed fibre is less air-permeable. Consequently, more fibres are transported to the areas with less fibres at the bottom of the chute. Furthermore, the air vents are segmentised and adjustable using servomotors, therefore allowing more or less fibres to accumulate in one section

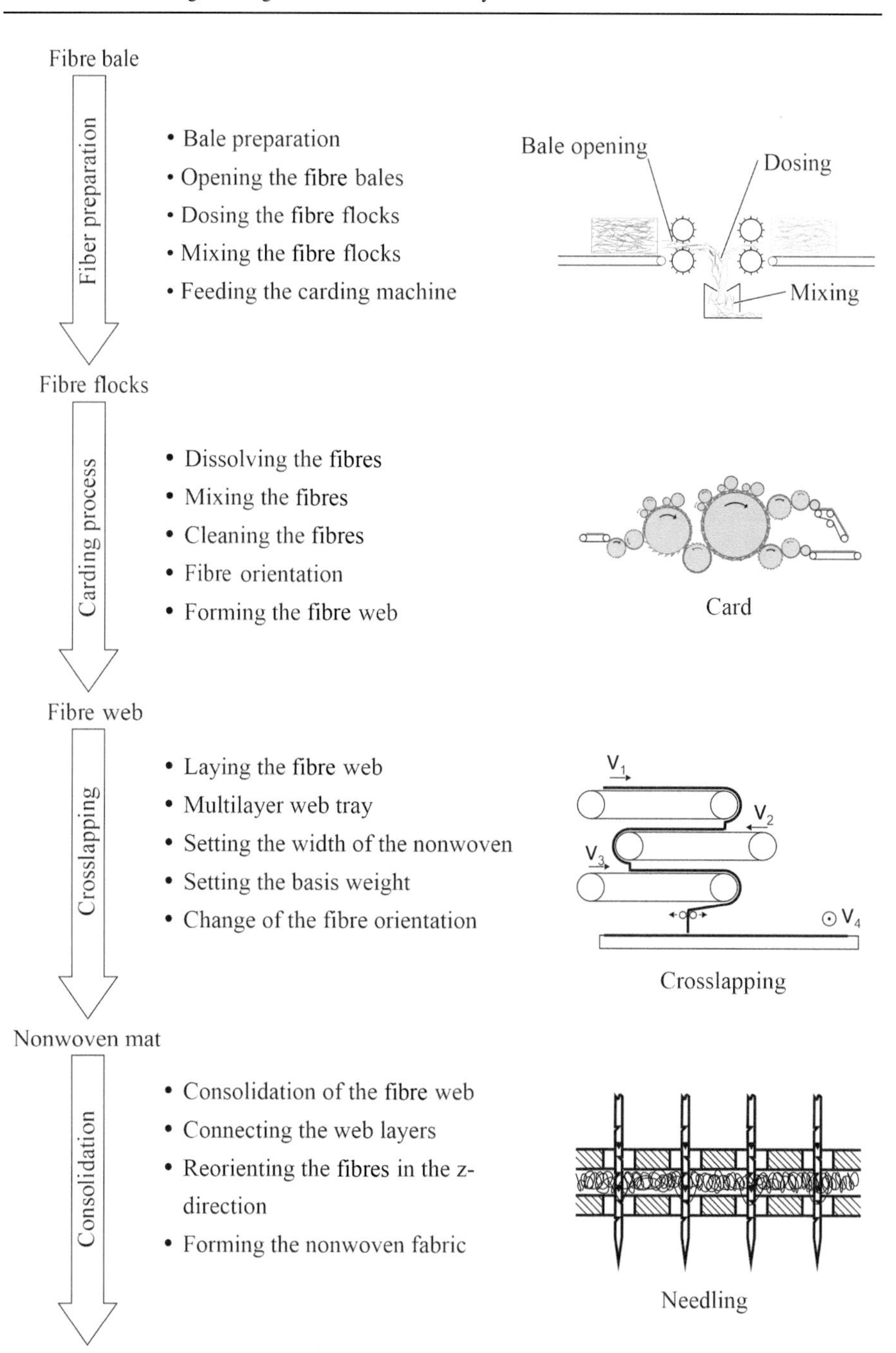


**Figure 5.7** Manufacturing process for carded nonwovens.
Based on Cloppenburg (2019).

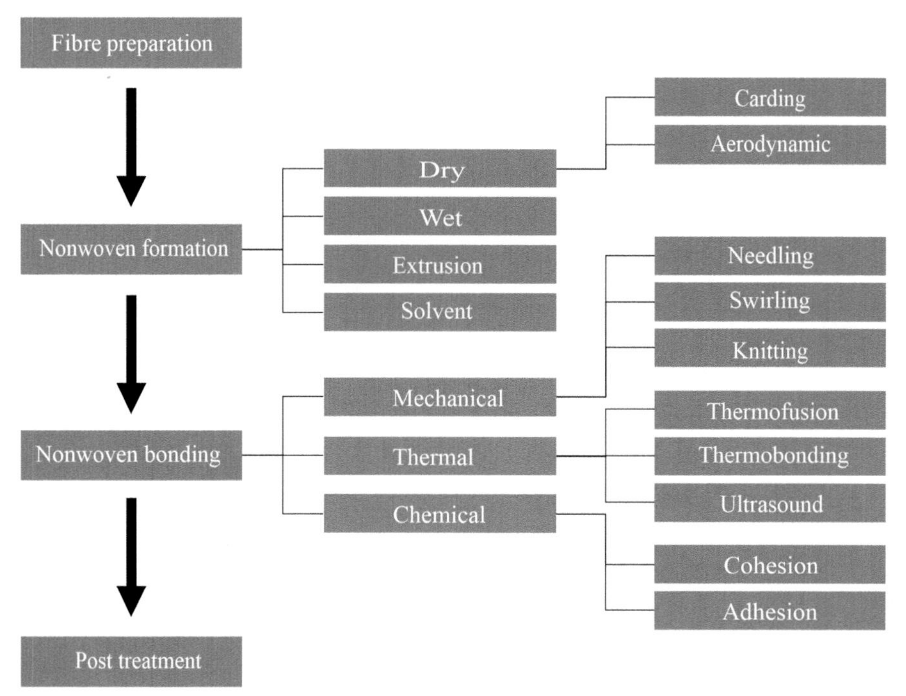


**Figure 5.8** Manufacturing processes for nonwovens.
Based on Cloppenburg (2019), Wiertz Fuchs (2012), Fuchs (2012), and Dilo et al. (2012).

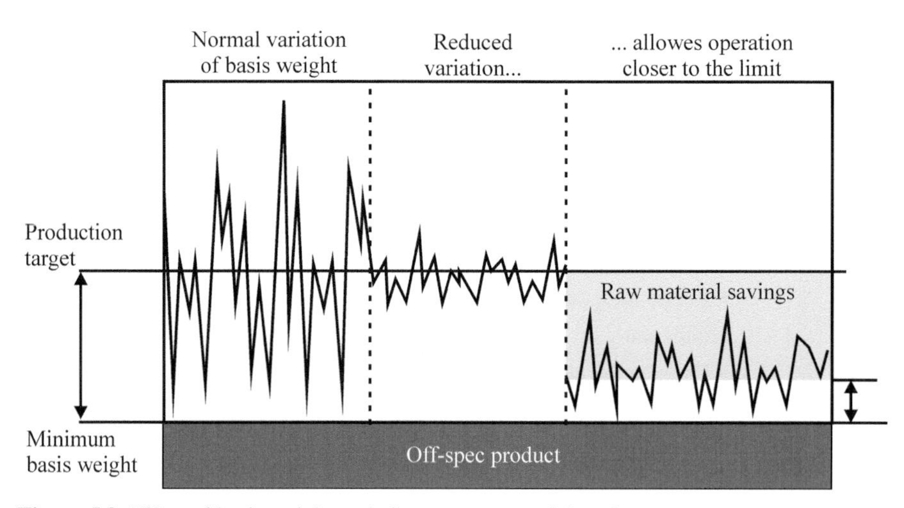


**Figure 5.9** Effect of basis weight variation on raw material savings.

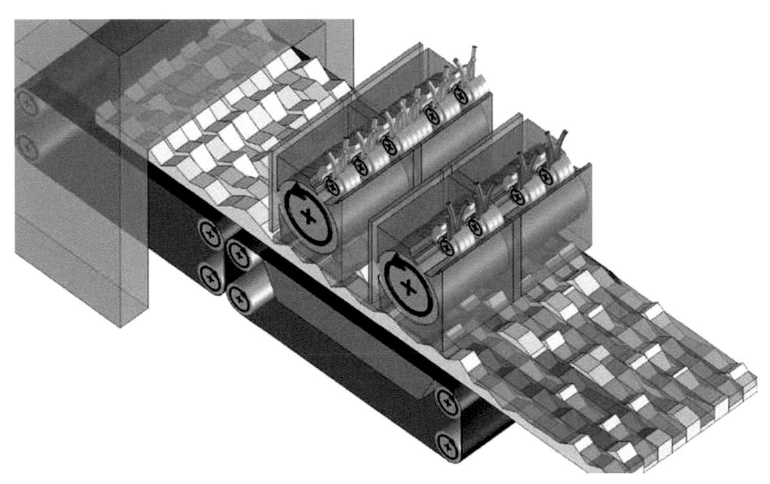

**Figure 5.10** IsoFeed system for mass control of the nonwoven card feed.
Courtesy of Dilo Systems (2019).

depending on the position. Sectional moulds at the bottom of the chute measure the actual density of the fibre web, which is processed to the card, delivering a control signal, which is controlled by the position of the air vents (Brydon and Pourmohammadi, 2007; Schlichter et al., 2012).

A different approach has been pursued by the Dilo Group. The Isofeed technology as shown in Fig. 5.10 does not aim to produce the fibre mat more evenly directly, but compensates for irregularities after the formation of the fibre mat (Dilo Systems, 2019; Dilo, 2022). For this purpose, individual aerodynamic web-forming units with a width of 33 mm are placed across the production width, which are fed with preneedled nonwoven strips or card sliver. Naturally occurring defects are detected by a basis weight measuring system and subsequently compensated for. This results in a very uniform feed web with coefficients of variation below 5% (Dilo Systems, 2019; Dilo, 2022).

To further increase the uniformity of the mass flow into the nonwoven card, inline weight measurement systems using radiometric systems or belt scales are used. Deviations from the desired working point are reduced by adjusting the speed of the card feed depending on the actual mass on the feed table and therefore delivering a uniform mass flow in the direction of production (MD: machine direction) (Brydon and Pourmohammadi, 2007).

In the further progress of crosslapping and consolidation, shrinkage and deformation occur to the produced fabric resulting in an uneven basis weight distribution in the CD direction. Especially the edges of the fabric are heavier than the middle (so-called 'smile-effect'). The machine suppliers developed different strategies to reduce the variation of the basis weight with closed-loop control systems. All systems have in common that the resulting base weight is measured at the end of the production line with a transversal moving scanner using gamma back-radiation, X-ray, near-

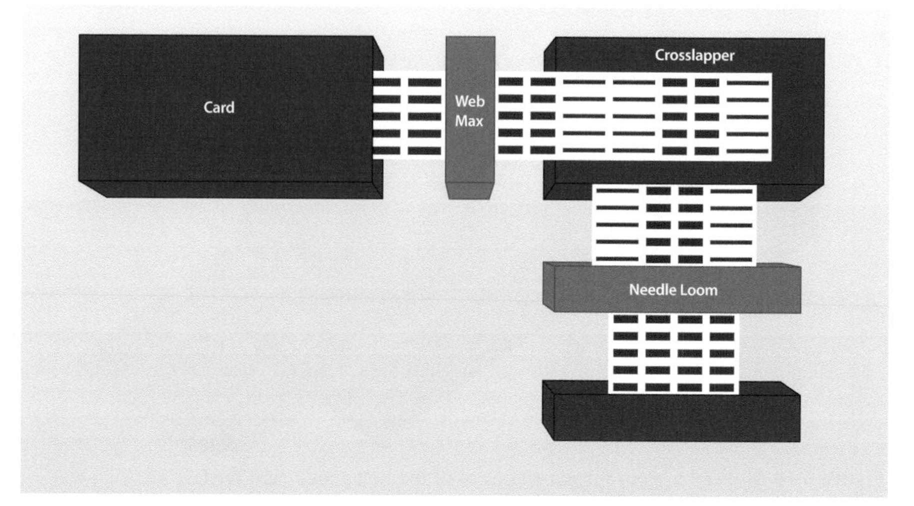


**Figure 5.11** Nonwoven line with integrated WebMax system.
Courtesy of Autefa Solutions.

infrared or beta-radiation transmission depending on the product and surrounding conditions (Brydon and Pourmohammadi, 2007)

The Andritz ProDyn System varies the speed of the card outlet depending on the weight distribution at the end of the line. For too heavy regions, the outlet speed is increased and therefore a web with lower basis weight is produced and vice versa (Andritz, 2022). A different approach has been pursued by the Dilo Group, Eberbach und Autefa Solutions with their CV1 and WebMax Systems shown in Fig. 5.11. In those systems draughting rollers between the card and the crosslapper produce thinner areas, which are laid in the edge areas of the crosslapped fabric. After shrinkage due to consolidation, the result is an even fabric with possible coefficients of variations of approximately 1% (Kunath, 2005; Autefa Solutions Germany GmbH, 2022a,b).

### *5.3.3.2 Energy efficiency*

Within a nonwoven line, the section of the fibre transport is crucial for the total energy demand of the whole production line. Within the fibre preparation part of the plant, 45%–50% of the energy demand derives from the fibre transport. To prevent clogging, the conveying fans are usually over-dimensioned by about 30% and are operated at maximum speed. Therefore, the correct setting of the fibre transport fans has high energy-saving potential. The possibilities for geometric optimisations of the fibre transport systems are limited. The highest potential for energy savings is therefore to run the conveying fans at the lowest possible operation speed (Möbitz, 2021).

Within a dissertation at RWTH-ITA, a complex measurement and control system for the fibre transport was developed. This system allows the operation of the fibre transport fans at the energetically most efficient operation point. A measurement system detects characteristic values of the fibre transport as the distribution factor and transport fluctuation. The behaviour of these characteristic values was examined under industrial conditions. Afterwards, a system model and state detection system were developed and used in a two-degrees-of-freedom control, as shown in Fig. 5.12. Under industrial conditions, an energy saving of approximately 60% can be achieved (Möbitz, 2021). A similar system was announced by the Dilo Group, which uses a sensor to control the fibre speed in the transport system and to prevent clogging of the transport pipes (Dilo Group, 2019).

Nonwoven processes can include drying processes to eliminate process water, which was either used for applying a finish onto the nonwoven or to consolidate a nonwoven using so-called hydro entanglement. The increase of energy efficiency of the drying process and the elimination of the wet process itself is therefore a constant subject of research and development for machine suppliers. A modern drying oven therefore uses up to six energy-efficiency modules to increase the energy efficiency of the drying process:

- High exhaust air saturation through counterflow principle
- Product preheating with saturated process air
- Fresh air preheating with exhaust air heat
- Preheating of the process air with residual heat of the product
- Process water recovery by condensation
- Solar-thermal preheating of the process air (Binnig, 2016)

Consolidation using hydroentanglement is an energy-intensive process as the water is shot at high pressure through the nonwoven material to create entanglement of the single fibres in the web. Especially for fine products in hygiene applications, this consolidation method is still the process of choice, as the product cannot become contaminated with machine oil or metal parts of broken needles. The machine supplier Autefa Solutions therefore developed and patented a new nozzle strip for hydroentanglement as shown in Fig. 5.13. Through the V-design of the nozzle strips, the distance between the nozzle and the fabric can be reduced from 25–40 to 10 mm. Therefore, less friction occurs between the air and the water jet. This reduces the energy demand by up to 30% compared to standard nozzle strips (Binnig, 2016; Hofmann, 2021).

Within a certain product range, needlepunching and hydroentanglement are equivalent as consolidation methods. For extremely fine products with low basis weight, hydroentanglement remains the consolidation method of choice. Because of the high energy demand of the hydroentanglement process and the resulting drying process, progress has been made to replace hydroentanglement for finer products using so-called micro-needling. Up to 8000 needles per metre and high stitching densities are used. The produced fabrics can match hydroentangled products regarding quality and productivity at a weight of 60 g/m$^2$. At the same time, the energy demand is significantly lower as no high-pressure pumps or drying equipment are needed (Dilo Systems, 2021).

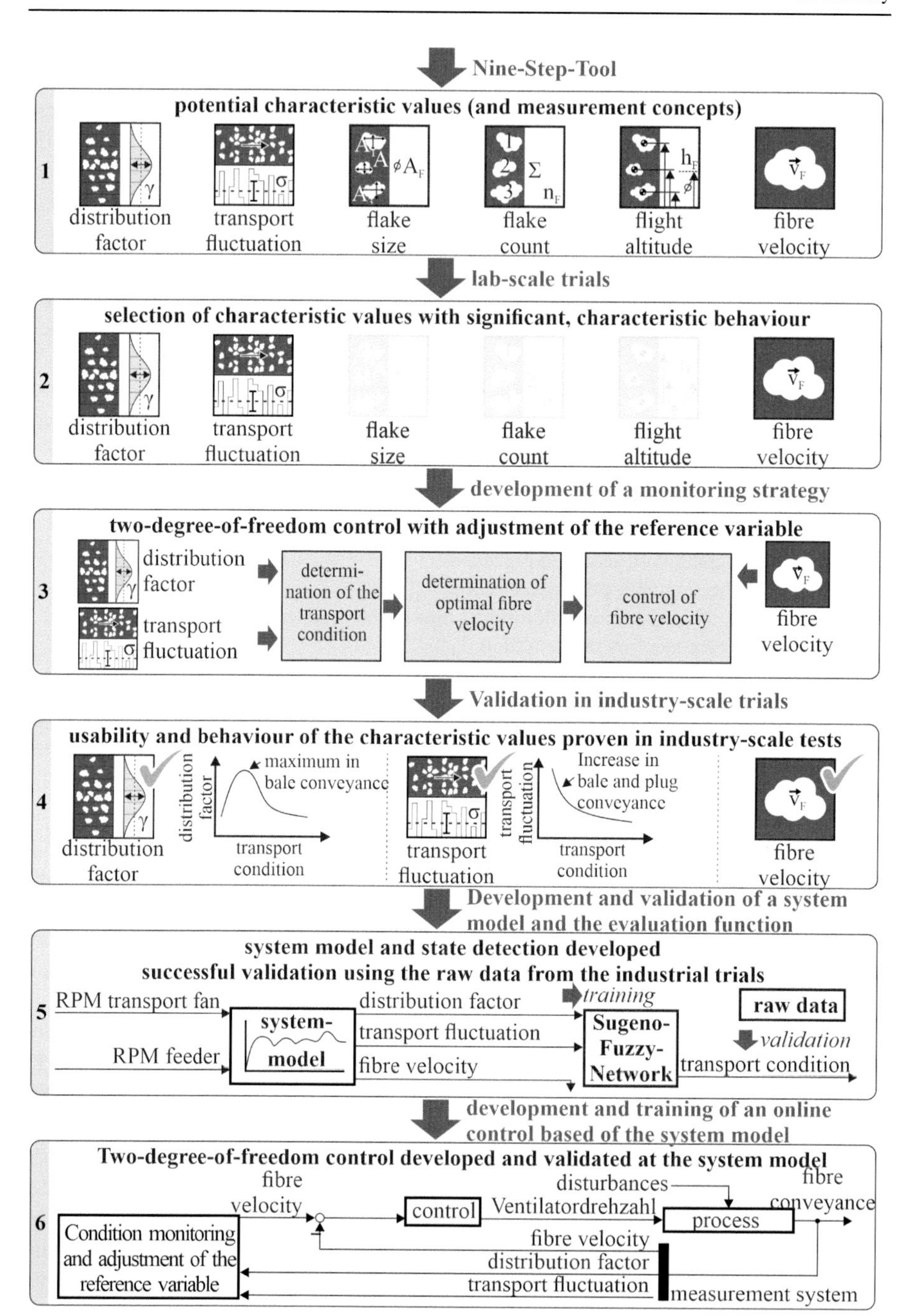


**Figure 5.12** Development of a control for energy-efficient fibre transport. Courtesy of Möbitz (2021).

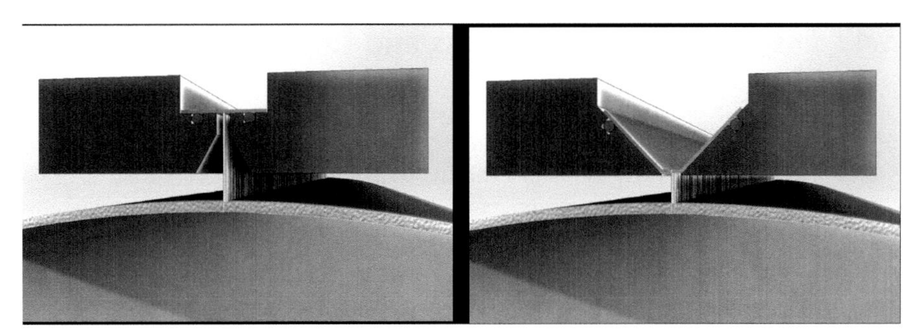

**Figure 5.13** Comparison of standard and V-Jet injectors.
Courtesy of Autefa Solutions.

### *5.3.3.3 Usage of data and digitalisation*

In the control of a nonwoven line the aggregation and analysis of production data has become increasingly important. The setup of nonwoven machines is highly complex and highly skilled long-time operators are less and less available due to demographic changes. In academic projects, applied artificial intelligence is used to optimise production in terms of quality and costs. As the optimisation routines aim to produce the desired product quality at the lowest possible cost, the production is also automatically improved in terms of energy efficiency. Those innovations therefore contribute to the sustainability-related key objectives A (reduce resource and energy consumption during production) and E (support skill-building for machine operators during use).

Within the project Easy Vlies 4.0, a measurement system for acquiring the produced quality of the carded web during the process was developed. Big data was collected during production on a pilot line and digital twins built from that data. The digital twins were used as simulation models in an optimisation routine for a setting aid of the nonwoven card. The setting aid considers not only the qualitative effects of settings but also the economic consequences. This prevents the selection of an uneconomic operation point in favour of better quality, which is not required by customers (Cloppenburg, 2019; Cloppenburg et al., 2021).

In the AutoNoM project, a holistic approach to machine learning-based optimisation of an entire nonwovens plant is being developed. During the implementation, both the technical and organisational hurdles are addressed. In addition to a central data lake, further measurement systems are developed to monitor the intermediate product qualities during the process. The collected data from the systems will be automatically time-corrected so that at any point in time a measuring point of the end product quality can be assigned to the actual settings, intermediate product qualities, disturbing influences and raw material qualities. With the help of models, production-related optimisation of the adjustable variables is achieved (Kins et al., 2021).

## 5.4 Environmental awareness and future perspectives

With the adoption of the EU Strategy for Sustainable and Circular Textiles and the corresponding legislation, pressure for change has been imposed on the textile industry, which manufactures in the EU or for the EU market. Beyond this legislative driver, there is a growing mass of consumers with enhanced awareness of sustainability-related aspects, especially among the younger generations of citizens. Whereas the resulting need of providing evidence for compliance with sustainability-related rules, norms and standards is mainly a concern for textile manufacturers, the machinery manufacturers have the role of providing their customers with what they need to enhance their sustainability in production – wherever machinery can have an influence on this performance.

On one hand, this potential of supporting the textile industry with sustainability-related innovation is connected with features of digitalisation (e.g., possibilities to link manufacturing characteristics and material information with a Digital Product Passport system). On the other hand, the optimisation of resource efficiency is as important as ever, especially in regions with rising energy costs and reduced access to water.

A significant challenge for machinery manufacturers will be their role in supporting textile manufacturers in building up trust in the processing of sustainable fibres. Yarns with high recycled content or which are biopolymers or which are based on natural fibres that are novel to the textile manufacturers are gaining importance. These novel yarns can cause costly delays in production, as suitable production parameters and even a re-design of textile products, which consider the limitations and boundary conditions of the novel material, need to be established.

Textile machinery manufacturers have the opportunity to support their customers on taking these challenging steps by offering assistance in the procurement of suitable auxiliary materials as well as in suggesting suitable processing parameters. In many cases, there will be a need to carry out pilot projects in which textile manufacturers and textile machinery manufacturers will learn and grow together.

## 5.5 Conclusion

The variety of concepts enabling a positive influence on the sustainability of textile manufacturing process chains in weaving, knitting and nonwovens is pertinent. As a selection of these concepts are presented and discussed, an understanding of the sustainability potentials and limitations of textile machinery is provided. Due to the abundant possibilities of process configurations, material and product portfolios as well as individual company strategies, the most suitable answer to the questions on the selection of textile machinery is customer-specific and unique for each individual company.

As with digitalisation concepts, there is rarely an 'off-the-shelf' solution, and textile manufacturers are in need of constantly seeking communication with technology providers in order to comply with the ever-changing legislative requirements and market needs to remain competitive. This communication is crucial also to the competitiveness of textile machinery manufacturers, as they need to understand the needs of their customers to implement their requirements into their own updated design guidelines for new generations of textile machinery. Accordingly, this importance of communication also applies to service providers maintaining, upgrading and 'retro-fitting' used textile machinery. One of the most striking challenges that technology providers and textile machinery manufacturers face is the fact that it is not always apparent to their customers what they actually need. In this sense, trade fairs like Techtextil, ITMA, India ITME, etc. offer a suitable platform to facilitate this necessary communication.

## References


Adanur, S., 2001. Handbook of Weaving. CRC Press, Boca Raton.

Albaugh, L., 2016. Languages for 3D Industrial Knitting Strange Loop Conference. https://www.youtube.com/watch?v=02h74L1PmaU&t=4s. (Accessed 13 April 2022).

Amos, D., 2010. Framework Knitters. http://www.nottsheritagegateway.org.uk/people/framework knitters.htm. (Accessed 13 April 2022).

Andersen, K., 1985. Frihavnen - Den Første Kongemoseboplads. Nationalmuseets Arbejdsmark. København.

Andritz, A.G., 2022. ProDyn System to Optimize Weight Profile and Save Fiber. Available at: https://www.andritz.com/products-en/group/nonwoven-textile/needlepunch/excelle-range/prodyn-forming-needlepunch-nonwoven-and-textile. (Accessed 13 April 2022).

Au, K.F., 2011. Advances in Knitting Technology. Verlag, 978-1-84569-372-5.

Autefa Solutions Germany GmbH, 2022a. Spunlace Lines. Autefa Solutions Germany, GmbH, Friedberg. Available at: https://www.autefa.com/fileadmin/user_upload/Nonwovens/Downloads/Autefa_Solutions_Spunlace_Lines.pdf. (Accessed 13 April 2022).

Autefa Solutions Germany GmbH, 2022b. Web Profile Control Technology WebMax. Autefa Solutions Germany, GmbH, Friedberg. Available at: https://www.autefa.com/nonwovens/product-range/web-forming/crosslapping/webmax. (Accessed 13 April 2022).

Barber, E.J.W., 1991. Prehistoric Textiles: The Development of Cloth in the Neolithic and Bronze Ages With Special Reference to the Aegean. Princeton University Press. https://doi.org/10.2307/j.ctv1xp9q1s.

Binnig, J., 2016. Energieeffiziente Herstellung von wasserstrahlverfestigten Vliesstoffen; 31. Hofer Vliesstofftage, 9.11.2016. Hof. Available at: www.hofer-vliesstofftage.de/vortraege/2016/2016-08-d.pdf. (Accessed 13 April 2022).

Britannica, 2023. https://www.britannica.com/technology/knitting. Accessed 15 March 2023.

Brydon, A.G., Pourmohammadi, A., 2007. Dry-laid web formation. In: Russell, S.J. (Ed.), Handbook of Nonwovens. Woodhead Publishing Ltd., Cambridge, UK, pp. 16–111.

Cherif, C. (Ed.), 2011. Textile Werkstoffe für den Leichtbau. Springer, Heidelberg.

Cloppenburg, F., 2019. Wirtschaftliche und technische Modellierung und Selbstoptimierung von Vliesstoffkrempeln. Shaker, Düren.

Cloppenburg, F., Krause, K., Hesseler, S., Peiner, C., Lechtaler, L., 2021. Case studies of modeling and simulation in textile engineering. In: Advances in Modeling and Simulation in Textile Engineering. Woodhead Publishing, Cambridge, UK, pp. 255–266.

de Araujo, M., Fangueiro, R., Hu, H., 2011. Weft-knitted structures for industrial applications. In: Au, K.F. (Ed.), Advances in Knitting Technology. https://doi.org/10.1533/9780857090621.2.136, 978-1-84569-372-5.

Dilo, J.P., 2022. 3D-Lofter – a new nonwoven technology by Dilo Group. Kohan Textile Journal. Available at. https://kohantextilejournal.com/3d-lofter-a-new-nonwoven-technology/. (Accessed 13 April 2022). Accessed.

Dilo Group, 2019. Dilo Group auf der ITMA 2019. Dilo Group, Eberbach. Available at: https://www.dilo.de/de/aktuelles/dilogroup-auf-der-itma-2019/. (Accessed 13 April 2022).

Dilo Systems, 2019. 3D-Lofter/IsoFeed. Dilo Systems, Eberbach. Available at: https://www.dilo.de/ecomaXL/files/3DLofter.pdf?download=1. (Accessed 13 April 2022).

Dilo Systems, 2021. Complete Production Lines for All Nonwovens Production Processes. Textile Technology. Available at: https://www.textiletechnology.net/technology/news/DiloSystems%20Complete%20production%20lines%20for%20all%20nonwovens%20production%20processes-30400. (Accessed 18 May 2022).

Dilo, J.P., Wizemann, G., Erth, H., Schreiber, J., Wegner, A., Münstermann, U., Möschler, W., Watzl, A., Pasternak, M., Schilde, W., Fuchs, H., Böttcher, P., Zäh, W., 2012. Vliesverfestigung. In: Fuchs, H., Albrecht, W. (Eds.), Vliesstoffe: Rohstoffe, Herstellung, Anwendung, Eigenschaften, Prüfung, second ed. Wiley-VCH, Weinheim, pp. 255–416.

EDANA, 2023. https://www.edana.org/nw-related-industry/what-are-nonwovens. Accessed 14 April 2023.

Elmogahzy, Y.E., 2019. Engineering Textiles, second ed. Woodhead Publishing, Cambridge, UK.

European Commission, 2022a. Commission Staff Working Document. Scenarios Towards Co-creation of a Transition Pathway for a More Resilient, Sustainable and Digital Textiles Ecosystem - SWD(2022) 105 Final. https://ec.europa.eu/docsroom/documents/49360.

European Commission, 2022b. Directorate-General for Environment: Communication from the Commission to the European Parliament, the Council, the European Economic and Social Committee and the Committee of the Regions EU Strategy for Sustainable and Circular Textiles (COM/2022/141 Final). https://eur-lex.europa.eu/legal-content/EN/TXT/HTML/?uri=CELEX:52022DC0141&from=EN.

FTR - Fachverband Textilrecycling: Geschichte des Textilrecyclings. https://www.bvse.de/themen/geschichte-des-textilrecycling.html. Accessed 13 April 2022.

Fuchs, H., 2012. Verfahrensübersicht über die Vliesstoffherstellung. In: Fuchs, H., Albrecht, W. (Eds.), Vliesstoffe: Rohstoffe, Herstellung, Anwendung, Eigenschaften, Prüfung, second ed. Wiley-VCH, Weinheim, pp. 121–122.

Hawkins, M., n.d. A Short History of Machine Knitting. https://knittinghistory.co.uk/resources/a-short-history-of-machine-knitting/#. Accessed 13 April 2022.

Hofmann, M., 2021. V-Jet Futura – the new spunlace generation. TextilPLUS 10/2021. Available at: https://www.autefa.com/fileadmin/user_upload/Nonwovens/Media/AUTEFA-Artikel-TextilPlus_09-10-2021-EN_211025-1.pdf. (Accessed 13 April 2022).

Huh, S.M., Kim, W.-J., 2022. Productivity optimisation for intarsia single-bed flat knitting machine using genetic algorithm. The Journal of The Textile Institute Band 113. H. 1, S. 33–44.

International textile Manufactures Federation, 2021. https://www.itmf.org/images/dl/press-releases/2021/ITMF-ITMSS-vol43-2020-Press-release.pdf.

Jeglinsky, V., Winkler, H., 2020. Untersuchung von Hindernissen zur Digitalisierung in der industriellen Produktion. BTU Cottbus, Senftenberg.

Jordan, J.V., Baukmann, K.-H., Warmer, J., Lopes Teixeira, J.M., Gries, T., 2018. Magnetisch geführter Schusseintrag für Webmaschinen auf Basis von Riemenantrieben. In:

Förderverein Cetex Chemnitzer Textilmaschinenentwicklung e.V (Ed.), Proceedings of 16th Chemnitzer Textiltechnik-Tagung : Technologievorsprung durch Textiltechnik. Förderverein Cetex Chemnitzer Textilmaschinenentwicklung e.V., Chemnitz, pp. 27–34, 28. & 29. May 2018.

Kins, R., Cloppenburg, F., Froese, T., Eibner, S., Pohlmeyer, A., Döringer, J., Gries, T., 2021. Vliesstoffe und künstliche Intelligenz: eine optimale Kombination? : AutoNoM: Automatisierte, KI-unterstützte On-Line-Optimierung der Vliesstoffproduktion. In: textile.4U : das Texdata International Magazin (3), pp. 30–31.

Kunath, P., 2005. Prof-Line System CV1 – Technik und Technologie zur Verbesserung der Vliesgleichmäßigkeit und zur Materialeinsparung; 20. Hofer Vliesstofftage, p. 9 (Hof).

Lin, J., Narayanan, V., McCann, J., 2018. Efficient transfer planning for flat knitting. In: Gershenfeld, N., Levin, D.I.W., MacCurdy, R. (Eds.), Proceedings of the 2nd ACM Symposium on Computational Fabrication. Cambridge Massachusetts, 17 06 2018 19 06.

Liu, Y., Hu, H., 2015. Three-dimensional knitted textiles. In: Chen, X. (Ed.), Advances in 3D Textiles. ISBN: 9781782422143.

McCann, J., Albaugh, L., Narayanan, V., Grow, A., Matusik, W., Mankoff, J., Hodgins, J., 2016. A Compiler for 3D Machine Knitting. https://doi.org/10.1145/2897824.2925940.

Möbitz, C., 2021. Steigerung der Energieeffizienz des pneumatischen Fasertransports. Shaker, Düren.

Matsumoto, E.A., 2021. Prescribing the Elasticity of Knits. stitch-by-stitch. https://www.on.kitp.ucsb.edu/online/films21/matsumoto/.

Popescu, M., Rippmann, M., Liew, A., Reiter, L., Flatt, R.J., Van Mele, T., Block, P., 2021. Structural design, digital fabrication and construction of the cable-net and knitted formwork. Structures 31. https://doi.org/10.1016/j.istruc.2020.02.013.

Ray, S.C., 2012. Fundamentals and Advances in Knitting Technology. Woodhead Publishing India, New Delhi.

Rezaey, A., Jordan, J., Gries, T., 2022. Leitfaden zur partizipativen Gestaltung von Smart Factory-Implementierungen. In: Stamm, P., Lundborg, M. (Eds.), Arbeit und Digitalisierung, Warum Digitalisierung besser mit einer partizipativen Arbeitsstruktur besser gelingt; Begleitforschung Mittelstand-Digital. WIK-Consult GmbH. https://www.mittelstand-digital.de/MD/Redaktion/DE/Publikationen/Arbeit-und-Digitalisierung-Studie.pdf?__blob=publicationFile&v=7.

Saggiomo, M., Wischnowski, M., Simonis, K., Gries, T., 2017. Automation in production of yarns, woven, and knitted fabrics. In: Nayak, R., Padhye, R. (Eds.), Automation in Garment Manufacturing. Woodhead Publishing, Cambridge, UK, pp. 49–74.

Schlichter, S., Rübenach, B., Morgner, J., Bernhardt, S., Kittelmann, W., Schäffler, M., Gulich, B., Krčma, R., Macková, I., Erth, H., Schilde, W., Blechschmidt, D., Dauner, M., Steinbach, U., 2012. Trockenverfahren. In: Fuchs, H., Albrecht, W. (Eds.), Vliesstoffe: Rohstoffe, Herstellung, Anwendung, Eigenschaften, Prüfung, second ed. Wiley-VCH, Weinheim, pp. 123–228.

Schulze, M., Ebert, J., 2002. Prozessmodell-basierte Präsentation von Produktionsfehler-Beschreibungen; Modellierung 2002, Modellierung in der Praxis - Modellierung für die Praxis, Arbeitstagung der GI, 25.-27. März 2002 in Tutzing, Deutschland, Proceedings, pp. S. 147–157. LNI P-12, GI 2002, ISBN 3-88579-342-3.

Spencer, D.J. (Ed.), 2001. Textiles, Knitting Technology, third ed. Woodhead Publishing. https://doi.org/10.1016/B978-1-85573-333-6.50004-0. ISBN 9781855733336.

Underwood, J., 2009. The Design of 3D Shape Knitted Preforms (Ph.D. dissertation). RMIT University, Australia.

United Nations Environment Programme, 2015. Sustainable Consumption and Production: A Handbook for Policy Makers. http://apps.unep.org/publications/index.php?option=com_pub&task=download&file=011712_en. ISBN: 978-92-807-3364-8.

Wiertz, P., Fuchs, H., 2012. Einführung. In: Fuchs, H., Albrecht, W. (Eds.), Vliesstoffe: Rohstoffe, Herstellung, Anwendung, Eigenschaften, Prüfung, second ed. Wiley-VCH, Weinheim, pp. 1−17.

Whitehouse, D., 2000. Woven Cloth Dates Back 27,000 Years. BBC News. http://news.bbc.co.uk/2/hi/science/nature/790569.stm. (Accessed 13 April 2022).

## Sources of further information

Autefa Solutions Germany GmbH, 2022c. Crosslapping. Autefa Solutions Germany, GmbH, Friedberg. Available at: https://www.autefa.com/fileadmin/user_upload/Nonwovens/Downloads/Autefa_Solutions_Crosslapping.pdf. (Accessed 13 April 2022).

Belcastro, S.-M., 2013. Adventures in mathematical knitting. American Scientist 101 (2). https://doi.org/10.1511/2013.101.124.

# Innovations in woven textiles

*Md. Saiful Hoque, Lauren M. Degenstein and Patricia I. Dolez*
Department of Human Ecology, University of Alberta, Edmonton, AB, Canada

## 6.1 Introduction

Textile fabrics are mainly flexible planar materials that can be produced from polymer solutions, natural fibres, filaments and yarns (Belal, 2018). The structures of textile fabrics have a very high length to thickness ratio. Weaving, knitting and nonwoven manufacturing techniques are prominent technologies for producing a variety of textile fabrics. However, the term 'weaving' is closely associated with the introduction of human civilisation and perhaps first comes to mind when discussing textile fabrics (Adanur, 2001). Historical evidence confirms the existence of weaving practices as early as 6000 years ago when Egyptians made woven fabrics for the first time. Furthermore, evidence of woven silk fabric production by the Chinese over 4000 years ago has been recorded also (Loar and Mohamed, 1982).

A woven fabric can be defined as the perpendicular interlacement of lengthwise yarns (warp) and widthwise yarns (weft) (Belal, 2018). Since the inception of weaving, various technologies have been developed for the interlacement of warp and weft yarns at right angles (Adanur, 2001). The variations in these yarns' interlacement form different woven fabric structures. These fabrics are primarily woven on looms, either by hand or automated processes, whereby the warp yarns are held taut in the machine while the weft yarns are inserted between the warp yarns.

As assumed from the definition of weaving, warp and weft yarns are the basic building blocks for manufacturing woven fabrics (Adanur, 2001). However, both warp and weft yarns obtained from the spinning machine are subjected to different processing techniques prior to weaving as they encounter different kinds of stress—strain conditions. For instance, warp yarns experience very high stress compared to weft yarns as they are held under tension on the loom, requiring chemical and physical processing to ensure smoothness and strength during weaving. On the other hand, yarn packages coming from spinning industries can be used as weft yarn without any further processing. However, in some instances, weft yarns must be physically prepared through winding processes.

The processes involved in fabric weaving, including various techniques/technologies, are a matter of concern when considering sustainability (Alkaya and Demirer, 2014). The textile industry, which includes the woven fabric manufacturing industry, pollutes the environment on a massive scale (Smith, 1994). As per the Ellen MacArthur Foundation (2017), the fashion industry is considered the second-largest polluting industry, consuming 93 billion cubic metres of water and dumping approximately half a million tons of microfibres into oceans, annually. Furthermore, textile industries are

**Sustainable Innovations in the Textile Industry. https://doi.org/10.1016/B978-0-323-90392-9.00019-7**

responsible for more carbon emissions than the cumulative emissions of maritime shipping and international flights. In particular, the chemical use and wastewater generation of several processing steps of textile raw materials – such as sizing, desizing, dyeing and finishing – are primary environmental concerns. Discharged effluent from textile industries poses an environmental hazard as it is generally alkaline, coloured, and has a high biological oxygen demand (BOD) and chemical oxygen demand (COD) (Ozturk et al., 2009). Apart from wastewater, solid waste and air pollutants are also generated by textile industries, further contributing to environmental deterioration.

In particular, the production of finished woven fabrics contributes to industry environmental concerns (Pattanayak, 2019). For instance, a high amount of electrical energy is consumed during the winding (0.1–0.4 kWh/kg), warping (0.1–0.3 kWh/kg) and sizing (0.05–0.08 kWh/kg) processes (Koç and Çinçik, 2010). Energy consumption during the weaving step is also high, which depends on the fibres, yarns, fabric construction and machines used. Approximately 1.7–4.2 kWh of energy is consumed to prepare and manufacture 1 kg of woven fabrics. A high amount of thermal energy (0.05–0.08 kWh/kg) is also further used during the sizing of warp yarns. Furthermore, the sizing and subsequent desizing processes involve high volumes of chemicals and water, which also poses an environmental concern. These points greatly contribute to the weaving industry's greenhouse gas (GHG) emissions. For instance, a report by Clothing Industry Training Authority (CITA) revealed that the weaving segment of the denim industry is responsible for more than half (52.32%) of GHG emissions from the whole industry (CITA, n.d.). In addition to high energy, water and chemical consumption, weaving industries also create sound pollution (90–115 dB from a shuttle loom) from several steps of the fabric manufacturing process (Pattanayak, 2019).

However, the industry is transitioning towards sustainable production approaches to mitigate the environmental impacts of textiles (Tobler-Rohr, 2011). These sustainable production approaches include reducing chemical, water and energy consumption during the manufacturing and processing of textile materials. Several techniques are in practice for implementing sustainable chemical and water consumption in textile industries, such as: (i) using synthetic sizing ingredients having low BOD; (ii) desizing with mineral acids; (iii) replacing chlorine bleach with ammonium salt or hydrogen peroxide bleaching technique; (iv) dyeing by incorporating pad-batch system; (v) pre-treatment and dyeing in one step; and (v) finishing without preservation compounds (Öner and Sahinbaskan, 2011; Hoque and Clarke, 2013). Moreover, sustainable innovations are improving the energy efficiency of the textile industry. For instance, supercritical dyeing and wet processing with ultrasonic and foam technology are promising techniques to reduce carbon emissions and increase the textile industries' energy efficiency (Hasanbeigi, 2010). Sustainable production approaches are welcomed in many textile manufacturing plants, including woven textile industries, as the positive effects of sustainable innovation are also realised in business performance (Zeng et al., 2010).

## 6.2 Historical aspects and current scenario

### 6.2.1 Historical perspectives of weaving

Following basketry and mat making, weaving emerged as an adaptive response to a changing environment; the migration of humans to new lands, the birth of agriculture, and the establishment of community settlements during the Neolithic period meant 'a change in lifestyle and household duties' (Broudy, 1979, p. 11). Items such as baskets, mats, coverings and carpets that were once coiled, twinned, wrapped and plaited were now woven using smooth fibres spun into yarns (Broudy, 1979). Prior to even the most rudimentary looms, textiles were woven by hand out of various materials available to weavers, dependent on their location (e.g., grasses, roots, straw, etc.), with stiff materials acting as the warp (Broudy, 1979). The availability of materials also affected the types of looms that were eventually developed; for instance, the Ojibway bag loom required only a simple cord from which to hang stiff warp yarns made of grass or hemp, while warp-weighted looms were required for softer yarns made of cotton (Broudy, 1979). Regardless of the loom type, the function of holding warp yarns in position to allow for the insertion of weft yarns at a right angle to the warp was consistent among these early looms (Broudy, 1979).

Aside from advancements in yarn spinning with the spindle, distaff and spinning wheel, little innovation occurred in woven fabric production until machinery from inventors such as John Kay (flying shuttle), James Hargreaves (spinning jenny), Richard Arkwright (water frame), Samuel Crompton (steam-powered mule) and Edmund Cartwright (steam-powered loom) advanced the speed and efficiency of spinning and weaving, while simultaneously lowering costs (Walton, 1912; Clair, 2018). These inventions incited a shift in the scale of woven textile production, moving from household craft, or 'cottage industry', towards the factory system we know today (Adanur, 2000, p. 2). The basic machinery principles of these early shuttle loom inventions still apply today, yet technological innovations have allowed for greater productivity and, thus, profit (Walton, 1912; Adanur, 2000). A general process flow chart of weaving is shown in Fig. 6.1.

In 1861, uniform requirements of the American Civil War prompted mass production and standardised garment sizing (Cole and Deihl, 2015). As Adanur (2000) describes, weaving efficiency was further improved with warp-tying and drawing-in automation in the early 1900s. In the latter half of the 20th century, synthetic fibres were invented alongside shuttle-less weft insertion methods such as projectile, rapier and air-jet machines. Taken together, these production advancements set the stage for the modern-day textile industry, eventually fostering an environment for fast fashion practices. Until recently, the goal of innovations in the woven textile sector has been to improve speed and productivity. However, as 'the filling insertion rates of today's conventional single phase weaving machines approach physical limits' (Adanur, 2000, p. 4) and as the environmental, social and economic impacts of boundless production are evident, the goals of the woven textile industry must move beyond profit-driven motives towards innovations with sustainability at the forefront.

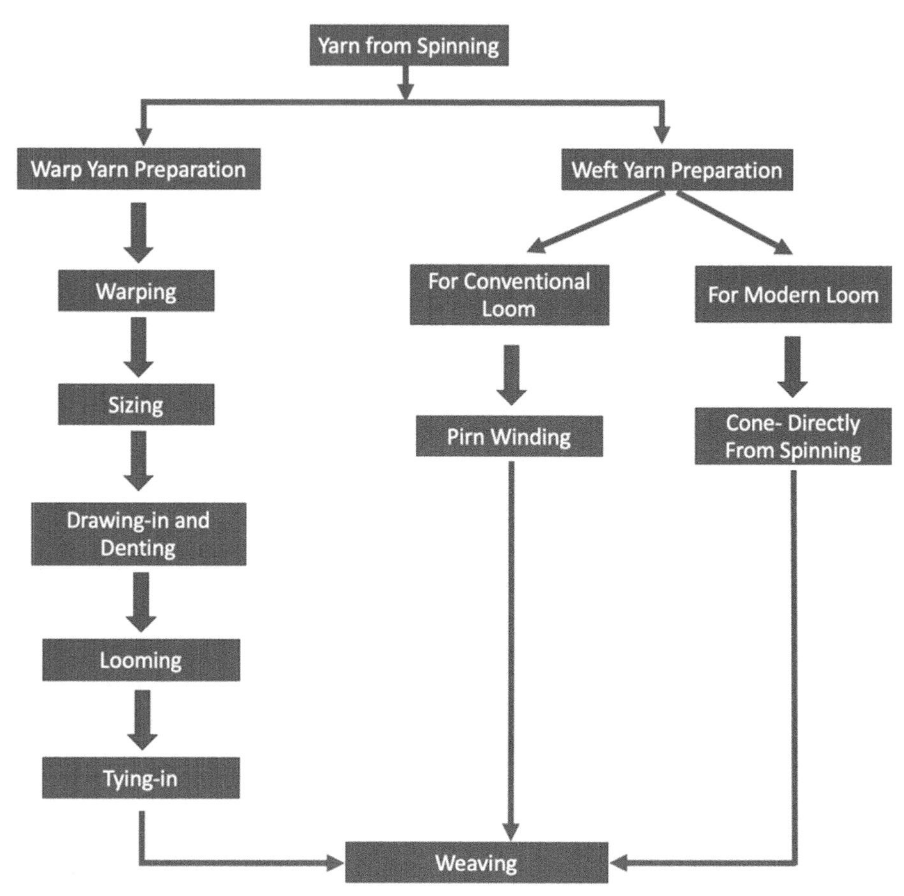


**Figure 6.1** Process flow chart of weaving.
Adapted from Belal (2018).

## 6.2.2 Sustainability of woven textile production

Greater recognition of the environmental and social impacts of the textile industry has driven efforts towards sustainable improvements throughout the value chain. As part of the fabric manufacturing stage, innovations in weaving are not exempt from these efforts. There are several ecological, social and economic consequences of current weaving practices that must be addressed to advance the sector towards greater sustainability.

### 6.2.2.1 Environmental impacts

The main environmental concerns of weaving are the energy requirements of each production step, chemical and water use in sizing processes, and waste generation from weaving processes or defects in the woven product itself (Koç and Çinçik, 2010;

Hasanbeigi and Berkeley, 2011; Hasanbeigi and Price, 2012; Saeidi et al., 2013; Hasanuzzaman and Bhar, 2016; Nayak et al., 2019; Pattanayak, 2019). Different loom actions within weaving consume varying amounts of energy with primary actions of shedding, picking, and beating up requiring more energy than secondary actions like letting off and taking up (Koç and Çinçik, 2010; Pattanayak, 2019). According to Pattanayak (2019), the actual process of weaving accounts for 13% of the energy consumed in fabric manufacturing. In Palamutcu's (2010) study of cotton fabric production in Turkey, weaving consumed 1.58–2.24 kWh of electrical energy for every kilogram of fabric produced. While looms utilise approximately 50%–60% of the energy required in a weaving plant, other energy-intensive processes in weaving are related to factory conditions such as room temperature, humidity, lighting, air compressors and sizing operations (Koç and Çinçik, 2010; Hasanbeigi and Berkeley, 2011; Hasanbeigi and Price, 2012). Weaving requires more resources than knitting due to pre- and post-treatment of yarns (e.g., sizing and desizing) in the weaving process (Nayak et al., 2019). Sizing and desizing steps demand large amounts of water and chemicals, and generate toxic effluent (Hasanuzzaman and Bhar, 2016). When this effluent evaporates, chemicals are released, contributing to air pollution (Hasanuzzaman and Bhar, 2016).

Historical woven fabrics were whole cloths that could be worn in different styles or configurations as demonstrated by traditional garments such as the Japanese kimono, Greek and Roman Chiton or Peplon, or the Indian sari (Senanayake and Gunasekara Hettiarachchige, 2020). The use of whole cloths meant that little yarn or fabric waste was produced as there was no selvage waste or need to cut the fabric into pattern pieces. The modern weaving process, however, generates both process and incidental waste (Goyal, 2021a). Process waste refers to any waste inherent to production operations and is therefore unavoidable, while incidental waste refers to all other avoidable waste such as faulty inputs and production inefficiencies (Goyal, 2021a). Goyal (2021a, p. 62) classifies this fabric waste as 'hard waste', whereby 'fibres are packed in a closed structure and need additional operations before reusing them'.

In other words, this fabric requires mechanical, chemical or thermal processing to be reused, whereas soft waste can be directly reused. Hard waste includes warp and weft yarn breaks, changing the warp yarn on the creel, residual yarn on weft or warp cones, and bobbin changes, amongst others (Goyal, 2021a). Additional waste is incurred from sizing, knotting, tying-in, cutting the false selvage, warp beam residual waste, trialling new weave structures on a loom (i.e., chindi waste) (Goyal, 2021a), and the eventual fabric off-cuts from garment manufacturing. Ecological and energy efficiency improvements of the weaving process can therefore be applied to weaving machinery, manufacturing conditions (e.g., room humidity and temperature), chemicals or additives in sizing/desizing, or in woven products themselves (Hasanbeigi and Price, 2012).

#### *6.2.2.2 Social impacts*

The main social impacts of woven textile production relate to the health and working conditions of weavers in both the handloom and power loom industry segments. These health impacts are a result of prolonged exposure to fibre dust, noise and ergonomic hazards, with the effects accumulating over time (Choobineh et al., 2004; Durlov et al., 2014; Kolgiri and Hiremath, 2018; Bori and Bhattacharyya, 2021). While handloom weaving is a source of economic security, social status and cultural preservation, weavers are often subject to environments with poor ventilation and ergonomic hazards, and lack the proper personal protective equipment (PPE) and training to mitigate these working hazards (Bori and Bhattacharyya, 2021). Since handloom weaving is a 'rural-based cottage industry' (Bori and Bhattacharyya, 2021, p. 623), weavers are responsible for preparing raw materials, carding and spinning fibres into yarns, dyeing, trimming threads, mending, and chemically finishing fabrics in addition to the weaving step (Durlov et al., 2014). These steps involve repetitive movements, prolonged sitting, working at a low height, awkward body positions, and exposure to chemicals and dust pollution (Durlov et al., 2014; Bori and Bhattacharyya, 2021). As a result, weavers develop a number of health problems including body pain and cramps, numbness or tingling of the extremities, fatigue, eye and skin irritation, headaches, cough, dizziness and dust allergies (Durlov et al., 2014; Bori and Bhattacharyya, 2021).

Workers in the power loom industry also experience health effects of weaving despite many processes being automated. While automation has reduced the physical strain on workers (Bedi, 2006), power loom weavers are exposed to hazards such as fibre dust, repetitive motion, noise and machinery vibrations (Nag et al., 2010; Hasanuzzaman and Bhar, 2016). In their study of 540 power loom employees in Solapur, India, Kolgiri and Hiremath (2018) found that workers suffered from respiratory complications (84.28%), musculoskeletal problems (73%) and issues with their eyesight (12%). Noise produced from high-speed weaving machines often exceeds the permissible noise levels [90 dB(A)] set by regulatory bodies such as the International Organisation for Standardisation (ISO) and the United States' Occupational Safety and Health Administration (OSHA) (Bedi, 2006; Hasanuzzaman and Bhar, 2016). As workers are exposed to these high levels of noise for 8 or more hours per day, prolonged noise pollution can lead to worker hearing loss (Bedi, 2006).

Addressing the social impacts of woven textile production is important for sustaining weavers' knowledge and skills, income and capacity for continued work (Bori and Bhattacharyya, 2021). Strategies such as updating machinery, implementing effective humidity controls and improving workplace ergonomics have been suggested to reduce health hazards and foster greater productivity, efficiency and quality of life of weavers (Durlov et al., 2014; Hasanuzzaman and Bhar, 2016). Furthermore, appropriate PPE and training are needed for workers to recognise and reduce exposure to workplace hazards.

#### *6.2.2.3 Economic impacts*

For many countries, weaving industries play an important economic role by providing employment and contributing to national exports (Hasanuzzaman and Bhar, 2016; Bori and Bhattacharyya, 2021). For instance, China, Pakistan and India each exported more than $1B (US) worth of woven cotton fabrics in 2020 alone (Statista, 2022). The physical properties of woven fabrics (e.g., tensile strength, air tightness, abrasion resistance) make these fabrics suitable for a variety of functions and applications such as medical, technical, fashion and home textiles. The versatility of woven fabrics demonstrates the importance of the weaving industry and how opportunities for innovations could benefit its many segments. Furthermore, the quality of woven fabrics is determined by several factors including the number of defects and uniformity of weave structure, contributing to the overall quality and performance of a textile product (Kadolph, 2011). Consequently, fabric defects and quality contribute to the cost of the woven product and influence customer satisfaction (Kadolph, 2011). These factors comprise some of the non-financial indicators of a firm's economic success.

Despite the profitability of the industry, scholars highlight changes to woven manufacturing processes that could ensure its long-term financial viability. These changes include improvements in the energy efficiency of plant operations and waste reduction initiatives (Hasanbeigi and Price, 2012; Goyal and Nayak, 2020). As Koç and Çinçik (2010) describe, the main costs of woven fabric production include raw and auxiliary materials, labour, energy and capital. Energy-saving practices such as transitioning to energy-efficient processes and machinery, process automation and optimisation, good in-house practices and reducing raw materials and resource waste would thus lower the costs of weaving. Furthermore, Hasanbeigi and Price (2012) suggest that '[k]now-how on energy-efficiency technologies and practices should, therefore, be prepared and disseminated to textile plants' so that firms are equipped to implement more sustainable practices.

## 6.3 Sustainability analysis of woven textiles

While the weaving stage of textile production does not account for a large proportion of the overall environmental, social or economic costs of manufacturing, improvements must be made at this stage of the value chain for industry sustainability. Silvestre and Ţîrcă (2019) propose a theoretical model for sustainable innovation that places equal emphasis on traditional innovation (profit driven), green innovation (environmentally driven) and social innovation (socially driven). Therefore, sustainable innovations go beyond green innovations and aim to balance the three pillars of sustainability – environment, social and economic – when creating novel, or improving upon, existing products and services. This theoretical model is reminiscent of Elkington's (1997) triple bottom line theory of sustainability, which considers people, planet and profit dimensions of a firm's products and activities. Frigelg et al. (2019) argue that the sustainable development of a company can be achieved through the connection of 'Environmental Integrity, Corporate Social Responsibility and Economic Prosperity'. Similarly, Annapoorani (2017, p. 59) refers to this as the 'sustainability triad'. Fig. 6.2 shows the environmental sustainability issues and innovations related to woven textiles.

Presently, no standardised practices exist for determining if a firm has fulfilled the three pillars of sustainability as each firm's environmental, social and economic concerns are dependent on their products and processes. The literature has emphasised the

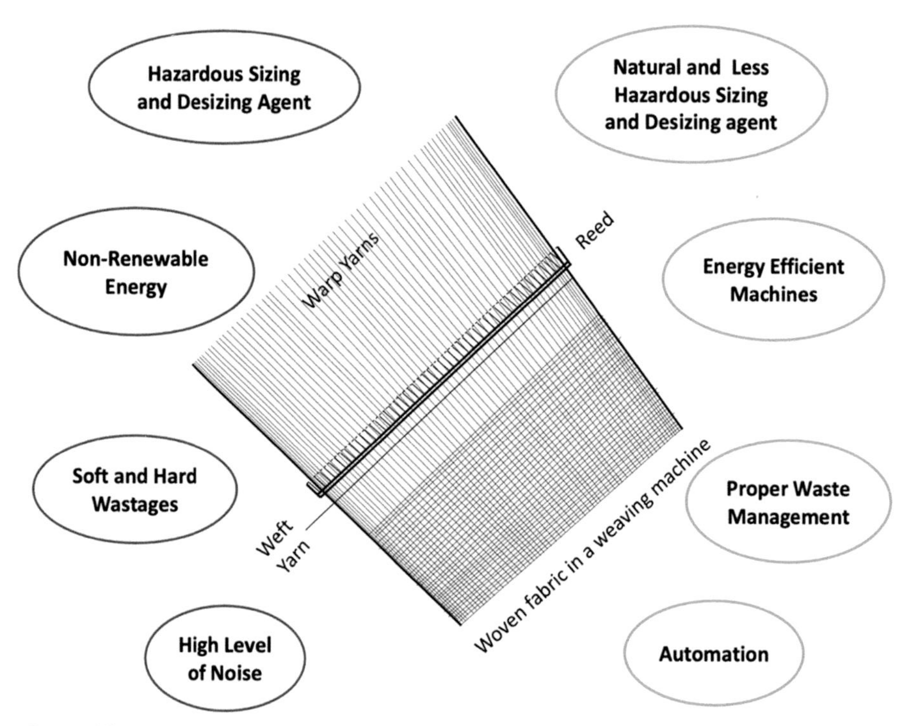


**Figure 6.2** Environmental sustainability issues (*red*) and innovations (*green*) related to woven textiles.

need to balance issues regarding people, planet and profit, yet few studies address all three aspects of sustainability (Abdul-Rashid et al., 2017; Wanniarachchi et al., 2020). Key performance indicators (KPIs) refer to 'the critical (key) indicators of progress toward an intended result. KPIs provide a focus for strategic and operational improvement, create an analytical basis for decision making and help focus attention on what matters most' (What is a Key Performance Indicator (KPI)?, n.d., para. 1). There are several tools that can measure these indicators at various levels, from broad (e.g., supply chain level) to narrow (e.g., individual firm level). However, many tools measure environmental, social or economic indicators separately, so firms may have to use multiple methods of measurement to gain a holistic view of its sustainability performance. Indeed, sustainability assessments in the textile industry often only account for one dimension of sustainability (Gbolarumi et al., 2021).

## 6.3.1 Environmental pillar

Improving the environmental sustainability of the woven textile industry means reducing natural resource use, decreasing all forms of pollution, prolonging the useful life of products and increasing the recyclability of fabrics (Frigelg et al., 2019). In addition, the incorporation of clean and renewable energy sources to reduce carbon emissions is key for sustainability (Abdul-Rashid et al., 2017). Environmental performance can therefore be measured by a firm's reduction in resource use, energy and waste generation using informal, holistic methods (e.g., waste diversion) or by adopting formalised methods. Methods for environmental sustainability evaluations in the textile industry include Life Cycle Assessments (LCA), ecological or environmental footprint analyses, eco-efficiency calculations, and the Higg Index (Luo et al., 2021).

According to the United States Environmental Protection Agency (2006), a Life Cycle Assessment 'is a technique for assessing the potential environmental aspects and potential aspects associated with a product (or service), by:

- Compiling an inventory of relevant inputs and outputs
- Evaluating the potential environmental impacts associated with those inputs and outputs
- Interpreting the results of the inventory and impact phases in relation to the objectives of the study' (United States Environmental Protection Agency, 2006, p. 2).

Various methods exist for conducting LCAs including EcoIndicator 95 (Goedkoop, 1995), EcoIndicator 99 (Goedkoop and Spriensma, 1999) and Centrum voor Milieukunde, Leiden (CML) (Heijungs et al., 1992). These methods differ in terms of process steps and impact categories (Tobler-Rohr, 2011). Additional LCA methods are reported in Muthu (2016). Tobler-Rohr (2011) presents an in-depth LCA of the textile value chain using ecological key figures (EKF) as an assessment method. Within this method, the impacts of stages in the value chain are calculated based on various impact categories such as greenhouse effect, acidification, carcinogenic substance release and pesticide use, with higher impact categories being represented by a higher number in the calculation. For weaving, the inputs and outputs are yarn and woven fabric, respectively, while impact categories include greenhouse effect from energy consumption and eutrophication due to sizing agents (Tobler-Rohr, 2011). LCAs can therefore be used to calculate the current impacts of products and processes, as a means for comparing products, and to identify higher impact categories needing to be addressed (Gonçalves and Silva, 2021).

Environmental or ecological footprints refer to the land and water requirements to supply resources to and absorb waste from humanity (Wackernagel and Rees, 1996). These quantitative tools can measure the footprints of different resource uses individually, such as carbon, energy or water footprints, although a complete ecological footprint estimation would consider all resources involved in a process (Costa et al., 2019). Butnariu and Avasilcai (2014) performed a case study estimating the energy, resource and waste footprints of the textile industry. The authors opted for this method of analysis as they argue that ecological footprint calculations are simple and easily understood indicators of sustainable development. Other studies examining the ecological footprints of the textile industry used similar categories to determine process inputs and outputs (Costa et al., 2019).

Eco-efficiency is a measurement of economic performance in relation to the environmental impacts of a process (Luo et al., 2021). This is calculated by dividing the economic value of a product by its environmental impact, which is based on factors such as resource use and greenhouse gas emissions (Luo et al., 2021). The intention of eco-efficiency calculations is to generate value with less inputs and emissions

(Shiwanthi et al., 2018). The Higg Index offers another holistic view of a company's performance as it accounts for quantitative and qualitative measurements of environmental and social indicators (Luo et al., 2021). Using Life Cycle Assessment data, the Higg Index is a tool to score the impacts of products, facilities and brands from the cradle to gate stages of the supply chain (Gonçalves and Silva, 2021).

Additional methods of environmental sustainability assessments include third-party certifications such as the Global Organic Textile Standard (GOTS), OEKO-TEX and Textile Exchange Standards, which require firms to adhere to certain production specifications (Amutha, 2017). Environmental performance evaluations (EPE) are another method of analysing environmental performance. In their case study of a Turkish woven fabric mill, Alkaya and Demirer (2014) conducted an environmental audit of the mill's resource use and waste generation. With the collected data, they performed an EPE to identify areas of impact and propose solutions to reduce these impacts. Based on the initial baseline data collection, it was determined that water use and waste, energy use, carbon emissions and salt consumption were areas in need of improvement. The strategies implemented by the mill, such as reusing cooling water and renovating their water softener, addressed these areas of concern. These strategies helped decrease water consumption (−40.2%), wastewater (−43.3%), energy use (−17.1%), carbon emissions (−13.5%) and salt consumption (−46.0%). Therefore, several tools are available for measuring environmental impacts, yet some may be more relevant than others for individual firms.

### *6.3.2 Social pillar*

The social pillar of sustainability refers to the 'capacity of a society, country, family and organization to function at a defined level of wellbeing and harmony for an indefinite period' (Gbolarumi et al., 2021, p. 3). Annapooranni (2017) discusses the different standards that should be followed to protect worker rights in the textile industry. Standards of the ILO address freedom of association, collective bargaining, child and forced labour, equality, wages and working hours, occupational health and safety, social security and maternity protection. As with environmental performance, there are several ways to analyse the social sustainability of a company and its processes. In a review of the literature, Köksal et al. (2017) found that indicators of textile supplier social performance included human rights, unfair wages, excessive work hours, child and/or forced labour, unhealthy or dangerous working conditions and discrimination. No studies in their review used diversity or the unethical treatment of animals as a social performance indicator, suggesting opportunities for further research related to these indicators.

While LCAs have traditionally applied to environmental impacts, the scope of LCAs has broadened to involve social impacts using Social Life Cycle Assessments (S-LCA) (Herrera Almanza and Corona, 2020). S-LCA measure the social impacts of products and processes over the lifetime of that item (Lenzo et al., 2017). S-LCA of the textile industry have been performed by Herrera Almanza and Corona (2020), Lenzo et al. (2017) and Zamani et al. (2018). One criticism of social assessments is that some metrics may be too narrow as they do not consider the broader societal impacts such as community progress or personal development (Luo et al., 2021). The Higg Index has expanded to include measurements of social sustainability such as employee treatment, wages and working conditions (Luo et al., 2021).

### 6.3.3 Economic pillar

Until recently, firms have only considered the economic bottom line of business – profit. Although financial success remains a foundational consideration of firms to remain in business, companies will also have to consider the social and environmental impacts of their products to appease changing consumer preferences and resource scarcity. To date, technical innovations in weaving have centred on increasing the speed of production by way of maximising the weft insertion rate (Adanur, 2001). However, there are other financial and non-financial indicators of economic performance besides productivity and manufacturing efficiency.

Economic indicators of business performance include factors such as material and labour efficiency, customer satisfaction and on-time delivery (Hourneaux et al., 2018). Zeng et al. (2010) separates business performance evaluation into financial indexes (profitability, net profit and return on equity) and non-financial indexes (market share, corporate reputation and shareholder confidence). These indicators can be measured using Life Cycle Costing or Balanced Score Card methods. Life Cycle Costing, or Economic Life Cycle Assessments (E-LCA), which evaluate the costs and benefits of a product over its lifetime, have not been frequently applied to the textile industry (Balanay and Halog, 2018). Kaplan and Norton's (1992) Balanced Score Card (BSC) measures both tangible and intangible business outcomes like customer perspectives (i.e., quality, performance, price), internal business perspectives, and learning and growth within a business in addition to financial performance (Kefe, 2019). At a broader level, traditional measurements of exports and contributions to gross domestic product can also be used to gauge the economic value of weaving industries.

### 6.3.4 Triple bottom line analysis of impacts

Literature on the sustainability of the textile industry reveals that analyses have mainly focused on the environmental dimension of the triple bottom line (Abdul-Rashid et al., 2017; Gbolarumi et al., 2021). Furthermore, it can be difficult to balance the three pillars of sustainability as there may be trade-offs (Luo et al., 2021). For instance, Weidner et al. (2021) describe how innovations beneficial for the environment may not be accessible to some due to higher costs of the product. Likewise, handloom weaving has social and environmental benefits, yet structural barriers such as a lack of training opportunities and government support for independent weavers hinder the economic viability of the craft (Wanniarachchi et al., 2020). However, other scholars reason that there are overlapping benefits when one or more sustainability pillars are addressed. Nidumolu et al. (2009) argue that incorporating environmental practices such as reducing energy use and waste can provide economic benefits for companies as they reduce energy costs and material inputs. In addition, Zeng et al. (2010) used structure equation modelling to investigate the association between cleaner production and business performance in the Chinese manufacturing industry and found there to be a significant positive relationship between the two. Goyal (2021a, p. 61) argues that 'proper waste management affects the quality, production efficiency, and profitability of the organization' as weaving waste can be reused or recycled, offsetting the costs of purchasing new materials. Therefore, implementing practices to address one pillar of sustainability can have unintentional benefits for the other pillars.

Sustainable supply chain management offers a management tool for the movement of information, capital and resources through the supply chain, while incorporating economic, social and environmental goals (Seuring and Müller, 2008). Kozlowski et al. (2014) examined the publicly disclosed sustainability indicators of apparel brands belonging to the Sustainable Apparel Coalition (SAC). They identified 87 indicators relating to the environment, social impacts and business innovation, with the greatest number of indicators relating to sustainable supply chain management. The authors found a lack of consistency in indicator reporting amongst brands, demonstrating the difficulty in comparing company sustainability performance as brands choose which information to disclose.

Lean production (LP) has been discussed as a manufacturing method to achieve a firm's triple bottom line. Lean production aims to eliminate physical waste (e.g., materials) and non-physical waste (e.g., time or labour of non-value-adding activities) from production processes (Raj et al., 2017). The idea here is that reducing materials and inputs will not only make processes more efficient but reduce costs as well. In an online survey of professionals working in the Indian apparel industry, Raj et al. (2017) found the top implemented lean production techniques to be quality management, bottleneck constraint removal and maintenance optimisation. Since these are issues dealt with in the weaving industry, it would be worthwhile for future research to examine how lean practices are applied to woven textile production and whether these practices aid in firms achieving the triple bottom line.

Complex and fragmented life cycle and production steps of textiles make it difficult to gather data. Separated locations of production hinder accurate or comparable data collection. Furthermore, the speed of the industry means that gathering timely data on processes and impacts is all the more complicated (Luo et al., 2021). Luo et al. (2021) propose a hybrid approach to sustainability analysis that considers quantitative and qualitative data and offers multi-criteria decision-making models to approach this. Researchers have used fuzzy or multi-criteria decision-making models to assess the sustainability performance of the textile industry (e.g., Charkha and Jaju, 2014; Acar et al., 2015; Wang et al., 2019).

Hourneaux et al. (2018) state that sustainable performance analyses should contain a minimum number of environmental, social and economic indicators to evaluate a firm's triple bottom line performance. Drawing on economic indicators from Henri (2009)'s Balanced Score Card and environmental and social indicators from the Global Reporting Initiative (GRI), the authors surveyed 149 companies to determine how they used the three types of indicators in their business (Hourneaux et al., 2018). Table 6.1 summarises the minimum indicators relevant for assessing the triple bottom line of a firm, according to the authors.

**Table 6.1** Indicators of economic, environmental and social dimensions of the triple bottom line.

| Economic dimension | Environmental dimension | Social dimension |
|---|---|---|
| • On-time delivery<br>• Number of customer complaints<br>• Survey of customer satisfaction<br>• Materials efficiency variance<br>• Rate of material scrap loss<br>• Labour efficiency variance | • Materials<br>• Energy<br>• Water<br>• Biodiversity<br>• Emissions, effluents and waste<br>• Environmental aspects of products and services<br>• Environmental compliance<br>• Transporting<br>• General environmental issues | • Labour/management relations<br>• Occupational health and safety<br>• Training and education<br>• Non-discrimination<br>• Freedom of association and collective bargaining<br>• Child labour<br>• Forced and compulsory labour<br>• Security practices<br>• Compliance |

Adapted from Houreaux et al. (2018) under the Creative Commons Attribution (CC BY 4.0) licence.

## 6.4 Innovations leading to sustainability in woven textiles

A paradigm shift towards sustainable innovation prevails in the textile manufacturing industry worldwide (Dolez and Vermeersch, 2018). In fact, stringent regulations for more sustainable practices, including removing toxic chemicals in textile processes, has steered the wheel towards sustainable action. Several lobby groups such as Zero Discharge of Hazardous Chemicals (ZDHC) and Ethical Fashion Forum are working continuously to prevent hazardous chemical use in textile manufacturing (Ethical Fashion Forum, 2015; ZHDC, 2021). For instance, ZDHC collaborates with textile and apparel brands, retailers and manufacturing plants to ensure sustainable chemical management. The ZDHC Roadmap to Zero Program released a Manufacturing Restricted Substance List (MRSL), which requires their collaborators to avoid certain substances when manufacturing goods (ZDHC, 2022).

The innovations leading to sustainability in woven textiles are mainly rooted in sustainable manufacturing processes. Sustainable manufacturing refers to the effective use of natural resources in the manufacturing process, satisfying economic, environmental and social aspects, and providing a better quality of life (Garetti and Taisch, 2012). In general, the manufacturing stage of products consumes an enormous amount of energy and generates hazardous wastes in solid, liquid and gas forms, thus creating a demand for sustainable manufacturing (Duflou et al., 2012). Likewise, sustainable manufacturing has been undertaken in woven textile manufacturing industries (Pattanayak, 2019).

### 6.4.1 Warping

'Good warping finishes fifty percent weaving' is a prevalent notion in weaving companies (Singh, 2014b, p.104). The warping process converts the spinning cone into a larger beam containing warp yarns required to weave fabrics. Warping is performed as per the final fabric construction requirements. For instance, the yarns are wrapped in the warping beam by following ends per inch (EPI), yarn linear density, final fabric length and width requirements. Considering the capacity of the industry and weaving requirements, warping can be performed either through direct/beam warping or sectional warping.

Direct or beam warping refers to the winding of yarns directly on a beam from cone packages placed on a creel (Adanur, 2001). This warping technique is widely used in the weaving industry that processes high-strength yarns, such as filament yarns, and does not require many warp ends to weave the fabric. Fig. 6.3 depicts a schematic diagram of direct/beam warping.

Sectional warping is also described as pattern, band and drum warping, where yarns are wrapped into the beam section-wise starting from the tapered end of the beam (Adanur, 2001). Sectional warping is performed when the industry has limited creel capacity, and warp yarn sizing is unnecessary (Singh, 2014b; Gandhi, 2019). In addition, sectional warping is also preferable for complex colour striped fabric constructions as it allows for warping of different-coloured yarns section-wise. Fig. 6.4 represents a schematic diagram of sectional warping.

Apart from direct and sectional warping, other warping techniques such as ball warping and draw warping are available for specific fabric constructions. For instance, ball warping is widely used for denim fabrics (Raina et al., 2015). Warping in the form of tow makes the rope dyeing process easier. On the other hand, draw warping is specially designed for warping thermoplastic yarns where heat setting is performed along with warping (Adanur, 2001). The inclusion of heat setting in the warping process ensures uniform stretching of yarns while improving the dyeability.

The warping stage of the weaving industry contributes to energy consumption, which eventually contributes to GHG emissions. It was reported for a denim yarn dyeing company that the warping section was responsible for 2.87% of its greenhouse gas (GHG) emissions (Cita, n.d.; Karthik and Murugan, 2017). Among these 2.87% GHG emissions, the warping machine contributed 1% of emissions. The rest of the

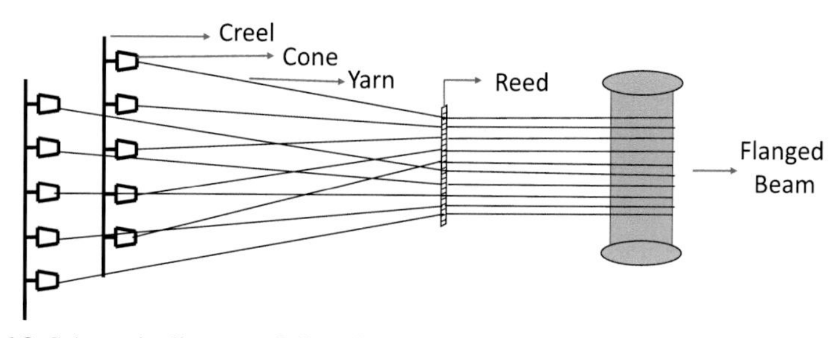


**Figure 6.3** Schematic diagram of direct/beam warping.

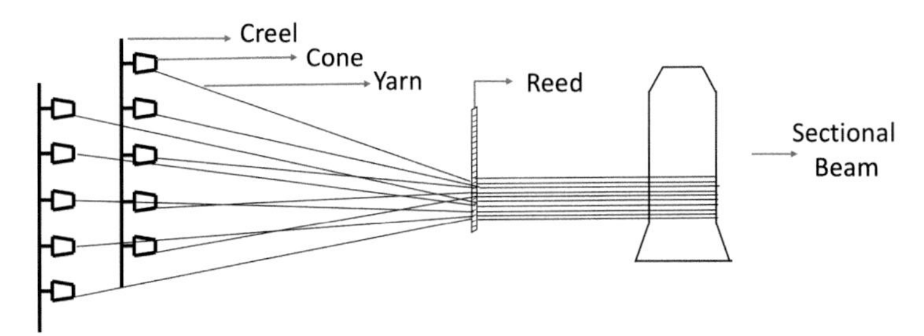


**Figure 6.4** Schematic diagram of sectional warping.

97.13% of GHG emission was attributed to the dyeing & sizing (94.50%), winding (1.92%) and packaging (0.71%). Although the energy consumption of warping machines is relatively lower, machine manufacturers and industry are working towards sustainable innovation to reduce energy consumption by warping machines and eventually to reduce GHG emissions.

To estimate the energy consumption of warping machines, researchers have investigated theoretical approaches and developed empirical models. For instance, Koç and Çinçik (2010) developed an equation (Eq. 6.1) for determining the energy consumption of warping machines where $EE_{WA}$ refers to electrical energy consumed (kWh), $t_{WA}$ indicates the warping machine operating time (hours), $E_{WA}$ specifies the power of the warping machine (kW) and $\eta_{EWA}$ refers to the energy efficiency (%) of the warping machine. This equation would give an idea of a warping machines' energy consumption prior to procurement so that an informed purchasing decision can be made.

$$EE_{WA} = t_{WA} \times E_{WA} \times \eta_{EWA} \tag{6.1}$$

Attempts have been made to modify both direct and sectional warping systems to increase warping efficiency (Thakkar and Bhattacharya, 2018). For instance, Hiroshi and Shozo (2007) patented a modification of a sectional warping machine that allows automatic switching of pattern warp yarns, effectively reducing creeling time. Fuhr (2010) recommended a modification in the flange of the beam for its automatic adjustment. Thakkar and Bhattacharya (2018) offered a novel design for combining both direct and sectional warping in one machine to reduce the types of warping machines used in the industry. They suggested using adjustable flanges in the warping beam to allow sectional beam adjustments, as necessary. However, their proposed design does not contain enough information for implementing the idea. Aside from machine modifications, researchers have also investigated the implementation of mathematical modelling to enable better control in warping machines. For example, Eren et al. (2016) developed a mathematical model that allows for a computer control system of the sectional warping machine. The authors suggested that their model eliminates the use of a sensor to measure the warp yarn length in the machine as the model precisely performs the warp yarn measurement.

The ISODIRECT direct warping machine developed by Karl Mayer Ltd. is an example of a sustainable innovation by the textile industry machine manufacturer (Textile Today, 2018). This machine was developed to provide better efficiency and reduce warping processes. In addition, this machine offers a smart reed system that allows automatic adjustment of the reed depending on the number of yarns and beam width, eventually reducing warping processing time.

### 6.4.2 Sizing

'Sizing is the heart of weaving' is an old aphorism in weaving industries (Goswami et al., 2004, p. 1). Sizing is an essential step in warp yarn processing as warp yarns are subjected to very severe conditions such as cyclic strain, flexing and abrasion during the weaving process (Singh, 2014b). Therefore, sizing can be considered a protective layer applied on warp yarns to provide the yarns with sufficient strength to withstand mechanical strain during fabric weaving. In addition, sizing also reduces the possibility of fluff binding among the warp yarns in the weaving machine, ensuring smooth and precise shedding.

In a sizing machine, warp beams are arranged in the creel in such a way as to allow a smooth withdrawal of warp yarns from the beam before moving to the next section of the machine – the size box (Singh, 2014a). The size box is considered the sizing machine's core as it contains the sizing solution. In conventional sizing machines, the size box contains an immersion roller that facilitates the immersion of warp yarns in the sizing liquor, whereas the size box of a modern sizing machine contains an immersion roller along with heating coils to properly maintain the temperature and viscosity of the size liquor. After passing through the size box, warp yarns go through the drying zone and wrap onto the sizing beam. A schematic diagram of the conventional sizing process is shown in Fig. 6.5.

Traditionally, starch and its derivatives have been used as a sizing agent of cellulosic yarns, whereas polyvinyl alcohols (PVA) and acrylics are preferable for synthetic yarn sizing (Goswami et al., 2004). Although starch demonstrates a well-performing, economical and biodegradable sizing for cotton fibres, synthetic sizing agents like PVA are often considered as the only solution for high-performance weaving machines. PVA is also preferred over starch as it allows easy desizing. However, using synthetic sizing is a great concern for environmental sustainability. For instance, the sizing and desizing process accounts for 40%–60% of the effluent load of the textile

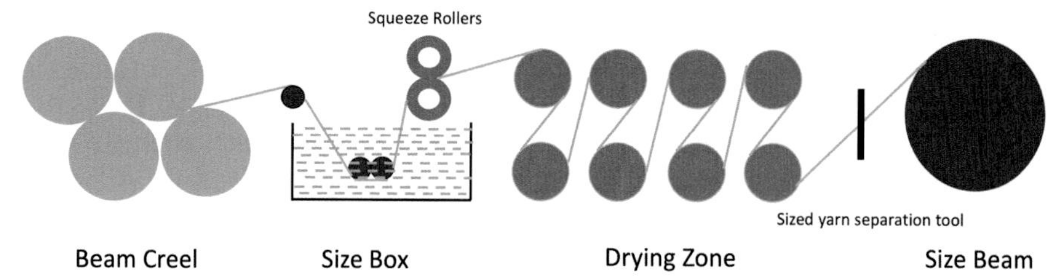


**Figure 6.5** Schematic diagram of the conventional sizing process.

industry (Reddy et al., 2014). In the case of PVA, it is responsible for 45% of the total BOD load and, more importantly, it resists general effluent treatments, eventually being discharged in water sources (Ren, 2000). To address the environmental issues of sizing, researchers are continuously working with alternative solutions such as developing environmentally friendly sizing and even size-free weaving techniques (Kabir and Haque, 2021).

Size-free weaving investigations started long ago. Gandhi (1975) explored the feasibility of weaving unsized warp yarn to manufacture plain-woven fabric. The author observed the formation of high hairiness on yarns during weaving, which led to the conclusion that size-free weaving is an impracticable process. However, their investigation was on single spun yarn, making it hard to perform size-free weaving. On the other hand, researchers reported the feasibility of size-free weaving for plied and folded or cord cotton yarns (Dumitras et al., 2004). In fact, the twists within plied yarns facilitate fibre protrusion on the surface of the yarn, leading to greater efficiency of size-free weaving for these yarns. However, there has not been much success with size-free weaving when considering the quality of the produced fabric. For instance, Sawhney (2008) reported the successful weaving of size-free cotton warp yarns while weaving a twill fabric on a modern high-speed weaving machine. Although it demonstrated the mechanical feasibility of weaving size-free cotton yarns, the fabric quality was not perfect as tiny fibrous balls were visible across the fabric. Researchers have also explored other techniques such as thermal treatment (Drexler and Tesoro, 2017), and graft copolymerisation (Bordes, 2007) of warp yarns to allow size-free weaving. Only a thermal treatment for 1.5 s at 250°C for three blended warp yarns (50/50, 65/35 and 80/20 cotton/polyester) was successful in imparting a sizing effect on the yarn without compromising yarn strength and elongation (Drexler and Tesoro, 2017). On the other hand, the graft polymerisation technique involved chemical treatment on the yarn surface while avoiding any size ingredients (Bordes, 2007). This researcher used a copolymer composed of three main components and found this technique efficient for performing size-free weaving.

Apart from the ongoing investigation into size-free weaving, there have been continuous attempts to incorporate environmentally friendly sizing agents in the sizing recipe. Traditionally, starch-based sizing ingredients, which are considered to be environmentally friendly, renewable and biodegradable, are used to size natural and manufactured fibres (Denny Farrow, 2008). However, some limitations of starch – such as moderate chemical and thermal resistance, excessive retrogradation phenomena, and compromised adhesion of starches to fibres – lead to difficulties in the post-processing of woven fabrics (Bismark et al., 2016). These drawbacks of pure natural starches led researchers to develop alternative environmentally sustainable sizing solutions, including using chemically modified starches. For example, Bismark and Zhu (2018) performed an amphipathic modification of starch to improve the adhesion properties of corn starch and applied it for sizing cotton warp yarns. They reported a considerable improvement in the adhesion of cotton yarns, leading to a robust tensile strength of the sized yarns. Also, they found a smooth desizing technique of the modified starch by oxidant desizing. However, chemically modified starch is still not ideal for industrial applications due to its high cost and concerns about biodegradability (Zhao et al., 2015).

The application of biodegradable sizing is essential when considering environmental sustainability. Animal and plant protein-derived sizing agents have been found to be the most effective biodegradable sizing agents (Stegmaier et al., 2008; Chen et al., 2013a,b; Reddy et al., 2014; Reddy and Yang, 2015; Yang et al., 2017). Among these, chitosan is one of the natural polymers gaining interest in the scientific community for sizing warp yarns. Stegmaier et al. (2008) reported both ecological and economic feasibility of modified chitosan as a potential sizing agent for sizing cotton yarns. Furthermore, they assessed the feasibility of applying modified chitosan in industrial applications and observed promising results. Interestingly, the application of modified chitosan avoids the need for desizing, and the properties of chitosan can be used effectively for other subsequent fabric treatment processes. Therefore, the authors used modified chitosan in place of pure chitosan to reduce the viscosity and cost ineffectiveness of chitosan sizing for industrial applications. This modification combined chitosan with cost-effective sizing agents such as starch and wax in an acid medium. Only a tiny addition of wax (0.05%) was found effective to reach the optimal viscosity of modified chitosan.

As a part of the continuous effort into ensuring sustainable practice in sizing, researchers have also explored different options for energy-efficient sizing. For instance, foam sizing, also known as 'short-liquor' sizing, reduces the use of water by 60%–80% in the sizing process compared to conventional sizing techniques (Drexler and Tesoro, 2017). The advantages of foam sizing include controlled size penetration into the warp yarns and low moisture pick-up, reducing the high energy requirements during drying, and facilitating a smooth desizing process. Accordingly, the energy efficiency of this technique has drawn attention for industrial application. However, it is crucial to mention that the successful implementation of the foam sizing technique depends on the chemical composition of the sizing ingredients. For instance, sizing ingredients that contain very low levels of foaming auxiliaries such as CMC (carboxymethyl-cellulose), PVA (polyvinyl alcohol) and acrylics are considered suitable for a foam sizing technique.

$CO_2$-based sizing approaches are considered another potential energy-saving sizing technique for weaving industries (Bowman et al., 1998). This technique eliminates the requirement of an aqueous sizing solution as pressurised liquid $CO_2$ is used as a solvent instead of water. Consequently, waste generation is reduced, and a considerable amount of energy is saved in the subsequent drying process of the sized yarns. Antony et al. (2018) demonstrated the environmental and economic feasibility of liquid and supercritical $CO_2$-based sizing techniques. Using $CO_2$-philes as the sizing agent and liquid and supercritical $CO_2$ as the sizing solvent, they applied the sizing recipe to cotton and polyester yarns. The authors studied three $CO_2$-philic sizing agents – sucrose octaacetate (SOA), α-D-glucose pentaacetate (AGLU) and polyethylene glycol (PEG). They found SOA to be the most suitable sizing agent for $CO_2$-based sizing as both the cotton and the polyester yarns sized with SOA showed improved mechanical performance alongside a desirable size coating on the yarns.

The sizing machine also plays an essential role in ensuring a cost-effective, efficient and sustainable sizing process (Seyam, 2019; Gandhi, 2020). Continuous developments have been seen in different sections of the sizing machine such as the pneumatic creel, temperature, squeeze pressure and stretch control of the size box; control in drying temperature and speed; improved command over the size; yarn moisture content, optimisation of beam pressure; and combined operations of sizing with other subsequent manufacturing processes. These developments have been mostly at the prototyping scale, which is essential to assessing the feasibility of new developments prior to the main production. For instance, Taiwan-based company CCI Tech Inc. developed a prototyping machine that can perform the combined actions of sizing and winding (CCI Tech Inc., n.d.; Seyam, 2019). This machine can run up to a sizing speed of 500 m/min and can control each spindle individually, leading to the possibility of sizing different yarns with different sizing recipes.

### 6.4.3 Weaving

Following sizing, warp and weft yarns are then woven into a fabric (Adanur, 2001). Weaving machines, otherwise known as looms, first came into existence under the name ‘pit loom’ (Gokarneshan, 2004). Since the handloom and power loom were developed, innovations have led to a variety of automated power looms available on the market. However, the basic principles and parts of a weaving machine or a loom are quite similar, as shown in Fig. 6.6. In a weaving machine, the warp yarns go through the heddle and reed of the machine. Heddle shafts are required for shed formation (shedding) by providing space for weft yarn insertion (picking). The number of heddle shafts required depends on the structure of the fabric. After the weft yarn insertion, the reed is used to compactly beat up (beating) the weft yarn into the fell

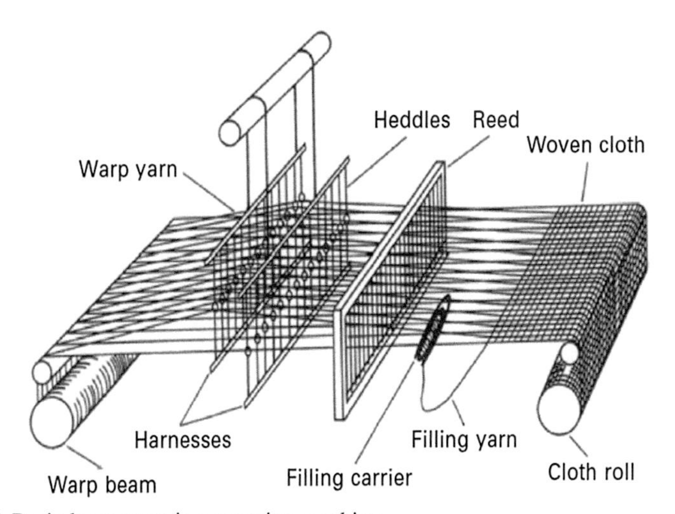


**Figure 6.6** Basic loom parts in a weaving machine.
Reprinted from Gandhi (2020), with permission from Elsevier.

of the woven cloth. Accordingly, three primary mechanisms – shedding, picking and beating – are required to complete one revolution of the loom, and this continuous mechanism in the weaving machine brings the final woven fabric in a roll form.

Most works regarding sustainable innovations in weaving machinery have attempted to reduce energy consumption. The first development to reduce weaving machine energy use was the transition from the shuttle loom to the era of the shuttle-less loom (Pattanayak, 2019). All shuttle-less looms available on the market such as rapier, projectile, air jet, water jet and multi-phase looms, are beneficial in terms of reduced pollution and high weft insertion rates, leading to less energy consumption compared to shuttle looms. Rapier looms are known for producing less noise pollution as these looms have comparatively less vibration during the weaving process. A wide variety of fabric constructions can be woven using a rapier loom. In particular, rapier looms are suitable for handling delicate weft yarns and complicated weave constructions.

On the other hand, projectile looms offer high production rates and great reliability. This type of loom is suitable for manufacturing high-width fabrics, though it has limitations in inserting a high number of coloured weft yarns and maintaining tension. Air jet looms have the highest weft insertion rate capacities and lowest production costs among the available shuttle-less looms, making them highly welcomed by weaving industries. Furthermore, air jet looms are also capable of weaving light- to medium-weight fabrics comprising both natural and man-made fibres. However, compressed air requirements in the weft insertion mechanism of an air jet loom make it a highly energy-consuming machine. This drawback of air jet looms resulted in the demand for the water jet loom. Like air jet looms, water jet looms have high weft insertion rates with comparatively low energy consumption. However, this loom is only suitable for weaving hydrophobic fibres as the weft insertion mechanism involves water. Multi-phase looms facilitate multi-phase weft insertion and are capable of high-weft insertion with less energy consumption (Hasanbeigi and Price, 2015). However, this loom has had a very low commercial adoption rate.

de Oliveira Neto et al. (2019) reported a comparative economic and environmental analysis of rapier and air jet looms. They conducted a case study of a denim fabric plant that had replaced their old rapier looms with new air jet looms to achieve better economic and environmental sustainability. The plant moved to air jet looms due to the low production cost, increased production efficiency and reduced environmental impact due to less fabric selvage waste (8 cm/insertion for the air jet loom compared to 14 cm/insertion for the rapier loom). From an economic standpoint, using an air jet loom could save the denim fabric plant up to $539,000 (USD) per year. In another study, Sanches et al. (2018) performed a comparative case study of woven denim fabric production (4,000,000 linear metres/month) by air jet and rapier looms. The same fabrics were produced on both looms with a 3/1 twill design; 10 Ne and 7 Ne warp and weft cotton yarns, respectively; 26/cm and 16.5/cm warp and weft yarns density in the fabric, respectively; and a 185 cm reed, with the aim of having a 167 cm finished fabric width. They found 91% and 89% efficiencies for air jet looms and rapier looms, respectively, whereas energy consumption of the air jet loom was 4 KWh/machine with an additional 7 KWh/machine for using compressed air for their weft insertion mechanism and 8 KWh/m for rapier looms. However, they reported that to produce the same style and amount of fabric (4,000,000 linear metres/month), air jet looms used 1,375,000 kWh total electricity, while 1,571,429 kWh total electricity was needed by the rapier looms. This finding led to the conclusion that air jet looms use less electrical resources than rapier looms. The economic evaluation of these two types of looms suggests that the air jet looms are more economical and can generate a monthly gain of $157,000 (USD) for producing the aforementioned type and quantity of denim fabric. For the environmental evaluation, the authors utilised the Material Input per Service Unit (MIPS). According to their calculation, they reported that the air jet loom could save 289 tons of material per month that would not be extracted or modified from the environment. The overall findings concluded that the air jet looms are more economical and environmentally sustainable than rapier looms. However, it is important to note that this research was conducted in only one type of fabric style, and therefore the findings cannot be generalised.

Despite the study by de Oliveira Neto et al. (2019), the use of electrical energy in air jet looms is still a concern for environmental sustainability. Therefore, Grassi et al. (2016) proposed a novel approach to modify the nozzle in the air jet weaving machine which can save up to 30% of the energy required for weft insertion. Using the convergent nozzle aerodynamic theory, the authors designed a high-volume low-pressure nozzle for weft insertion. The inner diameter (up to 5 mm) of the developed relay nozzle was comparatively larger, allowing it to carry larger volume flows while operating at a lower pressure. In fact, they demonstrated that their relay nozzle can facilitate the weft insertion across the warp at only 1 bar overpressure while the relay nozzles available in the market require up to 3–5 bar pressure. However, the authors still need to validate the concept in a commercial weaving process.

Weaving machine manufacturers are also coming up with innovations that foster sustainable development (Seyam, 2019). Itema, a weaving manufacturing company, introduced a new weaving machine called Discovery using a weft insertion mechanism they refer to as the 'Positive Flying Shuttle'. This weft insertion mechanism is similar to the projectile loom but differs in the acceleration mechanism of the projectile. In a projectile loom, the acceleration of the projectile comes from a picking arm whereas, in the Positive Flying Shuttle loom, the acceleration of the projectile comes from a picking bar. This picking bar contacts the projectile and moves it to a certain distance to give the projectile adequate energy to travel across the loom width, allowing the Positive Flying Shuttle loom to have greater control over the projectile than the traditional projectile loom. In addition, this mechanism allows a smoother motion and gradual acceleration of the projectile. Another example is the OptiMax-i sustainable innovation by Picanol. They introduced near fully digital weft insertion which reduces the amount of weft waste in the right-side selvage of the fabric (Picanol, 2021). In addition, the inclusion of the Electronic Right Gripper Opener (ERGO) in their machine facilitates control of the weft yarn length, thus reducing weft yarn waste. In addition, the use of a sumo drive concept in the machine design enables greater energy efficiency.

Many sustainable innovations can be observed in weaving sample prototyping machines (Seyam, 2019). For example, Taiwan-based CCI Tech Inc. came up with a sample prototyping machine called Kebalan that is capable of weaving 300 picks/minute. This is a record weft insertion speed in a sample prototyping machine, thus allowing for quick prototype manufacturing. Their other prototyping machine, Evergreen II, was developed by maintaining a compact structure so that it occupies less space in the weaving floor.

### 6.4.4 Desizing

Desizing refers to the complete removal of sizing from the sized yarns after weaving (Drexler and Tesoro, 2017). This step is essential before dyeing and finishing woven fabrics as sizing ingredients can react with soda during the dyeing process, leading to shade change in the fabric (Ul-Haq and Nasir, 2011). Ease of size removal, i.e., desizing, depends on the types of sizing ingredients used in this process (Adanur, 2001). Several factors such as the solubility of the film-forming polymer, viscosity of the solvated gel, dissolution rate, desizing agent concentration, mechanical action, desizing temperature, and required acidic and alkaline conditions influence the efficiency of the desizing process. Furthermore, different desizing methods exist, such as hydrolytic, oxidative, alkaline processes, solvent desizing, low-temperature plasma treatment, and desizing by thermochemical extraction and coronization.

An essential consideration in desizing is to be sure that the process does not hamper fabric properties (Ul-Haq and Nasir, 2011). For instance, cotton fabrics' acid desizing by dilute hydrochloric acid (HCl) or sulphuric acid ($H_2SO_4$) deteriorates fabric strength as the acid cannot differentiate between cellulose and starch. As an alternative, weak acids like hydrogen peroxide ($H_2O_2$) can be used to mitigate the effect of hydrolysis on cotton. However, using $H_2O_2$ in the desizing of starch leads to carbon dioxide ($CO_2$) formation and is a barrier to sustainable innovation in this sector. Therefore, enzymatic desizing is considered a potential solution to ensure environmental and economic sustainability (Aly et al., 2004). Moreover, enzymatic reactions are usually faster with the rare formation of toxic by-products (Gubitz and Cavaco-Paulo, 2003). Also, unlike the acid desizing process, enzymatic desizing by using $\alpha$-amylases can hydrolyse the starch only without affecting the cellulose fibre.

Researchers have also tried natural desizing agents for cotton fabrics (Hoque et al., 2018a,b). For example, Hoque et al. (2018b) investigated the use of soapnut as a desizing agent for denim fabrics, followed by washing with other agents such as lemon juice, tamarind powder, enzymes or calcium hypochlorite. They found that weight loss on naturally desized and washed denim fabrics is three times less than on synthetic chemical-washed denim fabrics. This finding indicates that the natural reagent did not affect the cellulose fibre. Also, they found higher fabric strength on naturally desized and washed specimens. However, an industrial feasibility study is needed before considering soapnut as a potential desizing agent for cotton-based fabrics on an industrial scale.

Another area of research attention is the integration of desizing with other wet processing stages. Ali et al. (2014) performed an experimental study to investigate the feasibility of integrated desizing, bleaching and reactive dyeing of cotton towel fabric. Their experimental procedure eliminated the discharge of desizing effluent and converted the effluent into $H_2O_2$, which facilitated the bleaching process. To achieve this, enzymatic (Dextrozyme DX) desizing was first performed in the cotton towel without any subsequent washing. Second, they used amyloglucosidase-pullanase and glucose oxidase (GOx) (*Aspergillus oryzae*) enzymes to generate $H_2O_2$ from the cotton towel desizing effluent, which was used in the bleaching stage. A catalase enzyme also treated the effluent of bleaching to get rid of the residual $H_2O_2$ from the water, and the treated water was used for the reactive dyeing process. Thus, they achieved up to 400% water-saving and 50% thermal energy-saving compared with the conventional method. Their findings suggest the process is highly economical while improving the environmental sustainability of the desizing stage.

### 6.4.5 Waste management

The concept of fast fashion has accelerated clothing consumption and generated substantial textile and resource waste (Bick et al., 2018). This high fabric production necessitates proper waste management in the textile production stage to ensure economic and environmental sustainability. Wastewater, fibre and fabric are the major sources of waste generation in textile manufacturing industries (Pensupa et al., 2017). Therefore, steps such as optimisation, reuse and recycling of fabric wastage, and proper maintenance of the effluent treatment plant (ETP) can help improve woven textile sustainability.

#### 6.4.5.1 Recycling and reuse of fabric and yarn waste

Selvage waste production is one of the limitations of shuttle-less looms (Akdeniz et al., 2017). Unlike shuttle looms, shuttle-less looms are unable to form a real selvage. Instead, they create an auxiliary or false selvage during the weaving process. This false selvage consists of five to nine warp yarns that remain at a certain distance from the last warp ends of the main fabric width. The formation of this false selvage leads to the consumption of more yarns (up to 7–9 cm of extra weft yarns at each side of the selvage) and as much as 9%–10% waste. However, recycling this selvage wastages can reduce landfill dumping while saving money on yarn costs. A garneting machine, or hard waste opener, can be used to break down the selvage to open the material to a fibre state, then spun to make new yarns. On the other hand, Qureshi et al. (2021) designed and developed a novel approach for the rapier loom that eliminates the need for false selvage. They proposed temporary gripping of the extra weft yarns rather than permanent gripping as is done in the regular process. This temporary gripping allowed for reuse of the gripper ends and eliminated the false selvage waste. They implemented the technique in an industrial setting and found the proposed design was feasible.

Different kinds of waste are also generated in the weaving industry, in addition to selvage waste. For instance, small amounts of yarns left on the cone during the warping process contribute to 0.5%–1% of weaving hard waste (Goyal, 2021b). However, nowadays weaving industries are trying to reuse the residual yarns from these cones by transferring them to large cones that can be used in the weaving process again. Nevertheless, these cones are usually used to produce low-grade fabrics as the cone contains a lot of knotting that affects the quality of the yarn. This process contributes to environmental sustainability but compromises industry profitability if the fabrics are sold at a lower price.

The sizing steps generate unsized yarn waste on the creel and sized yarn waste on the headstock sides (winding side of the machine) of a sizing machine (Goyal, 2021b). The total waste in the sizing step can be up to 0.7%, which usually varies from size to size depending on the yarn count processed. Unsized yarn waste is generated as some portion of warp yarn is kept, ensuring proper winding of warp on the weaving beam. Additionally, leftover yarns on the warp beam of the creel section contribute to unsized yarn waste. Sized waste is generated as a substantial amount of sized warp yarns are kept in the sizing machine to tie the yarns for the next sizing batch. However, sizing yarn waste is quite unavoidable and, therefore, reusing this waste is the next best option.

Knotting refers to the attachment when a new weaving beam is needed to produce the same woven fabric (Ghosh et al., 2014). Knotting is an automatic gaiting and drawing process that eliminates labour and time, effectively reducing manufacturing costs. However, the knotting process generates 0.3%–0.5% waste. Although this amount of waste can be considered tolerable, some modifications in the yarn process can reduce this waste. For instance, a taping material can be incorporated on the warp sheet that allows proper alignment of the yarn, reducing warp yarn damage before preparing it for the knotting process.

The solid waste generated in the weaving industry can also be reused to create value-added products or act as a source of energy (Pensupa et al., 2017). For example, Ryu et al. (2007) reported 16 MJ/kg energy content from cotton/polyester fabric waste. They found that the combustion of this type of fabric waste at 700°C for a few minutes can generate high heat energy. Campbell et al. (2000) co-combusted textile waste with coal in a circulating fluidised bed combustor (CFBC) to generate heat energy. They found that a 5% textile waste blend in the co-combustion process provided 122 MW energy in the CFBC power station, while increasing the textile waste to 20% in the blend gave a maximum of 24 MW energy. This finding indicates that incorporating textile waste with coal reduces the efficiency of the CFBC power station. However, they proved it economically sustainable when considering the textile waste disposal cost. Lokhande and Gotmare (1999) used loom waste (30% starch/70% cotton cellulose) to develop a highly absorbent graft co-polymer. Their developed product showed excellent water and saline solution absorbencies. In another study, Zheng et al. (2014) developed activated carbon fibre (ACF) using woven cotton waste. They found that their developed ACF was more efficient in treating oilfield wastewater than the commercially available ACF.

#### 6.4.5.2 Maintenance of effluent treatment plants

Waste treatment helps reduce the chemical concentration of wastewater before it is discharged into the environment (Pensupa et al., 2017). Composite woven textile industries with combined fabric weaving and finishing facilities maintain a functional ETP. The ETP is used to treat the effluent which mainly contains various chemicals, sizes, dyes and wastewater used in fabric wet processing stages (Ghaly et al., 2014). The effluent also contains different kinds of metals, such as chromium, arsenic, copper and zinc, that are detrimental to human and aqueous environments. The overall treatment of an ETP is divided into three steps: primary, secondary and tertiary treatments. The primary treatment process removes solids like oil, grease and gritty substances, while the secondary treatment step reduces BOD, phenol and residual oil from the primary treatment step and initiates the process of colour control in the wastewater. Finally, the tertiary step involves the removal of any residual contaminants from the first two steps. These treatments in an ETP can be done by various techniques, including physical, biological, chemical processes, or a combination of these. However, the biological process often has a prolonged retention time when the wastewater contains toxic metals as they inhibit the growth of micro-organisms and consequently delay the treatment process. In that scenario, a chemical process such as using an advanced oxidation treatment is more beneficial as this process can successfully treat most of the solid compounds found in textile effluent.

In recent years, nanotechnology has received high interest in textile wastewater treatment (Pervez et al., 2021). For instance, an iron (Fe) nanoparticle (NP)-loaded polyester fabric has been under development to remove dyes from textile effluents. Pervez et al. (2021) investigated the effectiveness of the Fe-NP-loaded polyester fabric on two dyes: 4-nitrophenol (4-NP) and methylene blue (MB). They obtained a conversion yield of 97.5% in 12 min for 4-NP and 98% in 16 min for MB dye. In another study, Dwivedi et al. (2020) used unzipped multi-walled carbon nanotube oxides (UMCNOs) to treat the cotton fabric and used it for wastewater treatment. They achieved 96.51% removal of MB dye from the wastewater. In addition, they also observed its effectiveness in removing organic diluents from the wastewater with 99% efficiency. These studies demonstrate the potential of nanotechnology and woven fabrics for improved textile wastewater treatment.

### 6.4.6 Production and logistics

Currently, industries are transforming towards the fourth industrial revolution (Industry 4.0), which necessitates adopting automation. The weaving industry is no exception to this. Most weaving companies operate with automatic power looms. Apart from automation in manufacturing, weaving companies are also taking advantage of automation in other sections of the industry, such as defect detection, machine downtime tracking, and monitoring of the fabric supply chain.

It is very unlikely to find a manufacturing process that is 100% free of defects (Jiang and Wong, 2018). Common defects in weaving include end out, oil spots and broken ends and picks. To be competitive and trustworthy in the market, manufacturers should maintain high-quality standards. Nateri et al. (2014) reported that fabric defects could be responsible for reducing the product price by up to 65%. Although the data are alarming, most weaving companies still depend on manual inspection for fabric defect detection. However, researchers are working on automatic defect detection systems, and some industries are transforming to automated processes to maintain high-quality standards.

Automatic defect detection works on the principle of artificial intelligence by using image processing and machine learning algorithms. Kopaczka et al. (2018) developed a unique approach for detecting faulty weft yarns during weaving on an air jet loom. They used computer vision techniques involving image acquisition and classification pipelines that precisely detect different defects based on machine learning methods. They observed excellent performance of their proposed automatic defect detection system while installed in a real production setting. Weninger et al. (2018) developed an automatic visual defect detection framework for plain woven fabrics that localises and tracks a single yarn defect in fabrics. They used fully conventional neural networks and float point analysis techniques to develop their automatic defect detection system. More information about the automatic fabric defect detection technique can be found in research by Rasheed et al. (2020) and Li et al. (2021), where they performed comprehensive reviews on state-of-the-art research on the topic.

Fabric weaving is a highly mechanical process that may be interrupted by sudden stoppage of the machine due to different kinds of mechanical, electrical or yarn-related faults (Haque et al., 2019). This sudden stoppage increases the fabric production time and can deteriorate fabric quality; thus, the operator needs to identify the loom issue as quick as possible. In this context, installing a downtime tracker for each weaving loom can aid in detecting the stopped machine and the reason for its interruption (Davey Textile Solutions Inc., 2021). Davey Textile Solution's custom-manufactured Internet of Things (IoT) device was installed for each loom on the production floor to collect data on loom stoppages. Instantly, the real-time data are stored in a cloud service that an electric device can monitor. This approach allows a weaving technician to monitor all the looms of a production floor from a single point, thus eliminating the need for human resources to troubleshoot the loom issue.

Finally, enterprise resource planning (ERP) is a supply chain management software that collects information from different industry sections and integrates it into a single database stored either locally or in the cloud (Ganesh et al., 2014). Weaving industries include different segments such as raw material collection, warping, sizing, desizing, finishing and packing of the fabric. An integrated system like ERP that comprises the real-time information of all segments in a database allows smooth operation of the industry. Surung et al. (2020) implemented ERP in a small weaving industry that previously managed data in a conventional way. They performed a user acceptance testing analysis after the implementation of ERP in that industry facility and observed very successful acceptance of the system by users. In the market, ERP software with different trade names such as Protex, Texplus, Inteos, Intex, Solinsyst and Loomdata are available to use for weaving applications (Thoney-Barletta, 2019).

## 6.5 Future perspectives

The textile industry, including in the weaving sector, has come a long way in terms of sustainability. Even though the situation is far from perfect, key players now understand that the topic cannot be swept under the rug. However, further progress in the textile industry is critical for global efforts towards sustainable development. Measurement is a key component for progress. The frequency with which progress is monitored has even been positively correlated with successfully achieving goals (Harkin et al., 2016). In the case of the textile industry, sustainability is an immense challenge considering the multitude of environmental, social and economic impacts of weaving activities. However, information obtained and models developed for other sectors of activity, which may display similar sustainability challenges, can guide weaving industry sustainability progress. For example, the minimum indicators of the economic, environmental and social dimensions of a firm's triple bottom line performance proposed by Hourneaux et al. (2018) (Table 6.1) can be adapted to the specificities of the textile industry and the woven fabric sector by defining the different aspects covered by each indicator in relation to the industry situation.

With consistent KPIs for sustainable performance/innovation, it is then possible to compare processes and results between different companies, locations and countries. Part of the solution may reside in initiatives launched over the last few years by standards organisations to promote and foster sustainable development. For instance, the International Organisation for Standardisation (ISO) adopted the London Declaration in 2021 'to combat climate change through standards' (ISO, 2021). The scope of the ISO's Technical Committee TC 38 on textiles also includes the development of terminology, definitions, test methods, and specifications related to ethics and environmental considerations in the textile supply chain (ISO, n.d.). Work Group 35 has been specially formed on the topic of environmental aspects. In the United States, the ASTM Committee D13 on Textiles comprises a subcommittee dedicated to textile sustainability (ASTM D13, 2022).

A second promising avenue of woven textile sustainability progress relies on new technologies. A first application can be to characterise the weaving process and identify areas of improvement. For example, sensor yarns have been developed and successfully used to get an insider view of the weaving process kinematics and locate high stress points on the yarns as they are processed in different areas of the loom (Boussu et al., 2016). The sensor yarn is produced by coating the weaving yarn with a thin layer of piezoresistive polymer, allowing the sensor yarn to become part of the weaving process and textile structure without affecting the in situ measurement it provides. In a study by Boussu et al. (2016), the sensor yarn allowed the authors to detect a difference in warp yarn tensioning in the whip roller area of a loom used to produce 3D woven fabrics. New technologies can also be used to automate some labour-intensive operations and increase overall weaving efficiency. Automatic defect detection is an example of such an improvement. The artificial intelligence applied to image analysis can provide powerful strategies towards economic sustainability to the textile industry. However, limitations outlined by Rasheed et al. (2020) and Li et al. (2021), including the lack of publicly available databases of fabric images, the need for the industry to upgrade its equipment and use of technology, and algorithm refinement, will have to be overcome first.

A third key component to weaving sustainability is collaboration within and along the value chain and between different disciplines. Many problems encountered in the weaving sector are multifaceted and multidisciplinary: an interdisciplinary approach is the most efficient way, and often the only one, to get to the best solution. For instance, when considering the case of sizing, chemistry and textile science come to mind first. The solution also needs input from textile engineers regarding the impact on/of the weaving equipment. In addition, mechanical engineers should be involved in the discussion as the sizing compound may affect the mechanical performance of the yarn, either permanently or in a temporary manner. Because the sizing compound/process may affect working conditions (e.g., the release of aerosols into the environment during the weaving process), it is critical that industrial engineers are at the design table too. Finally, considerations related to the potential impact of the process on the environment (e.g., through the release of its effluents) need to be kept in mind, with welcome insights from environmental scientists. However, interdisciplinary collaboration comes with challenges, such as being more complex and prone to communication failures (Bendix et al., 2017; Cairns et al., 2020). A good understanding of what interdisciplinary collaboration entails is thus critical to avoid pitfalls.

Another important dimension is collaboration along the textile value chain. The weaving sector is only one component of the puzzle that makes a finished product. Collaboration and concertation with those before and after weaving operations in the supply chain are essential to ensure that yarns are suitable for weaving and that woven fabrics are suitable for garment constructions or any other application they are intended for. For instance, Kim and Zorola (2018) reported how a Sustainable Supply Chain Network Innovation Model was developed and successfully implemented by TAL Apparel Group for recycled cotton clothing production. It included 'innovation co-creation' (Kim and Zorola, 2018, p.188) between yarn and garment manufacturers to ensure the woven textile product is fit for purpose/function and 'network of support' (Kim and Zorola, 2018, p. 189) from the various stakeholders along with their own networks. Collaboration and concertation with other potential partners within the textile industry and in other sectors can also offer unique opportunities for sustainable progress, such as selvage or other production waste being used by non-woven fabric manufacturers as raw materials and valuable compounds being extracted from desizing effluents. For instance, Ali et al. (2014) described how $H_2O_2$ generated from the desizing effluent of cotton towels can be used for the towel bleaching treatment. However, challenges exist as, depending on their position in the supply chain and how it may affect their business, companies may have differing perspectives on the importance of implementing sustainability practices [Hassini et al. (2012), as cited in Kim and Zorola (2018)].

The last level to be considered is within-industry collaboration. Partnering between actors in the weaving sector allows for strategy sharing to improve equipment and processes. However, it requires actors to go beyond the competitor-type mindset which has been the norm in the industry by acknowledging that any win is beneficial to everyone. Another positive outcome of collaborations within the weaving sector is related to cost and labour savings. As Kim and Zorola (2018) report, manufacturers are left to operate on a very small margin, which does not allow them to make the required investment for more sustainable practices. Therefore, approaching sustainability as a coordinated effort means that these investments are not up to individual firms alone, providing greater opportunities for more scalable and impactful innovations.

## 6.6 Conclusion

The weaving sector has an important role to play in making the textile industry more sustainable. An assessment of current woven textile production sustainability reveals several issues to tackle; in particular, intensive energy, chemical and water usage, generation of avoidable waste and poor working conditions are all areas of the sector in need of improvement. An analysis of the three pillars of sustainability identified environmental, social and economic opportunities for innovation. Environmental improvements relate to the reduction of use of natural resources and pollution, increases in product useful life and recyclability, and the incorporation of clean and renewable sources of energy; social improvements aim to better protect worker rights and create safe working conditions; while economic improvements consider financial and non-financial indexes beyond profit. However, sustainability in the weaving sector can only be achieved by considering these three pillars as a whole as their impacts are interconnected and entangled.

This chapter also reviews various innovations proposed and/or implemented for more sustainable woven textiles. These include new warping techniques for improved efficiency, size-free weaving, environmentally friendly/biodegradable sizing agents, more energy-efficient sizing, new weaving techniques and modifications in looms for lower energy consumption, and more environmentally friendly and less fibre-damaging desizing treatments. Progress has been achieved as well in terms of waste management, with the recycling and reuse of fabric and yarn wastage and improved practices for effluent treatment plant maintenance. Quality assurance in production and logistics has also benefitted from new technologies, specifically for defect detection.

Finally, three promising avenues to be explored for further progress have been described. The first relates to the need to measure progress in woven textile sustainability. Measurement strategies include taking advantage of data and models from other relevant sectors and partnering with standardisation organisations. A second promising avenue is harnessing new technologies to characterise the weaving process and identify areas of improvement as well as automate some labour-intensive operations and increase overall weaving process efficiency. The third key to progress in weaving sustainability relies on collaboration within and along the value chain and between different disciplines. By taking advantage of the different resources available and working as a team, the weaving sector has in its hands the tools to lead textile sustainability.

## References

Abdul-Rashid, S.H., et al., 2017. The impact of sustainable manufacturing practices on sustainability performance: empirical evidence from Malaysia. International Journal of Operations & Production Management 37 (2), 182–204. https://doi.org/10.1108/IJOPM-04-2015-0223.

Acar, E., Kiliç, M., Güner, M., 2015. Measurement of sustainability performance in textile industry by using a multi-criteria decision making method. Tekstil ve Konfeksiyon 25 (1), 3–9.

Adanur, S., 2000. Fabric formation by weaving. In: Handbook of Weaving. CRC Press, pp. 1–7. https://doi.org/10.4324/9780429135828-1.

Adanur, S., 2001. In: Adanur, S. (Ed.), Handbook of Weaving. Taylor & Francis.

Akdeniz, R., Ozek, H.Z., Durusoy, G., 2017. A study of selvedge waste length in rapier weaving by image analysis technique. TEKSTİL ve KONFEKSİYON 27 (1), 22–26.

Ali, S., et al., 2014. Integrated desizing–bleaching–reactive dyeing process for cotton towel using glucose oxidase enzyme. Journal of Cleaner Production 66, 562–567. https://doi.org/10.1016/J.JCLEPRO.2013.11.035.

Alkaya, E., Demirer, G.N., 2014. Sustainable textile production: a case study from a woven fabric manufacturing mill in Turkey. Journal of Cleaner Production 65, 595–603. https://doi.org/10.1016/j.jclepro.2013.07.008.

Aly, A.S., Moustafa, A.B., Hebeish, A., 2004. Bio-technological treatment of cellulosic textiles. Journal of Cleaner Production 12 (7), 697–705. https://doi.org/10.1016/S0959-6526(03)00074-X.

Amutha, K., 2017. Environmental impacts of denim. In: Sustainability in Denim. Woodhead Publishing.

Annapoorani, S.G., 2017. Social sustainability in textile industry. In: Muthu, S.S. (Ed.), Sustainability in the Textile Industry. Springer Nature, pp. 57–79.

Antony, A., et al., 2018. Sizing and desizing of cotton and polyester yarns using liquid and supercritical carbon dioxide with nonfluorous CO2-philes as size compounds. ACS Sustainable Chemistry and Engineering. American Chemical Society 6 (9), 12275–12280. https://doi.org/10.1021/ACSSUSCHEMENG.8B02699/SUPPL_FILE/SC8B02699_SI_001.PDF.

ASTM D13, 2022. Committee D13 on Textiles. American Society for Testing and Materials. Available at: https://www.astm.org/get-involved/technical-committees/committee-d13. (Accessed 11 May 2022).

Balanay, R., Halog, A., December 2018. Tools for circular economy: review and some potential applications for the Philippine textile industry. In: Circular Economy in Textiles and Apparel: Processing, Manufacturing, and Design, pp. 49–75. https://doi.org/10.1016/B978-0-08-102630-4.00003-0.

Bedi, R., 2006. Evaluation of occupational environment in two textile plants in Northern India with specific reference to noise. Industrial Health 44 (1), 112–116. https://doi.org/10.2486/indhealth.44.112.

Belal, S.A., 2018. Understanding Textile for a Merchandiser, second ed. BMN Foundation.

Bendix, R.F., Bizer, K., Noyes, D., 2017. Toward a sustainable interdisciplinarity: recommendations. In: Bendix, R.F., Bizer, K., Noyes, D. (Eds.), Sustaining Interdisciplinary Collaboration: A Guide for the Academy. University of Illinois Press, Urbana, pp. 94–108. Available at: https://www.jstor.org/stable/10.5406/j.ctt1n7qkfq.10?seq=1. (Accessed 11 May 2022).

Bick, R., Halsey, E., Ekenga, C.C., 2018. The global environmental injustice of fast fashion. Environmental Health: A Global Access Science Source 17 (1), 1–4. https://doi.org/10.1186/S12940-018-0433-7/PEER-REVIEW.

Bismark, S., Zhu, Z., 2018. Amphipathic starch with phosphate and octenylsuccinate substituents for strong adhesion to cotton in warp sizing. Fibers and Polymers 19 (9), 1850–1860. https://doi.org/10.1007/S12221-018-8058-6.

Bismark, S., Zhifeng, Z., Benjamin, T., 2016. Effects of differential degree of chemical modification on the properties of modified starches: sizing. The Journal of Adhesion 94 (2), 97–123. https://doi.org/10.1080/00218464.2016.1250629.

Bordes, B., 2007. Threads, Fibres and Filaments for Weaving without Sizing. United States.

Bori, G., Bhattacharyya, N., 2021. Factors affecting handloom weaving practices among women weavers of Assam. Economic Affairs 66 (4), 623–628. https://doi.org/10.46852/0424-2513.4.2021.15.

Boussu, F., et al., 2016. Fibrous sensors to help the monitoring of weaving process. In: Koncar, V. (Ed.), Smart Textiles and Their Applications. Woodhead Publishing, pp. 375–400. https://doi.org/10.1016/B978-0-08-100574-3.00017-5.

Bowman, L.E., et al., 1998. Advances in carbon dioxide based sizing and desizing. Textile Research Journal 68 (10), 732–738. https://doi.org/10.1177/004051759806801006.

Broudy, E., 1979. The Book of Looms: A History of the Handloom from Ancient Times to the Present. Van Nostrand Reinhold Co, New York.

Butnariu, A., Avasilcai, S., 2014. Research on the possibility to apply ecological footprint as environmental performance indicator for the textile industry. Procedia - Social and Behavioral Sciences 124 (March), 344–350. https://doi.org/10.1016/j.sbspro.2014.02.495.

Cairns, R., Hielscher, S., Light, A., 2020. Collaboration, creativity, conflict and chaos: doing interdisciplinary sustainability research. Sustainability Science 15 (6), 1711–1721. https://doi.org/10.1007/S11625-020-00784-Z/TABLES/1. Springer Japan.

CCI Tech Inc., n.d. Taroko - CCI Tech Inc., CCI Tech Inc. Available at: https://www.ccitk.com/product_detail.php?id=10. (Accessed 1 March 2022).

Campbell, P.E., McMullan, J.T., Williams, B.C., Aumann, F., 2000. Co-combustion of coal and textiles in a small-scale circulating fluidised bed boiler in Germany. Fuel Processing Technology 67 (2), 115–129.

Charkha, P.G., Jaju, S.B., 2014. Designing innovative framework for supply chain performance measurement in textile industry. International Journal of Logistics Systems and Management 18 (2), 216–230. https://doi.org/10.1504/IJLSM.2014.062327.

Chen, L., Reddy, N., Yang, Y., 2013a. Remediation of environmental pollution by substituting poly(vinyl alcohol) with biodegradable warp size from wheat gluten. Environmental Science and Technology 47 (9), 4505–4511. https://doi.org/10.1021/ES304429S.

Chen, L., Reddy, N., Yang, Y., 2013b. Soy proteins as environmentally friendly sizing agents to replace poly(vinyl alcohol). Environmental Science and Pollution Research 20 (9), 6085–6095. https://doi.org/10.1007/S11356-013-1601-5/FIGURES/10.

Choobineh, A., Shahnavaz, H., Lahmi, M., 2004. Major health risk factors in iranian hand-woven carpet industry. International Journal of Occupational Safety and Ergonomics 10 (1), 65–78. https://doi.org/10.1080/10803548.2004.11076596.

Clair, K.S., 2018. The Golden Thread: How Fabric Changed History. John Murray, London.

Cita, n.d. Apparel Supply Chain –Carbon Assessment about Jeans Producing. Available at: https://www.cita.org.hk/wp-content/uploads/2013/10/Fty_rpt_Eng_20130912.pdf. (Accessed 22 January 2022).

Cole, D.J., Deihl, N., 2015. The History of Modern Fashion: From 1850. Laurence King Publishing.

Costa, I., Martins, F.G., Alves, I., 2019. Ecological Footprint as a sustainability indicator to analyze energy consumption in a Portuguese textile facility. International Journal of Energy and Environmental Engineering 10 (4), 523–528. https://doi.org/10.1007/s40095-018-0268-6.

Davey Textile Solutions Inc, 2021. Case Study: How ShookIOT Downtime Tracker Enabled Davey Textile to Make Data-Driven Predictions and Decisions - Davey Textile Solutions - Custom Reflective Safety Trim. Davey Textile Solutions Inc. Blog. Available at: https://daveytextiles.com/blog/case-study-how-shookiot-downtime-tracker-enabled-davey-textile-to-make-data-driven-predictions-and-decisions/. (Accessed 5 April 2022).

de Oliveira Neto, G.C., et al., 2019. Cleaner production in the textile industry and its relationship to sustainable development goals. Journal of Cleaner Production 228, 1514–1525. https://doi.org/10.1016/J.JCLEPRO.2019.04.334.

Denny Farrow, F., 2008. Sizing: a review of the literature. Journal of the Textile Institute Transactions 14 (8), 254–263. https://doi.org/10.1080/19447022308661251.

Dolez, P.I., Vermeersch, O., 2018. Introduction to advanced characterization and testing of textiles. In: Dolez, P., Vermeersch, O., Izquierdo, V. (Eds.), Advanced Characterization and Testing of Textiles. Woodhead Publishing, pp. 3–21. https://doi.org/10.1016/B978-0-08-100453-1.00001-5.

Drexler, P.G., Tesoro, G.C., 2017. Materials and processes for textile warp sizing. In: Lewin, M., Sello, S.B. (Eds.), Handbook of Fiber Science and Technology: Volume I Chemical Processing of Fibers and Fabrics Fundamentals and Preparation: Part B, first ed. Routledge, New York, pp. 1–89. https://doi.org/10.1201/9780203719275-1/MATERIALS-PROCESSES-TEXTILE-WARP-SIZING-PETER-DREXLER-GIULIANA-TESORO.

Duflou, J.R., et al., 2012. Towards energy and resource efficient manufacturing: a processes and systems approach. CIRP Annals 61 (2), 587–609. https://doi.org/10.1016/J.CIRP.2012.05.002.

Dumitras, P.G., Sawhney, P.S., Bologa, M.K., 2004. Some new perspectives on improving weaving process... - electronic processing of materials. Operating Experience 40 (1), 82–87.

Durlov, S., et al., 2014. Prevalence of low back pain among handloom weavers in West Bengal, India. International Journal of Occupational and Environmental Health 20 (4), 333–339. https://doi.org/10.1179/2049396714Y.0000000082.

Dwivedi, P., Vijayakumar, R.P., Chaudhary, A.K., 2020. Synthesis of UMCNO-cotton fabric and its application in waste water treatment. Cellulose 27 (2), 969–980. https://doi.org/10.1007/S10570-019-02840-Z/FIGURES/13.

Elkington, J., 1997. Cannibals with Forks: The Triple Bottom Line of 21st Century Business. Capstone, Oxford.

Ellen MacArthur Foundation, 2017. A New Textiles Economy: Redesigning Fashion's Future. Available at: https://emf.thirdlight.com/link/2axvc7eob8zx-za4ule/@/preview/1?o. (Accessed 1 May 2022).

Eren, R., Suvari, F., Celik, O., 2016. Mathematical analysis of motion control in sectional warping machines. Textile Research Journal 88 (2), 133–143. https://doi.org/10.1177/0040517516676059.

Ethical Fashion Forum, 2015. The Issues: Chemicals, Common Objective. Available at: https://www.commonobjective.co/article/the-issues-chemicals#eff. (Accessed 16 January 2022).

Frigelg, E.L.C., Pereira, D.C., Curi, R.P., 2019. Sustainable innovation in the Brazilian textile industry. In: Stehr, C., Dziatzko, N., Struve, F. (Eds.), Corporate Social Responsibility in Brazil. Springer International Publishing, pp. 367–391. https://doi.org/10.1007/978-3-319-90605-8_18.

Fuhr, M., 2010. Warp Beam. https://patents.google.com/patent/EP2154277A1/en?oq=EP2154277A1.

Gandhi, K.L., 1975. A Comparison between Break-Spun and Ring-Spun Yarns with Particular Reference to Sizing and Weaving. PhD Thesis. The Victoria University of Manchester.

Gandhi, K.L., 2019. Yarn preparation for weaving: warping. In: Gandhi, K.L. (Ed.), Woven Textiles: Principles, Technologies and Applications, second ed. Woodhead Publishing, pp. 81–118. https://doi.org/10.1016/B978-0-08-102497-3.00003-9.

Gandhi, K.L., 2020. The fundamentals of weaving technology. In: Gandhi, K.L. (Ed.), Woven Textiles: Principles, Technologies and Applications, second ed. Woodhead Publishing, pp. 167–270. https://doi.org/10.1016/B978-0-08-102497-3.00005-2.

Ganesh, K., et al., 2014. Enterprise Resource Planning: Fundamentals of Design and Implementation. Springer International Publishing (Management for Professionals), Cham. https://doi.org/10.1007/978-3-319-05927-3.

Garetti, M., Taisch, M., 2012. Sustainable manufacturing: trends and research challenges. Production Planning & Control 23 (2–3), 83–104. https://doi.org/10.1080/09537287.2011.591619.

Gbolarumi, F.T., Wong, K.Y., Olohunde, S.T., 2021. Sustainability assessment in the textile and apparel industry: a review of recent studies. IOP Conference Series: Materials Science and Engineering 1051 (1), 012099. https://doi.org/10.1088/1757-899x/1051/1/012099.

Ghaly, A., et al., 2014. Production, characterization and treatment of textile effluents: a critical review. Chemical Engineering and Process Technology 5 (1), 1000182.

Ghosh, D.S.K., Gupta, M.K.R., Dey, M.C., 2014. Study on generation of knotting waste in weaving cotton fabric and its remedial measures. International Journal of Engineering Research and Technology 3 (6), 2185–2194. Available at: www.ijert.org. (Accessed 4 April 2022).

Global Reporting Initiative, 2008. Sustainability Reporting Guidelines. Available at: www.globalreporting.org.

Goedkoop, M., 1995. The Eco-Indicator 95: Final Report. NOH (National Reuse of Waste Research Programme), Amersfoort report 9523.

Goedkoop, M., Spriensma, R., 1999. The Eco-Indicator 99: A Damage Oriented Method for Life Cycle Impact Assessment (Methodology Report) (Amersfoort).

Gokarneshan, N., 2004. Fabric structure and design. First. In: Gokarneshan, N. (Ed.). New Age International (P) Ltd, New Delhi.

Gonçalves, A., Silva, C., 2021. Looking for sustainability scoring in apparel: a review on environmental footprint, social impacts and transparency. Energies 14 (11). https://doi.org/10.3390/en14113032.

Goswami, B.C., Anandjiwala, R.D., Hall, D.M., 2004. Textile Sizing. CRC Press, Taylor & Francis Group, New York.

Goyal, A., 2021a. Management of spinning and weaving wastes. In: Waste Management in the Fashion and Textile Industries. Elsevier Ltd. https://doi.org/10.1016/b978-0-12-818758-6.00003-x.

Goyal, A., 2021b. Management of spinning and weaving wastes. In: Nayak, R., Patnaik, A. (Eds.), Waste Management in the Fashion and Textile Industries, first ed. Woodhead Publishing, pp. 61–82. https://doi.org/10.1016/B978-0-12-818758-6.00003-X.

Goyal, A., Nayak, R., 2020. Sustainability in yarn manufacturing. In: Nayak, R. (Ed.), Sustainable Technologies for Fashion and Textiles. Woodhead Publishing, pp. 33–55.

Grassi, C., et al., 2016. Reducing environmental impact in air jet weaving technology. International Journal of Clothing Science & Technology 28 (3), 283–292. https://doi.org/10.1108/IJCST-03-2016-0037.

Gubitz, G.M., Cavaco-Paulo, A., 2003. In: Gubitz, G.M., Cavaco-Paulo, A. (Eds.), Textile Processing with Enzymes, first ed. Woodhead Publishing Limited, Cambridge, England.

Haque, M.M., et al., 2019. Study on loom stoppages in air jet weaving mill producing 100% cotton fabrics. Advance Research in Textile Engineering 4 (1).

Harkin, B., et al., 2016. Does monitoring goal progress promote goal attainment? A meta-analysis of the experimental evidence. Psychological Bulletin 142 (2), 198–229. https://doi.org/10.1037/BUL0000025.

Hasanbeigi, A., 2010. Energy-efficiency Improvement Opportunities for the Textile Industry. https://doi.org/10.2172/991751. Berkeley, CA.

Hasanbeigi, A., Berkeley, L., 2011. Energy Efficiency Technologies and Comparing the Energy Intensity in the Textile Industry, pp. 34–46.

Hasanbeigi, A., Price, L., 2012. A review of energy use and energy efficiency technologies for the textile industry. Renewable and Sustainable Energy Reviews 16 (6), 3648–3665. https://doi.org/10.1016/j.rser.2012.03.029.

Hasanbeigi, A., Price, L., 2015. A technical review of emerging technologies for energy and water efficiency and pollution reduction in the textile industry. Journal of Cleaner Production 95, 30–44. https://doi.org/10.1016/J.JCLEPRO.2015.02.079.

Hasanuzzaman, Bhar, C., 2016. Indian textile industry and its impact on the environment and health: a review. International Journal of Information Systems in the Service Sector 8 (4), 33–46. https://doi.org/10.4018/IJISSS.2016100103.

Hassini, E., Surti, C., Searcy, C., 2012. A literature review and a case study of sustainable supply chains with a focus on metrics. International Journal of Production Economics 140 (1), 69–82. https://doi.org/10.1016/J.IJPE.2012.01.042.

Heijungs, R., et al., 1992. Environmental Life Cycle Assessment of Products - Guide and Backgrounds (Leiden).

Henri, J.F., 2009. Taxonomy of performance measurement systems. Advances in Management Accounting 17, 247–288.

Herrera Almanza, A.M., Corona, B., 2020. Using social life cycle assessment to analyze the contribution of products to the sustainable development goals: a case study in the textile sector. The International Journal of Life Cycle Assessment 25 (9), 1833–1845. https://doi.org/10.1007/s11367-020-01789-7.

Hiroshi, M., Shozo, K., 2007. Warping System and Warping Method. Available at: https://scholar.google.com/scholar?hl=en&as_sdt=0%2C5&q=Hiroshi%2C+M.%2C+%26+Shozo+K.+%282007%29%2C+Warping+System+and+Warping+Method.+Patent+No.+JP4445437%2C+January+2007.&btnG=. (Accessed 24 January 2022).

Hoque, A., Clarke, A., 2013. Greening of industries in Bangladesh: pollution prevention practices. Journal of Cleaner Production 51, 47–56. https://doi.org/10.1016/J.JCLEPRO.2012.09.008.

Hoque, M.S., Rashid, M.A., et al., 2018. Alternative washing of cotton denim fabrics by natural agents. American Journal of Environmental Protection 7 (6), 79–83. https://doi.org/10.11648/j.ajep.20180706.12.

Hoque, M.S., Hossain, M.J., et al., 2018. Scope of dry wood and wood composite alternate to stone in case of acid wash on denim fabric. International Journal of Current Engineering and Technology 8 (02). https://doi.org/10.14741/ijcet/v.8.2.32.

Hourneaux, F., Gabriel, M.L. da S., Gallardo-Vázquez, D.A., 2018. Triple bottom line and sustainable performance measurement in industrial companies. Revista de Gestao 25 (4), 413–429. https://doi.org/10.1108/REGE-04-2018-0065.

ISO, 2021. ISO - The London Declaration, International Organization for Standardization. Available at: https://www.iso.org/ClimateAction/LondonDeclaration.html?utm_source=Members_2019-12-20&utm_campaign=571088869e-EMAIL_CAMPAIGN_2020_10_13_06_34_COPY_01&utm_medium=email&utm_term=0_64ffa93794-571088869e-223050729. (Accessed 11 May 2022).

ISO, n.d. ISO/TC 38 - Textiles, International Organization for Standardization. Available at: https://www.iso.org/committee/48148.html. (Accessed 11 May 2022).

Jiang, J.L., Wong, W.K., 2018. Fundamentals of common computer vision techniques for textile quality control. In: Wong, W.K. (Ed.), Applications of Computer Vision in Fashion and Textiles, first ed. Woodhead Publishing, pp. 3–15. https://doi.org/10.1016/B978-0-08-101217-8.00001-4.

Kabir, S.M.F., Haque, S., 2021. A mini review on the innovations in sizing of cotton. Journal of Natural Fibers 19 (87). https://doi.org/10.1080/15440478.2021.1941486.

Kadolph, S., 2011. Textiles. Prentice Hall, Upper Saddle River.

Kaplan, R., Norton, D., 1992. The balanced scorecard: measures that drive performance. Havard Business Review 70 (1), 71–79.

Karthik, T., Murugan, R., 2017. Carbon footprint in denim manufacturing. In: Muthu, S.S. (Ed.), Sustainability in Denim, first ed. Woodhead Publishing, pp. 125–159. https://doi.org/10.1016/B978-0-08-102043-2.00006-X.

Kefe, I., 2019. The determination of performance measures by using a balanced scorecard framework. Foundations of Management 11 (1), 43–56. https://doi.org/10.2478/fman-2019-0004.

Kim, J., Zorola, M., 2018. Sustainable innovation in the apparel supply chain: case study on TAL apparel limited. In: Chow, P.-S., et al. (Eds.), Contemporary Case Studies on Fashion Production, Marketing and Operations. Springer, Singapore, pp. 183–197. https://doi.org/10.1007/978-981-10-7007-5_11.

Koç, E., Çinçik, E., 2010. Analysis of energy consumption in woven fabric production. Fibres and Textiles in Eastern Europe 79 (2), 14–20.

Köksal, D., et al., 2017. Social sustainable supply chain management in the textile and apparel industry-a literature review. Sustainability 9 (1), 1–32. https://doi.org/10.3390/su9010100.

Kolgiri, S., Hiremath, R., 2018. Implementing sustainable ergonomics for power-loom textile workers. International Journal of Pharmacy and Pharmaceutical Sciences 10 (6), 108–112. https://doi.org/10.22159/ijpps.2018v10i6.26213.

Kopaczka, M., et al., 2018. Fully automatic faulty weft thread detection using a camera system and feature-based pattern recognition. In: Proceedings of the 7th International Conference on Pattern Recognition Applications and Methods (ICPRAM 2018), pp. 124–132. https://doi.org/10.5220/0006591301240132.

Kozlowski, A., Bardecki, M., Searcy, C., 2014. Environmental impacts in the fashion industry. The Journal of Corporate Citizenship 2012 (45), 16–36. https://doi.org/10.9774/gleaf.4700.2012.sp.00004.

Lenzo, P., et al., 2017. Social life cycle assessment in the textile sector: an Italian case study. Sustainability 9 (2092).

Li, C., et al., 2021. Fabric defect detection in textile manufacturing: a survey of the state of the art. Security and Communication Networks 2021. https://doi.org/10.1155/2021/9948808.

Loar, P.R., Mohamed, M.H., 1982. Weaving: Conversion of Yarn to Fabric. In: Loar, P.R., Mohamed, M.H. (Eds.). Merrow Publishing Co. Ltd, Durham, England.

Lokhande, H.T., Gotmare, V.D., 1999. Utilization of textile loomwaste as a highly absorbent polymer through graft co-polymerization. Bioresource Technology 68 (3), 283–286. https://doi.org/10.1016/S0960-8524(98)00148-5.

Luo, Y., et al., 2021. Environmental sustainability of textiles and apparel: a review of evaluation methods. Environmental Impact Assessment Review 86 (May 2020), 106497. https://doi.org/10.1016/j.eiar.2020.106497.

Muthu, S.S. (Ed.), 2016. Handbook of Life Cycle Assessment (LCA) of Textiles and Clothing. Woodhead Publishing.

Nag, A., Vyas, H., Nag, P.K., 2010. Gender differences, work stressors and musculoskeletal disorders in weaving industries. Industrial Health 48 (3), 339–348. https://doi.org/10.2486/indhealth.48.339.

Nateri, A.S., Ebrahimi, F., Sadeghzade, N., 2014. Evaluation of yarn defects by image processing technique. Optik Urban & Fischer 125 (20), 5998–6002. https://doi.org/10.1016/J.IJLEO.2014.06.095.

Nayak, R., Panwar, T., Van Thang Nguyen, L., 2019. Sustainability in fashion and textiles: a survey from developing country. In: Sustainable Technologies for Fashion and Textiles. Elsevier Ltd. https://doi.org/10.1016/B978-0-08-102867-4.00001-3.

Nidumolu, R., Prahalad, C.K., Rangaswami, M.R., 2009. Why sustainability is now the key driver of innovation. Harvard Business Review 87 (9), 56–64. https://doi.org/10.1109/emr.2015.7123233.

Öner, E., Sahinbaskan, B.Y., 2011. A new process of combined pretreatment and dyeing: rest. Journal of Cleaner Production 19 (14), 1668–1675. https://doi.org/10.1016/J.JCLEPRO.2011.05.008.

Ozturk, E., et al., 2009. A chemical substitution study for a wet processing textile mill in Turkey. Journal of Cleaner Production 17 (2), 239–247. https://doi.org/10.1016/J.JCLEPRO.2008.05.001.

Palamutcu, S., 2010. Electric energy consumption in the cotton textile processing stages. Energy 35 (7), 2945–2952. https://doi.org/10.1016/j.energy.2010.03.029.

Pattanayak, A.K., 2019. Sustainability in fabric manufacturing. In: Sustainable Technologies for Fashion and Textiles. Elsevier Ltd. https://doi.org/10.1016/B978-0-08-102867-4.00003-7.

Pensupa, N., et al., 2017. Recent trends in sustainable textile waste recycling methods: current situation and future prospects. In: Lin, C.S.K. (Ed.), Chemistry and Chemical Technologies in Waste Valorization. Topics in Current Chemistry Collections. Springer, Cham, pp. 189–228. https://doi.org/10.1007/978-3-319-90653-9_7.

Pervez, M.N., et al., 2021. Textile waste management and environmental concerns. In: Mondal, M.I.H. (Ed.), Fundamentals of Natural Fibres and Textiles. Woodhead Publishing, pp. 719–739. https://doi.org/10.1016/B978-0-12-821483-1.00002-4.

Picanol, 2021. OptiMax-i connect. Picanol. Available at: www.picanol.app. (Accessed 2 May 2022).

Qureshi, S.A., Ramesh Kumar, M., Prakash, C., 2021. Design and development of a novel mechanism for the removal of false selvedge and minimization of its associated yarn wastage in shuttleless looms. Autex Research Journal 21 (4), 352–362. https://doi.org/10.2478/AUT-2020-0031.

Raina, M.A., Gloy, Y.S., Gries, T., 2015. Weaving technologies for manufacturing denim. In: Paul, R. (Ed.), Denim: Manufacture, Finishing and Applications, first ed. Woodhead Publishing, pp. 159–187. https://doi.org/10.1016/B978-0-85709-843-6.00006-8.

Raj, D., et al., 2017. Implementation of lean production and environmental sustainability in the Indian apparel manufacturing industry: a way to reach the triple bottom line. International Journal of Fashion Design, Technology and Education 10 (3), 254–264. https://doi.org/10.1080/17543266.2017.1280091.

Rasheed, A., et al., 2020. Fabric defect detection using computer vision techniques: a comprehensive review. Mathematical Problems in Engineering 2020. https://doi.org/10.1155/2020/8189403.

Reddy, N., Yang, Y., 2015. Review: potential use of plant proteins and feather keratin as sizing agents for polyester-cotton. AATCC Journal of Research 2 (2), 20–27. https://doi.org/10.14504/AJR.2.2.3.

Reddy, N., et al., 2014. Reducing environmental pollution of the textile industry using keratin as alternative sizing agent to poly(vinyl alcohol). Journal of Cleaner Production 65, 561–567. https://doi.org/10.1016/J.JCLEPRO.2013.09.046.
Ren, X., 2000. Development of environmental performance indicators for textile process and product. Journal of Cleaner Production 8 (6), 473–481. https://doi.org/10.1016/S0959-6526(00)00017-2.
Ryu, C., et al., 2007. Combustion of textile residues in a packed bed. Experimental Thermal and Fluid Science 31 (8), 887–895. https://doi.org/10.1016/J.EXPTHERMFLUSCI.2006.09.004.
Saeidi, R.G., et al., 2013. An efficient DEA method for ranking woven fabric defects in textile manufacturing. International Journal of Advanced Manufacturing Technology 68 (1–4), 349–354. https://doi.org/10.1007/s00170-013-4732-4.
Sanches, A.G.O., Lucato, W.C., Oliveira Neto, G.C., 2018. Denim weaving technologies: environmental and economic evaluation. In: 7th Academic International Workshop Advances in Cleaner Production: 'Cleaner Production for Achieving Sustainable Development Goals.' Barranquilla – Colombia – June 21st and 22nd - 2018, pp. 1–9. Barranquilla, Colombia.
Sawhney, A.P.S., 2008. Size Free Weaving, Cotton Warp Yarn, High Speed Weaving Machine, Flexible Rapier Weaving Machine. FIBER2FASHION, pp. 2491–2496. Available at: https://www.fibre2fashion.com/industry-article/3272/sizefree-weaving-of-cotton-fabric-on-a-modern-highspeed-weaving-machine. (Accessed 26 February 2022).
Senanayake, R., Gunasekara Hettiarachchige, V., 2020. A zero-waste garment construction approach using an indigenous textile weaving craft. International Journal of Fashion Design, Technology and Education 13 (1), 101–109. https://doi.org/10.1080/17543266.2020.1725148.
Seuring, S., Müller, M., 2008. From a literature review to a conceptual framework for sustainable supply chain management. Journal of Cleaner Production 16 (15), 1699–1710. https://doi.org/10.1016/j.jclepro.2008.04.020.
Seyam, A.-F.M., 2019. Innovation in weaving at ITMA 2019. Journal of Textile and Apparel, Technology and Management 1–15. Special Issue-ITMA Show, Barcelona, Spain.
Shiwanthi, S., Lokupitiya, E., Peiris, S., 2018. Evaluation of the environmental and economic performances of three selected textile factories in Biyagama Export Processing Zone Sri Lanka. Environmental Development 27 (X), 70–82. https://doi.org/10.1016/j.envdev.2018.07.006.
Silvestre, B.S., Ţîrcă, D.M., 2019. Innovations for sustainable development: moving toward a sustainable future. Journal of Cleaner Production 208, 325–332. https://doi.org/10.1016/j.jclepro.2018.09.244.
Singh, M., 2014a. Yarn sizing. In: Singh, M.K. (Ed.), Industrial Practices in Weaving Preparatory, first ed. Woodhead Publishing India Pvt. Ltd, New Delhi, pp. 134–266.
Singh, M.K., 2014b. Warping. In: Singh, M.K. (Ed.), Industrial Practices in Weaving Preparatory, first ed. Woodhead Publishing India Pvt. Ltd, New Delhi, pp. 104–131.
Smith, B., 1994. Future pollution prevention opportunities and needs in the textile industry. In: Pojasek, B. (Ed.), Pollution Prevention Needs and Opportunities. Center for Hazardous Materials Research (May).
Statista, 2022. Leading Exporting Countries of Woven Cotton Fabrics Worldwide in 2020 (In Million U.S. Dollars).
Stegmaier, T., et al., 2008. Chitosan – a Sizing agent in fabric production – development and ecological evaluation. CLEAN – Soil, Air, Water 36 (3), 279–286. https://doi.org/10.1002/CLEN.200700013.

Surung, J.S., Bayupati, I.P.A., Putri, G.A.A., 2020. The implementation of ERP in supply chain management on conventional woven fabric business. International Journal of Information Engineering and Electronic Business 3, 8–18. https://doi.org/10.5815/ijieeb.2020.03.02.

Textile Today, 2018. Sustainable Warp Preparation-Economical and Good Cost: Benefit Ratio. Textile Today. Available at: https://www.textiletoday.com.bd/sustainable-warp-preparation-economical-and-good-cost-benefit-ratio/. (Accessed 22 January 2022).

Thakkar, A., Bhattacharya, S., 2018. Mechanical developments in design aspects of textile warping machines. In: International Conference on New Frontiers of Engineering, Science, Management and Humanities (ICNFESMH-2018). OM Institute of Technology & Management, Hisar, India, pp. 221–224.

Thoney-Barletta, K., 2019. Supply chain management software for textile networks at ITMA 2019. Journal of Textile and Apparel, Technology and Management 1–6.

Tobler-Rohr, M.I., 2011. Handbook of Sustainable Textile Production. In: Tobler-Rohr, M.I. (Ed.), first ed. Woodhead Publishing Limited, Cambridge.

Ul-Haq, N., Nasir, H., 2011. Cleaner production technologies in desizing of cotton fabric. The Journal of The Textile Institute 103 (3), 304–310. https://doi.org/10.1080/00405000.2011.570045.

United States Enviromental Protection Agency, 2006. Life Cycle Assessment: Principles and Practice. Available at: https://web.archive.org/web/20120916182225/http://www.epa.gov/nrmrl/std/lca/pdfs/chapter1_frontmatter_lca101.pdf.

Wackernagel, M., Rees, W., 1996. Our Ecological Footprint: Reducing Human Impact on the Earth. New Society Publishers, Philadelphia, PA and Gabriola Island, BC.

Walton, P., 1912. The Story of Textiles. John S. Lawrence, Boston.

Wang, C.N., Yang, C.Y., Cheng, H.C., 2019. A fuzzy multicriteria decision-making (MCDM) model for sustainable supplier evaluation and selection based on triple bottom line approaches in the garment industry. Processes 7 (7), 1–13. https://doi.org/10.3390/pr7070400.

Wanniarachchi, T., Dissanayake, K., Downs, C., 2020. Improving sustainability and encouraging innovation in traditional craft sectors: the case of the Sri Lankan handloom industry. Research Journal of Textile and Apparel 24 (2), 111–130. https://doi.org/10.1108/RJTA-09-2019-0041.

Weidner, K., Nakata, C., Zhu, Z., 2021. Sustainable innovation and the triple bottom-line: a market-based capabilities and stakeholder perspective. Journal of Marketing Theory and Practice 29 (2), 141–161. https://doi.org/10.1080/10696679.2020.1798253.

Weninger, L., Kopaczka, M., Merhof, D., 2018. Defect detection in plain weave fabrics by yarn tracking and fully convolutional networks. In: 2018 IEEE International Instrumentation and Measurement Technology Conference: Discovering New Horizons in Instrumentation and Measurement, Proceedings. Institute of Electrical and Electronics Engineers Inc., Houston, Texas, USA, pp. 1–6. https://doi.org/10.1109/I2MTC.2018.8409546.

What Is a Key Performance Indicator (KPI)?, n.d. Available at: https://kpi.org/KPI-Basics/KPI-Basics. (Accessed 13 April 2022).

Yang, M., et al., 2017. Biodegradable sizing agents from soy protein via controlled hydrolysis and dis-entanglement for remediation of textile effluents. Journal of Environmental Management 188, 26–31. https://doi.org/10.1016/J.JENVMAN.2016.11.066.

Zamani, B., et al., 2018. Hotspot identification in the clothing industry using social life cycle assessment—opportunities and challenges of input-output modelling. International Journal of Life Cycle Assessment 23 (3), 536–546. https://doi.org/10.1007/s11367-016-1113-x.

ZDHC, 2022. ZDHC Manufacturing Restricted Substances List (ZDHC MRSL) Version 2.0. Available at: https://mrsl.roadmaptozero.com/mrsl/MRSL2_0/export_pdf.php?complex=1&lang=1. (Accessed 16 January 2022).

Zeng, S.X., et al., 2010. Impact of cleaner production on business performance. Journal of Cleaner Production 18 (10–11), 975–983. https://doi.org/10.1016/J.JCLEPRO.2010.02.019.

Zhao, Y., et al., 2015. A sustainable slashing industry using biodegradable sizes from modified soy protein to replace petro-based poly(vinyl alcohol). Environmental Science and Technology 49 (4), 2391–2397. https://doi.org/10.1021/ES504988W/SUPPL_FILE/ES504988W_SI_001.PDF.

ZHDC, 2021. Roadmap to Zero. ZDHC Foundation. Available at: https://www.roadmaptozero.com/input. (Accessed 16 January 2022).

Zheng, J., Zhao, Q., Ye, Z., 2014. Preparation and characterization of activated carbon fiber (ACF) from cotton woven waste. Applied Surface Science 299, 86–91. https://doi.org/10.1016/J.APSUSC.2014.01.190.

# Innovations in knitted textiles

7

*Henning Löcken* [1] *and Mirela Blaga* [2]
[1]Institut für Textiltechnik of RWTH Aachen University, Aachen, Germany; [2]'Gheorghe Asachi' Technical University of Iasi, Faculty of Industrial Design and Business Management, Iasi, Romania

## 7.1 Introduction

Knitting, as a method of producing textiles by interlacing yarns, results in fabrics that are characterised by high structural elasticity and drapability, and are mainly used in clothing technology. Knitted fabrics are made with looping yarns, whereas woven fabrics are made by weaving of warp and weft yarns on a loom, and so they have a minimal amount of stretch in the resulting fabric. Knitted fabrics are generally preferred by fashion brands as they are highly versatile for making various types of apparel, from casual wear to sportswear and summer clothes to winter wear.

The main processes used to produce knitted fabrics are weft and warp knitting. Understanding the mechanics of stitch formation has contributed much to change the design of modern knitting machines and refine process parameters to make the most of the progress made so far. Warp knitting is much younger compared to weft knitting, but has quickly gained acceptance in the market, especially for high value-added structures, including technical textiles.

The market for fabrics produced with the two technologies is different: 60% of global sales are accounted for by weft knitting and the remaining 40% by warp knitting, but with an increasing trend in the near future. The knitting industry is capable of producing a wide range of technical textiles, which is reflected in the 12 application groups, which are: meditech, agrotech, mobiltech, protech, sportech, geotech, oekotech, hometech, buildtech, packtech, clothtech and indutech. As a result, about 80% of total knitted fabrics are currently treated as technical textiles, while the remaining 20% are used for apparel, fashion, and household items (EU Strategy for Sustainable and Circular Textiles, 2022).

## 7.2 Historical aspects and current scenario

Knitting is the second largest and most widespread technique for the production of fabrics after weaving, and arose from the need of people to make warm and soft products such as stockings, hats and gloves, which was not possible through weaving. In its long history, knitting has mainly produced two-dimensional fabrics, which were afterwards transformed into a final product by subsequent cutting and sewing or linking. More recently, with the development of advanced technologies, three-dimensional knitting

**Sustainable Innovations in the Textile Industry. https://doi.org/10.1016/B978-0-323-90392-9.00020-3**

has been used to produce lighter and softer fabrics by eliminating the process of sewing or linking. Newly developed automated industrial knitting machines are capable of producing advanced structures, including shaped and seamless fabrics, allowing faster processing and high potential for custom knitting (EU Strategy for Sustainable and Circular Textiles, 2022). The latest technology makes it possible to produce any type of textile product with a very high degree of accuracy, taking advantage of the flexibility of modern knitting machines with electronic control on the one hand, and the simultaneous development of new types of textile fibres on the other.

Currently, the knitting industry must follow the strategy set by the European Commission in March 2022, which aims to make textiles more durable, repairable, reusable and recyclable, and to promote innovation within the sector. The framework aims to promote the transformation of the textile sector so that by 2030 textile products placed on the market in the EU are durable, recyclable, made from recycled fibres, free of hazardous substances and produced in a way that respects social rights and the environment (Dissanayake and Perera, 2016).

To achieve this goal, all players in the knitting industry must be involved. Machine builders should produce knitting machines using environmentally friendly processes that have less impact on people and the environment. Product manufacturers need to revise their processes, production and consumption by selecting sustainable raw materials for knitwear production, minimising waste and using environmentally friendly chemicals. Wholesalers and retailers must promote the business of sustainably produced knitwear. Consumers should be adequately informed and make purchasing decisions based on the environmental impact of the product rather than the price (Babu et al., 2013).

## 7.3 Sustainable raw materials for knitted fabrics

An important parameter in the transformation of the textile industry towards a more sustainable future, in addition to innovations in processing technologies, is the use of sustainable yarns. Sustainable yarns, bio-based or recycled, are expected to reduce the carbon footprint and other impacts on the environment. Already in industrial use are, of course, conventional natural fibres such as wool and cotton, but also recycled polyester yarn from PET bottles. Above all, the use of natural fibres such as wool or cotton has always been pushed in knitting. However, synthetic fibres are also frequently used, especially for more technical applications such as medical textiles or sportswear. These fibres are made to a very large extent from petroleum and are therefore based on finite resources.

However, to achieve a more sustainable textile industry, many materials, especially petroleum-based yarns, need to be replaced with bio-based materials, and knitted fabrics need to be recycled, i.e., made recyclable, to enable a circular economy. In addition, the EU Textile Strategy decided to limit the use of recycled polyester fibres from PET bottles, as PET bottles already have a well-established circular economy in the beverage industry that should not be compromised by taking these resources out of circulation. The use of bottle regrind PET is therefore only an intermediate stage in the use of recycled fibres and does not offer a long-term perspective for a sustainable textile industry (Dissanayake and Perera, 2016).

Numerous research and development projects are currently underway to introduce new sustainable yarns. Bio-based synthetic fibres are being studied and developed using feedstocks from biomass such as algae or food waste, or recycled fibres (Knitting Views, 2022; Davis, 2021a). The ever-growing demand from customers for sustainable and environmentally friendly materials has motivated manufacturers to find solutions. RadianzaTM is one of the most environmentally friendly and sustainable fibres produced using gel dyeing technology, which allows far less resources to be used than the traditional dyeing process for yarns or fabrics and does not release unused dyes (Innova picks up Roica V550, 2022).

German yarn manufacturer Madeira has developed two sustainable yarns for use in knitting. Madeira's Polyneon Green yarn is made entirely from recycled polyester, and its strength and versatility make it suitable for workwear, sportswear, fashion and footwear. For Sensa Green, Madeira used Lenzing's Tencel lyocell fibres, which come from trees grown on non-agricultural land and which consume significantly less water than cotton. Closed-loop manufacturing recovers 99% of the organic solvents and water used in production. Spun and dyed exclusively in Germany, Sensa Green has a low carbon footprint due to shorter transport cycles and is certainly one of the most sustainable yarns on the market (World Textile Information Network, 2022a).

Sensil BioCare by Nilit is a high-quality, sustainable nylon 6.6 fibre enriched with special technology that helps reduce textile waste in the ocean and landfills by acting during and after the product's life cycle. When microplastics in Sensil BioCare garments are released during washing, they decompose much faster than conventional nylon 6.6 fibres, reducing textile waste. Roica V550 from Asahi Kasei is a sustainable stretch yarn that is degraded according to Hohenstein Environmental Certification without releasing harmful substances into the environment. Using these two yarns, Innova Fabrics, the Italian manufacturer of knitwear for apparel, underwear and sports, has strengthened its smart production with the introduction of the RF (Residual Free) line to reduce the impact of microplastic residues from the fashion industry (Davis, 2021b).

Agricultural wastes such as pineapple leaves and coconut shells are processed into new materials. Pineapple-based textiles are not new. What is new, however, is how pineapple waste can be processed on a large scale. The company Nextevo has developed a so-called ready-to-spin (RTS) fibre that entered trial production in September 2021. Similar to other plant fibres such as cotton, fabrics made from RTS pineapple leaf fibre (PALF) are smooth, soft and breathable. RTS PALF also has a high tensile strength and with its low crystalline structure, it has a high ability to absorb and retain dye. From a life cycle perspective, pineapple leaf waste is a viable alternative to cotton because the environmental impact in terms of carbon emissions, water consumption and freshwater ecotoxicity is much lower for pineapple leaves (13% of the total pineapple plant) than for cotton fibre (84% of the total cotton plant).

Southeast Asia is a vast agricultural region, the largest pineapple growing area in the world, and also an important growing area for rice, sugarcane, rubber, cassava and palm oil. There are so many ways to use these agricultural wastes that 50% of non-cotton fibres could be replaced by other natural fibres (Davis, 2021c).

Solvay, São Paulo has launched its first partially bio-based polyamide textile yarn, which can be used in a range of applications including sportswear, beachwear and underwear. Bio Amni is a polyamide that is produced entirely at the company's textile plant in Brazil. Solvay is thus following the growing global trend of demand for more sustainable textile products, particularly bio-based materials. In 2020, the company launched Amni Virus-Bac OFF, a functional polyamide that inhibits contamination between textiles and users, preventing the fabric transmitting viruses, including corona viruses, and bacteria (Ray, 2022).

Advanced materials specialist PrimaLoft won the Best Sustainable Textile Innovation award at the Drapers Sustainable Fashion Awards 2021 for PrimaLoft Bio, the first biodegradable, 100% recycled synthetic insulation and performance fabric often used in conjunction with knitwear. PrimaLoft Bio is made of high-performance fibres manufactured from 100% recycled plastic bottles that biodegrade in certain environments such as landfills, oceans and wastewaters. When exposed to these environments, naturally occurring microbes break down the fibres into their natural components without leaving harmful waste. Insulation and fabrics made from PrimaLoft organic fibres offer the same high-quality, durable performance as products made from pure polyester fibres – providing lightweight warmth and comfortable protection for a wide range of activities and weather conditions. PrimaLoft Bio fibres are renewable in a circular economy, as it is possible to rejuvenate the fibres into new high-performance material, over and over again, without reducing quality (http://www.ecolife.com/).

With these yarn developments, it is imperative to ensure good processability of the new materials on conventional textile machines. Only if a new yarn can be integrated into the existing process chain without major conversion measures or adjustments does this yarn have a chance of competing economically, despite all ecological impacts.

To ensure the processability of a newly developed yarn, it is necessary to conduct trials on industrial-scale production machines, especially to justify scale-up in the industry. Since only small quantities are usually available at the beginning of the development of new, sustainable raw materials, the knitting process is ideally suited to support the development. Even with small yarn quantities and without complex warp preparation, representative demonstrators can be produced on knitting machines, thus validating the applicability of novel materials in development.

## 7.4 Sustainability aspects of product development

The current context in knitting technology follows the general trend of overconsumption and shifting the production to other sectors, such as automotive, aerospace, footwear, medical, sports, construction, etc. The key players for the strategic development of sustainable products are the designers, yarn and machinery manufacturers, suppliers and retailers, consumers and policy makers. Solutions available to meet sustainability requirements include recycling and upcycling of raw materials and knitwear waste, sourcing of raw materials for knitwear, sustainable material design, and efficient production and finishing processes (EU Strategy for Sustainable and Circular Textiles, 2022).

A sustainable design considers environmental aspects such as limiting resource consumption through waste-free production, giving preference to renewable resources, and using recycled materials. It also focuses on waste prevention by minimising consumption, reusing when possible, and recycling when necessary. It will select non-toxic materials and those that contribute to human health and well-being. Last but not least, it will prioritise quality and durability over price (Ştefan et al., 2021). The energy required to produce knitwear is high, therefore, knitwear manufacturers have a great obligation to implement and ensure sustainability in all processes and stages of knitwear production as early as possible in order to remain at the forefront of supplying knitwear to the global market (Pavko Čuden, 2016).

The main developments focus on increasing the speed of knitting machines to improve efficiency. The introduction of robotics into the knitting process and the adaptation of knitting machines have allowed the production of unconventional materials and technical yarns. Technological processes have been merged to save production time, space, energy and labour. In the knitting industry, attention is paid to the recycling and upcycling of raw materials and knitting waste. The latest generation of knitting machines offers eco-friendly manufacturing processes with less impact on the environment and people. Sustainable finishing processes are also being developed to increase the value of the end products. The concept of complete garment or all-in-one knitting technology, which is produced on flat and circular knitting machines, can be seen in a continuous and shorter process, lower finishing costs, lower labour costs, fewer machines for production, faster response to the market and faster sampling (Pavko Čuden, 2020).

## 7.4.1 Weft knitting technology

### 7.4.1.1 Flat weft knitting

The weft knitting process enables the production of complex structures with low machine set-up time and minimal material input and is therefore already fundamentally suitable for sustainable production. With the fully-fashion and knit&wear approaches, many apparel textiles are already produced in this environmentally friendly way. However, the possibilities of knitting technology offer many more advantages for sustainable production than just the material-saving production of clothing.

With the integrated production of 3D knitted fabrics, it is possible in principle to reduce the amount of material used in knitting to a minimum, and not just for clothing. The 3D knitted fabrics are knitted directly in the desired shape, so that no further assembly steps are necessary. Since subsequent manufacturing is eliminated, there is no more waste and sewing threads are also saved. This approach makes it possible to produce almost no waste, with the exception of auxiliary yarns for a neat start to knitting. In flat knitting machines, the technique is ideal for technical textiles and other

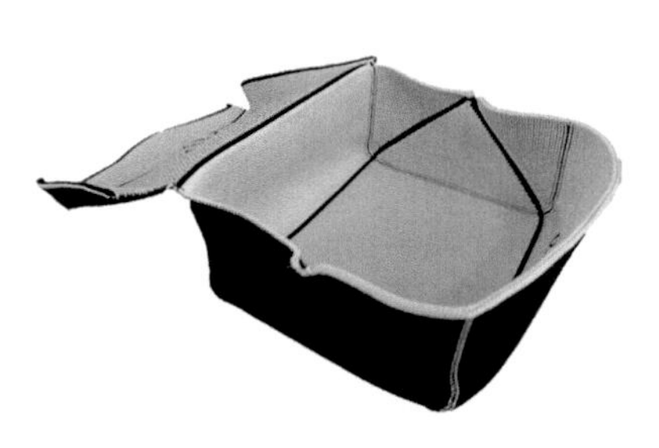

**Figure 7.1** 3D knitted textiles on flat bed knitting machines: Box (Stoll) car seat cover (Bache).

applications to avoid material waste during production. With 3D knitting, a wide variety of products such as car seats, medical functional fabrics or household applications can be produced in a single operation without waste. The production of 3D knitted fabrics is open to a wide range of shapes, from curves to the classic darts of the clothing industry as well as rectangular shapes (Gries et al., 2022; Bendt, 2016; Choi and Powell, 2005) (Fig. 7.1).

However, two challenges remain for broader application. First, product development and programming of 3D knitted fabrics is still very time-consuming and is still sometimes done with manual casting. A development towards automatic marker creation from CAD data, which is already being driven forward in many places, could increase the application possibilities of 3D flat knitting. On the other hand, there is still no industrial heat setting for 3D textiles. Especially in technical applications, e.g., in the automotive industry as seat covers, it is necessary to have the knitted fabrics heat-set after production in order to ensure the high requirements for shrinkage, resistance and dimensional stability (Liu et al., 2021).

Current commercially available heat-setting technologies do not allow 3D textiles to be set in a continuous process without a great deal of effort, often using calendar rolls or lateral clamping of the textiles. These technologies assume either a constant thickness or a constant width of the textile, which is not the case with 3D textiles. 3D knitting on flat knitting machines can save material and downstream process steps, but the production process in the knitting machine often takes a very long time due to the complex patterning and the many transfer sections required. Due to the costly production times and the lack of heat-setting, 3D knitwear is often not economically superior to classic cut-and-sew processes as long as ecological factors are not financially taken into account.

### *7.4.1.2 Circular weft knitting*

A newly developed approach to increasing production speed in 3D knitting is the transfer of 3D flat knits to circular knitting machines. Without the ability to transfer stitches, circular knitting machines are unable to achieve the same flexibility as current flat knitting machines, but with the newly developed 3D knitting pattern, it is still possible to create knitted 3D textiles on large circular knitting machines while taking advantage of

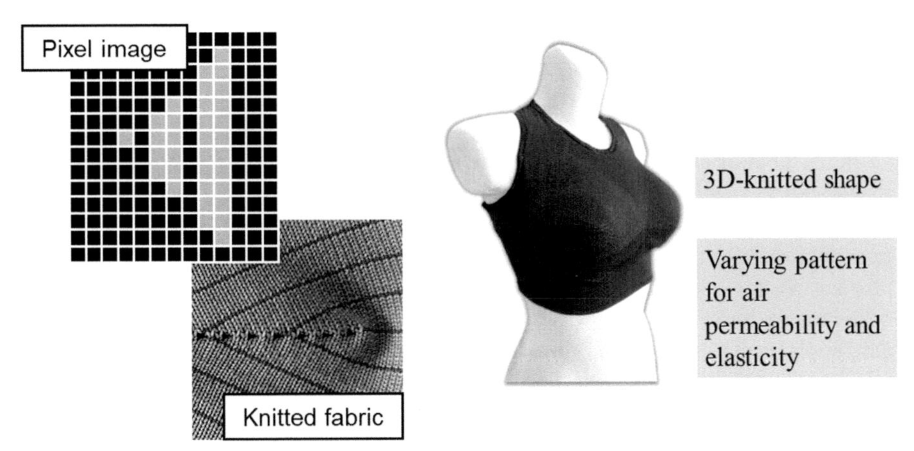


**Figure 7.2** Basic 3D knitting pattern on a large circular knitting machine (left) to produce sports bras (right).

the high productivity of large circular knitting machines. As examples of the suitability of circular knitting technology for the application of 3D knitting, sports bras and car seats have already been produced as demonstration products (Fig. 7.2).

The concept for producing 3D knitwear on large circular knitting machines is based on a new knitting pattern. With this new knitting pattern, it is possible to reduce the surface in certain local areas of the fabric, thus achieving a 3D shape analogous to the typical darts used in the cut & sew process. By integrating these 'sew-like' knitting patterns in various shapes and positions, a three-dimensional knitting form is created. The knitting pattern consists of floats and stitches alternating horizontally over the area to be reduced. Due to the continuous stitch wales of the floats, the corresponding needles are not moved in this area and thus hold the knitted fabric in position. The other needles continue to form stitches, but on the reverse side of the knitted fabric. In this way, the continuous movement of the circular knitting machines can be maintained and the knitting area can be reduced (Peiner et al., 2022).

To implement the 3D circular knitting pattern, an electronic needle control system, i.e., a Jacquard knitting machine, is required. When programming and developing the 3D geometries, the procedure is similar to that for 3D knitwear on flat knitting machines. First, the desired surface is casted, converted into a digital pattern and subsequently transformed into the knitting structure. In principle, 3D knitting is suitable for the production of various products. Due to the production speed advantages and a nevertheless individual design of the knitted structures, the technology is particularly suitable for use in mass customisation. For example, a sports bra can be knitted using the 3D large circular knitting technology in under 8 min with changing colours and knitting patterns without machine reconfiguration necessary.

In contrast to 3D knitting on flat knitting machines, however, 3D knitting does not produce a near-net-shaped knitted fabric. Although the 3D shape is completely present, the knitted fabric must be separated from the knitted tube after production. It also follows that 3D large circular knitting cannot be knitted in a zero-waste approach.

The potential of 3D large-diameter circular knitting therefore lies not primarily in material reduction, but in the saving of downstream process steps. For industrial use, however, digital product development must be further advanced and, analogous to 3D flat knitting, heat setting must be clarified (Simonis et al., 2017).

In the field of small-diameter circular knitting, Harry Lucas is a textile machinery manufacturer specialising in weft and warp knitting machines. The highly specialised machines of this traditional German company are exported worldwide for the production of fabrics for the medical, technical textiles and fashion markets. The company is also known for manufacturing circular knitting machines for low- and high-pressure hoses for the automotive industry, with about 80% of the world's production of radiator hoses made on their machines. The modular concept of its machines makes it possible to offer, within a very short time, a customised and performance-oriented solution to the specific requirements of the customer's production. The product range, which can be produced on machines with small circular diameter, includes simple cords with inlet, hoses with filling granules or recycled material, rubber hoses made of synthetic fibres, glass fibres, stainless steel, copper and carbon fibres. Single jersey fabrics can also be made with circular knits or plush fabrics used for pot cleaners. Other machines can be used to make bandages, cords and ribbons that are knit around a core. A range of machines have been designed for making knitted fabrics from synthetic or metallic threads. These can be used for filters, electronic shielding, vibration dampers and for the aerospace and defence industry (Maschinenfabrik Harry Lucas GmbH and Co. KG).

Tubes knitted on small-diameter circular knitting machines can be used to make circular knitted piezoelectric devices if they are made of piezoelectric fibres that have good mechanical properties required for use in knitting machines where mechanical tension must be applied during the knitting process. For the construction of the knitted wearable energy harvester device, woven conductive fabrics as inner and outer electrodes with thickness of 80 μm were embedded inside and outside of the circular knitted structure (Mokhtari et al., 2020).

The seamless circular knitting machines manufactured by Santoni represented a revolution in the world of textile machinery. Their use was initially limited to the production of underwear, but was then extended to the production of sportswear, beachwear, medical clothing, knitwear, sweaters and other types of outerwear. Their seamless knitting technology offers a unique combination of features such as comfort, fit, support, breathability, light weight, pleasant appearance and easy care. The technology offers the ability to design and manufacture innovative uppers efficiently and quickly, significantly reducing the overall time of the footwear production cycle and minimising production waste.

The constant research and development work and the resulting machine innovations involve the fusion of two concepts: the concept of hosiery and the concept of knitting technology have allowed Santoni machines to have both great potential and high performance, so that they can easily move from one sector of the textile field to another (West, 2020).

Santoni produced the innovative X MACHINE, a circular knitting machine for socks that enables a sustainable production process thanks to its record-breaking production times – from 5 to 7 min per piece. This machine produces seamless socks, eliminating the cutting and sewing of fabrics and the waste itself, and reducing completion times. The choice of patterns is wide, thanks to the intarsia concept for specific areas and different yarns and stitches arranged according to the design. The new Santoni X machine with automatic toe joint SbyS can process all types of fabrics, saving time, space and costs and giving products greater value and high-tech solutions.

In footwear, Santoni patented its innovative single-cylinder machine XT-MACHINE, which uses revolutionary intarsia technology that maps the foot for a perfect fit and shape and offers unlimited patterns and colour combinations. This machine can produce lightweight uppers with breathable mesh areas and eyelets ready for sole application in record time. This is the first machine in the world capable of knitting in terry cushions and transfer stitch points on the same course, due to the single-cylinder intarsia electronic with four reciprocating feeds and two selection points per feed and per rotating sense (Knitting Industry, 2022a).

### 7.4.2 Warp knitting technology

#### 7.4.2.1 Flat warp knitting

In the field of high-performance technical textiles, Karl Mayer manufactures biaxial and multiaxial warp knitting machines as well as warp knitting machines with and without weft insertion, on which functional articles can be produced for a wide range of applications. The company offers sustainable solutions for the composites industry, such as the production of fibre-reinforced plastics from natural fibres. The renewable raw materials are processed on weft and multiaxial warp knitting machines into reinforcement structures that open up completely new perspectives in terms of their environmental compatibility.

With another in-house development, 4D-KNIT, Karl Mayer has redefined the term spacer fabric, which stands for a completely new generation of spacer fabrics based on the basic structure of conventional 3D warp-knitted fabrics, but clearly distinguished from traditional constructions by their spacing (Fig. 7.3). This is something very special: the space between the top and bottom can be filled with voluminous yarns to create unprecedented effects in performance and design. Textile innovation 4D-KNIT offers opportunities for a better eco-balance of fashion items and outdoor clothing. The fabric for midlayer and softshell performance solutions impresses with

**Figure 7.3** 4D-KNIT structures by Karl Mayer (Karl Mayer Group).

a sophisticated structure that leads to significantly lower fibre release during washing compared to double-sided raised fleece fabrics. These knits can be produced on the double-needle bar Raschel machines, which don't produce a classic spacer fabric with monofilaments as spacers, but the space between the faces is filled with a bulky yarn. In addition, yarns with different shrinkage are processed in intelligent combinations on the front and back of the knitted fabric, and various laying techniques are used. In the finishing process, this results in high—low effects with differentiated properties. Voluminous fabrics with small and shallow reliefs or deep and voluminous shapes with different motifs are created. Mesh parts are used especially in functional clothing and footwear to provide breathability and a fashionable look (Davis, 2021d).

This is made possible by the ingenious bar arrangement and technical configuration in conjunction with Karl Mayer's proven, high-quality piezo Jacquard technology. The double-needle bar machines thus open up a new dimension and enable fabrics with a wide variety of formable patterns on both fabric faces. 4D-KNIT fabrics offer a new dimension of design possibilities. They are suitable for various applications such as automotive, footwear, home textiles, outerwear and mattresses.

Another attractive market is the innovative warp-knitted, seamless components for toes, fingers and even the head, which can be produced in a single operation and are used in functional sportswear, underwear, hosiery and fashion outerwear. Rascheltronic, the RSJ line from Karl Mayer, is a series of high-speed Jacquard Raschel machines for the production of Jacquard patterned stretch and non-stretch fabrics. The machines feature Jacquard patterning, which makes it possible to process an almost unlimited number of designs with the aid of an electronic guide bar. The computer controls permit rapid lapping changes and long-repeats on fabrics and also enable functional zones to be incorporated into the material. An example of a sports bra by Karl Mayer, which consists of various functional zones without disturbing seams, strong fabric zones for support, highly elastic zones for freedom of movement, and mesh structures for breathability, is shown in Fig. 7.4. For greater sustainability, the recycled PES and PA yarns can be processed on all standard Karl Mayer machines without any loss of speed or quality. In addition, knitting can be done without sizing. Fabric-like articles can be produced with less water, chemicals and energy than with weaving.

#### *7.4.2.2 Circular warp knitting*

Warp knitting is particularly well suited to the production of textile products for vascular applications (especially mitral heart valve replacements) because it can produce very thin, dense textile structures that prevent blood leakage around the valve. Porosity can be adjusted to recruit the desired cells according to size, creating specialised areas of tissue regeneration.

Cortland Biomedical designs and manufactures high-performance biomedical textile structures based on years of experience in medical textile technology, including circular knitting. In cardiovascular tissues, blood leakage inside the valve must be prevented, while ingrowth of native tissue outside is promoted. Knitted tissues are very pliable, allowing the implant to stretch and move with the body, reducing patient

**Figure 7.4** Sports bra from Karl Mayer (Karl Mayer Group).

discomfort and restoring natural mobility. Applications of knitted valve fabrics include vascular prostheses, haemostasis, cardiac assist devices and valve cuffs (Fig. 7.5).

Knitted fabrics are flexible enough to be inserted through a smaller catheter and expand within the vessels, allowing minimally invasive delivery methods without compromising mechanical integrity. This is particularly beneficial for patients with small vessels and for repairs in the three branches of the aortic arch, which has long been a challenge. Harry Lucas circular warp knitting machines available on the market are designed for the production of tubular nets for packaging or medical net dressings, tubular fabrics, trimmings and technical fabrics with or without inserts in diameters of 3–12 mm (Maschinenfabrik Harry Lucas GmbH and Co. KG).

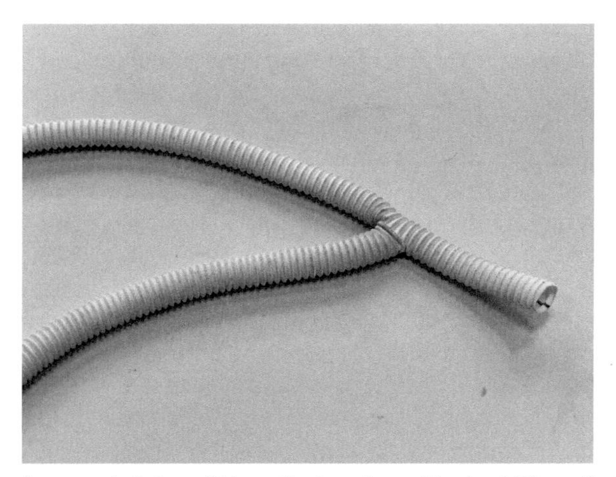

**Figure 7.5** Circular warp knitting of biomedical products (Cortland Biomedical).

## 7.5 Knitting challenges and digital solutions

Another sustainable approach to reducing the use of materials in knitting and especially in the development of textiles is digitalisation. Digitalisation and sustainability are the current trends in the knitting industry under the new concept of Industry 4.0. Known as 'using digital technologies to change a business model and create new revenue and value opportunities', digitalisation is gradually becoming a means not only to achieve sustainability (planet, profit, people) but also to reduce inventories, increase supply chain transparency, and customise production (World Textile Information Network, 2022b).

Knitting machine manufacturers are integrating their business and development strategies with Industry 4.0. The increasing use of digital devices inside and outside the industry means the right information is available for R&D, maintenance, operations, marketing and sales to make decisions and take action. In the case of knitting, the goal can be achieved through a direct and fast process from a digital sketch to prototyping (Pavko Čuden and Maity, 2022). The use of digitalisation helps knitwear manufacturers achieve four crucial parameters: greater transparency, shorter response times, shorter production cycles and higher productivity (Varshney, 2021).

KM.ON, the software specialist of the Karl Mayer Group, has acquired the German company Digitale Strickmanufaktur, which is developing a digital technology that connects retailers and producers via the cloud. Thanks to the network solutions, the necessary information about the machine is available, the customers have access to the current log file and are supported in their initial actions. Customers are provided with all the knowledge about warp knitting that is available at Karl Mayer. Basically, the aim is to help customers with special functions that ensure maximum productivity and reproducible quality, summarised under the umbrella term 'Smart Machine'. Digital solutions include recording machine performance and efficiency data, appropriate evaluation tools, and helpful features such as a machine alarm system. For example, the machine can automatically send a warning signal to the operator's cell phone to inform them when a warp beam needs to be changed soon (Michel, 2021).

The cloud solution developed could also play a role in industrial on-demand manufacturing in the future. More and more products are being ordered directly via online stores, e.g., from major brands or sporting goods manufacturers. With cloud technology, the order can be forwarded directly to the connected production facilities and the goods can be manufactured and shipped to the customer without any detours. This saves storage costs and reduces cost-intensive markdowns (Knitting Industry, 2022b).

While the most important topics in Industry 3.0 were hardware performance and networking, Industry 4.0 is primarily about data. The collection and processing of Big Data enables complete transparency. This transparency facilitates the prediction of events in areas such as maintenance, i.e., continuous monitoring can detect a rise in the operating temperature of a motor, and the correspondingly obtained data can then be used to localise and remedy the cause of the temperature rise before the motor fails. Predicting events also offers the possibility of automating certain actions, i.e., the speed of the machine could be automatically reduced until the engine problem is resolved (Michel, 2021).

As part of the knitelligence concept, STOLL by Karl Mayer has developed a new generation of knitting machines adapted to the specific needs of the digital world. Stoll's knitelligence technology is a modular system that combines all software solutions as well as knitwear production in a single platform and covers the entire value chain. Tailored to the specific needs of the digital world, this generation makes work easier through process automation, more transparency, shorter response times, shorter production cycles and higher productivity. Part of the new knitelligence machine generation is the new Multi Touch Panel. It is much more user-friendly thanks to its improved readability and brilliant surface. Faster response times, easier operation and other helpful functions make the new Multi Touch Panel a real relief. In addition, the extended knitting report is standard on every machine.

With this modern Internet of Things (IoT) technology, customers can not only make their processes much more efficient, but also respond much more flexibly to market requirements. In recent years, digitalisation processes have become increasingly automated. This leads from pattern creation to machine preparation, production planning and machine control, etc. STOLL has therefore developed STOLL-knitrobotic – part of the knitelligence software solution. STOLL-knitrobotic offers customers a reduction in workload through process automation, greater transparency, shorter response times, shorter production cycles and higher productivity. To take a step in this direction, the Karl Mayer Group has introduced a new design software k.innovation CREATE, which is one of the key modules of 'knitelligence' to simplify design development and speed up the programming process. The k.innovation CREATE PLUS combines an advanced programming system with a user interface for optimal use of knitting potential and a production planning system (PPS) for rational management of all processes (Varshney, 2020).

As a leading supplier of flat knitting technologies, Shima Seiki's goal has always been to produce machines and software best suited for the apparel and textile industries. According to the Japanese knitting machine manufacturer, factory automation has become a necessity in the textile industry, and interest in it has increased due to the COVID-19 pandemic (Varshney, 2021).

### 7.5.1 Virtual prototyping

There are virtual 3D patterning techniques with certain knitting machines from the leading manufacturers that are very useful, timely and competent in the planning, production and sale of knitwear with the help of high-resolution simulator software, which ultimately results in no waste, no leftovers and no inventory losses. It also eliminates the cost of raw materials and the time spent on the repeated production of prototype samples (Blaga, 2022).

Virtual prototyping is implemented by Shima Seiki with the help of the SDSONE APEX4 design system. As the centrepiece of the company's 'Total Fashion System' concept, SDS-ONE APEX4 comprehensively supports the entire production chain and integrates production into a smooth and efficient workflow – from yarn development, product planning and design to production and even sales promotion. Particularly effective is the way SDS-ONE APEX4 enhances the design evaluation process with its ultra-realistic simulation capability, with virtual samples minimising the need for actual sample production. When numerous variants must be evaluated before deciding on a final design, virtual product samples can be used to streamline the decision-making process by minimising the enormous time and cost normally associated with producing actual physical samples for each variant. The sustainability factor is undeniable, considering how much material is normally wasted in the sampling phase.

### 7.5.2 StoreFactory by adidas

Another digitalisation approach to shorten development times and enable local and on-demand production is the concept of in-store factories. Although such concepts exist for various products and associated production technologies, flat knitting in particular is ideal for such production concepts, as it is possible to produce off-the-rack textiles that do not require further cutting and sewing processes.

An example of an in-store factory is the StoreFactory, which was set up as a 3-month pop-up store by adidas in Berlin in 2017. In this factory, customers could design an individual sweater tailored to their body shape and have it knitted directly in the StoreFactory. To do this, customers were passed through various stations. After registration, various motifs were projected onto the upper body – a pattern for the sweatshirt was created by the movements of the customer's own body and their perception by sensors and selected by the customer. The data were used directly to create the appropriate knitting pattern and subsequently produce the sweatshirt on site. Although the StoreFactory only ran for 3 months as a pop-up store, the concept shows that it is possible to produce knitted textiles with flat knitting machines on-demand and locally on site (adidas AG and StoreFactory; Wolf, 2017).

## 7.6 Circularity of knitted garments

Another important point in the development of sustainable textiles is the establishment of a true circular economy. In a circular economy, the products and materials used are reused as much as possible and recycled at the end of their life cycle. In the case of clothing, the second-hand market is well known, but most textiles do not go through a closed loop. In particular, there is still much need for development in the recycling of used textiles, as currently only about 1% of all used textiles are sent to a recycling process (Chen et al., 2021).

To ensure a true circular economy, knitted textiles must be reused and the production waste generated at the end of use must be recycled to a large extent. Only if knitted fabrics are designed to be used in a true circular economy, will they be competitive in the future. An important aspect of enabling the circular economy is designing for recycling (Saha et al., 2021).

With design for recycling, the recyclability of knitted textiles is already considered and significantly increased during the development of knitted fabrics, with the three hierarchically organised approaches of avoidance, mono-material construction and preparation for separation. In order to significantly increase the recycling rate for textiles, the design of the textiles is crucial, in addition to suitable recycling plants and processes (Koszewska, 2018; Durham et al., 2015).

The first approach of design for recycling, avoidance, is already considered with material-saving production and reduction of production waste with 3D knitting and especially fully-fashion and knit&wear approaches, leading into a zero-waste production. This approach is particularly interesting for companies, as it directly reduces material costs. The second approach, mono-material construction, is much more difficult to implement, as many knitwear products use different yarns in one product, which are responsible for different functions. Important examples are the use of blends, e.g., polyester/cotton, or the usage of elastanc in knitted fabrics.

In order to realise a mono-material construction, either one of these functions must be dispensed with, or alternative approaches must be found to enable the necessary functions of the knitted fabric with only one yarn material. If a mono-material construction is not feasible for a product with certain necessary functions, the third approach of design for recycling may be helpful. With construction for separation the recyclability is supported, by enabling the separation of different materials into mono-material waste.

Another important part of opening up the recyclability of textiles is the transparent communication of the materials used across the various process and use stages. Only if at the end of use, directly before the textile is recycled, when it is still possible to understand what materials it is made of, recycling can be carried out effectively. Possible solutions for the transport of this crucial information for recycling include various textile tagging approaches that use directly knitted QR codes or integrated RFID chips (Durham et al., 2015; Roos et al., 2019).

### 7.6.1 Microfactory for recycling

One approach to improving the recycling of textiles is so-called microfactories. In these microfactories, used textiles are collected and shredded in a coherent process line and further processed into new yarns or even textiles (Deutsche Institute für Textil- und Faserforschung, 2020).

A pilot project for such a microfactory is the Loop system installed by H&M in Stockholm at the end of 2020. Here, customers can drop off their old textiles and have them 'transformed' into new ones. The old textiles are mechanically recycled. That is, the textiles are first shredded into individual fibres and spun back out into yarns. The recycled yarns are then used to make new textiles on flat knitting machines (Hennes & Mauritz, 2020). Because of the simple framework and the fact that only a few bobbins are needed, flat knitting allows direct production with the yarns that have just been recycled. Warp beam preparation and subsequent weaving or knitting have not yet been realised in microfactories.

### 7.6.2 SpeedFactory by adidas

A large part of global emissions, including from textiles, is caused by transport. Often, textiles traded in Europe are manufactured in Asia and then shipped by ocean-going freighters. Especially in the case of cotton, which is produced in the cotton belt worldwide, the production of textiles in Asia is a long diversion. Because of favourable wage structures and often lower environmental regulations, much of the textile industry has moved to Asia, as the favourable conditions even compensate for the more expensive transportation costs. When outsourcing, little or no thought has been given to the environmental impact of production in distant countries and the associated transportation.

With the new focus on sustainability, some sectors are trying to bring back local production to avoid the harmful greenhouse gases that result from long transportation. However, as local production must continue to compete with low-wage countries, production in high-wage countries is only conceivable with a higher degree of automation.

One approach to local production through networked and thus labour-saving processes was the so-called SpeedFactory at adidas AG, Herzogenaurach. In the SpeedFactory, local production of show uppers was developed in Germany. Together with partners like RWTH-ITA, Johnson Control, etc. the concept of digitalised, fully automated production was developed and built by adidas in Ansbach, near the adidas headquarters. In the SpeedFactory, the knitted fabrics for the shoe uppers were produced on large circular knitting machines and then cut by robots (European Commission, 2021).

### 7.6.3 Kniterate

Another approach to decentralising production by enabling local production is the Kniterate knitting machine (Fig. 7.6). Kniterate is a small and compact flatbed knitting machine produced by the start-up company of the same name. It is designed to enable the production of knitted garments in designer workshops, makerspaces, schools or at

**Figure 7.6** Kniterate flat knitting machine.

home. In addition, Kniterate's software approach aims to enable the development of new knitwear by makers without extensive training in knitwear design (Kniterate).

The Kniterate company is trying to bridge the gap between manual knitting and industrial knitting with a compact, low-cost knitting machine for home use. The machine has two needle beds, each with 252 needles, a working width of just over 90 cm, and is relatively compact, measuring 152 × 70 × 65 cm and weighing 210 kg (Kniterate).

Complexity and cost reduction are achieved by limiting variants, not using current machine technology and quantity of parts. The machine is offered exclusively in gauge E7 and with six yarn feeders. In contrast, knitting machines from established manufacturers are typically offered in a gauge range of E3–E18 and have up to 32 yarn guides. On the other hand, low-threshold service offers are available in the form of video tutorials and brochures that enable independent familiarisation, operation and maintenance of the machine.

A motorised knitting machine like the one from Kniterate for small workshops and even one's own living room is new. Through the company and the public attention and support, the control software of the knitting machine has also taken another step forward. Kniterate was also originally based on the 'Knitic' software and promises to be easy to use. Kniterate launched the venture in 2017 with a very successful crowd funding campaign and has since delivered several hundred machines. There are examples in the 'Antwerp Design Factory' and the 'weißensee kunsthochschule berlin' among others. It remains to be seen to what extent the Kniterate machine can close the gap between hand knitting and professional production in knitting factories in the future. However, the enormous response to the Kickstarter project suggests that the demand from designers or young companies is definitely there (Kniterate; FutureTEX, 2021).

## 7.7 Future perspectives

From the developments in textile machinery and new digital technologies, it can be concluded that digitalisation continues to be a driving force in optimising the industry towards its green goals. The sustainability of newly developed machinery and digital technologies are the latest trends of the textile industry. Closely related to the development of a sustainable industry is the diversification of production supported by digital technologies (Sterland, 2022). Driven by consumer demands for customisation and sustainability, new digitised production models have emerged as possible alternatives, including:

- On-demand manufacturing, where goods are produced when and how they are ordered, supported by investments in automation, web-to-print, and software technologies
- Manufacturing processes where factories operate with minimal or no human presence, enabled by full automation of processes
- A manufacturing process in which production from raw materials to finished product is carried out by only a few or a single production facility, which shortens the supply chain and improves sustainability
- Individual manufacturing processes, i.e., the production of an entire product, supported by advances in 3D knitting, weaving, and 3D printing technologies

These models are supported by digital technologies. Advances in hardware and software technologies revolve around automation, machines, additive manufacturing technologies, artificial intelligence (AI), machine learning (ML), machine-to-machine (M2M) communications, Internet of Things (IoT) technologies, the cloud, design platforms and 3D simulation.

The knitwear industry can make an important contribution to energy conservation, waste reduction and emission minimisation to meet environmental standards in the global scenario if all stakeholders in the knitwear industry are involved. From the manufacturers' point of view, selecting sustainable raw materials for knitwear production, using energy-efficient technologies, minimising emissions, using eco-friendly chemicals, and minimising waste are the concepts that need to be pursued. Wholesalers and retailers should promote the business of sustainably produced knitwear. Consumers are generally aware of sustainable products, but purchase decisions are still driven by price rather than the environmental impact of the product.

## 7.8 Conclusion

Recent innovations in textiles have much to do with the need to solve the problems associated with current overconsumption and related environmental issues. Consequently, the industry's current focus is on producing multifunctional, high-quality products with durable designs and longer life spans in line with the demand for sustainability in the textile and apparel sector. The knitting industry focuses on the reuse, recycling and recovery of raw materials and the reduction or even elimination of knitting waste.

A major contribution to this transformation is made by the existing design software solutions for knitting, which have reached an extremely high level of performance to meet the latest industry requirements and contribute to the digitalisation of the entire value chain in the knitting industry. In addition, the production of seamless knitted garments already contributes significantly to reducing waste, and fewer process steps are required in the production chain. Crucial to these developments is the responsible use of valuable natural resources. These achievements will not only have a significant impact on the knitting industry, but will also decisively shape the future of all business models.

The most current trend in the knitwear industry is the move towards a circular economy and Industry 4.0. To meet this trend, both machine builders and knitwear manufacturers have presented the adaptation of materials, equipment and production to sustainable concepts at recent trade shows, along with smart manufacturing, cloud computing, cyber production control and robotisation of knitting processes to meet market demands.

## References

Babu, R.P., O'Connor, K., Seeram, R., 2013. Current progress on bio-based polymers and their future trends. Progress in Biomaterials 8.

Bendt, E., 2016. Shape and Surface: the challenges and advantages of 3D techniques in innovative fashion, knitwear and product design. IOP Conference Series: Materials Science and Engineering 141.

Blaga, M., 2022. Use of CAD in knitted apparels. In: Advanced Knitting Technology. Elsevier, pp. 181–202.
Chen, X., Memon, H.A., Wang, Y., Marriam, I., Tebyetekerwa, M., 2021. Circular economy and sustainability of the clothing and textile industry. Materials Circular Economy 3, 1–9. https://doi.org/10.1007/s42824-021-00026-2.
Choi, W., Powell, N., 2005. Three dimensional seamless garment knitting on V-bed flat knitting machines. Journal of Textile and Apparel, Technology and Management 4, 1–33.
Davis, H., 2021a. Sustainable Threads From Madeira. https://www.knittingtradejournal.com/fibres-yarns-news/14698-sustainable-threads-from-madeira.
Davis, H., 2021b. Solvay Launches Bio Amni Sustainable Fibre. https://www.knittingtradejournal.com/fibres-yarns-news/14600-solvay-launches-bio-amni-sustainable-fibre.
Davis, H., 2021c. PrimaLoft Wins Coveted Sustainable Fashion Award. https://www.knittingtradejournal.com/fibres-yarns-news/14486-primaloft-wins-coveted-sustainable-fashion-award.
Davis, H., 2021d. Raschel Technology for 4D-Knits. https://www.knittingtradejournal.com/warp-knitting-news/14632-dissolving-design-boundaries-with-4d-knits.
Deutsche Institute für Textil- und Faserforschung, 2020. Microfactory: Virtuelle Bekleidungssimulation in 3D von Entwurf bis Produktion: Ein Meilenstein für die Modeindustrie auf dem Weg zu Industrie 4.0.
Dissanayake, G., Perera, S., 2016. New approaches to sustainable fibres. In: Sustainable Fibres for Fashion Industry, second ed. Springer Nature, pp. 1–12.
Durham, E., Hewitt, A., Bell, R., Russell, S., 2015. Technical design for recycling of clothing. In: Sustainable Apparel. Elsevier, pp. 187–198.
EU Strategy for Sustainable and Circular Textiles, 2022. Brussels.
European Commission, 2021. Data on the EU Textile Ecosystem and its Competitiveness: Request for Services 896/PP/2020/FC Implementing Framework Contract 575/PP/2016/FC. Final Report. Brussels.
FutureTEX, 2021. Craftbot 3D Drucker und Digitale Strickmaschine Kniterate am TPL. Weißensee Kunsthochschule Berlin. https://www.futuretex2020.de/aktuelles/meldungen/meldungen-detailseite/craftbot-3d-drucker-und-digitale-strickmaschine-kniterate-am-tpl-weissensee-kunsthochschule-berlin.
Gries, T., Bettermann, I., Blaurock, C., Bündgens, A., Dittel, G., Emonts, C., Gesché, V., Glimpel, N., Kolloch, M., Grigat, N., Löcken, H., Löwen, A., Jacobsen, J.-L., Kimm, M., Kelbel, H., Kröger, H., Kuo, K.-C., Peiner, C., Sackmann, J., Schwab, M., 2022. Aachen technology overview of 3D textile materials and recent innovation and applications. Applied Composite Materials 29, 43–64. https://doi.org/10.1007/s10443-022-10011-w.
Hennes, Mauritz, A.B., H and M, 2020. From Old to New with Looop: The World's First In-Store Recycling System Is Now Installed at H&M in Stockholm. Here's All You Need to Know. https://about.hm.com/news/general-news-2020/recycling-system–looop–helps-h-m-transform-unwanted-garments-i.html.
Innova picks up Roica V550 by new eco-collection. Knitting Views 17, 2022, 21.
Knitting Industry, 2022a. Circular Knitting: Latest Santoni Technologies on Show at ITM 2022. https://www.knittingindustry.com/circular-knitting/latest-santoni-technologies-on-show-at-itm-2022/.
Knitting Industry, 2022b. Flat Knitting: KM.ON Acquires Digitale Strickmanufaktur's Cloud Technology. https://www.knittingindustry.com/flat-knitting/kmon-acquires-digitale-strickmanufakturs-cloud-technology/.

Knitting Views, 2022. AW22/23 Collection Made of TAF's Radianza Launches by Full Thai Knitting. https://knittingviews.com/aw22-23-collection-made-of-tafs-radianza-launches-by-full-thai-knitting/.

Koszewska, M., 2018. Circular economy - challenges for the textile and clothing industry. Autex Research Journal 18, 337–347. https://doi.org/10.1515/aut-2018-0023.

Liu, S., Mao, S., Zhang, P., 2021. Toward on integrally-formed knitted fabrics used for automotive seat cover. Journal of Engineered Fibers and Fabrics 16, 1–12. https://doi.org/10.1177/15589250211006544.

Michel, S., 2021. The Digital Future of Warp Knitting: Interview with Christof Naier. https://textile-network.de/en/Business/The-digital-future-of-warp-knitting.

Mokhtari, F., Spinks, G.M., Fay, C., Cheng, Z., Raad, R., Xi, J., Foroughi, J., 2020. Wearable electronic textiles from nanostructured piezoelectric fibers. Advanced Materials Technologies 5, 1–15. https://doi.org/10.1002/admt.201900900.

Pavko Čuden, A., 2016. Modern knitting: sustainable orientation and circulation of innovations. Tekstilec 59, 63–75.

Pavko Čuden, A., 2020. Novelties in knitting. Tekstilec 63, 72–89.

Pavko Čuden, A., 2022. Recent developments in knitting technology. In: Maity, S. (Ed.), Advanced Knitting Technology. Elsevier, pp. 13–66.

Peiner, C., Löcken, H., Reinsch, L., Gries, T., 2022. 3D knitted preforms using large circular weft knitting machines. Applied Composite Materials 29, 273–288. https://doi.org/10.1007/s10443-021-09956-1.

Ray, S., 2022. Introduction to advances in knitting technology. In: Advanced Knitting Technology. Elsevier.

Roos, S., Sandin, G., Peters, G., Spak, B., Schwarz Bour, L., Perzon, E., Jönsson, C., 2019. Guidance for Fashion Companies on Design for Recycling. Unpublished. Mistra Future Fashion, Mölndal, Sweden.

Ştefan, E., Blaga, M., Penciuc, M., 2021. Digital Solutions for Education in Sustainable Knitting.

Saha, K., Dey, P.K., Papagiannaki, E., 2021. Implementing circular economy in the textile and clothing industry. Business Strategy and the Environment 30, 1497–1530. https://doi.org/10.1002/bse.2670.

Simonis, K., Gloy, Y.-S., Gries, T., 2017. IOP Conference Series: Materials Science and Engineering, pp. 1–6.

Sterland, H., 2022. IoT Technologies for Textile Manufacturing. Texprocess, p. 23.

Varshney, N., 2020. Future of Knitting Lies in Digitisation, Says STOLL. Attaining a Certain Level of Expertise in Knitting Technology in over 14 Decades Is Exemplary and STOLL is Proud of That. Here Are Glimpses of Our Interview with German Knitting Tech Giant. https://apparelresources.com/technology-news/manufacturing-tech/future-knitting-lies-digitisation-says-stoll/.

Varshney, N., 2021. Combating Knitting Challenges with Digitisation. https://apparelresources.com/technology-news/manufacturing-tech/combating-knitting-challenges-digitisation/.

West, A., 2020. Innovations in Knitting. https://www.textileworld.com/textile-world/features/2020/05/innovations-in-knitting-2/.

Wolf, S., 2017. Adidas Testet „Storefactory" in Berlin. https://www.textilwirtschaft.de/business/news/Unternehmen-Adidas-testet-Storefactory-in-Berlin-202646?crefresh=1.

World Textile Information Network, 2022a. Maximising South-East Asian Agricultural Waste. https://www.wtin.com/article/2022/january/030121/maximising-southeast-asian-agricultural-waste/?freeviewlinkid=134614.

World Textile Information Network, 2022b. Textile4.0.

# Sources of further information

Adidas AG, StoreFactory: Individuelle Produkte Gestalten und Direkt Produzieren.

Cortland Biomedical, Biomedical Textiles for Cardiovascular and Endovascular Applications. https://www.cortlandbiomedical.com/textile-solutions-cardiovascular-applications/.

Knitting Views knittingviews.com/wp-content/uploads/2022/06/KV_May_June_2022_Web.pdf.

Ecolife http://www.ecolife.com/.

Karl Mayer Group, Double Needle Bar Raschel Machines: 4D-Knit.Solutions. Introducing a New Generation of Warp Knit Fabric Using Karl Mayer Technology. https://www.karlmayer.com/en/products/warp-knitting-machines/double-needle-bar-raschel-machines/4d-knit-solutions/.

Kniterate, Kniterate: the Digital Knitting Machine: Knitting machines made for everyone. https://www.kniterate.com/about/.

Maschinenfabrik Harry Lucas GmbH & Co. KG, Harry Lucas - Modular Textile Machines for Growing Demands. https://www.lucas-elha.de/.

# Innovations in nonwoven textiles

*Mirza Mohammad Omar Khyum and Seshadri Ramkumar*
Department of Environmental Toxicology, The Institute of Environmental and Human Health, Texas Tech University, Lubbock, TX, United States

## 8.1 Introduction

The ASTM (Committee D-13 on Textile Materials) first described nonwoven materials as textile constructions comprising a web or mat composed of fibre that are bound together using a bonding substance (Buresh, 1962). During the 1970s, the ASTM definition of nonwoven materials stated that they are flat structures created through the bonding or interlocking of textile fibre, typically achieved through mechanical manipulation, thermal treatments, chemical or solvent actions, or a combination of these methods. This definition explicitly excluded paper, woven fabrics, knitted fabrics, tufted fabrics and materials made through wool or felting processes (ASTM, 1976). The ASTM definition of nonwoven structures remained focused on bonded fibrous webs and did not encompass structures that incorporated yarn. In 1992, ISO 90928 and CEN EN 29092:19929 defined nonwoven materials as sheets, webs or batts composed of fibre held together through friction, cohesion and/or adhesion, excluding paper, woven, knitted, tufted, stitch-bonded materials that incorporate binding yarns or filaments, and wet-milled felt products (Smith, 2000).

According to INDA (2007), nonwoven fabrics are characterised as planar and permeable structures that are formed by the interlocking of fibre or filaments, either directly from individual fibres or through the use of molten plastic or plastic film, without the intermediate step of converting fibre into yarns (Batra and Pourdeyhimi, 2012). In 2019, the International Organisation for Standardisation (ISO) 9092:2019 approved and published an updated and standardised definition of nonwoven, marking a significant milestone in promoting fair and equitable trade in this industry. The revised definition eliminates the reference to "textiles" that existed in the previous definition and introduces the concept of an engineered material. According to the new definition, nonwovens are characterised as engineered fibrous assemblies, primarily in a planar form, that has been deliberately imparted with a predetermined level of structural integrity through physical and/or chemical methods. It explicitly excludes weaving, knitting or papermaking processes from the definition (Innovation in Textiles, 2019).

Due to the widespread use of nonwovens and the diverse applications of the end products in different fields, sustainability is becoming a concern. Sustainability can be conceptualised as a metaphorical table that necessitates the presence of four distinct categories of "sustainability legs" to maintain stability. These categories include the elemental, environmental, energy and economic aspects. Consequently, sustainable manufacturing practices must encompass all four dimensions, often referred to as the "4Es".

**Sustainable Innovations in the Textile Industry. https://doi.org/10.1016/B978-0-323-90392-9.00012-4**

The concept of elemental sustainability encompasses the utilisation of raw materials with a reduced carbon footprint, thereby resulting in the production of environmentally friendly products. The utilisation of such materials results in reduced ecological impact, thereby facilitating the achievement of environmental sustainability. To accomplish this objective, it is essential to exercise control over energy consumption, thereby emphasising the critical importance of energy sustainability. The direct influence of energy sustainability on manufacturing and the economy is evident and can be attained through the implementation of conservation measures and the adoption of alternative energy sources. The economic viability of sustainability is important. Balanced approaches, commonly referred to as the 4Es four-legged table, have the potential to facilitate the creation of economically viable, environmentally sustainable, and technologically advanced products.

These cost-effective, sustainable and advanced products are crucial for driving economic growth while minimising the burden on nature. By incorporating energy sustainability measures into manufacturing processes, companies can reduce their environmental impacts and ensure long-term viability. This includes implementing conservation practices such as efficient resource management and waste reduction. Additionally, utilising alternate energy sources, such as solar or wind power, can further contribute to a more sustainable future. By focussing on the 4Es – energy sustainability, environmental stewardship, economic viability and advanced product development – we can create a harmonious balance that benefits both the planet and the economy. This holistic approach will pave the way for a brighter and more sustainable future for coming generations (Ramkumar, 2022a).

Nonwovens can achieve sustainability by incorporating raw materials with specific properties, including being compostable, derived from renewable sources, biodegradable, dispersible in water, free from plastic components, and containing a minimum of 15% recycled fibre content. Sustainability can also be achieved by incorporating raw materials with specific properties, including being compostable, derived from renewable sources, biodegradable, dispersible in water, free from plastic components, and containing a minimum of 15% recycled fibre content. By incorporating these specific properties into the raw materials, nonwoven materials can achieve sustainability. The use of compostable materials ensures that they can break down naturally, thereby reducing waste and environmental impacts. Derived from renewable sources, these materials are not only eco-friendly, but also help conserve our limited resources.

Additionally, biodegradability allows nonwovens to decompose without leaving harmful residues. Their ability to disperse in water further enhances their environmental friendliness as they do not contribute to water pollution. Furthermore, by being free from plastic components, nonwovens avoid the negative consequences of plastic waste. Finally, the inclusion of a minimum of 15% recycled fibre content shows their commitment to recycling and reducing the demand for virgin materials. Overall, by incorporating these specific properties into their raw materials, nonwovens have become a sustainable choice that promotes a healthier planet for future generations.

Although recycling of nonwoven materials presents various opportunities, it also entails a set of challenges that must be addressed. A notable challenge emerges within the automotive sector, wherein the disposal of end-of-life vehicles commonly involves the process of shredding. In the course of this shredding procedure, the need arises to segregate nonwoven substances. While it is possible to separate nonwoven materials, additional steps are often necessary to ensure their successful recycling. Therefore, it is imperative to develop a meticulously planned strategy that improves the ability to recycle.

The nonwoven industry thrives worldwide and has become a lucrative and multi-faceted investment sector. The performance is due to global economic trends that are on the rise and growing populations. The demand for nonwovens will continue to rise through 2025 in developed economies, while around one-third of emerging and developing economies will have similar trends. According to INDA and EDANA, there will be continual growth in the production of nonwovens worldwide. This increase will continue at an annual average rate of 2.7% and produce over 20.4 million metric tons in 2025.

In 2020, a significant portion of nonwoven manufacturing, comprising 88% of the overall tonnage, was accomplished using spunlaid and drylaid techniques. In the realm of drylaid production, a remarkable surge is anticipated in the near future. In the forthcoming quinquennium, a projection has been made indicating an anticipated average annual growth rate of 1.8% for the production of spunlaid materials on a global scale. It is projected that by 2025, the implementation of wetlaid technology will yield increased utilisation rates. It is anticipated that the absorbent hygiene sector will continue to be the largest market until 2025, when it will account for 21% of the total worldwide output. In addition, there is a growing demand for wipes and items that filter liquids because of the growing awareness of the significance of preserving clean surfaces, protecting the quality of the air we breathe, and guaranteeing the purity of the liquids we ingest.

In addition, it is anticipated that throughout the period of the forecast, the medical sector's share of output will revert to past levels, since there will be a decline in the need for protective medical apparel. At the same time, it is projected that the house building market and the automobile industry will witness an uptick in activity, both of which will contribute to their total production shares (INDA, 2022). With the ongoing trends, marketing, through its market-driven consumption-oriented practices, may have contributed to unsustainable production—consumption practices. Therefore, adopting sustainable marketing can help influence sustainable consumption through government intervention and corporate marketing strategy (Sheth and Parvatiyar, 2021).

## 8.2 Historical aspects and current scenario

The origins of nonwovens can be traced back to ancient times when natural fibres, such as flax, cotton and wool, were used for fabric production. The nonwoven industry, as it is known today, originated in the mid-20th century as a result of advancements in textile technology and the demand for novel materials with distinct characteristics. Large-scale industrial production began in 1942. In 1947, George Schroder introduced the first disposable diaper made from nonwoven fabric. Traditional textile manufacturing techniques, such as weaving and knitting, involve interlacing yarns to create fabrics. However, nonwovens are produced by bonding or interlocking fibres without weaving or knitting. In the 1940s, synthetic fibres such as polyester and polypropylene provided new possibilities for nonwoven production, offering improved strength, durability and cost-effectiveness compared to natural fibres. The first commercial nonwovens were primarily produced using dry-laid or air-laid techniques and were used for disposable products such as wipes, surgical masks and industrial filters (Batra and Pourdeyhimi, 2012).

Over time, nonwoven technology has evolved and diversified, with new manufacturing methods and processes being developed to enhance the performance and versatility of nonwoven fabrics. The spunbond process emerged in Europe and the United States in the late 1950s. Nonetheless, the comprehensive recognition of the commercial viability and efficacy of spunbond technology was only achieved during the period spanning from the mid-1960s to the early 1970s. Likewise, during the early part of the 1950s, the technology pertaining to meltblown nonwoven fabric was specifically devised to produce microfibre filters that could effectively capture radioactive particles present in higher regions of the atmosphere. Microfibres, alternatively referred to as superfine or meltblown fibres, generally have a diameter smaller than 10 μm (Malkan, 1995).

The industrial application of the spunlacing technique can be traced back to the early 1970s, when prominent nonwoven manufacturers such as DuPont and Chicopee took the lead in its implementation. Since the 1990s, this particular technique has witnessed substantial expansion and has emerged as a prominent and well-received nonmechanical bonding method for manufacturing a diverse array of nonwoven products. In contrast to the needle-punching technique, spunlacing employs high-pressure water jets to interlock the nonwoven webs. This method yielded denser webs with diverse surface textures and enhanced cleanliness. These benefits have resulted in the widespread application of spunlacing technology in the manufacture of nonwoven materials used in hygiene and wiping products. As a result, the majority of currently available spunlacing systems are designed specifically for the production of these delicate personal care items (Chen et al., 2008).

Electrospinning technology garnered considerable interest in the early 1990s, primarily owing to the increasing interest in nanotechnology. This methodology provides a wide range of options for producing fibres with diameters ranging from micrometres to nanometres that exhibit complex and distinct morphologies. Electrospinning has emerged as a prominent fabrication method for submicrometre/nanoscale fibres, offering scalability, cost-effectiveness and automation. Over the last 2 decades, it has gained significant traction in various fields, such as tissue engineering, drug delivery, biotechnology and environmental engineering, because of its intricate applications (Keirouz et al., 2023).

## 8.3 Selection of sustainable fibres

Over the past few years, there has been an increasing interest in investigating the possibilities of utilising natural fibres in a wide range of applications. Extensive research has been conducted on lignocellulosic fibres such as jute, hemp, flax, sisal and ramie to promote the adoption of sustainable nonwoven materials. Jute fibres, known for their coarse texture and mesh-like structure, are particularly suitable for needle-punched nonwovens due to their high moisture retention. Moreover, jute is cost-effective and widely available, making it popular in industries such as the automotive, pulp and paper, furniture, bedding and packaging sectors (Debnath, 2017).

Another natural fibre that has garnered attention is hemp fibre, which is a natural fibre known for its advantageous properties, including wet softness, strength, absorbency and wicking ability. Its wet softness ensures comfort, even when damp, making it suitable for products such as towels and wipes. The strength of the fibre allows nonwoven products to withstand usage without tearing or disintegration. This absorbency enables effective moisture absorption. Wicking properties remove moisture and promote dryness and comfort. Additionally, hemp fibre has a low carbon footprint owing to its fast growth, minimal water requirements, and reduced need for pesticides compared to crops such as cotton. Therefore, it is an environmentally sustainable choice. Moreover, hemp fibres are chemical-free, thus benefiting both the environment and human health. Due to its inherent valuable properties and notable sustainability advantages, hemp fibre presents an attractive alternative for the utilisation of nonwoven materials in various product applications. Similarly, sisal fibres have been extensively used in composite development (Neira and Marinho, 2009). Flax-based geotextiles offer excellent properties for civil engineering applications (Alimuzzaman et al., 2014). Additionally, cotton-based nonwovens have been developed for oil spill clean-up applications (Singh et al., 2014).

The utilisation of recycled natural fibres is of utmost importance for attaining circularity within the textile and fashion sectors. The procedure entails the collection of waste materials, either post-consumer or post-industrial, composed of natural fibres, such as cotton, linen or wool. These materials are then transformed into reusable products, including nonwovens. The initial step in the recycling process of natural fibres involves the gathering and categorisation of textile waste, which encompasses various sources, such as discarded clothing, textile remnants and residual materials from manufacturing. The materials that have been gathered undergo a series of processing stages, including sorting, cleaning and shredding. During shredding, the fibres undergo fragmentation, resulting in a reduction in their size into smaller pieces or even individual fibres. After processing, the fibres can be combined with virgin fibres or converted into novel textiles using various techniques, including spinning, carding, and needle-punching. Recycled natural fibre-based textiles have been applied in diverse sectors, such as apparel, home textiles and industrial composites. The exertion of recycled natural fibres facilitates the achievement of circularity by diminishing dependence on primary resources and redirecting textile waste away from disposal sites. The conservation of valuable resources, minimisation of energy consumption and reduction of greenhouse gas emissions can be achieved by prolonging the lifespan of natural fibres through recycling (Cestari et al., 2019; Zhao et al., 2022; Arafat and Uddin, 2022).

Moreover, from a sustainability point of view, it is important that the developed nonwovens are recyclable and reusable at the end of life. Therefore, in order to achieve a circular economy, the objective should be to encourage the utilisation and adaptation of nonwoven materials for alternative purposes, rather than their deposition in landfill sites. The flowchart presented in Fig. 8.1 provides a visual representation of the sustainable recycling process for natural fibre nonwovens, within the framework of a circular economy, emphasising the integration of waste-reducing measures in future manufacturing techniques.

Biofibres include the materials composed of lignocellulose and cellulose obtained from plant biomass, as shown in Fig. 8.2, as well as chemically regenerated cellulose derived from natural cellulose sources. In contrast, bio-based fibres are manufactured using polymers derived from renewable resources, such as polylactic acid (PLA), chitosan fibres, algae-based fibres and collagen fibres. Biobased fibres exhibit characteristics comparable to those of conventional synthetic fibres yet offer enhanced environmental sustainability. Biofibres and bio-based fibres have been widely utilised in the production of nonwoven goods, including wipes, diapers and personal hygiene products (Xu and Rowell, 2011; Pourmohammadi, 2013).

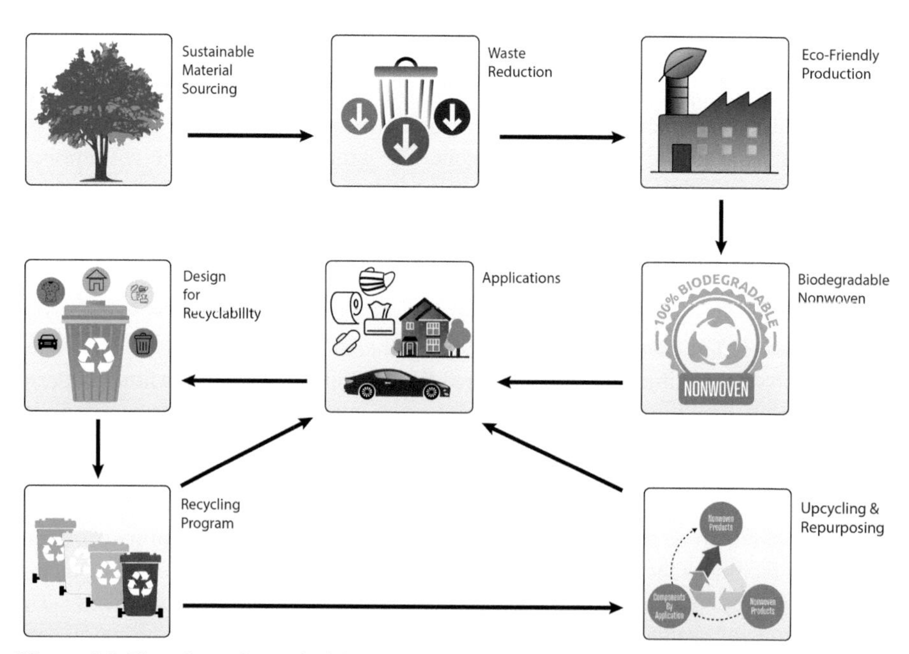


**Figure 8.1** Flowchart of sustainable recycling of natural fibre nonwovens.

**Figure 8.2** Sustainable fibres from plant biomass: (a) wood chips, (b) pulp extracted from the chips, (c) sustainably made fibres from the pulp.

## 8.4 Innovations leading to sustainable nonwoven products

### 8.4.1 Civil engineering applications

The utilisation of sustainable nonwovens in the field of civil engineering has drawn considerable interest recently, primarily owing to their capacity to tackle environmental issues and enhance the sustainability of construction methodologies. Nonwoven fabrics derived from recycled textile waste or natural fibres are viable substitutes for traditional materials in diverse civil engineering applications. Sustainable nonwovens have the potential to serve as reinforcements in concrete composites, thermal and acoustic insulators, erosion control geotextiles and other applications. Such nonwovens play a significant role in waste reduction, resource conservation and the promotion of a circular economy within the civil engineering sector through the repurposing of textile waste and utilisation of environmentally friendly materials.

The construction and building sectors are responsible for significant energy and material consumption, making it essential to adopt a circular economy model. In a circular green manufacturing process, textiles, which are frequently disposed of through thermal destruction, incineration or landfills, can be transformed into building materials.

The nonwoven polyester textiles were shredded and compressed into compact thermal-insulation boards. The results from this experiment demonstrated that the thermal conductivity coefficient of the created samples was similar to that of traditional materials commonly employed in construction. Furthermore, the introduction of waterglass facilitated the gradual spread of the flame throughout the entire product. Therefore, this thermal insulation could serve as a viable and environmentally friendly option for building construction (Antolinc and Filipic, 2021). Expandable graphite, an intumescent flame retardant, when incorporated into sustainable building insulation felt made of recycled cotton fibre, resulted in very good flame retardancy, without emitting any toxic gases (Paul et al., 2012).

Researchers have developed sustainable nonwovens with excellent insulating properties using hemp waste from the textile industry. The nonwovens showed lower thermal conductivity values than traditional materials, but had decreased mechanical properties and were inefficient against fungal proliferation. Despite their poor mechanical properties and inefficiency against fungal proliferation, sustainable nonwovens made from hemp waste still hold promise for various applications. With further research and development, it is possible to enhance their mechanical strength and address the issue of fungal resistance, making them viable alternatives to traditional materials in terms of insulation (Gutierrez-Moscardo et al., 2022).

The use of flax nonwovens as reinforcement in calcium aluminate cement composites can lead to sustainable construction materials with high strength. Utilising a calcium aluminate cement matrix can help alleviate fibre degradation following accelerated ageing, whereas treating the nonwoven fabrics did not have any impact on their mechanical performance or durability (Claramunt et al., 2018). The application of alkaline treatment not only enhanced the mechanical properties and durability of the composites based on calcium aluminate cement but also improved the adhesion between the flax nonwoven fabrics and the cement matrix. These findings suggest that sustainable surface treatments such as alkaline treatment can significantly enhance the performance of flax nonwoven fabrics as textile reinforcements in cement-based composites (Gonzalez-Lopez et al., 2020).

Researchers have combined coir fibre with hemp and flax to create a composite product with high-performance properties for industrial applications, resulting in improved mechanical properties compared with pure coir–polylactic acid composites (Ciccarelli et al., 2020). The incorporation of glass fibre preform and jute nonwoven fabric in hybrid composites resulted in notable enhancements in the tensile, flexural and izod impact strengths, surpassing the performance of composites reinforced with cotton webs. Moreover, wool waste fibre can be transformed into new suitable raw materials for building components, exhibiting favourable hygrothermal, acoustic and non-acoustic properties (Kamble and Behera, 2021; Rubino et al., 2021).

The use of nonwoven sisal fibre materials in buildings can contribute to energy efficiency and sustainability goals. Sisal fibre nonwoven materials possess remarkable thermal performance, low density, fire resistance and air permeability characteristics, making them extremely promising for thermal insulation in building applications. This material offers a substantial reduction in annual cooling and heating energy demands, positioning it as a competitive and effective solution for insulation. Additionally, sisal fibre nonwoven material is a sustainable and eco-friendly option, as it is derived from the agave sisalana plant, which requires minimal water and pesticides to grow. Its biodegradable nature makes it a desirable choice for those seeking environmentally friendly alternatives (Ouhaibi et al., 2022).

Geotextiles made from meandrically arranged Kemafil ropes produced from textile waste materials were used to protect and reinforce ditch banks in clay grounds. Geotextiles slow down water flow, prevent erosion and facilitate the growth of protective vegetation owing to their water-holding capacity and resistance to biodegradation (Broda et al., 2017). Innovative geotextiles made from coarse ropes filled with sheep wool and wrapped with woollen nonwovens were buried in the ground for erosion control, and their slow biodegradation maintained their protective potential for one vegetation session. These geotextiles offer a sustainable solution for erosion control, as they not only prevent soil erosion but also provide a natural habitat for microorganisms, promoting biodiversity. In addition, their use can help reduce the need for synthetic materials in construction projects, thereby contributing to a more eco-friendly approach (Broda et al., 2016)

The use of carboxymethyl cellulose/potassium nitrate solutions and citric acid as a cross-linker provides an effective method for enhancing the properties of nonwoven fabrics in agriculture. Precoating with polyvinyl alcohol further improved the performance of the fabric, leading to significant improvements in water absorption, water retention capacity and fertiliser release. These advancements have contributed to sustainable agricultural practices by promoting efficient water usage and nutrient delivery to plants (Ozen et al., 2018). Two textile waste nonwovens were tested for 3 months under accelerated weathering circumstances (ultraviolet light, moisture and heat) to determine whether they may serve as sustainable alternatives to plastic mulching films. This study aimed to assess the durability and degradation rate of textile waste nonwovens compared with traditional plastic mulching films. These findings provide valuable insights into the potential of nonwovens as eco-friendly alternatives to agricultural practices (Abidi et al., 2021). Agricultural practices generate large amounts of plastic waste, but sustainable textiles made from bio-based and eco-friendly polymers, such as polylactic acid, can be used as a support element and to replace conventional plastics. Experiments have shown that polylactic acid has biodegradation potential and can reduce pollution and consumption of agricultural inputs and increase the productivity and quality of crops (Stroe et al., 2021).

### 8.4.2 Healthcare, hygiene and safety applications

The development of sustainable nonwoven materials has opened up exciting possibilities in various fields ranging from flame retardancy and wound healing to temperature control and antimicrobial applications. By harnessing the unique properties of natural compounds, such as phytic acid and chitosan-based additives, researchers have successfully enhanced the flame resistance of poly(lactic acid) nonwoven fabrics and improved the mechanical strength and stability of meshes used in wound healing. In addition, there is promising evidence that mosquitoes can be repelled and disease transmission can be reduced through the use of plant-based spatial repellents incorporated into cotton-based substrates. These sustainable nonwoven materials not only offer improved performance but also contribute to environmental conservation owing to their biodegradability and nontoxic nature.

In the fields of hygiene and sustainability, innovative companies are pushing boundaries to create products that are both effective and environmentally friendly. Nice-Pak has significantly transformed the flushable wipes sector through the introduction of its innovative SecureFLUSH technology. The wipes, composed entirely of cellulose nonwoven material, exhibit exceptional strength and durability, while also demonstrating a disintegration rate five times faster than that of the leading brand of two-ply toilet paper. The key aspect of their design expertise lies in the utilisation of a specialised "lock and key" mechanism that incorporates plant-based fibres that efficiently disintegrate upon flushing, thereby promoting responsible plumbing practices and wastewater management. A noteworthy advancement in the field involves the development of Danufil QR, a type of viscose fibre that carries a positive charge. This innovation aims to hinder the binding of quaternary ammonium compounds (quats) to cellulose.

The utilisation of renewable cellulose fibres offers efficient disinfection, compatibility with various nonwoven technologies, and complete biodegradability. A company known as SHE has employed proprietary technology in the development of a product called Go! Pads. These pads were designed to include an absorbent core derived from agricultural waste sourced from banana plants, thereby obviating the need for chemical additives and superabsorbent polymers. Not only does this guarantee enhanced absorption, it also contributes to the development of a more sustainable and environmentally friendly product. Kudos has recently launched a disposable diaper that incorporates cotton, a material highly regarded by medical professionals, which is both carbon-negative and suitable for sensitive skin. In contrast to diapers manufactured using polymers derived from fossil fuels, Kudos diapers possess a composition that is four times more reliant on plant-based materials, rendering them a more environmentally sustainable option for parents who prioritise ecological considerations (INDA Innovation and Achievement Awards, n.d.).

Glatfelter, a prominent company specialising in sustainable solutions, has received recognition for its environmentally conscious cellulose-based products. The GlatClean disinfecting wipes, which utilise a resilient airlaid substrate, exhibit notable absorption properties, thereby presenting a viable and environmentally friendly option within the wipes industry. Furthermore, Glatfelter's GlatPure backsheet, which is employed in hygiene products, possesses desirable characteristics of breathability and impermeability. This sheet serves as a responsible disposal option upon reaching the end of its lifespan (2021–2022 Sustainability Report, 2022).

Collaboration between Coterie, a renowned brand in the baby care industry, and the VEOCEL brand has resulted in the development of a novel wipe that possesses two key characteristics: it is entirely derived from plants and is capable of undergoing biodegradation. The VEOCEL lyocell fibres utilised in these wipes have undergone comprehensive certification testing and have demonstrated complete compostability within a short span of a few weeks, even when subjected to challenging environmental conditions (Coterie, 2021).

Sero Hemp is a specific variety of hemp that has been cultivated and developed by Bast Fibre Technologies. Sero Hemp refers to a distinct cultivar of hemp that has undergone selective breeding to exhibit an elevated level of fibre content and superior fibre quality. It is widely recognised for its exceptional fibre attributes, including elongation, tensile strength and pliability, which have led to its utilisation in various sectors such as infant care, intimate care and personal care merchandise (Bast Fibre Technologies, 2023).

The utilisation of phytic acid, an inherent compound characterised by a substantial phosphorus composition, was employed to enhance the flame-retardant properties of the polylactic acid nonwoven fabric. The fabric that underwent treatment demonstrated favourable flame retardancy, with phytic acid serving as an environmentally friendly and efficient flame-retardant agent. Furthermore, phytic acid is biodegradable and nontoxic, making it an environmentally friendly choice for flame-retardant applications. Its ability to enhance the flame resistance of polylactic acid nonwoven fabrics opens new possibilities for the development of safer and more sustainable materials in various industries (Cheng et al., 2016).

Polyethylene terephthalate nonwoven fabric was successfully utilised for the immobilisation of glucose oxidase (GOx) enzyme through a combination of plasma treatment and chemical grafting using hyperbranched dendrimers. Modification of the polyester with amine terminal functional groups resulted in enhanced surface properties, enabling enzyme loading of up to 31% with an active immobilisation efficiency of 81%. The immobilised GOx enzyme displayed improved stability, retaining 50% of its initial activity even after six repeated uses, thus outperforming the unbound enzyme in terms of its longevity. The fibrous biocatalysts exhibited significant antibacterial efficacy against pathogenic bacterial strains when exposed to both oxygen and glucose, thereby offering greater potential for sustainable utilisation in industrial settings (Morshed et al., 2019a).

PHAs and chitin nanofibril (CN) and nanolignin (NL) complexes were incorporated into the meshes to enhance their mechanical strength and stability. Additionally, the bioactive factors loaded onto the meshes facilitate faster wound closure and promote tissue regeneration, further highlighting their potential for wound healing (Azimi et al., 2020).

Heat stress caused by helmet wear in tropical regions and hot climates can cause discomfort to motorcyclists. To control the temperature, researchers have explored the use of paraffinic Phase Change Materials (PCMs) as an interlayer between the helmet and scalp. These nontoxic, low-cost, lightweight and sustainable PCM nonwoven materials can reduce helmet temperature by 3.8°C, absorbing 17.8 W of heat. This novel approach presents a prospective means to enhance the comfort and safety of motorcyclists when exposed to high-temperature environments. By effectively regulating the temperature inside the helmet, the use of nonwoven PCM materials can prevent heat-related discomfort and potential health risks for riders (Sinnappoo et al., 2020).

Contamination of medical products, particularly nonwoven medical covers, poses a significant risk to patient well-being. To address this, a cost-effective and scalable approach was used to create a PLA/LMPLA-TPU/Triclosan nonwoven material. This innovative solution enhances the performance of the nonwoven fabric by improving its waterproofing, antimicrobial, mechanical and abrasion-resistant properties. The development of eco-friendly and sustainable nonwovens is important in the design of medical products. The use of PLA/LMPLA-TPU/Triclosan nonwoven materials in medical covers reduces the risk of contamination and ensures patient safety. Additionally, the eco-friendly nature of this material aligns with the growing demand for sustainable healthcare solutions (Zhang et al., 2021).

The demand for environmentally friendly PLA nonwovens has witnessed growth, but little attention has been given to flame-retardant PLA fabrics. To address this, a novel halogen-free self-intumescent polyelectrolyte, tris (hydroxymethyl)-aminomethane polyphosphate (APTris), was synthesised and utilised to enhance the fire resistance of nonwoven PLA fabrics. The resulting assessment of flammability demonstrated an improvement in the limiting oxygen index and a reduction in vertical burning damage. The dense char barrier created by APTris enhances thermal dilatability and flame retardancy, showing its potential as a sustainable and halogen-free treatment for commercially viable flame-retardant PLA nonwovens with reduced environmental impact (Wang et al., 2021).

Mosquito-borne arboviral infections and malaria are serious public health problems around the world. Plant-based spatial repellents applied to cotton substrates offer an excellent and environmentally friendly alternative to synthetic materials. These repellents, derived from plants such as citronella and lemongrass, have shown promising results in repelling mosquitoes and in reducing the risk of disease transmission. Additionally, the use of cotton-based substrates promotes sustainability and reduces reliance on harmful synthetic materials, making them a win–win solution for both human health and the environment (Hron et al., 2021).

Novel alginate nanofibres infused with oregano essential oil, known for their remarkable antimicrobial properties, have been developed for use in wound dressing and food packaging. Increasing the concentration of oregano essential oil from 2 to 3 wt.% resulted in enhanced antimicrobial efficacy against prevalent wound and foodborne pathogens. These nanofibres have the potential to significantly reduce the risk of infection and spoilage in both the medical and food industries. Their enhanced antimicrobial activity makes them promising sustainable candidates for future wound care and packaging materials (Lu et al., 2021).

A flame-retardant additive based on chitosan (NCS) was created by chemically modifying CS with silicon dioxide ($SiO_2$) via an ion interchange reaction. The introduction of NCS into vinyl ester/bamboo fibre (VE/BF) composites results in enhanced mechanical properties and flame retardancy. The synthesised NCS is well suited for producing sustainable flame-retardant composites using natural fibre (NF) for substructural components in engineering applications without compromising the mechanical properties. The incorporation of NCS not only improved the flame retardancy of the VE/BF composites but also contributed to their overall sustainability. The successful synthesis of NCS paves the way for the development of eco-friendly and fire-resistant NF composites, opening up diverse possibilities for their use in various engineering applications (Prabhakar et al., 2021).

Atmospheric plasma technology was used to prepare cotton-based nonwovens with one-way water transport. The cloth was treated with a small coating of polymerised hexamethyldisiloxane, making one side of the material superhydrophobic while leaving the other entirely wettable. Because of the asymmetry in its wettability, liquid water can flow in just one direction without impeding the passage of vapour or air. Hygiene, infection prevention and medical device applications can all benefit from the versatile moisture management capabilities of this eco-friendly cotton nonwoven fabric (Pu et al., 2021).

Hydrogel wound dressings have the advantages of absorbing high wound exudate, loading drugs and providing pain relief. Cellulose hydrogel nonwoven cotton composites were fabricated with improved moisture management, air permeability and fluid absorptive capacity for sustainable wound dressing application. These composites were developed by combining a cellulose hydrogel with nonwoven cotton, resulting in a highly absorbent and breathable material. Enhanced moisture management and fluid absorptive capacity make these dressings ideal for promoting faster wound healing and preventing infection (Ahmad et al., 2021).

An antimicrobial wound dressing was created by treating cellulosic nonwoven fabric with a combination of polyvinyl alcohol hydrogel, zinc oxide nanoparticles and mesoporous silica nanoparticles. This innovative nanocomposite exhibited exceptional antibacterial properties against *Staphylococcus aureus* and *Escherichia coli*, representing a significant advancement in the medical treatment of infected skin wounds. The inclusion of polyvinyl alcohol hydrogel in the fabric created a moist environment that facilitated wound healing by promoting cell migration and proliferation. Moreover, the presence of ZnO and mesoporous silica nanoparticles enhanced the antibacterial activity of the fabric by effectively suppressing bacterial growth and preventing infections (Nikdel et al., 2021).

Seaweed, particularly *Undaria pinnatifida*, has potential for use in the development of functional textiles for health and body care owing to its bioactive compounds. A flexible bi-laminate textile was successfully created by sandwiching fine dust extracted from *Undaria* seaweed between two layers of nonwoven fabric. This textile demonstrated excellent crease recovery properties, quickly regaining its original shape after folding or bending. Additionally, the fabric efficiently released mucilage, a viscous substance found in *Undaria*, which can offer various functional benefits in textile applications. Further research is needed for its incorporation into nonwoven fibre for dermatological use (Martinez and Becherucci, 2022).

A novel food packaging film with sustainable properties was developed by incorporating the N-halamine compound 1-chloro-2,2,5,5-tetramethyl-4-imidazolidinone (MC) into polylactic acid (PLA). The film successfully inhibited the process of PLA crystallisation, leading to enhanced physical characteristics. Furthermore, the film's antimicrobial properties effectively deactivated microorganisms, offering a potentially viable approach to prolonging the storage duration of perishable agricultural products. In addition, incorporation of the N-halamine compound into the film also provided a long-lasting antimicrobial effect, reducing the risk of foodborne illnesses. In addition, the inherent sustainability of polylactic acid renders it a viable environmentally conscious substitute for conventional materials used in food packaging (An et al., 2022).

A layer-by-layer approach was employed to coat nonwoven cotton fabrics with polysodium acrylate (PNaA) and poly 2-acrylamido-2-methylpropane sulphonic acid (PAMPS). Microwave techniques have been used to induce antibacterial activities. The efficacy of microwave treatment was observed in its ability to exhibit potent antibacterial properties while concurrently preserving a substantial level of viability in fibroblast cells. The dressings produced demonstrated the desired characteristics of strength, breathability, flexibility and wettability, making them highly viable candidates for use in wound dressings (Sadeghianmaryan et al., 2022).

## *8.4.3 Toxic decontamination*

The patented decontamination wipe (FibreTect), as shown in Fig. 8.3, consists of three layers: an outer layer, an adsorption layer and a next-to-skin layer. The outer layer's weight is reduced without compromising its strength, and its surface smoothness is improved by utilising a single or multiple punching technique that enhances interlocking. The adsorption layer incorporated fibres with diameters of approximately 500 nm or less. Similarly, the next-to-skin layer achieves reduced weight, improved strength and enhanced surface smoothness through multiple punching, enhancing interlocking (Ramkumar, 2005).

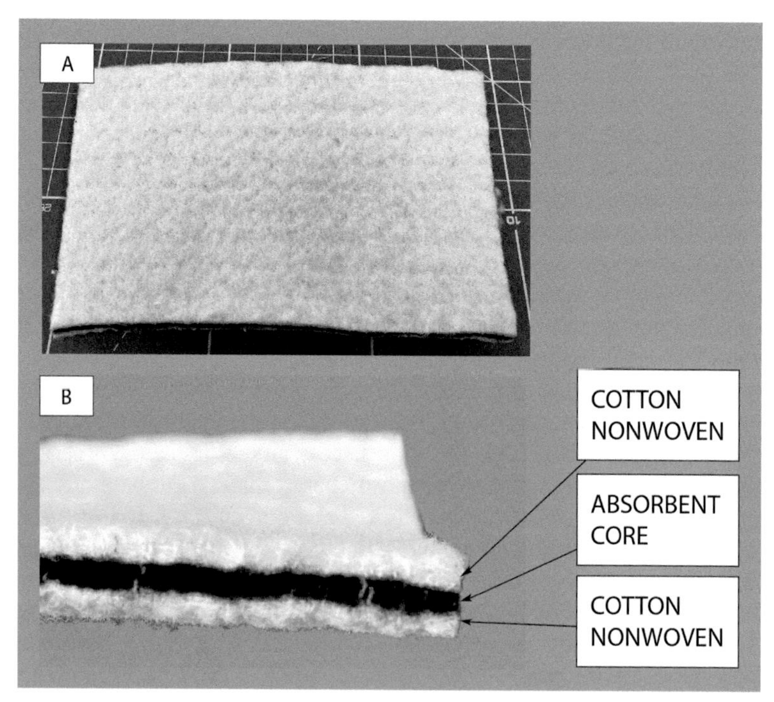


**Figure 8.3** Sustainable high-tech wipe (a) absorbent–adsorbent wipe, (b) cross-sectional view of the absorbent–adsorbent wipe.

The wipe is versatile, accommodating various fibres, including cotton, and is suitable for use on human skin and military equipment. This multilayered wipe has a unique fabric structure that enables it to effectively wipe away liquid and vapour toxins (Davis, 2009). The efficacy of this wipe was assessed in a comparative evaluation, wherein it was tested alongside alternative decontamination products against the presence of mustard gas and other hazardous chemical agents. The wipe, which consists of an activated carbon core enclosed between layers of absorbent polyester and cellulose, exhibited superior performance compared to 30 various decontamination products, including those presently employed in military decontamination kits (Dry Wipe Can Clean up Toxic Chemicals, 2008). The activated carbon core in the wipe is highly effective in adsorbing and trapping toxic chemicals, making it an ideal choice for decontamination purposes. Additionally, the absorbent polyester and cellulose layers ensure the efficient absorption of both liquid and vapour toxins, providing a comprehensive solution for decontamination needs.

The efficacy of the "Blot-Apply-Remove" technique, employing dry FibreTect wipes, in the removal of substantial quantities of hazardous substances has been demonstrated. The efficacy of FibreTect in removing harmful microorganisms, such as bacterial spores, through a dry wiping process was demonstrated in a study conducted by the U.S. Army. This approach offers notable benefits in regions characterised by extreme cold weather conditions, such as the Siachen glacier. In such environments, the freezing of liquids hampers the effectiveness of wet decontamination techniques. The findings of the study also indicated that the FibreTect dry wipe demonstrated a decontamination rate of up to 94.93% for *Bacillus atrophaeus* var. *globigii* (BG) spores (Ramkumar, 2022b). These findings highlight the potential of FibreTect as a reliable and effective solution for decontamination in extreme environments. The ability to remove toxic agents without the need for liquid-based methods could greatly benefit military personnel and first responders operating in cold weather regions.

### 8.4.4 Oil spill remediation

A novel and sustainable sorbent for oil spill cleanup was developed using raw, unprocessed cotton fibre. The sorbent, called cotton batt, exhibited a high oil sorption capacity, surpassing that of many commercial sorbents. This study investigated the fundamental mechanisms of oil sorption, such as adsorption, absorption and capillary action, using advanced microscopic techniques, as shown in Fig. 8.4. The characteristics of the cotton fibre, including fineness and maturity, were found to influence their oil sorption capacity.

Cotton batts made from immature and fine fibre showed a 7% higher oil sorption capacity than batts made from mature and coarser fibre. This environmentally friendly sorbent is easy to use and provides both a high oil sorption capacity and sustainability. This research focused on utilising raw unprocessed cotton and reducing the bulk density of the fibrous assembly through the carding process. The results demonstrate the superior oil sorption capacity of cotton batts, particularly those made from finer and immature fibre. The sorbent effectively absorbed oil without leaving visible oil sheets on the water surface. These findings highlight the potential of raw cotton batts as a

**Figure 8.4** Sustainable oil sorption mechanism: (a) visual representation of both bleached cotton (*left*) and raw cotton (*right*), (b) interplay between water and both bleached cotton (*left*) and raw cotton (*right*), (c) dynamic interaction between water and oil, (d and e) effective mechanism by which raw cotton accomplishes the separation of oil from water.

highly efficient and environmentally sustainable material for oil spill cleanup. Furthermore, the study revealed that the cotton batts maintained their structural integrity even after multiple oil sorption cycles, making them suitable for repeated use. This not only enhances their cost-effectiveness but also reduces the overall waste generated during oil spill cleanup operations (Singh et al., 2014).

In the field of environmental remediation, the development of effective and sustainable materials for oil spill cleanup and oily water treatment is of paramount importance. Researchers have explored various nonwoven fabric options derived from natural fibres, carbon fibre waste, and silkworm cocoon waste, with the aim of enhancing oil sorption capabilities and separation efficiencies. These innovative studies have yielded promising results, highlighting the potential of these materials to address the challenges of oil spillage and contaminated water. By improving fabric strength, achieving controllable wettability and employing surface modifications, these materials offer practical and environmentally friendly solutions for oil spill removal and water treatment.

In terms of oil sorption and retention capabilities, natural fibre nonwovens, such as milkweed and kapok, outperform cotton and polypropylene nonwoven fabrics. These natural fibres demonstrate higher capacities for absorbing oil and retaining it, and they also exhibit faster rates of oil sorption. However, their fabric strength was relatively low, highlighting the need for further research to enhance their strength and ensure sustainable oil spill removal. Strengthening the fabric of natural fibre nonwovens not only improves their effectiveness in oil spill cleanup but also enhances their durability and longevity (Renuka et al., 2016).

An innovative method was developed to attain adjustable wettability in polylactic acid nonwoven fabrics, leading to the creation of superwetted nonwoven fabrics. These fabrics possess excellent capabilities in absorbing and selectively separating oil and water, while also enabling the simultaneous photocatalytic degradation of water-soluble toxic organic pollutants. This advancement offers promising solutions for oily water treatment by allowing efficient oil and water separation through hydrophobic coating modifications. Furthermore, it addresses the removal of hazardous organic pollutants via photocatalysis, presenting a viable option for environmental clean-up efforts (Shi et al., 2018). Another method employed the treatment of a nonwoven mesh using a biomineralisation coating technique, which demonstrated strong oil/water separation capabilities. This environmentally friendly approach provides a long-lasting and sustainable solution for treating oily wastewater (Deng et al., 2019).

Differences in oil sorption properties have been observed in needle-punched nonwoven fabrics made from various types of natural fibres, including cotton, cotton waste, cotton/kapok blend and nettle fibre. The oil sorption capacities of cotton and cotton/kapok nonwovens were found to be higher than those of polypropylene, whereas nettle fibre demonstrated comparatively lower sorption capacities. Cotton/kapok nonwovens achieved over 95% recovery of diesel oil through compression, but their sorption capacity decreased with repetitive cycles owing to thickness loss. The results of this study indicate that nonwoven materials made from cotton and cotton/kapok blends exhibit promising characteristics for use as efficient and environmentally sustainable sorbents for oil. Further investigations could focus on mitigating the decrease in sorption capacity during repetitive cycles to enhance their practical applicability (Sinha et al., 2020).

To implement an ecologically sustainable approach, carbon fibre waste was employed to fabricate functional nonwoven textiles, which exhibited exceptional efficacy in the process of separating oil and water. The recycling procedure encompassed the application of fluorine-free coatings comprising polydimethylsiloxane (PDMS) and zeolite imidazole framework-8 (ZIF-8) to carbon fibre scraps. This treatment led to the creation of surfaces that exhibited exceptional hydrophobic and oleophilic characteristics. The application of this coating resulted in enhanced absorption capabilities for organic solvents and oils, thereby facilitating the effectiveness of gravity-driven and continuous oil–water separation procedures (Pakdel et al., 2021).

In one study, researchers examined the oil sorption properties of needle-punched nonwoven textiles produced from discarded silkworm cocoons. The results indicated that the tested fabrics demonstrated sorption capacities of 31.52 g/g for crude oil and

25.92 g/g for vegetable oil. The silk fabric exhibited hydrophobic and oleophilic characteristics, enabling its potential for multiple reuses with a minimum of five cycles. In addition, the fabric underwent complete biodegradation in soil within a 100-day timeframe, thereby emphasising its potential as an environmentally sustainable material for oil spill cleanup (Viju et al., 2022).

### 8.4.5 Energy materials

In the ever-evolving landscape of advanced materials and energy storage technologies, researchers and scientists continue to push innovation boundaries to create sustainable and high-performance solutions. From heat-resistant alginate separators for high-voltage lithium batteries to carbon-based materials with enhanced properties, and from piezoelectric generators to recycled polypropylene in microbial fuel cells, these advancements demonstrate the immense potential for sustainable and efficient energy-storage systems. Through their ingenuity and dedication, researchers are paving the way for a greener and more sustainable future.

Significant advancements have been achieved in the development of high-voltage lithium batteries through the introduction of sustainable and heat-resistant alginate nonwoven separators. These separators surpass commercially available polyolefin separators in terms of their mechanical properties, thermal stability and ionic conductivity. Notably, even at an elevated temperature of 55°C, the alginate separator retained 79.6% of its initial discharge capacity after 200 cycles, demonstrating exceptional stability throughout repeated use. The alginate separator has tremendous potential for high-voltage lithium batteries operating in elevated temperature environments, as it mitigates the risk of thermal runaway and ensures safer operation. In addition, its improved ionic conductivity facilitates faster charging and discharging rates, thereby enhancing the overall battery performance in high-voltage applications (Wen et al., 2017).

In the realm of electrochemical applications, carbon-based materials are highly sought after, owing to their robustness, electrical conductivity and corrosion resistance. Researchers have successfully enhanced the wettability of nonwoven carbon fibre felts by treating them with cold remote plasma and bio-functionalising them with glucose oxidase. This treatment results in increased enzymatic activity and opens the possibility of sustainable electrode usage. Modified carbon fibre felts are promising for various applications, including wearable electronics and large-area flexible devices (Kahoush et al., 2019).

Piezoelectric devices offer a promising alternative to fossil fuels for small electronic devices, because they can convert acoustic energy into electricity. Notably, researchers have utilised single-walled carbon nanotubes to create flexible and robust all-nonwoven polymer-based piezoelectric generators. These generators have extensive applications in various fields, such as wearable electronics and large-scale flexible devices (Hwang et al., 2019).

To address the challenges of supercapacitors, carbon nanofibre films derived from waste coal liquefaction have emerged as a solution. These films possess not only economic value but also effectively tackle the environmental pollution associated with coal waste. Furthermore, their unique properties enable enhanced energy storage and improved performance for various applications (Li et al., 2019).

A novel approach to increasing the lifetime of microbial fuel cells (MFCs) was introduced using recycled polypropylene nonwoven fabric. Reusing polypropylene as an extra layer or coating the ceramic surface are also approaches that have been investigated by researchers. The most notable achievement was observed in the PP/373 composite, which exhibited a remarkable 92% increase in power generation compared to unmodified 373. The effectiveness of PP coatings in preventing fouling depends on the properties of the ceramic material. This innovative approach presents a new method for improving the recycling process of polypropylene, a type of plastic that currently faces challenges owing to its low global recycling rates (Pasternak et al., 2021).

The energy generation and pollutant removal capabilities of osmotic microbial fuel cells (OsMFCs) utilising thin-film composites with embedded polyester screens (CTA-ES) and cellulose triacetate with cast nonwovens as intermediate membranes have been explored. The research findings indicate that OsMFCs utilising CTA-ESs as intermediate membranes exhibit the highest energy generation and pollutant removal efficiencies among the three configurations studied. This highlights the promising potential of CTA-ES membranes in enhancing the performance of osmotic fuel cells (Jiang et al., 2021).

The pursuit of environmentally friendly separators for lithium-ion batteries has led researchers to develop poly(hydroxybutyrate-co-hydroxyvalerate) (PHBV) membranes incorporated with cobalt ferrite fillers. These composite membranes demonstrate excellent cycling performance and are promising alternatives to synthetic polymers for battery applications. The utilisation of PHBV membranes not only reduces the environmental impact of lithium-ion batteries but also enhances their overall performance. Furthermore, the incorporation of cobalt ferrite fillers improves the battery stability and conductivity, establishing them as a viable and sustainable option for diverse battery applications (Barbosa et al., 2022).

Researchers have successfully created sustainable carbon fibre electrodes using lignin, a by-product of the pulp and paper industry, through electrospinning. The researchers conducted additional investigations to examine the influence of carbonisation temperature and oxygen plasma treatment on the electrochemical performance of supercapacitors. Electrospinning facilitates the production of highly porous carbon fibre electrodes with a substantial surface area, thereby improving their electrochemical performance. This study revealed that increasing the carbonisation temperature enhanced the conductivity and specific capacitance of the electrodes. Moreover, oxygen plasma treatment enhances electrode performance by increasing wettability and promoting better electrolyte penetration (Thielke et al., 2022).

### 8.4.6 Acoustic applications

Finding eco-friendly and novel ways to enhance sound insulation and absorption across a wide range of applications is a major focus of acoustic material research. This pursuit is driven by the desire to enhance acoustic comfort and experience in environments such as automobiles, buildings and home interiors. With the growing awareness of the environmental impact and the need for eco-friendly alternatives, researchers have turned their attention to natural fibre and recyclable materials as potential substitutes for traditional fibrous materials.

In order to reduce noise pollution in vehicles, researchers aimed to test the acoustic properties of nonwovens fabricated from a wide variety of natural fibres combined with PET and PP. Among the tested materials, the bamboo/PET/PP composite exhibited the highest absorption coefficient, indicating its suitability for noise control in vehicles. This finding suggests that the bamboo/PET/PP composite has the potential to significantly diminish the noise levels in automobiles, thus enhancing the overall driving experience. Further research and development of this composite material could lead to improved soundproofing solutions specifically designed for automotive applications (Islam and Mominul Alam, 2018).

Standardised characterisation of nanofibrous membranes prepared from recyclable materials has been studied. When applied to porous bulk materials, these membranes significantly improved their ability to absorb sound. As a result, they present a potential alternative to fibrous materials with a high carbon footprint, such as glass, minerals and ceramic fibrous materials (Ulrich and Arenas, 2020).

The Colombian agave species fique has been examined extensively due to its potential as a natural thermoacoustic material. Fique fibres were characterised, nonwoven samples were made, and their acoustic properties, flow resistivity, dynamic stiffness and thermal conductivity were examined. Compared to synthetic materials, fique nonwovens were shown to have qualities similar to those of thermal insulation, sound impact reduction and sound absorption at frequencies greater than 1000 Hz. These findings suggest that fique could replace synthetic materials in a number of technological settings because of their sustainability and reduced environmental impact. Further research could focus on optimising fique nonwovens for specific uses, such as in the construction and automotive industries, where thermal insulation and sound absorption play crucial roles (Gomez et al., 2020).

Automotive manufacturers are exploring sustainable and recyclable insulation materials to meet the regulatory requirements and customer preferences. Wool fibres, both waste and virgin, have emerged as cost-effective and eco-friendly alternatives to synthetic materials for sound and heat insulation in automobiles. They possess efficient insulation properties and biodegradability at the end of their lives. Bamboo is gaining attention as a promising sustainable material for automobile insulation. Bamboo offers similar advantages to wool, such as effective insulation properties and biodegradability, while also being lightweight and readily available. This makes it an appealing choice for automotive manufacturers seeking to achieve sustainability goals without compromising their performance (Cai et al., 2021).

Recycled nonwoven textiles like jute, polyester and hybrid jute–polyester were included as reinforcements to test the acoustic and thermal insulation properties of sustainable thermoplastic sandwich composite panels for building floor systems. Compression moulding with a variety of reinforcement to matrix ratios was used to make these composite panels. According to the results, the panels are more effective at dampening impact sounds at lower frequencies, while the efficacy of the jute and hybrid composites declines with increasing frequency. Better impact sound absorption was achieved by increasing the reinforcement to matrix ratio. When compared to other materials, polyester composites had the highest resistance and the lowest thermal conductivity (Aly et al., 2021). Hydrophilic nonwoven mats with optimal acoustic properties were produced through acetylation of windmill-palm fibre. Due to its ability to reduce noise, windmill palm fibre could be used in places like wall coverings and insulation (Chen et al., 2022).

The use of biduri fibre, which has a low density and is recognised for its hollow structure, allowed for the creation of a nonwoven fabric with remarkable high-frequency sound absorption qualities. The requirements for high- and low-frequency sound absorbers in the Indonesian National Standard 8443:2017 were met by nonwoven textiles having at least 80% biduri fibre. In addition, ISO 916754:1 categorised these textiles as class B sound absorbers (Judawisastra et al., 2022).

### 8.4.7 Filtration applications

In this rapidly advancing world, the need for innovative filtration and purification materials and technologies that promote sustainability and address environmental challenges is becoming increasingly crucial. Nonwoven fabrics have emerged as a versatile class of materials with a wide range of applications that offer unique properties and opportunities for sustainable solutions in various fields, including environmental remediation, water filtration, air purification and wastewater treatment. By harnessing the inherent advantages of nonwoven fabrics, such as their customisable structures, enhanced functionality and eco-friendly nature, these studies highlight promising advancements and exciting possibilities that lie ahead for these materials in creating a more sustainable future.

The incorporation of guanidine and amidoxime into polypropylene nonwoven fabric increases its selective adsorption capacity for uranium in seawater, presenting a highly promising solution for large-scale extraction. Moreover, the antifouling properties of the fabric effectively prevent biofouling and fouling by other organic and inorganic substances, thereby ensuring its long-term effectiveness. The fabric can be efficiently regenerated, allowing uranium recovery and reuse, making it a sustainable approach for extracting uranium from the sea (Zhang et al., 2018).

Using a novel strategy, zero-valent iron ($Fe^0 = ZVI$) particles were synthesised, immobilised and stabilised on a PET nonwoven membrane. When coupled with dendrimer/ex situ nonwovens, the resulting immobilised membrane showed outstanding efficacy in the Fenton-like degradation of malachite green dye, with quick colour removal (98% in 20 min). Moreover, the immobilised nonwovens maintained their efficacy even after many cycles, exhibiting exceptional reusability. These results demonstrate the promise of the dendrimer/ex situ nonwoven hybrid as a long-term strategy for purifying water of harmful contaminants (Morshed et al., 2019b).

The utilisation of membrane separation as a technology has gained momentum due to its energy efficiency and expanding market. However, sustainability concerns have emerged due to the employment of hazardous solvents and non-biodegradable polymers derived from petroleum. Sustainable and biodegradable nonwoven composite membrane supports made from bamboo fibre, polylactic acid and dimethyl carbonate have been created to address these issues. The bio-based membranes that were produced demonstrated a porous structure and tensile strength that were equivalent to those of traditional materials. The utilisation of bamboo led to heightened mechanical stability, reduced swelling and improved permeance. The utilisation of bamboo/polylactic acid membrane supports presents a viable and environmentally conscious substitute for conventional membrane-backing substances. This approach effectively eradicates the requirement for non-biodegradable polymers and harmful solvents (Le Phuong et al., 2019).

To develop effective and environmentally friendly solutions for removing dyestuffs from wastewater, the use of cationic nonwoven textiles has been investigated. Under optimal conditions, these textiles achieved an impressive removal rate of 99.47% and showed potential for industrial applications. Consequently, cationic nonwoven textiles have emerged as viable alternatives for colour removal, offering both effectiveness and environmental benefits (Demirel et al., 2021).

Although traditional nonwoven air filters have limitations in removing both particulate matter (PM) and harmful gases, a breakthrough has been made with the development of a high-efficiency integrated air filter. This filter incorporated keratin nanofibre and bicomponent microfibre nonwoven fibre with a bimodal structure. Notably, the KNF/BMF nonwovens demonstrated exceptional long-term filtration ability and high efficiency in removing PM as small as 0.26 μm, all while maintaining low airflow resistance. This innovative filter has immense potential for various applications in the multi-efficiency air filter industry (Hu et al., 2022).

A three-layer antibacterial cartridge system has been developed for the purpose of water filtration. This system utilises chitosan/poly(lactic acid) (CS/PLA) nanofibre and silver nanowires (AgNWs). The composite nanofibrous membrane consisting of CS/PLA/AgNWs, with a 3% AgNW loading, demonstrated complete antibacterial efficacy against both bacterial strains. Additionally, it effectively eliminated bacterial and heavy metal ion pollutants from pond water. The exceptional antibacterial performance and contaminant removal capabilities of the membrane make it a highly promising option for water-treatment systems. Furthermore, the utilisation of chitosan and polylactic acid ensures biocompatibility and environmental sustainability, further enhancing their viability for water filtration in diverse settings (Gadkari et al., 2022).

### 8.4.8 Conductive sensors

Technology is constantly evolving, and researchers are continuously pushing boundaries to develop innovative solutions. In one study, a breakthrough was achieved with the development of a conductive and superhydrophobic nonwoven fabric-based piezoresistive pressure sensor. This sensor represents a significant advancement in the field because it combines eco-friendly materials with exceptional performance. Owing to its high sensitivity, rapid response, stability and reusability, this sensor has a remarkable ability to detect a wide range of human motion. Moreover, the use of environmentally friendly materials in their fabrication caters to the increasing demand for sustainable and eco-friendly technologies.

The development of piezoresistive pressure sensors utilising eco-friendly materials has been achieved by employing conductive and superhydrophobic polylactic acid nonwoven fabric as the foundation. The sensor exhibits notable attributes such as heightened sensitivity, rapid response time, commendable stability and exceptional reusability. These qualities enable it to effectively identify diverse human movements, rendering it well-suited for applications in touch-sensitive displays, electronic skins, intelligent robotics and wearable devices. Furthermore, the use of eco-friendly materials in the fabrication of sensors aligns with the growing demand for sustainable and environmentally friendly technologies. This innovative sensor has the potential to revolutionise the field of human−machine interaction by enabling more intuitive and seamless interactions between humans and technology (Zeng et al., 2022).

## 8.5 Future perspectives

Nonwoven technology holds great potential for various industries as it addresses sustainability challenges by developing eco-friendly, biodegradable and renewable materials. Manufacturers will explore new fibres, such as recycled or bio-based materials, to minimise the environmental impact. Smart technologies will become more prevalent by embedding sensors, electronics or conductive fibres into nonwoven devices for applications in healthcare, sports and wearable devices. These materials will improve filtration capabilities, remove smaller particles and contaminants and enhance biodegradability. They will also play a vital role in biomedical and healthcare products, such as wound dressings, drug-delivery systems, tissue engineering and protective barriers.

Nonwoven materials may also be used in the energy sector, particularly in high-performance batteries and energy-storage devices. They can enable advancements in electric vehicles, renewable energy storage and portable electronics. Nonwoven manufacturing processes will become more flexible, allowing for customisation and on-demand production. Additive manufacturing techniques, such as 3D printing, can be integrated with nonwoven technology to create complex structures and tailored products. In addition, nonwoven materials can be seamlessly integrated into the Internet of Things (IoT) ecosystem, enabling communication and data exchange between interconnected devices, leading to smart textiles, smart packaging and other IoT-enabled applications.

## 8.6 Conclusion

The rise of sustainable nonwoven textiles represents a significant step towards addressing environmental concerns and promoting eco-conscious practices in various industries. By utilising natural and renewable fibres, such as cotton, jute and hemp, biofibre and bio-based fibres, and recycled natural fibres, sustainable nonwovens offer a range of desirable properties such as strength, absorbency and wicking capabilities. These materials have applications in diverse fields, including biocomposites, oil spill clean-up, civil engineering, flame retardancy, wound healing, personal hygiene, mosquito repellency, temperature control and antimicrobial applications. Additionally, sustainable nonwovens have been applied in energy materials, such as battery separators and conductive sensors in smart textiles.

In civil engineering, sustainable nonwovens contribute to waste reduction, resource conservation and the establishment of a circular economy. They can be employed as reinforcements in concrete composites, thermal and acoustic insulators and erosion control geotextiles. These applications do not only enhance sustainability in construction methodologies but also address environmental issues through their eco-friendly composition. Moreover, sustainable nonwoven materials have been utilised in the automotive industry for sound insulation and filtration.

This chapter serves as a valuable resource, providing insights into the sustainable aspects of nonwoven production and offering strategies to promote eco-conscious practices in the textile manufacturing industry. It is a call for researchers, industry professionals and policymakers to prioritise sustainable textile solutions, ultimately fostering a green future. By embracing sustainable nonwovens, a more environmentally friendly and socially responsible approach to textile manufacturing can be achieved.

## Acknowledgements

The authors would like to acknowledge Brad Thomas and Lori Gibler of the Department of Environmental Toxicology, Texas Tech University, for their contributions in editing and enhancing the quality of the figures used in this book chapter. Additionally, the authors would like to acknowledge Kenneth Kikanme for his contribution in finding scientific articles relevant to this book chapter.

## References

2021–2022 Sustainability Report, 2022. Glatfelter Engineered Materials. https://www.glatfelter.com/wp-content/uploads/GLT_2021-2022_ESG-Report-20221222.pdf. (Accessed 19 July 2023).

Abidi, H., Rana, S., Chaouch, W., Azouz, B., Aissa, I.B., Hassen, M.B., Fangueiro, R., 2021. Accelerated weathering of textile waste nonwovens used as sustainable agricultural mulching. Journal of Industrial Textiles 50 (7), 1079–1110.

Ahmad, F., Mushtaq, B., Butt, F.A., Zafar, M.S., Ahmad, S., Afzal, A., Ulker, Z., 2021. Synthesis and characterization of nonwoven cotton-reinforced cellulose hydrogel for wound dressings. Polymers 13 (23), 4098.

Alimuzzaman, S., Gong, R.H., Akonda, M., 2014. Biodegradability of nonwoven flax fiber reinforced polylactic acid biocomposites. Polymer Composites 35 (11), 2094–2102.

Aly, N.M., Seddeq, H.S., Elnagar, K., Hamouda, T., 2021. Acoustic and thermal performance of sustainable fibre reinforced thermoplastic composite panels for insulation in buildings. Journal of Building Engineering 40, 102747.

An, L., Hu, X., Perkins, P., Ren, T., 2022. A sustainable and antimicrobial food packaging film for potential application in fresh produce packaging. Frontiers in Nutrition 9, 924304.

Antolinc, D., Filipič, K.E., 2021. Recycling of nonwoven polyethylene terephthalate textile into thermal and acoustic insulation for more sustainable buildings. Polymers 13 (18), 3090.

Arafat, Y., Uddin, A.J., 2022. Recycled fibers from pre-and post-consumer textile waste as blend constituents in manufacturing 100% cotton yarns in ring spinning: a sustainable and eco-friendly approach. Heliyon 8 (11).

ASTM, 1976. Materials. Committee E-8 on Nomenclature and Definitions. In: Compilation of ASTM Standard Definitions. American Society for Testing and Materials.

Azimi, B., Thomas, L., Fusco, A., Kalaoglu-Altan, O.I., Basnett, P., Cinelli, P., Lazzeri, A., 2020. Electrosprayed chitin nanofibril/electrospun polyhydroxyalkanoate fibre mesh as functional nonwoven for skin application. Journal of Functional Biomaterials 11 (3), 62.

Barbosa, J.C., Correia, D.M., Fidalgo-Marijuan, A., Gonçalves, R., Fernandes, M., de Zea Bermudez, V., Costa, C.M., 2022. Sustainable lithium-ion battery separators based on poly (3-hydroxybutyrate-co-hydroxyvalerate) pristine and composite electrospun membranes. Energy Technology 10 (2), 2100761.

Bast Fibre Technologies, 2023. Our Hemp Fibre - Bast Fibre Technologies. https://www.bastfibretech.com/sero-hemp. (Accessed 19 July 2023).

Batra, S.K., Pourdeyhimi, B., 2012. Introduction to Nonwovens Technology. DEStech Publications, Inc.

Broda, J., Przybyło, S., Kobiela-Mendrek, K., Biniaś, D., Rom, M., Grzybowska-Pietras, J., Laszczak, R., 2016. Biodegradation of sheep wool geotextiles. International Biodeterioration & Biodegradation 115, 31–38.

Broda, J., Gawlowski, A., Laszczak, R., Mitka, A., Przybylo, S., Grzybowska-Pietras, J., Rom, M., 2017. Application of innovative meandrically arranged geotextiles for the protection of drainage ditches in the clay ground. Geotextiles and Geomembranes 45 (1), 45–53.

Buresh, F.M., 1962. Nonwoven Fabrics (No Title).

Cai, Z., Al Faruque, M.A., Kiziltas, A., Mielewski, D., Naebe, M., 2021. Sustainable lightweight insulation materials from textile-based waste for the automobile industry. Materials 14 (5), 1241.

Cestari, S.P., da Silva Freitas, D.D.F., Rodrigues, D.C., Mendes, L.C., 2019. Recycling processes and issues in natural fiber-reinforced polymer composites. In: Green Composites for Automotive Applications. Woodhead Publishing, pp. 285–299.

Chen, Y., Mueller, D.H., Niessen, K., Muessig, J., 2008. Spunlaced flax/polypropylene nonwoven as auto interior material: mechanical performance. Journal of Industrial Textiles 38 (1), 69–86.

Chen, C., Liu, Y., Wang, Z., Wang, G., Wang, X., 2022. Windmill palm waste fibre used as a sustainable nonwoven mat with acoustic properties. Fibre and Polymers 23 (10), 2960–2969.

Cheng, X.W., Guan, J.P., Tang, R.C., Liu, K.Q., 2016. Phytic acid as a bio-based phosphorus flame retardant for poly (lactic acid) nonwoven fabric. Journal of Cleaner Production 124, 114–119.

Ciccarelli, L., Cloppenburg, F., Ramaswamy, S., Lomov, S.V., Van Vuure, A., Vo Hong, N., Thomas, G., 2020. Sustainable composites: processing of coir fibres and application in hybrid-fibre composites. Journal of Composite Materials 54 (15), 1947–1960.

Claramunt, J., Fernandez-Carrasco, L., Ardanuy, M., 2018. Mechanical performance of flax nonwoven-calcium aluminate cement composites. In: Strain-Hardening Cement-Based Composites: SHCC4, vol 4. Springer Netherlands, pp. 375–382.

Coterie, May 26, 2021. Coterie Introduces Baby Wipes with Lenzing's VEOCEL$^{TM}$ Fibers to US Market. Lenzing – Innovative by Nature. https://www.lenzing.com/newsroom/press-releases/press-release/coterie-introduces-baby-wipes-with-veoceltm-fibers-to-us-market. (Accessed 19 July 2023).

Davis, J., May 7, 2009. Texas Tech Receives Patent for Decontamination Wipe Creation Process. Texas Tech Today. In: https://today.ttu.edu/posts/2009/05/texas-tech-receives-patent-for-decontamination-wipe-creation-process-2. (Accessed 15 July 2023).

Debnath, S., 2017. Sustainable production and application of natural fibre-based nonwoven. In: Sustainable Fibres and Textiles. Woodhead Publishing, pp. 367–391.

Demirel, T., Yavuz, Y., Ureyen, M.E., Koparal, A.S., 2021. Investigation of adsorption behavior of cationic nonwoven textiles as an alternative and environmentally friendly adsorbent to remove the reactive blue 21 dye from a model solution. Water Treatment 242, 304–315.

Deng, W., Li, C., Pan, F., Li, Y., 2019. Efficient oil/water separation by a durable underwater superoleophobic mesh membrane with TiO2 coating via biomineralization. Separation and Purification Technology 222, 35–44.

Dry Wipe Can Clean up Toxic Chemicals, 2008. NBC News. https://www.nbcnews.com/id/wbna28059956. (Accessed 15 July 2023).

Gadkari, R.R., Ali, S.W., Das, A., Alagirusamy, R., 2022. Silver nanowires embedded chitosan/poly-lactic acid electrospun nanocomposite web based nanofibrous multifunctional membrane for safe water purification. Advanced Sustainable Systems 6 (6), 2100360.

Gomez, T.S., Navacerrada, M.A., Díaz, C., Fernández-Morales, P., 2020. Fique fibres as a sustainable material for thermoacoustic conditioning. Applied Acoustics 164, 107240.

Gonzalez-Lopez, L., Claramunt, J., Hsieh, Y.L., Ventura, H., Ardanuy, M., 2020. Surface modification of flax nonwovens for the development of sustainable, high performance, and durable calcium aluminate cement composites. Composites Part B: Engineering 191, 107955.

Gutierrez-Moscardo, O., Canet, M., Gomez-Caturla, J., Lascano, D., Fages, E., Sanchez-Nacher, L., 2022. Sustainable materials with high insulation capacity obtained from wastes from hemp industry processed by wet-laid. Textile Research Journal 92 (7–8), 1098–1112.

Hron, R.J., Hinchliffe, D.J., Cintrón, M.S., Linthicum, K.J., Condon, B.D., 2021. Functional assessment of biodegradable cotton nonwoven substrates permeated with spatial insect repellants for disposable applications. Textile Research Journal 91 (13–14), 1578–1593.

Hu, Y., Ni, R., Lu, Q., Qiu, X., Ma, J., Wang, Y., Zhao, Y., 2022. Functionalized multi-effect air filters with bimodal fibrous structure prepared by direction growth of keratin nanofibre. Separation and Purification Technology 302, 122070.

Hwang, Y.J., Choi, S., Kim, H.S., 2019. Highly flexible all-nonwoven piezoelectric generators based on electrospun poly (vinylidene fluoride). Sensors and Actuators A: Physical 300, 111672.

INDA, 2022. Global Nonwoven Markets Report: A Comprehensive Survey and Outlook, 2020-2025. INDA.

INDA Innovation and Achievement Awards. (n.d.). Available online: https://www.inda.org/awards/. (Accessed 8 July 2023).

Innovation in Textiles, 2019. Global Nonwovens Association Leaders Meet at IDEA. Available online: https://www.innovationintextiles.com/global-nonwovens-association-leaders-meet-at-idea/. (Accessed 8 July 2023).

Islam, S., Mominul Alam, S.M., 2018. Investigation of the acoustic properties of needle punched nonwoven produced of blend with sustainable fibre. International Journal of Clothing Science & Technology 30 (3), 444–458.

Jiang, N., Huang, L., Huang, M., Cai, T., Song, J., Zheng, S., Chen, L., 2021. Electricity generation and pollutants removal of landfill leachate by osmotic microbial fuel cells with different forward osmosis membranes. Sustainable Environment Research 31 (1), 22.

Judawisastra, H., Sukmawati, A., Zaidi, S.Z.J., Harito, C., 2022. Sustainable sound absorber from nonwoven fabric of natural biduri fibre (calotropis gigantea) with polyester binder. Journal of Natural Fibre 19 (16), 12791–12804.

Kahoush, M., Behary, N., Cayla, A., Mutel, B., Guan, J., Nierstrasz, V., 2019. Surface modification of carbon felt by cold remote plasma for glucose oxidase enzyme immobilization. Applied Surface Science 476, 1016–1024.

Kamble, Z., Behera, B.K., 2021. Sustainable hybrid composites reinforced with textile waste for construction and building applications. Construction and Building Materials 284, 122800.

Keirouz, A., Wang, Z., Reddy, V.S., Nagy, Z.K., Vass, P., Buzgo, M., Radacsi, N., 2023. The history of electrospinning: past, present, and future developments. Advanced Materials Technologies, 2201723.

Le Phuong, H.A., Izzati Ayob, N.A., Blanford, C.F., Mohammad Rawi, N.F., Szekely, G., 2019. Nonwoven membrane supports from renewable resources: bamboo fibre reinforced poly (lactic acid) composites. ACS Sustainable Chemistry & Engineering 7 (13), 11885–11893.

Li, X., Tian, X., Yang, T., He, Y., Liu, W., Song, Y., Liu, Z., 2019. Coal liquefaction residues based carbon nanofibre film prepared by electrospinning: an effective approach to coal waste management. ACS Sustainable Chemistry & Engineering 7 (6), 5742–5750.

Lu, H., Butler, J.A., Britten, N.S., Venkatraman, P.D., Rahatekar, S.S., 2021. Natural antimicrobial nano composite fibres manufactured from a combination of alginate and oregano essential oil. Nanomaterials 11 (8), 2062.

Malkan, S.R., 1995. An overview of spunbonding and meltblowing technologies. Tappi Journal 78 (6), 185–190.

Martinez, M.A., Becherucci, M.E., 2022. Study of the potential use of the invasive marine algae Undaria pinnatifida in the preliminary development of a functional textile. Journal of Industrial Textiles 51 (5_Suppl. l), 8127S–8141S.

Morshed, M.N., Behary, N., Bouazizi, N., Guan, J., Chen, G., Nierstrasz, V., 2019a. Surface modification of polyester fabric using plasma-dendrimer for robust immobilization of glucose oxidase enzyme. Scientific Reports 9 (1), 15730.

Morshed, M.N., Bouazizi, N., Behary, N., Guan, J., Nierstrasz, V., 2019b. Stabilization of zero valent iron ($Fe^0$) on plasma/dendrimer functionalized polyester fabrics for Fenton-like removal of hazardous water pollutants. Chemical Engineering Journal 374, 658–673.

Neira, D.S.M., Marinho, G.S., 2009. Nonwoven sisal fibre as thermal insulator material. Journal of Natural Fibre 6 (2), 115–126.

Nikdel, M., Rajabinejad, H., Yaghoubi, H., Mikaeiliagah, E., Cella, M.A., Sadeghianmaryan, A., Ahmadi, A., 2021. Fabrication of cellulosic nonwoven material coated with polyvinyl alcohol and zinc oxide/mesoporous silica nanoparticles for wound dressing purposes with cephalexin delivery. ECS Journal of Solid State Science and Technology 10 (5), 057003.

Ouhaibi, S., Mrajji, O., El Wazna, M., Gounni, A., Belouaggadia, N., Ezzine, M., Cherkaoui, O., 2022. Sisal-fibre based thermal insulation for use in buildings. Advances in Building Energy Research 16 (4), 489–513.

Özen, İ., Okyay, G., Ulaş, A., 2018. Coating of nonwovens with potassium nitrate containing carboxymethyl cellulose for efficient water and fertilizer management. Cellulose 25, 1527–1538.

Pakdel, E., Wang, J., Varley, R., Wang, X., 2021. Recycled carbon fibre nonwoven functionalized with fluorine-free superhydrophobic PDMS/ZIF-8 coating for efficient oil-water separation. Journal of Environmental Chemical Engineering 9 (6), 106329.

Pasternak, G., Ormeno-Cano, N., Rutkowski, P., 2021. Recycled waste polypropylene composite ceramic membranes for extended lifetime of microbial fuel cells. Chemical Engineering Journal 425, 130707.

Paul, R., Brouta-Agnésa, M., Esteve, H., 2012. Chapter 10: Protective insulation materials. In: Bischof, S. (Ed.), Functional Protective Textiles, Published by Faculty of Textile Technology. University of Zagreb, Croatia, pp. 281–303.

Pourmohammadi, A., 2013. Nonwoven materials and joining techniques. In: Joining Textiles. Woodhead Publishing, pp. 565–581.

Prabhakar, M.N., Venakat Chalapathi, K., Atta Ur Rehman, S., Song, J.I., 2021. Effect of a synthesized chitosan flame retardant on the flammability, thermal properties, and mechanical properties of vinyl ester/bamboo nonwoven fibre composites. Cellulose 28, 11625–11643.

Pu, Y., Yang, J., Russell, S.J., Ning, X., 2021. Cotton nonwovens with unidirectional water-transport properties produced by atmospheric plasma deposition. Cellulose 28, 4427–4438.

Ramkumar, S., 2005. U.S. Patent Application No. 10/874,793.

Ramkumar, S., 2022a. Sustainability and Fibrous Substrates. Advanced Textiles Source.

Ramkumar, S., November 22, 2022b. Nonwoven Cotton Wipe Proven Innovation for Decontamination and Security. Cotton Grower. https://www.cottongrower.com/cotton-news/nonwoven-wipe-is-a-proven-innovation-for-decontamination-and-security/. (Accessed 15 July 2023).

Renuka, S., Rengasamy, R.S., Das, D., 2016. Studies on needle-punched natural and polypropylene fibre nonwovens as oil sorbents. Journal of Industrial Textiles 46 (4), 1121–1143.

Rubino, C., Aracil, M.B., Liuzzi, S., Stefanizzi, P., Martellotta, F., 2021. Wool waste used as sustainable nonwoven for building applications. Journal of Cleaner Production 278, 123905.

Sadeghianmaryan, A., Naghieh, S., Salimi, A., Kabiri, K., Cella, M.A., Ahmadi, A., 2022. Fabrication of cellulosic nonwoven-based wound dressings coated with CTAB-loaded double network PAMPS/PNaA hydrogels. Journal of Natural Fibre 19 (16), 12718–12735.

Sheth, J.N., Parvatiyar, A., 2021. Sustainable marketing: market-driving, not market-driven. Journal of Macromarketing 41 (1), 150–165.

Shi, J., Zhang, L., Xiao, P., Huang, Y., Chen, P., Wang, X., Chen, T., 2018. Biodegradable PLA nonwoven fabric with controllable wettability for efficient water purification and photocatalysis degradation. ACS Sustainable Chemistry & Engineering 6 (2), 2445–2452.

Singh, V., Jinka, S., Hake, K., Parameswaran, S., Kendall, R.J., Ramkumar, S., 2014. Novel natural sorbent for oil spill cleanup. Industrial & Engineering Chemistry Research 53 (30), 11954–11961.

Sinha, S.K., Kanagasabapathi, P., Maity, S., 2020. Performance of natural fibre nonwoven for oil sorption from sea water. Tekstilec 63 (1).

Sinnappoo, K., Nayak, R., Thompson, L., Padhye, R., 2020. Application of sustainable phase change materials in motorcycle helmet for heat-stress reduction. The Journal of the Textile Institute 111 (11), 1547–1555.

Smith, P.A., 2000. Technical fabric structures—3. Nonwoven fabrics. In: Handbook of Technical Textiles, 12, p. 130.
Stroe, C.E., Sârbu, T., Manea, V., Burnichi, F., Toma, D.M., Tudora, C.Ă.T.Ă.L.I.N.A., 2021. Study on soil burial biodegradation behaviour on polylactic acid nonwoven material as a replacement for petroleum agricultural plastics. Industria Textila 72 (4), 434—442.
Thielke, M.W., Lopez Guzman, S., Victoria Tafoya, J.P., García Tamayo, E., Castro Herazo, C.I., Hosseinaei, O., Sobrido, A.J., 2022. Full lignin-derived electrospun carbon materials as electrodes for supercapacitors. Frontiers in Materials 9, 859872.
Ulrich, T., Arenas, J.P., 2020. Sound absorption of sustainable polymer nanofibrous thin membranes bonded to a bulk porous material. Sustainability 12 (6), 2361.
Viju, S., Rengasamy, R.S., Thilagavathi, G., Singh, C.J., Mohamed, H.A.K., 2022. Sustainable development of needle punched nonwoven fabrics from silk worm cocoon waste for oil spill removal. Journal of Natural Fibre 19 (11), 4082—4092.
Wang, X., Wang, W., Wang, S., Yang, Y., Li, H., Sun, J., Zhang, S., 2021. Self-intumescent polyelectrolyte for flame retardant poly (lactic acid) nonwovens. Journal of Cleaner Production 282, 124497.
Wen, H., Zhang, J., Chai, J., Ma, J., Yue, L., Dong, T., Cui, G., 2017. Sustainable and superior heat-resistant alginate nonwoven separator of LiNi0. 5Mn1. 5O4/Li batteries operated at 55°C. ACS Applied Materials & Interfaces 9 (4), 3694—3701.
Xu, Y., Rowell, R.M., 2011. Biofibers. In: Sustainable Production of Fuels, Chemicals, and Fibers from Forest Biomass. American Chemical Society, pp. 323—365.
Zeng, Q., Lai, D., Ma, P., Lai, X., Zeng, X., Li, H., 2022. Fabrication of conductive and superhydrophobic poly (lactic acid) nonwoven fabric for human motion detection. Journal of Applied Polymer Science 139 (26), e52453.
Zhang, H., Zhang, L., Han, X., Kuang, L., Hua, D., 2018. Guanidine and amidoxime cofunctionalized polypropylene nonwoven fabric for potential uranium seawater extraction with antifouling property. Industrial & Engineering Chemistry Research 57 (5), 1662—1670.
Zhang, Y., Li, T.T., Shiu, B.C., Sun, F., Ren, H.T., Zhang, X., Lin, J.H., 2021. Eco-friendly versatile protective polyurethane/triclosan coated polylactic acid nonwovens for medical covers application. Journal of Cleaner Production 282, 124455.
Zhao, X., Copenhaver, K., Wang, L., Korey, M., Gardner, D.J., Li, K., Ozcan, S., 2022. Recycling of natural fiber composites: challenges and opportunities. Resources, Conservation and Recycling 177, 105962.

# Innovations in dyes and chemoinformatics approach

*Felix Y. Telegin* [1], *Jayesh V. Malanker* [2], *Jianhua Ran* [3] *and Nagaiyan Sekar* [4]
[1]G.A. Krestov Institute of Solution Chemistry, Russian Academy of Sciences, Ivanovo, Russia; [2]Malanker Enterprises, Bilimora, Gujarat, India; [3]State Key Laboratory of New Textile Materials and Advanced Processing Technologies, Wuhan Textile University, Wuhan, China; [4]Department of Speciality Chemicals Technology, Institute of Chemical Technology, Mumbai, Maharashtra, India

## 9.1 Introduction

In general, the textile industry relies heavily on several non-renewable resources including petroleum to produce synthetic fibres, fertilisers and pesticides to grow natural fibres such as cotton, and chemicals to produce dyes and finishes. Textiles are considered as the fourth highest pressure category for the use of primary raw materials and water, after food, housing and transport, and fifth for greenhouse gas (GHG) emissions (EU Circular Economy Action Plan). The textile industry has a significant impact on environment and is regarded as the second largest polluter of clean water (Brigden et al., 2012).

Due to the growing demand for clothing exacerbated by the proliferation of the fast fashion business model, a consistent throughput of natural resources is needed. With its low rates of use and low levels of recycling, the current linear system is the main cause of the massive and ever-expanding pressure on natural resources. The immense footprint of the industry extends beyond the use of raw materials. If the textile and fashion industry continues on its current path, by 2050 it could use more than 26% of the carbon budget associated with a 2°C global warming limit (Ellen Mac Arthur Foundation, 2017). Being one of the most polluting industries, textiles and clothing have a detrimental ecological footprint, caused by high energy, water and chemical use, the generation of textile waste, and coloured and chemical-loaded effluents (Niinimäki et al., 2020).

The textile colouration industry is one of the most polluting businesses in the world. Even though the industry has taken significant steps in developing sustainable manufacturing processes and materials (Nieminen et al., 2007), a large quantity of harmful and toxic chemicals is still used in the synthetic dye manufacture, and the coloured effluents from dyeing with synthetic colourants remains a serious environmental concern (Singh and Arora, 2011). Dyes are the most important additives of the textile fibres since the buying decision of a fabric is strongly affected by its visual appearance (Demirbilek and Sener, 2003).

**Sustainable Innovations in the Textile Industry. https://doi.org/10.1016/B978-0-323-90392-9.00008-2**

## 9.2 Historical aspects

Natural dyes have been used since time immemorial to colour the body, food, cave walls, textiles, leather, and everyday objects. The earliest known use of a dye was by Neanderthal man about 180,000 years ago, where they used red ochre (iron oxide), an inorganic pigment obtained from river beds. The first known use of an organic colourant was much later, around 4000 years ago, when the blue dye indigo was found in the wrappings of mummies in Egyptian tombs (Gordon and Gregory, 1983).

A large number of plant, animal, insect and mineral sources have been identified for the extraction of dyes and pigments. Dyed textiles found during archaeological excavations in different locations around the world provide evidence of the practice of dyeing in ancient civilisations. Plant-based dyes such as woad (*Isatis tinctoria*), indigo (*Indigofera tinctoria*), saffron and madder were grown commercially and were important trade goods in the economies of Asia, Africa and Europe.

As there is no universal classification, natural dyes are sorted in many different ways. It is a common procedure to group them in accordance with some specific characteristic, namely: origin, chemical constitution, application method, hue, etc. Another way to classify natural dyes is based upon their natural affinity to the fibre (Adeel et al., 2020; Gupta, 2020). Mordants are added to increase the rub fastness, washing fastness and light fastness of dyed goods, and traditionally metallic salts are widely used as they develop brilliant colours and enhance fastness properties due to metal complex formation (İşmal and Yıldırım, 2019; Saxena and Raja, 2014).

Since W.H. Perkin accidentally invented the synthetic colourant 'mauve' in 1856, synthetic dyes have been extensively employed in the colouring business for more than 160 years, while the use of natural dyes has fallen precipitously throughout the world. Currently, it is difficult to envision a time before bright, quick, affordable synthetic dyes and pigments that are of high quality were available. Even though it was only a few decades ago that textile industries turned to synthetic dyes, they were so successful that natural dyes now make up only about 1% of the total number of dyes used worldwide (Agarwal and Tiwari, 1989).

## 9.3 Impacts of dyes on health and the environment

The textile sector as a whole bears a significant share of the harmful impact on human health and the environment. Some of the dyes, along with a large number of industrial pollutants, are highly toxic and potentially carcinogenic (Sharma et al., 2018). Dyes have been shown to be hazardous to human health, the environment, and animals. As the discharge of azo dyes into waterways is hazardous to humans and the environment, a few synthetic dyes have been evaluated for potential toxicity. Research findings have revealed that these dyes are hazardous to a wide range of creatures, including aquatic animals (Young and Yu, 1997; Al-Amrani et al., 2022).

The most common hazard of powder dyes is respiratory problems due to the inhalation of dye particles, which can sometimes affect a person's immune system. In general, the diseases caused by textile dyes range from dermatitis to disorders of the central nervous system (Khan and Malik, 2018). The acute toxicity of textile dyes is caused by oral ingestion and inhalation, especially by exposure to dust (Clark, 2011), triggering irritation to the skin and eyes (Christie, 2007). The workers who produce or handle reactive dyes may be exposed to dermatitis, allergic conjunctivitis, rhinitis, occupational asthma or other allergic reactions. Some dyes reveal mutagenic potential also (Hunger, 2003).

Dyes exhibit high solubility in water, making it difficult to remove them by conventional methods (Hassan and Carr, 2018). The tendency to be recalcitrant in aerobic environments, especially in conventional treatment plants is responsible for bioaccumulating dyes in sediments and soil and transporting them to public water supply systems. Another risk involves combining dyes with intermediate synthetic compounds or their degradation products to generate other mutagenic and carcinogenic substances (Vikrant et al., 2018). The consequences of the xenobiotic and recalcitrant nature of dyes end up impacting the structure and functioning of ecosystems (Rawat et al., 2016).

## 9.4 Sustainable innovations in dyes

### *9.4.1 Natural dyes*

Natural dyes can exhibit better biodegradability and generally have greater compatibility with the environment. However, the limited range of natural colourants often fails in the desired colour intensity and light stability and they are not provided at an affordable cost (Cardon, 2007; Vankar, 2016). Currently, there is an increasing search and demand for renewable and sustainable resources and this includes natural dyes. Some types of natural dyes are already known and others remain to be discovered. In addition, there is concern about the toxicity of synthetic dyes, especially those used in the food industry (Freitas da Silva et al., 2010). The inherent antimicrobial activity of several natural dyed textiles is also known (Bautista et al., 2007; Singh et al., 2005). Other advantages include that most natural dyes have low toxicity, being widely used, even in food, and there are several health benefits (Cardarelli et al., 2008; Freitas da Silva et al., 2010).

The main challenge is the quality issues of natural dyes in textile colouration, especially the low colour intensity, thermal degradation and the low light fastness. The colours in textiles can degrade and fade over the course of time due to exposure to UV light and oxidation. As a solution to these problems, several mordants, which can bind both the fibre and natural dyes together by forming a complex, are traditionally used. Mordants based on heavy metals are actually creating a new problem, heavy metal pollution, instead of ecological dyeing with natural dyes (Manian et al., 2016), and the emergence of a new classification of natural mordants is inevitable. In one study, dyeing of ecru denim with onion extract using different natural mordants individually and in combinations was carried out (Deo and Paul, 2000, 2003). In another study, the wastes of sugarcane bagasse, wheat bran and rice husk were used for dye and mordant extraction (Nazir et al., 2022). Tartaric acid, tannic acid and several natural products are considered as natural mordants.

#### *9.4.1.1 Novel natural dyes*

Natural dyes derived from biodegradable plant sources may be a feasible substitute for synthetic colorants for usage in niche markets and non-textile spaces. Waste streams that are derived from agriculture or industry could be considered for the extraction of valuable components. One of the most sustainable sources of natural colorants could be by-products from the fruit and vegetable industries (Phan et al., 2021). Several natural dyes for textile dyeing can be extracted from food and beverage industry waste such as pressed berries, pressed grapes, waste and peels from vegetable processing. The dyes were extracted with boiling water and dyeing on wool yarn was performed (Bechtold et al., 2006).

Biotechnological methods of the production of indigo and quinonoid dyes are widely talked about in the literature. BioPolyCol is a project funded by the Bioeconomy International 2021 Call, Germany, with the main objective being to develop sustainable natural dyes with good colour intensity and high fastness for industrial dyeing of biopolymers. It aims to apply the untapped potential of Amazonian natural dyes derived from renewable plant resources, secondary crops and food industry residues from the Amazon region, without harming the environment (BioPolyCol, 2022). Colorifix is a UK-based company that sources and replicates true colours from nature, which are biologically produced. In contrast to conventional dyeing, involving chemicals and mordants, the engineered microorganisms of Colorifix, however, are able to concentrate the nutrient salts and metals that are already present in water to levels that facilitate this dye—fabric interaction with no added substances (ColoriFix, 2022).

#### *9.4.1.2 Algal dyes*

Currently, there is interest in the untapped potential of algal colours. Common algal pigments include phycoerythrin, phycocyanin, beta-carotene, chlorophyll and fucoxanthin (Raposo et al., 2013). The marine red alga *Porphyridium* spp. is known for its ability to synthesise a wide array of high-value compounds such as health-promoting pigments and poly-unsaturated fatty acids (PUFAs), as well as bioactive and thickening exopolysaccharides (EPS) (Dyes and Colorants from Algae). The *Porphyridium* biomass contains four major pigments under physiological growth conditions, namely the green chlorophyll *a*, red phycoerythin, and the blue phycocyanin and allophycocyanin (Kathiresan et al., 2007). Spira has recently developed an electric sky colour using spirulina algae containing phycocyanin and chlorophyll giving its blue-green tint, which has great potential as it is non-toxic and bright in colour (Sendilkumar, 2021). Greencolor is a Spanish project aiming to substitute synthetic dyes with algae-based alternatives in order to create more sustainable dyeing and printing processes which have lower socio-economic and environmental impacts (AITEX, 2020).

#### 9.4.1.3 Fungal pigments

Fungal pigments show excellent dyeability and colour fastness, and they are of great interest to the sustainable textile industry. They warrant production under controlled conditions, with no seasonal fluctuations, and are biodegradable (Mapari et al., 2010). Filamentous fungi produce pigments such as carotenoids, melanins, flavins, phenazines, quinones and monascins from different chemical classes (Dufossé et al., 2014). Most fungi produce pigments that are water-soluble and ideal for industrial production since they are easy to scale up in industrial fermenters and could be extracted easily without organic solvents on an industrial scale. Fungal pigments, due to their stability and consistency, have been reported for their use as sustainable alternatives to synthetic dyes in the textile industry (Robinson et al., 2012). In the FunColor project, funded in the Innovationsraum: BioTexFuture by BMBF, Germany, fungal pyomelanin is developed to replace the synthetic grey, brown and black dyes for textiles.

#### 9.4.1.4 Microbial pigments

A number of microorganisms such as *Rhodotorula*, *Sarcina*, *Cryptococcus*, *Monascus purpureus*, *Phaffia rhodozyma* and *Bacillus* sp. are known to produce pigments (Joshi et al., 2011). Bacterial secondary metabolites can be used as dyes for textile dyeing. Several bacterial strains are able to produce pigments or dyes capable of imparting colour to textile materials. Fermentation is a suitable process for bacterial pigment production. Bacterial pigment production is highly affected by the incubating temperature and pH (Mazotto et al., 2021).

Bacterial pigments usually have bright colours and also some special properties such as antimicrobial, antioxidant, UV protective, etc. These properties indicate that these pigments could be used as functional dyes for different textile materials, as they offer a range of potential applications in addition to their colouration properties (Kramar and Kostic, 2022).

### 9.4.2 Synthetic dyes

Numerous synthetic dyes are extensively used in the textile colouration industry, which have negative impacts on the environment and are particularly linked to harmful effects on human health (Shahid et al., 2013). The main sources of this problem are the use of hazardous chemicals to fix colourants on textiles and the use of non-biodegradable petroleum-based dyes for textiles, as well as the large environmental leakage of these dyes and fixatives. As a result, natural dyes have returned, at least to some extent, into fashion as a suitable co-partner or alternative to synthetic colours (Mirjalili et al., 2011; Yusuf et al., 2015).

Given the size of the textile colouration business, natural dyes cannot develop into a viable alternative until the economy is transformed into one that is mostly based on agriculture. Even with an understanding of agriculture-based taxonomy, one still needs to put a lot of effort into sustainable isolation methods, colour consistency and standardisation. The textile and apparel industry as a whole is currently looking for sustainable colouring techniques. The colourant industry could survive if the objectives listed below are met:

- Colours that can completely absorb into or interact with textile materials are developed
- High tinctorial strength dyes are created to reduce the amount of dye needed
- Dyes are produced using environmentally friendly techniques, such as biotechnological methods
- Dyes that can be quickly biodegrade
- Dyes that are applicable in digital printing
- Dyes that do not participate in any photochemical process that could harm the skin

Around the world, these areas are the subjects of direct and/or indirect colour chemistry research from different perspectives. The developments that are likely to take this to the sustainable growth level are discussed further.

#### 9.4.2.1 Diazotisation and coupling reaction

The most common type of dyes manufactured are azo dyes. Despite the fact that there are other techniques, diazotisation and coupling processes are most often employed in the manufacture of azo dyes (Merino, 2011). Although this sector has seen some notable advancements, organic chemists may find it helpful to look for mild conditions for azo dye preparations. The diazotisation–azo coupling reaction has been the main focus of the most significant techniques for making azo dyes. Their incompatibility with the environment is the fundamental constraint of this procedure. Since using potent liquid acids for diazotisation reactions – such as sulphuric acid and hydrochloric acid – damages the environment permanently, their use runs counter to sustainable chemistry principles. These liquid acids also need to undergo special neutralisation processing, which includes expensive and ineffective catalyst separation from homogeneous reaction mixtures and ultimately produces non-recyclable waste. The use of recyclable strong solid acids as substitutes for non-recyclable liquid acid catalysts has been encouraged by the need for a 'sustainable' approach to chemical processing (Clark, 2002; Okuhara, 2002; Su, 2013; Wilson and Clark, 2000).

There are growing economic, environmental and safety problems about weakly basic amines that are diazotised using sodium nitrite in an industrial process. This process uses excessive concentrations of concentrated sulphuric acid as a solvent and generates effluent with a high acid content. When sodium nitrite solution is added, concentrated sulphuric acid produces a strong dilution heat effect that calls for effective cooling of the reaction mixture to regulate the process temperature. Due to the aforementioned, it is difficult to prepare aryl diazonium salts with two or three potent electron-withdrawing functional groups, which are crucial for producing deep shade dispersing azo dyes with excellent colour fastnesses. The use of counterions, such as tetrafluoroborate, hexafluorophosphates or disulphonimide, can significantly boost the stability of diazonium salts (Roglans et al., 2006).

Tang and colleagues reported that diazotisation by tertbutyl nitrite in the presence of excessive quantities of 1,5-naphthalenedisulphonic acid as a proton donor and stabiliser of diazonium salts allowed them to synthesise stable solid diazonium salts of weakly basic amines. The obtained compounds had mixes of mono- and diaryldiazonium 1,5-naphthalenedisulphonates in a ratio that could not be predicted or controlled, while having an outstanding stability profile. This approach is less appealing due to the relatively high cost of 1,5-naphthalenesulphonic acid and the unclear final composition, despite the mild reaction conditions used and the use of ethyl acetate as the reaction solvent at room temperature (Qiu et al., 2017).

Arenediazonium tosylates are reported to have excellent high thermal stability. The developed method uses less expensive p-toluenesulphonic acid to diazotise anilines, but it also uses polymer supported nitrite (Filimonov et al., 2008). In response to the high demand for a sustainable method of producing isolable and non-hazardous arenediazonium salts, the two methods described above have been combined to create a truly practical protocol for the synthesis of arenediazonium tosylates that takes place in a green reaction solvent and at room temperature using commercially available reagents (Khaligh, 2020; Li and Jia, 2018; Mihelač et al., 2021; Oger et al., 2014). Another novel technique is using nanomagnetic supported sulphonic acid for the diazotisation and diazo coupling process of aromatic amines. After the reaction is finished, the catalyst may be easily magnetically separated, which removes the need for catalyst filtration (Koukabi et al., 2016).

#### *9.4.2.2 Completely absorbing dyes*

In order to have dyes that can be completely absorbed into or interact with textile materials, reactive dyes with high fixing potential need to be produced. Heterocyclic disazo reactive dyes that have been synthesised, demonstrated good substantivity due to their reactive vinylsulphone derivatives and decreased degree of sulphonation. Higher exhaustion and overall fixation yield were consequently attained on all textiles (Youssef et al., 2014) (Scheme 9.1).

In theory, any dyestuff with a primary aromatic amino group can be converted into a reactive dye for cellulosic fibres using a unique reactive system based on diazonium ions. The dyeing procedure is quick and efficient. These dyes will be less expensive to produce and use less raw materials (Bhate et al., 2017). It has been found that using Lanasol or Remazol reactive dyes on wool in a medium of sea water gives better results in terms of better exhaustion (Broadbent et al., 2018).

As the typical water-based dyeing systems require a lot of water and energy and generate a lot of waste, which raises the cost of treatment, they have a negative

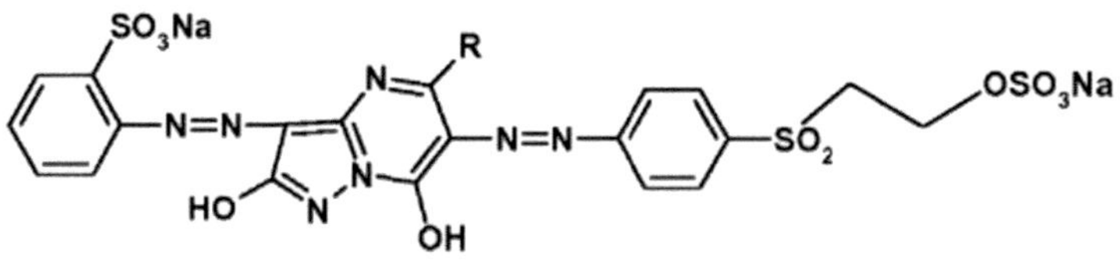

1, R = CH3
2, R = NH2

**Scheme 9.1** Heterocyclic disazo reactive dyes.

influence on the environment and ecology. As an alternative, a number of non-nucleophilic solvent assisted dyeing systems are developed (Tang and Kan, 2020). Quantitative fixation is a well-known property of polymeric reactive dyes. As far as the application characteristics are concerned, they may be sustainable (Lv et al., 2017; Mao et al., 2018; Shan et al., 2020). Reactive dyeing of silk using the Mannich method has been shown to be superior, with good exhaustion (Guo et al., 2020). Higher fixation efficiencies from carboxylate-activated reactive dyes have the potential to enhance reactive dyeing procedures (Lewis et al., 2022).

### 9.4.2.3 High tinctorial strength dyes

Several high tinctorial strength dyes are created to reduce the amount of dye needed. Because chromophores are present in azomethine dyes' structures, these dyes typically exhibit spectacular colour and brightness qualities. When colouring blend materials, azomethine dichloro-s-triazinyl reactive disperse dyes with antibacterial and UV protection qualities exhibit outstanding exhaustion and fixation (Ibrahim et al., 2021) (Scheme 9.2).

Fluorescence and dye absorptivity are frequently linked to brightness. Novel fluorescent heterocyclic acid dyes have been prepared by diazotising various sulphonic acid-based amines and coupling with 4-hydroxyl-1-methyl-2(1H)-quinolone (Shinde and Sekar, 2019) (Scheme 9.3). An effective technique for creating high-performance acid dyestuffs is to introduce rigid quinoidal heterocyclic structures with high coplanarity as the chromophore backbone (Cai et al., 2020, 2021). Novel multifunctional dyes with antibacterial and UV protection capabilities that included a benzophenone as a core component have also been created (Bait et al., 2020).

**Scheme 9.2** Azomethine dichloro-s-triazinyl dyes.

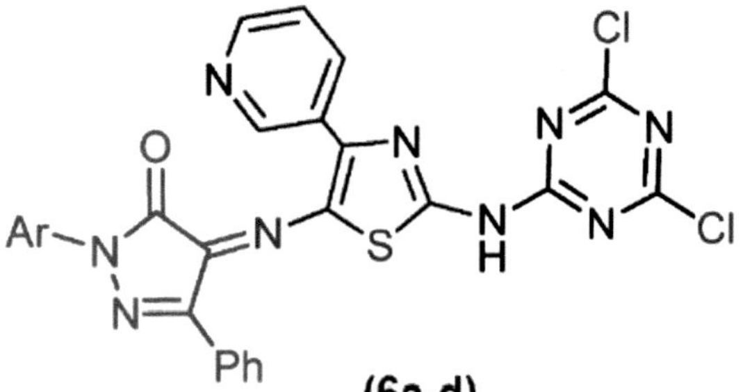

**(6a-d)**

6a; Ar = $C_6H_4(Cl)$-2
6b; Ar = $C_6H_4(NO_2)$-4
6c; Ar = $C_6H_3(NO_2)$-2,4
6d; Ar = $C_6H_2(Cl)$-2,4,6

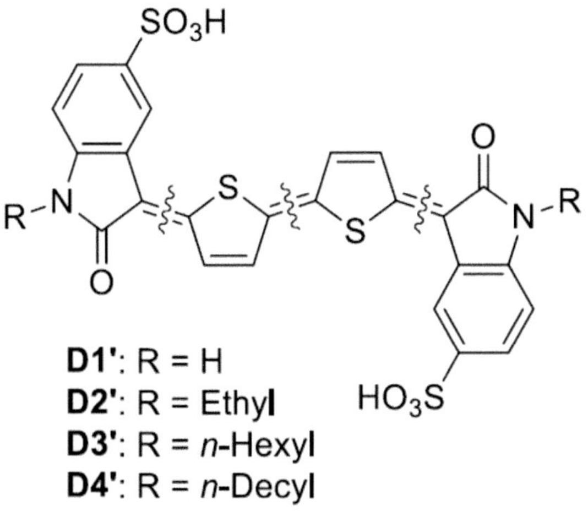

$SO_3H$ $HO_3S$

**D1'**: R = H
**D2'**: R = Ethyl
**D3'**: R = *n*-Hexyl
**D4'**: R = *n*-Decyl

**Scheme 9.3** Fluorescent heterocyclic acid dyes.

### *9.4.2.4 Dyes for digital printing*

The dyeing industry uses a lot of energy and creates huge pollution. The colouration business, which needs numerous dyeing auxiliary chemicals, large amounts of industrial water and high thermal energy, must build anti-pollution facilities to combat severe air and water pollution. A sustainable solution is the conversion of traditional colouring to digital textile printing (DTP). The practise of printing a pattern or an image directly on fabric using graphic design is known as digital textile printing. Depending on the type of textiles, water-soluble and solvent soluble inks for DTP are applied in various ways. While it is possible to evaluate existing dyes for DTP compatibility, new dyes are also developed specifically for this function (Dawson and Hawkyard, 2000).

Water-soluble inks are currently becoming more popular because of their broad colour spectrum, high colour intensity and great jetting stability. High solubility, high thermal and photo stability, high colour purity, high productivity and good dyeing qualities are the primary requirements for DTP inks, and hence for the dyes used in such inks. DTP inks in particular should be continuously released into the fabric depending on the electric signals of the printer head without any clogging. They are made with colourants, vehicles and additives (Park et al., 2012). Perylene-based acid dyes are produced with great thermal, photo and tinctorial stability for use as ink colourants in high-speed DTP (Choi et al., 2019).

### *9.4.2.5 Functional dyes for one-step dyeing and finishing*

Dyeing with functional dyes with incorporated properties such as antimicrobial, antifungal, UV protection, hydrophobicity, mosquito repellence, etc. can result in one-step dyeing and finishing. Such simultaneous dyeing and functional finishing is an innovative step towards achieving sustainability in the most polluting sectors of the textile industry. An antimicrobial reactive dye was prepared by reacting the dye with 4-

amino-2,2,6,6-tetramethylpiperidine (Jiang et al., 2017). When a reactive dye was reacted with chitosan, the dye showed very efficient antibacterial activity due to the presence of chitosan in the dye structure (Tang et al., 2016). Quinazolinone-based mono-chlorotriazine reactive dye showed both antibacterial and antifungal activities (Patel and Patel, 2015). A novel antibacterial disperse dye based on the combination of two antibacterial moieties, 4-hydroxy coumarin and 4-(1,3-dimethyl-2-phenyl-3-oxo pyrazolyl), has good potential in developing antibacterial polyester textiles (Sahoo et al., 2015). Pyrazolopyrimidine azo disperse dyes can also develop good antibacterial properties in dyed textiles (Khattab et al., 2017).

UV-protective acid dyes for silk were synthesised by incorporating 2-hydroxy-4-methoxybenzophenone, a UV absorber (Wang et al., 2011). Benzophenone-based disperse dye imparted excellent UV protection to polyamide and polyester fabrics (Shinde et al., 2021). A natural resource, areca nut was utilised for the preparation of UV-protective azo disperse dye (Pawar et al., 2018). Functional anthraquinone-based dyes are synthesised for the development of superhydrophobic cotton (Salabert et al., 2015). Mosquito-repellent polyester was developed by dyeing with functional 4-hydroxy coumarin-based disperse dye (Singh and Sheikh, 2022). Multifunctional azoic dyes were prepared by in situ azo coupling, using modified DEET (DEET-NH2) for imparting mosquito repellency and antibacterial activity to cotton (Teli and Chavan, 2018). Thus the incorporation of different functional molecules in the dye structures can lead to one-step dyeing and multifunctional finishing with good fastness properties.

### 9.4.3 Structural colours

As it is difficult for synthetic dyes to meet all the sustainability requirements, a new type of colouration is highly desired. In nature, there are three major sources of colour: pigments, structural colours and bioluminescence. Structural colour is the colour produced by micro-/nano-structures, and is bright and lively (Sun et al., 2013). It is generated by the interaction between the micro-/nano-structures of the material surface and the incident light, which is closely related to the sizes and periods of the structures. Based on different nano arrangements, the structures can be roughly divided into photonic crystals and non-photonic crystals (Li et al., 2023; Fenzl et al., 2014). Under the illumination of white light, light of a specific colour can be scattered on the surface of materials (Dumanli and Savin, 2016).

The most common mechanisms of structural colours are film interference, diffraction grating, scattering and photonic crystals. Biological colours are mainly derived from film interference, which includes thin-film and multi-film interference. The diffraction grating mechanism is found in the mollusc *Haliotis glabra* and the *Hibiscus trionum* flower. Coherent scattering includes colours produced by brilliant iridescent butterfly wing scales and feather barbules, such as the peacock's tail feathers. Some examples of colours produced by photonic crystal structures include opal in beetles and iridescent spines in the sea mouse (Sun et al., 2013).

#### 9.4.3.1 Structural colouration of textiles

Compared with conventional textile colouration with dyes and pigments, structural coloured fabrics have attracted broad attention due to their advantages of eco-friendliness, brilliant colours and anti-fading properties (Han et al., 2021). Structural colouration of textiles is possible by thin-film deposition, self-assembly, printing and micro-/nano-structuring.

Magnetron sputtering technology is commonly used for thin-film deposition. Metal or non-metal films can be deposited on textiles such as polyester, cotton, etc. as long as the appropriate sputtering process is selected (Liu et al., 2011). Self-assembly is the process of realising structural sequences of different scales without any direct external effects, and colloidal self-assembly is the most common method. Under gravity, colloidal microspheres can be deposited on the textile surfaces, forming periodic photonic crystal structures spontaneously after the volatilisation of solution, which is called the gravitational sedimentation method (Li et al., 2019). Polydopamine (PDA)-based 3D colloidal photonic balls and fibres have been developed, where the structural colour was bright for the shell of PDA. This could absorb light and so is suited for the development of structural coloured inks (Kohri et al., 2018).

Printing methods are carried out mainly through screen printing, spray guns or atomisers. Large-scale production is not an issue for screen printing. Vivid non-iridescent structural colours with good mechanical stability on white fabric have been developed by a rapid screen printing method. The multicolour pattern output was feasible as long as the printing process was repeated. Colours obtained were highly visible on different substrates. Furthermore, washing and rubbing tests proved that the structural and mechanical stability were enhanced to a great extent because of the addition of polyacrylate (Zhou et al., 2020). Multiple colour outputs could be obtained on the fabrics at the same time, so that fascinating patterns could be printed. The colours of the fabrics could be adjusted by changing the diameters of the microspheres. Due to the adhesive capacity of polyacrylate, 3D colloidal crystals were closely combined with yarn, and even after repeated washing and friction, the colouring film retained good integrity. This method has been proved to have good prospects for the colouration industry (Meng et al., 2017).

In order to create micro-/nano-structuring on fibres, micro-/nano rough surfaces are either added to fibres, or a porous amorphous structure is kept on or inside them. These two methods control the physical interaction between periodic structures and light by combining absorption, reflection, refraction, transmission, interference, scattering and diffraction to obtain the structural colours. Various techniques have been developed, including moulding and embossing, mask deposition, optical lithography, electron beam lithography and scanning probe technology, for patterning on nanometre to millimetre length scales (Li et al., 2023).

The most investigated structural colour on fabrics originated from a band gap of multilayered photonic crystals or amorphous photonic structures. However, limited by the nature of the colour generation mechanism and a multilayered structure, it is challenging to achieve structural coloured fabrics with brilliant non-iridescent colours and high fastness (Han et al., 2021). Non-iridescent structural colour coatings were successfully constructed on silk by combining an amorphous photonic structure (APS) with polymers as adhesives (Meng et al., 2019).

In another study, structural coloured fabrics with vivid non-iridescent bright colours and high colour fastness were fabricated by spray coating $Cu_2O$ spheres on a fabric to form disordered $Cu_2O$ thin layers. The colour of the fabric can be simply tuned by changing the particle size of $Cu_2O$ spheres. According to colour fastness test standards, the rubbing and light fastness of the structural colour on fabric can reach excellent levels that can withstand rubbing, photobleaching, washing, rinsing, kneading, stretching and other external mechanical forces. This colouring method has very good potential for textile colouration (Han et al., 2021).

## 9.5 Chemoinformatics approach for sustainable innovations

Modern aspects of sustainable innovation in dyes for textile materials are based on the design of new chemical structures of dyes that satisfy several requirements including dye affinity, fastness of dyeing to physicochemical treatments, colour strength and the spectral properties of dyes. Besides this, at the present time, there are some new requirements such as chemical, photochemical and biochemical elimination as well as toxic properties of dyes and their carcinogenic activity.

The study of structure–property relationships was essential for textile chemistry research in previous years. This was based on a detailed study of physicochemical aspects of textile colouration and properties of dyes adsorbed by the fibres (Burkinshaw, 2016; Burkinshaw, 2021a,b; Giles et al., 1978; Johnson, 1989; Peters, 1975; Shore, 2002; Telegin, 2009; Vickerstaff, 1954). Currently, QSPR/QSAR (quantitative structure–property/activity relationships) research and chemoinformatics analysis cover almost every fundamental and applied field in chemical studies.

Presently, 27,000 individual products under 13,000 generic names are incorporated in the Colour Index (SDC, AATCC). Much information regarding the properties of commercial textile dyes for all technical groups of dyes is provided on the World Dye Variety website (World Dye Variety, 2022). Water-soluble dyes include about 50% of the total amount of dyes. The Max Weaver Dye Library (Kuenemann et al., 2017) at Eastman Kodak Company represents a collection of 98,000 vials of custom-made and largely sparingly water-soluble dyes. As a part of this collection temporary, water-soluble, hair dyes are collected and analysed in the research (Williams et al., 2018a,b). Some results of recent research on the bioelimination of large groups of commercial acid, direct and reactive dyes (Churchley et al., 2000, a,b) are suitable for discussion of dye affinity to cellulose. Chemometric analysis of different classes of dyes was performed in a series of research: acid dyes for silk (Giorgi et al., 1993, 1994; Giorgi and Carpignano, 1996), acid dyes for wool and nylon (Giorgi et al., 1997) and disperse dyes for synthetic fibres (Giorgi et al., 1998; Kats et al., 1988; Kats and Krichevskii, 1979). One of the early examples of software for QSPR analysis was SPARC software which is widely used up to now for predicting ionisation constants pKa and hydrophobicity LogP of organic compounds and, as an example, azo dyes and related compounds (Hilal et al., 1994).

The problem of QSPR analysis of dye affinity to textile fibres and physicochemical properties of dyeing became a key point of the development of new ideas on the eve of 2000. The fundamental contribution to the problem had been done by the research of S. Timofei and co-authors (Timofei et al., 2000). The review cited above makes reference to several papers of their research published since 1994 based on the application of the comparative molecular field analysis method. The beginning of their research started directly with vat, disperse, acid and direct dyes. Later, the authors continued modification of the method for acid dyes in more recent research (Funar-Timofei et al., 2012). Their results and a database for anionic dyes are used in the research (Polanski et al., 2003, 2004; Wang et al., 2014; Zhokhova et al., 2005). Recent studies (Kumar and Kumar, 2020; Yu et al., 2018) continue the development of the chemoinformatics approach based on the experimental data for the affinity of anionic dyes to cellulose fibres collected in the research (Funar-Timofei et al., 2012; Timofei et al., 2000) mentioned above.

A wide variety of empirical properties of water-soluble dyes are covered by chemoinformatics research, for example, tinctorial properties of acid dyes on cellulose fibres in domestic washing of mixed cellulose/polyamide/wool materials (Oakes and Dixon, 2003a), photodegradation (Zhang and Zhang, 2020), and catalytic elimination (Xu et al., 2016) of dyes in the wastewater, behaviour in the advanced oxidation process (Awfa et al., 2021; Li et al., 2013), ecotoxicity of dyes (Funar-Timofei and Ilia Gheorghe, 2020; Roy, 2020; Umbuzeiro et al., 2019), etc.

Recent advances in QSPR methods have been analysed in reviews (Cherkasov et al., 2014; Katritzky et al., 1997; Muratov et al., 2020). Particular problems of chemoinformatics of dyes are also reviewed (Luan et al., 2013; Toropova et al., 2022; Yu and Wang, 2021). The analysis of the chemical structure and property relationships usually applies multiple linear regression models or neuro-models based on chemoinformatics software applied for structure—property analysis of chromophores and other organic compounds:

- SPARC, by L. A. Carreira et al., ARChem, USA, 1994 (Hilal et al., 2003)
- CODESSA, by A. R. Katritzky, M. Karelson, R. Petrukhin, University of Florida USA, 2001—5 (CODESSA PRO: COmprehensive DEscriptors for Structural and Statistical Analysis, 2001—2005)
- DRAGON, by Kode Chemoinformatics, R. Todecini et al., Pisa, Italy, 1994 (DRAGON, 1994)
- NASAWIN, by I. I. Baskin et al., Moscow State University, Russia, 1995 (Baskin et al., 1999; Baskin et al., 1997a,b; Baskin and Varnek, 2008)
- ISIDA, by Solov'ev, Frumkin Institute of Physical Chemistry, RAS, 2000 Russia (Solov'ev)
- CORAL, Mario Negri Institute, E. Benfenati, A. A. Toropov, A. P. Toropova, Italy, 2010 (CORAL software/databases, 2010—2018)
- OCHEM, I. I. Tetko et al., International project, 2011 (Sushko et al., 2011)

Different kinds of descriptors are represented by physicochemical parameters, functional groups, topology and geometry of the molecules, fragments of a different kind, etc. These descriptors are collected and the databases are arranged through the use of chemoinformatics software, for example, ChemAxon (ChemAxon).

One of the advanced approaches realised in NASAWIN and ISIDA based on fragment descriptors generated by the use of defragmentation of the molecule on substructures of chains, branches, and cycles of atoms was applied in the research (Baskin et al., 1999; Baskin et al., 1997a,b; Baskin and Varnek, 2008; Marcou et al., 2012; Solov'ev et al., 2000; Telegin and Koifman, 2016; Telegin and Marfin, 2022) for diverse organic compounds. This approach was applied for the structure and properties analysis of textile dyes (Ran et al., 2022; Telegin et al., 2016; Telegin et al., 2016; Zhokhova et al., 2005).

Another modern approach to chemoinformatics analysis CORAL explores SMILES-based (simplified molecular-input line system) representation of the molecules typically used in chemoinformatics software mentioned above. Recent publications of such an analysis cover different problems of chemistry, for example Nesmerak et al. (2013) and Toropov et al. (2020), as well as the chemistry of dyes (Toropova et al., 2022).

In contrast to the above approaches for chemoinformatics, one of the simple methods is based on the application of a correlation analysis of technical properties of dyes with calculated physicochemical parameters reflecting correspondent quantities in a series of dyes. Systematic use of such an approach performed in the research (Telegin, 2009; Telegin et al., 2013) for discussion of the dye affinity, wash fastness, and light fastness is similar to the classical application of Hammett substituent constants in dye chemistry.

### 9.5.1 Simple correlation analysis of technical properties of dyes

The general approach of a simple correlation analysis is based on the theoretical evaluation of amphiphilic and electrophilic properties of dyes by the use of empirical methods of molecular modelling of organic molecules realised in chemical software JChem for Office (ChemAxon) or other similar programs. Amphiphilic properties are characterised by the log of partition coefficient of a substance between immiscible solvents, such as water and n-octanol, LogP. Electronic properties are characterised by HOMO and LUMO energies, respectively, the highest occupied and lowest unoccupied molecular orbitals. According to the physicochemical meaning, LogP value is proportional to the thermodynamic energy of dye transfer from one solution to another. Correspondent values of HOMO and LUMO energies are equivalent to an energy of ionisation and affinity to the electron of the dye molecule. In other words, HOMO energy characterises the ability of dyes for accepting electrons in oxidation reactions and LUMO energy characterise the ability of dyes to donate electrons in reduction reactions.

### 9.5.2 Wash and light fastness of disperse dyes on synthetic fibres

Disperse dyes are the most suitable class of dyes for such analysis as far as the above definition of hydrophobicity is true for nonionic substances. Several examples listed below demonstrate the applicability of the parameter LogP in the studies of dye structure–property relationships describing the behaviour of disperse dyes in different dyeing systems. For example, the results of studies of adsorption of azobenzene dyes containing amido groups, and naphthylamine derivatives on polyester fibres and films in a water dyebath collected from the literature demonstrate that evaluated parameters of LogP with experimental partition coefficients between polyester and dyebath in logarithmic scale exhibits rather good correlation (Telegin et al., 2008, 2013). The affinity of dyes to textile materials controls several properties of dyeings, the most important of which is wash fastness. Correlation analysis of individual series of pyridine derivatives and azobenzene derivatives with heteroaryl residues (Telegin et al., 2008, 2013) exhibit good correlations between parameter LogP and dye affinity as well as with wash fastness on acetate, nylon and silk fibres.

It is well known that the light fastness of dyes is controlled by redox properties of dyes and their photo-destruction follows an oxidation or reduction mechanism, therefore correlation of light fastness with oxidation and reduction potentials of dye molecules is quite expected. Analysis of the results for light fastness of a series of disperse dyes (Telegin et al., 2008, 2013) demonstrates a typical correlation of this quantity with HOMO energy, characterising the photo-oxidation mechanism of azo-group destruction on polyester. The results of thiophene derivatives exhibit a correlation of light fastness with LUMO energy. In this case, the destruction of dye molecules follows a reduction mechanism due to the sulphur atom in the thiophene moiety.

### 9.5.3 Affinity, wash and light fastness of acid and direct dyes

Acid and direct dyes as anionic water-soluble dyes play an important role in fundamental research on colour-fading properties (Chudgar and Oakes, 2003; Oakes et al., 1998; Oakes, 2002, 2003) and adsorption on fibrous materials (Burkinshaw, 2021a,b; Burkinshaw and Salihu, 2019a,b,c,d; Oakes and Dixon, 2003a,b). Reactive dyes are a special class of anionic dyes due to their irreversible chemical fixation on the fibres, therefore their wash fastness properties could not be analysed within a single concept for anionic dyes. Studies of the substantivity of the hydrolysed form of reactive dyes (Ferus-Comelo, 2013) do not provide information about the chemical structure of dyes. On the other hand, fundamental research on the colour-fading properties of reactive dyes by peroxide (Bredereck and Schumacher, 1993a) and the light fastness of dyeings (Bredereck and Schumacher, 1993b) does not contain sufficient information for the application of statistical tools of chemoinformatics.

It is common to discuss the properties of acid and reactive dyes from the point of view of solubility in water. However, analysis of some known results demonstrates the role of hydrophobic properties of water-soluble dyes in their physicochemical and technical properties (Telegin and Zarubina, 2004). For example, the set of commercial acid azo dyes shown in Scheme 9.4a and b demonstrate amphiphilicity LogP within −2 to 6 and a number of sulphonic groups from 1 to 4. According to the analysis of experimental data (Telegin and Zarubina, 2004), in the region of LogP above −2, the rate of dyeing of wool fibres is increased while levelling and migration factors are decreased with increasing LogP. Fig. 9.1 shows an illustration of the variation of dye exhaustion and light fastness of dyeings on wool.

The data reported by Carpignano et al. (Carpignano et al., 1983) for a model monosulphonic azo 4-aminobenzene acid dye shown in Scheme 9.5a–c indicate a good correlation between LogP and wash fastness on wool represented in Fig. 9.2. Several commercial acid red dyes (World Dye Variety, 2022) shown in Scheme 9.6a and b were analysed and the correlation of LUMO energy with the light fastness is

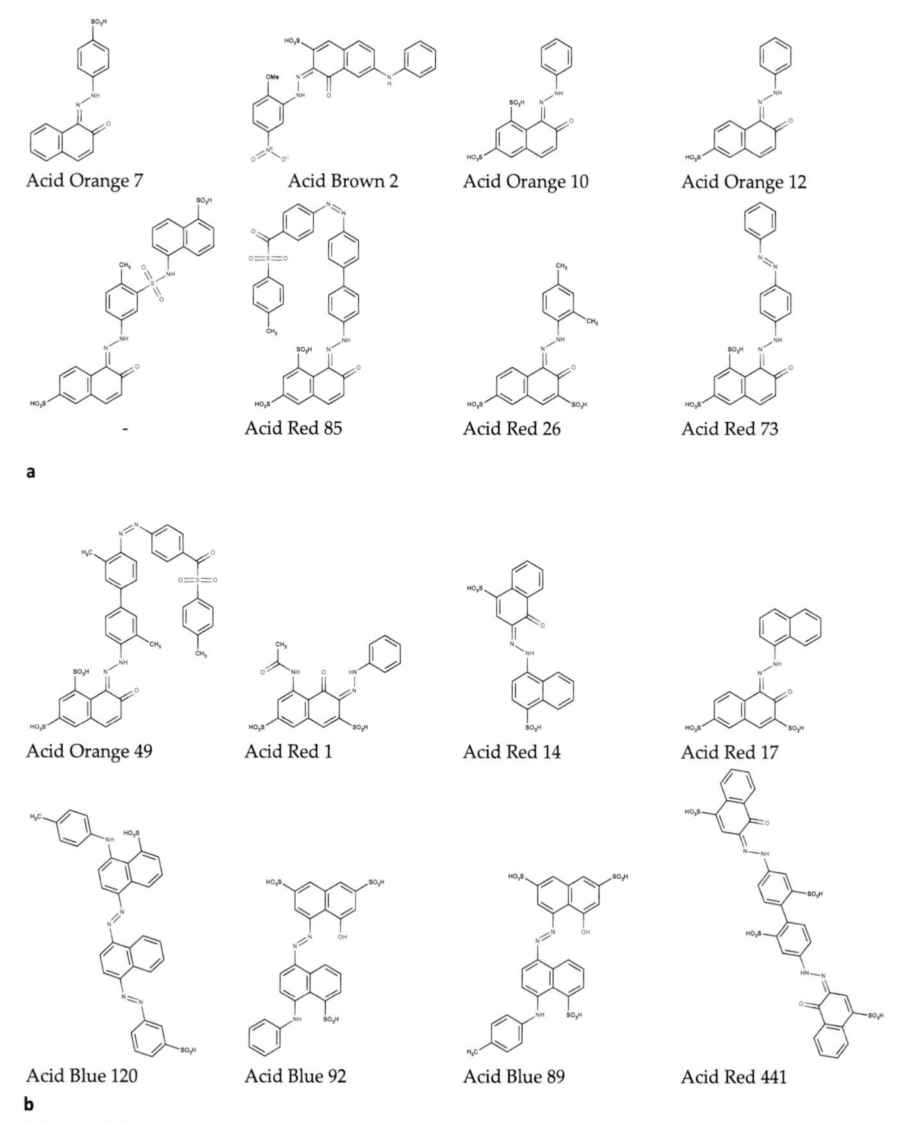


**Scheme 9.4** A set of commercial acid azo dyes.

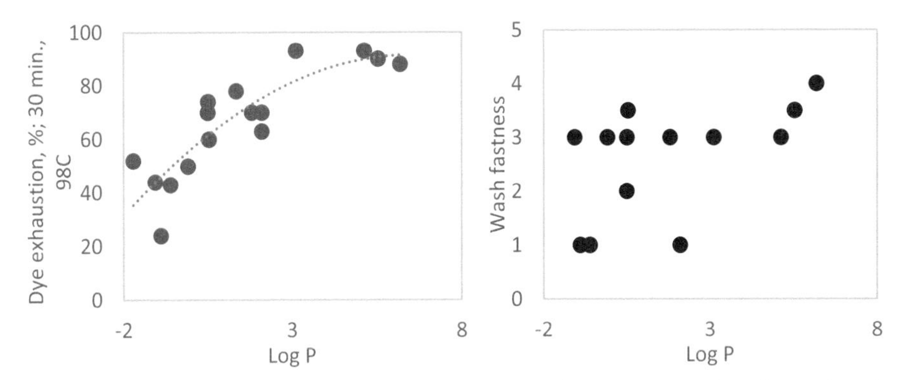


**Figure 9.1** Correlation between acid azo dye hydrophobicity LogP, dye exhaustion and wash fastness on wool.

demonstrated in Fig. 9.3, characterising their photo reduction mechanism of azo group destruction.

Studies of the modified non-mutagenic benzidine direct dyes (Gong et al., 2002) exhibit opposite technical property behaviours. Simple correlations shown in Fig. 9.4 show that in a series of 13 direct dyes (Scheme 9.7a–m) derived from 3,3′disubstituted benzidenes, exhaustion is decreased with the increase of hydrophobicity of compounds. The light fastness of dyes is decreased in the same direction. The different behaviours of acid dyes are explained by the role of hydrophobic interactions of acid dyes with wool and hydrogen bonding in the adsorption of direct dyes on cotton. Interestingly, dyes structures with high affinity to cellulose predicted in the research (Toropova et al., 2022) bear additional hydrophilic groups.

Studies of sorption of acid dyes on cotton are important for clarifying the mechanism of the undesirable colouring of fibres in domestic washing. From a theoretical viewpoint, such research is valuable for understanding the driving force of dye affinity to cotton. A set of model acid dyes, derivatives of $\gamma$, R and H-acid, shown in Scheme 9.8a and b (Funar-Timofei et al., 2012), provide an example of simple compounds for such research. The evaluated dye affinity of the compounds exhibits a correlation with calculated values of dye hydrophobicity LogP and dye solubility LogS, shown in Fig. 9.5. Comparison of Figs 9.4 and 9.5 demonstrates the role of dye hydrophobicity in the adsorption mechanism by cotton, i.e., acid dyes increase their affinity (or dyebath exhaustion) with the increase of their hydrophobicity within $-5$ to $+3$ or decrease of solubility in solution, while the exhaustion of direct dyes is increased at low hydrophobicity within from 3 to 9 due to hydrophilic character of cotton fibres.

### *9.5.3.1 Comparative analysis of fragment descriptors of regression models*

Comparative analysis of fragment descriptors of regression models can be applied on different kinds of fibres and colour fastness tests. Application of chemoinformatics

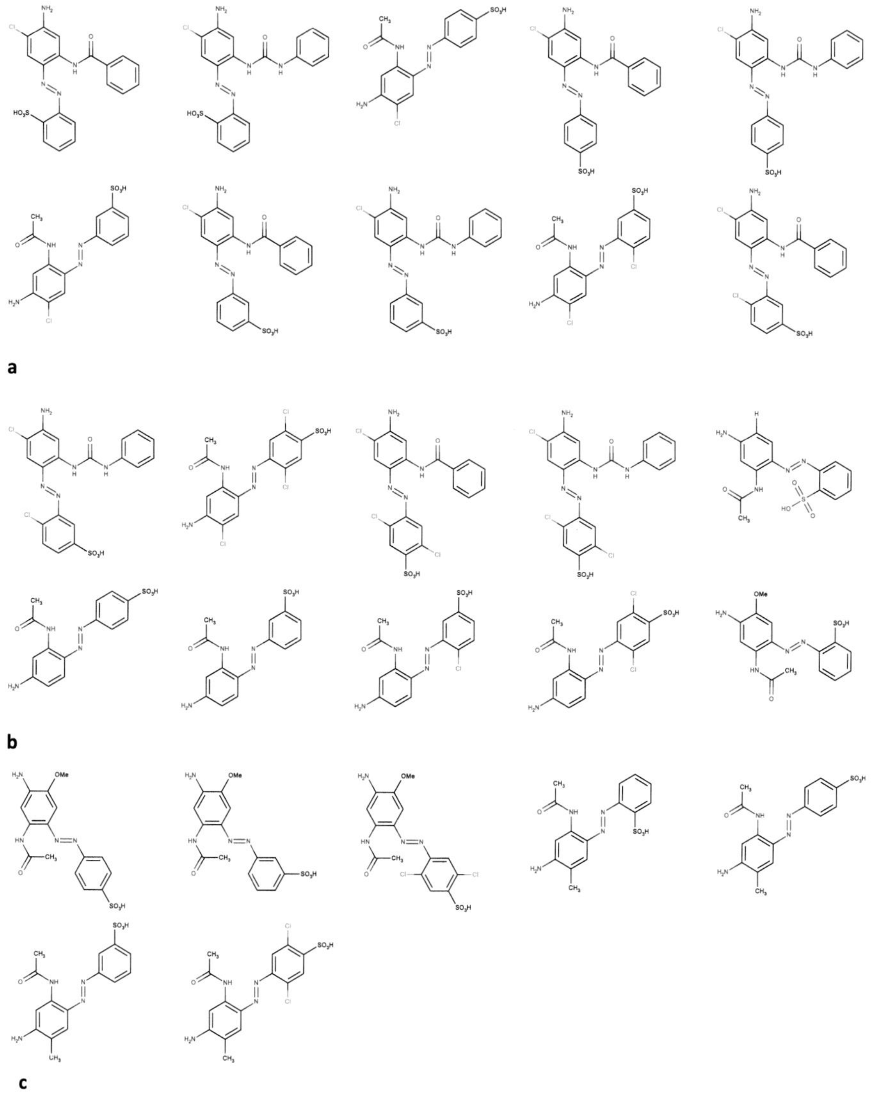

**Scheme 9.5** A set of model monosulphonic azo benzene dyes.

tools for analysis of large databases of dyes performed in the research (Ran et al., 2022) demonstrated the efficiency of the method for analysis of dye chemical structure–property relationships in several case studies: light fastness of acid dyes on wool and acid dyes on polyamide, the sensitivity of acid dyes on wool to oxygen bleaching,

Wash fastness
4
3
2
1
1
3
5
7
Log P

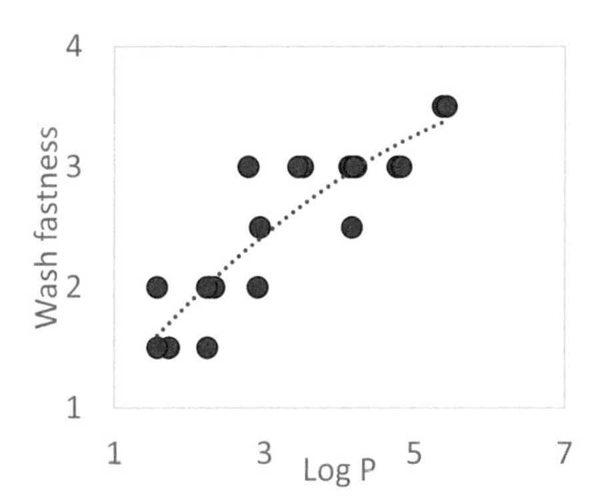

**Figure 9.2** Correlation between acid azo dyes hydrophobicity LogP and their wash fastness on wool.

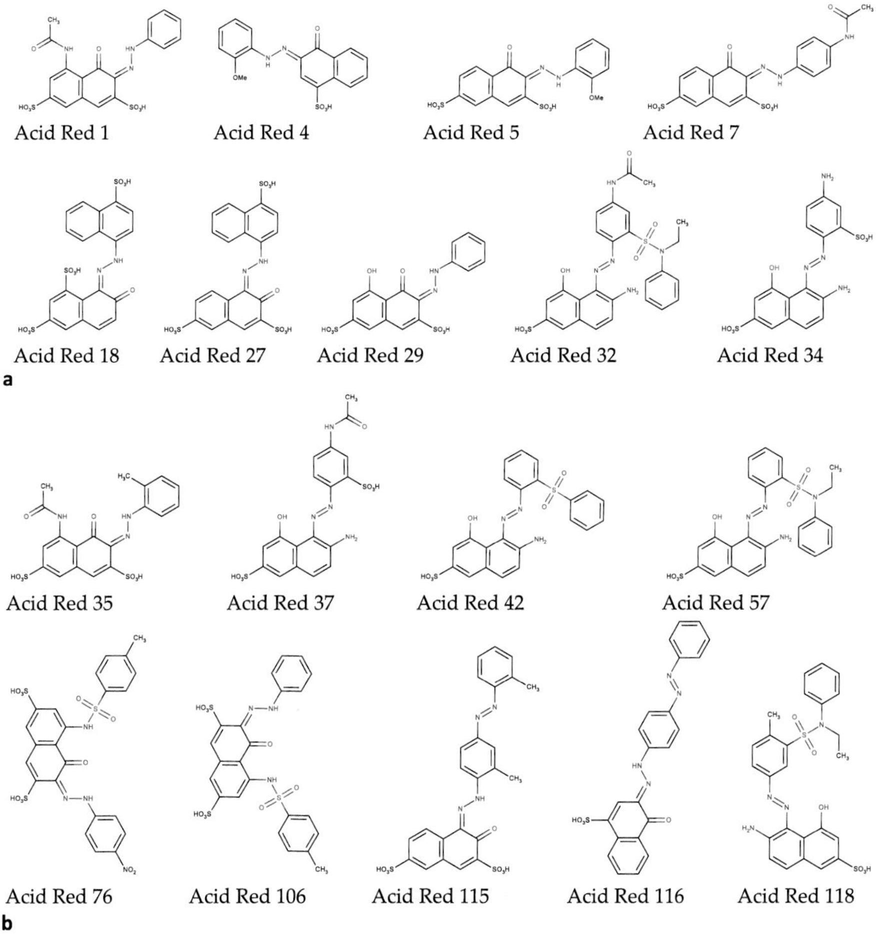

Acid Red 1
Acid Red 4
Acid Red 5
Acid Red 7
Acid Red 18
Acid Red 27
Acid Red 29
Acid Red 32
Acid Red 34
a
Acid Red 35
Acid Red 37
Acid Red 42
Acid Red 57
Acid Red 76
Acid Red 106
Acid Red 115
Acid Red 116
Acid Red 118
b

**Scheme 9.6** A set of commercial red acid dyes.

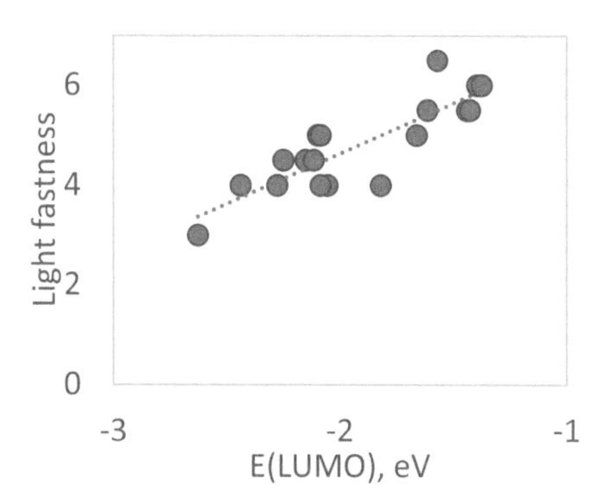


**Figure 9.3** Correlation between LUMO energy and light fastness of commercial red acid dyes on wool.

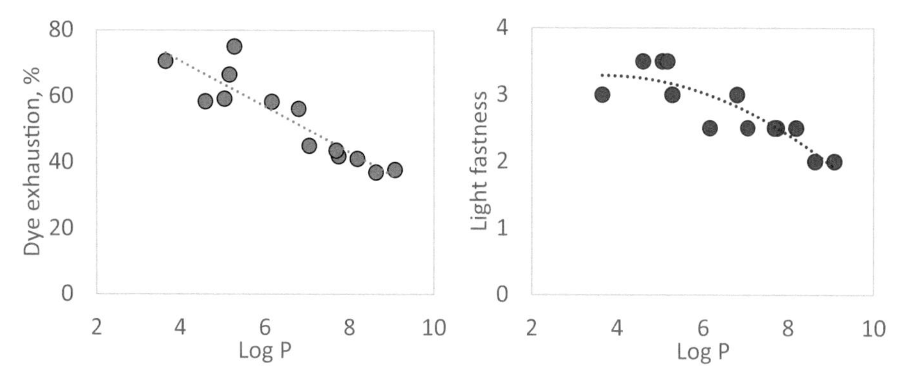


**Figure 9.4** Correlation of hydrophobicity LogP with dye exhaustion and light fastness of direct dyes on cotton.

wash fastness of acid dyes on wool, adsorption of direct dyes on cotton, and photodestruction of azo dyes in solution. The fragment approach of QSPR depicts several substructural descriptors reflecting the mechanism of destruction and dye—fibre interaction. Experimental data for chemoinformatics analysis of colour fastness properties of textiles are provided in various sources of information. References for colour fastness of wool, polyamide and cotton fibres dyed with acid and direct dyes are reported in the above-mentioned research (Ran et al., 2022).

Combination of the models for various fibres and colour fastness tests is of interest for comparative analysis of the physicochemical mechanism of dye destruction and their interaction with the fibre. Tables 9.1—9.4 demonstrate several molecular fragments of similar chemical nature, indicating the coinciding physicochemical routes of dye destruction and adsorption regardless of the nature of the fibre.

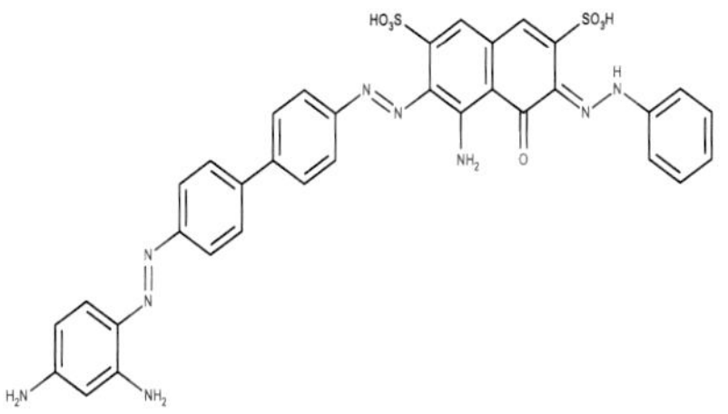

a – Direct Blue 38

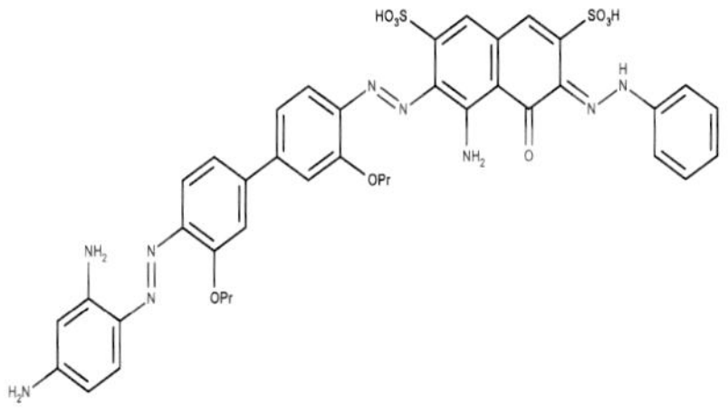

b

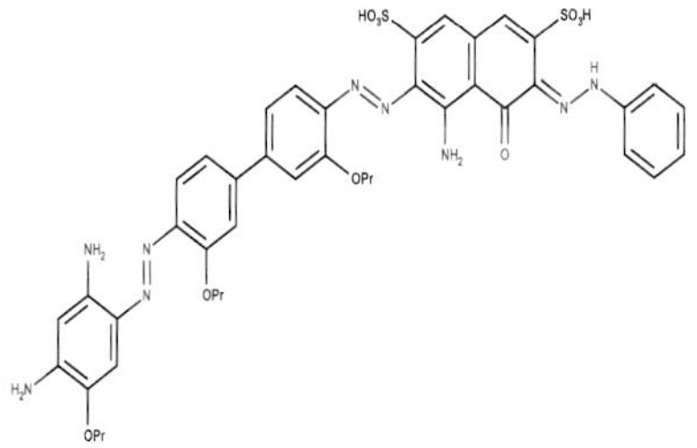

c

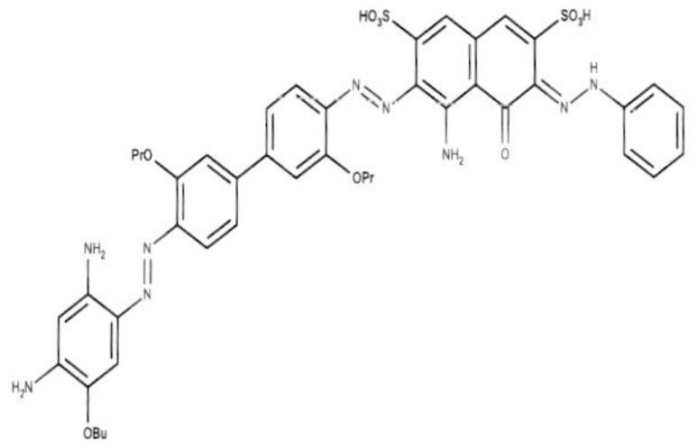

d

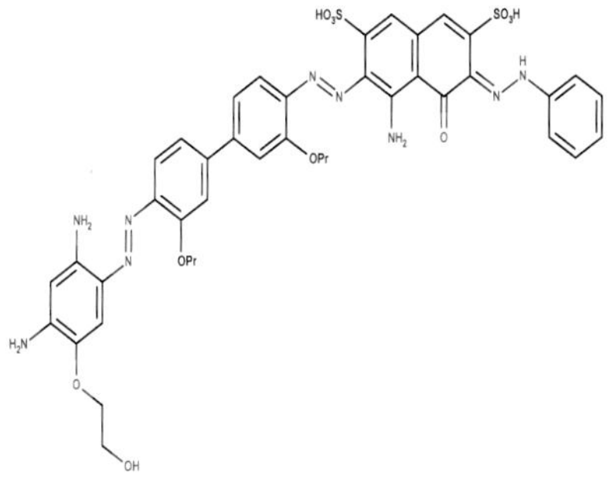

e

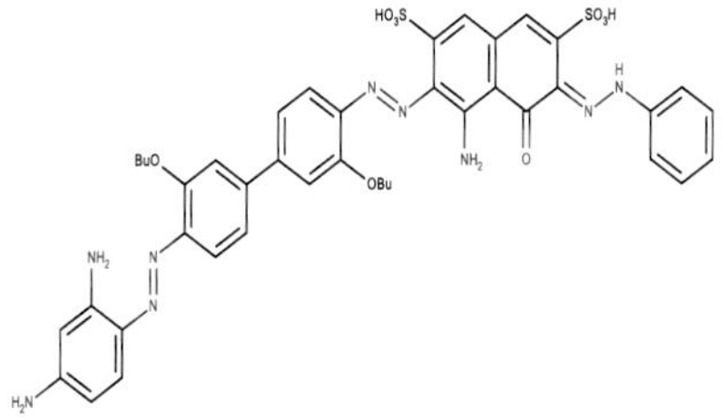

f

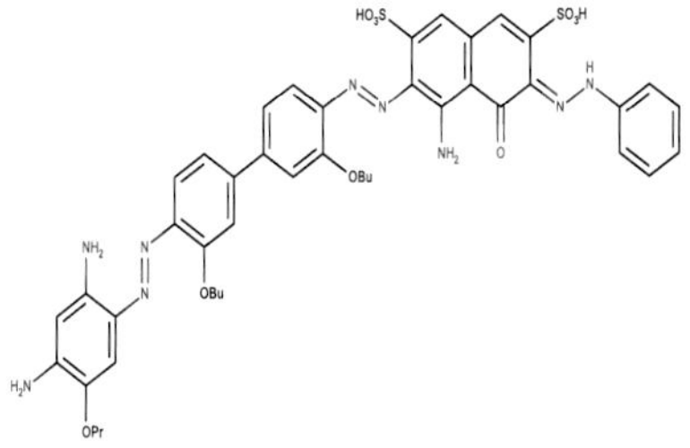

g

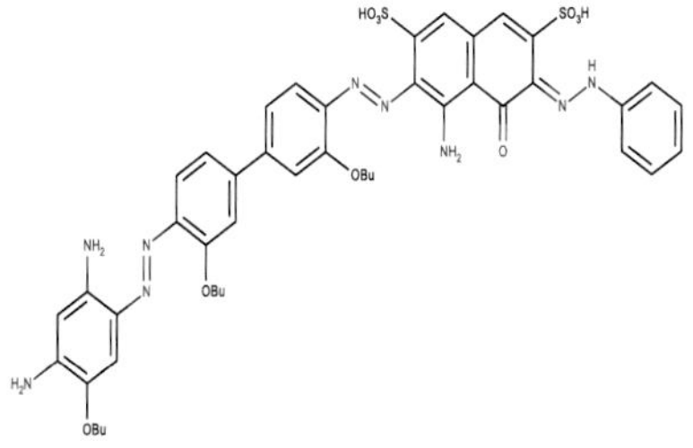

h

**Scheme 9.7** (a-d) Mutagenic dye Direct Blue 38 and a set of model non-mutagenic direct dyes. (e-m) A set of model non-mutagenic direct dyes.

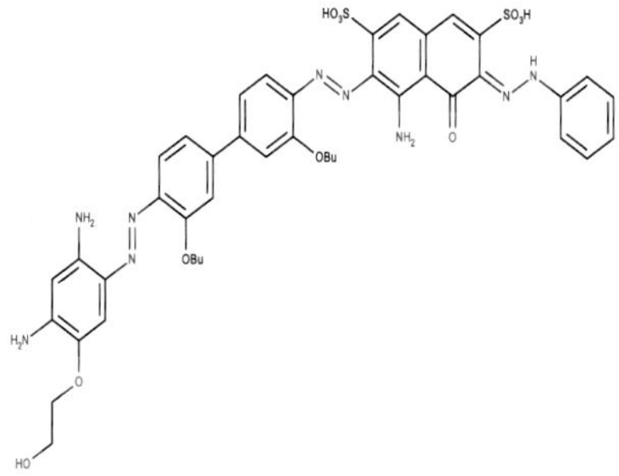

i

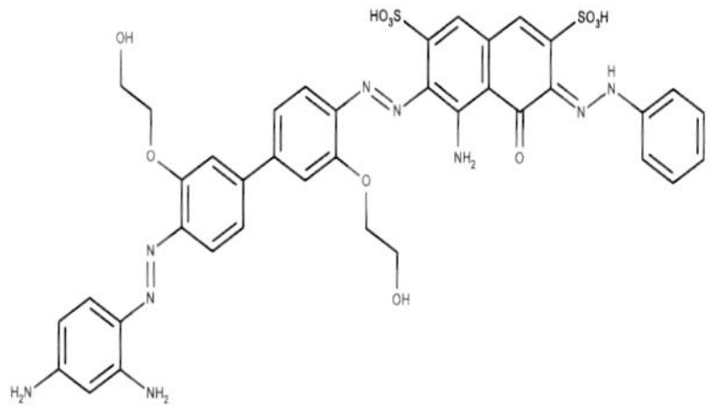

j

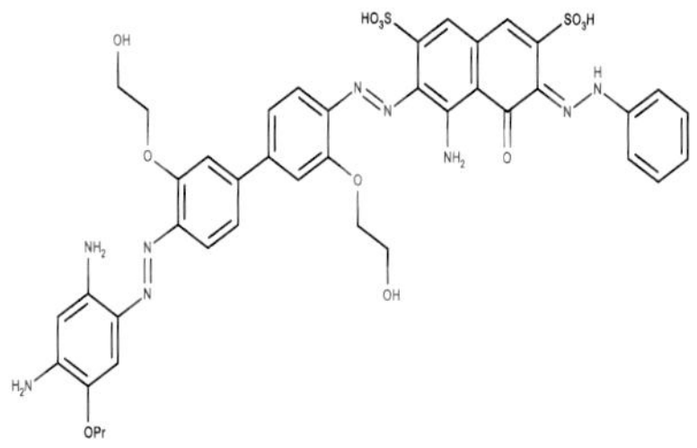

k

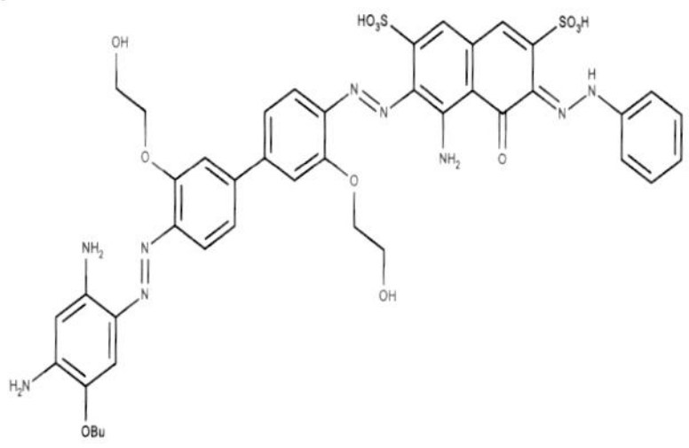

L

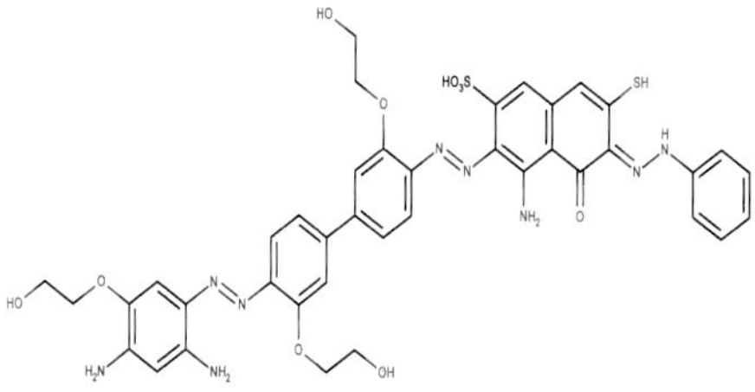

m

**Scheme 9.7** cont'd.

The negative (destructive) effect on dye chromophore observed in the tests for light fastness of dyeings on wool and polyamide, as well as the sensitivity of dyeings on wool for oxygen bleaching, are controlled by the fragments shown in Table 9.1: primary or substituted amino groups; azo-bond as a part of a chain of conjugated double bonds with primary or substituted amino groups; chain of aromatic carbon atoms; and the nitrogen atom of azo-bond.

Positive (stabilising) effect on the dye chromophore observed in the tests for the light fastness of dyes on wool and polyamide fibres, as well as the sensitivity of dyes on wool to oxygen bleaching is explained by the fragments shown in Table 9.2

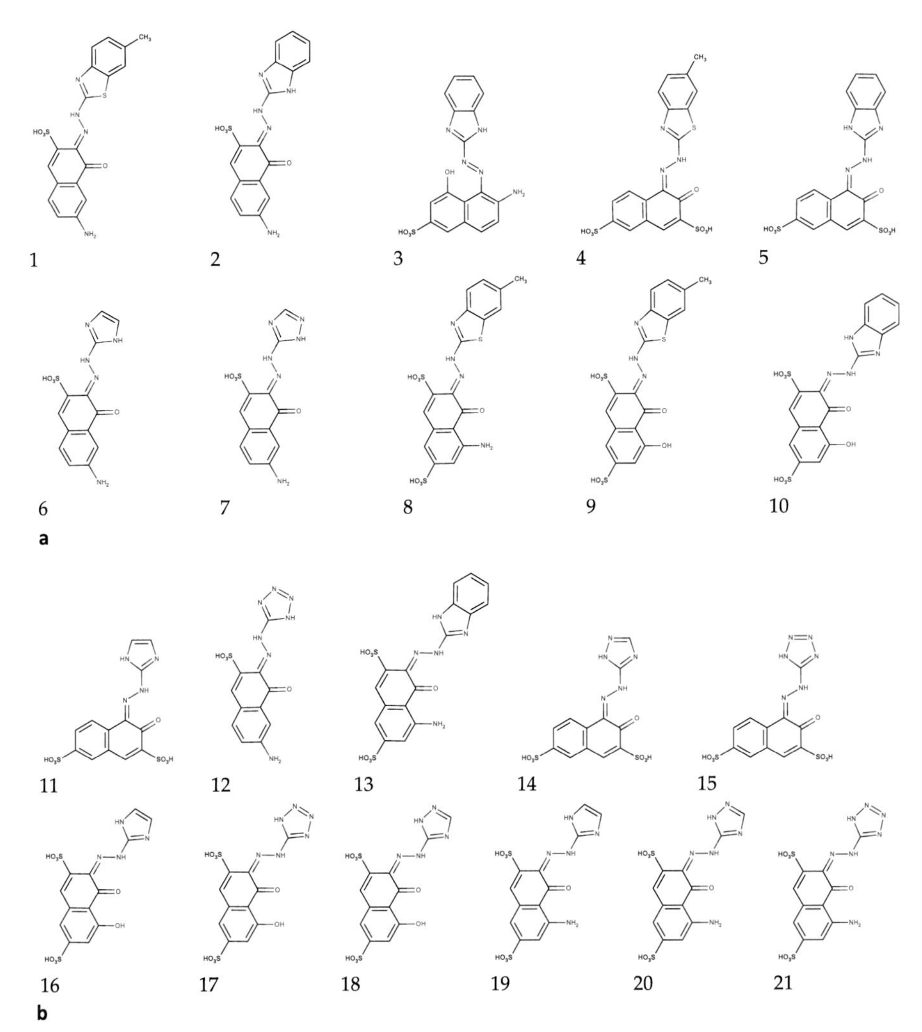


**Scheme 9.8** A set of model acid dyes.

azo group as a part of a chain of conjugated double bonds; azo group as a part of a chain of conjugated double bonds and sulphonic group; azo-bond as a part of a chain of conjugated double bonds and carbamide group.

The fragment shown in Table 9.3 has a positive effect on the wash fastness of dyeings on wool and adsorption on cotton fibres: two azo groups as a part of a chain of

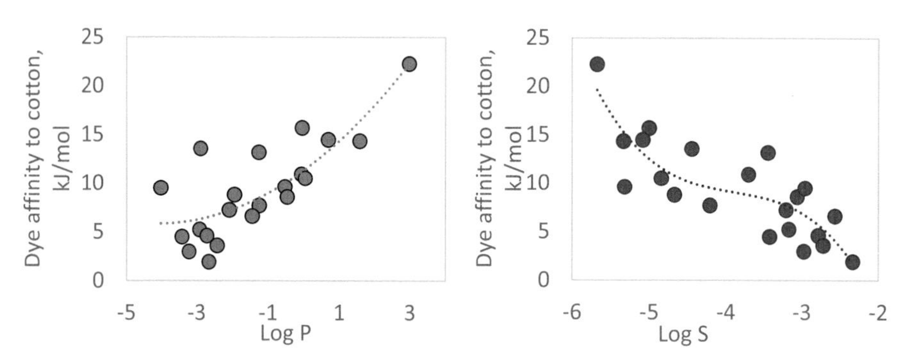


**Figure 9.5** Relationship between acid dye hydrophobicity LogP or dye solubility LogS and affinity to cotton.

**Table 9.1** Comparison of fragments responsible for the destruction of dyes in light fastness (LF) and sensitivity of dyeings to oxygen bleaching (OB) tests.

| Primary or substituted amino group | Azo-bond and the primary or substituted amino group | Aromatic chain and nitrogen of azo group |
|---|---|---|
| LF-W-3 | | LF-W-10 |
| OB-W-3 | OB-W-4 | |
| LF-PA-2 | LF-PA-8 | LF-PA-6 |

**Table 9.2** Comparison of fragments responsible for stabilisation of dyes in light fastness (LF) and sensitivity of dyeings to oxygen bleaching (OB) tests.

| **Azo group in a chain of conjugated double bonds** | **Azo group in a chain of conjugated double bonds and sulphonic group** | **Azo-bond in a chain of conjugated double bonds and carbamide group** |
|---|---|---|
| OB-W-9 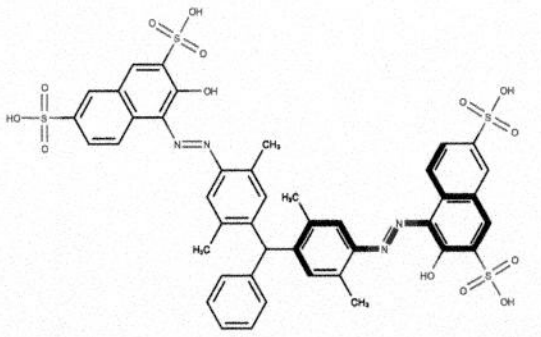 | LF-W-7 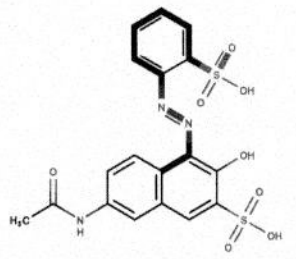 | |
| LF-PA-5 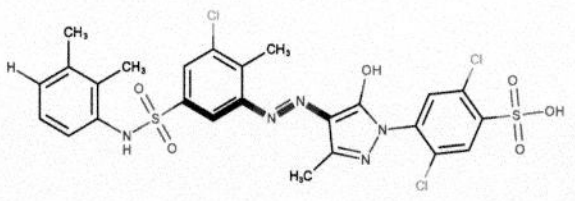 | LF-PA-9 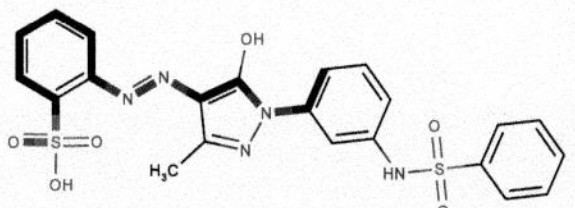 | LF-PA-10 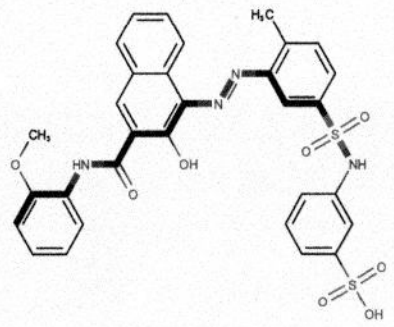 |

conjugated double bonds; chain of conjugated double bonds containing an azo group and hydrophobic substituent. Both chain fragments play the role of bulky hydrophobic fragments. In contrast to the above, the fragments presented in Table 9.4 reduce the wash fastness of acid dyes on wool and adsorption of direct dyes on cotton: an aromatic chain with a terminal sulfonic group; the azo group as a part of a chain of conjugated double bonds with hydrophilic substituents -$NH_2$ or -OH.

**Table 9.3** Comparison of fragments responsible for the positive impact on wash fastness (WF) and adsorption (A) of dyes on cotton.

| Two azo-bonds in a chain of conjugated double bonds | Azo-bond in a chain of conjugated double bonds and terminal hydrophobic terminal group |
|---|---|
| WF-W-5 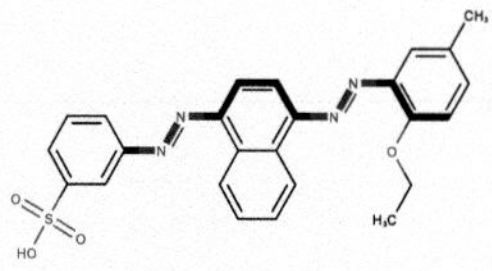 | WF-W-6 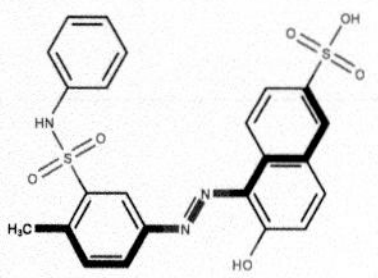 |
| A-C-8 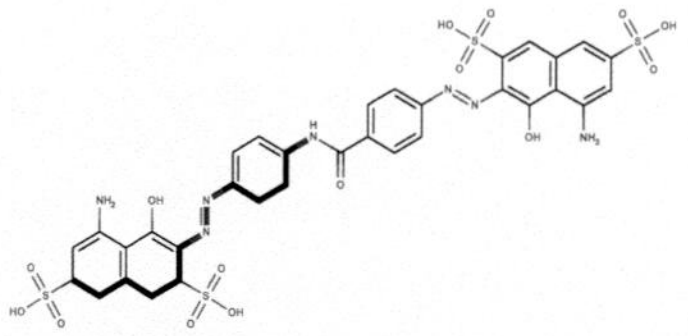 | |

**Table 9.4** Comparison of fragments responsible for the negative impact on wash fastness (WF) and adsorption (A) of dyes on cotton.

| Sulphonic group | Azo-bond and hydrophilic terminal group |
|---|---|
| WF-W-3 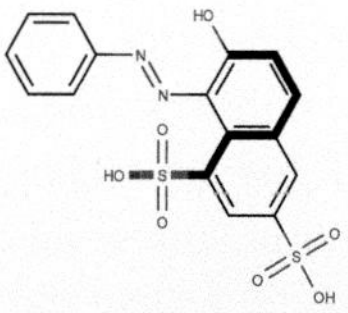 | WF-W-7 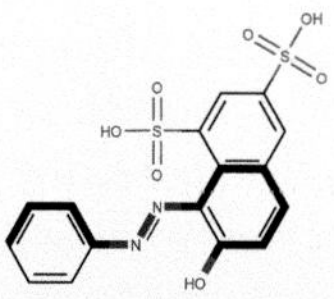 |
| A-C-10 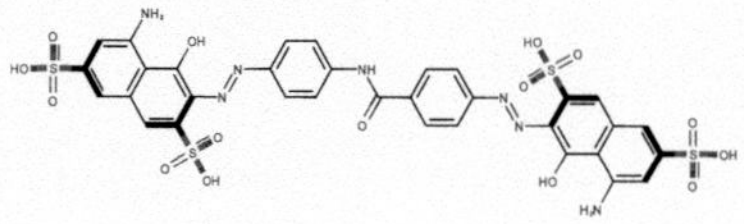 | A-C-9 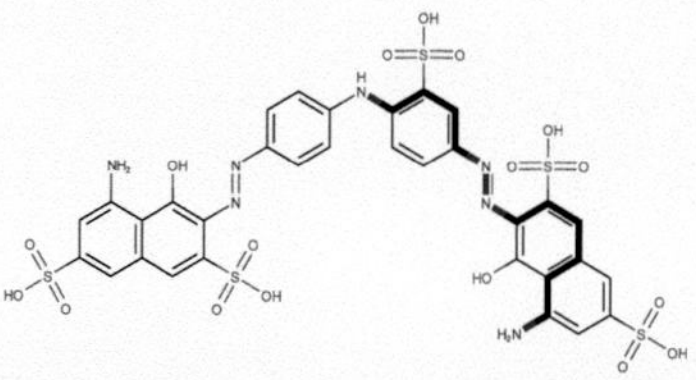 |

## 9.6 Future perspectives

In future, new sources of natural dyes, and natural and ecological mordants, will be explored extensively, particularly for niche and high-end markets. Synthetic dyes are here to stay, but the future molecules and the production processes will be sustainable and environmentally friendly, using green chemistry and the chemoinformatics approach. Oxygenated heterocycles are known to biodegrade well, and therefore the development of new dyes bearing oxygenated heterocycles may ensure sustainable colorants. Digital colouration using high tinctorial power dyes will lead the way to zero effluent discharge. Multifunctional dyes with inherent functional properties can lead to sustainable one-step dyeing and finishing. Structural colours will also be regarded as an ideal alternative to dyes and pigments. Therefore, the time will soon arrive when the textile colouration industry will no longer be considered as a heavily polluting industry.

## 9.7 Conclusion

Colour has been cherished by civilization since the dawn of time and has an impact on everyone in some way. Natural dyes were used to dye textiles until the 19th century, when synthetic dyes replaced them. There are several sustainable innovations taking place in the textile colouration industry. The current focus on the bioeconomy, sustainability and circularity is bringing back natural dyes for special applications and niche markets. In the case of synthetic dyes, green chemistry is being pursued aiming at increasing the dye affinity, which can in turn result in reduced effluents. The incorporation of different functional molecules in the dye structures can lead to sustainable single step dyeing and multifunctional finishing. Moreover, the chemoinformatics approach is leading sustainable innovations in dyes. Structural colours that are generated by the interaction between the micro-/nano-structures of the material surface and the incident light can lead to totally sustainable colouration.

## Acknowledgements

Felix Y. Telegin would like to acknowledge financial support from the Ministry of Science and Higher Education of the Russian Federation (No. 075-15-2021-579).

## References

Adeel, S., Salman, M., Zahoor, A.F., Usama, M., Amin, N., 2020. An insight into herbal-based natural dyes. In: Recycling From Waste in Fashion and Textiles, pp. 423–456.

Agarwal, O.,P., Tiwari, R., 1989. Mineral pigments of India. In: Compendium of the National Convention of Natural Dyes'. National Handloom Development Corporation, Lucknow, Jaipur.

AITEX, 2020. AlgaeL-based Dyes; A New Alternative for Sustainable Development in the Textile Industry. Available at: https://www.aitex.es/colorantes-procedentes-de-algas-greencolor/?lang=en. (Accessed 20 March 2023).

Al-Amrani, W.A., Hanafiah, M.A.K.M., Mohammed, A.-H.A., 2022. A Comprehensive Review of Anionic Azo Dyes Adsorption on Surface-Functionalised Silicas. Springer, Heidelberg.

Awfa, D., Ateia, M., Mendoza, D., Yoshimura, C., 2021. Application of quantitative structure−property relationship predictive models to water treatment: a critical review. ACS ES&T Water 1 (3), 498−517.

Bait, S., Shinde, S., Adivarekar, R., Sekar, N., 2020. Multifunctional properties of benzophenone based acid dyes: synthesis, spectral properties and computational study. Dyes and Pigments 180, 108420. https://doi.org/10.1016/j.dyepig.2020.108420.

Baskin, I., Varnek, A., 2008. Fragment descriptors in SAR/QSAR/QSPR studies, molecular similarity analysis and in virtual screening. In: Varnek, A., Tropsha, A. (Eds.), Chemoinformatics Approaches to Virtual Screening. RSC Publishing, Cambridge, pp. 1−43.

Baskin, I.I., Ait, A.O., Halberstam, N.M., Palyulin, V.A., Alfimov, M.V., Zefirov, N.S., 1997a. Application of methodology of artificial neural networks for predicting the properties of sophisticated molecular systems: prediction of the long-wave absorption band position for symmetric cyanine dyes. Russian Chemical Bulletin, International Edition 357 (1−3), 353−355.

Baskin, I.I., Keschtova, S.V., Palyulin, V.A., Zefirov, N.S., 1999. Combining molecular modelling with the use of artificial neural networks as an approach to predicting substituent constants and bioactivity. In: Gundertofte, K., Jørgensen, F.S. (Eds.), Molecular Modeling Nd Prediction of Bioactivity.

Baskin, I.I., Palyulin, V.A., Zefirov, N.S., 1997b. A neural device for searching direct correlations between structures and properties of chemical compounds. Journal of Chemical Information and Computer Sciences 37 (4), 715−721.

Bautista, L., Paul, R., Mota, J., La Varga, M. de, Garrido-Franco, M., Aubouy, L., Briz, A., Pérez de la Ossa, P., 2007. Dyeing of low-pressure oxygen plasma treated wool fabrics with antibacterial natural dyes. International Dyer 192 (10), 14.

Bechtold, T., Mussak, R., Mahmud-Ali, A., Ganglberger, E., Geissler, S., 2006. Extraction of natural dyes for textile dyeing from coloured plant wastes released from the food and beverage industry. Journal of the Science of Food and Agriculture 86 (2), 233−242.

Bhate, P.M., Devi, R.V., Dugane, R., Hande, P.R., Shaikh, L., Vaidya, S., Masand, S., 2017. A novel reactive dye system based on diazonium salts. Dyes and Pigments 145, 208−215. https://doi.org/10.1016/j.dyepig.2017.06.007.

BioPolyCol, Development of Amazonian Bio-dyes from Renewable Resources for Industrial Dyeing of Biopolymers, 2022. Available at: https://www.ita.rwth-aachen.de/global/show_document.asp?id=aaaaaaaabsaerre (Accessed 20 March 2023).

Bredereck, K., Schumacher, C., 1993a. Structure reactivity correlations of azo reactive dyes based on H-acid. III. Dye degradation by peroxide. Dyes and Pigments 23 (2), 121−133.

Bredereck, K., Schumacher, C., 1993b. Structure reactivity correlations of azo reactive dyes based on H-acid. IV. Investigations into the light fastness in the dry state, in the wet state, and in presence of perspiration. Dyes and Pigments 23 (2), 135−147.

Brigden, K., Labunska, I., Johnston, P., Santillo, D., 2012. Organic Chemical and Heavy Metal Contaminants from Communal Wastewater Treatment Plants with Links to Textile Manufacturing, and in River Water Impacted by Wastewater from a Textile Dye Manufacturing Facility, in China. Greenpeace Research Laboratories. Available at: https://www.greenpeace.to/greenpeace/wp-content/uploads/2012/12/TechnicalReport-07-2012.pdf. (Accessed 6 March 2023).

Broadbent, P.J., Carr, C.M., Rigout, M., Kistamah, N., Choolun, J., Radhakeesoon, C.L., Uddin, M.A., 2018. Investigation into the dyeing of wool with Lanasol and Remazol reactive dyes in seawater. Coloration Technology 134 (2), 156–161.
Burkinshaw, S.M., 2016. Physico-chemical Aspects of Textile Coloration. John Wiley & Sons, Ltd, Chichester, UK, SDC.
Burkinshaw, S.M., 2021a. The role of inorganic electrolyte (salt) in cellulosic fibre dyeing: Part 1 fundamental aspects. Coloration Technology 137 (5), 421–444.
Burkinshaw, S.M., 2021b. The role of inorganic electrolyte (salt) in cellulosic fibre dyeing: Part 2 theories of how inorganic electrolyte promotes dye uptake. Coloration Technology 137 (6), 547–586.
Burkinshaw, S.M., Salihu, G., 2019a. The role of auxiliaries in the immersion dyeing of textile fibres part 2: analysis of conventional models that describe the manner by which inorganic electrolytes promote direct dye uptake on cellulosic fibres. Dyes and Pigments 161, 531–545.
Burkinshaw, S.M., Salihu, G., 2019b. The role of auxiliaries in the immersion dyeing of textile fibres: Part 3 theoretical model to describe the role of inorganic electrolytes used in dyeing cellulosic fibres with direct dyes. Dyes and Pigments 161, 546–564.
Burkinshaw, S.M., Salihu, G., 2019c. The role of auxiliaries in the immersion dyeing of textile fibres: Part 4 theoretical model to describe the role of liquor ratio in dyeing cellulosic fibres with direct dyes in the absence and presence of inorganic electrolyte. Dyes and Pigments 161, 565–580.
Burkinshaw, S.M., Salihu, G., 2019d. The role of auxiliaries in the immersion dyeing of textile fibres: Part 5 practical aspects of the role of inorganic electrolytes in dyeing cellulosic fibres with direct dyes. Dyes and Pigments 161, 581–594.
Cai, J., Jiang, H., Chen, W., Cui, Z., 2020. Design, synthesis, characterization of water-soluble indophenine dyes and their application for dyeing of wool, silk and nylon fabrics. Dyes and Pigments 179, 108385. https://doi.org/10.1016/j.dyepig.2020.108385.
Cai, J., Jiang, H., Cui, Z., Chen, W., 2021. N-Sulphonatoalkyl indophenine derivatives: design, synthesis and dyeing properties on wool, silk and nylon fabrics. Coloration Technology 137 (2), 181–192.
Cardarelli, C.R., Benassi, M.d.T., Mercadante, A.Z., 2008. Characterization of different annatto extracts based on antioxidant and colour properties. LWT - Food Science and Technology 41 (9), 1689–1693.
Cardon, D., 2007. Natural dyes - sources, tradition, technology and science. In: Vankar, P.S., National Institute of Industrial Research (Eds.), Handbook on Natural Dyes for Industrial Applications. National Institute of Industrial Research, Delhi (India).
Carpignano, R., Barni, E., Di Modica, G., Grecu, R., Bottaccio, G., 1983. Quantitative relationships between chemical structure and technical properties of 4-aminoazobenzene sulphonic acid dyes. Dyes and Pigments 4 (3), 195–211.
ChemAxon: Free Academic License for JChem. Available at: www.chemaxon.com. (Accessed 12 January 2017).
Cherkasov, A., Muratov, E.N., Fourches, D., Varnek, A., Baskin, I.I., Cronin, M., Dearden, J., Gramatica, P., Martin, Y.C., Todeschini, R., Consonni, V., Kuz'min, V.E., Cramer, R., Benigni, R., Yang, C., Rathman, J., Terfloth, L., Gasteiger, J., Richard, A., Tropsha, A., 2014. QSAR modeling: where have you been? Where are you going to? Journal of Medicinal Chemistry 57 (12), 4977–5010.
Choi, S., Cho, K.H., Namgoong, J.W., Kim, J.Y., Yoo, E.S., Lee, W., Jung, J.W., Choi, J., 2019. The synthesis and characterisation of the perylene acid dye inks for digital textile printing. Dyes and Pigments 163, 381–392. https://doi.org/10.1016/j.dyepig.2018.12.002.

Christie, R.M., 2007. Environmental Aspects of Textile Dyeing. Elsevier.

Chudgar, R.J., Oakes, J., 2003. Dyes, azo. In: Kirk-Othmer Encyclopedia of Chemical Technology. John Wiley & Sons, Inc.

Churchley, J., Greaves, A., Hutchings, M., Phillips, D.A.S., Taylor, J.A., 2000. A chemometric approach to understanding the bioelimination of anionic, water-soluble dyes by a biomass - Part 2: acid dyes. Coloration Technology 116 (7−8), 222−228.

Churchley, J.H., Greaves, A.J., Hutchings, M.G., Phillips, D.A.S., Taylor, J.A., 2000a. A chemometric approach to understanding the bioelimination of anionic, water-soluble dyes by a biomass - Part 3: direct dyes. Coloration Technology 116 (9), 279−284.

Churchley, J.H., Greaves, A.J., Hutchings, M.G., Phillips, D.A.S., Taylor, J.A., 2000b. A chemometric approach to understanding the bioelimination of anionic, water-soluble dyes by a biomass - Part 4: reactive dyes. Coloration Technology 116 (10), 323−329.

Clark, J.H., 2002. Solid acids for green chemistry. Accounts of Chemical Research 35 (9), 791−797.

Clark, M. (Ed.), 2011. Handbook of Textile and Industrial dyeing:Principles, Processes and Types of Dyes. Elsevier.

CODESSA PRO: COmprehensive DEscriptors for Structural and Statistical Analysis (2001-2005). Available at: www.codessa-pro.com. (Accessed 29 April 2022).

ColoriFix ColoriFix, 2022. Available at: https://colorifix.com/. (Accessed 6 March 2023).

CORAL software/databases (2010-2018). Available at: www.insilico.eu/coral/. (Accessed 29 April 2022).

Dawson, T.L., Hawkyard, C.J., 2000. A new millennium of textile printing. Review of Progress in Coloration and Related Topics 30 (1), 7−20.

Demirbilek, O., Sener, B., 2003. Product design, semantics and emotional response. Ergonomics 46 (13−14), 1346−1360.

Deo, H.T., Paul, R., 2000. Dyeing of ecru denim with onion extract using natural mordant combinations. Indian Journal of Fibre & Textile Research 25 (2), 152.

Deo, H.T., Paul, R., 2003. Eco-friendly mordant for natural dyeing of denim. International Dyer 188 (11), 49.

Dragon, 1994. Available at: https://chm.kode-solutions.net/products_dragon.php. (Accessed 29 April 2022).

Dufossé, L., Fouillaud, M., Caro, Y., Mapari, S.A.S., Sutthiwong, N., 2014. Filamentous fungi are large scale producers of pigments and colorants for the food industry. Current Opinion in Biotechnology 26, 56−61. https://doi.org/10.1016/j.copbio.2013.09.007.

Dumanli, A.G., Savin, T., 2016. Chemical Society Reviews 45, 6698−6724.

Dyes and Colorants from Algae. Available at: http://www.seacolors.eu/images/dyes_and_colourants_from_algae.pdf. (Accessed 20 March 2023).

Ellen Mac Arthur Foundation, 2017. Fashion and the Circular Economy. Available at: https://archive.ellenmacarthurfoundation.org/explore/fashion-and-the-circular-economy. (Accessed 20 March 2023).

EU Circular Economy Action Plan. https://ec.europa.eu/environment/pdf/circular-economy/new_circular_economy_action_plan.pdf. (Accessed 10 April 2023).

Fenzl, C., Hirsch, T., Wolfbeis, O.S., 2014. Angewandte Chemie International Edition 53, 3318−3335.

Ferus-Comelo, M., 2013. An analysis of the substantivity of hydrolysed reactive dyes and its implication for rinsing processes. Coloration Technology 129 (1), 24−31.

Filimonov, V.D., Trusova, M., Postnikov, P., Krasnokutskaya, E.A., Lee, Y.M., Hwang, H.Y., Kim, H., Chi, K.-W., 2008. Unusually stable, versatile, and pure arenediazonium tosylates:

their preparation, structures, and synthetic applicability. Organic Letters 10 (18), 3961–3964.

Freitas da Silva, Jane, G., Beltro Lessa Constant, P., Wilane de Figueiredo, R., Medeiros Moura, S., 2010. Formulacao e estabilidade de corantes de antocianinas extraidas das cascas de jabuticaba (Myrciaria ssp.). Alimentos e Nutricao (Brazilian Journal of Food and Nutrition) 21, 429. Available at: https://link.gale.com/apps/doc/A245953143/AONE?u=googlescholar&sid=bookmark-AONE&xid=7d53cc9d.

Funar-Timofei, S., Fabian, W.M., Kurunczi, L., Goodarzi, M., Ali, S.T., Heyden, Y.V., 2012. Modelling heterocyclic azo dye affinities for cellulose fibres by computational approaches. Dyes and Pigments 94 (2), 278–289.

Funar-Timofei, S., Gheorghe, I., 2020. In: Roy, K. (Ed.), QSAR Modeling of Dye Ecotoxicity, Ecotoxicological QSARs. Springer US, New York, NY, pp. 405–436.

Giles, C.H., Duff, D.G., Sinclair, R.S., 1978. Relation between the molecular structure of dyes and their technical properties: chapter VII. In: Venkataraman, K. (Ed.), The Chemistry of Synthetic Dyes. Academic Press, New York-London, pp. 279–329.

Giorgi, M. R. de, Carpignano, R., Crisponi, G., 1997. Structure optimization in a series of acid dyes for wool and nylon. Dyes and Pigments 34 (1), 1–12.

Giorgi, M. R. de, Carpignano, R., Scano, P., 1994. Structure optimization in a series of dyes for wool and cotton. A chemometric approach. Dyes and Pigments 26 (3), 175–189.

Giorgi, M. R. de, Cerniani, A., Carpignano, R., Savarino, P., 1993. Design of high fastness acid dyes for silk: a chemometric approach. Journal of the Society of Dyers and Colourists 109 (12), 405–410.

Giorgi, M. de, Carpignano, R., 1996. Design of dyes of high technical properties for silk by a chemometric approach. Dyes and Pigments 30 (1), 79–88.

Giorgi, M. de, Carpignano, R., Cerniani, A., 1998. Structure optimization in a series of thiadiazole disperse dyes using a chemometric approach. Dyes and Pigments 37 (2), 187–196.

Gong, G., Gao, X., Wang, J., Zhao, D., Freeman, H.S., 2002. Trisazo Direct Black dyes based on nonmutagenic 3,3′-disubstituted benzidines. Dyes and Pigments 53 (2), 109–117. https://doi.org/10.1016/S0143-7208(02)00010-4 [Online].

Gordon, P.F., Gregory, P., 1983. Organic Chemistry in Colour. Springer-Verlag, Berlin.

Guo, Q., Chen, W., Cui, Z., Jiang, H., 2020. Reactive dyeing of silk using commercial acid dyes based on a three-component Mannich-type reaction. Coloration Technology 136 (4), 336–345.

Gupta, V.K., 2020. Fundamentals of natural dyes and its application on textile substrates. In: Kumar Samanta, A., Awwad, N., Majdooa Algarni, H. (Eds.), Chemistry and Technology of Natural and Synthetic Dyes and Pigments. IntechOpen.

Han, Y., Meng, Z., Wu, Y., Zhang, S., Wu, S., 2021. Structural colored fabrics with brilliant colors, low angle dependence, and high color fastness based on the mie scattering of Cu2O spheres. ACS Applied Materials & Interfaces 13 (48), 57796–57802. https://doi.org/10.1021/acsami.1c17288.

Hassan, M.M., Carr, C.M., 2018. A critical review on recentadvancements of the removal of reactive dyes from dyehouseeffluent by ion-exchange adsorbents. Chemosphere 209 (1), 201–219.

Hilal, S.H., Carreira, L.A., Baughman, G.L., Karickhoff, S.W., Melton, C.M., 1994. Estimation of ionization constants of azo dyes and related aromatic amines: environmental implication. Journal of Physical Organic Chemistry 7 (3), 122–141.

Hilal, S.H., Karichhoff, S.W., Careira, L.A., 2003. Prediction of Chemical Reactivity Parameters and Physical Properties of Organic Compounds from Molecular Structure Using SPARC. Research Triangle Park, NC 27711.

Hunger, K., 2003. Industrial Dyes: Chemistry, Properties and Appli-Cations. Willey-VCH, Weinheim.

Ibrahim, S.A., Rizk, H.F., Aboul-Magd, D.S., Ragab, A., 2021. Design, synthesis of new magenta dyestuffs based on thiazole azomethine disperse reactive dyes with antibacterial potential on both dyes and gamma-irradiated dyed fabric. Dyes and Pigments 193, 109504. https://doi.org/10.1016/j.dyepig.2021.109504.

İşmal, Ö.E., Yıldırım, L., 2019. Metal mordants and biomordants. In: The Impact and Prospects of Green Chemistry for Textile Technology. Elsevier, pp. 57–82.

Jiang, Z., Peng, W., Li, R., Huang, D., Ren, X., Huang, T.S., 2017. Synthesis and application to cellulose of reactive dye precursor of anti-bacterial N-halamine, Color. The Tech 133, 376–381.

Johnson, A. (Ed.), 1989. The Theory of Coloration of Textiles, second ed. Society of Dyers and Colourists, Bradford, UK.

Joshi, V.K., Devender, A., Neerja, S.R., December 2011. Indian Journal of Natural Products and Resources 2 (4), 421–427.

Kathiresan, S., Sarada, R., Bhattacharya, S., Ravishankar, G.A., 2007. Culture media optimization for growth and phycoerythrin production from Porphyridium purpureum. Biotechnology and Bioengineering 96 (3), 456–463.

Katritzky, A.R., Karelson, M., Lobanov, V.S., 1997. QSPR as a means of predicting and understanding chemical and physical properties in terms of structure. Pure and Applied Chemistry 69, 2.

Kats, M.D., Krichevskii, G.E., 1979. Mathematical model for the relationship between lightfastness and chemical structure of monoazo disperse dyes. Izvestiya vuzov Technology text prom (5), 60–63.

Kats, M.D., Lysun, N.V., Mostoslavskaya, E.I., Krichevskii, G.E., 1988. Studies of the relationships between chemical structure of diperse monoazo dyes and their light protecting properties on polyamide fibres. Zhurnal Prikladnoi Khimii (5), 1196–1199.

Khaligh, G.N., 2020. Recent advances and applications of tert-butyl nitrite (TBN) in organic synthesis. Mini-Reviews in Organic Chemistry 17 (1), 3–25. https://doi.org/10.2174/1570193X15666181029141019.

Khattab, T.A., Elnagdi, M.H., Haggaga, K.M., Abdelrahmana, A.A., Aly, S.A., 2017. Green synthesis, printing performance, and antibacterial activity of disperse dyes incorporating arylazopyrazolopyrimidines. AATCC Journal of Research 4, 1–8.

Khan, S., Malik, A., 2018. Toxicity evaluation of textile effluentsand role of native soil bacterium in biodegradation of a textiledye. Environmental Science and Pollution Research International 25 (5), 4446–4458.

Kohri, M., Yanagimoto, K., Kawamura, A., Hamada, K., Imai, Y., Watanabe, T., Ono, T., Taniguchi, T., Kishikawa, K., 2018. ACS Applied Materials & Interfaces 10, 7640–7648.

Koukabi, N., Otokesh, S., Kolvari, E., Amoozadeh, A., 2016. Convenient and rapid diazotization and diazo coupling reaction via aryl diazonium nanomagnetic sulfate under solvent-free conditions at room temperature. Dyes and Pigments 124, 12–17. https://doi.org/10.1016/j.dyepig.2015.03.041.

Kramar, A., Kostic, M.M., 2022. Bacterial secondary metabolites as biopigments for textile dyeing. Textiles 2, 252–264. https://doi.org/10.3390/textiles2020013.

Kuenemann, M.A., Szymczyk, M., Chen, Y., Sultana, N., Hinks, D., Freeman, H.S., Williams, A.J., Fourches, D., Vinueza, N.R., 2017. Weaver's historic accessible collection of synthetic dyes: a cheminformatics analysis. Chemical Science 8 (6), 4334–4339.

Kumar, P., Kumar, A., 2020. In silico enhancement of azo dye adsorption affinity for cellulose fibre through mechanistic interpretation under guidance of QSPR models using Monte

Carlo method with index of ideality correlation. SAR and QSAR in Environmental Research 31 (9), 697–715.

Lewis, D.M., Broadbent, P.J., Carr, C.M., He, W.D., 2022. Investigation into the reaction of reactive dyes with carboxylate salts and the application of carboxylate-modified reactive dyes to cotton. Coloration Technology 138 (1), 58–70.

Li, B., Dong, Y., Ding, Z., 2013. Heterogeneous Fenton degradation of azo dyes catalyzed by modified polyacrylonitrile fiber Fe complexes: QSPR (quantitative structure property relationship) study. Journal of Environmental Sciences 25 (7), 1469–1476.

Liu, J., Kou-Bing, C., Hwang, J.F., Lee, M.H., 2011. The Journal of Industrial Textiles 41, 123–141.

Li, P., Jia, X., 2018. tert-Butyl nitrite (TBN) as a versatile reagent in organic synthesis. Synthesis 50 (04), 711–722.

Li, R., Zhang, S., Zhang, R., 2023. Recent progress in artificial structural colors and their applications in fibers and textiles. Chemistry of Materials 1–17. https://doi.org/10.1002/cmtd.202200081. Wiley-VCH GmbH.

Li, W., Ji, W., Sun, H., Lan, D., Wang, Y., 2019. Langmuir 35, 113–119.

Luan, F., Xu, X., Liu, H., Cordeiro, M.N.D.S., 2013. Review of quantitative structure-activity/property relationship studies of dyes: recent advances and perspectives. Coloration Technology 129 (3), 173–186.

Lv, D., Cui, J., Wang, Y., Zhu, G., Zhang, M., Li, X., 2017. Synthesis and color properties of novel polymeric dyes based on grafting of anthraquinone derivatives onto O-carboxymethyl chitosan. RSC Advances 7 (53), 33494–33501.

Manian, A.P., Paul, R., Bechtold, T., 2016. Metal mordanting in dyeing with natural colourants. Coloration Technology 132 (2), 107–113.

Mao, H., Lin, L., Ma, Z., Wang, C., 2018. Dual-responsive cellulose fabric based on reversible acidichromic and photoisomeric polymeric dye containing pendant azobenzene. Sensors and Actuators B: Chemical 266, 195–203. https://doi.org/10.1016/j.snb.2018.02.131.

Mapari, S.A.S., Thrane, U., Meyer, A.S., 2010. Fungal polyketide azaphilone pigments as future natural food colorants? Trends in Biotechnology 28, 300–307. https://doi.org/10.1016/j.tibtech.2010.03.004.

Marcou, G., Horvath, D., Solov'ev, V., Arrault, A., Vayer, P., Varnek, A., 2012. Interpretability of SAR/QSAR models of any complexity by atomic contributions. Molecular Informatics 31 (9), 639–642.

Mazotto, A.M., De Ramos Silva, J., De Brito, L.A.A., Rocha, N.U., De Souza Soares, A., 2021. How can microbiology help to improve sustainability in the fashion industry? Environmental Technology & Innovation 23, 101760.

Meng, F., Umair, M.M., Iqbal, K., Jin, X., Zhang, S., Tang, B., 2019. Rapid fabrication of noniridescent structural color coatings with high color visibility, good structural stability, and self-healing properties. ACS Applied Materials & Interfaces 11, 13022–13028.

Meng, Y., Tang, B., Ju, B., Wu, S., Zhang, S., 2017. ACS Applied Materials & Interfaces 9, 3024–3029.

Merino, E., 2011. Synthesis of azobenzenes: the coloured pieces of molecular materials. Chemical Society Reviews 40 (7), 3835–3853.

Mihelač, M., Siljanovska, A., Košmrlj, J., 2021. A convenient approach to arenediazonium tosylates. Dyes and Pigments 184, 108726. https://doi.org/10.1016/j.dyepig.2020.108726.

Mirjalili, M., Nazarpoor, K., Karimi, L., 2011. Eco-friendly dyeing of wool using natural dye from weld as co-partner with synthetic dye. Journal of Cleaner Production 19 (9), 1045–1051. https://doi.org/10.1016/j.jclepro.2011.02.001.

Muratov, E.N., Bajorath, J., Sheridan, R.P., Tetko, I.V., Filimonov, D., Poroikov, V., Oprea, T.I., Baskin, I.I., Varnek, A., Roitberg, A., Isayev, O., Curtarolo, S., Fourches, D., Cohen, Y., Aspuru-Guzik, A., Winkler, D.A., Agrafiotis, D., Cherkasov, A., Tropsha, A., 2020. QSAR without borders. Chemical Society Reviews 49 (11), 3525−3564.

Nazir, F., Siddique, A., Nazir, A., Javed, S., Hussain, T., Abid, S., 2022. Eco-friendly dyeing of cotton using waste-derived natural dyes and mordants. Coloration Technology 138 (6), 684−692.

Nesmerak, K., Toropov, A.A., Toropova, A.P., Kohoutova, P., Waisser, K., 2013. SMILES-based quantitative structure-property relationships for half-wave potential of N-benzylsalicylthioamides. European Journal of Medicinal Chemistry 67, 111−114.

Nieminen, E., Linke, M., Tobler, M., Beke, B.V., 2007. EU COST Action 628: life cycle assessment (LCA) of textile products, eco-efficiency and definition of best available technology (BAT) of textile processing. Journal of Cleaner Production 15 (13−14), 1259−1270.

Niinimäki, K., Peters, G., Dahlbo, H., Perry, P., Rissanen, T., Gwilt, A., 2020. The environmental price of fast fashion. Nature Reviews Earth & Environment 1 (4), 189−200. https://doi.org/10.1038/s43017-020-0039-9.

Oakes, J., 2002. Principles of colour loss. Part 1: mechanisms of oxidation of model azo dyes by detergent bleaches. Review of Progress in Coloration and Related Topics 32 (1), 63−79.

Oakes, J., 2003. Principles of colour loss. Part 2: degradation of azo dyes by electron transfer, catalysis and radical routes. Review of Progress in Coloration and Related Topics 33 (1), 72−84.

Oakes, J., Dixon, S., 2003a. Adsorption of dyes to cotton and inhibition by polymers. Coloration Technology 119 (3), 140−149.

Oakes, J., Dixon, S., 2003b. Adsorption of dyes to cotton and inhibition by surfactants, polymers and surfactant−polymer mixtures. Coloration Technology 119 (6), 315−323.

Oakes, J., Gratton, P., Clark, R., Wilkes, I., 1998. Kinetic investigation of the oxidation of substituted arylazonaphthol dyes by hydrogen peroxide in alkaline solution. Journal of the Chemical Society, Perkin Transactions 2 (12), 2569−2576.

Oger, N., d'Halluin, M., Le Grognec, E., Felpin, F.-X., 2014. Using aryl diazonium salts in palladium-catalyzed reactions under safer conditions. Organic Process Research & Development 18 (12), 1786−1801.

Okuhara, T., 2002. Water-tolerant solid acid catalysts. Chemical Reviews 102 (10), 3641−3665.

Park, J.-Y., Hirata, Y., Hamada, K., 2012. Relationship between the dye/additive interaction and inkjet ink droplet formation. Dyes and Pigments 95 (3), 502−511. https://doi.org/10.1016/j.dyepig.2012.04.019.

Pawar, A.B., More, S.P., Adivarekar, R.V., 2018. Dyeing of polyester and nylon with semi-synthetic azo dye by chemical modification of natural source areca nut. Natural Products and Bioprospecting 8, 23−29.

Patel, D.R., Patel, K.C., 2015. Synthesis, characterization and in vitro antimicrobial screening of some new MCT reactive dyes bearing nitro quinazolinone moiety. Journal of Saudi Chemical Society 19, 347−359.

Peters, R.H., 1975. Textile Chemistry: The Physical Chemistry of Dyeing. Elsevier Sci. Publ. Co, Amsterdam-Oxford-New York.

Phan, K., Raes, K., van Speybroeck, V., Roosen, M., Clerck, K. de, Meester, S. de, 2021. Non-food applications of natural dyes extracted from agro-food residues: a critical review. Journal of Cleaner Production 301, 126920. https://doi.org/10.1016/j.jclepro.2021.126920.

Polanski, J., Gieleciak, R., Wyszomirski, M., 2003. Comparative molecular surface analysis (CoMSA) for modeling dye-fiber affinities of the azo and anthraquinone dyes. Journal of Chemical Information and Computer Sciences 43 (6), 1754–1762.

Polanski, J., Gieleciak, R., Wyszomirski, M., 2004. Mapping dye pharmacophores by the comparative molecular surface analysis (CoMSA): application to heterocyclic monoazo dyes. Dyes and Pigments 62 (1), 61–76.

Qiu, J., Tang, B., Ju, B., Xu, Y., Zhang, S., 2017. Stable diazonium salts of weakly basic amines—convenient reagents for synthesis of disperse azo dyes. Dyes and Pigments 136, 63–69.

Ran, J., Pryazhnikova, V.G., Telegin, F.Y., 2022. Chemoinformatics analysis of the colour fastness properties of acid and direct dyes in textile coloration. Colorants 1, 280–297.

Raposo, M.F.de J., de Morais, R.M.S.C., Bernardo de Morais, A.M.M., 2013. Bioactivity and applications of sulphated polysaccharides from marine microalgae. Marine Drugs 11 (1), 233–252.

Rawat, D., Mishra, V., Sharma, R.S., 2016. Detoxification of azodyes in the context of environmental processes. Chemosphere 155, 591–605.

Robinson, S.C., Tudor, D., Cooper, P.A., 2012. Utilizing pigment-producing fungi to add commercial value to American beech (Fagus grandifolia). Applied Microbiology and Biotechnology 93, 1041–1048. https://doi.org/10.1007/s00253-011-3576-9.

Roglans, A., Pla-Quintana, A., Moreno-Mañas, M., 2006. Diazonium salts as substrates in palladium-catalyzed cross-coupling reactions. Chemical Reviews 106 (11), 4622–4643.

Roy, K. (Ed.), 2020. Ecotoxicological QSARs. Springer US, New York, NY.

Salabert, J., Sebastián, R.M., Vallribera, A., 2015. Anthraquinone dyes for superhydrophobic cotton. Chemical Communications 51, 14251–14254.

Sahoo, J., Mekap, S.K., Kumar, P.S., 2015. Synthesis, spectral characterization of some new 3-heteroaryl azo 4-hydroxy coumarin derivatives and their antimicrobial evaluation. Journal of Taibah University for Science 9, 187–195.

Saxena, S., Raja, A.S.M., 2014. Natural dyes: sources, chemistry, application and sustainability issues. In: Muthu, S.S. (Ed.), *Roadmap to Sustainable Textiles and Clothing,* Singapore. Springer Singapore, pp. 37–80.

SDC, AATCC *Colour Index™ Online*. Available at: https://colour-index.com/about. (Accessed 29 April 2022).

Sendilkumar, D., 2021. The Rise of Algae-Based Dyes. Available at: https://www.spirainc.com/news/the-rise-of-algae-based-dyes. (Accessed 20 March 2023).

Shahid, M., Shahid-ul-Islam, Mohammad, F., 2013. Recent advancements in natural dye applications: a review. Journal of Cleaner Production 53, 310–331. https://doi.org/10.1016/j.jclepro.2013.03.031.

Shan, B., Tang, B., Zhang, S., Tan, W., 2020. Synthesis and dyeing properties of polyvinylamine dyes for cotton. Coloration Technology 136 (3), 288–295.

Sharma, B., Dangi, A.K., Shukla, P., 2018. Contemporary enzymebased technologies for bioremediation: a review. Journal of Environmental Management 210, 10–22.

Shinde, S., Bait, S.P., Adivarekar, R., Nethi, N.S., 2021. Benzophenone based disperse dyes for UV protective clothing: synthesis, comparative study of UPF, light fastness and dyeing properties and computational study. Journal of the Textile Institute 112, 71–84.

Shinde, S., Sekar, N., 2019. Synthesis, spectroscopic characteristics, dyeing performance and TD-DFT study of quinolone based red emitting acid azo dyes. Dyes and Pigments 168, 12–27. https://doi.org/10.1016/j.dyepig.2019.04.028.

Shore, J. (Ed.), 2002. Colorants and Auxiliaries: Organic Chemistry and Application Properties. Colorants, vol. 1. Society of Dyers & Colourists. Available at: http://www.lob.de/cgi-bin/work/suche2?titnr=236940463&flag=citavi.

Singh, A., Sheikh, J., 2022. Synthesis of novel, coumarin-based, mosquito repellent-cum-multifunctional azo disperse dye for functional dyeing of polyester. Journal of the Textile Institute 1–11. https://doi.org/10.1080/00405000.2022.2109103.

Singh, K., Arora, S., 2011. Removal of synthetic textile dyes from wastewaters: a critical review on present treatment technologies. Critical Reviews in Environmental Science and Technology 41 (9), 807–878.

Singh, R., Jain, A., Panwar, S., Gupta, D., Khare, S.K., 2005. Antimicrobial activity of some natural dyes. Dyes and Pigments 66 (2), 99–102. https://doi.org/10.1016/j.dyepig.2004.09.005.

Solov'ev, V.P. ISIDA/QSPR. Available at: https://vpsolovev.ru/programs/isidaqspr/. (Accessed 21 December 2022).

Solov'ev, V.P., Varnek, A., Wipff, G., 2000. Modeling of ion complexation and extraction using substructural molecular fragments. Journal of Chemical Information and Computer Sciences 40 (3), 847–858.

Su, C.-H., 2013. Recoverable and reusable hydrochloric acid used as a homogeneous catalyst for biodiesel production. Applied Energy 104, 503–509. https://doi.org/10.1016/j.apenergy.2012.11.026.

Sun, J., Bhushan, B., Tonga, J., 2013. Structural coloration in nature. RSC Advances (35).

Sushko, I., Novotarskyi, S., Körner, R., Pandey, A.K., Rupp, M., Teetz, W., Brandmaier, S., Abdelaziz, A., Prokopenko, V.V., Tanchuk, V.Y., Todeschini, R., Varnek, A., Marcou, G., Ertl, P., Potemkin, V., Grishina, M., Gasteiger, J., Schwab, C., Baskin, I.I., Palyulin, V.A., Radchenko, E.V., Welsh, W.J., Kholodovych, V., Chekmarev, D., Cherkasov, A., Aires-de-Sousa, J., Zhang, Q.-Y., Bender, A., Nigsch, F., Patiny, L., Williams, A., Tkachenko, V., Tetko, I.V., 2011. Online chemical modeling environment (OCHEM): web platform for data storage, model development and publishing of chemical information. Journal of Computer-Aided Molecular Design 25 (6), 533–554.

Tang, A.Y., Kan, C., 2020. Non-aqueous dyeing of cotton fibre with reactive dyes: a review. Coloration Technology 136 (3), 214–223.

Tang, R., Yu, Z., Zhang, Y., Qi, C., 2016. Synthesis, characterization, and properties of antibacterial dye based on chitosan. Cellulose 23, 1741–1749.

Telegin, F., Shushina, I., Ran, J., Biba, Y., Mikhaylov, A., Priazhnikova, V., 2013. Structure – property relationships for dyes of different nature. Advanced Materials Research 821–822, 488–492.

Telegin, F.Y., 2009. Structure and properties of dyes in theory and practice of coloration. Design, Materials, Technology 11 (4), 163–167.

Telegin, F.Y., Khaylenko, E.S., Telegin, P.F., 2008. Quantitative relationships for design of disperse dyes of high technical properties. In: Proceedings of the 21th IFATCC Congress, 6-9 May. Barcelona, Spain.

Telegin, F.Y., Koifman, O.I., 2016. Fragment approach in QSPR analysis of organic molecules for aqueous redox flow batteries. In: ACS Publications Symposium in Partnership with ICCAS. Innovation in Molecular Science, pp. 111–112. Beijing, October 23-25. Beijing, China.

Telegin, F.Y., Marfin, Y.S., 2022. Polarity and structure of BODIPYs: a semiempirical and chemoinformation analysis. Russian Journal of Inorganic Chemistry 67 (3), 362–374.

Telegin, F.Y., Priazhnikova, V.G., Ran, J., 2016a. Fragment approach in QSPR analysis of anionic dyes. In: ACS Publications Symposium in Partnership with ICCAS. Innovation in Molecular Science, pp. 181–182. Beijing, October 23-25. Beijing, China.

Telegin, F.Y., Ran, J., Morshed, M., Pervez, M.N., Sun, L., Zhang, C., Priazhinikova, V.G., 2016b. Structure and properties of dyes in coloration of textiles. Application of fragment approach. Key Engineering Materials 703, 261–266.

Telegin, F.Y., Zarubina, N.P., 2004. Relationshipts between structure of acid azo dyes and their behavior in wool dyeing. Izvestiya Vysshikh Uchebnykh Zavedenii, Khimiya i Khimicheskaya Tekhnologiya 47 (8), 42–47 (in Russian).

Teli, M.D., Chavan, P.P., 2018. Dyeing of cotton fabric for improved mosquito repellency. Journal of the Textile Institute 109, 427–434.

Timofei, S., Schmidt, W., Kurunczi, L., Simon, Z., 2000. A review of QSAR for dye affinity for cellulose fibres. Dyes and Pigments 47 (1), 5–16.

Toropov, A.A., Toropova, A.P., Marzo, M., Benfenati, E., 2020. Use of the index of ideality of correlation to improve aquatic solubility model. Journal of Molecular Graphics and Modelling 96, 107525.

Toropova, A.P., Toropov, A.A., Roncaglioni, A., Benfenati, E., 2022. Monte Carlo technique to study the adsorption affinity of azo dyes by applying new statistical criteria of the predictive potential. SAR and QSAR in Environmental Research 33 (8), 621–630.

Umbuzeiro, G.D.A., Albuquerque, A.F., Vacchi, F.I., Szymczyk, M., Sui, X., Aalizadeh, R., Ohe, P. C. von der, Thomaidis, N.S., Vinueza, N.R., Freeman, H.S., 2019. Towards a reliable prediction of the aquatic toxicity of dyes. Environmental Sciences Europe 31 (1).

Vankar, P.S. (Ed.), 2016. Handbook on Natural Dyes for Industrial Applications (Extraction of Dyestuff from Flowers,Leaves, Vegetables), second ed. NIIR Project Consultancy Services.

Vickerstaff, T., 1954. The Physical Chemistry of Dyeing, second ed. Oliver & Boyd, London-Edinburgh.

Vikrant, K., Giri, B.S., Raza, N., Roy, K., Kim, K.H., Rai, B.N., et al., 2018. Recent advancements in bioremediation of dye: currentstatus and challenges. Bioresource Technology 253, 355–367.

Wang, X., Sun, Y., Wu, L., Gu, S., Liu, R., Liu, L., Liu, X., Xu, J., 2014. Quantitative structure–affinity relationship study of azo dyes for cellulose fibers by multiple linear regression and artificial neural network. Chemometrics and Intelligent Laboratory Systems 134, 1–9.

Wang, Y., Tang, B., Ma, W., Zhang, S., 2011. Synthesis and UV-protective properties of monoazo acid dyes based on 2-hydroxy-4-methoxybenzophenone. Procedia Engineering 18, 162–167.

Williams, T.N., Kuenemann, M.A., Van Den Driessche, George, A., Williams, A.J., Fourches, D., Freeman, H.S., 2018a. Toward the rational design of sustainable hair dyes Using cheminformatics approaches: step 1. Database development and analysis. ACS Sustainable Chemistry & Engineering 6 (2), 2344–2352.

Williams, T.N., Van Den Driessche, George, A., Valery, A.R.B., Fourches, D., Freeman, H.S., 2018b. Toward the rational design of sustainable hair dyes using cheminformatics approaches: step 2. Identification of hair dye substance database analogs in the Max Weaver dye library. ACS Sustainable Chemistry & Engineering 6 (11), 14248–14256.

Wilson, K., Clark, J.H., 2000. Solid acids and their use as environmentally friendly catalysts in organic synthesis. Pure and Applied Chemistry 72 (7), 1313–1319.

World Dye Variety, 2022. Available at: http://www.worlddyevariety.com/. (Accessed 29 April 2022).

Xu, Y., Chen, X., Li, Y., Ge, F., Zhu, R., 2016. Quantitative structure–property relationship (QSPR) study for the degradation of dye wastewater by Mo–Zn–Al–O catalyst. Journal of Molecular Liquids 215, 461–466.

Young, L., Yu, J., 1997. Ligninase-catalysed decolorization of synthetic dyes. Water Research 31, 1187–1193. https://doi.org/10.1016/S0043-1354(96)00380-6.

Youssef, Y.A., Kamel, M.M., Taher, M.S., Ali, N.F., Abd El Megiede, Saadia, A., 2014. Synthesis and application of disazo reactive dyes derived from sulfatoethylsulfone pyrazolo [1,5-a]pyrimidine derivatives. Journal of Saudi Chemical Society 18 (3), 220–226. https://doi.org/10.1016/j.jscs.2011.06.015.

Yu, S., Zhou, Q., Zhang, X., Jia, S., Gan, Y., Zhang, Y., Shi, J., Yuan, J., 2018. Hologram quantitative structure–activity relationship and topomer comparative molecular-field analysis to predict the affinities of azo dyes for cellulose fibers. Dyes and Pigments 153, 35–43.

Yu, X., Wang, H., 2021. Support vector machine classification model for color fastness to ironing of vat dyes. Textile Research Journal 91 (15–16), 1889–1899.

Yusuf, M., Shahid, M., Khan, M.I., Khan, S.A., Khan, M.A., Mohammad, F., 2015. Dyeing studies with henna and madder: a research on effect of tin (II) chloride mordant. Journal of Saudi Chemical Society 19 (1), 64–72. https://doi.org/10.1016/j.jscs.2011.12.020.

Zhang, G., Zhang, S., 2020. Quantitative structure-activity relationship in the photodegradation of azo dyes. Journal of Environmental Sciences 90, 41–50.

Zhokhova, N.I., Baskin, I.I., Palyulin, V.A., Zefirov, A.N., Zefirov, N.S., 2005. A study of the affinity of dyes for cellulose fiber within the framework of a fragment approach in QSPR. Russian Journal of Applied Chemistry 78 (6), 1013–1017.

Zhou, C., Qi, Y., Zhang, S., Niu, W., Ma, W., Wu, S., Tang, B., 2020. Dyes and Pigments 176.

# Speciality chemicals, enzymes and finishes

# 10

*Babita U. Chaudhary, Srishti Tewari and Ravindra D. Kale*
Department of Fibres and Textile Processing Technology, Institute of Chemical Technology, Mumbai, Maharashtra, India

## 10.1 Introduction

Innovation has been a pivotal driver for the advancement of the textile industry. Progressing from natural dyes, mordants, and starch, to highly successful synthetic dyes and organic sizes, the textile industry has seen it all. Since its origin, the textile auxiliaries or speciality chemicals have evolved along with the changing needs of mankind. Today is an era of greater functionality and less environmental damage, hence the textile auxiliaries or speciality chemicals being developed and utilised are focused on being functional as well as sustainable.

Speciality chemicals in textiles are chemicals that impart a unique effect or property to textiles that are used in small quantities but have a high value. Various chemicals are used in numerous textile processes during the conversion of fibre into a finished fabric. According to the report published by Grand View Research, the worldwide textile chemicals market was valued at USD 23.62 billion in 2018, and it was predicted to increase at a CAGR of 4.5% between 2019 and 2025. High demand from the fast-growing apparel industry is responsible for this huge development.

The 21st century is the century of introspection of human activities and their impacts on the environment. There is an increasing focus of the global market on biodegradable products, reduction in utilisation of hazardous chemicals, and development of sustainable practices. Amongst all the sectors of the economy, textiles is the second-largest contributor to global pollution. The majority of the chemical utilisation occurs in the wet finishing process, which further leads to higher water utilisation. In order to address this issue, a plethora of approaches have been implemented, however chemical consumption has not be eliminated from the textile processing industry. For the last three to four decades, researchers have been directing most of their efforts into minimising the use of chemicals and looking for cleaner substitutes for the chemicals used in the wet processing of textiles. Concerning sustainable practices, enzymes have great potential to replace the conventional catalysts in the textile industry.

The word 'enzyme' was first used by the German physiologist Wilhelm Kühne in 1878 when he was describing the ability of yeast to produce alcohol from sugars, and it is derived from the Greek words en (meaning 'within') and zume (meaning 'yeast'). Enzymes are biological catalysts (also known as biocatalysts) that speed up

**Sustainable Innovations in the Textile Industry. https://doi.org/10.1016/B978-0-323-90392-9.00011-2**

biochemical reactions in living organisms. Similarly to chemical catalysts, enzymes are only required in very low concentrations, and they speed up reactions without themselves being consumed during the reaction.

Amino acid-based enzymes are globular proteins that range in size from less than 100 to more than 2000 amino acid residues. These amino acids can be arranged as one or more polypeptide chains that are folded and bent to form a specific three-dimensional structure, incorporating a small area known as the active site, where the substrate binds. The active site may well involve only a small number (less than 10) of the constituent amino acids. It is the shape and charge properties of the active site that enable it to bind to a single type of substrate molecule so that the enzyme can demonstrate considerable specificity in its catalytic activity.

The hypothesis that enzyme specificity results from the complementary nature of the sub-state and its active site was first proposed by the German chemist Emil Fischer in 1894, and became known as Fischer's 'lock and key hypothesis', whereby only a key of the correct size and shape (the substrate) fits into the keyhole (the active site) of the lock (the enzyme).

Enzymes can work at atmospheric pressure and in mild conditions with respect to temperature and acidity (pH). Most enzymes function optimally at a temperature of 30–70°C and at pH values which are near the neutral point (pH 7). Enzyme processes are potentially energy-saving and save investing in special equipment resistant to heat, pressure, or corrosion. Due to their efficiency, specific action, the mild conditions in which they work, and their high biodegradability, enzymes are very well suited for a wide range of industrial applications.

Normally, enzymes have been found to work on natural as well as synthetic substrates. Fruits, cereals, milk, fats, cotton, leather, and wood are some typical candidates for enzymatic conversion in the industry. Enzymes are used in the textile industry because they accelerate reactions, act only on specific substrates, operate under mild conditions, are safe and easy to control, can replace harsh chemicals, and are biologically degradable.

## 10.2 Historical aspects and current scenario

Since the advent of the cotton gin in the 18th century, the textile industry has progressed significantly. From ancient times to the present, textile production methods have evolved, and the materials available have influenced how people carried their belongings, dressed, and decorated their surroundings. Fabric manufacture was mechanised throughout the Industrial Revolution, with machines powered by waterwheels and steam engines (History of Clothing and Textiles – Wikipedia, n.d.). Production evolved from small-scale cottage operations to large-scale assembly-line operations. Clothing, on the other hand, was still created entirely by hand. Sewing machines revolutionised garment production in the 19th century (Parker, 1984).

The 20th century was marked by new applications for textiles as well as inventions in synthetic fibres and computerised manufacturing control systems. With increasing demand and modern machinery becoming available, it was possible for manufacturers to make their products with desirable attributes, such as enhanced strength, flexibility, or durability. Mechanical solutions, such as various weaving and knitting patterns, fibre alterations, and textile finishing (textiles) can all be used to achieve these features. Stains, fires, wrinkles, and microbiological life have all proven possible to resist since the 1960s. Dye technology advancements made it possible to colour natural and synthetic textiles that were previously impossible to dye (Kadolph and Marcketti, 2016).

However, with the advancements and increase in demand and supply, and advancements in technology, there has been a rise in fast fashion at affordable prices. Fast fashion, according to critics, adds to pollution, waste, and deliberate obsolescence because of the low-cost materials and manufacturing methods it employs. Due to ever-changing fashion there is a race among brands to provide the latest trend at affordable rates which in turn puts pressure on manufacturers. The manufacturer for profit uses cheap and hazardous chemicals for manufacturing. This leads to various problems as shown in Fig. 10.1. The badly manufactured clothing does not age properly, yet

**Figure 10.1** Problems faced due to the non-sustainable growth in the textile industry.

cannot be recycled because it is mostly made of synthetics (over 60%). As a result, when discarded it rots for years in landfills (Schlossberg, 2019).

As well as being highly potent catalysts, enzymes also possess remarkable specificity in that they generally catalyse the conversion of only one type (or at most a range of similar types) of substrate molecules into product molecules. Some enzymes demonstrate group specificity. For example, alkaline phosphatase can remove a phosphate group from a variety of substrates. Other enzymes demonstrate much higher specificity, which is described as absolute specificity. For example, glucose oxidase shows almost total specificity for its substrate, β-D-glucose, and virtually no activity with any other monosaccharides.

The technology for producing and using commercially important enzyme products combines the disciplines of microbiology, genetics, biochemistry, and engineering, which have been developed and matured through time both singly and in an interactive manner. The organism is preferred which provides high yields of enzymes in the shortest possible fermentation time. The production strains used in industry are normally modified by genetic manipulation to enable high levels of production.

## 10.3 Impacts on health and the environment

The fashion sector, according to the Ellen MacArthur Foundation, is responsible for 10% of global $CO_2$ emissions, which is more than all international flights and shipping combined. This is largely due to the lengthy transportation routes that textiles travel from manufacturing to purchase and disposal (A New Textiles Economy: Redesigning Fashion's Future, 2017).

A deadly combination of chemicals is used to dye cheap clothing. Denim manufacture is the world's second-largest polluter of fresh water. Over 2.5 billion gallons of waste from the textile sector have contaminated 70% of Asia's lakes and rivers, resulting in a catastrophic ecological and public health problem (Ribeiro et al., 2020).

Another problem is that polyester sheds microfibres with each wash cycle. These trace elements end up in the water supply. They increase the amount of plastic in the water and damage marine food cycles there. According to studies, these microscopic fibres end up in the intestines of animals and even make their way onto our plates in the form of seafood (Webber, 2018). The textile industry has been a major contributor to water pollution and generates a large quantity of toxic waste in the form of heavy metals, dyes, detergents, etc. Enzyme technology tackles these challenges head-on as the enzymes are biodegradable, require a lower temperature to function, and can be reused. Enzymes tend to replace harsh chemicals and also improve the quality of the product (Kumar et al., 2021).

# 10.4 Innovations leading to sustainable textile processes

## 10.4.1 *Textile pretreatments*

### 10.4.1.1 *Sizing and desizing*

#### Speciality chemicals

Lubricants are commonly used with warp sizes in fabric production, with polyamide filament fabrics being an exception. Warp sizes are used to protect yarns during weaving and to improve the weaving performance of filament and spun yarns. Textile lubricants impart essential properties for wet and dry spinning, such as low friction and reduced fibre breakage to synthetic and natural fibres. Sizing formulation mostly comprises of binder/adhesive, lubricant and in some cases, a small number of other additives are added like weighting agent, softeners and antiseptic agents. Sizing agents are primarily classified as natural materials and their derivatives like starch, carboxymethyl cellulose, pectin, and synthetic polymers like polyvinyl alcohol (PVA), polyester, polyacrylates. With the introduction of new spinning and weaving technology and various high-speed shuttleless looms, a mixture of natural and synthetic sizes is used (Goswami et al., 2004).

Among various textile processes, sizing and desizing account for about 40%–60% of the effluent load in the textile industry (Hebeish et al., 2006). Although PVA is more expensive compared to starch, it is preferred due to its easy desizeability and excellent sizing performance. However, PVA accounts for 45% of the total BOD load and does not degrade in effluent treatment plants, and is present in the water when released from the effluent plant. Attempts to recover and reuse PVA and other sizing agents and limit their release into the environment have been technically challenging and/or economically unviable (Sarkar et al., 2012).

To overcome problems caused by PVA many companies are coming up with new sizing formulations, such as SIZOL AQ from Zydex which is a homogeneous chemically modified starch for partially replacing additives used for spun yarn desizing. The product works well with polyester, cotton, and viscose fibres (Zydex Industries, n.d.). Sicoflex ES by Indokem is one shot agent for sizing warp yarns for high-speed weaving which can be removed easily using hot water without using an enzyme (Welcome to Indokem, n.d.).

Various works have been carried out on substituting the PVA or reducing the sizing load. Reddy et al. (2014) have used keratin from chicken feathers to develop a warp sizing agent for polyester (PE) and polyester/cotton (P/C) blend. They studied the COD levels of both PVA and keratin in activated sludge and it was found that there was a substantial decrease in the case of keratin, whereas there was a negligible decrease for PVA (Reddy et al., 2014). Lihong et al. used soy protein to develop a sizing agent and found that a 5-day BOD/COD ratio of 0.57 compared to 0.01 for PVA indicating soy protein sizes was easily biodegradable in activated sludge. Also, a PE and P/C blend showed improvement in strength and abrasion resistance compared to commercially available PVA-based sizes (Chen et al., 2013). Zhao et al. used soymeal to fabricate biodegradable and low-cost warp sizes containing natural plasticisers and tackifiers. Compared to triethanolamine plasticised SPI sizes, soymeal has greater flexibility, adhesion to yarn, and abrasion resistance (Zhao et al., 2016). Zhang et al. synthesised carboxymethylated corn starch (CMCS) via a microwave-assisted method using corn starch as raw material. CMCS as a sizer was used on cotton yarn and showed some performances close to those of PVA (Zhang et al., 2015).

#### Enzymatic desizing

Many natural and chemical formulations have been utilised to size yarns ranging from starch to CMC. Starch and its derivatives are most commonly used because of their excellent film-forming capacity, availability, and relatively low cost (Feitkenhauer et al., 2003). After weaving, the sizing agent and natural non-cellulosic materials present in the cotton need to be removed to prepare the fabric for dyeing and finishing.

Desizing was conventionally carried out by treatment of greige fabric with acid, alkali, or oxidising agents accompanied by high temperatures. However, the chemical treatment was not efficient as it did not completely degrade starch, leading to imperfections in dyeing, and also resulted in degradation of the cotton fibre, destroying its natural, soft feel.

Amylase is a hydrolytic enzyme that catalyses the breakdown of dietary starch into short-chain sugars, dextrin, and maltose. An amylase enzyme can be used for desizing processes at low temperature (30−60°C) and the optimum pH is 5.5−6.5 (Cavaco-Paulo and Gübitz, 2003). Nowadays, amylases are commercialised and preferred for desizing due to their high efficiency and specificity, completely removing the size without any harmful effects on the fabric (Etters and Annis, 1998). The starch is randomly cleaved into water-soluble dextrins that can then be removed by washing. This also reduces the environmental load as the discharge of waste chemicals is eliminated with the use of enzymes (Araújo et al., 2008).

### *10.4.1.2 Scouring and bleaching*

Greige or untreated cotton contains various non-cellulosic impurities, such as waxes, pectins, hemicelluloses, and mineral salts, present in the cuticle and primary cell wall of the fibre (Etters, 1999). These impurities are responsible for the hydrophobic properties of raw cotton and impede dyeing and finishing (Mojsov et al., 2020). Scouring is therefore a pretreatment process to remove the impurities to prepare the fabric for further wet processing. Along with the removal of non-cellulosic impurities, scouring also improves the wettability of the fabric. Strong alkali treatment is preferred to scour the natural fibres. Although cotton has a good affinity for the alkali, scouring chemicals also affect the cellulose negatively, leading to a reduction in strength and hence a loss of fabric weight. Furthermore, the resulting wastewater has a high COD (chemical oxygen demand), BOD (biological oxygen demand), and salt content (Buschle-Diller et al., 2016).

Bleaching is the process of removing the natural pigment or colour of the fibres and is done to achieve a whiter fabric which further ensures level dyeing. The traditional bleaching recipe included chlorine-containing oxidising compounds but today the most common industrial bleaching agent is hydrogen peroxide. Radical reactions of bleaching agents with the fibres lead to a decrease in the degree of polymerisation and, thus, lead to severe damage to the fibres and hence fabric. A good amount of chemical preparation for bleaching also renders huge quantities of rinse water to be treated.

## Innovations in scouring and bleaching

Scouring is done to remove natural impurities present in fabric such as waxes, pectin, etc. and bleaching is done when either a white finish fabric is desired or lighter shades of fabric are required. Earlier both were done separately but nowadays the process is combined so as to reduce the environmental load through a decrease in chemical consumption and less water usage. Earlier sodium hypochlorite and hydrogen peroxide were commonly used bleaching agents; however, many toxic by-products are formed and large amounts of water and energy are required to remove these chemicals, hence work is been done to substitute them or develop a new method of bleaching. Zhou et al. used glycerol triacetate as an activator for hydrogen peroxide bleaching at a lower temperature as an alternative to alkaline high-temperature bleaching (Zhou et al., 2021).

## Speciality chemicals

UNIV-AIO, a product from Sarex, India, serves as a one-stop pretreatment solution. It is a low foaming product for cotton scouring and bleaching. It requires only peroxide and does not require the use of caustic during the bleaching process. It is environmentally friendly, having low COD and TDS (Sarex, n.d.). IndiDye has developed a clay bleach under the trade name Clay White. In this product, natural mineral-rich clay is used to bleach fibres. It uses less water as compared to the traditional chlorine-based bleaching and bleached fibres are free from hypochlorous acid and other hazardous chemicals (Why IndiDye – IndiDye, n.d.).

## Enzymatic scouring

Cellulase and pectinase are combinedly used for bio-scouring. The two enzymes function synergistically to accomplish the scouring of cotton. The pectinase destroys the cotton cuticle structure by digesting the pectin and breaking the linkage between the cuticle and the body of the cotton fibre, whereas cellulase destroys the cuticle structure by digesting the primary wall of cellulose immediately under the cuticle of cotton (Mojsov, 2011).

Biological oxygen demand (BOD) and chemical oxygen demand (COD) of the enzymatic scouring process are 20%–45% as compared to the alkaline scouring (100%). Total dissolved solids (TDS) of the enzymatic scouring process are 20%–50% as compared to alkaline scouring at 100%. As a bonus, the handle of bio-scoured fabric is softer than that of the chemically scoured fabrics. Enzymatic scouring effectively scours fabric without negatively affecting the fabric or the environment. It also aims to minimise health risks by not exposing operators to aggressive chemicals.

### Enzymatic bleaching

Amyloglucosidases, pectinases, and glucose oxidases are compatible with the bleaching of cotton based on their active pH and temperature range (Jhatial et al., 2020). The generation of peracids is the only limitation of enzymatic bleaching. To overcome the peracids, the laccase/mediator system is efficient as it targets only coloured substances. Laccases are multi-copper-containing oxidoreductase enzymes capable of oxidising phenols and aromatic amines, reducing the molecular oxygen of water. Laccase in combination with redox mediators is used in textiles. Until recently, the laccases in the textile industry were potentially used to decolourise dye effluents. However, researchers have concluded that laccases are also used to bleach textiles, modify fabric surfaces, and in the colouration of cotton (Madhu and Chakraborty, 2017).

## 10.4.2 Dyeing and printing

### 10.4.2.1 Dyeing

Dyeing is the most important process in textile processing. The wastewater from textile industrial plants is regarded as the dirtiest of all mechanical areas, both in terms of quantity and the composition of effluents released. According to the United Nations Environment Programme (UNEP), it takes 3781 L of water to make a pair of jeans, from cotton production to final product delivery to the store. This equates to approximately 33.4 kg of carbon equivalent emissions (UNEP, 2018). Conventional bulk dyeing and wet processing of garments consume a lot of water and, according to the World Bank, account for up to 20% of all industrial water pollution. Untreated effluent, laden with toxic chemicals, is dumped into local streams, putting vulnerable populations at risk in countries where water is already scarce and of poor quality (UNEP, 2018). To address these concerns, new methods of colouring textiles with little to no water and chemicals are being researched and developed. Some of the game-changing solutions that are currently available, as well as what is expected in the future, are discussed further and illustrated in Fig. 10.2.

#### Dope, spin or solution dyeing

For synthetic fabrics, several companies colour the fibres in the manufacturing state, thus reducing the usage of various chemical additives and water. Synthetic fibres can be coloured in a one-step dope dyeing process that introduces pigment, in the form of colour chips, to the liquid polymer solution prior to extrusion into fibres. This means that the pigment is incorporated into the fibre, resulting in superior colour fastness and consistency. We are SpinDye has pioneered such dyeing methods, uses recycled synthetic fibres for manufacturing clothes, and has partnerships with various brands. They claim to have reduced water usage by 75% in the entire production chain process. Patagonia, Inc., an outdoor clothing US-based firm has started solution dyeing for almost all of its fabrics required for luggage, fleece, and shell (Solution Dyeing, n.d.). Various other companies such as Croozer, Burton, have also adopted the dope dyeing method for their products (Solution Dyeing, n.d.; The Advantages of Solution Dyed Fabrics, 2020). Another novel technology involves the solubilisation of a water-

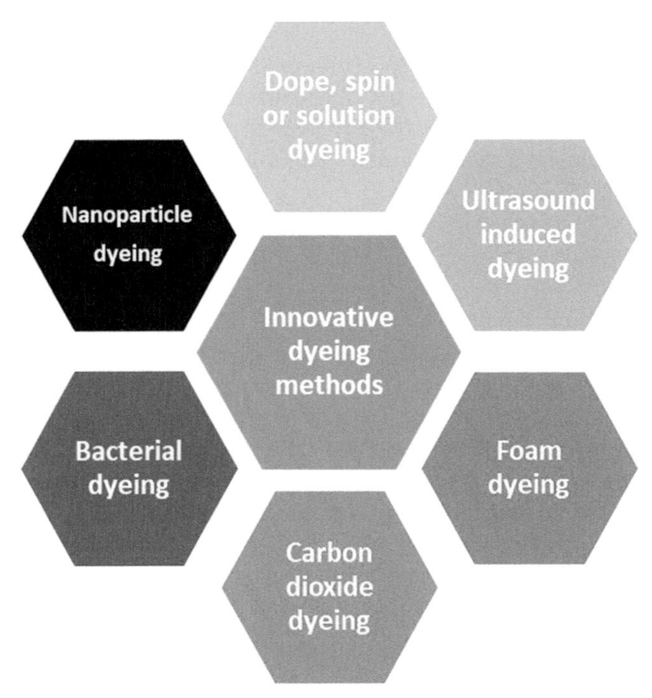


**Figure 10.2** Various innovative dyeing methods.

insoluble natural dye in oil in water microemulsions. The dyeing behaviour on wool was studied and the mechanism elucidated (Paul et al., 2005).

## Ultrasound-induced dyeing

Due to increased awareness of the environmental and health hazards of synthetic dyes, much research has been conducted to increase the dyeability of fabrics using an ultrasound technique in conjunction with natural dyes extracted from various sources such as coconut coir extract, *Eclipta alba* (Vankar et al., 2007; Adeel et al., 2020). IndiDye is one such company that uses ultrasonic technology along with its patented natural dyes and auxiliaries with a high level of fastness without using any mordants, salts, or hazardous chemicals. As per the company's website data, the life cycle assessment (LCA) of their dyes shows up to a 98% water reduction as compared to conventional dyes and dyeing can be done in shorter times and at lower temperatures (Why IndiDye – IndiDye, n.d.).

## Foam dyeing

Foam dyeing is a water-saving, environmentally friendly technology that is becoming increasingly popular around the world, particularly for textiles. Research work has been done for using foam dyeing for various fabrics such as for cotton fabric using reactive dye (Yu et al., 2014) and pigment (Shang et al., 2011), on polyamide fabrics with acid dye (Shang et al., 2011; Wang et al., 2019) to reduce the usage of energy and water. It is gaining popularity mainly for denim dyeing. Gaston Mills, Indigo Mill Designs, and Tejidos Royo, a Spanish textile mill, have collaborated to transform the denim industry with a dry indigo dyeing technique. The patented IndigoZERO process, which was introduced in 2018, uses foam to dye the yarn, with air transporting the dye onto the fabric instead of water. IndigoZERO uses 99% less water than traditional rope dyeing, allows for small lot sizes, and is performed in a machine one-quarter the size of a typical indigo dyeing machine.

### Supercritical $CO_2$

A lot of research work has been done to use $CO_2$ for the dyeing of various fabrics along with imparting various value-added properties. Polyester was dyed using supercritical $CO_2$ with photochromic dyes to develop UV-sensing smart textile (Abate et al., 2020). The Dutch company, DyeCoo, uses $CO_2$ to dye synthetic fabrics on an industrial scale. In the dye vessel, the $CO_2$ is heated and heavily compressed to reach a supercritical state between a gas and a liquid, which makes it very solvent. This allows the dye to dissolve easily and penetrate deep into the fibre. The efficient colour absorption and short batch cycles cut energy usage by half, what is more, the $CO_2$ used is reclaimed from existing industrial processes, and up to 95% is recycled (Dyecoo, n.d.).

### Bacterial pigments

London Biodesign Lab, Faber Futures, as well as the Vienna Textile Lab and the research project Living Colour, have obtained colour from naturally pigment-producing bacteria to dye both natural and synthetic fabrics. The process requires very little water, low temperatures, no chemicals or fixatives, and does not generate any polluting by-products. Growing bacteria is also far less resource-intensive than cultivating dye plants, while the biodegradable pigment is friendly to humans and the environment, even possessing antimicrobial and UV-protection properties. Scaling things up, French start-up Pili and Colorfix in the UK have bio-engineered bacteria to carry colour molecules, which then undergo fermentation to produce large quantities of pigment.

### Algal colours

Many researchers are exploring the usage of photosynthetic organisms for colouration of textiles. Algae are abundantly available organisms with different coloured pigments and can act as a safe and sustainable choice for the environment. Three major categories of coloured pigments occur in algae, i.e., chlorophyll which yields green, carotenoids yield red, orange, yellow, amber or brown colour, and phycobilins yield red or blue colour (Campos et al., 2001). Blue-green algae were used for extracting blue phycocyanin dye which was applied on cotton and silk (Baek et al., 2022). Brown and red algae were used for dyeing of cotton fabric using mordants and revealed a difference in colour with pre- and post-mordant techniques (Azeem et al., 2019). Green algae were used to dye wool fibres using a microwave technique and the effect of chitosan and tannic acid on dye strength was also studied (El-Khatib, 2016). Green algae, *Cladophora glomerata* L., were used for dyeing and it was found that 2% NaOH was the best medium for colour extraction and gave excellent shade at 75°C for 45 min

(Mir et al., 2019). Berlin design studio, Blond and Bieber, use algae to create colourful dyes for textile printing. They have created an analogue textile printer that produces its own dye using different types of algae. The prints are photo-sensitive and change tone over time when exposed to sunlight, creating a 'biodynamic colour palette' (Blond and Bieber's algae textile dyes won the Lodz Design Festival prize, date not known).

#### Nanoparticles

Nanoparticles are used in dyes to reduce the amount of harmful salt and water in the dye solution, which helps to protect the environment. Various nanoparticles have been used for dyeing purposes. Silica nanoparticles have been used as filler to absorb reactive dye particles and form coloured nanoparticles which were used for the dyeing of cotton, thus eliminating the use of a high amount of salt required during dyeing (Sampaio et al., 2011). For high-end dyeing of cotton and wool, even gold nanoparticles were used as stable colourfast colourants. Because of their great stability, long shelf-life, usage of aqueous solvent, a huge number of N—H donor groups, high positive charge (III), and prospective applications, cobalt (III) nanoparticles are utilised in dyes (Nazeer et al., 2018). A sono-chemical synthesis approach was used to create Co (III) nanocomplexes, $[Co(NH_3)_6]Cl_3.2H_2O$, $[Co(en)_3]Cl_3.3H_2O$, and $[Co(dien)_2]Cl_3.3.5H_2O$, and their antibacterial activity and textile dyeing behaviour investigated. It was found that the padding and curing procedure was appropriate for dyeing Co (III) colours on cotton garments and that the dye was also efficient against *Bacillus subtilis* bacteria and *Alternaria alternata* fungi (Bala et al., 2017).

### *10.4.2.2 Printing*

Printing, like dyeing, is a process for applying colour to a substrate. However, instead of colouring the whole substrate (cloth, carpet, or yarn) as in dyeing, print colour is applied only to defined areas to obtain the desired pattern. Various speciality chemicals like wetting agents, binders, carriers, oxidising or reducing agents, antifoaming agents, hygroscopic agents, thickeners, etc., are used in different styles of printing.

A novel aqueous binder of polyurethane acrylate based on either polyethylene glycol or glycerol ethoxylate-co-propoxylate having zero volatile organic compounds was used for preparing printing paste for screen printing of all types of textile fabrics using pigment dyes. The highest colour strength was obtained and fastness properties range between good and excellent for samples printed using polyurethane acrylate based on glycerol ethoxylate-co-propoxylate as a binder, irrespective of the type of fabric used (Elmolla and Schneider, 2006). A cheap natural thickener from *Amaranthus paniculates* starch was used for printing of cotton using vat and indigosol dyes instead of maize starch (Teli et al., 2007). Fenugreek gum was isolated from fenugreek seeds and subjected to cyanoethylation, which converts it into a water-soluble product that resists fermentation when stored. Furthermore, the results show that cyanoethylated fenugreek gum can be used safely as a thickener in printing transfer paper. The latter produces printed goods with a soft handle and excellent colour fastness (El-Kashouti, et al., 2004). Altraplex AMK/CB by L.N Chemical Industries is an anti-migrating agent for pigments (L.N. Chemical Industries, n.d.).

### 10.4.3 Textile finishes

#### 10.4.3.1 Enzymatic finishes

##### Enzymatic wash

Stone washing is a process that was adapted in the denim sector as a result of continuous laundering and a faded look. The traditional denim jeans had a dark blue appearance. With prolonged wear and repeated laundering, faded denim was more attractive and more acceptable by wearers. In the last 2 decades of the 20th century, the process of denim manufacturing was revolutionised by stone washing. This included removing the indigo from the selective surface of the denim fabric using pumice stone.

The pumice stones were a viable option for the initial days, but with increased demand for denim, the utilisation of pumice stone was not that efficient, as 1–2 kg of pumice stone is required to stone wash one pair of jeans for the desired effect. Also, the sludge generated due to the stone washing gets deposited in an effluent tank which requires extra efforts for separating the sludge from the effluent before effluent treatment.

The application of pumice stones has now been replaced with cellulases. Cellulase acts on the cell wall of the fibre and hence disrupts the yarn surface and leads to the loosening of indigo dye from the fabric surface and causes fading. A vast number of cellulases are available to obtain the desired appearance that is functional over a wide range of temperatures from 30 to 60°C. Based on pH, cellulases are classified as acidic (pH 4.5–5.5), basic (pH 9–10), or neutral (pH 6.6–7). Laccase has been effective in reducing the backstaining in denim washing with a mediator as well as without a mediator (Campos et al., 2001; Paul and Naik, 1997).

##### Biopolishing

The fibre of cotton is of short staple length and hence the fabric has a fuzzy appearance which is responsible for pilling on abrasion. Conventionally, softening is done for cotton fabrics to get rid of the fuzzy appearance. The disadvantage of chemical softening is the greasy feel of the fabric even after complete washing of the fabric.

Biopolishing eliminates the microfibrils from the surface of cellulosic fabrics, which in turn reduces the fuzziness and pilling tendency. This enhancement of surface properties is brought about by cellulases which hydrolyse the microfibrils protruding from the fabric surface (Cavaco-Paulo et al., 1996). Biopolishing also renders a cleaner surface to the fabric, hence adding a cool and soft feel to the fabric, however excessive lustre is obtained which is a drawback. The fabric obtained after biopolishing is non-greasy however.

Commercially available cellulases for biopolishing are a mixture of endoglucononases, exoglucanases, and cellobiases, which has the capability to modify cellulosic fibres in a controlled and desired manner (Madhu and Chakraborty, 2017). Acid cellulases which are enriched with endogluconase are well suited for biopolishing of cellulosic fibre. Cellulase treatment has also been reported to increase softness and enhance the post-dyeing and resin finishing of the fabric. Swelling of cotton was used as a tool to enhance the response of cellulase enzymes. It was found that the accessibility of enzymes to difficult-to-reach crystalline regions of cellulose was increased by the action of intracrystalline swelling agents, thus achieving the desired effect, with a lower quantity of enzymes (Paul and Teli, 2010). In another research work, the biosoftening of polyester/cotton blended fabric using cutinase and cellulose was studied (Mccloskey and Jump, 2005).

#### *10.4.3.2 Flame-retardant finishes*

Flame retardants (FR) are chemicals applied to fabrics to inhibit or suppress the combustion process. Flame-retardant finishes are important for reducing or avoiding damage to people and property caused by fires triggered by a variety of man-made and natural factors. They interfere with combustion at various stages of the process, e.g., during heating, decomposition, and ignition of flame spread (Fibre2fashion.com, n.d.). Generally, flame retardants can be divided into three major categories:

**Category I:** Gas-phase FR that prevent combustion by the reduction of heat in a gaseous phase by scavenging reactive free radicals, for example, halogen and phosphorus-based FR.
**Category II:** Endothermic FR which function in both gas phase and condensed phase by releasing non-flammable gases ($H_2O$, $CO_2$) which dilute the fuel and cool the polymer, for example, carbonates of Ca and Mg and metal hydroxides.
**Category III:** Char forming FR that acts in condensed phase by preventing fuel release, and therefore thermal insulation for underlying polymer takes place. For example, intumescents that swell as a result of heat exposure, thus reducing flame retardancy and novel flame retardants, such as expandable graphite and nanocomposite.

Halogen-based flame retardants were most commonly used due to their cost, availability, and extensive experience with usage of this class of additives. Halogen-based flame retardants interfere with the combustion cycle primarily through a free radical scavenging mechanism inhibiting flame propagation in the vapour phase.

##### Innovations in flame-retardants

Tetrabromobisphenol-A (TBBPA), polybromodiphenylether, and chlorinated paraffins have been high-production halogen-based flame retardants for many years (Papaspyrides and Kiliaris, 2014). Brominated flame retardants (BFRs) have been found in plants and wildlife throughout the food chain, as well as in human tissues, blood serum, and breast milk from exposed occupational populations (individuals working in the production of BFRs or the production, recycling, or disposal of BFR-containing products) and in general populations (Darnerud, 2003; Segev et al., 2009).

With halogenated FRs being banned, focus turned to different type of FRs such as phosphorus-based, nitrogen-based, various nanocomposites, inorganic compounds, etc. Phosphorus-based flame retardants come in a variety of forms, including inorganic and organic. Inorganic examples include ammonium phosphates, ammonium polyphosphates, and red phosphorus. Organic examples include organophosphates, organophosphonates, polyphosphonates, and hybrid metal phosphonate salts (Morgan, 2019).

Nitrogen-based flame retardants are generally used in combination with other flame retardants. Many companies are coming up with various FRs which are less harmful to the environment. Pyrovatex CP New from Huntsman International LLC (dialkylphosphonocarboxylic acid amide) and Saraflam CWF of Sarex Chemicals (organophosphorous compound) are examples of durable flame retardants for cellulosic fibre fabrics or fabric blends where the synthetic component does not exceed 25%. Pekoflam PES.CN liq c (organic phosphonate esters) of Archroma is a durable flame retardant for textiles made from polyester and polyamide fibres (Mowbray, 2015a). Eccoshield PPE-4X – ORCO product of Organic Dyes and Pigments LLC is a nitrogenous/phosphate condensate flame-retardant finishing product for polypropylene, polyester, other synthetic fabrics, and cellulosics (Organic Dyes and Pigments, n.d.). Expandable graphite is an intumescent flame-retardant agent, which is of great interest. An industrial research study established that expandable graphite can be used as an alternative to conventional flame-retardant agents, with the main advantage of not emitting any toxic or harmful fumes (Paul et al., 2012).

### *10.4.3.3 Easy care finishes*

#### Innovations in easy care finishes

Easy care finish or durable press finish is done for fabrics mainly with high cellulosic content. This finish generally provides antishrinkage properties, and improves dry and wet wrinkle recovery, i.e., crease resistance. Earlier formaldehyde-based finishing agents included urea formaldehyde or melamine formaldehyde, and non-formaldehyde products, such as 1,2,3,4-butane tetracarboxylic acid (BTCA), citric acid (Harifi and Montazer, 2012), etc. The most important and widely used chemical is DMDHEU and its derivatives because of their stability and durability. However, formaldehyde poses a significant risk and causes severe health issues such as teary eyes, breathing difficulties, and headache, and contact with the skin can lead to allergic reaction or eczema. Therefore, formaldehyde-free compounds like polycarboxylic acids, ionic cross-linking agents, and aqueous ionic polyurethane have been used.

Polycarboxylic acid should contain at least three carboxylic acids in order to be effective as a durable finish. As a result, the most common polycarboxylic acids are BTCA, citric acid, succinic acid, and malic acid. Maleic acid, as well as itaconic acid and sucrose acid, have also been used (Yang et al., 1998; Dehabadi et al., 2013; Patil and Netravali, 2020). Ahmed et al. used citric acid along with fibroin solution for a crease-resistant finish on cotton fabric (Ahmed et al., 2021). Various ionic cross-linking agents have also been used for a durable finishing (Hashem et al., 2003; Wijesena et al., 2014). These methods work on the principle of imparting ionic nature to the cotton fabric which can then absorb opposite-charged polyelectrolyte and form a cross-linking net (Dehabadi et al., 2013).

Finotex developed 'Finopret ZF' and 'Finopret ZFR', a zero-formaldehyde cross-linking agent with no built-in catalyst for a wash and wear anticrease finish (Fineotex, n.d.). Rucon FNF by Rudolf is a formaldehyde-free cross-linking agent for cellulosic fibres or their blends with synthetic fibres (Rudolf, n.d.).

#### 10.4.3.4 Antimicrobial finishes

Traditionally antimicrobial finishing was mainly done for surgical clothes, undergarments, luxury fabrics, baby wear, etc. However, with rising consumer awareness and with the increase in health concerns due to the COVID-19 pandemic, there has been a sharp rise in demand and supply for antimicrobial finishes in textiles. The antimicrobial agents either kill or inhibit the growth of microorganisms, and help in reducing their spread. Various compounds are used for imparting antimicrobial properties.

Metals and metal salts are among the commercially widely used compounds for imparting antimicrobial properties. Various metals such as copper, zinc, cobalt, silver, and gold are used in textiles. Silver, by far, is the most widely used in the textile industry. For synthetic fibres, silver particles are incorporated into the polymer before extrusion (Maleknia et al., 2015), or via electrospinning they are deposited onto the fibre (Hong et al., 2006). For cotton, it might be first treated with ligands such as succinic acid anhydride, tannic acid, or chelating agents such as EDTA, so as to enhance the subsequent absorption of metallic salts (Hipler et al., 2006; Purwar and Joshi, 2004). However, such methods are not eco-friendly and hence not commercially used.

##### Innovations in antimicrobial finishes

Recent breakthroughs in technology have overcome cost, environmental, and technical challenges associated with producing some metal-treated textiles on a commercial scale. As a result, silver is now used in a large number of commercial antimicrobial synthetic fibres and yarns. For example, Thomson Research Associates manufactures UltraFresh and Silpure products. The silver is in the form of ultra-fine metallic particles and is primarily applied to polyester fabrics at the finishing stage. Milliken has developed a silver-based antimicrobial agent, AlphaSan, which is a zirconium phosphate-based ceramic ion-exchange resin containing silver which is added during the extrusion process of synthetic fibres. AlphaSan is being used by a number of companies to produce antimicrobial textiles, for example, the polyester and nylon yarn by O'Mara (MicroFresh and SoleFresh) and the polyester yarn by Sinterama (GuardYarn). Sanitized PL 19-30, a water-based product, is being added to its product line by Sanitized AG, a specialist in hygiene function and material protection for plastics and textiles. It provides complete defence against fungi, moulds, algae, and bacteria. This water-based product is made only with active components that comply with Biocidal Products Regulation (BPR) (Sanitized, n.d.). Nano-Care has developed Liquid Guard antimicrobial coating that is long-acting as a permanent surface disinfectant and provides protection against a wide range of microorganisms and also acts as a mildew and odour inhibitor (Nano-Care, n.d.).

Quaternary ammonium compounds are composed of a hydrophobic alkyl chain and a hydrophilic counterpart. These compounds are active against a wide range of microorganisms, such as Gram-positive and Gram-negative bacteria, fungi, and certain types of viruses. In the textile industry, the compounds containing long alkyl chains (12−18 carbon atoms) are most used, mainly for cotton, polyester, nylon, and wool (Simoncic and Tomsic, 2010; Gao and Cranston, 2008). The antimicrobial action of these compounds depends on the alkyl chain length, the presence of the per fluorinated group, and the cationic ammonium group's number in the molecules. Different antimicrobial compounds used are given in Table 10.1.

#### 10.4.3.5 Water- and oil-repellent finishes

Fluorocarbons are used in water- and stain-repellent fabrics since they are known to provide durable water and oil repellence (DWOR). Fluorocarbon compounds are known for their ability to repel both water and oil. Finishes based on C8 fluorocarbons (structures with eight carbon atoms) were commonly utilised in the past. However, there have been concerns raised about these C8 fluorocarbons, particularly PFOA (perfluorooctanoic acid or pentadecafluorooctanoic acid) and PFOS (perfluorooctane sulphonate or heptadecafluoro-1-octanesulfonic acid). Because of its CMR (carcinogenic, mutagenic, or toxic for reproduction) and PBT (persistent, bioaccumulative, and toxic) qualities, PFOA has been designated as a chemical of extremely high concern.

C6 or C4 fluorocarbon compounds, or even fluorine-free water repellents, are increasingly being used to replace C8 fluorocarbon chemistry. New commercial DWOR finishes based on short-chain fluorocarbons (C6 or C4 fluorocarbon chemistry), hybrid systems, or fluorine-free finishes are currently on the market. Daikin Chemicals have UNIDYNE XF Series, a non-fluorinated repellent finish for a wide array of textile substrates that is made from greater than 50% bio-based materials (UNIDYNE XF Series for Textiles | Fluorochemicals | Daikin Global, n.d.). BioTex have a series of fluorine-free alternatives for good water and oil repellency such as Bioguard ZO, Bioguard ZOP, and Bioguard Zero. BioGuard ZERO is made up of 68% smart raw materials, including recycled stearyl polymers, which contributes to reduction in $CO_2$ footprint ("BiotexTM | Fluorine−Free Water Repellency ZERO Flouro + PFOA," n.d.). Maflon has a series of Hydrosin fluorine-free polymers that have been designed specifically for textile applications. They provide treated materials with long-lasting water repellence (DWR) and stain release properties. They are free of PFC, PFOA, PFOS, APEO, and organotin compounds (Maflon, n.d.). Rudolf's RUCO-GUARD and RUCOSTAR fluorocarbon finishing products are examples of C6 fluorocarbon finishing materials. The BARRIER series is based on fluorine-free dendrimer and 3D hyperbranched coral-like polymer technology, as well as clean C6- and C8-based fluorine, and was developed jointly by Rudolf and HeiQ (Rudolf, n.d.).

Water, oil, and stain repellence are all available in the BARRIER collection for outdoor use. The fluorocarbon-based TUBIGUARD 42-AT by CHT is a water-repellent finish for PA raincoats, while the fluorocarbon-based TUBIGUARD SRO provides dirt release. Thor's QUECOPHOB nano-dispersions are fluorocarbon-based nano-dispersions with water and dirt repellence and self-cleaning qualities. The LE5 variant is ideal for clothes and has a strong wash resistance. Archroma's NUVA range includes a variety of C6 water- and oil-repellent compounds. NANOFICS is a water- and oil-repellent coating (C6-based) developed by Europlasma for textile fabrics and apparel. The application is carried out using a low-pressure plasma method that uses only a few chemicals and no water (Mowbray, 2015).

**Table 10.1** Various chemicals used for antimicrobial finishing.

| Compounds used | Chemical structure/examples | Fibres applied on | Action mode |
| --- | --- | --- | --- |
| Metal and metallic salts | $TiO_2$ AND $ZnO_2$ | Cotton, wool, polyester, nylon | Generate reactive oxygen species, damaging cellular proteins, lipids, and DNA (Maleknia et al., 2015; Hong et al., 2006) |
| QAC | 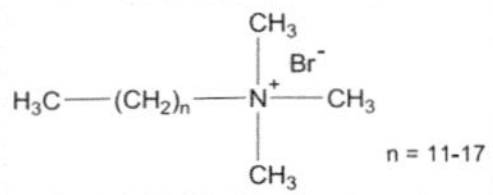 $H_3C-(CH_2)_n-N^+(CH_3)_2-CH_3$ $Br^-$, n = 11-17<br>(Example: Monoquaternary ammonium salt: alkyl trimethylammonium bromide) | Cotton, wool, polyester, nylon | Damage cell membranes; denature proteins; inhibit DNA production, avoiding multiplication (Simoncic and Tomsic, 2010; Gao and Cranston, 2008) |
| Chitosan | 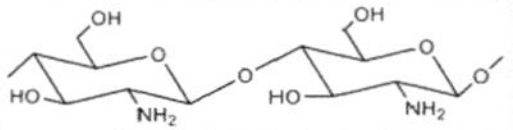 | Cotton, wool, polyester | Low Mw: inhibits synthesis of mRNA, preventing protein synthesis<br>High Mw: causes leakage of intracellular substances or blocks the transport of essential solutes into the cell (Simoncic and Tomsic, 2010) |
| PHMB | 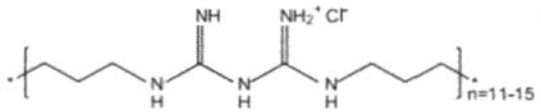 NH; $NH_2^+Cl^-$; n=11-15 | Cotton, polyester, nylon | Interacts with membrane phospholipids, resulting in their disruption and the lethal leakage of cytoplasmic materials (Simoncic and Tomsic, 2010) |
| N-halamines | 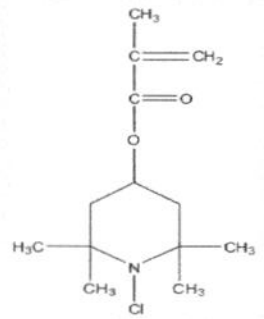 | Cotton, wool, polyester, nylon | Precludes the cell enzymatic and metabolic processes, causing consequent microorganism destruction (Gao and Cranston, 2008) |

#### 10.4.3.6 Antistatic finishes

Static electricity can cause many processing problems for textile materials, especially those made from hydrophobic synthetic fibres. In most dry textile processes, fibres and fabrics move at high speed over various surfaces, which can generate electrostatic charging from frictional forces. This electrical charge can cause fibres and yarns to repel each other, leading to ballooning and difficulty in handling. Also, due to static build-up on upholstery and carpets, a small electric shock can be experienced, and also garments cling to the body if there is static build-up.

Estofeel (Conc.), a product from Sarex is a hydrophilising agent for synthetic fibres that helps polyester, polyamide, and other synthetic fibres to overcome the static charge they generate (Sarex, n.d.). Internal antistatic agents such as LOXIOL by Emery Oleochemicals come under green polymer additives, and are intended to be incorporated into the plastics material in the compound or masterbatch. After the thermoplastic conversion, such as extrusion, injection moulding, calendaring, or compression moulding, the additive migrates to the plastics' surface to avoid the build-up of static charge (Emery Oleochemicals, n.d.). HT Fine Chemical Co. has developed the Antistatic Agent KD which is environmentally friendly and has no APEO or NPEO in it, and meets the environment standards of the UK, EU, and USA (HT Fine Chemical Co. Ltd, n.d.).

### 10.4.4 Enzymes as a tool for textile modification

Of more than 7000 enzymes known to date, there is a meagre number that are commonly used in textile industry processes. The principal enzymes applied in the textile industry are hydrolases and oxidoreductases. The group of hydrolases includes amylases, cellulases, proteases, pectinases, and lipases/esterases.

Enzymes have been classified into six classes by the International Commission on Enzymes (EC) established in 1956 by the International Union of Biochemistry (IUB) in consultation with the International Union of Pure and Applied Chemistry (IUPAC). The enzymes have been systematically categorised based on the function they perform or the reaction they catalyse (Mojsov, 2011). The EC classification system is divided into six categories of basic functions (Table 10.2).

**Table 10.2** Classification of enzymes.

| Reference code | Class | Function |
|---|---|---|
| EC1 | Oxidoreductases | Catalyse oxidation/reduction reactions |
| EC2 | Transferases | Transfer a functional group |
| EC3 | Hydrolases | Catalyse the hydrolysis of various bonds |
| EC4 | Lyases | Cleave various bonds by means other than hydrolysis and oxidation |
| EC5 | Isomerases | Catalyse isomerisation changes within a single molecule. |
| EC6 | Ligases | Join two molecules with covalent bonds. |

Enzymes have been also used to treat other natural fibres. These include sericinases for enzymatic degumming of silk (Gulrajani, 1992), proteases for felt-free-finishing of wool, and xylanases and cellulases for softening of jute (Kundu et al., 2016). Amongst synthetics, esterases have been studied for partial hydrolysis of the surface of synthetics fibres, increasing their hydrophilic property. Presently, the textile industry is reaping a good share of dividends from enzymes. These processes and finishes include fading of denim and non-denim, bio-scouring, bio-polishing, wool finishing, peroxide removal, and decolourisation of dyestuff.

#### 10.4.4.1 Modification of synthetic fibres

Synthetic fibres represent almost 50% of the worldwide textile fibre market. Polyethylene terephthalate (PET), polyamide (PA), and polyacrylonitrile (PAN) fibres show excellent features such as good strength, high chemical resistance, low abrasion, and shrinkage properties. However, synthetic fibres share common disadvantages, such as high hydrophobicity and crystallinity, which affect not only wearing comfort (making these fibres less suitable to be in contact with human skin), but also the processing of fibres, impeding the application of finishing compounds and colouring agents.

Consistent efforts are being made to achieve modification of synthetic fibres. The surface modifications in synthetic fibres are required to render them with texture and functionality imitating natural fibres. Enzymatic modification by hydrolysis tends to increase hydrophilicity, improve dyeing, and reduce the build-up of static charge. Some of these modifications have been commercialised to a limited extent but still, the application of enzymes for surface modification is at the experimentation stage. Hydrolase is a prominent class of enzymes which finds its application in synthetic fibre modification. Cutinase, esterase, and lipase have been found to be suitable for the modification of polyester.

PET was modified using esterase and increased hydrophilicity was observed after the enzymatic treatment (Fischer-Colbrie et al., 2009). An attempt was made to modify the PET yarns by esterase derived from actinomycetes. The dyeing of enzyme-modified PET with reactive dye signified the increased number of hydroxyl groups (Alisch et al., 2009). Lipases have been utilised for manufacturing bifunctional polyesters, coating cellulose by PET, and biocompatible sorbitol-loaded polyesters (Guebitz and Cavaco-Paulo, 2008). Table 10.3 shows enzyme modifications of different synthetic fibres.

#### 10.4.4.2 Clean up of bleach bath

In the textile industry, one of the necessary preliminary processes, bleaching of fabrics, is performed with $H_2O_2$ after desizing and scouring. For neutralising the effluent, a considerable amount of chemical treatment and huge volumes of water were previously required. To mitigate this situation, nowadays, enzymes are being used to reduce the bleach bath. This eliminates the need for a reducing agent and minimises the need for rinse water, resulting in less polluted wastewater and lower water consumption. The higher cost of enzymes for the degradation of hydrogen peroxide in bleaching

**Table 10.3** Enzyme modifications of synthetic fibres.

| S. no. | Enzyme | Source | Substrate | Attribute obtained | References |
|---|---|---|---|---|---|
| 1. | Cutinase | *Aspergillus oryzae*, *Humicola insolens*, *Penicillium citrinum*, *Fusarium solani*, *Thermobifida fusca*, *Thermobifida cellulolytica*, *F. oxysporum* | PET | Hydrolysis of PET | Cavaco-Paulo and Gübitz (2003), Kanelli et al. (2017), Silva and Cavaco-Paulo (2008) |
| 2. | Lipase | *Humicola* sp., *Candida antarctica*, *Thermomyces lanuginosus*, *Burkholderia* spp., *Triticum aestivum*, *Rhizopus delemar* | PET | Hydrolysis | Walter et al. (1995), Cavaco-Paulo and Gübitz (2003) |
| 3. | Laccase | Different sources | PET | Oxidative degradation without cleaving the polymer | Miettinen-Oinonen et al. (2004) |
| 4. | Laccase with mediator | Different sources | Nylon 66 | Increase the hydrophilicity | Miettinen-Oinonen et al. (2004), Silva and Cavaco-Paulo (2008) |
| 5. | Protease | *Beauveria* sp., an amidase from *Nocardia* sp. and a cutinase from *F. solani pisi* | Nylon 66 | Increase the dye bath exhaustion | Parvinzadeh et al. (2009) |
| 6. | Protease and lipase | Different sources | Nylon 6 | Higher dye exhaustion | Kiumarsi and Parvinzadeh (2010), Parvinzadeh et al. (2009) |
| 7. | Protease | *Bacillus* | Nylon | Improve hydrophilicity and cationic dye affinity without hampering the mechanical strength | Begum et al. (2018), El-Bendary et al. (2012) |
| 8. | Nitrilase | Different sources | PAN | Increase hydrophobicity | Cavaco-Paulo and Gübitz (2003), Silva & Cavaco-Paulo (2008) |
| 9. | Cutinase and lipase | Different sources | PAN | Hydrolyse the vinyl acetate moiety | Agrawal et al. (2011), Cavaco-Paulo and Gübitz (2003), Silva and Cavaco-Paulo (2008) |

effluents could be reduced by immobilisation of enzymes, which will not only allow the recovery of enzymes, but also reutilise treated bleaching effluents for dyeing (Fruhwirth et al., 2002).

The enzyme specifically breaks down hydrogen peroxide into non-active oxygen and water under mild temperature conditions. The activity of the enzyme is 10 $KCIU/_g$. One KCIU activity unit (Kilo Catalase International Unit) is the amount of enzyme that breaks down 1 millimole of hydrogen peroxide per minute under standard conditions (25°C, pH 7.0, 10 mmol $H_2O_2$). Under industrial conditions, Terminox Ultra (Novozymes) takes 10−15 min to break down the hydrogen peroxide completely and then the dye can be added (Kannan and Nithyanandan, 2008).

### *10.4.4.3 Antifelting finish of wool*

Raw wool is hydrophobic due to the epicuticle surface membranes containing fatty acids and hydrophobic impurities like wax and grease. Harsh chemicals are commonly used for their removal, such as alkaline scouring using sodium carbonate, pretreatment using potassium permanganate, sodium sulphite, or hydrogen peroxide. Wool fabric tends to felt and shrinks on wet processing. The shrinkage behaviour of wool can be regulated by various chemical means.

The most successful commercial shrink-resistant process available is the chlorine-Hercosett process developed more than 30 years ago by Heiz. Although this is a beneficial method (good antifelt effect, low damage, and low weight loss) there are some important drawbacks (limited durability, poor handling quality, yellowing of fibres, difficulties in dyeing, and environmental impact of the release of absorbable organic halogens) (Julià et al., 2000).

Papain has also been used to develop 'shrink-proof' wool. This method is comprised of partial hydrolysis of the scale tips. This treatment gave wool a silky lustre and added to its value. However, this process was abandoned due to economic reasons (Araújo et al., 2008). More recently, and mainly for environmental reasons, proteases of the subtilisin type have been studied as an alternative to chemical pretreatment of wool. Several studies have reported that pretreatment of wool fibres with proteases improved antishrinkage properties, removed impurities, and increased subsequent dyeing affinity (Parvinzadeh, 2007).

The size of the wool fibre is small and hence the enzyme can penetrate the fibre cortex, which destroys the inner parts of the wool structure. Several reports have shown that increasing enzyme size by chemical cross-linking with glutaraldehyde or by the attachment of synthetic polymers like polyethylene glycol can reduce enzyme penetration and the consequent reduction of strength and weight loss (Schroeder et al., 2006). Some of these processes have been tested on an industrial process scale (Shen et al., 2007).

Pretreatment of wool fibres with hydrogen peroxide, at alkaline pH in the presence of high concentrations of salts, also targets enzymatic activity on the outer surface of the wool, by improving the susceptibility of the cuticle to proteolytic degradation (Lenting et al., 2006). Wool pretreated with biosurfactant surfactin resulted in the increase of antifelting of wool by protease without the loss of tensile strength (Iglesias et al., 2019).

#### 10.4.4.4 Treatment of leather for dyeability

Treatment of leather with transglutaminase (TG), together with keratin or casein, has a beneficial effect on the subsequent dyeing and colour properties of leather (Collighan et al., 2002). The application of TG for leather treatment seems to be a promising strategy but is still at the research level. Products in the family effectively break down and remove starch-based sizing agents. They are available in a range of concentrations. Oxidative desizing methods can negatively affect cellulose. Products only target the starch, and so do not impact the fabric strength. Coenzymatic processes tend to show better results for leather treatment in comparison to chemical processes (De Souza and Gutterres, 2011). Different commercial enzymes available in the market are shown in Table 10.4.

**Table 10.4** Commercial enzyme products available in the market.

| S. no. | Commercial product | Main enzyme | Application |
|---|---|---|---|
| 1. | Terminox | Catalase | • Safe dyeing<br>• Batch-to-batch reproducibility<br>• Saves water, energy, and time<br>• Higher production throughput<br>• Flexible implementation<br>• Reduces effluent output or load |
| 2. | Bioprep | Pectinases | • Eliminates high-temperature alkaline scouring<br>• Saves energy<br>• aves water<br>• Reduces weight loss and maintains strength<br>• Reduces COD and salt loads in wastewater<br>• Simplifies the process for higher mill capacity<br>• Safer to use |
| 3. | Cellusoft | Cellulase | • Suitable for cotton biopolishing<br>• Upgrades fabric quality<br>• Gives a durable soft, clean, and bright surface<br>• Reduces weight and strength loss<br>• Eliminates undyed spots on garments<br>• Available in wide pH and temperature ranges<br>• Can be used before, during, and after dyeing |

**Table 10.4 Continued**

| S. no. | Commercial product | Main enzyme | Application |
|---|---|---|---|
| 4. | Denilite | Laccase | • Creates innovative fashions and looks<br>• Improves worker safety<br>• A more sustainable solution<br>• Easy to use<br>• Enhances abrasion effect<br>• Excellent anti-back staining properties<br>• No damage to stretch fibres<br>• A safer solution for lightweight fabrics |
| 5. | Aquazym | Amylase | • Improves fabric quality<br>• Improves wettability<br>• Available in wide pH and temperature ranges<br>• Reproducible performance<br>• Improves worker safety<br>• A more sustainable solution |

Based on data collected from Enzymes for Textiles | Novozymes (n.d.).

## 10.5 Environmental awareness and future perspectives

In the global market, the textile sector has been under growing pressure to meet demanding social and environmental standards. The public has become more conscious of global environmental and social issues, which has been aided by increased consumer pressure. Several worldwide customers have established their corporate ethical codes, forcing non-compliant suppliers to reconsider these norms in their operations. To ensure environmentally friendly chemicals are used and that there is a reduction in the carbon footprint, various certifications are been issued like Oeko Tex, Asthma and allergy friendly, Certification Program Nordic Swan Eco-label, Global Organic Textile Standard, etc.

The technological advances in the textile industry were earlier directed at maximising production to cater to the needs of the growing population and being able to become cost-effective for affordability. The present diegesis of society is of mindful production. With the awareness of the environmental issues and the Sustainable Development Goals (SDGs) in place, and various international treaties and pacts, it has now become imperative to be more watchful and conscious about what is being produced, how it is to be utilised, and later the methods of disposal. Three decades ago, life cycle analysis (LCA) was a novel concept but now it is referred to for every new product which is being developed and rolled out into the market for consumption.

The compliances and certifications which used to be the resplendent names and processes to attract consumers have now become the starting point for all textile processing units and companies. Parameters like BOD, COD, and discharge of hazardous chemicals in the effluent, are considered symbolic of cleaner textile processing by the supply chain as well as the end customer. Presently, several organisations and independent entities have bloomed to cater to these upcoming demands of the textile industry. The trend of specialised testing for developing textile products has begun to gain customer attention and the end consumers are not only demanding environmentally safe and sustainable products but are also willing to pay a higher price for them. The same trend is forecast to follow when every textile product will come not only with a price tag but also a sustainable tag constituting all the parameters that might evolve to regulate the eco-friendly textile processes.

## 10.6 Conclusion

This chapter has discussed the important speciality chemicals and sustainable innovations around them. Finishes for flame retardance and water and oil repellence are the hot topics for current textile needs. These finishes utilise halogenated compounds, however, due to the increasing health hazards of fluorocarbons, nitrogen- and phosphate-based finishes are being encouraged. Similarly, formaldehyde-based formulations are also being replaced by carboxylic acids in easy-care textile finishes. Antimicrobial finishes are the newest yet most exploited finish in present-day textiles. Not only natural products but metal nanoparticles have been utilised to impart the finish.

Enzyme technology has been used in the textile industry for a long time, but it is being popularised these days owing to the sustainable focus on textile processing. For almost every textile pretreatment, there is an enzyme alternative. From desizing to degumming and scouring to denim wash, enzymes have made their mark. Cutinase, pectinase, and cellulase are some of the commonly used enzymes by the textile industry. Enzyme technology has surely contributed to cleaner textile processing. Present-day trends in textile processing are paving the way for more sustainable textile materials, which is furthering the path to achieving the SDGs and protecting the environment.

## References

A New Textiles Economy: Redesigning Fashion's Future, 2017. Ellen MacArthur Foundation.

Abate, M.T., Seipel, S., Yu, J., Viková, M., Vik, M., Ferri, A., Guan, J., Chen, G., Nierstrasz, V., 2020. Supercritical $CO_2$ dyeing of polyester fabric with photochromic dyes to fabricate UV sensing smart textiles. Dyes and Pigments 183, 108671. https://doi.org/10.1016/j.dyepig.2020.108671.

Adeel, S., Kiran, S., Habib, N., Hassan, A., Kamal, S., Qayyum, M.A., Tariq, K., 2020. Sustainable ultrasonic dyeing of wool using coconut coir extract. Textile Research Journal 90 (7–8), 744–756. https://doi.org/10.1177/0040517519878795.

Agrawal, P., Parvinzadeh Gashti, M., Willoughby, J., 2011. Surface and Bulk Modification of Synthetic Textiles to Improve Dyeability.

Ahmed, M., Sukumar, N., Gideon, R.K., 2021. Crease resistance finishing optimization of citric acid and fibroin solution for cotton fabrics. Journal of Natural Fibers 18 (2), 297–307. https://doi.org/10.1080/15440478.2019.1623740.

Alisch, M., Feuerhack, A., Müller, H., Mensak, B., Andreaus, J., Zimmermann, W., 2009. Biocatalytic modification of polyethylene terephthalate fibres by esterases from actinomycete isolates. Biocatalysis and Biotransformation 22 (5–6), 347–351. https://doi.org/10.1080/10242420400025877.

Araújo, R., Casal, M., Cavaco-Paulo, A., 2008. Application of enzymes for textile fibres processing. Biocatalysis and Biotransformation 26 (5), 332–349. https://doi.org/10.1080/10242420802390457.

Azeem, M., et al., 2019. Harnessing natural colorants from algal species for fabric dyeing: a sustainable eco-friendly approach for textile processing. Journal of Applied Phycology 31 (6), 3941–3948. https://doi.org/10.1007/s10811-019-01848-z.

Baek, N.W., Zhang, X., Lou, J.F., Fan, X.R., 2022. Dyeing fabrics with a colorant extracted from blue-green algae. AATCC Journal of Research 9 (5), 223–230. https://doi.org/10.1177/24723444221103673.

Bala, R., et al., 2017. Sonochemical synthesis, characterization, antimicrobial activity and textile dyeing behavior of nano-sized cobalt(III) complexes. Ultrasonics Sonochemistry 35, 294–303. https://doi.org/10.1016/j.ultsonch.2016.10.005.

Begum, S., Wu, J., Takawira, C.M., Wang, J., 2018. Surface modification of polyamide 6,6 fabrics with an alkaline protease. Subtilisin 11 (1), 64–74. https://doi.org/10.1177/155892501601100110.

BiotexTM | Fluorine-free Water Repellency ZERO Flouro + PFOA, n.d. BiotexTM Malaysia. Retrieved May 25, 2023, from https://www.biotex-malaysia.com/portfolio/fluorine-free-water-repellency/.

Buschle-Diller, G., el Mogahzy, Y., Inglesby, M.K., Zeronian, S.H., 2016. Effects of scouring with enzymes, organic solvents, and caustic soda on the properties of hydrogen peroxide bleached cotton yarn. Textile Research Journal 68 (12), 920–929. https://doi.org/10.1177/004051759806801207.

Campos, R., Kandelbauer, A., Robra, K.H., Cavaco-Paulo, A., Gübitz, G.M., 2001. Indigo degradation with purified laccases from Trametes hirsuta and Sclerotium rolfsii. Journal of Biotechnology 89 (2–3), 131–139. https://doi.org/10.1016/S0168-1656(01)00303-0.

Cavaco-Paulo, A., Gübitz, G., 2003. Textile Processing with Enzymes, pp. 1–228.

Cavaco-Paulo, A., Almeida, L., Bishop, D., 1996. Effects of agitation and endoglucanase pretreatment on the hydrolysis of cotton fabrics by a total cellulase. Textile Research Journal 66 (5), 287–294. https://doi.org/10.1177/004051759606600501.

Chen, L., Reddy, N., Yang, Y., 2013. Soy proteins as environmentally friendly sizing agents to replace poly(vinyl alcohol). Environmental Science and Pollution Research 20 (9), 6085–6095. https://doi.org/10.1007/s11356-013-1601-5.

Collighan, R., Cortez, J., Griffin, M., 2002. The biotechnological applications of transglutaminases. Minerva Biotecnologica 14 (2), 143–148. https://doi.org/10.3390/biom3040870.

Darnerud, P.O., 2003. Toxic effects of brominated flame retardants in man and in wildlife. Environment International 29 (6), 841–853. https://doi.org/10.1016/S0160-4120(03)00107-7.

De Souza, F.R., Gutterres, M., 2011. Application of enzymes in leather processing: a comparison between chemical and coenzymatic processes. 31st IULTCS Congress 29 (03), 473–481. www.abeq.org.br/bjche.

Dehabadi, V.A., Buschmann, H.J., Gutmann, J.S., 2013. Durable press finishing of cotton fabrics: an overview. Textile Research Journal 83 (18), 1974–1995. https://doi.org/10.1177/0040517513483857.

Dyecoo, n.d. $CO_2$ Dyeing.

El-Bendary, M.A., Abo El-Ola, M., Moharam, M.E., 2012. Enzymatic surface hydrolysis of polyamide fabric by protease enzyme and its production. Indian Journal of Fibre & Textile Research 37, 273–279.

El-Kashouti, M.A., El-Molla, M., Salem, T.S., Halwagy, A., 2004. Chemical modification of fenugreek gum via cyanoethylation, and its utilization as thickener in textile printing. Egyptian Journal of Chemistry 47, 163–181.

El-Khatib, E.M., 2016. Enhancing dyeing of Wool Fibers with colorant pigment extracted from green algae. Available at: www.jocpr.com. (Accessed 13 February 2023).

Elmolla, M., Schneider, R., 2006. Development of ecofriendly binders for pigment printing of all types of textile fabrics. Dyes and Pigments 71 (2), 130–137. https://doi.org/10.1016/j.dyepig.2005.06.017.

Emery Oleochemicals, n.d. Green Polymer Additives | Emery Oleochemicals. Retrieved February 16, 2023, from https://greenpolymeradditives.emeryoleo.com/additives/?https://www.emeryoleo.com/contact-us.php?_ga=2.264184844.1040244861.1541924961846402291.1541924961&gclid=Cj0KCQiAorKfBhC0ARIsAHDzslsX0xvSuR6m0mTulzWbUgi66Df-pgkUor0tyG6L0-1H9PVuYiDlzq4aAjCLEALw_wcB.

Enzymes for Textiles | Novozymes, n.d. Retrieved November 8, 2021, from https://biosolutions.novozymes.com/en/textiles.

Etters, J.N., 1999. Cotton preparation with alkaline pectinase: an environmental advance. Textile Chemist and Colorist and American Dyestuff Reporter 1 (3), 33–36.

Etters, J., Annis, P.A., 1998. Textile Enzyme Use: A Developing Technology (Undefined).

Feitkenhauer, H., Fischer, D., Fäh, D., 2003. Microbial desizing using starch as model compound: enzyme properties and desizing efficiency. Biotechnology Progress 19 (3), 874–879. https://doi.org/10.1021/BP025790Q.

Fibre2fashion.com, n.d. Flame Retardant Finishes—Free Technical Textile Industry.

Fineotex. https://fineotex.com/anticrease-resins/. (Accessed 26 December 2021).

Fischer-Colbrie, G., Heumann, S., Liebminger, S., Almansa, E., Cavaco-Paulo, A., Guebitz, G.M., 2009. New enzymes with potential for PET surface modification. Biocatalysis and Biotransformation 22 (5–6), 341–346. https://doi.org/10.1080/10242420400024565.

Fruhwirth, G., Paar, A., Gudelj, M., Cavaco-Paulo, A., Robra, K.H., Gübitz, G., 2002. An immobilised catalase peroxidase from the alkalothermophilic Bacillus SF for the treatment of textile-bleaching effluents. Applied Microbiology and Biotechnology 60 (3), 313–319. https://doi.org/10.1007/S00253-002-1127-0.

Gao, Y., Cranston, R., 2008. Recent advances in antimicrobial treatments of textiles. Textile Research Journal 78 (1), 60–72. https://doi.org/10.1177/0040517507082332.

Goswami, B.C., Anandjiwala, R.D., Hall, D., 2004. Textile Sizing. CRC Press. https://doi.org/10.1201/9780203913543.

Guebitz, G.M., Cavaco-Paulo, A., 2008. Enzymes go big: surface hydrolysis and functionalisation of synthetic polymers. Trends in Biotechnology 26 (1), 32–38. https://doi.org/10.1016/J.TIBTECH.2007.10.003.

Gulrajani, M.L., 1992. Degumming of silk. Review of Progress in Coloration and Related Topics 22 (1), 79–89. https://doi.org/10.1111/J.1478-4408.1992.TB00091.X.

Harifi, T., Montazer, M., 2012. Past, present and future prospects of cotton cross-linking: new insight into nano particles. Carbohydrate Polymers 88 (4), 1125–1140. https://doi.org/10.1016/j.carbpol.2012.02.017.

Hashem, M., Hauser, P., Smith, B., 2003. Wrinkle recovery for cellulosic fabric by means of ionic crosslinking. Textile Research Journal 73 (9), 762–766. https://doi.org/10.1177/004051750307300903.

Hebeish, A., Higazy, A., El-Shafei, A., 2006. New sizing agents and flocculants derived from chitosan. Starch/Staerke 58 (8), 401–410. https://doi.org/10.1002/star.200500460.

Hipler, U.C., Elsner, P., Fluhr, J.W., 2006. Antifungal and antibacterial properties of a silver-loaded cellulosic fiber. Journal of Biomedical Materials Research—Part B Applied Biomaterials 77 (1), 156–163. https://doi.org/10.1002/jbm.b.30413.

History of Clothing and Textiles—Wikipedia, n.d.

Hong, K.H., et al., 2006. Preparation of antimicrobial poly(vinyl alcohol) nanofibers containing silver nanoparticles. Journal of Polymer Science Part B: Polymer Physics 44 (17), 2468–2474. https://doi.org/10.1002/polb.20913.

HT Fine Chemical Co. Ltd., n.d. Antistatic Agent for Textile. Retrieved February 20, 2023, from https://www.htfine-chem.com/product/antisatic-agent-kd.html.

Iglesias, M.S., Sequeiros, C., García, S., Olivera, N.L., 2019. Eco-friendly anti-felting treatment of wool top based on biosurfactant and enzymes. Journal of Cleaner Production 220, 846–852. https://doi.org/10.1016/J.JCLEPRO.2019.02.165.

Jhatial, A.K., Yesuf, H.M., Wagaye, B.T., 2020. Pretreatment of Cotton, pp. 333–353. https://doi.org/10.1007/978-981-15-9169-3_13.

Julià, M.R., Pascual, E., Erra, P., 2000. Influence of the molecular mass of chitosan on shrink-resistance and dyeing properties of chitosan-treated wool. Journal of the Society of Dyers and Colourists 116 (2), 62–67. https://doi.org/10.1111/J.1478-4408.2000.TB00023.X.

Kadolph, S.J., Marcketti, S.B., 2016. Textiles, twelfth ed. Pearson.

Kanelli, M., Vasilakos, S., Ladas, S., Symianakis, E., Christakopoulos, P., Topakas, E., 2017. Surface modification of polyamide 6.6 fibers by enzymatic hydrolysis. Process Biochemistry 59, 97–103. https://doi.org/10.1016/J.PROCBIO.2016.06.022.

Kannan, M.S.S., Nithyanandan, R., 2008. Enzymatic Application for Bleach Cleanup | Processing, Dyeing & Finishing | Features | The ITJ. https://indiantextilejournal.com/articles/FAdetails.asp?id=853.

Kiumarsi, A., Parvinzadeh, M., 2010. Enzymatic hydrolysis of nylon 6 fiber using lipolytic enzyme. Journal of Applied Polymer Science 116 (6), 3140–3147. https://doi.org/10.1002/APP.31756.

Kumar, D., Bhardwaj, R., Jassal, S., Goyal, T., Khullar, A., Gupta, N., 2021. Application of enzymes for an eco-friendly approach to textile processing. Environmental Science and Pollution Research. https://doi.org/10.1007/S11356-021-16764-4.

Kundu, A.B., Ghosh, B.S., Chakrabarti, S.K., Ghosh, B.L., 2016. Enhanced bleaching and softening of jute by pretreatment with polysaccharide degrading enzymes. Textile Research Journal 61 (12), 720–723. https://doi.org/10.1177/004051759106101204.

Lenting, H.B.M., Schroeder, M., Guebitz, G.M., Cavaco-Paulo, A., Shen, J., 2006. New enzyme-based process direction to prevent wool shrinking without substantial tensile strength loss. Biotechnology Letters 28 (10), 711–716. https://doi.org/10.1007/S10529-006-9048-0.

L.N. Chemical Industries, Retrieved April 25, 2023 from https://www.lnchemicals.com/dyeing-printing-auxiliaries.html.

Madhu, A., Chakraborty, J.N., 2017. Developments in application of enzymes for textile processing. Journal of Cleaner Production 145, 114–133. https://doi.org/10.1016/J.JCLEPRO.2017.01.013.

Maflon. Textile fluorine free water repellence & crosslinkers - Maflon S.p.A. https://www.maflon.com/index.php?option=com_content&view=article&id=61&Itemid=321. (Accessed 20 February 2023).

Maleknia, L., Rashidi, A.S., Ghamsari, N.A., 2015. Preparation and characterization of nylon 6/silver nanocomposite fibers for permanent antibacterial effect. Oriental Journal of Chemistry 31 (1), 257–262. https://doi.org/10.13005/ojc/310128.

Mccloskey, S.G., Jump, J.M., 2005. Bio-polishing of polyester and polyester/cotton fabric. Textile Research Journal 75 (6), 480–484. https://doi.org/10.1177/0040517505053846.

Miettinen-Oinonen, A., Pere, J., Lantto, R., Kruus, K., Puolakka, A., Buchert, J., 2004. Modification of Textile Fibres by Laccase. https://doi.org/10.2/JQUERY.MIN.JS.

Mir, R.A., et al., 2019. Green algae, Cladophora glomerata L.—based natural colorants: dyeing optimization and mordanting for textile processing. Journal of Applied Phycology 31 (4), 2541–2546. https://doi.org/10.1007/S10811-018-1717-6/FIGURES/4.

Mojsov, K., 2011. I-17 application of enzymes in the textile industry: a review primena na enzimi vo tekstilnata industrija: pregled. In: II International Congress "Engineering, Ecology and Materials in the Processing Industry", p. 78.

Mojsov, K., Janevski, A., Andronikov, D., Jordeva, S., Golomeova, S., Gaber, S., 2020. Enzymatic treatments for cotton. Tekstilna Industrija 68 (2), 12–17. https://doi.org/10.5937/TEKSTIND2002011M.

Morgan, A.B., 2019. The future of flame retardant polymers—unmet needs and likely new approaches. Polymer Reviews 59 (1), 25–54. https://doi.org/10.1080/15583724.2018.1454948.

Mowbray, J., 2015. Archroma Flame Retardant Certified by Oeko-Tex. Ecotextile News.

Nano-Care, Retrieved April 25, 2023 from https://nano-care.com/antimicrobial-coating/.

Nazeer, A.A., Dhandapani, S., Vijaykumar, S.D., 2018. Advanced technologies for coloration and finishing using nanotechnology. In: Handbook of Renewable Materials for Coloration and Finishing. John Wiley & Sons, Inc., Hoboken, NJ, pp. 473–500. https://doi.org/10.1002/9781119407850.ch17.

Organic Dyes and Pigments. Technical data sheet of Eccoshield PPE-4X. https://www.organicdye.com/wp-content/uploads/2015/06/ECCOSHIELD®-PPE-4X-Tech-Bulletin.pdf.

Papaspyrides, C.D., Kiliaris, P. (Eds.), 2014. Polymer green flame retardants. Elsevier.

Parker, R.E., 1984. History of the textile industry committee. In: IEEE Transactions on Industry Applications. https://doi.org/10.1109/TIA.1984.4504531.

Parvinzadeh, M., 2007. Effect of proteolytic enzyme on dyeing of wool with madder. Enzyme and Microbial Technology 40 (7), 1719–1722. https://doi.org/10.1016/J.ENZMICTEC.2006.10.026.

Parvinzadeh, M., Assefipour, R., Kiumarsi, A., 2009. Biohydrolysis of nylon 6,6 fiber with different proteolytic enzymes. Polymer Degradation and Stability 94 (8), 1197–1205. https://doi.org/10.1016/J.POLYMDEGRADSTAB.2009.04.017.

Patil, N.V., Netravali, A.N., 2020. Multifunctional sucrose acid as a 'green' crosslinker for wrinkle-free cotton fabrics. Cellulose 27 (9), 5407–5420. https://doi.org/10.1007/s10570-020-03130-9.

Paul, R., Naik, S.R., 1997. Stoneless stone washing—an innovative concept in denim washings. Textile Dyer and Printer 30 (11), 13.

Paul, R., Teli, M.D., 2010. Swelling of cotton as a tool to enhance the response of cellulase enzymes. Coloration Technology 126 (6), 325–329. https://doi.org/10.1111/j.1478-4408.2010.00264.x.

Paul, R., Solans, C., Erra, P., 2005. Study of a Natural dye solubilization in o/w microemulsions and its dyeing behavior. Colloids and Surfaces 253, 175.

Paul, R., Brouta-Agnésa, M., Esteve, H., 2012. In: Bischof, S. (Ed.), Chapter 10: Protective Insulation Materials in Functional Protective Textiles. Faculty of Textile Technology, University of Zagreb, Croatia, pp. 281–303.

Purwar, R., Joshi, M., 2004. Recent developments in antimicrobial finishing of textiles—a review. AATCC Review 4 (3), 22–26.

Reddy, N., Chen, L., Zhang, Y., Yang, Y., 2014. Reducing environmental pollution of the textile industry using keratin as alternative sizing agent to poly(vinyl alcohol). Journal of Cleaner Production 65, 561–567. https://doi.org/10.1016/j.jclepro.2013.09.046.

Ribeiro, F., Okoffo, E.D., O'Brien, J.W., Fraissinet-Tachet, S., O'Brien, S., Gallen, M., Samanipour, S., Kaserzon, S., Mueller, J.F., Galloway, T., Thomas, K.V., 2020. Quantitative analysis of selected plastics in high-commercial-value Australian seafood by pyrolysis gas chromatography mass spectrometry. Environmental Science and Technology 54 (15), 9408–9417. https://doi.org/10.1021/acs.est.0c02337.

Rudolf Group. Textile auxiliaries. https://www.rudolf.de/en/products/textile-auxiliaries/. (Accessed 12 April 2022).

Sampaio, S., Martins, C., Gomes, J.R., 2011. Colored nanoparticles for ecological dyeing of cellulosic fibres. Advanced Materials Research 332–334, 1136–1139. https://doi.org/10.4028/www.scientific.net/AMR.332-334.1136.

Sanitized, A.G., Retrieved April 25, 2023 from https://www.sanitized.com.

Sarex, n.d. Retrieved April 20, 2023 from https://www.sarex.com/textile/.

Sarkar, A., Sarkar, D., Gupta, M., Bhattacharjee, C., 2012. Recovery of polyvinyl alcohol from desizing wastewater using a novel high-shear ultrafiltration module. Clean—Soil, Air, Water 40 (8), 830–837. https://doi.org/10.1002/clen.201100527.

Schlossberg, T., 2019. How Fast Fashion Is Destroying the Planet. The New York Times, New York Times.

Schroeder, M., Lenting, H.B.M., Kandelbauer, A., Silva, C.J.S.M., Cavaco-Paulo, A., Gübitz, G.M., 2006. Restricting detergent protease action to surface of protein fibres by chemical modification. Applied Microbiology and Biotechnology 72 (4), 738–744. https://doi.org/10.1007/S00253-006-0352-3.

Segev, O., Kushmaro, A., Brenner, A., 2009. Environmental impact of flame retardants (persistence and biodegradability). International Journal of Environmental Research and Public Health 6 (2), 478–491. https://doi.org/10.3390/ijerph6020478.

Shang, S., Hu, E., Poon, P., Jiang, S., Kan, C.W., Koo, R., 2011. Foam dyeing for developing the wash-out effect on cotton knitted fabrics with pigment. Research Journal of Textile and Apparel 15 (1), 44–51. https://doi.org/10.1108/RJTA-15-01-2011-B005.

Shen, J., Rushforth, M., Cavaco-Paulo, A., Guebitz, G., Lenting, H., 2007. Development and industrialisation of enzymatic shrink-resist process based on modified proteases for wool machine washability. Enzyme and Microbial Technology 40 (7), 1656–1661. https://doi.org/10.1016/J.ENZMICTEC.2006.07.034.

Silva, C., Cavaco-Paulo, A., 2008. Biotransformations in synthetic fibres. Biocatalysis and Biotransformation 26 (5), 350–356. https://doi.org/10.1080/10242420802357845.

Simoncic, B., Tomsic, B., 2010. Structures of novel antimicrobial agents for textiles—a review. Textile Research Journal 80 (16), 1721–1737. https://doi.org/10.1177/0040517510363193.

Solution dyeing: Fabrics dyed for sustainability—in both senses of the word I Croozer, n.d. Retrieved May 25, 2023, from https://www.croozer.com/en/sustainable-dyeing-method-for-bicycle-trailer-textiles.

Teli, M.D., Shanbag, V., Dhande, S.S., Singhal, R.S., 2007. Rheological properties of Amaranthus paniculates (Rajgeera) starch vis-à-vis Maize starch. Carbohydrate Polymers 69 (1), 116–122. https://doi.org/10.1016/j.carbpol.2006.09.008.

The advantages of solution dyed fabrics, April 27, 2020. Burton Snowboards. http://www.burton.com/blogs/the-burton-blog/what-is-solution-dye/.

UNEP, 2018. Cleaning up Couture: What's in Your Jeans? UNEP.ORG.

UNIDYNE XF Series for Textiles I Fluorochemicals I Daikin Global, n.d. Retrieved May 25, 2023, from https://www.daikinchemicals.com/solutions/products/water-and-oil-repellents/unidyne-non-fluorinated-textiles.html.

Vankar, P.S., Shanker, R., Srivastava, J., 2007. Ultrasonic dyeing of cotton fabric with aqueous extract of Eclipta alba. Dyes and Pigments 72 (1), 33–37. https://doi.org/10.1016/j.dyepig.2005.07.013.

Walter, T., Augusta, J., Müller, R.J., Widdecke, H., Klein, J., 1995. Enzymatic degradation of a model polyester by lipase from Rhizopus delemar. Enzyme and Microbial Technology 17 (3), 218–224. https://doi.org/10.1016/0141-0229(94)00007-E.

Wang, Q., Zhou, W., Du, S., Xiao, P., Zhao, Y., Yang, X., Zhang, M., Chang, Y., Cui, S., 2019. Application of foam dyeing technology on ultra-fine polyamide filament fabrics with acid dye. Textile Research Journal 89 (23–24), 4808–4816. https://doi.org/10.1177/0040517519839377.

Webber, K., 2018. The Environmental and Human Cost of Making a Pair of Jeans. EcoWatch.

Welcome to Indokem, n.d.

Why IndiDye® – IndiDye®, n.d.

Wijesena, R.N., et al., 2014. Side selective surface modification of chitin nanofibers on anionically modified cotton fabrics. Carbohydrate Polymers 109, 56–63. https://doi.org/10.1016/j.carbpol.2014.03.035.

Yang, C.Q., et al., 1998. Nonformaldehyde durable press finishing of cotton fabrics by combining citric acid with polymers of maleic acid. Textile Research Journal 68 (6), 457–464. https://doi.org/10.1177/004051759806800611.

Yu, H., Wang, Y., Zhong, Y., Mao, Z., Tan, S., 2014. Foam properties and application in dyeing cotton fabrics with reactive dyes. Coloration Technology 130 (4), 266–272. https://doi.org/10.1111/cote.12088.

Zhang, H., Wang, J.K., Liu, W.J., Li, F.Y., 2015. Microwave-assisted synthesis, characterization, and textile sizing property of carboxymethyl corn starch. Fibers and Polymers 16 (11), 2308–2317. https://doi.org/10.1007/s12221-015-5321-y.

Zhao, Y., Xu, H., Mu, B., Xu, L., Hogan, R., Yang, Y., 2016. Functions of soymeal compositions in textile sizing. Industrial Crops and Products 89, 455–464. https://doi.org/10.1016/j.indcrop.2016.05.047.

Zhou, Y., Zheng, G., Zhang, J., Wang, Q., Zhou, M., Yu, Y., Wang, P., 2021. An eco-friendly approach to low-temperature and near-neutral bleaching of cotton knitted fabrics using glycerol triacetate as an activator. Cellulose 28 (12), 8129–8138. https://doi.org/10.1007/s10570-021-04030-2.

Zydex Industries, n.d. Technical Data Sheet.

# Innovations in textile wet processing machinery

*Zafar Juraev*[1], *Mukhitdin Sattarov*[1] *and Roshan Paul*[2]
[1]Department of Metrology, Standardization and Product Quality Management, Andijan Machine-Building Institute, Andijan, Uzbekistan; [2]Institut für Textiltechnik der RWTH Aachen University, Aachen, Germany

## 11.1 Introduction

The European Union (EU) strategy for sustainable and circular textiles addresses the production and consumption of textiles while recognising the importance of the textiles sector. It implements the commitments of European initiatives such as the green deal and circular economy action plan. This strategy looks at the entire life cycle of textile products and proposes coordinated actions to change how to manufacture and consume textiles (European Commission, 2022).

The statistics of the world chemical fibre industry in 1975 showed that globally, about 24 million metric tons of textile fibres were produced. By 2020, that number nearly quintupled to over 108 million metric tons. The production of natural fibres such as cotton or wool was 27.4 million metric tons, whereas man-made fibres accounted for the remaining 80.9 million tons. Chemical fibres include synthetic fibres such as polyester or polyamide and man-made cellulose fibres such as viscose. The production of man-made fibres surpassed that of cotton in the mid-1990s and has more than doubled in the past few decades. In 2020, synthetic fibres accounted for approximately 62% of global fibre production. Polyester alone accounted for 52% of the market, whereas polyamide and other synthetics accounted for 5% and 5.2%, respectively. Other synthetic fibres include acrylic, elastane, and polypropylene.

Although the need for the more responsible consumption of resources is becoming widely recognised, forecasts point to a significant increase in fibre production in coming years. Global fibre production is expected to reach 156 million metric tons by 2030, up 43% from 2020. Therefore, in 2030, per capita production will be approximately 17 kg per person. The industry's steady growth can be attributed to the constant demand for research and development as well as the willingness of the textile industry to innovate and seize the opportunities created by global trends and changes in consumer behaviour (Statista, 2022).

All of these textile fibres undergo lengthy manufacturing and technological processes before they are ready for clothing. They are first converted into textiles, and move on to wet processing techniques such as pretreatments (desizing, scouring, and bleaching), colouration (dyeing or printing) and finishing. Wet processing of textiles is used in a broader sense and includes not only fabric dyeing and finishing but

**Sustainable Innovations in the Textile Industry. https://doi.org/10.1016/B978-0-323-90392-9.00018-5**

also yarn dyeing and finishing. Functional finishing of textiles is mainly attributed to chemical finishing – or wet finishing – because it is usually employed to provide textiles with new functional properties. However, functionality is achieved through chemical finishes, but it could also combine chemical, mechanical and biotechnological finishes as well as fabric characteristics. For example, the yarn type and fabric structure are also important for finishing. Mechanical finishing is based on the physical means of changing the appearance, surface and dimensional characteristics of textiles, such as calendering, embossing, suede, emery, brushing, napping, heat setting and Sanforizing. Biotechnological finishes are mostly enzyme-based and also use natural products (Paul, 2014).

Wet processing of textiles is critical to improving the performance and handle of textiles. These processes involve different types of wet processing machinery. Hundreds of large, medium and small companies produce wet processing machinery worldwide. The most significant manufacturers of wet processing machinery are traditionally Western European firms from Germany, Switzerland, the Netherlands and Italy, as well as Japanese and US firms. In the past few decades, machinery manufacturers have offered no revolutionary solutions, but the equipment is being innovated to save chemical, materials, water and electricity, and improve the quality of products and reproducibility of results, as well as minimise harm to the environment.

## 11.2 Historical aspects and current scenario

The history of the appearance of decorated clothing goes back to the origins of the world's civilisations. According to archaeological research, human decoration appeared much earlier than clothing. To stand out from the environment as well as perform religious rites, primitive people painted their bodies with colour obtained from minerals, a mixture of plant sap and clay, and applied tattoos and scars to the body. With the advent of clothing, and later with the invention of weaving, exotic patterns and ornaments were transferred to fabric. Art historians note that ancient drawings on fabrics had a simple, primitive character. The subsequent development of humanity led to the emergence of more complex and undoubtedly artistically valuable ornaments, which have survived in some national costumes.

An analysis of historical literature suggests that one of the oldest ways to decorate clothes is to draw a pattern with fabric paint. In China, for example, a pattern was applied to silk fabrics with a brush, and in the countries of North Africa, the 'nodular' method of colouring materials was widely used. The earliest mention of drawing a pattern on fabric using printed boards (cliches) dates back to 6000–6500 BCE. During archaeological excavations near the settlement of Chatal Hiiyiik, located in the territory of modern Turkey, clay stamps were found for applying images to fabrics. In India, the art of colouring fabrics using the backup printing method was born. Since the time of the traveller Marco Polo, Indian printed fabrics have become known across the world. In the countries such as Malaysia, Indonesia and Myanmar, the printing of fabrics using the techniques of cold and hot batik is widespread. This method of designing fabrics is used by several designers from different countries to develop exclusive coloured products.

Historically, two methods of colouring fabric are most common. The first is plain dyeing, in which fabric is immersed in the colouring pigment. The second method is printing, which uses patterns and ornaments prepared in advance by designers on the surface of the fabric. It was necessary to perform more than 30 operations, and the whole process took around 4 months. Most ancient textile colouration techniques are veritable art forms that combine culture, tradition and nature.

From a manual or artistic technique, colouration further progressed rapidly and led to the development of the first prototypes of wet processing machinery. Block printing technique begins with the creation of a stamp from a block of wood by hand carving the intended design. The stamp is then dipped into ink. Immediately afterward, it is pressed firmly against the fabric, stamp by stamp. Different stamps can be used on the same textile, and several layers of dyes can be added. About 4500 years ago, the technique of printing on textiles and paper using woodcuts flourished in China. Several centuries later, Indian artisans studied and perfected this technique and have continued to pass it on for generations.

Shibori is another printing technique used in Japan since the 8th century, although it appears to have originated in China many centuries ago. Shibori is a term that combines several methods of dyeing fabric by winding, sewing, folding or shrinking. Knot dyeing is considered one of the most ancient ways of ornamenting fabrics. Shibori uses a thread to pinch and sew undyed fabrics so that it does not absorb dye, and the pattern on the fabric appears only after dyeing. Batik is another ancient dyeing technique that uses wax to create intricate designs. Hot wax is applied with brushes or poured on certain parts of the undyed fabric by craftsmen who paint amazing patterns, and then the wax is removed with boiling water.

In the 18th century, wooden prototypes of early textile processing machinery further progressed to metal, when an expert from Switzerland named Oberkampf produced the idea of using metal instead of wooden heels. Engraved boards had an in-depth pattern. This made it possible to improve the quality of the image applied to the fabric. Another leap was made with the invention of the cylindrical shaft, and then the printing press itself. During the Industrial Revolution, which spanned from about 1760 to 1840, the transition of textile industry to new manufacturing processes started. Textile machines for spinning and weaving were pioneers that paved the way to wet processing machines for pretreatments, dyeing, printing and finishing. It took several years for wet processing machines to evolve and to reach the current stage. The latest innovations in textile machinery are exhibited in International Textile Machinery Association (ITMA) (ITMA, 2023).

## 11.3 Innovations in wet processing machinery

Fabrics made on a loom are harsh, with an ugly appearance, and they contain various impurities and contaminants. These fabrics are not suitable for manufacturing garments, and they require wet processing. Wet processing is a technological process that ennobles fabrics, improves their quality and gives them a marketable appearance. Wet processing includes three main steps: pretreatment, dyeing or printing and finishing. Each stage consists of several physical, mechanical and chemical operations. Fig. 11.1 depicts the main operations involved in textile wet processing.

After dyeing and finishing, the fabric is subjected to sorting, the quality control and determination of the grade of the products. The quality of fabrics is determined by carefully reviewing its appearance with a strict consideration of all raw materials and weaving defects, as well as dyeing, printing and finishing defects. The fastness

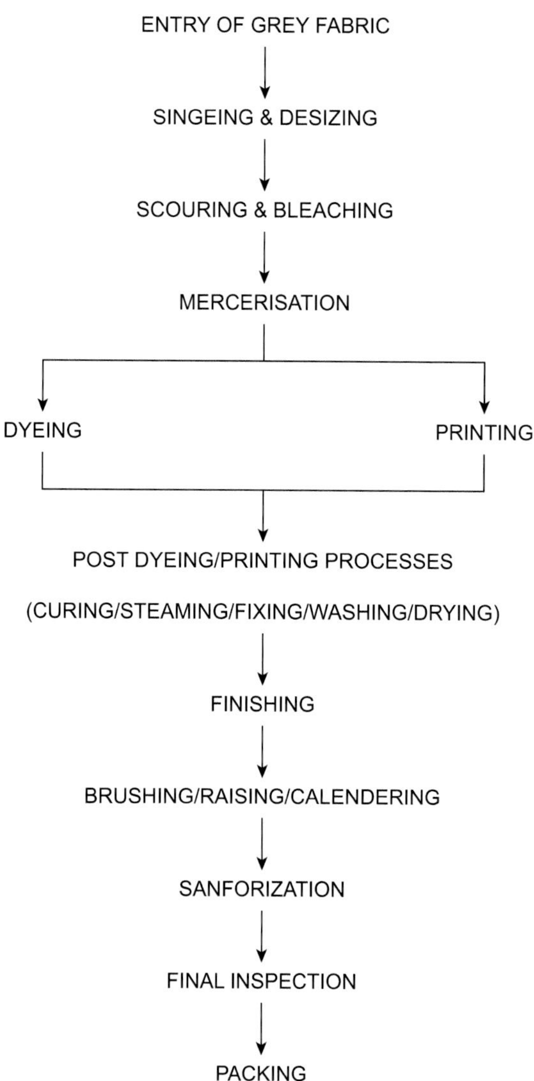


**Figure 11.1** Operations involved in textile wet processing.

of dyeing and printing, the width and density of the fabric, and other indicators provided for by relevant technical documents are determined. After sorting, rolls of fabric go to the warehouse. From the warehouse, the fabric is shipped to consumers for self-delivery or is delivered by the company's transport to places specified by the buyer.

## 11.3.1 Pretreatment machines

### 11.3.1.1 Singeing

On the front surface of the textile fabric, protruding tips of fibres inevitably appear. Singeing is removes randomly protruding fibres from the front surface of the fabric by burning to give it a more even and shiny appearance. To burn protruding fibres requires energy. The supplied energy must be enough to burn only protruding fibres while keeping firmly bound fibres intact. This is achieved by enabling the time of contact between the singeing flame and the fabric to be a mere fraction of a second. Because the temperature of flames is too high (around 1300°C), any regulation of this temperature is out of the question. However, metering and control of the thermal energy of the flame are essential for safe but effective burning off.

The most widely used machines are gas-fired. They enable fabric to be singed from two sides, ensuring the burning of protruding fibres on the surface and within the thickness of the fabric, with a high speed of fabric movement, a high coefficient of automation and an efficiency factor of at least 0.95. Burning machines operate both individually and as part of lines that provide singeing, impregnation with a process solution and rolling into a roll to keep it cold. The principle of a gas singeing machine is shown in Fig. 11.2 (Fibre2fashion, 2022).

There are several well established manufacturers of singeing machines. The components of the most common gas-burning equipment from Osthoff-Senge GmbH,

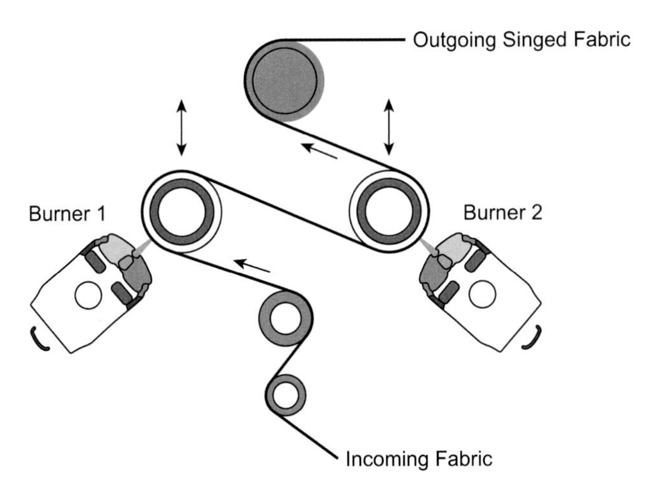


**Figure 11.2** Principle of gas singeing machine.
From Fibre2fashion. (2022).

Germany are a filling device, a brush and fluff chamber, a burning chamber, a washing and impregnation chamber, a squeezing device, a rolling device and an ageing station (Osthoff-Senge, 2023). The Wakayama machine from Japan includes a roller with a tray compensator, a lint chamber, a burning chamber, a spark arresting chamber, two-shaft padding, a roller compensator, and a rolling device with a tray compensator (Wakayama, 2023). The singeing line of the Intes, Italy includes a rolling device, a lint chamber, a burning chamber, an impregnation bath with a squeezing device, a rolling device and an ageing station. The singeing machine of Küsters, Germany includes a filling device, a brush and fluff cleaning chamber, a burning chamber, a spark chamber, a bathtub with a wringer and an optional roller-type device. Table 11.1 lists the characteristics of different singeing machines available on the market.

### *11.3.1.2 Desizing*

Desizing is first chemical pretreatment process. Before weaving, it is mandatory to size warp yarns because single-ply spun yarn has poor weavability. In sizing, a thin film of starch or other sizing agents, binders and softeners is applied to the warp yarn surface.

**Table 11.1** Characteristics of different singeing machines.

| Technical specifications | Osthoff-Senge (Germany) | Wakayama (Japan) | Intec (Italy) | Küsters (Germany) |
|---|---|---|---|---|
| Working width, mm | 2000 | 1800 | 2600 | 1800 |
| Mass of processed fabric, g/m$^2$ | No more than 500 | 70–350 | 100–350 | No more than 400 |
| Fabric movement speed, m/min | 10–140 | 20–200 | 20–90 | 8–180 |
| Installed power of electric motors, kW | 26.4 | 27.1 | 8.0 | 8.0 |
| Several burners, pieces | 4 | 4 | 2 | 2 |
| Number of clothes in dressing, pieces | 1 | 1 | 1 | 1 |
| Type of singeing | Double-sided | Double-sided | Double-sided | Double-sided |
| Bath volume, m$^3$ | 0.8 | 0.1 | 1.3 | 0.2 |
| Max flow: | | | | |
| - Water, m$^3$/h | – | 0.34 | – | 4.2 |
| - Steam, t/h | – | 0.12 | – | 0.1 |
| Consumption per 1000 m of fabric: | | | | |
| - Water, m$^3$/h | 0.05 | 0.056 | 0.05 | – |
| - Steam, t/h | 5.9 | 10.0 | 5.2 | – |
| Overall dimensions, mm: | | | | |
| - Length | 8000 | 15,000 | 11,170 | 19,620 |
| - Width | 4500 | 3900 | 4200 | 2860 |
| - Height | 3200 | 4100 | 3900 | 5100 |

When the fabric is dyed, problems including poor absorbency arise owing to size present on the fabric. Thus, the applied size has to be removed from woven fabric for further wet processing such as bleaching, dyeing, printing and finishing.

Desizing of cotton and other fabrics can be done by physical or chemical methods or their combination. The most important methods of desizing are rot steeping, acid steeping and enzymatic desizing. Industrially, desizing can be carried out in a jigger machine, which is a batch process. This is a simple method in which fabric from one roll is processed in a bath and rewound on another roll. Continuous desizing is carried out on machines such as the Winch and J-box. In the Winch, grey fabric is first passed through hot water and then through a desizing solution at 50 to 60°C. For the J-box, fabric is padded with a desizing solution and fed through an open-width J-box kept at 80 to 90°C with a dwell time of 25–40 min. After-wash of fabrics is carried out at 95 to 100°C, and then they are rinsed thoroughly with hot water.

### *11.3.1.3 Scouring*

Scouring is a pretreatment process in which natural impurities such as oil and wax as well as added impurities are completely removed from natural fibres. All textile materials in their natural form are known as greige or grey materials. They have a natural colour, smell and contaminants, which are unsuitable for clothing materials (Hall, 1969). The purpose of scouring is to improve the wettability and permeability of the fabric. It is an important process before dyeing and finishing. There are two types of methods for scouring: batch and continuous. The batch or discontinuous process is mainly carried out in a kier boiling machine or in jigger and winch dyeing machines.

A kier boiler is a cylindrical, long, mild steel or cast-iron container fitted with two tubular perforations (a disc with several holes). One is on the ground and the other is on the top. These discs are linked to the upper compartment with tubes that bring the liquor. Steam is carried through the centre of the bay. Thus, liquor pipes are surrounded by steam that heats them. In the batch scouring process, fabric is packed into machine and stored in a tubular shape. A circular tube to the fabric pumps warm liquor and sprays it, and the liquor goes through the packed fabric slowly and is collected on the opposite side of the kier. The liquor is then pumped to the heater again using a centrifugal pump, and this cycle is repeated. After scouring at 80°C, water is used to remove material impurities from the textile. Afterward, it is neutralised with 0.1% acetic acid and cold cleaning is done (Dspattextile, 2023). Fig. 11.3 shows the technical details of a kier boiler machine, and Fig. 11.4 that of a J-box machine.

The continuous process is usually done in a J-box machine. It is called a J-box because the scouring vessel looks like the letter '*J*'. It is mainly a steel chute with a large capacity for fabrics. Compared with a kier, fabric is supplied from one end and pulled from the other. The J-box's internal side is polished and insulated to minimise thermal losses. Desizing, scouring and bleaching can be performed at once in this machine.

The J-box can be separated into four parts: saturation, preheater, J-box and washing unit. The saturation is made without sodium hydroxide, but with a scouring recipe. Fabric is then into the solution via the guide roller at a temperature of approximately

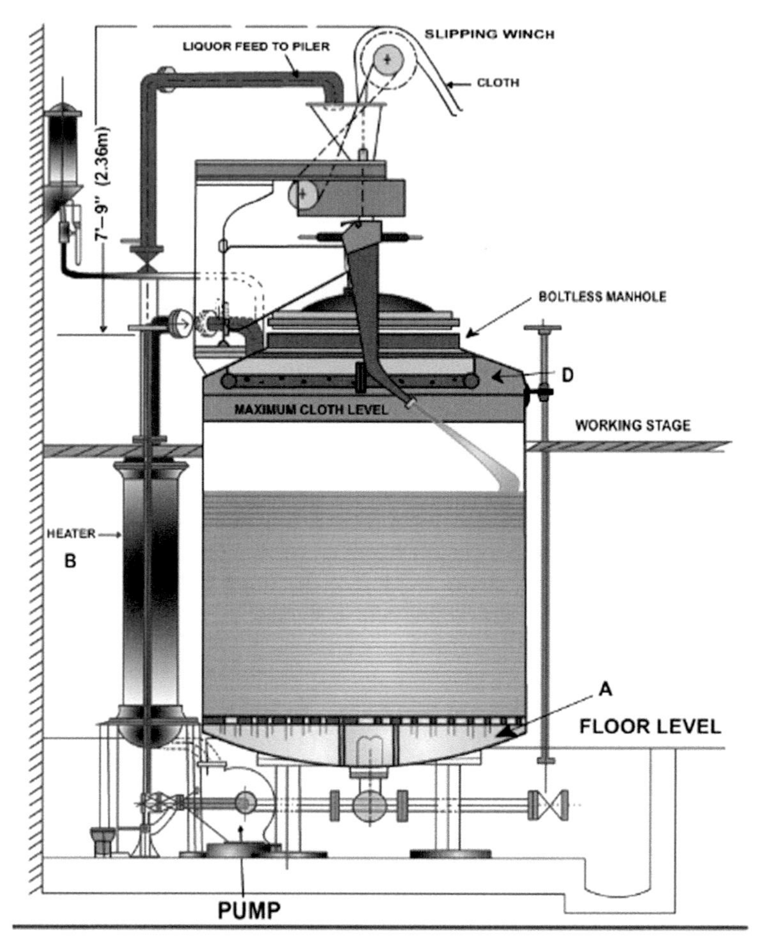


**Figure 11.3** Technical details of kier boiler machine.

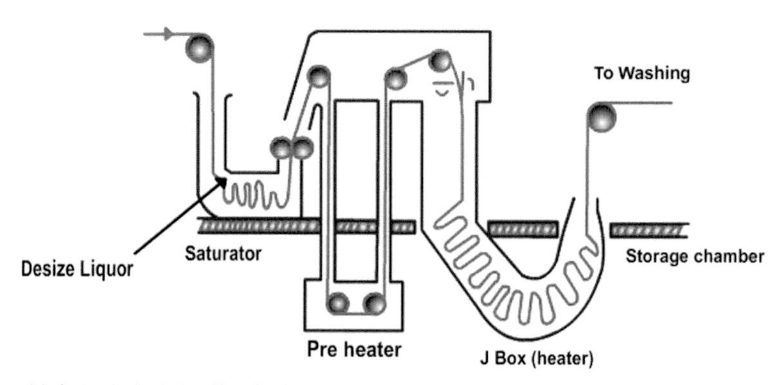


**Figure 11.4** Technical details of J-box machine.

**Table 11.2** Comparison of discontinuous and continuous scouring processes.

| No. | Discontinuous process | Continuous process |
|---|---|---|
| 1 | In this process, fabric is scoured in rope form | In this process, fabric is scoured in an open-width form |
| 2 | This process is suitable for a small order | This process is unsuitable for a small order |
| 3 | Uneconomical process for a large order | Economical process for a large order |
| 4 | More time is required than for a continuous process | Less time-consuming process |

80°C. The material is transferred to the J-box during pre-heating at 110 to 120°C. Sodium hydroxide solution from the J-box is maintained, and this solution is applied on the fabric at approximately 100°C. During this process, sodium hydroxide responds to and removes impurities in the fabric in the J-box. Water-soluble impurities or products are removed in the washing unit. The material is first cleaned in warm water and then dried in cold water (Img, 2023). The advantages and disadvantages of discontinuous and continuous scouring processes are listed in Table 11.2.

#### *11.3.1.4 Bleaching*

Bleaching is the last textile pretreatment process. Its purpose is to remove natural colour for subsequent steps such as dyeing or printing or to achieve a full white colour (Hummel, 1898). If the scoured fabric contains natural colours and pigments, the brightness of the fabric may be affected and result in uneven dyeing. Pastel shades do not appear prominently on the fabric if the fabric has traces of natural colours and pigments. The main objective of bleaching is to whiten the cotton fabric, yarn and fibres. In the textile industry, bleaching is carried out in a continuous bleaching range, in which it is possible to combine all three pretreatment processes such as desizing, scouring and bleaching. This minimises energy consumption and reduces the number of operations.

There are several innovative concepts for pad-steam bleaching. In the Scout pad-steam bleaching range developed by Erbatech GmbH, Germany, continuous bleaching can be done. In their Galaxy model, continuous pad-steam bleaching is also possible in a tubular form and it is possible to achieve high production and outstanding fabric quality with low processing costs (Erbatech, 2022). Shree Shakti Krupa Industries, India offers an innovative continuous bleaching range. In this machine, fabric is treated with a bleaching chemical and then is steamed. This helps to increase the whiteness of the fabric so that it is prepared for better dyeing and finishing (Shree Shakti Krupa, 2022).

#### *11.3.1.5 Mercerisation*

Mercerisation is the optional short-term treatment of cotton fabric with a concentrated solution of sodium hydroxide under tension in the cold, followed by rinsing with hot and cold water. It enhances the fabric's lustre and dyeability with reactive and direct dyes, preventing fading, increasing strength, and giving the material a silky sheen. Mercerisation is also believed to improve the shrinkage of fabrics.

The process was developed in 1844 by John Mercer of Great Harwood, Lancashire, United Kingdom, who swelled cotton fibres with sodium hydroxide. The process did not become popular until it was developed in its current form in 1890. Cotton was stretched to avoid shrinking (Schmidt, 1938), and the originator of the idea, Horace Lowe, found that the fibres took on a sheen (Cook, 1984).

For classical mercerisation, two types of machinery are used: chain lines and roller machines. Chain lines are mainly offered by Japanese companies Wakayama and Kyoto, and roller machines are mainly from Unitechna in Germany, Benninger in Switzerland, Textima in Germany, Wakayama in Japan and EFI Mezzera Mercerising in Italy. The most important technical characteristics of roller mercerising machines are listed in Table 11.3.

Disadvantages of chain lines include lower productivity than roller lines, work on only one canvas and the possibility of mechanical damage to the fabric (breakage at the edges). There could also be random releases of edges from the double grips, causing uneven fabric in width, and possible unevenness in the density of fabric because of its more efficient expansion in the region of edges. Advantages include the possibility of processing dense fabrics, the higher quality of mercerisation owing to warp and weft tension, and less shrinkage across the width of the fabric.

Compared with chain lines, roller mercerising machines cause less damage to the fabric, there is no breakage on edges, and they offer higher productivity owing to processing in several canvases. Moreover, the maintenance and repair of machines are easier, and they work in silent mode. Disadvantages of roller machines include a significant reduction in the width of the fabric owing to tension on the warp. Also, the effect of mercerisation is worse (quality of gloss) than that achieved on chain lines (Holme, 2016; Ferro et al., 2020).

#### *11.3.1.6 Washing*

Prewashing and washing remove surface contaminants and residual chemicals from fabrics before and after wet processing. Ben-Wash from Benninger AG, Switzerland is the complete range of all washing processes, with excellent washing performance. The innovative machine works with minimal water consumption, high energy efficiency and up to 50% water and steam savings compared with exhaust processes. The company has also innovated its Fortracta prewasher to maximise the removal of surface contaminants. The vertical counterflow device optimises the use of water and energy. Trikoflex, Benninger's low-tension washer, is designed for stretch knits and wrinkle-prone fabrics. Improved washing efficiency is provided by the unique grooved drum surface.

The Fortracta washer has been specially designed to remove surface dirt and high concentrations of chemicals. The fabric is passed up through two narrow channels. Like a waterfall, the washing solution flows downwards under the

**Table 11.3** Technical characteristics of roller mercerising machines.

| | | Benninger (Switzerland) | | Textima (Germany) | |
|---|---|---|---|---|---|
| **Technical specifications** | **Unitechna (Germany)** | **Option 1** | **Option 2** | **6241 model** | **6243 model** |
| Working width L, mm | 1800–2400 | 1800–2500 | 1200–3400 | 2240 | 1800–2600 |
| Mass of processed fabric, g/m$^2$ | Up to 500 | Up to 210 | Up to 300 | Up to 210 | Up to 250 |
| Fabric movement speed, m/min | 20–100 | 36–72 | 20–120 | 20–60 | 20–100 |
| Installed power of electric motors, kW | – | 53 | 85–90 | 50.95 | 18.8 |
| Working volume of alkaline bath, m$^3$ | 2.0–2.4 | 1.0 | 0.3 | 3.0 | 3.0 |
| Steam consumption: | | | | | |
| -kg/1000 m | 1900–2090 | 450 | 315 | 400–520 | 280 |
| -kg/kg of fabric | 3.0 | – | – | – | – |
| Water consumption: | | | | | |
| -m$^3$/kg fabric | 17.5 | – | – | – | – |
| -m$^3$/1000 m | 2.63–2.8 | 2.7 | 1.9 | 2.83 | 2.0 |
| Consumption of electricity: | | | | | |
| -kWh/1000 m | 17.5 | – | – | – | – |
| -kWh/kg fabric | 0.19 | – | – | – | – |

influence of gravity. Owing to the counterflow principle, a high flushing effect is created in the channel, and the unit is also self-cleaning. Separation of the solution takes place with the help of a scraper shaft as well as flat nozzles at the outlet of the Fortracta compartment.

The prewashed article can then be passed through a conventional roller machine with no problems. Because the chemical content and mud ballast have been reduced by 50%, rubber rollers can be used again. Main highlights of the machine are the high washing performance, high level of uniformity and reproducibility, minimum water consumption, high energy efficiency and up to 50% savings in water and steam (Benninger, 2022). Ultrasonic washing is an upcoming innovative washing technology that leads to efficient washing with low energy consumption.

### 11.3.2 Dyeing machines

Colour is the result when white light from a source such as the sun reflects off dye on a surface. The colour it reflects is determined by the structure of the molecule, specifically by parts of the chromogen molecule called the chromophore group (Collier, 1970). Two processes are used to apply colour: dyeing and printing. Dyeing is the application of dyes or pigments to textile materials such as fibres, yarns and fabrics to achieve the desired colour fastness. Dye molecules attach to fibre by absorption, diffusion or binding. Temperature and time are important controlling factors. In printing, colour is applied only to a specific area with desired patterns.

#### 11.3.2.1 Yarn dyeing

In yarn dyeing, skeins, bundles and spaces are used. When skeins are dyed, the yarn is wound loosely into skeins and then dyed. The dye penetrates the yarn well, but the process is slow and comparatively expensive. Yarn wound on perforated spools in the package are dyed in a pressurised tank. The process is comparatively faster, but colour uniformity may not be as good as with skein-dyed yarn. When the bundle is dyed, the perforated bundle of the base is used. Space dyeing is used to produce yarns of different colours. As a rule, yarn dyeing provides adequate colour absorption and penetration of most materials. Thick and heavily twisted yarn may not have good dye penetration. This process is commonly used when yarns of different colours are used to create fabrics (e.g., plaids, iridescent fabrics).

In the industry, package dyeing machines are widely used, in which the yarn is wound into suitable packages such as cones and mounted onto perforated spindles, and then dyeing is carried out by forcing liquor through the package. The Pulsar dyeing system for packages, tops and warp beams from Italian company Loris Bellini Srl is equipped with a circulation pump and a hydraulic circuit that can achieve 70% energy savings while using 30% fewer chemicals and water at a liquid-to-mass ratio of 3.8:1. Pulsar is the result of an innovative but simple idea aimed at redefining the standard concept of yarn dyeing in the package.

Because of its innovative concept, Pulsar ensures cost-savings as well as an overall improvement in the quality of the final dyed product. The aspect of quality, confirmed by extensive laboratory tests and by end users who may personally have direct evidence, explains the reduced installed power and renewed hydraulic circuit. Owing to the combination of these aspects, the dye liquor has an increased affinity for the treated fibre. Moreover, a mixer (thermocolourmix) installed in the lower portion of the main kier provides significant savings to the complete dyeing process, such as 70% less electric energy and nearly 30% less water (Liquor Ratio [LR] 1:4). Even for steam, which is required in smaller quantities owing to the reduced level of water, Pulsar brings savings of approximately 20% over a traditional system (Loris Bellini, 2022). Table 11.4 lists the main technical characteristics and innovations of the Pulsar dyeing machine.

**Table 11.4** Technical characteristics of pulsar dyeing machine.

| Ordinary yarn dyeing system | | | Pulsar yarn dyeing system | | |
|---|---|---|---|---|---|
| **Machine model** | **Installed capacity (kg)** | **Installed power on main circulation pump (kW)** | **Machine model** | **Installed capacity (kg)** | **Installed power on main circulation pump (kW)** |
| 680/1965 | 100 | 18.5 | 680/1965 PD | 100 | 3 |
| 840/1965 | 150 | 22 | 840/1965 PD | 150 | 4 |
| 1040/ 1965 | 250 | 30 | 1040/ 1965 PD | 250 | 5.5 |
| 1280/ 1965 | 400 | 45 | 1280/ 1965 PD | 400 | 11 |
| 1400/ 1965 | 500 | 55 | 1400/ 1965 PD | 500 | 11 |
| 1600/ 1965 | 700 | 55 | 1600/ 1965 PD | 700 | 15 |
| 2000/ 1965 | 1000 | 110 | 2000/ 1965 PD | 1000 | 30 |

#### *11.3.2.2 Fabric dyeing*

Solid colour dyeing is a method for dyeing a fabric containing two or more types of fibres or yarns in the same shade to achieve the appearance of a solid colour fabric (Hoechst, 1999). Fabrics can be dyed in a single or multistep process. Union dyeing is used to dye single-colour blends and combinations of fabrics commonly used for clothing and household items. Several innovations are taking place in the dyeing machinery sector aiming to reduce water and energy consumption and effluents.

The Italian company Brazzoli SpA offers Innowash, an advanced washing process for its Ecologic Plus II forward flow dyeing machine, producing a liquid-to-product ratio of 3.8:1. The required amount of washing is determined automatically to minimise water consumption. An evolution of the previous Ecologic, this version controls and automatically operates all machine parameters as well as dyeing phases, optimising consumption and costs and standardising dyeing results in different product batches. Table 11.5 lists the main specifications of the Ecologic Plus II progressive flow machine.

The automatic control and management system of the washing process keeps the water consumption ratio at the lowest possible level. The optimised performance system is in its third version with important mechanical and software innovations, to optimise the washing process even more as well as simplify maintenance (Brazzoli, 2022).

The Italian company Flainox Srl developed an innovative garment dyeing machine, NRG-DL. It processes garments at a low liquor-to-product ratio of 5:1 and

**Table 11.5** Specifications of ecologic plus II machine.

| Specification | Unit | Light shade | Medium shades | Dark shades |
|---|---|---|---|---|
| | | Basis knit cotton 100% (reactive process) | | |
| | | Fabric absorption: 200% | | |
| Water | L/kg | 20–22 | 25–26 | 28–32 |
| Steam | kg steam/kg fabric | 1.1 | 1.2 | 1.7 |
| Electric energy | kW | 0.11 | 0.12 | 0.13 |
| Process time (including loading and unloading) | h and min | 4:20 | 4:40 | 5:00 |
| | | Final liquor ratio: 3.8:1 | | |

can measure energy, water and chemical consumption in real-time, optimising the process with a minimal use of energy, water and chemicals. The company also introduced a colouring system designed for use with natural dyes. AOM/C-WOOL is a recirculating extraction–dye–dye bath system in which the selected plant material is extracted just before application. After the fabric has been dyed, the dye bath is reclaimed for later use, which is a typical practice with natural dyes. The machine is claimed to reduce the carbon footprint of the manufacturing plant by 50% (Flainox, 2023). The main characteristics of the NRG-DL garment dyeing machine are listed in Table 11.6.

### *11.3.3 Printing machines*

In practice, not all fabrics are dyed in one colour. To manufacture many products, it is necessary to apply a pattern to the surface so that only that part of the fabric is dyed, as in printing. Textile printing is also considered localised dyeing. This is the application of colour in the form of a paste or ink on the surface of the fabric according to a given pattern. It is also possible to print patterns on already dyed fabric. In properly printed fabrics, colour bonds to the fibre to resist washing and rubbing.

There are several printing methods, such as stamp printing, transfer printing, sublimation printing and screen printing. Screen printing with flat screens uses a stencil and a nylon mesh to create the print design, and a waterproof material is used to block out the designs. When the screen is flooded with ink, the design appears. In a basic operation, flat screen and rotary screen printing machines are similar. Rotary screen printing is a continuous method in which a perforated cylindrical screen applies a colourant made from either pigment or dye. The screens rotate in contact with the

**Table 11.6** Characteristics of NRG-DL garment dyeing machine.

| **Nominal load** | **kg** | **60** | **90** | **180** | **240** |
|---|---|---|---|---|---|
| Number of compartments | n | 3 | 3 | 3 | 3 |
| Usable drum volume | $m^3$ | 1 | 1.5 | 3 | 4 |
| Average power consumption | kW/h | 6 | 8 | 14 | 18 |
| Maximum speed | Rpm | 550 | 550 | 450 | 450 |
| Machine width and depth | mm | 2600<br>1670 | 2700<br>2070 | 2970<br>2430 | 2970<br>2930 |
| Total height | mm | 2350 | 2350 | 2700 | 2700 |
| Weight of machine and accessories | kg | 2200 | 2800 | 4700 | 5500 |
| Maximum working temperature | °C | 98/Optional 105 | 98/Optional 105 | 98/Optional 105 | 98/Optional 105 |

substrate and the print paste is fed from inside the screens. The paste is forced out from the screen by means of a metal squeegee blade. Rotary screen printing is the most widely used type in the industry. It is a highly successful technique, but it leads to the huge waste of dyes and chemicals (Magictextiles, 2023).

Rotascreen textile printing machines from Zimmer, Austria offer a cost-effective method for high-quality mass production with remarkably high printing speeds of up to 120 m/min. They are based on pioneering technologies such as the magnet system plus roll rod, which combines the advantages of the magnet system with the conventional blade technique. They guarantee uniform application over the complete working width with no left-centre-right variations (Zimmer, 2022).

The EFI Reggiani ecoTERRA pigment solution offered by Electronics for Imaging, Inc is designed for streamlined and greener textile printing. It is an all-in-one solution for water-based pigment printing that requires no ancillary equipment for pretreatment and posttreatment. The technology results in a drastic reduction in energy and water consumption in the overall process for a more sustainable direct-to-textile printing experience. The ecoTERRA ink range features seven colours for an expanded colour gamut. An enhanced polymerisation and finishing unit for the machine gives the fabric a softer hand feel, delivering a performance in line with the most stringent textile industry requirements (Efi, 2023).

Digital printing is a sustainable innovation in printing. It is an easy-to-use method for printing designs onto fabric using a computer system employing modified or specialised inkjet technology. The global digital textile printing market is valued at approximately US $2.7 billion and is expected to have a phenomenal compound annual growth rate of 16.3%, reaching US $8 billion by the end of 2029. This has forced many to rethink their approach to using resources that were once abundant but are now scarce and expensive. This is evident in three main areas:

**Water savings:** Dye- and pigment-based printing systems are the biggest positive contributors, with savings of 70–80 L/m of print. For example, if these technologies were adopted worldwide, it would result in a potential saving of over 2 trillion L of water per year, or roughly double the United Kingdom's annual water consumption.

**Energy savings:** On average, a digital printer will consume about 0.14 kW/m of printed material, whereas a conventional rotary screen printer will consume an average of 0.46 kW/m of printed material. This represents a potential energy savings of over 63%. If adopted globally, it would reduce the textile industry's electricity needs by over 900 GW, three times the United Kingdom's electricity consumption.

**Pollution reduction:** The textile printing industry results in a huge volume of polluted water that is dumped into local aquifers. Conventional screen printing is an intensive process. Ink dispersions used are measured in litres, as opposed to the millilitres used in digital printing. The disposal of waste in the form of unused dyes, wash screens and effluent for sewage washing results in millions of litres of contaminated wastewater that must be managed every year (McKeegan, 2022).

Mimaki Engineering introduced preprinting and postprinting equipment such as the TR300-1850C coater and the TR300-1850S steamer, which allows direct printing on textiles and the transfer of paper on the same machine. Taking an integrated approach to digital fabric production, the company also offers TS55-1800, which provides uninterrupted performance and offers the best value for money. It is suitable for producing interior fabrics, sports and fashion wear as well as soft signs. Mimaki UCJV300-160 is an inkjet-integrated UV printer/cutter for roll-to-roll printing with white ink support. This versatile printing system provides unique quad-layer printing capabilities for advanced backlit materials, as well as five-layer printing, which prints different patterns on both sides of the material at the same time (Mimaki, 2022).

The Swift-Jet from Technijet, United Kingdom uses a spray coating system that is claimed to save 50% of the volume of water and energy used in precoating fabric for digital textile printing. This machine prepares reactive, pigment and acid ink sets using constant coating employing a contactless technology as an in-line or stand-alone process. This printer brings cost savings and flexibility and enhances the scope of market opportunities. Main advantages include multiple and consistent results from 10 to 1000 m, print result standards equivalent to conventional processes, energy consumption reduced by 50% and chemical consumption reduced by 80% (Technijet, 2022).

SPGPrints, Austria, which specialises in digital and screen printing solutions, developed three digital printing systems called Jasmine, Rose and Magnolia. Roll-to-roll printer Rose is for printing on sublimation paper, with a maximum capacity of up to 720 $m^2/h$. Jasmine and Magnolia print directly on fabric. The former is designed for maximum productivity, whereas the latter has 1200 dpi print resolution, and Archer+ technology raises the bar for print quality (SPGPrints, 2022).

Another generation of high-speed combination-processing machines for single-stage printing is available, such as the EFI Reggiani Bolt hybrid machine, with innovative printheads. The printer can achieve cruising print speeds of up to 90 m/min (over 8000 $m^2$/h) with a resolution of 600 × 600 dpi. A greyscale with a variable droplet size of 5−30 pL enables of the highest complexity to be printed (Efi, 2022). Optimum Digital Planet, Turkey developed the Picasso Carpet printer, designed for direct digital printing on polyester carpets at a speed of 120 $m^2$/h, and the Nirvana Belt textile printer. The latter prints on cotton, viscose, polyester and other materials with various types of ink, providing a speed of 800 $m^2$/h with a single pass. The Nirvana Belt eliminates wasted water and the need for additional machinery by preventing the prepress and postpress processes. With this technology, which can print on both cotton and polyester fabrics, the production capacity increases and water consumption is reduced, saving labour and time. With no pretreatment, costs are decreased by 50% in this waterless printing, which also reduces energy and chemical costs by 80%. Because the chemicals usually applied during pretreatment are not used, and no extra energy, time and cost are spent on drying, the process minimises damage to the environment (Optimumdigital, 2023).

Zimmer, Austria developed an innovative product for the digital printing of technical textiles, particularly for military camouflage. The Colaris camouflage printing system offers perfect colour adjustment with additional infrared reflectance control by imitating specific reflectance characteristics. This digital printing solution matches the requirements for most internationally operating forces and security agencies. Another development by the company is the Colaris digital inkjet printing line for printing on woolen and polyamide moquette fabrics used for the interior decoration of vehicles such as buses, subways, trains and aeroplanes (Zimmer, 2023).

## 11.3.4 Finishing machines

Finishing is the final stage in the wet processing of textile materials. It is a series of processes that bleached, dyed, printed and some grey fabrics undergo before being placed on the market. A fabric can gain significant added value through the application of one or more finishing processes. The purpose of textile finishing is to adapt textiles to their intended purpose or end use and/or to improve the performance properties of the fabric, facilitating further cutting and sewing operations in the garment industry. Finishing includes mechanical, chemical and biotechnological processes (Paul, 2014).

### 11.3.4.1 Mechanical finishing

#### Napping

Napping is an important mechanical textile finish and the oldest one. A wide variety of fabrics can be produced using this process, including blankets, flannelettes and industrial fabrics. Napping lifts from the main part of the fabric a layer of fibres protruding from the surface, called a pile. The formation of lint on the fabric results in a lofty handle and can soften the weave or pattern and colour of the fabric (Britannica, 2022). There are two types of napping machines: card-making and teaser. The speed of the wire card machine ranges from 11 to 14 m/min, which is 20% to 30% faster than the teaser machine. That is why the card lifting machine is widely used.

## Shearing

Shearing is a type of mechanical finishing in which the appearance of a fabric is improved by trimming the loops or embossing the surface to a uniform and even height. It may also be used to create stripes and other patterns by varying the surface height. Shearing is most commonly used to make woollens and worsted materials. A shearing machine can cut the loop or pile to a desired level, and the machine may have a spiral blade similar to a grass clipper (Elsasser and Sharp 2022; Joseph et al., 1992).

## Peach effect

The peach skin effect on fabric is a mechanical process comparable to napping, but it is gentle. The peach effect is achieved by lightly sanding the fabric, in which the fabric is passed through abrasive rollers on a special machine. A peach finish is also possible with the help of certain chemicals, or with a wash. After the technology is applied, the fabric acquires properties such as a velvety and soft surface, a comfortable feeling when it is worn, increased wear resistance and the form stability of clothing.

## Calendering

Calendering is an important mechanical finishing process in which the goals are to compress the fabric, reduce its thickness, improve the opacity and give the fabric a smooth, silky finish. The breathability of the fabric can also be reduced by changing its porosity to give the fabric a different degree of sheen and reduce yarn slippage. In calendering, fabric is passed between heated rollers to obtain smooth, polished or embossed effects, depending on the surface properties of the roller and the relative speeds (Collier, 1970). An ordinary calendar consists of a series of hard and soft (elastic) rollers arranged in a certain order. The hard metal bowl is made from hardened iron, cast iron or steel. The soft roller for a calendar can be divided into a wool paper roll and a cotton pulp paper roll. The sequence of rollers is such that no two rigid rollers are in contact with each other. Pressure can be applied using complex levers and weights, or hydraulic pressure can be used. The pressure and heat used in calendering depend on the type of finish required. Guoguang Group, China offers as series of fabric calendar machines including a friction calendar, a pressure and heated roller calendar, a three-roll calendar and a four-roll calendar (Guoguang, 2023).

## Sanforization

Sanforization is mechanical shrinkage, in which fabric is forced to contract in width and/or length, and residual tendency to shrink after subsequent laundering is minimal (Collier, 1970). Shrinkage is a change in the dimensions of the fabric along the length and width after washing, use and exposure. Shrinkage is of two types: minus or plus. In minus shrinkage, there is a noticeable size reduction; any noticeable increase in size is known as plus shrinkage or expansion.

Sanforization is the heat treatment of cotton or any other fabric to prevent possible shrinkage after the finished product is washed, or to minimise it. This process is named after its creator, Sanford Lockwood Clott, who patented it in 1930. During sanforization, fabric presoaked in water is pulled under a load between dense rubber rollers of the machine. The rollers strongly compress the threads, but because of the elastic properties of rubber, the threads do not break. Then, fabric processed by the rollers is dried on a preheated cylinder. In another version of the same technology, the fabric may not be wetted, but simply doused with hot steam. In any case, the goal of sanforization is to remove tension from the cotton threads. After washing, the threads will no longer critically shrink in the direction the fibres are woven (Choudhury, 2017). The Clip Type mercerising range from Kyoto Machinery, Japan offers a high speed operation with a practical operation speed up to 200 m/min (Kyoto, 2023). The innovative design of the mercerising system from Prabhat Textile Corp, India prevents water vapour from dripping, and the fully sealed steam cover design prevents steam leakage. The vibration mechanism of this machine helps it maintain fabric fluffiness. It offers A long working life, A low production cost and minimal maintenance charges (Prabhat, 2023).

#### *11.3.4.2 Chemical finishing*

Chemical finishing can be carried out in many pretreatment and dyeing machines. chemical finishing processes pose a greater environmental hazard than mechanical ones owing to their higher potential for air emissions and effluents. Also referred to as wet finishing, chemical finishing involves processes that change the chemical composition of fabrics and/or the surfaces to which they are applied (Paul, 2014).

Fabric finishing is an energy-intensive process. There are two energy-saving machine development strategies: a technology that skips one or more production steps to shorten the finishing process, and the use of latest-technology components such as motors, electronic or intelligent controls, which can significantly reduce energy consumption. Optimising finishing processes is another way to save energy. The highest potential for energy saving would be to reduce the use of steam, water or electricity in the different processing stages, and to adopt technical solutions to limit or partially recover the energy consumed (Cafaggi, 2023).

Given the increasing level of automation and remote services in the textile fabric finishing industry, intensive relationships among the finishers, machine manufacturers, chemical suppliers and other technology providers is the key to a sustainable future. Modern machine concepts with intelligent control systems are significantly reducing the consumption of water, chemicals and energy consumption. The use of these machines leads to the massive reduction of the environmental footprint compared with older models (Brückner, 2023).

### *11.3.5 Steam generator*

Steaming is an important process in textile wet processing such as in pretreatment, dyeing and finishing, in which steam enables fast energy transfer into the fibre. Textile finishing processes such as forming, embossing, drying or dampening use large amounts of hot air steam at 121 to 260°C. Industrial steam generators are widely used for these purposes, which are sealed boilers made of durable metal. A heating element is placed inside, which brings water to a boiling point. A steam generator is a special machine designed to produce water vapour with a pressure above atmospheric pressure. As a rule, the generation of water vapour in the apparatus occurs as a result of the working medium, such as water, being heated.

According to the type of heating element, there are different types of steam generators. In gas steam generators, the energy of gas combustion is used to heat water. A steam boiler of this type has a higher performance than do electrical appliances, but it is less than a diesel one. Diesel steam generators have high working power and provide maximum performance: steam volume and temperature. Solid fuel or biofuel generators work on the energy of combusting coal, production waste and other solid fuels. Electric steam generators are more often used together with a thermoelectric type of heater, and less often with an electrode type. Sensors monitor the flow of water and regulate heating processes. The structure of a typical steam generator is shown in Fig. 11.5.

A sustainable innovation in this sector is the induction steam generator. The uniqueness of this electronic equipment is that it produces steam through the most efficient method of energy conversion, induction. Water is heated by the boiler body, which is the role of the core of the induction coil. When an electric current passes through the winding, the temperature of the boiler walls reaches 200 to 220°C. As a result, water is heated. The efficiency of induction steam generators is 97% to 98%, and the cost of electricity is two to three times lower than for conventional ones.

In addition to cost-effectiveness, advantages of the equipment are stability and trouble-free operation. It is available in automatic and manual modes. The equipment does not require warming up, which saves time and energy. It takes no more than 10 min from the moment the induction steam generator is started to when steam is produced. In the latest innovations of induction steam generators, the process of steam formation occurs using a combination of water, current and a magneto dynamic plasma discharge—low-temperature plasma, which converts energy from one type to another with the highest possible efficiency. The use of low-temperature plasma makes it

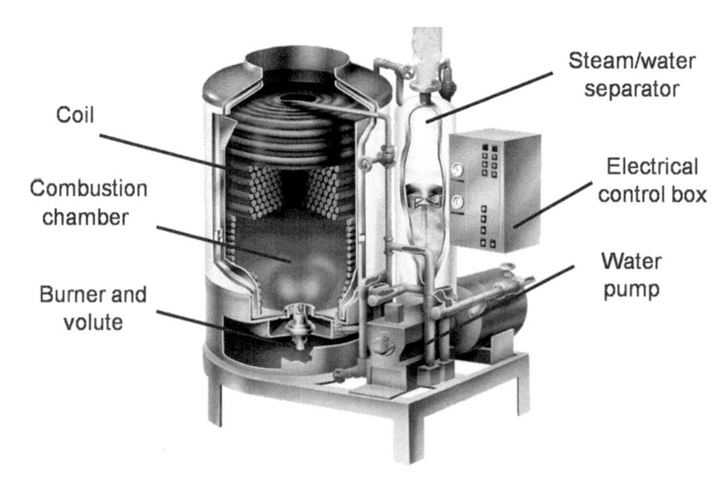


**Figure 11.5** Structure of a steam generator.
From Claytonsteam (2023).

possible to obtain such high temperatures that could not previously be achieved. In these steam generators, the thermal effect on the heated substance is combined with the effect of a magnetic field. The magnetic field does not transfer active energy, but under its influence, atoms of the substance are excited, which facilitates the transfer of active energy from the heat source to the heated medium. Accordingly, high efficiency of thermal energy transfer is ensured (Claytonsteam, 2023; Miuraboiler, 2023; Veit, 2023).

## 11.4 Innovations leading to sustainability and energy efficiency

The EU strategy for sustainable textiles aims for a shift to a climate-neutral circular economy in which textile products are designed to be more durable, reusable, repairable, recyclable and energy-efficient (European Commission, 2022). Such initiatives have led to the development of several innovative and emerging wet processing technologies and machines able to provide economic and environmental benefits such as reduced energy, water and chemical use. Textile wet processing is considered to be the most delicate link in the textile chain, and so machinery innovations can have a big role in achieving textile sustainability.

### *11.4.1 Enzymatic technologies*

A sustainable alternative to conventional textile pretreatment and finishing processes is the biotechnological application involving enzymes. Enzymatic processes can be carried out in existing wet processing machines in the industry, and no special machines are required. The use of enzymes in textile processing can solve important issues such as the development of cleaner, softer, more environmentally friendly and nonaggressive textile production. From the economic point of view, there is reduced energy and the cost of chemicals. A more complete processing of low-quality natural fibres such as coarse wool fibres, short flax fibres and other fibrous materials can also be achieved.

Enzymatic scouring has several ecological benefits compared with alkaline scouring, because there is a 20% to 50% reduction in rinse water consumption. There is also a 20% to 40% reduction in biochemical oxygen demand and chemical oxygen demand loads in wastewater. The enzymatic removal of residual postbleaching hydrogen peroxide also has several environmental benefits. There are energy and water savings because peroxidases have no negative influence on downstream dyeing, so the liquor does not have to be drained after enzymatic treatment and before dyeing. There is complete biodegradability of peroxidases and a reduction in the rinse steps. It also eliminates wastewater pollution from chemical reducing agents used in conventional processes. Enzymes are also widely used to finish natural fibres such as cotton and wool, as well as synthetic ones such as polyester, to polish fibre surfaces and impart hydrophilicity (Paul, 2014; Canal et al., 2004).

### 11.4.2 Ultrasonic dyeing

The use of ultrasound in the textile industry is a sustainable advance in which ultrasonic waves are used as sources of energy in the wet processing of textiles. When ultrasonic waves are absorbed by a liquid system, cavitation occurs, which is the alternating formation of waves, vibrations and the collapse of tiny bubbles or cavities. During each wave cycle, compression or rarefaction occurs. During the rarefaction part of the wave cycle, dissolved gas molecules act as nuclei to form cavities, which can expand relatively slowly to a diameter of up to 0.1 cm and then rapidly collapse during the compression part of the cycle (Saravan, 2006).

Ultrasonic cavitation accelerates the speed of dyeing and increases the adsorption of the dye into the fabric. A typical dyeing process involves the use of chemicals and thermal energy, which can be reduced with ultrasonic energy. Ultrasonic vibrations contribute to the dispersion of dyes in solutions and favour emulsification. The use of ultrasonic technologies can make it possible to intensify processes such as the bleaching, dyeing and finishing of textile materials, the laundering of contaminants and the degreasing of materials. It also improves some properties of natural and synthetic fibres. Ultrasonic wet processing is gaining popularity in the textile industry because it achieves comparable effects with a lower temperature and lower chemical concentrations. It also has several other benefits, such as shorter cycle times and a reduced consumption of dyes and chemicals, which allows for a 20% to 30% reduction in effluents (Warmoeskerken et al., 2002).

Several ultrasonic machines are available at the laboratory or pilot scale, but industrial machines are still under development. However, the machine described in patent CN102535064B of Hangzhou Intelligent Dyeing and Finishing Equipment Co Ltd, China is an industrial-scale machine that is relatively inexpensive and may be suitable for dyeing and finishing various fabrics. The ultrasonic frequency and intensity are easily adjusted so that dye uptake can be improved, dyes are saved and the dyeing time is shortened. The requirements of energy conservation, reduced emissions, a low carbon footprint and environment friendliness are also met (Hangzhou, 2023).

Sonotronic GmbH, Germany developed an ultrasonic washer for textiles. Ultrasound at 20 kHz acts directly on the fabric and mechanically beats substances to be dissolved or removed from the fabric. The main feature of this ultrasonic washer is that excess substances are loosened from the fabric mechanically as well as by cavitation (Sonotronic, 2023).

### 11.4.3 Supercritical $CO_2$ dyeing

The use of water as a solvent for dyes and chemicals is mainly due to its abundant availability and low cost. The problem associated with using water is the formation of wastewater. After each stage, an additional stage is required to dry the fabrics. The amount of energy used to remove water is also enormous, which adds to problems for processors by making wet processing the weakest link in the entire textile chain. To eliminate the shortcomings of wet processing, it is proposed to replace water as a solvating medium with some gases. Dyes require high pressure and temperature to dissolve. Of all the gases that can be converted to supercritical fluids, $CO_2$ is the most versatile and widely used (Joshi et al., 2006). Supercritical CO2 dyeing needs much research to make it commercially viable. It has been well-established on a laboratory and pilot scale, and dyeing with this system has proven successful for both synthetic and natural fibres. Industrial-scale integration may require further research and innovation.

DyeCoo is a pioneer in this field. Its $CO_2$-based technology uses reclaimed $CO_2$ as a dyeing medium in a closed loop. Under pressure, $CO_2$ becomes supercritical (SC–$CO_2$), and in this state, $CO_2$ has an exceedingly high solvent power, which allows the dye to dissolve easily. Owing to the high permeability, dyes easily and deeply penetrate into the fibres, creating bright colours. The technology uses $CO_2$, a by-product of the industry, as the dye medium in a closed system, and around 95% of $CO_2$ is recycled in the machine for reuse.

$CO_2$ dyeing does not require the addition of process chemicals to dissolve dyes. It has high solvent power, which allows the dye to dissolve easily. The technology uses 100% pure dyes, with an absorption of more than 98%, and there is no waste. Because there are no process chemicals and water, there is no need for wastewater treatment. Efficient colour absorption and short batch cycles make this technology energy-efficient. The investment cost of $CO_2$ dyeing equipment is high compared with traditional dyeing equipment, whereas operating and environmental costs are lower. Short batch cycles, the efficient use of dyes and no wastewater treatment contribute to a significant reduction in operating costs. All of this contributes to a significant reduction in operating costs (DyeCoo, 2023).

Hisaka Works, Ltd, Japan is a producer of supercritical $CO_2$ dyeing machines. Their supercritical fluid processing machine reuses $CO_2$ from liquid to supercritical, gas and then liquid in a closed system, and is effective in reducing the environmental impact without using water. The dye is dissolved in the $CO_2$ under supercritical conditions, and the fabric is dyed with the supercritical fluid in the tank. After the machine is opened to remove fabrics, $CO_2$ returning to the gas at the time of opening is pressurised and compressed in the next process, and then is recovered and stored as liquid $CO_2$ (Hisaka, 2023).

### 11.4.4 Foam dyeing and finishing

Foam is the dispersion of gas in a liquid. The wet processing of textiles, which uses air in the form of dispersion foam, is called foam dyeing or foam finishing. In foam dyeing technologies, which is carried out at an exceptionally low bath modulus, a reduction in water consumption up to 30%, energy up to 40%, and dyes up to 15%, as well as significantly less pollution from industrial waste are achieved. In foam printing processes, first of all, the possibility of reducing the consumption of thickener is achieved. The reduction of thickener for reactive dyes is around 65%, for disperse dyes it is 60% and for pigments it is 50%.

By foaming a concentrated finishing solution, its volume is greatly increased, minimising the problem of uniform distribution throughout the textiles. The foam is dispensed onto the fabric to ensure that no excess liquid needs to be removed and recycled. The essence of foam finishing textiles is to replace most of the liquid in the finishing media with air. As a result, on average, the moisture content of the treated material is reduced by three to four times. Accordingly, energy consumption to remove liquid from it during heat treatment processes is also reduced. In finishing, promising results can be obtained when processing textile materials with high-expansion foams with a low content of an aqueous phase, which makes it possible to increase the speed of operation of tumble dryers by 1.5 to 2.0 times with a decrease in the heat treatment temperature by 15 to 20°C (Cotton Incorporated, 2011).

Prism Textile Machinery, India is a manufacturer of foam dyeing and finishing machinery. The machine is user-friendly and offers high productivity with a working width of 1800−3200 mm and a speed of 40 meter per minute (mpm) (Prism, 2023). Foammix 600 from Rollmac, Italy is a foam-generating machine for the high production of fine foam with high homogenous characteristics for textile finishing (Rollmac, 2023). Table 11.7 lists the comparative indicators of fabric finishing by water and foam technologies.

### 11.4.5 Plasma technology

Plasma technology is considered an interesting and future-oriented process owing to its environmental acceptability and the wide range of applications. It is a dry process that produces no wastewater or chemical emissions. Thus, the methods are economical and reduce the environmental impact caused by the chemical textile industry. Gas-discharge plasma is an ionised gas containing a large number of charged ion particles, radicals and so on (Crespo et al., 2009).

The combination of particles of various chemical activities, energy and penetrating abilities makes low-temperature plasma a powerful tool for modifying the properties of textile materials, merging elements of both chemical and physical factors of influence. Low-temperature plasma is widely used to solve specific production problems in the finishing of textiles, leather and fur materials. External parameters of low-temperature plasma include the type of plasma-forming gas, its pressure and flow rate, discharge current as well as the geometry of the plasma-chemical reactor and

**Table 11.7** Comparison of fabric finishing by water and foam technologies.

| Index | Water technology | Foam technology |
|---|---|---|
| Surface density of fabric, g/m$^2$ | 800 | 800 |
| Moisture absorption during impregnation, % | 400 | 40 |
| Specific costs for heating fabric from 20 to 100°C | 1250 | 240 |
| Moisture content of fabric after processing in trough, % | 75 | 45 |
| Amount of heat for drying fabric to 10% moisture content, kJ | 1440 | 530 |
| Energy consumption, kW/h | 440 | 150 |

its structural materials that are in contact with the discharge zone. Depending on the processing conditions that occur under the influence of electric current and its discharges, electromagnetic field and plasma jet, it is possible to modify the surfaces of almost any textile material. During plasma treatments, only the surface of the textile material is modified, that too a very thin near-surface layer, the thickness of which is around 1000 A°. The main mass of the material remains unchanged, so the mechanical, physicochemical and electrophysical properties are preserved. Usually $O_2$, $N_2$, $CO_2$, Ar, $NH_3$ and air are used as plasma-forming gases (Bautista et al., 2010; Crespo et al., 2009).

The full-scale industrialisation of plasma treatment in textiles is still in progress, but there are several laboratory- and pilot-scale developments. Several functional finishes such as hydrophilicity and hydrophobicity are created on textiles using plasma technology (Crespo et al., 2009). Temperature could be an important parameter during atmospheric pressure plasma treatments on polyester fabrics to generate different functional groups (Bautista et al., 2010). In other work, wool fabric was treated with low-pressure plasma and dyed with commercially available natural colorants, which resulted in high dye uptake and better fastness properties (Paul et al., 2012).

Several companies offer plasma treatment machines for textile processing. Diener Electronic GmbH, Germany offers both low-pressure plasma and atmospheric pressure plasma machines (Diener, 2023). Openair-Plasma pretreatment from Plasmatreat, Germany significantly improves the wettability of fibres and yarns, enabling even solvent-free dyes to bond well and durably. Textile surfaces can be provided with hydrophobic and dirt-repellent finishes (Plasmatreat, 2023). Another manufacturer of plasma treatment systems for cleaning, activation and coating is Henniker Plasma, Germany (Henniker, 2023). PLAtex cold atmospheric plasma from Grinp Srl, Italy can functionalise fabric using a specific combination of gases and power that makes the fibre receptive to chemicals used in the cavitational impregnation unit of the machine. These plasma machines are available at an industrial scale and are able to treat textiles up to 4 m wide (Grinp, 2023). Table 11.8 shows a comparative description of conventional water-based processes and plasma processes.

**Table 11.8** Conventional and plasma processing of textiles.

| Index | Conventional treatment | Plasma treatment |
|---|---|---|
| Medium | Water (or organic solution) | Dry |
| Chemicals | Large quantities | Gases in limited quantities |
| Types of reactions | Simple and well-studied | Complex and little studied |
| Energy consumption | High | Low for atmospheric pressure plasma, higher for vacuum plasma |
| Temperature | High | Room |
| Treatment type | Bulk or surface treatment (>100 nm), frequent change in bulk properties | Bulk modification of capillary-porous materials (<10 to 60 nm) |

Continued

**Table 11.8 Continued**

| Index | Conventional treatment | Plasma treatment |
|---|---|---|
| Substrates | Different textile materials require different processes and equipment | The same equipment can be used for all textile materials under different process parameters |
| Process type | Batch or continuous | Batch (vacuum plasma), batch or continuous (atmospheric pressure plasma) |
| Equipment | Simple and well-known | Complex, evolving |
| Required qualifications | Standard | High specialisation |
| Waste treatment costs | High | Low |

## 11.5 Future perspectives

Sustainability in the textile sector suggests the use of green clothing and other textile products. Many governmental agencies, nonprofit organisations and innovative start-up companies are producing innovative programs and ideas to promote sustainability and responsible consumption in textiles. A circular economy or circularity is the key to achieving textile sustainability. It involves sharing, leasing, reusing, repairing, refurbishing and recycling existing materials and products for as long as possible. This aims to tackle global challenges such as climate change, biodiversity loss and waste disposal (Goyal, 2022).

To achieve this goal, enlightened consumers should opt for fabrics from recycled materials or innovative natural and bio-based fibres. Creating more value for safety, comfort, health, ecology and nature, including the creation of smart textiles, can provide adaptive functions to changes in the environment and human needs. The modern textile processing scenario requires the conservation of energy or the use of a small amount of energy. The use of ultrasonic waves, microwave dyeing, plasma technology and supercritical carbon dioxide dyeing are some revolutionary and futuristic paths to improve the wet processing of textiles. The sustainability of textile production is impossible without further introducing resource-saving and energy-efficient wet processing machinery, which also reduces the effluent load. Several sustainable innovations are occurring in the wet processing machinery field. International machinery exhibitions such as ITMA, India International Textile Machinery Exhibition (India ITME) are relevant in this sense.

## 11.6 Conclusion

Sustainability, climate awareness and social responsibility have become important components of textile production and business worldwide. Natural resources are scarce and their judicious use is imperative. The shift to greener alternatives can essentially reduce the carbon footprint of the textile industry. Another major solution is the circular economy, which emphasises reducing and recycling waste textile materials. The global development goal is technological innovations in the field of eco-textiles aimed at solving environmental problems.

The wet processing of textiles has countless steps leading to a finished product. Each step has several complex variables, and each batch is like a new one, which depends on a well-trained workforce rather than innovative machines and technology. Developments in machinery are taking place at a fast pace to satisfy users with a quality product at a competitive price. Although the main principle of development is to meet the needs of users, it will also increase cost-competitiveness and overall sustainability.

Main trends in textile wet processing include the increased efficiency and productivity of wet processing machinery and the efficient use of energy and water as well as the creation of environmentally friendly technologies. This paradigm is integrates the design, technology and production based on automatic machines with intelligent programme control and the use of electronic business tools. As a result, it is possible to systematise the main focuses of textile sustainability by saving raw materials, energy and time, and increasing productivity, by innovations in wet processing machinery.

## References


Bautista, L., Paul, R., Mota, J., Crespo, C., De la Varga, M., Aubouy, L., Pavan, C., 2010. Effect of temperature during atmospheric pressure plasma treatments on polyester fabrics. In: Paper Presented in AUTEX 2010 World Textile Conference, pp. 21–23. Vilnius, Lithuania.

Benninger, A.G., 2022. Available at: https://benningergroup.com/textile-finishing/overview. (Accessed 22 August 2022).

Brazzoli SpA, 2022. Available at: https://www.brazzoli.it. (Accessed 22 August 2022).

Britannica, 2022. Available at: https://www.britannica.com/technology/napping. (Accessed 25 May 2022).

Brückner, R., 2023. Sustainability and Digitalisation in Textile Finishing. The Textile Magazine. Available at: https://itma.com/blogs/automation-digital-future/2021/finishing/sustainability-and-digitalisation-in-textile-fin. (Accessed 5 February 2023).

Cafaggi, G., 2023. Finishing With Environmental Friendliness in Mind. ITMAConnect Virtual Visitor. Available at: https://itma.com/blogs/Expert-Insights-(1)/Jan-2023/Finishing/Finishing-with-environmental-friendliness-in-mind. (Accessed 5 February 2023).

Canal, J.M., Navarro, A., Calafell, M., Rodriguez, C., Caballero, G., Vega, B., Canal, C., Paul, R., 2004. Effect of various bio-scouring systems on the accessibility of dyes into cotton. Coloration Technolology 120 (6), 311. https://doi.org/10.1111/j.1478-4408.2004.tb00236.x.

Choudhury, A.K.R., 2017. Principles of Textile Finishing. Woodhead Publishing Ltd., Cambridge, UK, p. 55.

Claytonsteam, 2023. Available at: https://claytonsteam.com/en-EN/products/steam-generators. . (Accessed 15 April 2023).

Collier, A.M., 1970. Handbook of Textiles. Pergamon Press, Oxford, UK, p. 160.
Cook, J.G., 1984. Handbook of Textile Fibres: Volume I: Natural Fibres. Woodhead Publishing Ltd., Cambridge, UK.
Cotton Incorporated, 2011. Available at: https://www.cottoninc.com/wp-content/uploads/2017/12/TRI-4009-Foam-Applications-on-Textiles.pdf. (Accessed 20 March 2023).
Crespo, L., Bautista, L., Mota, J., Paul, R., Garrido-Franco, M., García-Montaño, J., De la Fuente, M., Amantia, D., Aubouy, L., De la Varga, M., Detrell, A., 2009. New textile finishes using plasma technology. Revista de Quimica Textil 192, 36.
Diener Electronic GmbH, 2023. Available at: https://www.plasma.com/en. (Accessed 20 March 2023).
Dspattextile, 2023. Available at: https://www.dspattextile.com/2022/06/kiers-in-scouring-processing.html#more. (Accessed 20 March 2023).
DyeCoo, 2023. Available at: https://goexplorer.org/wp-content/uploads/2018/06/Fashion_2015_DyeCooTextileSystems1.jpg. (Accessed 4 March 2023).
Efi, 2022. Available at: https://www.efi.com/products/inkjet-printing-and-proofing/reggiani-textile/reggiani-digital-industrial-printers/digital-printers-for-direct-printing-on-fabric/efi-reggiani-bolt/overview/. (Accessed 22 August 2022).
Efi, 2023. Available at: https://go.efi.com/en-ecoTERRA-Solution. (Accessed 22 May 2023).
Elsasser, V.H., Sharp, J.R., 2022. Textiles: Concepts and Principles. Fairchild Books, New York, p. 197. ISBN: 9781501366550.
Erbatech GmbH, 2022. Available at: https://www.erbatech.com/en/continuous-bleaching. (Accessed 15 July 2022).
European Commission, 2022. Available at: https://environment.ec.europa.eu/strategy/textiles-strategy_en. (Accessed 15 April 2022).
Ferro, M., Mannu, A., Panzeri, W., Theeuwen, C.H.J., Mele, A., 2020. An integrated approach to optimizing cellulose mercerization. Polymers 12 (7), 1559. https://doi.org/10.3390/polym12071559.
Fibre2fashion, 2022. Available at: https://www.fibre2fashion.com/industry-article/3474/singeing-fundamentals. (Accessed 15 April 2022).
Flainox, 2023. Available at: https://www.flainox.com/en/nrg-dl-rotary-machine. (Accessed 22 April 2023).
Goyal, P., 2022. Significance of Sustainability in Textiles. Textile World (January/February).
Grinp srl, 2023. Available at: https://www.grinp.com/textile. (Accessed 20 March 2023).
Guoguang Group, 2023. Available at: http://www.calenderingmachine.com/1-1-fabric-calendering-machine.html. (Accessed 20 March 2023).
Hall, A.J., 1969. Standard Handbook of Textiles. Heywood Books. ISBN 13: 9780592063287.
Hangzhou, 2023. CN102535064B Available at: https://patents.google.com/patent/CN102535064B/en. (Accessed 4 March 2023).
Henniker Plasma, 2023. Available at: https://www.henniker-plasma.de/en. (Accessed 20 March 2023).
Hisaka, 2023. Available at: https://www.hisaka.co.jp/english/textile/product/product10.html. (Accessed 4 March 2023).
Hoechst, 1999. Dictionary of Fibre and Textile Technology. Kosa, p. 258. ISBN 13: 9780967007106.
Holme, I., 2016. Coloration of technical textiles. In: Horrocks, A.R., Anand, S.C. (Eds.), Handbook of Technical Textiles. Woodhead Publishing Ltd., Cambridge, UK, pp. 231–284.
Hummel, J.J., 1898. Dyeing of Textile Fabrics. Harvard University, London, New York, p. 86. Cassell and Company, Limited.

Img, 2023. Available at: https://img.directindustry.com.ru/images_di/photo-g/54005-10354174.webp. (Accessed 20 March 2023).
ITMA, 2023. Available at: https://itma.com. (Accessed 4 March 2023).
Joseph, M.L., Hudson, P.B., Clapp, A.C., Kness, D., 1992. Joseph's Introductory Textile Science. Holt Rinehart and Winston, p. 392. ISBN 13: 9780030507236.
Joshi, A.S., Malik, T., Parmar, S., 2006. Polyester Dyeing in Supercritical Carbon Dioxide. Asian Dyer, pp. 51–54.
Kyoto Machinery, 2023. Available at: https://kyoto-machinery.com/pdf/clipTypeMercerinzingRange_en-A4.pdf. (Accessed 4 March 2023).
Loris Bellini, 2022. Available at: https://www.lorisbellini.com/en/index.php/products/tintura/. (Accessed 22 August 2022).
Magictextiles, 2023. Available at: https://www.magictextiles.co.uk/rotary-screen-printing. (Accessed 22 May 2023).
McKeegan, D., 2022. How Digital Technology is Democratising the textile industry. Available at: https://www.fespa.com/en/news-media/features/how-digital-technology-is-democratising-the-textile-industry. (Accessed 22 August 2022).
Mimaki, 2022. Available at: https://mimaki.com. (Accessed 30 July 2022).
Miuraboiler, 2023. Available at: https://miuraboiler.com/steam-boiler-guide-for-textile-manufacturing-processes. (Accessed 15 April 2023).
Optimumdigital, 2023. Available at: https://www.optimumdigital.com/en/products/textile-printer/picasso-carpet/i-2592. (Accessed 15 April 2023).
Osthoff-Senge, 2023. Available at: https://www.osthoff-senge.com/?page_id=2088. (Accessed 25 July 2023).
Paul, R. (Ed.), 2014. Functional Finishes for Textiles: Improving Comfort, Performance and Protection. Woodhead Publishing Ltd., Cambridge, UK. ISBN: 9780857098399.
Paul, R., Erra, P., Molina, R., 2012. Dyeing of plasma treated wool with commercially available natural colorants. International Dyer 197 (10), 32.
Plasmatreat, 2023. Available at: https://www.plasmatreat.com/en. (Accessed 20 March 2023).
Prabhat Textile Corporation, 2023. https://www.prabhattextiles.com/sanforizing-machine-6021917.html. (Accessed 4 March 2023).
Prism, 2023. Available at: https://www.prismtextilemachinery.com/fabric-finishing-machine-with-foam-or-wet-technology.html. (Accessed 10 May 2023).
Rollmac, 2023. Available at: https://www.rollmac.it/a_21_EN_98_1.html. (Accessed 10 May 2023).
Saravan, D., 2006. Ultrasonic textile processing—update. Colourage 53 (4), 111–116.
Schmidt, O., 1938. Great Soviet Encyclopedia, vol 39.
Shree Shakti Krupa Industries, 2022. Available at: https://www.shreeshaktikrupa.com/continuous-bleaching-range. (Accessed 15 July 2022).
Sonotronic GmbH, 2023. Available at: https://sonotronic.de/en/technologies/ultrasonic/ultrasonic-washing. (Accessed 4 March 2023).
SPGPrints, 2022. Available at: https://www.spgprints.com. (Accessed 10 April 2022).
Statista, 2022. Production Volume of Textile Fibres Worldwide 1975-2020. https://www.statista.com/statistics/263154/worldwide-production-volume-of-textile-fibres-since-1975. (Accessed 10 April 2022).
Technijet, 2022. Available at: https://www.nessancleary.co.uk/technijet-shows-textile-treatment-system. (Accessed 15 July 2022).
Veit, 2023. Available at: https://www.veit.de/en/products/cleaning-germ-and-virus-protection/steam-generator-with-steam-applicator-sg-67-66mf. (Accessed 15 April 2023).

Wakayama Iron Works Ltd, 2023. US4125921A, Apparatus for gas-singeing knitted fabrics. Available at: https://patents.google.com/patent/US4125921A/en. (Accessed 25 July 2023).
Warmoeskerken, M., et al., 2002. Laundry process intensification by ultrasound. Colloids and Surfaces A: Physicochemical and Engineering Aspects 210 (2), 277–285.
Zimmer, 2022. Available at: https://www.zimmer-klagenfurt.com/en/content/rotaryscreen-printing. (Accessed 10 April 2022).
Zimmer, 2023. Available at: https://www.zimmer-kufstein.com/en/download/file/fid/7490. (Accessed 22 May 2023).

# Innovations in textile pretreatments

12

*Thiago Felix dos Santos* [1], *Caroliny Minely da Silva Santos* [1], *Jose Heriberto Oliveira do Nascimento* [1] *and Roshan Paul* [2]
[1]Micro and Nanotechnologies Innovation Research Group, Federal University of Rio Grande do Norte, Natal, Rio Grande do Norte, Brazil; [2]Institut für Textiltechnik der RWTH Aachen University, Aachen, Germany

## 12.1 Introduction

The textile industry is process intensive and technologically complex. The word textile means to weave, and it encompasses a set of industrial processes that use a variety of natural, regenerated and synthetic fibres such as cotton, wool, silk, jute, viscose, nylon, polyester and acrylic to produce various fabrics. Natural textiles such as cotton, wool and silk undergo different pretreatments to make them ready for further processes such as dyeing and finishing. These pretreatment processes can help eliminate sizing agents, oils, waxes, seed hulls, soils, pectin and other impurities, while improving whiteness and the feel of the fabric.

The textile industry makes a significant contribution to many national economies, encompassing both small- and large-scale operations around the world (Ghaly et al., 2013; Khan and Malik, 2014; Dey and Islam, 2015; Bhatia, 2017; Sarwar and Khan, 2022). Despite the economic importance of the textile industry, the growing demand for products has proportionally increased the generation of wastewater and gaseous emissions. The textile industry is considered as an industry with the highest water consumption and high water pollution in the world. With the exponential increase in the global population and innovations in chemistry during the past century, chemical production has grown enormously (Karthik and Gopalakrishnan, 2014). Textile production, which started with the first industrial revolution in Europe, gradually shifted to developing countries because of the low labour cost and relaxed regulations, and it led to the pollution of natural water bodies in these countries. This has ultimately caused a major pollution problem in the world. It puts both human health and the environment at risk because it has the potential to contaminate soil, air and water (Karthik and Gopalakrishnan, 2014; Khan and Malik, 2014; Roy Choudhury, 2014; Nimkar, 2018; Sivaram et al., 2019; Subramanian et al., 2022).

In addition to the large volume of water required for various unit operations, including pretreatments, there is energy consumption and the use of chemicals in various processes. The textile industry uses more than 8000 chemicals in its supply chain that can contaminate natural water bodies owing to indiscriminate disposal and improper handling. With a population of seven billion people and an average clothing

**Sustainable Innovations in the Textile Industry. https://doi.org/10.1016/B978-0-323-90392-9.00010-0**

consumption of 7–13 kg/person, the use of chemicals is on the order of 5 billion kg. The misuse of water consumption and the lack of treatment of textile effluents make water resources increasingly scarce and increase the rates of water pollution, whereas the global industrial demand for water is increasing in the economic scenario. The contradiction between these trends encourages industrial manufacturers to adopt cleaner production technologies to save water consumption and reduce pollution (Hussain and Wahab, 2018; Samanta et al., 2019; Mamun et al., 2022; Papamichael et al., 2022).

In the dyeing process, washing coloured fabrics can release 10%–50% of dye-related chemicals into the environment. Ineffective dyeing and other finishing processes may be responsible for the discharge of around 200,000 tons of dye into the environment, in addition to microplastics derived from textile fibres present in the effluents. About 40% of dyes used globally contain organically bound chlorine, a known carcinogen. The textile effluent has its effect on the environment, because it drastically reduces the concentration of oxygen owing to the presence of hydrosulphides and blocks the passage of light through the body of water, harming the aquatic ecosystem. Chemicals, on the other hand, can evaporate and be absorbed through breathing, or find an entry point through the skin and cause allergic reactions or harm to children even before birth (Joshi et al., 2004; Pereira and Alves, 2012; Khan and Malik, 2014; Bhatia, 2017; Chen et al., 2017; Nimkar, 2018; Sarwar and Khan, 2022).

They can also cause changes in the physiology and biochemical mechanisms of animals, resulting in the impairment of important functions such as breathing, osmoregulation, reproduction and mortality. Heavy metals present in textile industry effluents are not biodegradable but accumulate in primary organs of the body. Over time, the organs become infected, leading to various symptoms of disease. Thus, untreated or incompletely treated textile effluents can be harmful to aquatic and terrestrial life, adversely affecting the natural ecosystem and causing long-term health effects (Joshi and Santani, 2012; Khan and Malik, 2014; Bhatia, 2017; Chen et al., 2017; Samchetshabam Gita and Choudhury, 2017; Nimkar, 2018; Sarwar and Khan, 2022).

## 12.2 Historical aspects of pretreatments

In textiles, pretreatment are a process used to remove impurities from fibres or fabric. Such processes are particularly relevant for cotton and include desizing, scouring, mercerisation and bleaching, which, apart from removing impurities, make the fabric hydrophilic and thus ready for further wet processing such as dyeing, printing or finishing. Each process requires different types of textile auxiliary chemicals along with water, from which unexhausted chemicals are drained in the effluent stream.

The concept of pretreatment can be traced back to ancient times, when various methods were used to prepare materials for further processing or use. For example, in ancient Egypt, papyrus was soaked in water before it was written on, to improve ink absorption and prevent ink from bleeding. Similarly, in ancient China, silk was boiled in water and treated with natural substances such as alum and vinegar to improve its strength and dye uptake.

In modern times, the use of pretreatment methods has become increasingly important in industrial applications to improve the performance of materials and enhance product quality. In the early 20th century, chemical pretreatments such as acid and chromate coatings were developed to improve the adhesion of paint and other coatings to metals. These methods used toxic chemicals and were associated with environmental and safety concerns. In the 1960s, the concept of surface modification by plasma treatment was introduced, which involved exposing materials to low-pressure plasma to modify their surface properties. This method was highly effective in improving adhesion, wettability and other surface properties of materials. It quickly gained popularity in various industrial applications, including textiles.

In the following decades, other pretreatment methods such as enzymatic treatment, ultraviolet (UV) treatment, sonication and microwave treatment were developed, offering new ways to modify the surface properties of materials. These methods are often more environmentally friendly and safer to use than traditional chemical treatments. Today, pretreatment methods continue to have a critical role in processing natural textiles such as cotton, wool, silk and bast fibres. They are used to improve the performance of textile materials and enhance the functionality of end products. As new textile materials and technologies are developed, the need for effective pretreatment methods will continue to grow.

## 12.3 Textile wet processing and pollution

The textile industry is complex with regard to the varieties of products, processes and raw materials. For fabric to have added value, it is important for it to go through finishing stages. Thus, fabrics are bleached, dyed and printed, making the products suitable for an intended purpose or end use and/or improving the maintainability of the fabric. Wet chemical processes such as desizing, scouring, cleaning, mercerisation, bleaching, dyeing, printing and finishing consume large volumes of fresh water. They also discharge large volumes of effluents that are usually intensely coloured, with a high concentration of organic compounds and large variations in composition (Imran et al., 2015). Within these baths, various chemical products such as dyes and auxiliaries are used to impart the desired quality to the fabrics, but they can cause environmental problems unless they are properly treated before disposal (Yusuf et al., 2017). Industrial wastewater is highly alkaline in nature and contains a high concentration of biological oxygen demand, chemical oxygen demand, total dissolved solids (TDS) and alkalinity (Imran et al., 2015; Bhatia, 2017; Chen et al., 2017; Mondal et al., 2017).

Desizing is the first pretreatment process for natural textiles such as cotton. It removes sizes from fabric and is carried out to facilitate subsequent treatments. Scouring is an important pretreatment process performed on cotton to remove impurities such as natural wax, pectins and nonfibrous matter with a wetting agent and sodium hydroxide. For raw wool, scouring or washing is the first pretreatment process to remove grease, suint, dirt, dust and sand. It is carried out using detergent and sometimes an alkali. Natural silk also undergoes pretreatments before dyeing and printing, because it is necessary to remove sericin partially or completely, as well as natural oils and organic impurities. Degumming removes sericin from fibroin, in which the long protein molecule of sericin is broken down into smaller fractions that are easily dispersed or solubilised in hot water.

Mercerisation is a treatment for cotton fabrics and yarns that gives the fabric or yarn a glossy appearance and resistance and increases dye absorption. During mercerisation, fabric is treated with concentrated sodium hydroxide solution to provide better fibre swelling. This results in increased mechanical resistance and affinity for the dye. Bleaching of raw cotton improves whiteness by removing the natural colour and remaining impurities. The desired degree of whitening is determined by the required whiteness and absorbency. Bleaching of wool is normally carried out with hydrogen peroxide to develop white wool. Another bleaching agent is sodium percarbonate, which can achieve higher whiteness and brilliant colour with minimal fibre property changes (Vujasinović et al., 2023). The silk also contains natural colouring matter, tinted with yellow or brown pigments. These natural colouring matters are bleached or removed by hydrogen peroxide bleaching to produce a pure white material. Table 12.1 compares desizing, scouring, mercerising, cleaning and bleaching processes.

Another important chemical process is dyeing, which colours textile materials such as fibres and threads (Tzanov et al., 2002; Babu et al., 2007; Ul-Haq and Nasir, 2011; Mondal et al., 2017). Three main pollutants found in textile effluents are colour, dissolved solids and toxic metals. The presence of colour in wastewater is a main problem of the textile industry. It is easily visible to the human eye even at low concentrations of dye. Most dyes are not easily degradable. Solid particles can be found with the use of inorganic sodium salts (NaCl and $Na_2SO_4$), directly increasing the level of TDS in the effluent, which forms the largest fraction of total solids and is not removable by conventional treatments.

The presence of toxic metals in textile wastewater is another major problem and can be encountered with the use of chemicals such as sodium hydroxide, sodium carbonate and other salts or in dyes (such as metallised mordant dyes). In addition, owing to the

**Table 12.1** Comparison of desizing, scouring, mercerising, cleaning and bleaching.

| Process | Applications |
|---|---|
| Desizing | Removal of sizing agents from textiles for improved dyeing and finishing; used to produce denim, cotton and other textiles |
| Scouring | Removal of natural wax, grease, pectins and dirt from raw textiles to improve absorbency; used to produce cotton, wool and silk textiles |
| Mercerising | Cotton fabric treatment to improve fabric strength, lustre, and dye uptake; used to produce high-end apparel, home textiles and industrial fabrics |
| Cleaning | Removal of dirt, stains and contaminants from textiles materials; used to produce apparel and technical textiles |
| Bleaching | Whitening of textiles materials; used to produce apparel, home textiles and industrial fabrics |

competitiveness of textile industry production, the increased use of synthetic dye combinations has contributed to dyeing effluents, generating an even greater volume of effluents. The dye can remain in the environment for a long time because it has high thermal photostability and resists biodegradation (Mondal et al., 2017; Periyasamy and Militky, 2020). Fig. 12.1 shows the environmental impact created by the textile industry.

Textile chemical processes are indispensable for obtaining value-added textile products, but it is necessary to use numerous chemical products in both pretreatment and the dyeing, printing and finishing of these materials. In addition to using huge volumes of water, all of these processes are major polluters, generating a high volume of effluent. With the increase in industrialisation and urbanisation, this problem tends to aggravate the demand for water. A forecast carried out by the Federation of Indian Chambers of Commerce and Industry Water Mission demonstrated that the demand for water for the industrial sector will probably increase by 8.5%, and 10.1% will be taken from fresh water in 2025 to 2050. Simple and sustainable reuse and recycling strategies are being researched to evaluate ways to reduce water consumption in wet textile processing in general and pretreatment in particular, which involves desizing, scouring, mercerising, cleaning and bleaching processes.

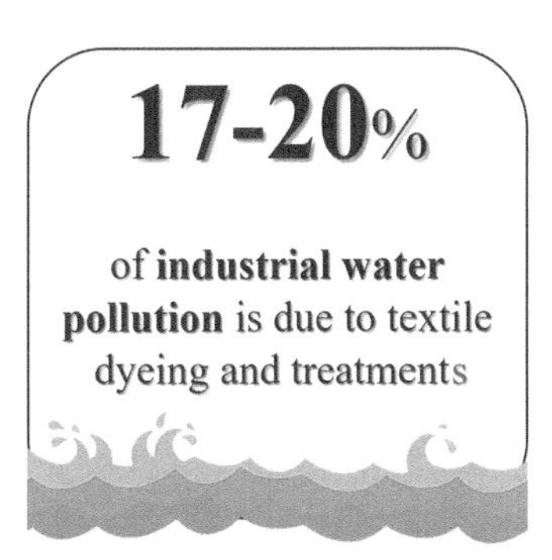


**Figure 12.1** Impact of textile industry on water and environment.

There are also studies involving the treatment and reuse of water from these processes (Wang et al., 2011; Bhatia, 2017; Harane and Adivarekar, 2017; Periyasamy and Militky, 2020; Wang et al., 2022). Thus, it is clear that the textile industry is one of the biggest contributors to environmental threats worldwide, producing 60 billion kg of fabric annually and using up to 34 trillion litres of water (Periyasamy and Militky, 2020). The release of textile pretreatment and dye effluents into sea and river water is destructive to aquatic life and other organisms. Therefore, it is important to study and raise awareness about alternative processes that reduce polluting loads (Periyasamy and Militky, 2020; Stone et al., 2020).

## 12.4 Impacts on health and environment

The textile industry is an important industry that generates a large amount of industrial effluents every year. Thus, it is well-known worldwide as the main source of water pollution that is harmful to sea life and the environment (Boström and Micheletti, 2016; Manzoor and Sharma, 2020; Islam et al., 2022). The fundamental strength of the textile industry flows from its strong production base of a wide range of fibres and yarns, from natural resources such as cotton, wool, silk, jute and flax to synthetic and man-made fibres such as polyester, viscose, nylon and acrylic (Hassaan and Nemr, 2017). With the growing demand for textile products, textile factories and their effluents have been increasing proportionally, causing a major pollution problem for future generations and in the postindustrialisation world. Among the many chemicals present in textile wastewater, dyes are considered important pollutants because when they are used in the textile industry, they cause environmental and health problems (Djehaf et al., 2017; Lellis et al., 2019; Markandeya et al., 2022). The environmental costs of developing textile materials are shown Fig. 12.2. The environmental impacts of pretreatment processes such as desizing, scouring, mercerising, cleaning and bleaching are compared in Table 12.2.

There is a high global consumption of chemical products to supply the demand of a world population of almost eight billion inhabitants. Many of the world's

**Figure 12.2** Environmental costs of textile materials.

**Table 12.2** Environmental impacts of desizing, scouring, mercerising, cleaning and bleaching processes.

| Process | Environmental impact | Potential solutions |
|---|---|---|
| Desizing | High water use and waste | Use of biotechnology (enzymes) or nanotechnology for reduced water use and waste reduction; recycling and reuse of water |
| Scouring | Use of harsh chemicals, high water use | Use of enzymes in bioscouring for reduced water use, biodegradable effluents and waste reduction |
| Mercerising | High energy use, chemicals and waste | Use of continuous mercerisation or ultrasonic-assisted mercerisation for reduced energy use and waste reduction; use of new mercerising agents for reduced chemical use and waste reduction |
| Cleaning | High water and energy use, use of harsh chemicals | Use of plasma technology, ultrasound or ozone treatment for reduced water and energy use; use of environmentally friendly cleaning agents |
| Bleaching | Use of harsh chemicals, high water use and waste | Use of hydrogen peroxide or ozone for reduced chemical use and waste; use of electrochemical or photocatalytic bleaching for reduced water use and waste reduction |

environmental problems associated with the textile industry are associated with water pollution caused by the release of untreated effluents and those owing to the use of toxic chemicals, especially during textile pretreatment (Azanaw et al., 2022). Textile effluent is of great environmental concern because it decreases photosynthetic activity, which is harmful to the aquatic ecosystem and drastically decreases the oxygen concentration, damaging flora and fauna (Islam, 2020; Meghwal et al., 2020; Serrano-Martínez et al., 2020; Dragun et al., 2022; Kumar, 2022). Because of this heavy pollution, the normal functioning of cells is disturbed, which can cause physiologic alterations and changes in the biochemical mechanisms of the animals, resulting in the regeneration of important functions such as respiration, reproduction and even mortality (Khan and Malik, 2014; Naidu et al., 2021).

Thus, it is important to reduce the quantity, complexity and colour of textile effluents. Untreated textile effluents carry major environmental hazards. Several health problems are associated with chemicals used in the textile industry (Khan and Malik, 2014; Mondal et al., 2017; Mani and Bharagava, 2018). Heavy metals (Cr, As, Cu and Zn) present in textile effluents are not biodegradable. Therefore, they accumulate in the body's primary organs, and over time the organs become infected, leading to various disease symptoms (Khan and Malik, 2014; Briffa et al., 2020; Swarnkumar Reddy and Osborne, 2020; Afzaal et al., 2022; Mitra et al., 2022; Singh et al., 2022).

## 12.5 Sustainable innovations in textile pretreatments

The wet pretreatment of textile materials consumes large amounts of electricity, fuel and water (Prabaharan and Rao, 2001; Eren and Ozturk, 2011). Therefore, greenhouse gas emissions and contaminated effluents are an environmental problem now and for future generations (Sevimli and Sarikaya, 2002; Hasanbeigi and Price, 2015). Commercial demand for textile products, particularly natural ones such as cotton, has increased dramatically. A few decades ago, these textile products benefitted by wet pretreatment processes that had a profound impact on the modern world and human lifestyle (Karthik and Gopalakrishnan, 2014). Since ancient times, the environmental impact and long-term depletion of natural resources during the use of textile products has been a significant challenge for humans (Verma et al., 2021). There is a great need to innovate sustainably in pretreatment processes. Fig. 12.3 depicts expectations from the innovations in textile pretreatment processes.

Over time, the textile industry has evolved from conventional wet pretreatment processes to a generation of technologies that provide innovative pretreatment methods as well as textile materials with high sustainability and a low impact on natural resources. Sustainable innovations in textile pretreatments are leading to eco-friendliness, and energy and resource efficiency. The old and conventional processes of textile pretreatment assumed a bold system with less or no water in the process, resulting in improved production with a minimal chance of damage to the environment, natural resources and humanity (Kabir et al., 2019). In the current scenario, promoting these pretreatment techniques in the textile industry is significant challenging when it comes to concern for the environment. Table 12.3 shows the status of applications of different sustainable technologies on an industrial scale.

The growing consumption of natural textile materials leads to a failure to choose sustainable textile pretreatment processes. Together with the cost of these technologies, different sustainable options need to be found (Juanga-Labayen et al., 2022; Lara et al., 2022). Using sustainable or green textile pretreatment processes reduces

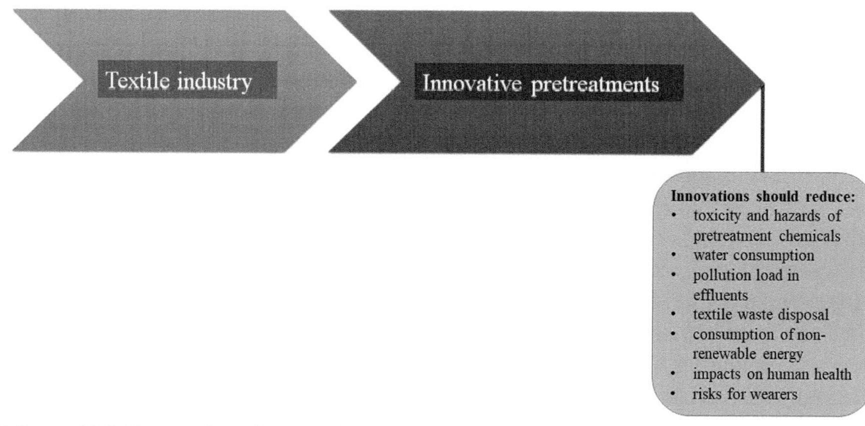


**Figure 12.3** Expectations from textile pretreatment innovations.

**Table 12.3** Status of applications of sustainable technologies on an industrial scale.

| No. | Type of technology | Application status on an industrial scale |
|---|---|---|
| 1 | Enzymatic | In advanced stage, but depends on application field |
| 2 | Ultrasonic | Pilot plant/industrial scale |
| 3 | Ozonisation | Under development |
| 4 | Plasma | Pilot plant/industrial scale |
| 5 | Microwave energy | Under development |
| 6 | Nanotechnology | Pilot plant/industrial scale |

the high levels of pollution and contamination in the textile industry and guarantees a future for the next generations and the conservation of natural resources for the future. Thus, there is a focus on developing eco-friendly textile pretreatment techniques (Biji et al., 2015; Verma et al., 2021). Several sustainable textile pretreatment techniques are depicted in Fig. 12.4. By applying eco-friendly concepts, textile pretreatment can be improved in several ways, such as:

- Substitution of unsustainable textile materials and chemicals by greener ones
- Elimination or reduction of the use of toxic chemicals in textile pretreatments
- Reduction in the use of water and chemicals, or the recycling of them during textile pretreatment
- Reduction of consumption of energy and fuel in pretreatments

Generally, the wet processes are composed chemical, fixation, washing and drying stages. The aim of pretreatment is to improve the appearance, texture or performance of natural textile materials such as cotton. Newer sustainable pretreatment methods

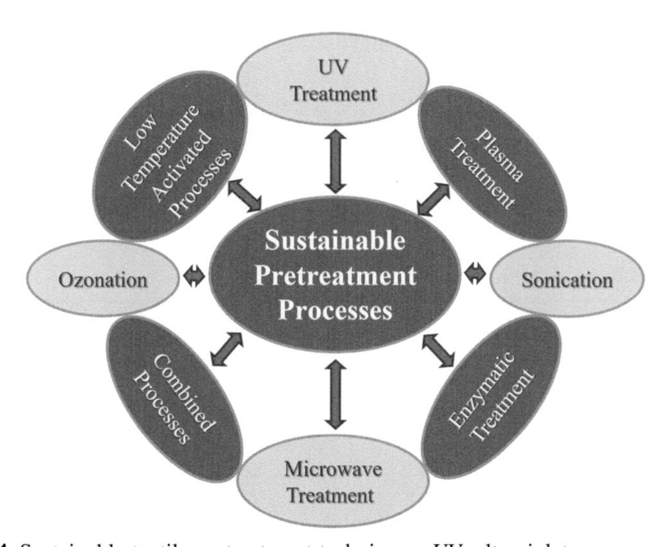


**Figure 12.4** Sustainable textile pretreatment techniques. *UV*, ultraviolet.

consider all possibilities that will lessen the effects of pollution and save water and/or energy. These sustainable pretreatment methods aim to:

- Use environmentally friendly chemicals
- Optimise processes (water and time savings in all possible areas, using less chemicals and working at lower temperatures)
- Reuse water from pretreatment or other stages of the textile chain

Several novel sustainable technologies can lead to ecologically correct production processes in pretreating natural textile materials.

### 12.5.1 Enzymatic technology

The most important green technology in textile pretreatment processes is the use of enzymes, because they are biological catalysts. Enzymes were discovered in the second half of the 19th century. Since then, they have been widely used in various textile industrial processes, mainly to pretreat cotton fabrics (Kumar et al., 2021). Energy savings and water conservation are some promising advantages of enzymatic pretreatments. They also omit the use of substances that are dangerous and harmful to humans and natural resources (Hoque et al., 2021). Enzymes are usually globular proteins. Like other proteins, they consist of long linear chains of amino acids that fold together to produce a complex three-dimensional protein composed of polypeptide chains. However, they are biocatalysts composed of metabolic products of living organisms acquired from bacterial derivatives (Subash and Perumalsamy, 2022). These are extremely efficient and highly specific biocatalysts. In some commercial sources, enzymes are obtained from three primary resources: animal tissue, plants and microbes.

Enzymes generally used in the textile chain are amylase, pectinase, cellulase, laccase and catalase, which are employed to remove starch, degrade lignin, remove fabric fuzz and degrade excess fabric bleach and hydrogen peroxide. Table 12.4 lists different enzymes used in cotton pretreatment. Natural enzymes are not readily available in sufficient quantities for use in the textile industry. Enzymes are relatively fragile substances and are susceptible to denaturation (i.e., degradation owing to temperature, ionising radiation, light, acids, alkalis and biological factors), and thus become inactive. They are manufactured largely by fermentation, a technique that has been well-

**Table 12.4** Different enzymes used in textile processing.

| Enzyme type | Substrate | Use at an industrial scale |
|---|---|---|
| Amylase | Starch | Desizing |
| Pectinase | Pectin | Bioscouring replacing sodium hydroxide |
| Cellulase | Cellulose | Stone wash – biopolishing – biofinishing |
| Laccase | Indigo dye | Improved appearance of denim clothes |
| Catalase | Peroxides | Peroxide removal in dyeing processes |

known for over 3 millennia (Choudhury, 2014). Microorganisms are important producers of enzymes for the textile industry that are used in sustainable textile pretreatment (Kabir and Koh, 2021).

Enzymes are increasingly being used in pretreatment to improve the efficiency and sustainability of various industrial processes, particularly in the field of natural textiles. In a study, the effect of four different enzymatic bioscouring systems was compared against the accessibility of dyes into cotton. The dyeability of reactive, cationic and acid dyes was studied, and each enzymatic system provided its own end-groups, resulting in differences in the response of the fibre to dyeing. Bioscouring reduced the pollution level of the effluent (Canal et al., 2004). Another study on coarse wool fibres investigated changes in fibre properties during bioscouring with an enzyme complex. This bio-innovative scouring resulted in excellent fibre properties (Vujasinović et al., 2023). There are several advantages and some disadvantages to using enzymes in pretreatments.

#### 12.5.1.1 Advantages

- Selectivity: Enzymes are highly selective in nature, so they can target specific components in the biomass matrix without affecting other components. For example, specific enzymes can selectively break down lignin, hemicellulose or cellulose, which are the main constituents of biomass.
- Mild conditions: Enzymes work under mild reaction conditions, such as lower temperatures and pressures, and at neutral pH. This reduces the energy consumption of the process and minimises the formation of unwanted by-products.
- Eco-friendly: Enzymes are derived from natural sources and are biodegradable, which makes them environmentally friendly. Moreover, they can be recycled and reused in multiple cycles, further reducing the environmental impact.
- Improved efficiency: They can improve the efficiency of various industrial processes by increasing the yield of the desired products, reducing the reaction time and improving the overall process economics.

#### 12.5.1.2 Disadvantages

- Cost: Enzymes can be expensive to produce, which can increase the overall cost of pretreatment. However, advances in enzyme production technology and the use of recombinant DNA technology are reducing the cost of enzyme production.
- Limited stability: They can be sensitive to environmental factors such as temperature, pH and inhibitors, which can reduce their stability and activity over time. This limits their application in certain industrial processes that require long-term stability.
- Substrate specificity: Enzymes are highly substrate-specific, so they work on only certain types of biomass or specific chemical structures. This limits their application in certain industrial processes that require a broad range of substrate compatibility.
- Regulatory requirements: The use of enzymes in industrial processes may be subject to regulatory requirements and safety assessments, which can increase the complexity and cost of the process.

### 12.5.2 Plasma technology

Plasma technology is a relatively new approach in textile pretreatment that involves using a high-energy gas plasma to modify the surface properties of materials. Plasma discharges are formed when a substance in the gaseous state held at high energy results in the release of the outer electron of an atom, and that released electron becomes a free negative charge and the atom becomes a free positive charge. Plasma consists of an electronic soup defined as a partially ionised gas with an equal proportion of positive and negative charges under extreme conditions of pressure and temperature (dos Santos et al., 2014). Plasma gas ionisation can be induced as an ecological pretreatment for desizing, bleaching and so forth. Ionisation can be carried out using various methods such as dielectric barrier discharge, atmospheric pressure plasma technique, corona and glow discharge.

Plasma has in its composition species of radicals, ions, electrons, UV radiation and other molecules that react and superficially modify the properties of the textile material. This sustainable textile pretreatment technology is mainly used to induce surface modifications at nanometric level and improve the properties of both natural and synthetic textile materials, increase dyeing rates, and improve the colour, diffusion and adhesion of dyes, coatings and so on. The operating principle of this technology is interesting. The filamentary or diffuse discharges act on the textile substrate without altering its bulk properties, because modifications and treatments occur on a micrometric and nanometric scale (Okuno et al., 1992). The textile material is placed between parallel plates and the plasma is excited. Particles are generated and then interact with the surface of the textile material. Then, a thin film of nanometre size is formed on the surface of the material and the surface is structured with new functional groups (Cai et al., 2003; Gorjanc et al., 2013). Plasma is a dry technique that can be used for the pretreatment and surface modification of different textiles without generating effluents (Paul et al., 2001). In a study, both low-pressure and atmospheric plasma treatments were used to improve the wettability, mechanical and durability properties of polyester fabrics (Bautista et al., 2007, 2010). The main uses of plasma pretreatment are:

- Deposit and coating: to add materials as thin layers on the surface of textiles
- Add effects: the activation of textile surface can be a temporary effect to modify surface energy
- Remove effects: cleaning textile, desizing, etching to modify topography of the surface and sterilisation from contamination
- Textile surface to functionalise: a permanent effect that includes introducing new chemical groups to the surface

There are advantages and disadvantages of using plasma technology in textile pretreatments.

#### 12.5.2.1 Advantages

- High efficiency: Plasma technology is a dry process and can modify the surface properties of materials quickly and efficiently without using chemicals, solvents or solvents. This makes it a highly efficient pretreatment method that can improve the adhesion, wettability and biocompatibility of materials.

- Versatility: This technology can be applied to a wide range of materials, including textiles, metals, ceramics and composites. This makes it a versatile pretreatment method that can be used in various industrial applications.
- Selectivity: It can be tuned to modify specific surface properties of materials, such as surface roughness, surface energy and chemical composition. This makes it a selective pretreatment method that can be tailored to meet specific application requirements.
- Environmental sustainability: It is an environmentally friendly pretreatment method that does not use chemicals or solvents. Moreover, it can be operated at low temperatures and pressures, which reduces the energy consumption of the process.

#### 12.5.2.2 Disadvantages

- Equipment cost: Plasma technology requires specialised equipment that can be expensive to purchase and maintain. Moreover, the complexity of the equipment may require specialised training for operators.
- Safety concerns: It uses high-energy gas plasmas that can pose safety risks, such as electrical shock, gas leaks and ozone generation. This requires strict safety protocols to be put in place to ensure operator safety.
- Surface degradation: Plasma can cause the surface degradation of materials if not properly controlled. This can result in reduced mechanical strength, increased surface roughness and reduced durability.
- Limited scalability: This technology may not be scalable to larger volumes or higher production rates. This can limit its application in certain industrial processes that require a high throughput.

### 12.5.3 Ultraviolet treatments

UV treatment is commonly combined with the use of plasma. UV can be employed for surface engineering on various textile substrates, avoiding the use of water as a processing medium. Thus, it can replace conventional pretreatments such as scouring and improve absorption, hydrophilicity, oleophilicity, dyeing, printing, antistatic and antifelting properties. UV treatment can be applied to only one side of the sample surface and can use less processing time owing to the exclusion of multiple-step operations and the partial reduction in effluent treatment. Therefore, it requires a minimum amount of chemicals and energy and lowers the cost of the final product. UV treatment also contributes to reducing the dyeing time and temperature, aids in the dye bath without compromising fastness properties and contributes to similar or better depths of shade compared with untreated samples.

Durable flame retardant surface treatments using a patented UV and atmospheric plasma facility were reported. They offer a way to bond flame retardant precursor species directly to fibres, which can be used before plasma/UV exposure or in the plasma/UV reaction zone itself. Thus, a series of wet processing cycles are eliminated from the process (Horrocks et al., 2018).

Oxidation processes using UV are also useful for treating fully or partially degrading recalcitrant and/or toxic industrial pollutants. Heterogeneous UV-$TiO_2$ photocatalysis can be used to remove organic pollutants from textile effluents, and is being widely studied and commercialised in many developing countries (Olmez-Hanci et al., 2011; Samanta et al., 2014; Al-Mamun et al., 2019; Horrocks et al., 2018). UV treatment can be used as a sustainable pretreatment method to expose textile materials to radiation to modify their surface properties. There are several advantages and some disadvantages of using UV treatments in pretreatments.

#### 12.5.3.1 Advantages

- High efficiency: UV treatments can modify the surface properties of materials quickly and efficiently without using chemicals or solvents. This makes it a highly efficient pretreatment method to improve the adhesion, wettability and biocompatibility of materials.
- Versatility: It can be applied to a wide range of materials, including polymers, metals, ceramics and composites. This makes it a versatile pretreatment method that can be used in various industrial applications.
- Selectivity: It can be tuned to modify specific surface properties of materials, such as the surface roughness, surface energy and chemical composition. This makes it a selective pretreatment method that can be tailored to meet specific application requirements.
- Environmental sustainability: It is an environmentally friendly pretreatment method that does not use chemicals or solvents. Moreover, it can be operated at room temperature and ambient pressure, which reduces the energy consumption of the process.

#### 12.5.3.2 Disadvantages

- Limited penetration depth: UV treatments can modify the surface properties of materials only to a certain depth, typically up to a few microns. This limits their application in certain industrial processes that require deeper modifications.
- Surface degradation: It can cause the surface degradation of materials if not properly controlled. This can result in reduced mechanical strength, increased surface roughness and reduced durability.
- Substrate specificity: It may be effective on only certain types of materials or specific chemical structures. This limits its application in industrial processes that require a broad range of substrate compatibility.
- Safety concerns: It involves exposure to radiation, which can pose safety risks such as to the skin and eyes. This requires strict safety protocols to be put in place to ensure operator safety.

### 12.5.4 Microwave technology

Microwave radiation can replace conventional heating methods. It provides fast and effective uniform volume heating and heats all particles simultaneously. It also helps in various chemical reactions such as the synchronisation and synthesis of organic compounds and polymers. In addition, microwave radiation increases the dispersion of organic molecules in polymers, which form additional pores and increase the rate of dye fixation in the textile material. Several researchers have used microwave radiation to dye various fibres, achieving a high exhaustion rate, increased dyeing rate, high dye diffusion and excellent colour fastness properties. The influence of exhaust dyeing of knitted poly(butylene terephthalate) fabrics with disperse dyes by microwave dielectric heating was investigated. Researchers studied whether this technique could increase dyeability with faster processing and whether dyeing was obtained with sufficient wet fastness. It was found that microwave thermal dyeing increased dye exhaustion and fixation, with good colour fastness and dyeing repeatability achieved with short dye bath heating times (Öner et al., 2013; Hashem et al., 2014; Amesimeku et al., 2021).

Microwave technology is a relatively new approach to pretreatment that involves microwave radiation to modify the surface properties of materials. In a study, microwaves were used to pretreat cotton fabrics to reduce the pretreatment time and use chemicals and water (Hashem et al., 2014). There are some advantages and disadvantages of using microwave technology in pretreatments.

#### 12.5.4.1 Advantages

- High efficiency: Microwave technology can modify the surface properties of materials quickly and efficiently without using chemicals or solvents. This makes it a highly efficient pretreatment method that can improve the adhesion, wettability and biocompatibility of materials.
- Versatility: It can be applied to a wide range of materials, including textiles, polymers, metals, ceramics and composites. This makes it a versatile pretreatment method that can be used in various industrial applications.
- Selectivity: It can be tuned to modify specific surface properties of materials, such as surface roughness, surface energy and chemical composition. This makes it a selective pretreatment method that can be tailored to meet specific application requirements.
- Scalability: This technology can be easily scaled up to larger volumes or higher production rates. This makes it suitable for industrial processes that require a high throughput.

#### 12.5.4.2 Disadvantages

- Equipment cost: Microwave technology requires specialised equipment that can be expensive to purchase and maintain. Moreover, the complexity of the equipment may require specialised training for operators.
- Limited penetration depth: It can modify the surface properties of materials only to a certain depth, typically up to a few millimetres. This limits its application in certain industrial processes that require deeper modifications.
- Substrate specificity: This technology may be effective on only certain types of materials or specific chemical structures. This limits its application in certain industrial processes that require a broad range of substrate compatibility.
- Safety concerns: This involves exposure to microwave radiation, which can pose safety risks, such as thermal burns and electric shock. This requires strict safety protocols to be put in place to ensure operator safety.

### 12.5.5 Sonication

Ultrasonic waves (USW) are sound vibrations with a frequency greater than 17 kHz, which is outside the audible range for humans. When USW propagate, particles in the electrical medium vibrate, transferring energy through the propagation medium. The perceptible effects of ultrasound result from the way sound propagates, which can occur in two ways: by longitudinal waves (particles vibrate parallel to the direction of propagation of the wave) and by transverse waves (particles vibrate perpendicular to the direction of propagation of the wave). Low pressure in the dilution region can lead to the formation of cavities or bubbles, which are called cavitation and are responsible for most of the physical and chemical effects observed in solid−liquid or liquid−liquid systems. The sonication technique can be used both to synthesise nanomaterials and for applications in polymeric composites, functional textiles, coating nanoparticles on fabric surfaces and cleaning textile materials.

The use of ultrasonic energy in the pectinase enzymatic scouring of cotton fabrics reduces processing time, making it more competitive than conventional alkaline scouring, and significantly improving pectinase efficiency, but without affecting the fabric tensile strength. Ultrasonic treatment can also improve the textile dyeing of the natural dyes *Acacia catechu* and *Tectona grandis*, because it promotes greater and faster absorption of the dye in cotton fabric pr-treated with enzymes. Associated with the use of mordants, there is a significant improvement in dyeing results, such as resistance to washing and light. In research, sonication contributed to the formation of a thick and strongly aggregated $TiO_2$ nanocoating, also obtaining the effective result of washing fastness and promoting better permeability relative to water vapour owing to the acceleration of fluid flow within the internal structure of a fibre and the swelling of the textile substrate owing to acoustic cavitation (Yachmenev et al., 2001; Vankar and Shanker, 2008; Elshemy et al., 2019; Noman and Petru, 2020). Sonication is a pretreatment method that exposes materials to high-frequency sound waves to modify their surface properties. There are some advantages and disadvantages of using sonication in pretreatments.

#### 12.5.5.1 Advantages

- High efficiency: Sonication can modify the surface properties of materials quickly and efficiently without using chemicals or solvents. This makes it a highly efficient pretreatment method that can improve the adhesion, wettability and biocompatibility of materials.
- Versatility: It can be applied to a wide range of materials, including polymers, metals, ceramics and composites. This makes it a versatile pretreatment method that can be used in various industrial applications.
- Selectivity: It can be tuned to modify specific surface properties of materials, such as surface roughness, surface energy and chemical composition. This makes it a selective pretreatment method that can be tailored to meet specific application requirements.
- Scalability: It can be easily scaled up to larger volumes or higher production rates. This makes it suitable for industrial processes that require a high throughput.

#### 12.5.5.2 Disadvantages

- Limited penetration depth: Sonication can modify the surface properties of materials only to a certain depth, typically up to a few microns. This limits its application in certain industrial processes that require deeper modifications.
- Surface degradation: It can cause the surface degradation of materials if not properly controlled. This can result in reduced mechanical strength, increased surface roughness and reduced durability.
- Substrate specificity: It may be effective on only certain types of materials or specific chemical structures. This limits its application in certain industrial processes that require a broad range of substrate compatibility.
- Equipment cost: Sonication requires specialised equipment that can be expensive to purchase and maintain. Moreover, the complexity of the equipment may require specialised training for operators.
- Safety concerns: It involves exposure to high-frequency sound waves, which can pose safety risks, such as hearing loss and tissue damage. This requires strict safety protocols to be put in place to ensure operator safety.

### 12.5.6 Ozonisation technology

Textile pretreatments with ozone is a clean technology that proposes a solution to environmental pollution during conventional pretreatment processes (Prabaharan et al., 2000; Prabaharan and Rao, 2003). Ozone is a strong oxidising agent and can be produced synthetically, just as it is naturally available in the atmosphere. Ozonation technology is a sustainable process that greatly reduces process time, water, chemical and energy consumption and the amount of industrial effluents. Therefore, it can be considered an innovative, simple, clean and environmentally friendly pretreatment process. Ozone gas is spread and applied over the textile materials at a controlled speed (Ab Rasid et al., 2021).

Ozonation can be used as an eco-friendly and alternative pretreatment method of bleaching compared with conventional methods (Pai et al., 2020; Gottschalk et al., 2009). Ozone generated in the equipment can provide a whitening effect, commercially fading the colour. That is, it is like a washing machine without water (Tarhan and Sarıışık, 2009; Kamppuri and Mahmood, 2019). There are several advantages and disadvantages of ozonisation technology.

#### 12.5.6.1 Advantages

- Its combination with clean technologies such as UV, plasma and ultrasound promotes even higher performance
- Improve the dyeability of fibres, particularly fibres of natural origin
- No need to store chemicals compared with other conventional pretreatment methods
- It is a disinfectant treatment of various textile materials (sheets, bandages, hydrophilic cotton, gauze, handkerchiefs, etc) owing to the intrinsic properties of ozone

- Lower consumption of water and chemicals and loss of time in the ozonation process compared with conventional wet pretreatment processes
- No hazardous waste owing to decomposition of ozone into oxygen
- Nonhalogenated organic compounds in wastewater
- Higher degree of whiteness than conventional bleaching processes (Eren and Eren, 2017; Eren and Yetişir, 2018).

#### 12.6.2.2 Disadvantages

- High capital investment for new configurations of industrial clean textile pretreatment machines
- Prevention of problems with yellowing requiring posttreatment (catalase, reducing washing, etc)
- Difficulty of implementing ozone gas in industrial textile machines because of structural adequacy of textile processing machine for wet pretreatment processes
- High level of corrosion in metallic parts of industrial processing machine because of high potential of ozone oxidation, except for stainless steel
- High loss of strength in textile materials owing to use without a rich understanding of effects and dosages of ozone
- Unsuitable for storage owing to quick decomposition of ozone
- Irreparable damage to plastic parts of industrial textile pretreatment machines because of high oxidation potential of ozone
- Flammability and explosivity of ozone (Eren and Eren, 2017; Muthu, 2018).

### 12.5.7 Low-temperature processes

Low-temperature processes in textile production refer to processes that operate at lower temperatures than conventional ones. These processes offer several advantages in terms of energy savings, reduced fabric damage, sustainability and product quality. Cotton fabric is traditionally scoured with alkali and bleached with alkaline hydrogen peroxide ($H_2O_2$) at high temperatures. It can provide cotton fabric good absorbency and whiteness, but it has negative aspects such as high energy and water consumption and damage to the fibres. However, well-established possibilities exist to achieve the low-temperature scouring of cotton with enzymes.

To bleach cotton, it can employ a low-temperature process using catalysed peroxide systems formed by a transition metal complex in the hydrogen peroxide solution. The peroxide systems are activated by a bleach activator in the hydrogen peroxide solution, in which the bleach activator is converted to a peracid by reaction with hydrogen peroxide, which enables bleaching at low temperatures. Potassium persulphate is used as a catalyst for the decomposition of hydrogen peroxide. A Fourier transform infrared analysis indicated that the potassium persulphate/hydrogen peroxide bleaching system decreased carbonyl groups, generating bleaching and brightness in the cotton samples, and requiring 1.96 times less energy than the conventional process.

A substance similar to nonanoyloxybenzene sulphonate was also studied as a low-temperature bleach, the sodium activator 4-(2-decanoyloxyethoxycarbonyloxy) benzene sulphonate. It is effective for the peroxide bleaching of cotton fabrics at low temperatures. Compared with the performance of this bleach against conventional peroxide bleaching, activated bleaching provided cotton fabrics with competitive whiteness, wettability and dyeing (Zeng and Tang, 2015; Inamdar et al., 2017; Yu et al., 2017). There are several advantages and disadvantages of low-temperature processes.

#### *12.5.7.1 Advantages*

- Energy savings: Low-temperature processes require less energy to operate, which can lead to significant cost savings and a reduced environmental impact.
- Reduced fabric damage: These processes can reduce the risk of fabric damage, shrinkage and colour fading compared with high-temperature processes.
- Improved sustainability: They can reduce water use and the release of harmful chemicals, leading to a more sustainable textile production process.
- Enhanced product quality: They can improve the quality of textile products by reducing defects, improving colour consistency and enhancing fabric properties such as softness and hand feel.

#### *12.5.7.2 Disadvantages*

- Longer processing times: Low-temperature processes can take longer to complete than high-temperature processes, which can increase production time and reduce efficiency.
- Limited applicability: They may not be suitable for all textile types and end uses, because some fabrics may require higher temperatures to achieve the desired outcomes.
- Equipment limitations: These processes may require specialised equipment or modifications to existing equipment, which can increase the cost of implementation.
- Lack of standardisation: They are not yet standardised, which can make it difficult to compare results among different processes and manufacturers.

### *12.5.8 Combined processes*

Combined processes are based on choosing methods that complement each other and on criteria that depend on the requirement, such as cost and effectiveness. When different processes are combined, there could also be a synergistic effect on the overall process efficiency. Combined processes in textile pretreatment use multiple techniques or treatments to achieve a desired outcome in a faster or more efficient way.

A study combined the effects of ultrasound and bioscouring on cotton using alkaline pectinase. Two different sources of ultrasound, an ultrasonic bath and an ultrasonic homogeniser, were applied, along with the enzymatic scouring process. Results indicated a clear increase in the efficiency of the enzymatic scouring process when the ultrasonic homogeniser was used. The wettability time and wicking distance after the ultrasonic homogeniser-assisted bioscouring were much better than for enzymatic scouring without the use of ultrasound energy. Ultrasonic bath-assisted enzymatic bioscouring gave slightly better results than enzymatic scouring without sonication (Aksel Eren and Erismis, 2013). The possibility of applying ultrasound in the enzymatic scouring of wool with proteases was studied. Ultrasound did not impair the specific activity of enzymes, but it led to an increase in their effect on the surface of wool fibres and lowered the quantity of effluents (Rositza et al., 2011).

Nanophotoscouring and nanophotobleaching of raw cellulosic fabric were carried out using nano-$TiO_2$ under two different light exposures, UV rays and daylight. The desized cotton fabrics were treated in an ultrasonic bath containing a colloidal aqueous solution of nano-$TiO_2$, citric acid and sodium hypophosphite. Incorporating citric acid in the treatment bath enhanced the treatment durability against washing and created a durable hydrophilic white cotton fabric even after several successive washings. This combined process can be considered a replacement for conventional scouring and bleaching processes on cotton fabric (Montazer and Morshedi, 2012).

Apart from textile pretreatments, combined processes may be applied in other areas of textiles, such as the effluent treatment of wastewater from pretreatments and other wet processes. A study was carried out on the effects of an economical and efficient treatment of textile effluents for colour removal by the combined process of bentonite adsorption followed by electroflotation. The quality of the treated effluent met the requirements of the water disposal standards (Djehaf et al., 2017).

Another study showed the feasibility of treating real textile effluents by a combined process of electrocoagulation (using aluminium electrodes), chemical coagulation (using polyaluminium chloride as a coagulant) and adsorption process (using pistachio shell ash). The experiments demonstrated that the combined processes were superior to these processes alone for the removal of organic and inorganic compounds from real textile effluents (Bazrafshan et al., 2016). There are some advantages and disadvantages of combined processes in textile pretreatment.

#### 12.5.8.1 Advantages

- Improved efficiency: Combined processes can increase the efficiency of textile pretreatment by reducing the number of steps required, reducing the time and cost involved in the process.
- Reduced environmental impact: These processes can reduce the use of chemicals, water and energy, leading to a lower environmental impact.
- Improved quality: They can lead to improved fabric quality by reducing defects, improving colour consistency and enhancing fabric properties such as strength, dye uptake and shrinkage resistance.
- Flexibility: Combined processes can be tailored to specific fabric types and end uses, enabling greater flexibility in textile production.

#### 12.5.8.2 Disadvantages

- Complex process: Combined processes can be complex and require specialised equipment and expertise, which can increase the cost of implementation and operation.

- Risk of damage: They can increase the risk of fabric damage if not properly optimised or if incompatible treatments are used together.
- Difficult to optimise: Optimising these processes can be challenging because of interactions between different treatments and the need to balance the desired outcomes with process efficiency and cost.
- Lack of standardisation: Combined processes are not yet standardised, which can make it difficult to compare results among different processes and manufacturers.

Combined processes in textile pretreatment offer several advantages but require a careful consideration of specific textile product and process requirements to ensure optimal results. Nanotechnology is also an upcoming field in textile pretreatments, particularly combined with other sustainable processes. Table 12.5 lists applications of nanotechnology in textile pretreatment processes.

**Table 12.5** Use of nanotechnology in textile pretreatment processes.

| Process | Use of nanotechnology |
| --- | --- |
| Desizing | Development of nanoscale enzyme systems for improved efficiency and reduced environmental impact in desizing process |
| Scouring | Use of nano-$TiO_2$ for photocatalytic decomposition of natural impurities on cotton |
| Mercerising | Use of nanocellulose for improved fabric properties and reduced environmental impact in mercerisation process |
| Cleaning | Use of nanoscale cleaning agents and coatings for improved cleaning efficiency and reduced environmental impact in cleaning process |
| Bleaching | Use of nanocatalysts for improved efficiency and reduced environmental impact in bleaching process |

Table 12.6 lists advantages and disadvantages of sustainable textile pretreatment technologies.

**Table 12.6** Sustainable textile pretreatment technologies.

| Process | Technology | Advantages | Disadvantages |
| --- | --- | --- | --- |
| Desizing | Enzymes | Reduced environmental impact, cost savings | Further research needed for large-scale use |
| | Nanotechnology | Improved efficiency | Further research needed for large-scale use |
| | Microwave technology | Reduced process time, improved efficiency | Further research needed for large-scale use |

*Continued*

**Table 12.6 Continued**

| Process | Technology | Advantages | Disadvantages |
|---|---|---|---|
| Scouring | Enzymes | Reduced environmental impact, cost savings | No disadvantages, already in large-scale use |
| | Nanotechnology | Improved efficiency | Further research needed for large-scale use |
| | Ultrasound-assisted bioscouring | Improved efficiency | Further research needed for large-scale use |
| Mercerising | Continuous mercerisation | Improved efficiency, reduced environmental impact | Equipment limitations, high initial cost |
| | Ultrasonic-assisted mercerisation | Improved efficiency, reduced environmental impact | Equipment limitations, high initial cost |
| | New mercerising agents | Improved fabric properties | Limited applicability, lack of standardisation |
| Cleaning | Plasma technology | Reduced environmental impact, reduced water use | Equipment limitations, high initial cost |
| | Ultrasound | Improved cleaning efficiency | Equipment limitations, high initial cost |
| | Ozone treatment | Reduced environmental impact, reduced water use | Equipment limitations, high initial cost |
| Bleaching | Hydrogen peroxide | Reduced environmental impact, reduced fabric damage | Limited effectiveness on certain fabric types |
| | Ozone | Reduced environmental impact, reduced fabric damage | Equipment limitations, lack of standardisation |
| | Electrochemical bleaching | Reduced water use, reduced environmental impact | Equipment limitations, high initial cost |

## 12.6 Environmental awareness and future perspectives

Conventional technologies for textile pretreatment have prevailed in the world since ancient times and have the same principle in all parts of the textile chain, such as spinning, weaving, knitting, dyeing, finishing and some other techniques that have previously been relevant to the textile industry. Currently, sustainable or green technologies in industrial scale textile pretreatments have gained much more attention owing to their ecologically correct approach, economic viability and the nonbankruptcy of natural resources for the next generations. This is because environmentally friendly technology, or more specifically sustainable technology, within the textile chain has become a promising trend for the future of the textile industry.

In almost all cases of pollution, the fundamental problem of the textile industry is the low efficiency of textile pretreatments and a poor understanding of the life cycle of the auxiliary chemicals involved. Textile wet pretreatment in the textile industry continues to grow rapidly, and it is clear that there is still room for the textile industry in the medium and long term to make its pretreatment processes and all of the chemistry involved green. Therefore, application of the principles of green chemistry and other aspects of sustainable technologies will increasingly lead the textile industry toward production models that are more compatible with the environment, the preservation of natural resources and human health.

## 12.7 Conclusion

An efficient and effective pretreatment is the key to making subsequent processes such as dyeing and finishing defect-free. Adopting sustainable processes and improving conventional textile pretreatment techniques are the future trends of the world textile industry and reduce impacts on the environment. Developments in sustainable or clean technologies in pretreatments can lead to ecologically correct textile production processes. Desizing is a critical pretreatment process. Innovations in this area have led to improved efficiency, reduced environmental impact and cost savings. The use of enzymes, nanotechnology and microwave technology have shown promising results in improving desizing. In the case of scouring, enzymatic treatments are available at an industrial scale, and other techniques such as ultrasound-assisted bioscouring and nanotechnology are making inroads.

Mercerising is an important process that enhances the strength, lustre and dye uptake of cotton fibres. Advances in mercerising include the use of continuous mercerisation, ultrasonic-assisted mercerisation and the use of new mercerising agents. These techniques have shown promising results in improving the efficiency and effectiveness of mercerising and have the potential to revolutionise the industry. Proper cleaning is essential for maintaining the quality and longevity of textile products. Advances in cleaning techniques, such as the use of plasma technology, ultrasound and ozone treatment, have shown promise in improving cleaning efficiency, reducing the environmental impact, and decreasing water use.

Bleaching is an essential process in textile production. Innovations have led to improvements in efficiency, cost-effectiveness and a reduced environmental impact. Novel bleaching agents such as hydrogen peroxide and ozone have shown promise in improving bleaching while reducing the use of harmful chemicals. In addition, methods such as electrochemical bleaching and photocatalytic bleaching have the potential to revolutionise the industry by reducing water use and the environmental impact. There is still scope for further innovations. Thus, further research is needed to optimise all of these techniques and ensure their effectiveness on an industrial scale.

## Acknowledgements

The authors from the Federal University of Rio Grande do Norte acknowledge the Carrefour group, and the Graduate Program in Chemical Engineering at the Federal University of Rio Grande do Norte (PPGEQ/UFRN), for financial and technical support.


## References


Ab Rasid, N.S., et al., 2021. Recent advances in green pre-treatment methods of lignocellulosic biomass for enhanced biofuel production. Journal of Cleaner Production 321, 129038. https://doi.org/10.1016/j.jclepro.2021.129038.

Afzaal, M., et al., 2022. Heavy metals contamination in water, sediments and fish of freshwater ecosystems in Pakistan. Water Practice and Technology 17 (5), 1253–1272. https://doi.org/10.2166/wpt.2022.039.

Aksel Eren, H., Erismis, B., 2013. Ultrasound-assisted bioscouring of cotton. Coloration Technolology 129, 360–366. https://doi.org/10.1111/cote.12035.

Al-Mamun, M.R., et al., 2019. Photocatalytic activity improvement and application of UV-$TiO_2$ photocatalysis in textile wastewater treatment: a review. Journal of Environmental Chemical Engineering 7 (5), 103248. https://doi.org/10.1016/j.jece.2019.103248.

Amesimeku, J., et al., 2021. Dyeing properties of meta-aramid fabric dyed with basic dye using ultrasonic-microwave irradiation. Journal of Cleaner Production 285, 124844. https://doi.org/10.1016/j.jclepro.2020.124844.

Azanaw, A., et al., 2022. Textile effluent treatment methods and eco-friendly resolution of textile wastewater. Case Studies in Chemical and Environmental Engineering 6, 100230. https://doi.org/10.1016/j.cscee.2022.100230.

Babu, B.R., et al., 2007. An overview of wastes produced during cotton textile processing and efluent treatment methods. The Journal of Cotton Science 11, 110–122.

Bautista, L., Briz, A., Aubouy, L., De la Varga, M., Garrido-Franco, M., Paul, R., 2007. Wettability, mechanical and durability properties improvements in textiles using low-pressure and atmospheric plasma treatments. In: Poster presented in International Textile Conference 2007, Aachen, Germany, 29–30 November.

Bautista, L., Paul, R., Mota, J., Crespo, C., De la Varga, M., Aubouy, L., Pavan, C., 2010. Effect of temperature during atmospheric pressure plasma treatments on polyester fabrics. In: Paper presented in AUTEX 2010 World Textile Conference, Vilnius, Lithuania, 21–23 June.

Bazrafshan, E., Alipour, M.R., Mahvi, A.H., 2016. Textile wastewater treatment by application of combined chemical coagulation, electrocoagulation, and adsorption processes. Desalination and Water Treatment 57 (20), 9203–9215. https://doi.org/10.1080/19443994.2015.1027960.

Bhatia, S.C., 2017. In: Devraj, S. (Ed.), Pollution Control in Textile Industry. WPI Publishing. https://doi.org/10.1201/9781315148588.

Biji, K.B., et al., 2015. Smart packaging systems for food applications: a review. Journal of Food Science and Technology 52 (10), 6125–6135. https://doi.org/10.1007/s13197-015-1766-7.

Boström, M., Micheletti, M., 2016. Introducing the sustainability challenge of textiles and clothing. Journal of Consumer Policy 39 (4), 367–375. https://doi.org/10.1007/s10603-016-9336-6.

Briffa, J., Sinagra, E., Blundell, R., 2020. Heavy metal pollution in the environment and their toxicological effects on humans. Heliyon 6 (9), e04691. https://doi.org/10.1016/j.heliyon.2020.e04691.

Cai, Z., et al., 2003. Effect of atmospheric plasma treatment on desizing of PVA on cotton. Textile Research Journal 73 (8), 670–674. https://doi.org/10.1177/004051750307300

Canal, J.M., Navarro, A., Calafell, M., Rodriguez, C., Caballero, G., Vega, B., Canal, C., Paul, R., 2004. Effect of various bio-scouring systems on the accessibility of dyes into cotton. Coloration Technolology 120 (6), 311. https://doi.org/10.1111/j.1478-4408.2004.tb00236.x.

Chen, L., et al., 2017. A process-level water conservation and pollution control performance evaluation tool of cleaner production technology in textile industry. Journal of Cleaner Production 143, 1137–1143. https://doi.org/10.1016/j.jclepro.2016.12.006.

Choudhury, A.K.R., 2014. Sustainable textile wet processing: applications of enzymes. In: Muthu, S.S. (Ed.), Roadmap to Sustainable Textiles and Clothing. Textile Sc. Springer, Singapore, pp. 203–238. https://doi.org/10.1007/978-981-287-065-0_7.

Dey, S., Islam, A., 2015. A review on textile wastewater characterization in Bangladesh. Resources and Environment 5 (1), 15–44. https://doi.org/10.5923/j.re.20150501.03.

Djehaf, K., et al., 2017. Textile wastewater in Tlemcen (Western Algeria): impact, treatment by combined process. Chemistry International 3 (4), 414–419.

Dragun, Z., et al., 2022. Yesterday's contamination–a problem of today? the case study of discontinued historical contamination of the Mrežnica River (Croatia). Science of The Total Environment 848, 157775. https://doi.org/10.1016/j.scitotenv.2022.157775.

Elshemy, N.S., et al., 2019. Optimization and characterization of prepared nano-disperse dyes via a sonication process and their application in textile dyeing and printing. Fibres and Polymers 20 (12), 2540–2549. https://doi.org/10.1007/s12221-019-9135-1.

Eren, H.A., Eren, S., 2017. Ozone bleaching of cellulose. IOP Conference Series: Materials Science and Engineering 254 (8), 082009. https://doi.org/10.1088/1757-899X/254/8/082009.

Eren, H.A., Ozturk, D., 2011. The evaluation of ozonation as an environmentally friendly alternative for cotton preparation. Textile Research Journal 81 (5), 512–519.

Eren, S., Yetişir, İ., 2018. Ozone bleaching of woven cotton fabric. Pamukkale University Journal of Engineering Sciences 24 (7), 1245–1248. https://doi.org/10.5505/pajes.2017.82231.

Ghaly, A.E., Ananthashankar, R., Alhattab, M., Ramakrishnan, V.V., 2013. Production, characterization and treatment of textile effluents: a critical review. Journal of Chemical Engineering and Process Technology 05 (01). https://doi.org/10.4172/2157-7048.1000182.

Gorjanc, M., et al., 2013. Multifunctional textiles – modification by plasma, dyeing and nanoparticles. In: Eco-Friendly Textile Dyeing and Finishing. InTech. https://doi.org/10.5772/53376.

Gottschalk, C., Libra, J.A., Saupe, A. (Eds.), 2009. Ozonation of Water and Waste Water. Wiley. https://doi.org/10.1002/9783527628926.

Harane, R.S., Adivarekar, R.V., 2017. Sustainable processes for pre-treatment of cotton fabric. Textiles and Clothing Sustainability 2 (1), 2. https://doi.org/10.1186/s40689-016-0012-7.

Hasanbeigi, A., Price, L., 2015. A technical review of emerging technologies for energy and water efficiency and pollution reduction in the textile industry. Journal of Cleaner Production 95, 30–44. https://doi.org/10.1016/j.jclepro.2015.02.079.

Hashem, M., et al., 2014. New prospects in pretreatment of cotton fabrics using microwave heating. Carbohydrate Polymers 103, 385–391. https://doi.org/10.1016/j.carbpol.2013.11.064.

Hassaan, M.A., Nemr, A. El, 2017. Health and environmental impacts of dyes: mini review. American Journal of Environmental Science and Engineering 1 (3), 64–67. https://doi.org/10.11648/j.ajese.20170103.11.

Hoque, M.T., Mazumder, N.-U.-S., Islam, M.T., 2021. Enzymatic wet processing. In: Rather, L.J., Shabbir, M., Haji, A. (Eds.), Sustainable Practices in the Textile Industry. Wiley, pp. 87–110.

Horrocks, A., et al., 2018. Environmentally sustainable flame retardant surface treatments for textiles: the potential of a novel atmospheric plasma/UV laser technology. Fibres 6 (2), 31. https://doi.org/10.3390/fib6020031.

Hussain, T., Wahab, A., 2018. A critical review of the current water conservation practices in textile wet processing. Journal of Cleaner Production 198, 806–819. https://doi.org/10.1016/j.jclepro.2018.07.051.

Imran, M.A., et al., 2015. Sustainable and economical one-step desizing, scouring and bleaching method for industrial scale pretreatment of woven fabrics. Journal of Cleaner Production 108, 494–502. https://doi.org/10.1016/j.jclepro.2015.08.073.

Inamdar, U.Y., et al., 2017. Low-temperature bleaching of cotton fabric by activated peroxide system. Emerging Materials Research 6 (2), 387–395. https://doi.org/10.1680/jemmr.16.00148.

Islam, S., 2020. A study on the solutions of environment pollution and worker's health problems caused by textile manufacturing operations. Biomedical Journal of Scientific and Technical Research 28 (4). https://doi.org/10.26717/BJSTR.2020.28.004692.

Islam, T., et al., 2022. Impact of textile dyes on health and ecosystem: a review of structure, causes, and potential solutions. Environmental Science and Pollution Research 30 (4), 9207–9242. https://doi.org/10.1007/s11356-022-24398-3.

Joshi, V.J., Santani, D.D., 2012. Physicochemical characterization and heavy metal concentration in effluent of textile industry. Universal Journal of Environmental Research and Technology 2 (2), 93–96.

Joshi, M., Bansal, R., Purwar, R., 2004. Colour removal from textile effluents. Indian Journal of Fibre and Textile Research 29, 239–259.

Juanga-Labayen, J.P., Labayen, I.V., Yuan, Q., 2022. A review on textile recycling practices and challenges. Textiles 2, 174–188. https://doi.org/10.3390/textiles2010010.

Kabir, S.M.M., Koh, J., 2021. Sustainable textile processing by enzyme application. In: Mendes, K., Sousa, R. de, Mielke, K. (Eds.), Biodegradation Technology of Organic and Inorganic Pollutants. IntechOpen, pp. 1–27. https://doi.org/10.5772/intechopen.97198.

Kabir, S.M.F., et al., 2019. Sustainability assessment of cotton-based textile wet processing. Clean Technology 1, 232–246. https://doi.org/10.3390/cleantechnol1010016.

Kamppuri, T., Mahmood, S., 2019. Finishing of denim fabrics with ozone in water. Journal of Textile Engineering and Fashion Technology 5 (2). https://doi.org/10.15406/jteft.2019.05.00189.

Karthik, T., Gopalakrishnan, D., 2014. Environmental analysis of textile value chain: an overview. In: Roadmap to Sustainable Textiles and Clothing. Textile Sc, pp. 153–188. https://doi.org/10.1007/978-981-287-110-7_6.

Khan, S., Malik, A., 2014. Environmental and health effects of textile industry wastewater. In: Environmental Deterioration and Human Health. Springer Netherlands, Dordrecht, pp. 55–71. https://doi.org/10.1007/978-94-007-7890-0_4.

Kumar, A., 2022. Impact of textile wastewater on water quality. Central Asian Journal of Medical and Natural Sciences 3 (3), 11.

Kumar, D., et al., 2021. Application of enzymes for an eco-friendly approach to textile processing. Environmental Science and Pollution Research International 30 (28), 1–11. https://doi.org/10.1007/s11356-021-16764-4.

Lara, L., Cabral, I., Cunha, J., 2022. Ecological approaches to textile dyeing: a review. Sustainability 14, 1–17.

Lellis, B., et al., 2019. Effects of textile dyes on health and the environment and bioremediation potential of living organisms. Biotechnology Research and Innovation 3 (2), 275–290. https://doi.org/10.1016/j.biori.2019.09.001.

Mamun, A. Al, et al., 2022. An assessment of energy and groundwater consumption of textile dyeing mills in Bangladesh and minimization of environmental impacts via long-term key performance indicators (KPI) baseline. Textiles 2 (4), 511–523. https://doi.org/10.3390/textiles2040029.

Mani, S., Bharagava, R.N., 2018. Textile industry wastewater: textile industry wastewater environmental and health hazards and treatment approaches. In: Recent Advances in Environmental Management, first ed. CRC Press, Boca Raton, p. 546.

Manzoor, J., Sharma, M., 2020. Impact of Textile Dyes on Human Health and Environment, pp. 162–169. https://doi.org/10.4018/978-1-7998-0311-9.ch008.

Markandeya, Mohan, D., Shukla, S.P., 2022. Hazardous consequences of textile mill effluents on soil and their remediation approaches. Cleaner Engineering and Technology 7, 100434. https://doi.org/10.1016/j.clet.2022.100434.

Meghwal, K., et al., 2020. Effect of Dyes on Water Chemistry, Soil Quality, and Biological Properties of Water, pp. 90–114. https://doi.org/10.4018/978-1-7998-0311-9.ch005.

Mitra, S., et al., 2022. Impact of heavy metals on the environment and human health: novel therapeutic insights to counter the toxicity. Journal of King Saud University Science 34 (3), 101865. https://doi.org/10.1016/j.jksus.2022.101865.

Mondal, P., Bose, D., Baksi, S., 2017. Study of environmental issues in textile industries and recent wastewater treatment technology. World Scientific News 61 (2), 98.

Montazer, M., Morshedi, S., 2012. Nano photo scouring and nano photo bleaching of raw cellulosic fabric using nano $TiO_2$. International Journal of Biological Macromolecules 50 (4), 1018–1025. https://doi.org/10.1016/j.ijbiomac.2012.02.018.

Muthu, S.S. (Ed.), 2018. Sustainable Innovations in Textile Chemical Processes. Springer Singapore (Textile Science and Clothing Technology), Singapore. https://doi.org/10.1007/978-981-10-8491-1.

Naidu, R., et al., 2021. Chemical pollution: a growing peril and potential catastrophic risk to humanity. Environment International 156, 106616. https://doi.org/10.1016/j.envint.2021.106616.

Nimkar, U., 2018. Sustainable chemistry: a solution to the textile industry in a developing world. Current Opinion in Green and Sustainable Chemistry 9, 13–17. https://doi.org/10.1016/j.cogsc.2017.11.002.

Noman, M.T., Petru, M., 2020. Effect of sonication and nano $TiO_2$ on thermophysiological comfort properties of woven fabrics. ACS Omega 5 (20), 11481–11490. https://doi.org/10.1021/acsomega.0c00572.

Okuno, T., Yasuda, T., Yasuda, H., 1992. Effect of crystallinity of PET and nylon 66 fibres on plasma etching and dyeability characteristics. Textile Research Journal 62 (8), 474–480. https://doi.org/10.1177/004051759206200807.

Olmez-Hanci, T., Arslan-Alaton, I., Basar, G., 2011. Multivariate analysis of anionic, cationic and nonionic textile surfactant degradation with the $H_2O_2$/UV-C process by using the capabilities of response surface methodology. Journal of Hazardous Materials 185 (1), 193–203. https://doi.org/10.1016/j.jhazmat.2010.09.018.

Öner, E., Büyükakinci, Y., Sökmen, N., 2013. Microwave-assisted dyeing of poly(butylene terephthalate) fabrics with disperse dyes. Coloration Technology 129 (2), 125–130. https://doi.org/10.1111/cote.12014.

Pai, R.S., Doraswami, U.R., Belino, N.J.R., Paul, R., 2020. Ozone in textile bleaching: roadblocks and path forward. AATCC Review 20 (6), 34–36. https://doi.org/10.14504/ar.20.6.3.

Papamichael, I., et al., 2022. Building a new mind set in tomorrow fashion development through circular strategy models in the framework of waste management. Current Opinion in Green and Sustainable Chemistry 36, 100638. https://doi.org/10.1016/j.cogsc.2022.100638.

Paul, R., Pardeshi, P.D., Manjrekar, S.G., 2001. Plasma treatment - an effective tool for polymer surface modification. Synthetic Fibres 30 (4), 5.

Pereira, L., Alves, M., 2012. Dyes—environmental impact and remediation. In: Environmental Protection Strategies for Sustainable Development. Springer Netherlands, Dordrecht, pp. 111−162. https://doi.org/10.1007/978-94-007-1591-2_4.

Periyasamy, A.P., Militky, J., 2020. Sustainability in Textile Dyeing: Recent Developments, pp. 37−79. https://doi.org/10.1007/978-3-030-38545-3_2.

Prabaharan, M., Rao, J.V., 2001. Study on ozone bleaching of cotton fabric − process optimisation, dyeing and finishing properties. Coloration Technology 117 (2), 98−103. https://doi.org/10.1111/j.1478-4408.2001.tb00342.x.

Prabaharan, M., Rao, J.V., 2003. Combined desizing, scouring and bleaching of cotton using ozone. Indian Journal of Fibre and Textile Research 28, 437−443.

Prabaharan, M., et al., 2000. A study on the advanced oxidation of a cotton fabric by ozone. Coloration Technology 116 (3), 83−86. https://doi.org/10.1111/j.1478-4408.2000.tb00024.x.

Reddy, S., Osborne, W.J., 2020. Heavy metal determination and aquatic toxicity evaluation of textile dyes and effluents using Artemia salina. Biocatalysis and Agricultural Biotechnology 25, 101574. https://doi.org/10.1016/j.bcab.2020.101574.

Rositza, B., Dancho, Y., Lubov, Y., 2011. Enzyme assisted ultrasound scouring of raw wool fibres. Journal of Biomaterials and Nanobiotechnology 2 (1), 65−70. https://doi.org/10.4236/jbnb.2011.21009.

Roy Choudhury, A.K., 2014. Environmental impacts of the textile industry and its assessment through life cycle assessment. In: Roadmap to Sustainable Textiles and Clothing. Textile Sc, pp. 1−39. https://doi.org/10.1007/978-981-287-110-7_1.

Samanta, K.K., Basak, S., Chattopadhyay, S.K., 2014. Environment-friendly textile processing using plasma and UV treatment. In: Technologies, and Processing Methods, pp. 161−201. https://doi.org/10.1007/978-981-287-065-0_6.

Samanta, K.K., et al., 2019. Water consumption in textile processing and sustainable approaches for its conservation. In: Water in Textiles and Fashion. Elsevier, pp. 41−59. https://doi.org/10.1016/B978-0-08-102633-5.00003-8.

Samchetshabam Gita, A.H., Choudhury, T.G., 2017. Impact of textile dyes waste on aquatic environments and its treatment. Environment and Ecology 35 (3C), 2349−2353.

dos Santos, T.F., et al., 2014. Functionalization of 100% polyester fabrics through dielectric barrierdischarge. In: ABM Proceedings. Editora Blucher, São Paulo, pp. 6774−6781. https://doi.org/10.5151/1516-392X-24456.

Sarwar, T., Khan, S., 2022. Textile Industry: Pollution Health Risks and Toxicity, pp. 1−28. https://doi.org/10.1007/978-981-19-2832-1_1.

Serrano-Martínez, A., et al., 2020. Degradation and toxicity evaluation of azo dye Direct red 83:1 by an advanced oxidation process driven by pulsed light. Journal of Water Process Engineering 37, 101530. https://doi.org/10.1016/j.jwpe.2020.101530.

Sevimli, M.F., Sarikaya, H.Z., 2002. Ozone treatment of textile effluents and dyes: effect of applied ozone dose, pH and dye concentration. Journal of Chemical Technology & Biotechnology 77 (7), 842−850. https://doi.org/10.1002/jctb.644.

Singh, A., et al., 2022. Heavy metal contamination of water and their toxic effect on living organisms. In: The Toxicity of Environmental Pollutants. IntechOpen. https://doi.org/10.5772/intechopen.105075.

Sivaram, N.M., Gopal, P.M., Barik, D., 2019. Toxic waste from textile industries. In: Energy from Toxic Organic Waste for Heat and Power Generation. Elsevier, pp. 43–54. https://doi.org/10.1016/B978-0-08-102528-4.00004-3.

Stone, C., et al., 2020. Natural or synthetic – how global trends in textile usage threaten freshwater environments. Science of The Total Environment 718, 134689. https://doi.org/10.1016/j.scitotenv.2019.134689.

Subash, M. c, Perumalsamy, M., 2022. Green degumming of banana pseudostem fibres for yarn manufacturing in textile industries. In: Biomass Conversation and Biorefinary, pp. 1–10. https://doi.org/10.1007/s13399-022-02850-1.

Subramanian, K., et al., 2022. An overview of cotton and polyester, and their blended waste textile valorisation to value-added products: a circular economy approach – research trends, opportunities and challenges. Critical Reviews in Environmental Science and Technology 52 (21), 3921–3942. https://doi.org/10.1080/10643389.2021.1966254.

Tarhan, M., Sarıışık, M., 2009. A comparison among performance characteristics of various denim fading processes. Textile Research Journal 79 (4), 301–309. https://doi.org/10.1177/00405175080908.

Tzanov, T., et al., 2002. Hydrogen peroxide generation with immobilized glucose oxidase for textile bleaching. Journal of Biotechnology 93 (1), 87–94. https://doi.org/10.1016/S0168-1656(01)00386-8.

Ul-Haq, N., Nasir, H., 2011. Cleaner production technologies in desizing of cotton fabric. Journal of the Textile Institute 103, 304–310. https://doi.org/10.1080/00405000.2011.570045.

Vankar, P.S., Shanker, R., 2008. Ecofriendly ultrasonic natural dyeing of cotton fabric with enzyme pretreatments. Desalination 230 (1–3), 62–69. https://doi.org/10.1016/j.desal.2007.11.016.

Verma, M.K., et al., 2021. Trends in packaging material for food products: historical background, current scenario, and future prospects. Journal of Food Science and Technology 58 (11), 4069–4082. https://doi.org/10.1007/s13197-021-04964-2.

Vujasinović, E., Tarbuk, A., Pušić, T., Dekanić, T., 2023. Bio-innovative pretreatment of coarse wool fibers. Processes 11, 103. https://doi.org/10.3390/pr11010103.

Wang, Z., et al., 2011. Textile dyeing effluent treatment. In: Advances in the Treatment of Textile Effluents, pp. 91–116.

Wang, X., Jiang, J., Gao, W., 2022. Reviewing textile wastewater produced by industries: characteristics, environmental impacts, and treatment strategies. Water Science and Technology 85 (7), 2076–2096. https://doi.org/10.2166/wst.2022.088.

Yachmenev, V.G., Bertoniere, N.R., Blanchard, E.J., 2001. Effect of sonication on cotton preparation with alkaline pectinase. Textile Research Journal 71 (6), 527–533. https://doi.org/10.1177/004051750107100610.

Yu, J., et al., 2017. Pilot-plant investigation on low-temperature bleaching of cotton fabric with TBCC-activated peroxide system. Cellulose 24 (6), 2647–2655. https://doi.org/10.1007/s10570-017-1276-z.

Yusuf, M., Shabbir, M., Mohammad, F., 2017. Natural colorants: historical, processing and sustainable prospects. Natural Products and Bioprospecting 7 (1), 123–145. https://doi.org/10.1007/s13659-017-0119-9.

Zeng, H., Tang, R.-C., 2015. Application of a novel bleach activator to low temperature bleaching of raw cotton fabrics. The Journal of The Textile Institute 106 (8), 807–813. https://doi.org/10.1080/00405000.2014.945764.

# Dyeing, printing and digital colouration

*Avinash Pradip Manian, Thomas Bechtold and Tung Pham*
Research Institute for Textile Chemistry and Textile Physics, University Innsbruck, Dornbirn, Austria

## 13.1 Introduction

The total production of textiles for clothing and footwear is expected to grow to 102 million tons by 2030. The vast majority of products is disposed by incineration or landfill, which leads to a considerable environmental impact and a substantial contribution to global warming. These facts brought the European Commission to formulate a general strategy to convert the European textile and garment industry into a more sustainable and circular textile economy (European Commission, 2022).

The sustainable textile approach requires a consideration of all stakeholders who are participating along the full life cycle of a textile product. In the 1990s, health concerns about textiles led to the development of the first eco-labels, which collected requirements for textile products to avoid the presence of hazardous substances in textiles. As examples, the presence of maximum allowable concentration of amines, the concentration of formaldehyde and the concentration of heavy metals were restricted. Nevertheless, the presence of residual chemicals still led to allergic reactions owing to individual level sensitivity and a predisposition to sensitisation (Aerts et al., 2014). A considerable number of labels have been presented to producers and customers since then. Moreover, besides the primary focus on restricted substances present in goods, an increasing number of aspects of production, such as water consumption, the use of best available techniques, as well as the working conditions of employees have been included.

Aspects of ecological production require the integration of new elements such as a design for recycling, sorting and separation (Nørup et al., 2019) in addition to more localised approaches to improve production techniques to reduce their contribution to global warming (de Oliveira Neto et al., 2019). These activities should cover the full lifetime of a product in a coordinated approach. The most relevant stages include the substitution of virgin fibre material by recycled products, the development and integration of design for recycling into all stages of technical and mechanical processing of textiles, and garment design for the longer duration of use and repair. Optimised textile design and better labelling to facilitate the sorting and disassembly or separation of wasted textile products as well as efficient polymer recycling techniques are important. The intelligent use of dyeing and printing technologies and the adaptation of currently used processes and chemicals to the requirements of a circular economy will require a reconsideration all stages of production.

**Sustainable Innovations in the Textile Industry. https://doi.org/10.1016/B978-0-323-90392-9.00009-4**

## 13.2 Historical aspects and current scenario

Historically, movement toward the higher sustainability of processes and products in textile processing was initiated by different driving forces. Each step led to the adaptation of processes, energy, water and chemical consumption, mostly as part of incremental development. In the 1980s, the recovery of chemicals was mainly driven by cost and ecological aspects were of minor relevance. As an example, the recovery of spent sodium hydroxide solution from mercerisation processes by evaporation was introduced into dye houses to reduce costs (Bechtold et al., 1985). Energy consumption was reduced by installing large heat exchangers, which enabled energy recovery in the form of hot water.

The treatment of wastewater was seen as an unwanted additional requirement that generated additional costs. However legal requirements for effluents released from dye houses were set up, such as to reduce pollution in the lake Bodensee, which serves as a drinking water reservoir for a large part of southern Germany (Österreich, 2000). Process optimisations led to a reduction in water consumption in exhaust processes through machinery that was able to operate at a low liquor ratio (mass of goods in kilograms per volume of dye bath in liter). The optimised flow regime in continuous processes also brought savings in water and energy.

A good overview of technical processes to reduce emissions in textile production was published by Cotton, Inc (Cotton Incorporated, 2009). There are advantages and limitations of existing technologies. For instance, low liquor exhaust dyeing, cationisation, size recovery, membrane filtration are discussed, and an estimation of costs and return on investment is given.

Digital colouration by inkjet printing reached technical scale production in the early 2000s. The use of digital printing techniques enabled new approaches in design, pattern construction and colour resolution (Riisberg, 2007). High production volumes are processed by rotary screen printing whereas inkjet printing concentrates on specialised products and lower batch sizes. High flexibility with regard to the patterns of inkjet printing offers opportunities for mass customisation, but flexibility is limited in the dye systems used.

More integrated approaches to process and product use were highlighted by the concepts of life cycle analysis (LCA) (Shen, 2011; Velden et al., 2014; Cotton Incorporated, 2009) and cradle-to-cradle (C2C) (Wolford, 2022). LCA analysis quantifies the overall environmental impact of a product, whereas in C2C analysis, products are based on a regenerative design and must fit into a biological or technical cycle. The discussion about the persistence of fluorocarbon-based water repellent finishes in the environment clearly addressed the need to consider the behaviour of products at the end of their use, as well as chemicals and additives released with them into the environment during use or disposal.

In the 2000s new keywords appeared in searches, such as biobased and renewable, water footprint, $CO_2$ footprint, global warming, microplastic, circularity of products and design for recycling. The appearance of these keywords indicated the end of localised process and product optimisations and the need for integrated approaches that consider the production of raw material, processing into textile products, consumer use and disposal or recycling into new products at the same time (Dahlbo et al., 2017). The development of closed loop strategies suddenly included consumer behaviour, such as washing and drying procedures as well as the collection of postconsumer waste and separation into different fractions for fibre and polymer recycling (Schmutz et al., 2021).

# 13.3 Circular economy in textiles

## *13.3.1 Fibres in textile circularity*

The landscape of clothing manufacture and use has changed dramatically over the past few decades. Globally, the production of clothing has doubled but average use has been reduced by about a third (Circular Fibers Initiative, 2017). Some discarded clothing is resold as second-hand goods, but the major proportion is incinerated or disposed in landfills, and there is little recycling of the materials. In Europe, around 4.3 million tons of textiles waste is disposed of annually owing to the lack of viable recycling strategies (Briga-Sá et al., 2013). Fibre production consumes significant amounts of raw materials, energy and water. Therefore, the current system wastes resources and is detrimental to the environment. A proposed remedy, termed the circular economy, envisages that 'clothes, fabric, and fibres are kept at their highest value during use, and re-enter the economy after use, never ending up as waste' (Circular Fibers Initiative, 2017). This vision is endorsed by political entities such as the European Union, which has legislated that by 2025, all member states should put in place systems to collect waste textiles separately and ensure they are not incinerated or landfilled (European Environment Agency, 2019).

The concept of the circular economy requires polymers from waste clothing to be recycled into the manufacture of new textiles and clothing. Cellulose, polyester (PES) and polyamide (PA) represent most polymer types found in textiles. Work on the chemical recycling of cellulose from textile wastes is based on the use of amine oxide solvents such as *N*-methylmorpholine *N*-oxide (NMMO) (Brinks et al., 2018; Negulescu et al., 1998) or ionic liquids (IL) (De Silva et al., 2014; Haslinger et al., 2019a,b). For example, Sixta et al. used 1,5-diazabicyclo[4.3.0]non-5-ene acetate as a direct solvent to dissolve the cotton part selectively in cotton PES-blended textile waste (Haslinger et al., 2019a). The cellulose solution was then spun to new man-made fibres, including the remaining dyestuff from textile waste (Haslinger et al., 2019b). The strategy for chemical recycling of PES focuses on depolymerising PES, yielding its precursor monomers (e.g. terephthalic acid and ethylene glycol), through processes such as thermal and catalytic decomposition (Kenny et al., 2008; Du et al., 2016) hydrolysis (Paszun and Spychaj, 1997) or methanolysis (Genta et al., 2005).

Different strategies have also been proposed for PAs 6 (PA6) and 66 (PA66). Those for PA6 include hydrolysis (Hocker et al., 2014; Bockhorn et al., 2001) using supercritical fluids (Chen et al., 2010; Kamimura et al., 2014) and using ILs under microwave irradiation (Kamimura et al., 2019) via ring-closing reactions to produce *N*-acetylcaprolactam (Alberti et al., 2019). For PA66, acidic hydrolysis (Patil and Madhamshettiwar, 2014), supercritical water (Zhao et al., 2018) and glycolysis using ethylene glycol and triethylenetetramine as decomposing agents (Datta et al., 2018) have been investigated.

The depolymerisation strategy works best for single polymer synthetic products, because depolymerising polymer mixes entails extensive purification steps to separate and isolate the different monomers before repolymerisation can proceed. If a synthetic polymer is mixed with nonsynthetics such as wool or cellulose, depolymerisation treatment of the mixture leads to the total loss of nonsynthetic fibres, because there is no known method to repolymerise glucose to cellulose fibres or reassemble amino acids into wool fibres. However, often clothing textiles are not single-component products but mixtures or blends of different polymer types, such as PES with cotton or wool with PA. Effective recycling requires the complete separation and recovery of all polymers from blends without damage. That is difficult to achieve with mechanical processes because components in blends are intimately mixed. Thus, the selective dissolution of individual polymers is an attractive option. However, it must be ensured that the solvent used does not damage either the polymer dissolved out or the undissolved polymer(s) left behind.

The approach of nondestructive dissolution and recovery of PA fibres was developed, and investigations to dissolve out and recover PA by the controlled complexation and decomplexation of PA without degradation was reported (Rietzler et al., 2019). The solvent employed in the investigations was a mixture of $CaCl_2$, ethanol and water (CEW). Depending on the relative molar fractions of its components, the mixture exerts a swelling effect or dissolves PA (Rietzler et al., 2018a,b). The dissolution and swelling action was previously investigated (Rietzler et al., 2018a, 2019) and researchers confirmed that the solvent does not cause the molecular degradation of PA (e.g. chain scission, lowering the molecular weight). Apart from the ability of CEW to dissolve PA, another important aspect for the possible application of this solvent system for recovering PA from textile waste is quality of the polymer structure of the dissolved and reprecipitated polymer (in the ideal case, free of $Ca^{2+}$). This can be achieved only through proper decomplexation. Table 13.1 lists different dissolution technologies of major polymer components in textiles with respect to recycling mixed textile waste. Classical organic solvents are excluded from the comparison because in most cases they are unsuitable for textile recycling and are difficult to handle in large-scale operations.

Apart from NMMO and ILs for cellulosic-based textiles, other technologies for PES and PA are only partly suitable for application to recycling mixed textile waste. Although depolymerisation (e.g. of PA6) has reached an industrial scale, and chemically recycled PA6 fibre is commercially available (e.g. Econyl), the separation of individual monomeric forms from a mix of polymer types to the required purity levels for repolymerisation is significantly challenging. Furthermore, polymerisation has to be considered an additional important cost factor in the recycling of mixed textile waste.

**Table 13.1** Different technologies with regard to chemical recycling of mixed textile waste.

| Concept | Fibre type | Status | Challenge |
|---|---|---|---|
| *N*-methylmorpholine *N*-oxide (Brinks et al., 2018; Negulescu et al., 1998) | Cellulose | Industrial scale, good separation of cellulose from mixed textiles | Influence of contaminants on fibre spinning process |
| Ionic liquids (De Silva et al., 2014; Haslinger et al., 2019a,b; Kamimura et al., 2019) | Cellulose, polyamide | Well-developed, good separation of cellulose from mixed textiles | Influence of contaminants on fibre spinning process of cellulose fibre, complicated separation and purification of monomer for repolymerisation of PA |
| Hydrolysis (Hocker et al., 2014; Paszun and Spychaj, 1997; Patil and Madhamshettiwar, 2014) | Polyester, polyamide | Well-developed, industrial scale polyamide (PA6) fibre available | Clean (single) feedstock required, complicated separation and purification of monomer units |
| Supercritical fluids (Chen et al., 2010; Genta et al., 2005; Kamimura et al., 2014; Zhao et al., 2018) | Polyester, polyamide | Under development, industrial example available for polyethylene terephthalate | Process design and conditions, clean (single) feedstock required, complicated separation and purification of monomer units |
| $CaCl_2$, ethanol and water nondestructive (Rietzler et al., 2018a, 2019, 2021) | Polyamide | Under development | Process design, optimisation and upscaling |
| Solvent mixture (Vonbrül et al., 2022) | Polyurethane | Under development | Process design, optimisation and upscaling |

Considering only PA, the global production of PA fibres in 2018 for the textile and carpet sectors amounted to about 5.4 million metric tons (Textile exchange, Preferred Fiber and Materials Market Report 2019). PA fibres are predominantly found in the textile (75% to 85%) and carpet sectors (15% to 25%) (Wesołowski, 2016). Apart from products such as stockings and carpets, for which PAs constitute the major proportion or a significant proportion of the material composition, it is common to find PAs in minor proportions (15% to 20%), such as in uniforms and socks (Kothari, 2008). Assuming that half of the PA fibres produced were employed in PA-only products, and the other half in products in which PA constituted 50% of the mass, the total amount of PA-containing textiles manufactured in 2018 may have been about 10.8 million metric tons. According to this projected volume, PA-containing textile stream can be considered a huge feedstock source not only for new PA fibrous materials, but also to recover other fibre parts such as PES, cellulose and wool.

Fig. 13.1 illustrates the material flow when nondestructive separation is applied to recycling PA-containing mixed textile waste. Apart from textiles as second-hand goods, the process includes first sorting used textiles into materials containing PA and those containing no PA fibres, and further sorting PA-containing materials into different fractions with different second fibre types such as PES, cellulose and wool. The process can be done using analytical methods such as spectroscopy (near infrared or Raman) and melting point indication (Wang, 2010). Nondestructive separation through controlled complexation or decomplexation separates PA material from other

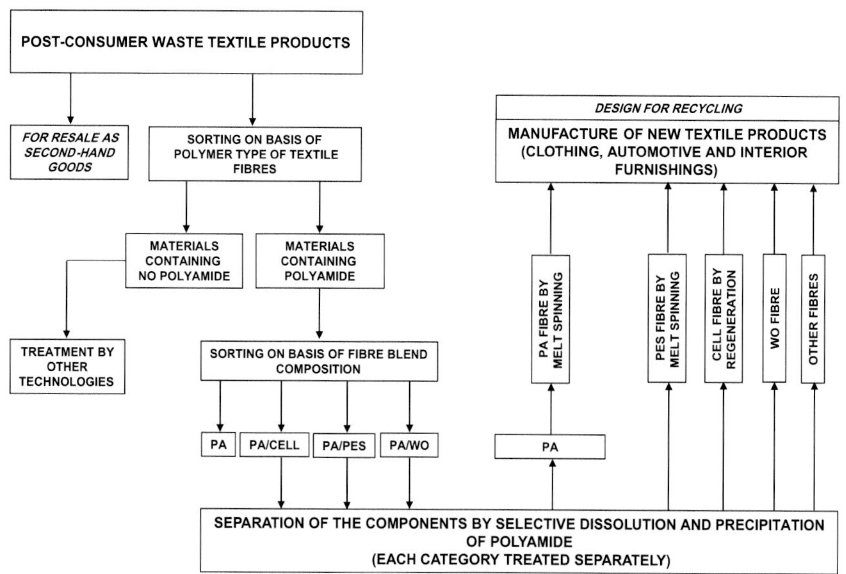


a) Textiles containing PUR will be sorted as per the main components, e.g. PA/PUR will be sorted into the PA category, PA/Wo/PUR will be sorted into the PA/Wo category, etc.

b) PA = polyamide, Cell = Cellulose, PES = polyester, Wo = Wool, PUR = polyurethane

c) Life cycle assessment (LCA) studies will be performed of the whole sequence.

d) There will be engagement with stakeholders at each stage.

**Figure 13.1** Representation of material flow by nondestructive separation of polyamide with regard to recycling of mixed textile waste.

fibres. Whereas isolated PA material can be further melt processed into new PA fibre, other parts of mixed textile waste can be used directly (e.g. wool in the case of PA–wool mixtures for new wool-based yarn) or converted into new fibres (e.g. PES fibre by melt spinning or cellulose fibre by solution spinning) (NMMO or IL) (Table 13.1), depending on the composition of the textile waste that is used.

Polyesterurethane and polyetherurethane (Spandex and Lycra) are fibre components frequently used in textiles to provide highly elastic properties. These polymers are often combined with other synthetic fibres such as PES and PA. The polyurethane component should be removed before the polymer recycling of the main textile constituent can be considered. Many solvent-based systems use toxic dimethylformamide (DMF) or DMF-containing solvents, which leads to expensive installations on the technical scale. The dissolution of polyurethane fibres using solvent mixtures of low toxicity was reported, which solves this critical step of fibre polymer recycling (Vonbrül et al., 2022). Looking toward future textile circularity, the concept of designing for recycling can be incorporated as an integrated part of the circle to reduce the number of different components in textiles, easing a later separation step.

## 13.3.2 Dyes and pigments in textile circularity

Besides challenges that need to be solved in fibre separation and recovery for reuse, approaches to minimise the environmental impact of textile dyeing are urgently needed to reduce the generation of further effluents such as wastewater from dyeing (Circular Fibers Initiative, 2017). Globally, wastewater from dyeing is still a major issue, especially in developing countries, because it is often discharged directly into the environment with no treatment step. As an estimation, up to 2,00,000 tons of dyes is lost into the wastewater stream every year (Drumond Chequer et al., 2013).

Some efforts have been made to cope with the retention of colour during the reuse of textiles. As an example, textile waste with different colours were blended, giving a broad spectrum of yarns (Esteve-Turrillas and de la Guardia, 2017). However, the approach is limited because mainly preconsumer waste with defined colours is used, whereas only small amounts of postconsumer textile can be added to yarn production (Esteve-Turrillas and de la Guardia, 2017). The main challenges to using postconsumer textile waste are colour change and degradation by UV irradiation, abrasion at use and laundering (Palme et al., 2014; Wedin et al., 2019).

In the most cases, textile waste is dyed, such as PES with disperse dyes and PA with acidic dye and metal complex dye, whereas polypropylene (PP) is dyed en masse using pigment master batches during fibre spinning. Thus, dyes and pigments are making the separation and recycling of textile even more complicated. However, the efficient removal of dyes and pigments from textiles is needed to boost the industrialisation of waste textile recycling (Li et al., 2020). The associated high cost of removing dyes and pigments is therefore a limiting factor of the large-scale reuse of dyed textile waste (Leal Filho et al., 2019; Lyndsay McGregor, 2015). Thus, only colourless plastic bottles are reused to produce fibres on a large scale (Tournier et al., 2020). Other challenges are the possible bleaching of dyes during fibre separation from the environment and the undefined colour of recovered fibrous materials.

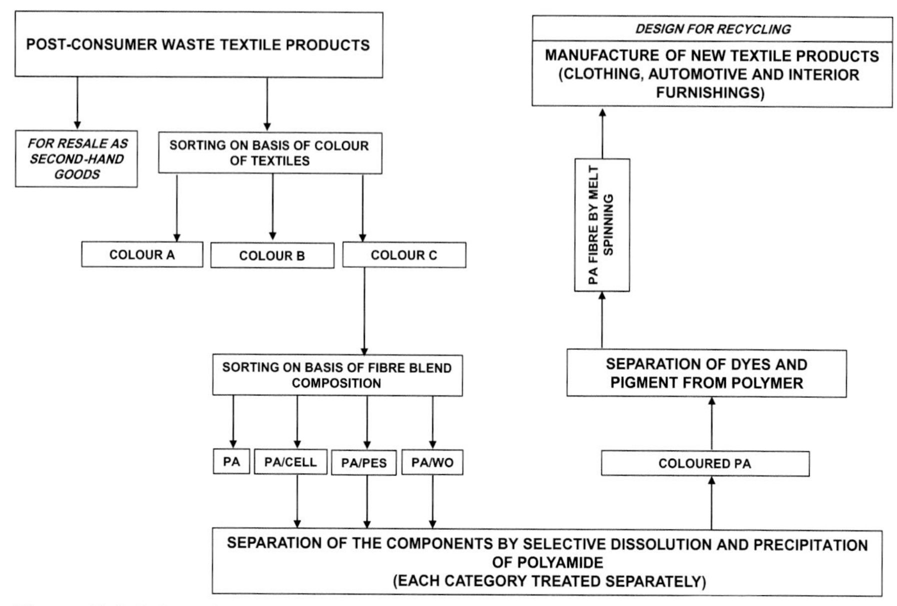


**Figure 13.2** Schematic view of ideal separation and recovery process of dyed polyamide (PA) fibres from textile waste. *PES*, polyester; *WO*, wool.

Thus, the ideal process of separating and recovering dyed polymer fibres must cover the steps of mechanical sorting, dissolution or depolymerisation, separation and recovery of dyes and pigments, recovery of polymer or purification of monomer followed by polymerisation (Thiounn and Smith, 2020). Fig. 13.2 shows such an idealised process for dyed PA fibres.

To decolour textile fibres, two approaches are often applied: dye destruction and dye extraction. However, the balance between efficient dye removal and polymer damage remains a major issue for both approaches. The destruction of dyes by oxidation and photodegradation degrades polymers and changes the dyeability of regenerated fibres (Cristóvão et al., 2009; Maryan et al., 2013; Qian et al., 2017). On the other hand, the extraction of dyes cannot completely remove colours from fibres (Groves et al., 2016). The main reasons for those drawbacks are the strong interaction between dyes and fibres and the low chemical potential of dyes in polymers (Cheng et al., 1991). Thus, much effort has been put into evaluating solvent systems with suitable dye solubility. Furthermore, harsh extraction conditions, such as high temperatures and long extraction times, often lead to the degradation of fibre polymers (Yang et al., 2001).

Mu and Yang (Mu and Yang, 2022) successfully demonstrated the complete removal of disperse dyes, acid dyes and direct dyes from PES, PA and cotton fibres, respectively, by minimising the fibre density by suitable solvents and temperatures. Mixtures of ethylene carbonate and tetramethyl urea with a mass ratio of 7:3 was used for PES with a disperse dye, dimethyl sulfoxide (DMSO) and water with a mass ratio of 95:5

was used for PA with acid dye and DMSO was used to separate direct dye from cotton. The authors reported that after the removal, the chemical structures of dyes and average molecular weight of polymers did not change (Mu and Yang, 2022).

## 13.4 Innovations leading to sustainability and resource efficiency

Previously, improvements in sustainability, energy and resource efficiency were straightforward and incremental. Individual localised developments were highlighted and discussed as the best available technology (BAT). Awareness about interconnections among all stages of the life cycle of a product led to questions about how to assess the BAT and to a request to consider all stages of a product life cycle in parallel, which made it more difficult to compare certain technologies. Instead, in this chapter, representative cases of technologies are discussed as examples of concepts that may be compatible with future integrative concepts for textile production.

### *13.4.1 Mass colouration*

Mass colouration, spin-dyeing or dope dyeing is 'a method of colouring manufactured fibres by incorporation of the colourant in the spinning composition before extrusion into filaments' (AATCC, 2000). Thus, it is limited to manufactured fibres. There are disadvantages to mass colouration: the process economics is improved with an increase in the lot size and thus unsuitable for small fibre lots, and because fibres cannot easily be redyed, there is reduced flexibility in adapting products in response to changes in fashion requirements. Nevertheless, the method offers ecological benefits, such as greater consistency in the uniformity of colouration. Moreover, coloured fibres generally exhibit excellent light fastness and excellent use of dyestuff (i.e. minimal loss of unfixed dye). Also, no wastewater is generated. A predominant area of its application is in colouring fibres that are otherwise difficult to dye, such as polypropylene, but the ecological benefits make it an attractive prospect even for fibre types such as PES and cellulosics, which can be coloured with standard dyeing techniques.

A common method of mass colouration is through the dispersion of insoluble pigment particles in polymer dopes (melts or solutions). Care is required to ensure that particulates do not interfere with fibre spinning or negatively affect the mechanical properties of fibres that are produced. Thus, the particle fineness and dispersion stability as well as stability to the temperature of polymer melts or the solvent in polymer solutions are critical factors. Pigment dispersion may be introduced and dispersed in the bulk of the polymer dope, or smaller volumes of master batches may be formulated, which are polymer dopes pigmented at high concentrations, to be blended with the bulk of the polymer dope. An alternative to insoluble pigments are solvent dyes dissolved in a solvent compatible with the polymer dope. Other alternatives being investigated are adding colourants before polymerisation or to applying them to the formed polymer chips or granules (Roy Choudhury, 2011).

The mass colouration of regenerated cellulosics produced via viscose and lyocell processes presents specific challenges because of the strong reducing or oxidizing agents used in formulations of the polymer dope. For the viscose process in which cellulose is dissolved in a medium of strong reducing power (carbon disulphide and sodium hydroxide), vat dyes were investigated for dope dyeing. The addition of a reduced vat dye to the spinning dope (I. G. Farbenindustrie A.G., 1937; Kline and Helm, 1939) or the addition of the parent form of the dye to reduce it in the dope by itself (Lockhart, 1932) or with the help of additional reducing agents such as sodium hydrosulphite (Batt, 1961) were investigated. Other options include dispersing the vat dye in the spinning dope as a pigment, spinning the fibre, and subjecting it to reducing conditions to solubilise the pigments (I. G. Farbenindustrie A.G., 1936; Maloney, 1967; Ruesch and Schmidt, 1936). In these techniques, the oxidation of the vat dye back to its parent form is achieved by the final treatment of fibres with oxidizing agents. Colouration was also achieved by adding vat acids or the ester derivatives of leuco compounds of vat dyes to the polymer dope (Dosne, 1936; Lutgerhorst, 1956).

The addition of dyes dissolved in polar, water-miscible solvents and the dissolution of dyes directly in the spinning dope were also attempted (Ciba Ltd., 1968, 1966, 1965; Phrix-Werke, 1965; Riehen and Reinach, 1971a,b; Soc. pour l'ind. chim. a Bale., 1941; Wegmann and Booker, 1966). The mass colouration of cellulose regenerated via the lyocell process was attempted by adding inorganic pigments containing minimal amounts of heavy metals (Bartsch and Ruef, 1999), or by dyeing cellulose pulp with a vat dye before dissolution in the NMMO (Manian et al., 2007). Fig. 13.3 shows the mass colouration of lyocell type fibres with C.I. Vat Green 1.

### 13.4.2 Denim and jeans

A substantial amount of the yearly production of cotton (27 Mt) is used to produce several billion denim products (jeans). The main dye used for denim production is indigo, with an estimated annual production of 70,000 tons (Paul et al., 2021). The huge amount of production thus makes a substantial environmental impact. A simplified overview of the circular concept of denim production is given in Fig. 13.4.

The main fibre used to produce denim is cotton. However, the worldwide production of cotton is based on the substantial consumption of fertilizers and pesticides as well as high water consumption for irrigation (Paul, 2015; Imran et al., 2019). Indigo is the main dye used to produce denim. The dye requires reduction (reduction potential of approximately −700 mV vs. (Ag/AgCl, 3 M KCl) reference) to become soluble in the alkaline dye bath (optimum pH of 11.5−12). Furthermore, reducing agents are required to maintain a reduced state in the dye bath during continuous dyeing. A substantial reduction of nearly 50% of the total consumption of reducing agents may be achieved using a prereduced indigo solution (hydrogenation) or the direct cathodic reduction of indigo (Blackburn et al., 2009). Both techniques reduce the dye in an alkaline solution without forming byproducts from the use of reducing agents. In case of hydrogenation, the removal of unwanted

**Figure 13.3** Mass colouration of lyocell-type fibres through dissolution and regeneration of coloured pulp (C.I. Vat Green 1).

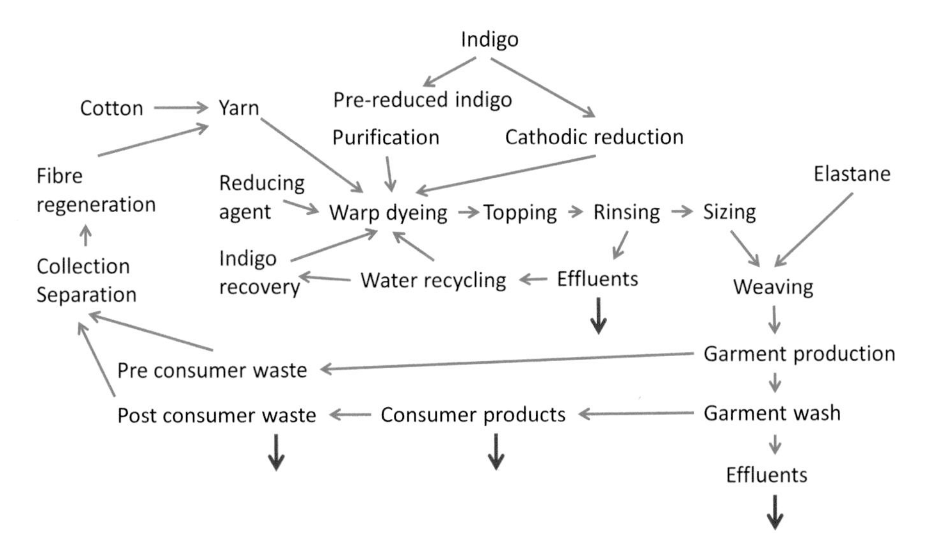


**Figure 13.4** Positions of textile chemical operations in the concept of circularity. *Red arrows* indicate main sites of waste formation.

byproducts from the synthesis of indigo (aniline and N-methyl-aniline) can also be achieved.

Indigo is dyed through the repetitive immersion of the warp yarn into the dye bath and then air oxidation. This leads to the considerable consumption of a reducing agent, which is necessary to maintain the indigo in the dye bath in its reduced form and compensate for the continuous intake of oxygen. The standard process is based on the use of sodium hydrosulphite ($Na_2S_2O_4$), which irreversibly forms sulphite ($SO_3^{2-}$) as a consequent product of the reaction with indigo and sulphate ($SO_4^{2-}$) in a later reaction with air oxygen. As a result, indigo dyeing units release considerable concentrations of sulphate into wastewater. Representative concentrations of sulphate in wastewater range from 600 to 1600 mg/L and are far beyond the legal limit for release of 200 mg/L sulphate into a communal wastewater treatment plant (Abdelileh et al., 2020).

Recycling of indigo by membrane filtration using membranes with an appropriate cutoff can be achieved, and technical installations have been proposed (Buscio et al., 2015). The separation of indigo from wastewater may also be achieved using selective sorption and desorption processes (Wambuguh and Chianelli, 2008). A barrier to achieving a technical-scale installation results from the low cost of indigo, which makes the economic value of dyestuff recycling low. In addition, standardisation of the regenerated dye is required and the presence of sulphur dyes from bottoming and topping and impurities released from cotton have to be considered. The load of sulphate in the wastewater hinders the recirculation of water owing to the accumulation of salt in the dye bath, which changes the uptake of indigo and results in variable dyeing results. Thus, the future concept of a sustainable process should fulfil several requirements:

- Recycling of water and indigo should be possible
- The release of chemicals and subsequent products from reducing agents that cannot be eliminated in a biological stage should be minimised.

Two concepts have the potential to reduce the chemical load in wastewater:

- The use of organic reducing agents
- Indirect electrochemical reduction

With the use of reducing sugars for dye bath reduction, subsequent biodegradable products are released into wastewater. However, higher amounts of the biobased reducing agent are required, and both pH and the dyeing temperature are higher than the recommended optimum range of pH 11.5 and 25 to 30°C owing to the slower buildup of reducing conditions (Saikhao et al., 2018). Adaptation of the dye bath composition to higher indigo concentrations could compensate for the lower affinity of the leuco-di-anion of indigo.

As an advantage, an organic reducing system is compatible with process conditions used for sulphur topping. Thus, dye bath transfer into the topping bath will not cause problems, and the further use of unconsumed reducing sugars is possible. The comparable higher concentrations of sugar used in the dye bath will also lead to a higher organic load in the wash water. However, the reuse of indigo containing concentrated wash water to prepare the dye bath or for prewetting would be possible. Reducing organic substances present in the wash water will still be valuable for dyeing, which is not possible in the case of dithionite. For hydrosulphite, the salt load must be eliminated (e.g. by reverse osmosis) before water recycling is possible. Thus, the installation of dyeing processes using sugar-based reducing agents is significantly valuable when more closed water and chemical cycles are intended.

Another possible technique for reducing dye is to use electrochemical processes. The use of an iron complex for the indirect cathodic reduction of indigo could replace nonregenerable reducing agents (Blackburn et al., 2009). As chemicals are transported into the wash water, the reconcentration and reuse of all valuable constituents of the iron complex, dispersed indigo and water might be achieved, minimising the release of chemicals and water from the dyeing process (Abdelileh et al., 2020). Although large-scale tests of electrochemical indigo dyeing have been undertaken, the high investment costs have hindered introducing it into regular production.

A completely different approach to recycling indigo-dyed cellulose fibre-based textiles was reported. When the cellulose polymer is recycled by producing viscose fibres, indigo dye present in used denim can also be recycled. The viscose process is tolerant to the addition of a reduced indigo solution, and thus dope-dyed indigo fibres can be obtained. That study presents an approach to the combined recycling of cellulose material and textile dye (Manian et al., 2022).

Ecru denim fabrics were dyed with several natural dyes using natural mordants without the use of metallic mordant, which resulted in sustainable denim garments (Deo and Paul, 2000, 2004a). Furthermore, sustainable indigo dyeing methods for ecru denim fabrics were developed by a standing bath technique (Deo and Paul, 2004b) and the reuse of indigo vats without replenishment (Deo and Paul, 2004c). Both techniques resulted in varying shades on denim fabrics with almost complete exhaustion of the dye baths.

### *13.4.3 Garment wash*

A considerable part of the overall environmental load during the production of denim results from garment washing. Garment processing is an essential part of the development of different looks resulting from intense mechanical and chemical treatment of the garment. Depending on the depth of colour and the desired effects in the final product, a major part of indigo is removed and substantial amounts of indigo and indigo-dyed fibre residue are found in wastewater. Traditional denim washing formulas can use 70 L of water, 150 g of chemicals and 1 kWh of energy per pair of jeans, which is substantial compared with water and chemical consumption in dyeing (Khan and Jintun, 2021).

A huge number of processes have been applied. Major processes use oxidants such as $NaOCl$ and $KMnO_4$ followed by treatment with reducing agents. More sustainable processes used different technologies, all of which employ a low liquor ratio or process the garment in a wet or dry state based on enzyme treatments (cellulases and laccases), ozone treatment, plasma processes and laser treatment (Khan and Jintun, 2021).

Two general principles are followed in these processes: weakening and removing fibres from the surface, and the oxidative destruction of indigo dye. In any case, a certain amount of washing is required to remove subsequent products from the treatment. A reduced number of treatment baths leads to reduced water consumption, energy consumption and $CO_2$ equivalents (Nielsen, 2012). A comparison of the ecological profile of different processes also has to include the balance of resources for the generation of products used. For example, in case of enzymes, their biotechnological production should be included.

### 13.4.4 Sulphur dyeing without reducing agents

Based on the annual production volume, C.I. Sulphur Black 1 belongs to the most important dyes. The dye offers high tinctorial strength at a comparable low cost. Black denim is a major field of application. The addition of reducing agents to the dye bath is required to break down the high-molecular weight dye molecule into soluble reduced fragments, which adsorb onto the fibre and recombine to the final black dye. Because of the lower redox potential required to reduce the disulphide bonds and the quinoneimine groups in the dye molecule and the higher temperature and pH value upon application, less powerful reducing agents can be applied. The former use of sulphides and polysulphides has been replaced by organic reducing agents such as glucose, which has led to a substantial reduction in the sulphide content of the wastewater. The accumulation of subsequent products from the use of reducing agents in the dye bath prevents the development of circular concepts. However, when sulphur black is used for topping or bottoming, the use of a common reducing agent both for sulphur black and indigo reduction may be a step ahead to facilitate the recycling of effluents.

The ability of the polymeric dye molecule to undergo several reduction steps enables dye molecules to serve as soluble electron transfer for cathodic dye reduction by themselves (Fig. 13.5) (Bechtold et al., 2000). The introduction of such processes could reduce the consumption of dyestuff in coloration of black denim substantially during the preparation of the liquid formulation of the dye from a crude filter cake and during the application of the dye, which is a prerequisite for the dye bath and water recycling (Bechtold et al., 2008). Full-scale tests of the technology on a technical scale dyeing range have demonstrated the proper function of the concept.

In any case, driving forces to introduce electrochemical processes as a prerequisite for regenerating the dye bath in sulphur dyeing will not be based exclusively on the expected cost savings. They will stem from the need to reduce the chemical load into the wastewater and improve the ecological profile of sulphur dyeing. The introduction of cathodic dye reduction could lead to a substantial contribution.

In another approach, the use of selected blue sulphur dyes was introduced to denim production to replace complicated dyeing with indigo. In addition to replacing indigo, the use of sulphur dye-based concepts widens the range of colours available for denim production (Archroma, 2023a).

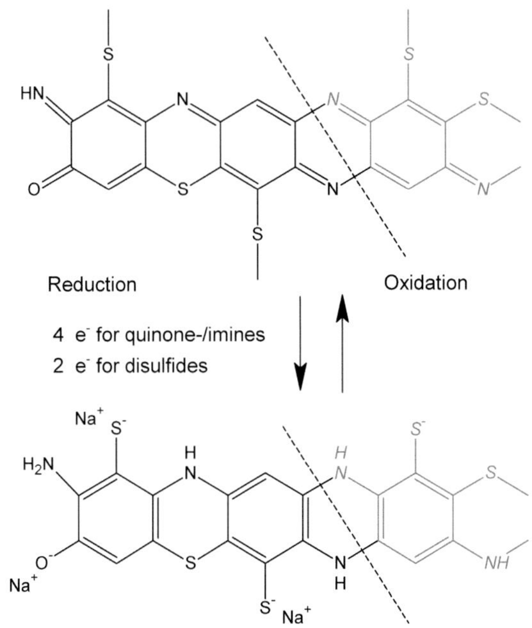

**Figure 13.5** Multielectron reduction of C.I. Sulphur Black 1.

## *13.4.5 Reactive dyes*

Reactive dyes represent the largest group of dyes used for cellulose fibre-based textiles. Among the most important advantages of this group of dyes are the wide range of brilliant colours available, which offer the maximum colouristic potential, a reasonable level of fastness properties and straightforward application delivering reliable results at acceptable costs. For their application, only inexpensive chemicals (alkali and salt) are required to form covalent binding between the reactive anchor of the dye and the hydroxyl groups of cellulose.

After extensive use in many different processes, the application of reactive dyes requires a reconsideration of the aspects of emissions and recycling of chemicals and water. The application of reactive dyes can follow highly efficient processes such as cold pad batch dyeing, in which a large amount of dyestuff fixation is achieved at a short liquor ratio using a comparably low amount of chemicals, energy and water. Also, in the reactive dyeing of wool highly efficient processes are used, which enable good bath exhaustion, a high degree of dyestuff fixation and the minimal use of chemicals to adjust the pH.

A different situation is observed in exhaust dyeing processes for cellulose textiles, which is another major field of reactive dye application. Increasing concerns arise from these characteristics:

- Comparable low affinity of dyes to cellulose fibres limits bath exhaustion. Thus, high amounts of salt (e.g. 50 g/L NaCl) are added to the dye bath to improve dye fixation.

- Hydrolysis of the reactive anchor in the alkaline bath leads to the significant loss of dye, which is no longer able to form a covalent linkage to the cellulose.

Typical liquor ratios in exhaust dyeing range from 1:5 to 1:10. Thus, when a salt concentration of 50 g/L NaCl is applied, a total of 0.5−1 kg of NaCl is spent per kilogram of dyed goods. Costs for salt are low, but there has been increased criticism addressing the release into the environment of such high amounts of salt after a single use. The situation becomes even worse when fibre blends (e.g. 50% cotton, 50% PES) are dyed, in which longer effective liquor ratios result from the presence of other fibres.

The limited dyestuff fixation also leads to the formation of highly coloured wastewater, which contains dyes with low biodegradability. Hydrolysis of the anchor functionality hinders the recovery and reuse of the dyes. Concepts for improving the ecological profile of reactive dyeing have to address two major principles:

- High dyestuff fixation and minimised losses in hydrolysed reactive dyes
- A reduction in overall salt consumption by increasing dye sorption and fixation or by reducing the liquor ratio

Approaches to reduce the liquor ratio have reached technical limits with regard to the movement of goods in jet dyeing as well as the solubility limits of dyes. Special techniques that use an extremely low liquor ratio (e.g. spray application) were introduced for particular applications such as in overdyeing processes for denim production (Russ, 2022).

A promising system to improve dyestuff fixation and reduce the salt load in wastewater from reactive dyeing modifies the surface charge of cellulose fibres by introducing cationic groups. The cationisation of cellulose fibres has been widely investigated as a process to increase the sorption of anionic dyes and improve the exhaustion and binding of reactive dyes (Roy Choudhury, 2014). Through the introduction of cationic sites by the covalent linkage of quaternary ammonium groups, the negative surface charge of cellulose is reduced and even inverted to the opposite sign. Zeta-potential measurements indicate that changing the surface charge and functionalising cellulose fibres with positively charged groups imparts them with ion-exchange properties (Fig. 13.6).

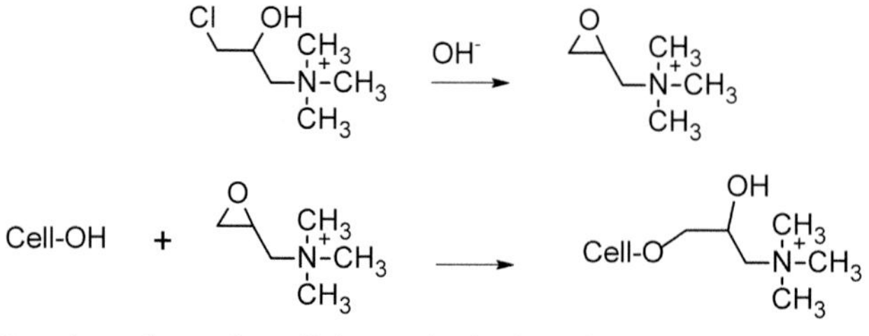

**Figure 13.6** Reaction scheme for cellulose cationisation with 3-chloro-2-hydroxypropyl-N,N,N-trimethylammonium chloride (CHPTAC).

The dyestuff fixation reaction is proposed via a two-step reaction. First, the anionic reactive dye adsorbs via coulombic forces to cationic sites. In the second step, a covalent bond between dye and fibre is formed (Pruś et al., 2022). As a result of cationisation and dye sorption, a high local concentration of dye is achieved in a short distance to the fibre surface, which consequently increases the reaction efficiency between the dye and fibre.

The introduction of cationic groups can be achieved via a chemical reaction with a reactive quaternary ammonium salt such as CHPTAC (Zhang et al., 2021). A major drawback of chemical cationisation is the low efficiency of the reaction between the cationisation agent and cellulose. Although literature reports argue that cationisation treatments are more ecological, the low reaction efficiencies lead to significant losses in chemicals. Typical formulas for the cationisation of cellulose in exhaust processes use 20−30 g/L of a cationisation agent at a liquor ratio of 1:5 or higher, which finally corresponds to 100−150 g of chemicals per kilograms of textiles. In addition, several treatment baths have to be used for the treatment.

A generally more technical problem to be solved for wider application is the uniformity of the modified fibre, which will be marked during the dyeing step, leading to visible variations in colour. Similar to cationic dyeing of polyacrylics, the competition of different dyes during the phase of dye sorption and fixation has to be considered and an appropriate selection of dyes is required.

Promising alternatives to the covalent modification of fibres with cationic anchors use the sorption of cationic polymers on the fibre surface (Correia et al., 2021). These polymers (e.g. cationic starch) then take on the function of the positively charged anchor. Such concepts have the potential for substantially higher efficiency with regard to chemical consumption (Zhang et al., 2005).

The cationic modification of cellulose fibres was also reported in combination with dyeing using other dye classes, such as direct dyes, anionic dyes or vat dyes (Abdelileh et al., 2019; Sutlović et al., 2021). In another approach to salt-free reactive dyeing, the chemical consumption of cationisation could be substantially reduced. It was demonstrated that a low degree of cationisation is sufficient to increase reactive dye exhaustion to the level obtained by adding 50 g/L NaCl (Setthayanond et al., 2023).

The future potential of modifying cationic cellulose fibre for a higher degree of efficiency in dyeing with reactive dyes is high. However, a clear assessment and comparison of overall chemical consumption, water use and other relevant ecological parameters is required to highlight achievements or at least discuss the need for the further optimisation of the balance of resources. Moreover, significant improvements in the results of the subsequent dyeing step with regard to uniformity and shade reproducibility are required to achieve the wider application of this promising technology.

### 13.4.6 Natural colourants

Before the development of synthetic dyes, natural colourants were used to produce coloured textiles. Synthetic dyes offered substantially better performance at lower costs, which enabled access to coloured garments for people with lower incomes. Growing awareness about health hazards as well as concerns about the lack of sustainability for synthetic colourants led to discussions about the possible use of natural colourants in textile dyeing.

Activities continue in the field of dyeing with natural colourants. Major actors are positioned in the field of traditional handicraft dyeing and small dye gardening. In addition, a limited number of larger activities are ongoing to grow dye plants and apply them in textile dyeing. As a consequence of growing interest in natural dyes, dye manufacturers have begun to offer a small range of nature-based dyes (Archroma, 2023b).

The use of biobased colourants for dyeing may contribute to making textile products more sustainable. However, critical aspects have to be considered to establish a product portfolio of natural dyes as well as dyed products. Simply replacing synthetic dyes with natural colourants is insufficient, but any alternative must be competitive in its ecological profile with the best available dyeing processes used with synthetic dyes (Bechtold et al., 2003). All stages of a product life cycle have to be considered for a concept able to substitute synthetic colourants in selected fields of application:

- Plant sources: The direct farming of dye plants should be kept to a minimum because competition with food production must be avoided. Inputs for fertiliser, crop seeding, growing and harvesting contribute to the overall balance of resources of the dyes. The use of byproducts from food (e.g. pressed berries, onion peels) and wood processing (e.g. bark extracted for tannins) is preferable because this adds value to process streams (Bechtold et al., 2007; Fitz-Binder and Bechtold, 2019).
- The extraction of natural colourants from plant material has to be performed with a strict consideration of the resources used, because a low content of dye in the plant will lead to a substantial amount of extracted residues. Thus, the use of solvents and processes with a high energy demand (e.g. evaporation of extract) has to be critically considered (Thomas Bechtold et al., 2008). The addition of chemicals during extraction leads to a high amount of chemically contaminated plant residue, which hinders their use for animal feed or composting.
- The application of dyes must be performed with strict control of all chemicals and additives, including mordanting procedures and the minimised consumption of water and energy. (Manian et al., 2016) Comparisons of results from natural dyeing should be made with the best available synthetic dye processes, to highlight achievements with regard to resource consumption, $CO_2$ and the water footprint.

For the successful introduction of natural colourants into commercial production, the correct positioning of the product is required. Limitations remain with regard to fastness properties (light fastness and wash fastness) and the limited colouristic range. A novel technology for natural dyeing of wool was developed, which involves dye solubilisation in oil—water microemulsions. The dyeing method resulted in good fastness properties (Paul et al., 2008). Another possible but rarely discussed strategy for widening the colouristic range of natural dyeing is the use of hybrid concepts, in which natural colourants combined with selected synthetic dyes.

In addition to their biobased origin, a major advantage of natural colourants comes from their biodegradability, which results from their sourcing. This property can be a valuable advantage for products:

- That are designed for a limited and comparably short period of use
- Made from biodegradable fibres (cellulose, wool and other biodegradable polymers)
- For which biodegradation and composting are the planned routes for disposal.

When such a product has been dyed with synthetic dyes, these dyes will be released during biodegradation of the polymer matrix. The release of biodegradable natural colourants during degradation of the fibre polymer then contributes to the overall ecological profile.

When circularity and textile recycling are intended, the fate of dye during recycling of the fibre polymer must be considered. Many times, dyeing with natural dyes results in limited wash fastness. In particular, when complexing agents are applied during washing, removal of the mordant leads to the destruction of the metal complex and release of the dye into the dye bath. This weakness of natural dyeing could also be used as a strength during fibre recycling, because lock–unlock systems could be designed to permit the simple and efficient decolourisation of fibres during recycling (Fig. 13.7).

## 13.4.7 Printing

### 13.4.7.1 Thickeners

Printing patterns and items on textiles and garments contributes to the overall design of a product. Thus, a wide range of techniques and methods are used to achieve an enormous variety of effects. A number of specialised techniques produce a fashionable or traditional look, such as three-dimensional foam print, flock print, transfer of films, silicone prints and batik colouration. After a short period of wider application for

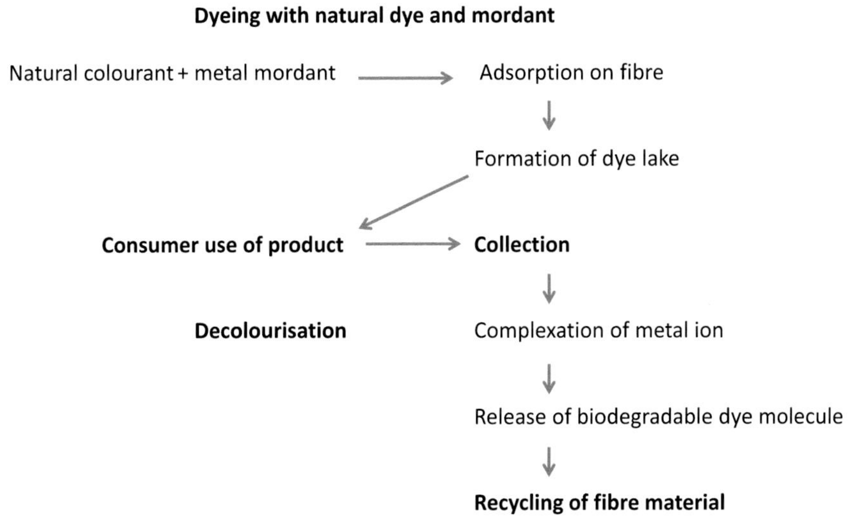

**Dyeing with natural dye and mordant**

Natural colourant + metal mordant → Adsorption on fibre

↓

Formation of dye lake

**Consumer use of product** → **Collection**

↓

**Decolourisation** Complexation of metal ion

↓

Release of biodegradable dye molecule

↓

**Recycling of fibre material**

**Figure 13.7** Lock–unlock systems with mordant dyeing for improved fibre decolourisation of collected postconsumer waste.

fashionable products, many techniques disappear or remain as low-volume products for niche markets. The short life of use for many techniques makes optimisation with regard to ecological aspects more difficult. Thus, in this chapter sustainability aspects are discussed for major printing techniques (screen printing and inkjet printing) and strategies for the technical design of specialised prints.

An assessment of the environmental profile of screen printing and inkjet printing for printing on cotton textiles led to remarkable results (Kujanpää and Nors, 2014). The cradle-to-gate greenhouse gas emissions for the production of 750 $m^2$ fabric as a functional unit were estimated with 390 kg $CO_2$eq for inkjet-printed cotton fabric, which is substantially low compared with screen printing (648 kg $CO_2$eq). A difference in processing between the techniques arises from fabric pretreatment for inkjet printing, which accounts for 23% of total kg $CO_2$eq, whereas for screen printing the test accounts for 22% of total kg $CO_2$eq for the process. However, when the production of cotton is included in the calculation, the values increase to 4112 and 4384 kg $CO_2$eq. for inkjet and screen printing, respectively. This indicates the high impact of cotton production on the overall production of greenhouse gas in terms of kg $CO_2$eq.

The impact in terms of kg $CO_2$eq for printed textiles also depends on the lifetime of the products. When higher colour stability of screen printed fabric leads to a 10% longer lifetime of the products, the overall impact in terms of kg $CO_2$eq favours screen-printed fabric (Kujanpää and Nors, 2014). A comparison of screen printing and inkjet printing with regard to water consumption (water withdrawal, water consumption and water volume required to dilute the pollutants) indicated substantially higher water consumption using the screen printing process (Kujanpää and Nors, 2014). However, energy consumption per metre of printed fabric of inkjet printing is higher than in screen printing, because the production speed of inkjet printing is comparably low and the energy consumption of the machinery is high. In addition, a high amount of dye is required in inkjet printing to achieve a good colour effect (Chen et al., 2015).

Screen printing with natural colourants has a long tradition and may contribute to more eco-friendly printing in the future (Yildirim et al., 2020). However, the processes and chemicals used require a careful consideration to achieve progress compared with the state of the art.

#### *13.4.7.2 Screen printing*

Screen printing is an efficient process for larger production volumes. At the end of a batch, a considerable amount of paste remains in the screen and tubes. The high concentration of chemicals, dyes and thickeners in printing pastes makes the collection and disposal of pastes obligatory. The biodegradability of synthetic dyes and many thickeners is limited, and the high dye content prevents the recycling of the paste (Xu et al., 2018).

For improved sustainability of the process, both the thickeners and dyes used in the printing paste should be fully biodegradable (Hassabo et al., 2021; Saad et al., 2021). Another important approach to reducing the environmental impact of textile chemicals was presented in a study using fruit waste to extract a thickener (pectin) (Ebrahim et al., 2023).

An example of research activity to improve sustainability in pigment printing is a project at the University Innsbruck using natural pigments and suitable biopolymers in pigment printing. The use of natural colourants in textile printing has a long tradition, but the limited range of colours makes introducing natural colourants into screen printing challenging (Kavyashree, 2020). Printing of leuco-indigo on silk and cotton was investigated for the use of natural indigo in screen printing (Mongkholrattanasit et al., 2018; Mongkholrattanasit et al., 2018). Natural dyes are collected in form of dye lakes, which form the basis for dye pigments (Leitner et al., 2012; Mahmud-Ali et al., 2012). The application of a natural dye extract from *Terminbalia chebula* (C.I. Natural Brown 6), *Rubia cordifolia* (C.I. Natural Red 16) and *Curcuma tinctoria* (C.I. Natural Yellow 3) combined with metallic mordants ($Al_2(SO_4)_3$, $FeSO_4$, $CuSO_4$) and *Butea monosperma* flowers was also reported as examples of using plant-based dyes in screen printing (Patel and Kanade, 2019; Babel and Gupta, 2015). However, the metallic mordants can lead to heavy metal pollution. The use of biopolymer-based binder formulations containing only biodegradable components was investigated to circumvent the separate disposal of residual printing paste and deliver a product in which the print is biodegradable.

When a pigment binder is applied in screen printing, the lower flexibility of the textile leads to stiffer handling. A binder-free formulation of a screen printing paste uses coloured anionic nanospheres consisting of anionic poly(styrene-methyl methacrylate)-acrylic acid and absorbed cationic dyes such as Rhodamine B (Li et al., 2023).

When fully biodegradable formulations are targeted for printing pastes, the use of preservatives against microbial growth in the paste has to be considered. Unwanted microbial growth can lead to degradation of the thickeners and reduced viscosity of the paste, as well as to the anaerobic decolourisation of azo dyes. However, the presence of preserving agents in the paste affects the biodegradation of the collected residual pastes and concentrates.

#### *13.4.7.3 Digital colouration*

Requirements for inks that perform well on technically demanding printing heads requires a careful selection of all constituents. Thus, synthetic dyes are used. Two groups of inks for digital or inkjet printing can be distinguished: 'dye' inks containing, for example, acid dyes, reactive dyes, disperse dyes and 'pigment' inks, which contain dispersed dye pigments. To optimise the ecological profile of inkjet printing, several aspects have to be considered (Tawiah et al., 2016):

- Optimisation of the composition of the ink reduces the consumption of additives (surfactants, solvents, viscosity controllers and humectants). Water-based inks reduce the release of volatile organic compounds into the air.
- Adequate pretreatment of the substrate increases dye fixation.
- The selection of appropriate dye systems enables more rapid dye fixation with less energy (e.g., lower demand for steaming).

In particular, to print on PES fabric, savings in the consumption of resources and a lower environmental impact are expected benefits of an optimised ink formulation (Gao et al., 2020). More efficient printing on PES fibres was achieved by adding polyvinylpyrrolidone to the ink. Microstructuring of the aqueous ink led to high sharpness of printed contours and improved colour depth and colour fastness (Kim et al., 2021).

Pretreatment of fabric with chemicals and polymers binds the printed liquid and achieves a sharp contour of the print, also contributing to the overall environmental burden. The amount and type of polymer used to pretreat the fabric is a critical issue for two reasons. The overall amount of polymer used will contribute to the organic load released when the print is washed. Thus, the more efficient the polymer works, the lower the chemical oxygen demand is in the wash water. The polymer also influences dye fixation, and thus it is a critical factor for overall dye consumption, dye fixation and colour depth (Liang et al., 2021; Kaimouz et al., 2010).

Urea-free inkjet printing with reactive dyes has been reported using the eco-steam process for reactive dye fixation. In this process, steam with 80% relative humidity enables a high fixation yield and good fastness properties while employing a reduced amount of chemicals (Wang et al., 2017). Activation of the fibre surface through plasma treatment leads to higher fibre reactivity and thus improved dye fixation in the subsequent inkjet printing. Therefore, there are environmental benefits from a higher colour yield and less colour loss in the wastewater (Wang et al., 2022). Also, improved results on printing on silk were obtained with plasma pretreatment. The time between plasma processing and printing should be no longer than 24 h because the level of activation decreases with time (Zhang et al., 2018). Plasma pretreatment on cotton fabric increased hydrophilicity and fibre roughness, which lead to improved dye penetration into the fabric and the better capture of ink on the fabric surface (Pransilp et al., 2016).

Scientific approaches address different aspects of inkjet printing, such as the use of the more efficient synthesis of reactive dyes (Faisal and Lin, 2020). The use of natural dyes in inkjet printing was also studied (Savvidis et al., 2013). Low-molecular weight natural dyes have also been successfully used for transfer printing on PES fabric (Yousef et al., 2021).

A substantial reduction in the amount of reactive dye present in printing effluents as well as a reduction in sodium carbonate and urea consumption may be achieved by replacing conventional reactive dyes with reactive dye-copolymer nanospheres. In a study, C.I. Reactive Red 218 was adsorbed on cationic poly(styrene-butyl acrylate-vinylbenzyl trimethylammonium chloride) nanospheres, which then were used in inkjet printing (Song et al., 2020). The digital printing of denim textiles may also open a resource-saving approach to producing denim garments. The reproduction of traditional denim by printing might significantly reduce water, energy and chemical use during indigo dyeing and garment washing. However, differences in dye penetration and colour saturation between classically ring-dyed denim and inkjet-printed products still limit the applicability of the technique (Wang et al., 2021).

In the sublimation printing of PES-based textiles, the transfer of dye from printed paper to the fibre by sublimation is used. By the direct inkjet printing of sublimation dyes on plasma-pretreated substrates, the use of transfer paper may be avoided, leading to a substantial reduction in such waste. However, the possible need to wash and dry the printed fabric has to be considered (Alihoseini et al., 2021).

A field of application for inkjet printing emerged from the need for miniaturised electronic circuits. Conductive lines can be formed through electroless copper deposition on silver seed, generated by the thermal decomposition of silver citrate on inkjet-printed lines (Biermaier et al., 2022, 2023). On synthetic fibres (i.e. PA), improved adhesion of the electroless deposited Cu-layer was achieved by plasma pretreatment or Ca complexation-based surface modification (Gleissner et al., 2022). In another approach, conductive lines were formed by the decomposition of inkjet-printed metal-organic layers (Kim et al., 2021).

## 13.5 Environmental awareness and future perspectives

The need to reduce the environmental impact of textiles has initiated a thorough reconsideration of all processes, techniques, chemicals, colourants and materials currently used. Optimisation of inputs in textile and garment production with regard to the biodegradability of outputs is insufficient to develop appropriate solutions for more sustainable production. Circularity requires the design of processes and products that include all stages of a product's lifetime. As a result, recycling of a polymer material will also determine the selection of dyes as well as the dyeing and printing techniques.

For synthetic fibres, reuse probably will include thermoplastic reshaping without preceding decolourisation. Efficient sorting into fractions of uniform composition with regard to batches of similar colour will lead to coloured regenerated fibres. Thus, the selection of dyes for synthetic fibres will have to consider the thermal stability of a colourant to avoid the uncontrolled decomposition of dyes during the melt extrusion of regenerates. The colour of the raw material will also define the possible range of colour that can be achieved in the next product cycle.

Cellulose-based fibres probably will be regenerated by dissolution and fibre regeneration processes using viscose fibre technology, the lyocell process or ILs. Decolourisation of collected cellulose fibres would require substantial amounts of chemicals, energy and water. Thus, the compatibility of dyes with the fibre regeneration process will be an imperative (Manian et al., 2022).

The implementation of circularity will lead to similar requirements for dye selection in textile printing. For example, pigment dye printing with binder systems may be considered critically compared with techniques based on fibre dyeing, such as disperse dyeing of PES or reactive or vat dyeing of cellulosics, because the binder polymer will be a source of complications in fibre-to-fibre recycling. Digital colouration based on high tinctorial power inks will take the lead in the future. The simultaneous incorporation of all aspects of the product's lifetime and recycling will form the biggest challenge for future textile technologists and textile chemists, who will have to expand their field of expertise from process optimisation to product design for recycling.

## 13.6 Conclusion

Textile dyeing and finishing as well as printing processes will have to be reconsidered with regard to their compatibility with requirements defined by the demand for the circularity of products. Technical equipment and machinery probably will not change dramatically; however, the dyes and chemicals that are used will need to be deeply reconsidered. Originally designed for optimum performance and durability, all chemicals used in textile dyeing and printing will be grouped into two classes: chemicals and auxiliaries released with processing baths, and the chemicals, dyes and polymers that will be part of the product. Printing techniques with a reduced consumption of chemicals, especially inkjet printing, have a high potential to take an important role in colouring products with a low consumption of chemicals and dyes.

An important aspect to increase the duration of product use will be to increase the quality of dyeing and products in terms of fastness and appearance. Whereas the optimisation of production costs has been the major focus for decades, the durability of a product will favour higher quality dyeing in the future. All chemicals released with processing waste in an ideal case should be recyclable with reasonable effort, or at least should be fully biodegradable. Arguments that a chemical or dye is removed in the communal wastewater treatment plant through adsorption in sewage sludge will be insufficient. Full mineralisation through biological degradation should be the goal.

Chemicals, dyes and polymers to be delivered with the product must fit into the polymer and fibre regeneration process, or biodegradation may be the foreseen pathway for disposal. Thus, the use of biodegradable polymers for fibre production will also require the use of biodegradable dyes and finishes. Recycling coloured material will define new specifications for dyes and pigments. Instead of the maximum performance at the lowest cost, products used in future textile dyeing operations will be selected with regard to their behaviour in planned recycling.

## Acknowledgements

The authors gratefully acknowledge financial support to the Competence Centers for Excellent Technologies (COMET) project 'Textile Competence Center Vorarlberg 2 – FFG 882502', funded within COMET by BMK and BMDW, as well as cofinancing from the Federal Province of Vorarlberg.

# References

AATCC, 2000. A glossary of AATCC standard terminology. In: AATCC Technical Manual, p. 399.

Abdelileh, M., Manian, A.P., Rhomberg, D., Ben Ticha, M., Meksi, N., Aguiló-Aguayo, N., Bechtold, T., 2020. Calcium-iron-D-gluconate complexes for the indirect cathodic reduction of indigo in denim dyeing: a greener alternative to non-regenerable chemicals. Journal of Cleaner Production 266, 121753.

Abdelileh, M., Ticha, M.B., Moussa, I., Meksi, N., 2019. Pretreatment optimization process of cotton to overcome the limits of its dyeability with indigo carmine. Chemical Industry and Chemical Engineering Quarterly 25, 277–288.

Aerts, O., Duchateau, N., Lambert, J., Bechtold, T., 2014. Sodium metabisulfite in blue jeans: an unexpected cause of textile contact dermatitis. Contact Dermatitis 70.

Alberti, C., Figueira, R., Hofmann, M., Koschke, S., Enthaler, S., 2019. Chemical recycling of end-of-life polyamide 6 via ring closing depolymerization. ChemistrySelect 4, 12638–12642.

Alihoseini, M.R., Khani, M.R., Jalili, M., Shokri, B., 2021. Direct sublimation inkjet printing as a new environmentally friendly approach for printing on polyester textiles. Progress in Color, Colorants and Coatings 14, 129–138.

Archroma, 2023a. Advanced Denim. https://www.archroma.com/solutions/coloration-denim-casual-wear. (Accessed 20 March 2023).

Archroma, 2023b. Earth Colours. https://www.archroma.com/innovations/earth-colors-by-archroma. (Accessed 23 March 2023).

Babel, S., Gupta, R., 2015. Screen printing on silk fabric using natural dye and natural thickening agent. Journal of Textile Science & Engineering 06, 1–3.

Bartsch, P., Ruef, H., 1999. Method for Producing Cellulosic Moulded Bodies. WO 99/46434.

Batt, I.P., 1961. Process for Producing Colored Pellicular Gel Structures of Regenerated Cellulose. US 3005723.

Bechtold, T., Berktold, F., Turcanu, A., 2000. The redox behaviour of CI Sulphur Black 1 - a basis for improved understanding of sulphur dyeing. Journal of the Society of Dyers and Colourists 116, 215–221.

Bechtold, T., Burtscher, E., Sejkora, G., Bobleter, O., 1985. Modern methdos of lye recovery. International Textile Bulletin Dye 31, 5–26. English Ed.

Bechtold, T., Mahmud-Ali, A., Ganglberger, E., Geissler, S., 2008. Efficient processing of raw material defines the ecological position of natural dyes in textile production. International Journal of Environment and Waste Management 2, 215–232.

Bechtold, T., Mahmud-Ali, A., Mussak, R., 2007. Anthocyanin dyes extracted from grape pomace for the purpose of textile dyeing. Journal of the Science of Food and Agriculture 87, 2589–2595.

Bechtold, T., Turcanu, A., Ganglberger, E., Geissler, S., 2003. Natural dyes in modern textile dyehouses - how to combine experiences of two centuries to meet the demands of the future? Journal of Cleaner Production 11, 499–509.

Bechtold, T., Turcanu, A., Schrott, W., 2008. Pilot-scale electrolyser for the cathodic reduction of oxidised C.I. Sulphur Black 1. Dyes and Pigments 77, 502–509.

Biermaier, C., Gleisner, C., Bechtold, T., Pham, T., 2022. Localised catalyst printing for flexible conductive lines by electroless copper deposition on textiles. In: FLEPS 2022 - IEEE

International Conference on Flexible and Printable Sensors and Systems, Proceedings, pp. 2–5.

Biermaier, C., Gleißner, C., Bechtold, T., Pham, T., 2023. The role of citrate in heterogeneous silver metal catalyst formation : a mechanistic consideration. Arabian Journal of Chemistry, 104803.

Blackburn, R.S., Bechtold, T., John, P., 2009. The development of indigo reduction methods and pre-reduced indigo products. Coloration Technology 125, 193–207.

Bockhorn, H., Donner, S., Gernsbeck, M., Hornung, A., Hornung, U., 2001. Pyrolysis of polyamide 6 under catalytic conditions and its application to reutilization of carpets. Journal of Analytical and Applied Pyrolysis 58, 79–94.

Briga-Sá, A., Nascimento, D., Teixeira, N., Pinto, J., Caldeira, F., Varum, H., Paiva, A., 2013. Textile waste as an alternative thermal insulation building material solution. Construction and Building Materials 38, 155–160.

Brinks, G.J., Bouwhuis, G.H., Agrawal, P.B., Gooijer, H., 2018. Processing of Cotton-Polyester Waste Textile - WO2014081291. WO2014081291.

Buscio, V., Crespi, M., Gutiérrez-Bouzán, C., 2015. Sustainable dyeing of denim using indigo dye recovered with polyvinylidene difluoride ultrafiltration membranes. Journal of Cleaner Production 91, 201–207.

Chen, J., Li, Z., Jin, L., Ni, P., Liu, G., He, H., Zhang, J., Dong, J., Ruan, R., 2010. Catalytic hydrothermal depolymerization of nylon 6. Journal of Material Cycles and Waste Management 12, 321–325.

Chen, L., Ding, X., Wu, X., 2015. Water Management Tool of Industrial Products: a case study of screen printing fabric and digital printing fabric. Ecological Indicators 58, 86–94.

Cheng, J., Wanogho, S.O., Watson, N.D., Caddy, B., 1991. The extraction and classification of dyes from cotton fibres using different solvent systems. Journal of the Forensic Science Society 31, 31–40.

Ciba Ltd., 1965. Dyeing of Regenerated Cellulosic Fibers and Films. FR 1417575.

Ciba Ltd., 1966. Transparent Colored Regenerated Cellulose. NL 6514672.

Ciba Ltd., 1968. Process for the Preparation of Transparent Colored Shaped Articles of Regenerated Cellulose with the Aid of Organic Dyestuffs of Low Solubility in Water. GB 1128158.

Circular Fibers Initiative, 2017. A New Textiles Economy: Redesigning Fashion's Future. Ellen MacArthur Found. http://www.ellenmacarthurfoundation.org/publications.

Correia, J., Oliveira, F.R., de Cássia Siqueira Curto Valle, R., Valle, J.A.B., 2021. Preparation of cationic cotton through reaction with different polyelectrolytes. Cellulose 28, 11679–11700.

Cotton Incorporated, 2009. A World of Ideas - Technologies for Sustainable Cotton Textile Manufacturing. Cotton Incorporated, Cary, USA.

Cristóvão, R.O., Tavares, A.P.M., Ferreira, L.A., Loureiro, J.M., Boaventura, R.A.R., Macedo, E.A., 2009. Modeling the discoloration of a mixture of reactive textile dyes by commercial laccase. Bioresource Technology 100, 1094–1099.

Dahlbo, H., Aalto, K., Eskelinen, H., Salmenperä, H., 2017. Increasing textile circulation—consequences and requirements. Sustainable Production and Consumption 9.

Datta, J., Błażek, K., Włoch, M., Bukowski, R., 2018. A new approach to chemical recycling of polyamide 6.6 and synthesis of polyurethanes with recovered intermediates. Journal of Polymers and the Environment 26, 4415–4429.

de Oliveira Neto, G.C., Correia, J.M.F., Silva, P.C., de Oliveira Sanches, A.G., Lucato, W.C., 2019. Cleaner production in the textile industry and its relationship to sustainable

development goals. Journal of cleaner production 228, 1514–1525. https://doi.org/10.1016/J.JCLEPRO.2019.04.334.
De Silva, R., Wang, X., Byrne, N., 2014. Recycling textiles: the use of ionic liquids in the separation of cotton polyester blends. RSC Advances 4, 29094–29098.
Deo, H.T., Paul, R., 2000. Dyeing of ecru denim with onion extract using natural mordant combinations. Indian Journal of Fibre & Textile Research 25 (2), 152.
Deo, H.T., Paul, R., 2004a. Natural dyeing of denim with eco-friendly mordant. International Textile Bullein 50 (5), 66.
Deo, H.T., Paul, R., 2004b. Standing bath technique for indigo denim dyeing. International Textile Bulletin 50 (2), 66.
Deo, H.T., Paul, R., 2004c. Reuse of indigo vats for shade development. International Dyer 189 (2), 14.
Dosne, H., 1936. Colored Cellulose Material. US 2041907.
Drumond Chequer, F.M., de Oliveira, G.A.R., Anastacio Ferraz, E.R., Carvalho, J., Boldrin Zanoni, M.V., de Oliveir, D.P., 2013. Textile dyes: dyeing process and environmental impact. In: Eco-Friendly Textile Dyeing and Finishing. InTech.
Du, S., Valla, J.A., Parnas, R.S., Bollas, G.M., 2016. Conversion of polyethylene terephthalate based waste carpet to benzene-rich oils through thermal, catalytic, and catalytic steam pyrolysis. ACS Sustainable Chemistry & Engineering 4, 2852–2860.
Ebrahim, S.A., Othman, H.A., Mosaad, M.M., Hassabo, A.G., 2023. Eco-friendly natural thickener (pectin) extracted from fruit peels for valuable utilization in textile printing as a thickening agent. Textiles 3, 26–49.
Esteve-Turrillas, F.A., de la Guardia, M., 2017. Environmental impact of Recover cotton in textile industry. Resources, Conservation and Recycling 116.
European Commission, 2022. EU Strategy for Sustainable and Circular Textiles. COM(2022), 141(final), 30.33.2022.
European Environment Agency, 2019. Textiles in Europe's Circular Economy.
Faisal, S., Lin, L., 2020. Green synthesis of reactive dye for ink-jet printing. Coloration Technology 136, 110–119.
Fitz-Binder, C., Bechtold, T., 2019. Extraction of polyphenolic substances from bark as natural colorants for wool dyeing. Coloration Technology 135, 32–39.
Gao, C., Zhang, Z., Xing, T., Hou, X., Chen, G., 2020. Controlling the micro-structure of disperse water-based inks for ink-jet printing. Journal of Molecular Liquids 297, 111783.
Genta, M., Iwaya, T., Sasaki, M., Goto, M., Hirose, T., 2005. Depolymerization mechanism of poly(ethylene terephthalate) in supercritical methanol. Industrial & Engineering Chemistry Research 44, 3894–3900.
Gleissner, C., Biermaier, C., Bechtold, T., Pham, T., 2022. Complexation-mediated surface modification of polyamide-66 textile to enhance electroless copper deposition. Materials Chemistry and Physics 288, 126383.
Groves, E., Palenik, C.S., Palenik, S., 2016. A survey of extraction solvents in the forensic analysis of textile dyes. Forensic Science International 268, 139–144.
Haslinger, S., Hummel, M., Anghelescu-Hakala, A., Määttänen, M., Sixta, H., 2019a. Upcycling of cotton polyester blended textile waste to new man-made cellulose fibers. Waste Management 97, 88–96.
Haslinger, S., Wang, Y., Rissanen, M., Lossa, M.B., Tanttu, M., Ilen, E., Määttänen, M., Harlin, A., Hummel, M., Sixta, H., 2019b. Recycling of vat and reactive dyed textile waste to new colored man-made cellulose fibers. Green Chemistry 21.
Hassabo, A., Othman, H., Ebrahim, S., 2021. Natural thickener in textile printing (A Mini Review). Journal of Textiles, Coloration and Polymer Science 18, 55–64.

Hocker, S., Rhudy, A.K., Ginsburg, G., Kranbuehl, D.E., 2014. Polyamide hydrolysis accelerated by small weak organic acids. Polymer 55, 5057–5064.

I. G. Farbenindustrie A.G., 1936. Process for the Manufacture of Dyed Filaments and Films. GB 448447.

I. G. Farbenindustrie A.G., 1937. Process for the Manufacture of Dyed Filaments and Films. GB 465606.

Imran, M.A., Ali, A., Ashfaq, M., Hassan, S., Culas, R., Ma, C., 2019. Impact of climate smart agriculture (CSA) through sustainable irrigation management on resource use efficiency: a sustainable production alternative for cotton. Land Use Policy 88, 104113.

Kaimouz, A.W., Wardman, R.H., Christie, R.M., 2010. The inkjet printing process for Lyocell and cotton fibres. Part 1: the significance of ink-jet chemicals and their relationship with colour strength, absorbed dye fixation and ink penetration. Dyes and Pigments 84, 79–87.

Kamimura, A., Ikeda, K., Suzuki, S., Kato, K., Akinari, Y., Sugimoto, T., Kashiwagi, K., Kaiso, K., Matsumoto, H., Yoshimoto, M., 2014. Efficient conversion of polyamides to ω-hydroxyalkanoic acids: a new method for chemical recycling of waste plastics. ChemSusChem 7, 2473–2477.

Kamimura, A., Shiramatsu, Y., Kawamoto, T., 2019. Depolymerization of polyamide 6 in hydrophilic ionic liquids. Green Energy & Environment 4, 166–170.

Kavyashree, M., 2020. Printing of textiles using natural dyes: a global sustainable approach. In: Samanta, A.K., Awwad, N., Algarni, H.M. (Eds.), Chemistry and Technology of Natural and Synthetic Dyes and Pigments. IntechOpen, Londonn, pp. 195–203.

Kenny, S.T., Runic, J.N., Kaminsky, W., Woods, T., Babu, R.P., Keely, C.M., Blau, W., O'Connor, K.E., 2008. Up-cycling of PET (polyethylene terephthalate) to the biodegradable plastic PHA (polyhydroxyalkanoate). Environmental Science & Technology 42, 7696–7701.

Khan, M.K.R., Jintun, S., 2021. Sustainability issues of various denim washing methods. Textile and Leather Review 4, 96–110.

Kim, I., Ju, B., Zhou, Y., Li, B.M., Jur, J.S., 2021. Microstructures in all-inkjet-printed textile capacitors with bilayer interfaces of polymer dielectrics and metal-organic decomposition silver electrodes. ACS Applied Materials & Interfaces 13, 24081–24094.

Kline, H.B., Helm, E.B., 1939. Manufacture of Artificial Silk. US 2143883.

Kothari, V.K., 2008. 14 - polyester and polyamide fibres – apparel applications. In: Deopura, B.L., Alagirusamy, R., Joshi, M., Gupta, B. (Eds.), Polyesters and Polyamides. Woodhead Publishing, pp. 419–440.

Kujanpää, M., Nors, M., 2014. Environmental Performance of Future Digital Textile Printing, pp. 1–30.

Leal Filho, W., Ellams, D., Han, S., Tyler, D., Boiten, V.J., Paço, A., Moora, H., Balogun, A.-L., 2019. A review of the socio-economic advantages of textile recycling. Journal of Cleaner Production 218, 10–20.

Leitner, P., Fitz-Binder, C., Mahmud-Ali, A., Bechtold, T., 2012. Production of a concentrated natural dye from Canadian Goldenrod (Solidago canadensis) extracts. Dyes and Pigments 93, 1416–1421.

Li, B., Wang, J., Luo, Z., Wang, J., Cai, Z., Ge, F., 2023. Colloids and Surfaces A: Physicochemical and Engineering Aspects Facile and Binder-free Fabrication of Deep Colors on Cotton Fabrics with Hand-Feel Enhancement via Screen Printing, p. 665.

Li, M., Lu, J., Li, X., Ge, M., Li, Y., 2020. Removal of disperse dye from alcoholysis products of waste PET fabrics by nitric acid-modified activated carbon as an adsorbent: kinetic and thermodynamic studies. Textile Research Journal 90, 2058–2069.

Liang, Y., Liu, X., Fang, K., An, F., Li, C., Liu, H., Qiao, X., Zhang, S., 2021. Construction of new surface on linen fabric by hydroxyethyl cellulose for improving inkjet printing performance of reactive dyes. Progress in Organic Coatings 154, 106179.
Lockhart, G.R., 1932. Manufacture of Rayon. US 1865701.
Lutgerhorst, A.G., 1956. Spundyed Rayon. US 2738252.
McGregor, L., 2015. Are Closed Loop Textiles the Future of Fashion?.
Mahmud-Ali, A., Fitz-Binder, C., Bechtold, T., 2012. Aluminium based dye lakes from plant extracts for textile coloration. Dyes and Pigments 94, 533–540.
Maloney, M.A., 1967. Mass Coloring of Regenerated Cellulose with Vat Dyes. DE 1253864.
Manian, A.P., Müller, S., Braun, D.E., Pham, T., Bechtold, T., 2022. Dope dyeing of regenerated cellulose fibres with leucoindigo as base for circularity of denim. Polymers 14.
Manian, A.P., Paul, R., Bechtold, T., 2016. Metal mordanting in dyeing with natural colourants. Coloration Technology 132, 107–113.
Manian, A.P., Ruef, H., Bechtold, T., 2007. Spun-dyed lyocell. Dyes and Pigments 74.
Maryan, A.S., Montazer, M., Harifi, T., 2013. One step synthesis of silver nanoparticles and discoloration of blue cotton denim garment in alkali media. Journal of Polymer Research 20, 189.
Mongkholrattanasit, R., Klaichoi, C., Sasithorn, N., Changmuang, W., Manarungwit, K., Maha-In, K., Ruenma, P., Boonkerd, N., Sangaphat, N., Pangsai, M., 2018. Screen printing on silk fabric using natural indigo. Vlakna a Textil 25, 51–56.
Mu, B., Yang, Y., 2022. Complete separation of colorants from polymeric materials for cost-effective recycling of waste textiles. Chemical Engineering Journal 427, 131570.
Negulescu, B.Z.I., Kwon, H., Collier, B.J., Collier, J.R., Pendse, A., 1998. Recycling cotton from cotton/polyester fabrics. Textile Chemist and Colorist 30, 31–35.
Nielsen, A.M., 2012. Combined denim washing process: save time, energy and water without sacrificing quality. International Dyer 197, 16–18.
Nørup, N., Pihl, K., Damgaard, A., Scheutz, C., 2019. Evaluation of a European textile sorting centre: material flow analysis and life cycle inventory. Resources, Conservation and Recycling 143, 310–319. https://doi.org/10.1016/J.RESCONREC.2019.01.010.
Österreich, R., 2000. BGBL. 215, Begrenzung von Abwasseremissionen aus der Herstellung von Textil-, Leder- und Papierhilfsmitteln. Abwasseremissionsverordnung.
Palme, A., Idström, A., Nordstierna, L., Brelid, H., 2014. Chemical and ultrastructural changes in cotton cellulose induced by laundering and textile use. Cellulose 21, 4681–4691.
Paszun, D., Spychaj, T., 1997. Chemical recycling of poly(ethylene terephthalate). Industrial & Engineering Chemistry Research 36, 1373–1383.
Patel, B., Kanade, P., 2019. Sustainable dyeing and printing with natural colours vis-à-vis preparation of hygienic viscose rayon fabric. Sustainable Materials and Technologies 22, e00116.
Patil, D.B., Madhamshettiwar, S.V., 2014. Kinetics and thermodynamic studies of depolymerization of nylon waste by hydrolysis reaction. Journal of Applied Chemistry 2014, 1–8.
Paul, R. (Ed.), 2015. Denim: Manufacture, Finishing and Applications. Woodhead Publishing Ltd (Elsevier), Cambridge, UK. ISBN: 9780857098436.
Paul, R., Blackburn, R.S., Bechtold, T., 2021. Indigo and indigo colorants. Ullmann's Encyclopedia of Industrial Chemistry 1–16.
Paul, R., Solans, C., Erra, P., 2008. Mechanism involved in the dyeing of wool with an oil-in-water microemulsion system. Journal of Applied Polymer Science 110, 156.
Phrix-Werke, A.-G., 1965. Spun-dyed Regenerated Cellulose. NL 6407087.

Pransilp, P., Pruettiphap, M., Bhanthumnavin, W., Paosawatyanyong, B., Kiatkamjornwong, S., 2016. Surface modification of cotton fabrics by gas plasmas for color strength and adhesion by inkjet ink printing. Applied Surface Science 364, 208–220.

Pruś, S., Kulpiński, P., Matyjas-Zgondek, E., Wojciechowski, K., 2022. Eco-friendly dyeing of cationised cotton with reactive dyes: mechanism of bonding reactive dyes with CHPTAC cationised cellulose. Cellulose 29 (7), 4167–4182.

Qian, H.-F., Zhao, X.-L., Dai, Y., Huang, W., 2017. Visualized fabric discoloration of bi-heterocyclic hydrazone dyes. Dyes and Pigments 143, 223–231.

Riehen, W.M., Reinach, F.S., 1971a. Sparingly Soluble Organic Dyestuffs. US 3620788.

Riehen, W.M., Reinach, F.S., 1971b. Difficultly Soluble Organic Dye Compositions for Dyeing Transparent, Shaped, Regenerated Cellulose Bodies. DE 1806199.

Rietzler, B., Bechtold, T., Pham, T., 2018a. Controlled surface modification of polyamide 6.6 fibres using CaCl2/H2O/EtOH solutions. Polymers 10, 207.

Rietzler, B., Bechtold, T., Pham, T., 2019. Spatial structure investigation of porous shell layer formed by swelling of PA66 fibers in CaCl2/H2O/EtOH mixtures. Langmuir 35, 4902–4908.

Rietzler, B., Manian, A.P., Rhomberg, D., Bechtold, T., Pham, T., 2021. Investigation of the decomplexation of polyamide/CaCl2 complex toward a green, nondestructive recovery of polyamide from textile waste. Journal of Applied Polymer Science 138, 51170.

Rietzler, B., Pham, T., Bechtold, T., 2018b. Polyamide Fibre - WO2018172453. WO20 18172453.

Riisberg, V., 2007. Digital tools and textile printing. In: Paper presented at Dressing Rooms: Current Perspectives on Fashion and Textiles, Oslo, Norway. https://adk.elsevierpure.com/ws/portalfiles/portal/106922/Digital_Tools_and_Textile_Printing.pdf. Accessed 29 January 2024.

Roy Choudhury, A.K., 2011. Dyeing of synthetic fibres. In: Handbook of Textile and Industrial Dyeing. Elsevier, pp. 40–128.

Roy Choudhury, A.K., 2014. Coloration of cationized cellulosic fibers— a review. AATCC Journal of Research 1, 11–19.

Ruesch, R., Schmidt, H., 1936. Preparation of Dyed Filaments and Films. US 2043069.

Russ, C., 2022. Zwei-Bad Kontinuefärbeverfahren für Kurzmetragen mit Reaktivfarbstoffen. Textilveredlung 19–22.

Saad, F., Mohamed, A.L., Mosaad, M., Othman, H.A., Hassabo, A.G., 2021. Enhancing the rheological properties of aloe vera polysaccharide gel for use as an eco-friendly thickening agent in textile printing paste. Carbohydrate Polymer Technologies and Applications 2, 100132.

Saikhao, L., Setthayanond, J., Karpkird, T., Bechtold, T., Suwanruji, P., 2018. Green reducing agents for indigo dyeing on cotton fabrics. Journal of Cleaner Production 197, 106–113.

Savvidis, G., Zarkogianni, M., Karanikas, E., Lazaridis, N., Nikolaidis, N., Tsatsaroni, E., 2013. Digital and conventional printing and dyeing with the natural dye annatto: optimisation and standardisation processes to meet future demands. Coloration Technology 129, 55–63.

Schmutz, M., Hischier, R., Som, C., 2021. Factors allowing users to influence the environmental performance of their t-shirt. Sustainable Times 13, 1–16.

Setthayanond, J., Netzer, F., Seemork, K., Suwanruji, P., Bechtold, T., Pham, T., Manian, A.P., 2023. Low - level cationisation of cotton opens a chemical saving route to salt free reactive dyeing. Cellulose 30 (7), 4697–4711.

Shen, L., 2011. Bio-based and Recycled Polymers for Cleaner Production an Assessment of Plastics and Fibres. Utrecht University.

Soc. pour l'ind. chim. a Bale, 1941. Pigment-containing Spinning Masses. CH 212386.

Song, Y., Fang, K., Bukhari, M.N., Ren, Y., Zhang, K., Tang, Z., 2020. Green and efficient inkjet printing of cotton fabrics using reactive Dye@Copolymer nanospheres. ACS Applied Materials & Interfaces 12, 45281–45295.

Sutlović, A., Glogar, M.I., Čorak, I., Tarbuk, A., 2021. Trichromatic vat dyeing of cationized cotton. Materials 14, 1–17.

Tawiah, B., Howard, E.K., Asinyo, B.K., 2016. The chemistry of inkjet inks for digital textile printing. BEST 4, 61–78.

Textile Exchange, 2019. Preferred Fiber and Materials Market Report 2019.

Thiounn, T., Smith, R.C., 2020. Advances and approaches for chemical recycling of plastic waste. Journal of Polymer Science 58, 1347–1364.

Tournier, V., Topham, C.M., Gilles, A., David, B., Folgoas, C., Moya-Leclair, E., Kamionka, E., Desrousseaux, M.-L., Texier, H., Gavalda, S., Cot, M., Guémard, E., Dalibey, M., Nomme, J., Cioci, G., Barbe, S., Chateau, M., André, I., Duquesne, S., Marty, A., 2020. An engineered PET depolymerase to break down and recycle plastic bottles. Nature 580, 216–219.

Velden, N.M. Van Der, Patel, M.K., Vogtländer, J.G., 2014. LCA benchmarking study on textiles made of cotton , polyester , nylon , acryl , or elastane. International Journal of Life Cycle Assessment 19, 331–356.

Vonbrül, L., Cordin, M., Bechtold, T., Pham, T., 2022. Method for Separating Polyurethane from a Textile, p. 7. EP22198684.

Wambuguh, D., Chianelli, R.R., 2008. Indigo dye waste recovery from blue denim textile effluent: a by-product synergy approach. New Journal of Chemistry 32, 2189–2194.

Wang, L., Yan, K., Hu, C., 2017. Cleaner production of inkjet printed cotton fabrics using a urea-free ecosteam process. Journal of Cleaner Production 143, 1215–1220.

Wang, M., Parrillo-Chapman, L., Rothenberg, L., Liu, Y., Liu, J., 2021. Digital textile ink-jet printing innovation: development and evaluation of digital denim technology. International Conference on Digital Printing Technologies 53–64. 2021-Octob.

Wang, M., Shi, F., Zhao, H., Sun, F., Fang, K., Miao, D., Zhao, Z., Xie, R., Chen, W., 2022. The enhancement of wool reactive dyes ink-jet printing through air plasma pretreatment. Journal of Cleaner Production 362, 132333.

Wang, Y., 2010. Fiber and textile waste utilization. Waste and Biomass Valorization 1, 135–143.

Wedin, H., Lopes, M., Sixta, H., Hummel, M., 2019. Evaluation of post-consumer cellulosic textile waste for chemical recycling based on cellulose degree of polymerization and molar mass distribution. Textile Research Journal 89, 5067–5075.

Wegmann, J., Booker, C., 1966. Colored Viscose Dope. DE 1220964.

Wesołowski, K.P., 2016. The polyamide market. Fibres Text. East. Europe 24, 12–18.

Wolford, 2022. C2C. Wolford - Cradle to CradleTM Certified Collection. https://company.wolford.com/de/nachhaltigkeit/cradle-to-cradle-certified-collection/. (Accessed 29 April 2022).

Xu, H., Yang, B., Liu, Y., Li, F., Shen, C., Ma, C., Tian, Q., Song, X., Sand, W., 2018. Recent advances in anaerobic biological processes for textile printing and dyeing wastewater treatment: a mini-review. World Journal of Microbiology and Biotechnology 34, 1–9.

Yang, Y., Plischke, L.W., Mclellan, G.R., Dickerson, J.L., 2001. Process for Separating Polyamide from Colorant. EP1085043A1.

Yildirim, F.F., Yavas, A., Avinc, O., 2020. Printing with sustainable natural dyes and pigments. In: Muthu, S.S., Gardetti, M.A. (Eds.), Sustainability in the Textile and Apparel Industries : Production Process Sustainability. Springer International Publishing, Cham, pp. 1–35.

Yousef, M., Othman, H., Hassabo, A., 2021. A recent study for printing polyester fabric with different techniques. Journal of Textiles, Coloration and Polymer Science 18, 247–252.

Zhang, C., Wang, L., Yu, M., Qu, L., Men, Y., Zhang, X., 2018. Surface processing and ageing behavior of silk fabrics treated with atmospheric-pressure plasma for pigment-based ink-jet printing. Applied Surface Science 434, 198–203.

Zhang, S., Ma, W., Ju, B., Dang, N., Zhang, M., Wu, S., Yang, J., 2005. Continuous dyeing of cationised cotton with reactive dyes. Coloration Technology 121, 183–186.

Zhang, T., Zhang, S., Qian, W., He, J., Dong, X., 2021. Reactive dyeing of cationized cotton fabric: the effect of cationization level. ACS Sustainable Chemistry & Engineering 9, 12355–12364.

Zhao, X., Zhan, L., Xie, B., Gao, B., 2018. Products derived from waste plastics (PC, HIPS, ABS, PP and PA6) via hydrothermal treatment: characterization and potential applications. Chemosphere 207, 742–752.

# Functional finishing and smart coating

*Nuno Belino* [1], *Roshan Paul* [2], *Prakash Pardeshi* [3] *and Rakesh Seth* [4]
[1]Universidade da Beira Interior, Covilhã, Portugal; [2]Institut für Textiltechnik der RWTH Aachen University, Aachen, Germany; [3]Associated Chemical Corporation, Mumbai, Maharashtra, India; [4]Om Tex Chem Pvt. Ltd., Thane West, Maharashtra, India

## 14.1 Introduction

The textile industry is constantly striving for innovative production techniques to improve product quality, and it is also important that these functional products are developed in an environmentally friendly way. Besides the traditional role of dressing people, textiles are capable of providing wear comfort and protection in dangerous environments. The most important requirements for protective wear are barrier effectiveness and thermo-physiological comfort for the wearer. Textile finishing consists of mechanical, chemical and biotechnological processes, which are used dependent on the types and end uses of the textile fabric. The functional finishes can be applied to apparel fabrics, household textiles and technical textiles to increase their appeal to the consumer and to stimulate growth in niche markets.

In the textile industry, finishing is usually carried out in the final stage of textile processing, as a result of which they gain several functional characteristics. A wide variety of finishing chemicals are available in the market that meet or exceed the expectations of consumers. Novel finishes providing high value addition to apparel fabrics are greatly appreciated by a more demanding consumer market. It is widely perceived that the end uses for technical textiles will continue to increase every year and the modification of commodity fibre and fabric properties by innovative finishes can be a cheaper route to high performance than using a high-cost fibre with inherently built-in performance properties.

Apart from the conventional finishing chemicals, a wide variety of sustainable and bio-based chemicals are being developed that can provide functional properties to textiles. The future trend in functional finishing is to develop multifunctional textiles, which are highly efficient, durable, cost-effective and manufactured in an environmentally sustainable manner. Smart coatings based on bifunctional anchor peptides are offering a sustainable and smart alternative to develop functional coatings on textiles.

**Sustainable Innovations in the Textile Industry. https://doi.org/10.1016/B978-0-323-90392-9.00014-8**

## 14.2 Historical aspects and current scenario

Finishing is an important step in the textile manufacturing chain, which aims to convert textile fabric into a final product that can be used as technical textiles or fashion apparel. Both mechanical and chemical finishes are classical finishes and these two types are, in fact, complementary. Mechanical finishing involves the application of physical principles such as friction, tension, temperature, pressure, etc. The chemical finishing is imparted by means of chemicals of different origins and a textile can receive new properties otherwise impossible to obtain with mechanical means. Industrial biotechnology has already made inroads into textile processing and it shows that social, environmental and economic benefits go hand-in-hand with applications of this technology in textiles.

Due to globalisation and free trade, the textile and apparel market has now become consumer-driven, and new demands for multiple functionalities are ever increasing. In this highly competitive scenario, manufacturers are trying to offer high-quality functional textile materials at a cost-effective price. The functional textiles have opened a plethora of applications in diverse fields and the utilisation of these functional textiles with enhanced properties is expanding rapidly.

The textile manufacturing processes, particularly wet processing, including finishing, have already created severe environmental damage. Huge quantity of chemicals, water and energy are consumed in the finishing process, and also wasted as solid waste and effluents. Consequently, a number of regulations and initiatives have been introduced in the textile industry to enhance the environment-friendly nature of textile processing. Enlightened consumers are also demanding a sustainable transformation of the textile and fashion industry. This has led to a sustainable trend in the textile manufacturing for developing environmentally friendly textile products that use less energy and water and fewer chemicals. Functional textile finishes are expected to provide the desired performance with minimum pollution.

## 14.3 Sustainable innovations in functional finishes

### *14.3.1 Ballistic and stab protection finishes*

Different types of threats to civilian and military life producing personal injuries or death are mainly stabbing, shooting, cutting, shrapnel, etc. The main textiles for protection against weapons and other impacts are body armours. The incorporation of rigid materials such as foils, metal, ceramic or composite plates, makes them heavy and inflexible and, therefore, uncomfortable to the wearer (Bautista, 2014; Egres et al., 2004). The most important factors related to ballistic fabric properties are the technical qualities of the fibre, type of fabric, thickness, structure, strength and strain.

Several chemical finishes and fibre additives are currently applied to complement ballistic and stab resistance of multilayered protective textiles, mainly based on Kevlar, and/or to confer extra properties onto them (Bautista, 2014). Shear thickening fluids

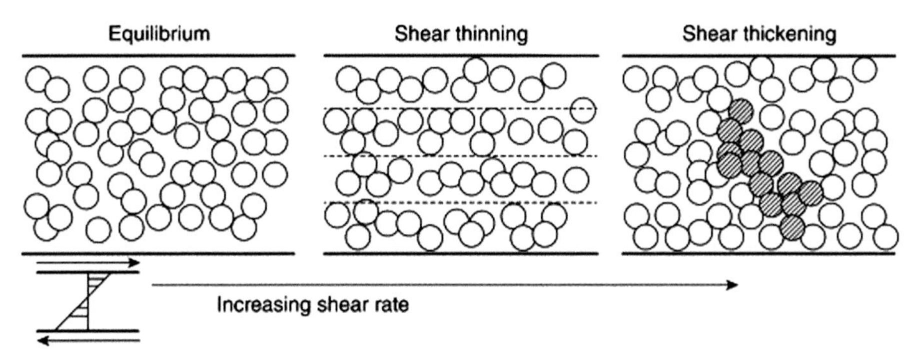


**Figure 14.1** Schematic representation of shear thickening fluids.
Based on Wagner and Brady (2009). Copyright 2009, American Institute of Physics.

are the most important and effective finishing chemical applied to achieve ballistic and stab resistance on protective textiles (Srivastava et al., 2012). Fig. 14.1 shows a schematic representation of shear thickening fluids.

Finishing with shear thickening fluids can lead to excellent puncture and cut protection, and also at a reduced cost, leading to lightweight and effective stab protective textiles (Smith, 1999). A shear thickening fluid was made by dispersing silica nanoparticles in polyethylene glycol and the ballistic performance of impregnated Kevlar fabric was studied. The impact resistance performance of the fabric was significantly enhanced. The design strategy for optimising weight and ballistic performance of soft body armour reinforced with shear thickening fluid was studied and it was found that shear thickening fluid impregnated panels were found to stop the impacting bullet earlier than a neat panel (Bajya et al., 2020). The ballistic performance of armours composed of a polyurea elastomer/Kevlar fabric composite and a shear thickening fluid structure was studied. It was found that a high-strength composite laminate using the best polyurea/Kevlar plates combined with the shear thickening fluid structure was significantly lighter and thinner than the conventional Kevlar laminate (Chang et al., 2021).

It is important to separate ballistic impacts from stab threats. Bullet impact takes place at a very high speed (300−1000 m/s), while stabbing takes place at a relatively low speed. Energy absorption required for textiles as protection from knife threats is generally higher than that required for puncture threats (Egres et al., 2004). Some other alternatives to shear thickening fluids include ceramic or metallic coatings applied onto textiles by thermal spray and silicon-based dilatant powders applied by coating or powder spray guns. Other fibre additives and functional finishes are also applied to protective textiles to complement ballistic and stab resistance and to confer multifunctional properties (Bautista, 2014).

Moreover, natural fibre-reinforced composites can be considered for the second layer of a multilayered body armour system owing to their good performance associated with other advantages such as being lighter, cheaper and environmentally friendly. Several such natural fibre composites, particularly with ceramic coatings, contribute towards their sustainability as their ballistic performance is comparable to that of aramid when used in body armour. Ramie fibre was reported to be efficient in ballistic applications as a bulletproof panel which can withstand ballistic impacts (Marsyahyo et al., 2009). Composites reinforced with curaua fibres are potentially able to replace synthetic fibre fabrics as a second layer in multilayered armour systems for personal protection (de Oliveira Braga et al., 2017).

### 14.3.2 Chemical and biological protection finishes

Chemical threat agents are compounds or substances that can be produced naturally or synthetically. These agents have the ability to cause significant morbidity or mortality in humans as well as other organisms. However, chemical weapons are designed or modified to maximise exposure through a range of delivery methodologies. Chemical terrorism is the deliberate use of chemical threats or weapon agents for terrorism (Turaga et al., 2014).

A range of multifunctionalities can be imparted to textile products as a means of offering enhanced levels of protection from chemical agents (Schreuder-Gibson et al., 2003). The growing concern over exposure to chemical threats and weapons has led researchers to develop effective countermeasures to deliver maximum protection for initial responders, military personnel and civilians (Gugliuzza and Drioli, 2013). The US Naval Research Laboratory (UNRL) has developed a chemical gradient on graphene to attract or repel water or nerve agent simulants. Chemical gradients created in the control set-up either repel or absorb dimethyl-methylphosphonate, depending upon the gas used in the process. It has been reported that the chemical gradient created on graphene substrate using oxygen gas moved the nerve agent simulant towards increasing oxygen concentrations (Hernandez et al., 2013).

Biological terrorism is an emerging concern across the globe that demands significant attention and necessitates the development of biological protective finishes. It is defined as the calculated use of microorganisms or toxins produced by living organisms. Chitosan is a sustainable polycationic molecule that has been used as a protective finish because of its proven antimicrobial activity. Quaternary ammonium salts, either in solution or immobilised on to suitable substrate, are basically absorbed by the cell membranes because of their positive charge. Once inside the membrane, they interact with the lipid and protein components of the membrane. This eventually results in the leakage of cell wall components and cell death (Elshafei and Zanfaly, 2011).

A few naturally occurring enzymes have been identified for their potential to be used as biological protective finishes. Researchers also have investigated the ability of Elafin for its use in countering bioterrorism. Elafin is a protein that interrupts catastrophic organ and tissue injury by destroying the enzymes responsible for it in the events of major trauma and infection. It is also known for its antibacterial activity (Turaga et al., 2014).

Protection against both chemical warfare agents and biological threats are developed using metal-organic frameworks (MOF) as active layers. These MOF/fibre composite systems could be adapted for use under realistic operating conditions to protect civilians, military personnel and first responders from harmful pathogens and chemical warfare agents. The strategies adopted for the fabrication of MOF/fibre composites include the in situ hydrothermal growth of MOFs on fibre surfaces, dip-coating of fibres into suspensions of pre-synthesised MOF particles, and a biosynthesis method that incorporates preformed MOF particles into nanofibre networks generated via fermentation (Ma et al., 2023).

### 14.3.3 Antiviral finishes

Antiviral textiles can effectively inhibit the spread of viruses and significantly reduce the risk of cross-infection and re-infection to protect human health and safety. The COVID-19 pandemic due to SARS-CoV-2 virus has affected almost all countries around the world. In high-risk areas, disposable surgical masks are commonly used by patients, doctors and even the general public. Antiviral textiles have become an important area of protective textiles, particularly in the field of health and hygiene. During the COVID-19 outbreak, they became an essential part of daily life. A representation of the COVID-19-causing virus, SARS-CoV-2, is shown in Fig. 14.2.

Carrington Textiles, UK, introduced an antiviral finish for woven textiles that could deactivate over 99% of enveloped viruses, with good wash durability. An antiviral finish is applied as a coating that is pressed and cured on the fabric at the end of the manufacturing process. This finish works as a barrier to capture viruses before they have the chance of passing through the pores of the fabric and deactivates them (Carrington Textiles, 2023). Viroff-tex, an antiviral treatment made by Klopman, Germany, is a powerful tool against COVID-19 virus. This finish is capable of protecting garments from enveloped viruses (Klopman, 2023). Viroblock NPJ03 is an intelligent technology from HeiQ, UK, that is added to the fabric during the final stage of the textile manufacturing process. Viroblock NPJ03 has been tested and found to be 99.99% effective in 30 min against SARS-CoV-2 virus (HeiQ, 2023).

In recent years, researchers have studied various sustainable antiviral materials, which can prevent the spread and reproduction of viruses by killing and reducing their attachment. Due to health and safety awareness in recent years, natural antiviral substances have received increased attention. These antiviral materials have a number of advantages, such as wide availability, safety and low toxicity. The most commonly used method is to apply natural antiviral materials and certain metal nano antiviral materials on textiles by impregnation and padding.

Commonly used natural antiviral materials can be divided into plants, animals and microbial extracts. Exopolysaccharides (EPS) synthesised by eukaryotes and

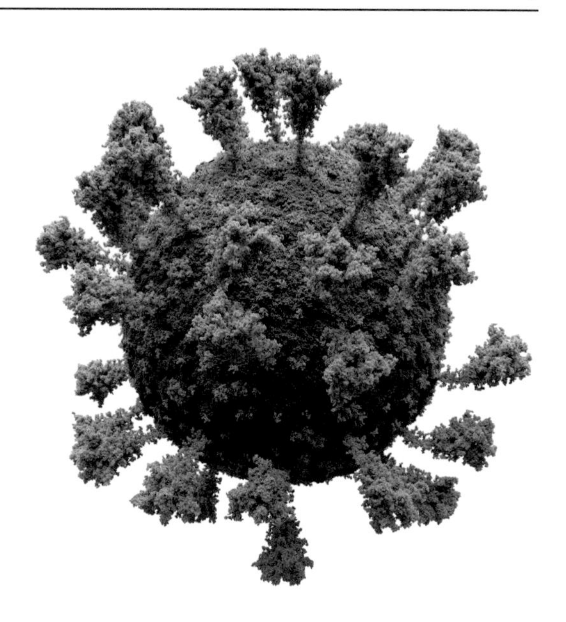

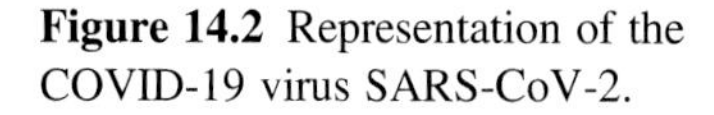

**Figure 14.2** Representation of the COVID-19 virus SARS-CoV-2.

prokaryotes, especially the sulphated polysaccharides such as dextran, have antiviral effects against enveloped viruses. They interfere with viral absorption and penetration into host cells, as well as inhibiting the reverse transcriptase activity of various retroviruses (Baba et al., 1988).

Propolis or bee glue is a resinous mixture that honey bees produce by mixing saliva and beeswax with exudate gathered from tree buds, sap flows or other botanical sources. It has a wide range of biological and pharmacological activities, as well as antibacterial and antiviral effects (Sforcin and Bankova, 2011). The mechanism of antiviral properties of propolis is related to viral inhibition to a certain extent. It can promote reverse transcriptase and virus proliferation inhibition. The antiviral activity of propolis may be related to the effect of flavonoids, which play a critical role in the antiviral process (Huleihel and Isanu, 2001).

Rather than using only the physical protection methods such as masks that create a virus barrier, the prospective of incorporating germicidal far UV-C in face masks and PPE was proposed as a sustainable method. Far UV-C can eliminate viruses in a short time before they reach inside human body. It has been established that the direct exposure of far UV-C in the 207–222 nm range is not hazardous to humans and so it is quite promising. They are safe and germicidal even at a very low intensity and a short exposure time (Khandual et al., 2021).

### 14.3.4 Antimicrobial finishes

Synthetic antimicrobial agents are very effective against a wide range of microbes and show good durability on textiles. However, the drawbacks in their application include the associated side effects, water pollution and action on non-target microorganisms. Hence, there is a demand for eco-friendly and natural antimicrobial agents. Natural products such as chitosan (El-Tahlawy et al., 2005; Zhang et al., 2003) and natural dyes (Singh et al., 2005; Gupta et al., 2004) for antimicrobial finishing are sustainable alternatives. Other herbal products such as aloe vera, tea tree oil, eucalyptus oil and azuki beans (*Vigna angularis*) can also be used for this purpose (Joshi et al., 2009; Ammayappan and Moses, 2009).

Chitosan is a natural antimicrobial agent and its activity depends on various intrinsic and extrinsic factors such as molecular weight, degree of deacetylation, pH and temperature. In addition to its antimicrobial activity, chitosan is nontoxic, biocompatible and biodegradable. Although the mechanism of antimicrobial activity of chitosan is not fully known, it is generally accepted that the primary amine groups provide positive charges, which interact with negatively charged residues on the surface of the microbes. This interaction extensively changes the cell surface and cell permeability, leading to leakage of intermolecular substances. Fig. 14.3 shows the chemical structure of chitosan.

The antimicrobial activity of several natural dyes against Gram-positive and Gram-negative bacteria species has been investigated. It was found that the antimicrobial efficacy of the dyes depends on their chemical structure, especially the functional groups present. The presence of tannins, which are naturally occurring polyphenols in the natural dyes contributed towards the antimicrobial activity. The use of natural mordants during dyeing of natural dyes increased the antimicrobial activity and the durability of the dyed samples (Gupta et al., 2004).

The antimicrobial activity of aloe vera is exploited for medical textile applications such as wound dressing, suture and bioactive textiles. The use of peroxyacids such as peroxyacetic acid as a sustainable antimicrobial finish has been reported (Dettenkofer and Block, 2005). In deactivating the microbes, peroxyacids are converted to carboxylic acid, which can be regenerated through the reaction with an oxidant such as hydrogen peroxide (Huang and Gang, 2003).

The antimicrobial characteristics of algae have been explored by many researchers (Quinto et al., 2019). The antimicrobial substances in macro-algae (*Himanthalia elongata*) and micro-algae (*Synechocystis* spp.) were explored and it was found that the extracts from both possessed antimicrobial and antioxidant activities in the case of *E. coli* and *S. aureus* (Herrero et al., 2013). Mushrooms possess various antimicrobial and antioxidant properties (Gyawali and Ibrahim, 2014). Extracts from wild

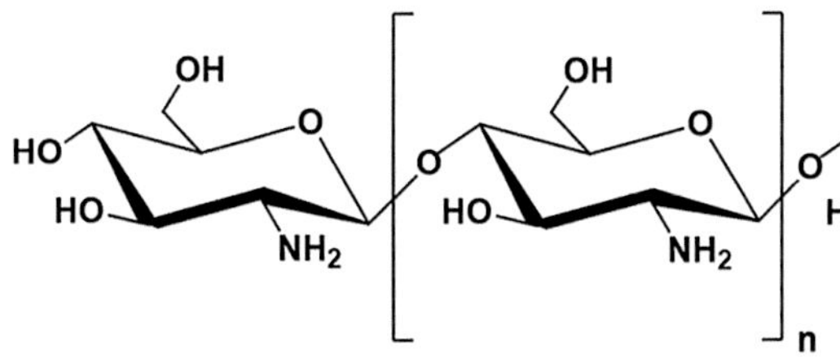

**Figure 14.3** Chemical structure of chitosan.

*Laetiporus sulphureus* fruiting bodies have been found to show antimicrobial effects against various microbial pathogens such as *C. albicans*, *C. parapsilosis*, *S. aureus*, *Enterococcus faecalis*, etc. Edible mushroom extracts from Aphyllophorales (Ramesh and Pattar, 2010), Agaricus (Öztürk et al., 2011), *Armillaria mellea*, *Paxillus involutus*, etc. also have been found to show antimicrobial activity.

### 14.3.5 Superabsorbent finishes

Superabsorbent finishes are essential tools for engineering textiles to develop materials with advanced properties, specific technical requirements and added functionalities for applications in areas such as biomedicine, hygiene, sanitation, agriculture, geotextiles and protective clothing. The basic categories of superabsorbents include hydrogels as well as composites of various polymers and inorganic materials. Hydrogels are three-dimensional polymeric or colloidal networks that are insoluble in water but can absorb large amounts of it in their structure, resulting in extensive swelling. In most cases, hydrogels have stimuli-responsive properties, i.e., they swell and shrink by absorbing and expelling water, respectively, depending on the ambient conditions of pH, temperature, light, ionic strength, magnetic field, etc. (Glampedaki et al., 2012a; Tokarev and Minko, 2009).

Single-use, disposable hygiene and sanitary products significantly contribute to plastic waste, creating environmental concerns. The technological progress and the need for alternative eco-friendly and inexpensive resources have put forward the use of natural and biodegradable polymers in superabsorbents. Polysaccharides such as starch, chitosan and guar gum are sustainable alternatives for synthetic superabsorbents (Glampedaki et al., 2012b). Superabsorbent finishes can be applied on fibre and textile surfaces using most conventional finishing techniques, which can be roughly divided into pre-textile and post-textile manufacturing.

Several fully biological superabsorbent polymers have been developed as eco-sustainable products with a low environmental impact. Biodegradable, compostable superabsorbent powder shows higher absorption and retention values and, due to its organic components, its application is safe in the food, medical, hygiene and sanitary sectors (Magicsrl, 2023). Biobased superabsorbent polymers have been designed and synthesised using biological cross-linker and backbone components to create a hydrogel system, which absorbs water into the polymeric matrix. The hydrogels are synthesised using chitosan and sodium alginate as the backbone foundation and genipin as the cross-linker, which are all commonly found in nature. Through chemical ratio alteration, including the cross-linker to backbone ratio, the superabsorbent polymers successfully absorb and retain water (Draney and Bates, 2022).

A series of hyperbranched polyether cross-linkers have been designed and employed as effective cross-linkers to create feasible, high-performance superabsorbent hydrogels that are environmentally friendly. The hydrogel materials achieved greater strength compared with traditional hydrogels by using hyperbranched polyether to create a homogeneous and strongly cross-linked polymer network. In addition, it was determined that the addition of modified sodium lignosulphonate improved several characteristics of traditional hydrogels, including the absorption and fixation of water, and that the developed hydrogels have good antibacterial properties (Junhao et al., 2023).

### 14.3.6 Flame-retardant finishes

Flame-retardant textiles are textiles or textile-based materials that can inhibit or resist the spread of fire. The use of highly efficient halogenated flame-retardant systems is being restricted due to the environmental concerns and it has given impetus to the development of novel phosphorus flame retardants. Compared to others, phosphorus-based flame retardants are found to generate less toxic gases and smoke during combustion (Lu and Hamerton, 2002). A novel durable polymeric organophosphorus flame retardant for finishing of cotton fabrics has been developed. This polymeric compound can be synthesised in a two-step procedure starting from $POCl_3$, pentaerythritol and bio-based tartaric acid (Liu et al., 2012). Nitrogen-based flame retardants are also becoming more popular as they are considered to be environment-friendly and non-toxic substitutes for halogenated ones.

A halogen-free hydroxy-functional organophosphorus oligomer (HFOP) flame-retardant agent, which is usually used in the finishing of polyurethane foams and has not been used for textile finishing, was evaluated on cotton. This flame-retardant agent does not have a functional group that can bind on cotton cellulose. Therefore, it was necessary to use cross-linking agents such as citric acid and appropriate catalysts such as $(Na_4)_2HPO_4$, which are able to react with both the flame-retardant agent and cotton, so that the system can become durable to multiple laundering cycles. The treated cotton fabric after curing in a microwave oven showed flame resistance after five home laundering cycles (Bischof et al., 2012).

Expandable graphite, an intumescent flame retardant, has high potential to be used as a flame retardant for insulation materials, particularly for those are made up of natural fibrous materials (cellulose, bast fibres, etc.). Flame retardant was applied on a nonwoven felt containing a majority of recycled cotton fibre, without using any non-eco-friendly binders such as epoxy or phenolic resins. The study established that expandable graphite can be used as an alternative to conventional flame-retardant agents, with the main advantage of not emitting any toxic or harmful fumes (Paul et al., 2012).

Carbon nanotubes and graphene can also be used for the surface modification of textiles for imparting flame retardance. Aqueous graphene dispersions are applied on different textile surfaces as a permanent coating. Graphene modifications such as graphene oxide and multi-layer graphene are used, as they have different properties. The extremely wafer-thin graphene coating produced a physical barrier that holds off heat and gases (Paul, 2016; Woelfling et al., 2016; Hohenstein Institute, 2023). A schematic representation of graphene is shown in Fig. 14.4.

**Figure 14.4** Schematic representation of graphene.

Plasma technology has opened the way for nanocoatings on textile surfaces with a heat-shielding effect. The simultaneous argon plasma-induced grafting and polymerisation of a series of flame-retardant monomers such as phosphate, phosphonate and phosphoramidate acrylate monomers was attempted on cotton (Tsafack and Levalois-Grützmacher, 2006a) and polyacrylonitrile (Tsafack and Levalois-Grützmacher, 2006b) fabrics.

Among the techniques used to impart flame retardancy to fabrics that are considered to have a negligible environmental impact, layer-by-layer (LbL) assembly is highly relevant. A multilayer thin film less than one micron thick can be obtained with this deposition method. In combination with, or instead of electrostatic attraction, an LbL assembly can also be obtained through covalent (Sun et al., 2000) or hydrogen bonds (Lv et al., 2009).

### 14.3.7 Hydrophobic and oleophobic finishes

Hydrophobic and oleophobic finishes are applied to textiles to enhance their properties and performance, and thereby to increase the value of textile products. Coatings based on per- and polyfluoroalkyl substances (PFAS) are widely used, as they are highly efficient. They are stable under intense heat, often have surfactant properties and possess functions such as very high water and oil repellence. However, these substances are a main concern because of their high persistence, their degradation products and their impacts on human health and the environment. Some PFAS are banned at the EU level and several restrictions exist in a number of countries (ECHA, 2023). Fig. 14.5 shows the structures of perfluorooctansulphonic acid (PFOS) and perfluorooctanoic acid (PFOA).

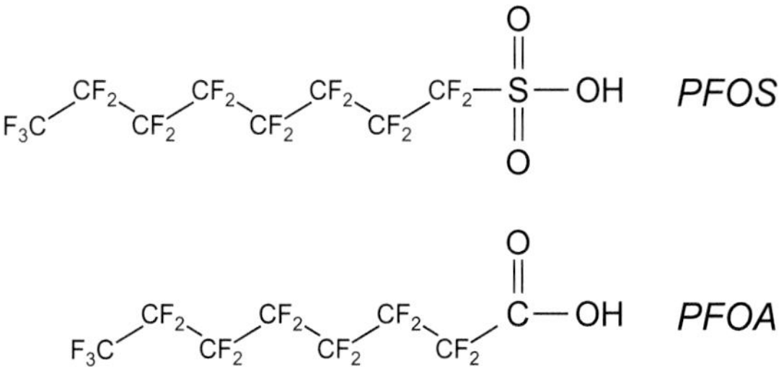

PFOS

PFOA

**Figure 14.5** Structures of perfluorooctansulphonic acid and perfluorooctanoic acid.

In the textile sector, C6-based PFAS is traditionally used for hydrophobic, stain-resistant as well as oleophobic coatings. There are some natural alternatives for hydrophobic properties, such as bio-based waxes or paraffins, but a commercially successful fluor-free oleophobic coating is yet to be developed. Nanotechnology-based solutions such as silica nanoparticles, calcium carbonate, carbon nanotubes, etc. are used for creating nanostructured hydrophobic surfaces on textiles.

Typically, modified $SiO_2$ particles are used as a sol–gel system aimed at hydrophobicity. These silica particles can be prepared by acidic or alkaline hydrolysis of tetraethoxysilane and are meta-stable in solution. This solution is often called colloidal silica or silica nanosol. By addition of hydrophobic modified silanes like hexadecyltriethoxysilane, hydrophobic groups can be covalently bonded to the silica particles of the nanosol and fixed on the textile surface. Non-fluorinated hydrophobic monomeric additives such as alkyltrialkoxysilanes can also be used in combination with a sol–gel process. For these additives, the gained hydrophobic effect greatly depends on the length of the alkyl chain of the used silane additive (Mahltig, 2014).

Silicones are very promising, and other novel materials include hydrophobins, dendrimers and vitrimers. Hydrophobins are spherical proteins with diameters of a few nanometres having a hydrophobic half and a hydrophilic half. They naturally occur on the top of mushroom heads to preventing soiling of the mushroom and applying a certain water repellency (Wohlleben et al., 2010). Depending on the properties of the used textile, the hydrophobic or hydrophilic part of the hydrophobin is placed on the fibre and the part with opposite properties is directed to the air leading to new textile properties. The hydrophobic part of hydrophobin is then orientated at the solid/air–interface leading to the hydrophobic modification of cotton (Opwis and Gutmann, 2011).

Bionics is making a sustainable breakthrough into textile finishing technologies. This is the application of biological methods and systems found in nature to design and develop engineering systems and technologies. Bionics is the science of constructing artificial systems that have some of the characteristics of living systems. Rudolf GmbH, Germany, has developed Bionic-Finish Eco, a bionics-based technology built upon dendrimers, leading to a non-fluorinated and highly durable water-repellent finish for high-performance applications. Dendrimers are highly branched polymers, also called star-like polymers. They are known for their well-defined regular structure, built up in several generations starting from a core and containing a surface with a high density of functional groups. The number of end groups increases exponentially with the number of generations of the dendrimer. These multifunctional branches can interact among themselves, co-crystallise, and self-organise into highly ordered, multi-component systems. They optimise the whole structure, attach to the textiles and create water repellence (Fischer and Vögtle, 1999; Rudolf GmbH, 2023).

Vitrimers, a subset of covalent adaptable networks, are a class of polymer material capable of self-healing and shape reprocessing at temperatures above their topology freezing temperature, where dynamic covalent bond exchange reactions dominate. Their properties include self-healing, malleability, orthogonal processability and multiple shape memory. Particularly, epoxy vitrimer composites are attractive as a potential industrial material due to their self-healing properties and enhanced thermomechanical performance (Hubbard et al., 2022).

Vitrimer design strategies can pave the way for the development of next-generation circular materials. Vitrimeric materials have attracted great attention as an interesting class of renewable plastic due to their potential to exhibit strength, durability and chemical resistance approaching that of traditional thermosets, while exhibiting end-of-life recyclability. This is due to their chemical structure, as vitrimers possess dynamic covalent cross-links, which impart stability while being reprocessable (Zheng et al., 2021). Ultra-thin self-healing vitrimer coatings can provide durable hydrophobicity. Polydimethylsiloxane vitrimer thin film deposited through dip-coating on a variety of substrates showed excellent hydrophobicity and optical transparency (Ma et al., 2021).

### *14.3.8 Ultraviolet protection finishes*

Textiles and apparel offer only limited UV light protection to the wearers and so ultraviolet protection finishes are applied to enhance the protection. The ultraviolet protection factor (UPF) of textiles is influenced by fibre contents, fabric construction, cover factor, fabric colour and the UV-protective finishes. Several UV finishes have been used in order to reduce the damaging effects on textile materials and health risks of UV radiation. The UV absorber molecules should be colourless or almost colourless compounds having high absorption coefficients in the UV range of 290−400 nm spectra. In order to offer effective protection against UV radiation, the finish molecules must quickly transform the absorbed energy into less harmful vibrational energy before reaching the surrounding substrate and exhibit good photostability (Kim, 2014).

There are both organic and inorganic UV blockers. The organic blockers are also known as UV absorbers as they absorb UV rays, whereas the inorganic blockers efficiently scatter both UVA and UVB rays, which are the main cause of skin cancer. Compared with organic UV absorbers, inorganic blockers such as nanoparticles are preferred due to their properties such as non-toxicity, chemical stability under UV radiation, etc.

In sol−gel finishing of fabrics, cotton was coated with $TiO_2$ nanoparticles formed by the hydrolysis and condensation reaction of titanium isopropoxide. A low initial concentration resulted in developing a uniform nanoparticle coating on the fabric surface. On the other hand, the fabric with the higher initial concentration did not show saturation of the nanoparticles, showing that further loading could be possible. Therefore, it is possible to substitute traditional UV finishing methods with the products and/or processes based on nanotechnology to obtain high UV protection. The developed process can be adapted easily to the existing machinery in a textile industry (Paul et al., 2010). Untreated white cotton fabric and the fabric coated with titanium dioxide sol−gel is shown in Fig. 14.6.

The UPF values of textiles can also be improved in a sustainable way by means of mechanical treatments such as calendaring, by changing the cover factor. This technique has proved successful, as the UPF value of different fabrics after only one pass through the calendar led to increases of 200% or more in the level of protection offered (Bernhard et al., 2022).

### 14.3.9 Insect-repellent finishes

Insect-repellent textiles have the characteristic of protecting humans from insects and particularly mosquitoes in areas that are their habitats and which are prone to vector-borne infectious diseases. Mosquitoes are the vectors for malaria (*Anopheles stephensi*), dengue fever (*Aedes aegypti*) and many other infectious diseases including yellow fever, chikungunya, filariasis, etc. It is a well-known fact that malaria and dengue fever are amongst the most lethal diseases in the world. Malaria is a vector-borne infectious disease caused by protozoan parasites, whereas dengue is a vector-borne disease caused by viruses.

Well-established insect-repellent finishes include synthetic chemicals such as N,N-diethyl-meta-toluamide (DEET), picaridin, permethrin, etc., which are not eco-friendly. Plant-based insect repellents have been used for generations in traditional practice as personal protection measure against mosquitoes. Essential oils and extracts belonging to the plants in the *Citronella* genus, mainly *Cymbopogon nardus*, are commonly used as ingredients of plant-based and sustainable mosquito repellents. When applied, citronella, which contains citronellal, citronellol, geraniol, citral, α pinene and limonene, is a very effective dose as DEET (Curtis et al., 1987), but many of these essential oils rapidly evaporate, causing loss of efficacy and leaving the user unprotected. However, by mixing citronella with a large molecule like vanillin

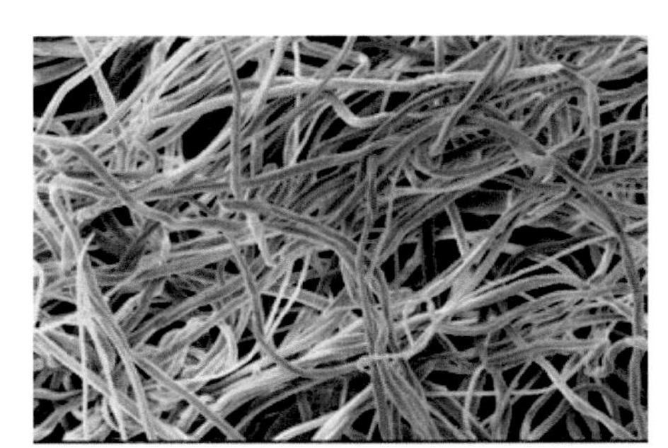

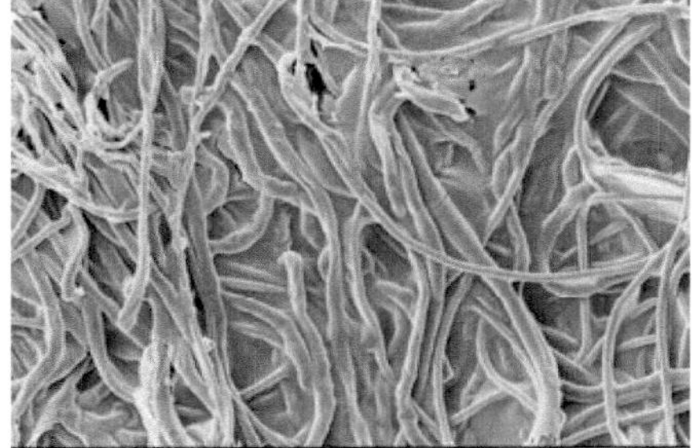

**Figure 14.6** Cotton fabric and fabric coated with titanium dioxide sol−gel.

(5%) the protection time can be considerably prolonged by reducing the release rate of the volatile oils (Tawatsin et al., 2001).

The use of microencapsulation technologies to enhance the performance of natural repellents makes plant oils like citronella a more viable option for making long-lasting repellents (Moore et al., 2007). *Azadirachta indica*, commonly known as neem, is widely considered as a natural alternative to DEET. Neem contains several active constituents called limonoids and the major components include azadirachtin, salannin, nimbin and meliantriol. Neem components show multiple effects against insects including mosquitoes, houseflies, cockroaches, etc. (Schmutterer, 1990).

*Corymbia citriodora*, also known as lemon eucalyptus, offers a potential natural repellent extracted from its leaves. The essential oil of lemon eucalyptus, which contains p-menthane-3,8-diol, shows pesticidal properties and is considered as an alternative to synthetic mosquito repellents. The mode of its action is most likely by masking the environmental cues that mosquitoes use to locate their target. It is believed that p-menthane-3,8-diol does not repel insects, but simply masks or confuses the attractive signals that humans emit so that mosquitoes are unable to locate them (Nasci et al., 2012; Drapeau et al., 2011).

A biodegradable and sustainable antimalarial technology has been developed through the functionalisation of PLA filaments. Knitted textiles were developed from PLA fibres, which are embedded with internally developed microcapsules containing a novel natural repellent agent, Schinus molle, added during the extrusion process. The mosquito repellency and wash resistance were found to be better than for DEET (Pinheiro et al., 2017, 2020).

### 14.3.10 Self-cleaning finishes

There are two major approaches, namely superhydrophobicity and photocatalysis, that can be used to impart self-cleaning properties to textile surfaces. Research on superhydrophobic surfaces has largely been based on mimicking nature, where over 200 species have been found to have the proper combination of surface chemistry and morphology that allows them to stay clean. Researchers have primarily followed two techniques to fabricate superhydrophobic surfaces: making a rough surface from a low surface energy material or modifying a rough surface using a material of low surface energy (Ma and Hill, 2006).

Silane-based sol–gel treatments have been used widely to create superhydrophobic surfaces on textiles. Biomimetic superhydrophobic cotton textiles have been developed. Silica particles coated with amine groups were generated in situ and covalently bonded to cotton by a one- or two-step reaction. The amine groups were reacted with mono-epoxy-functionalised polydimethylsiloxane to hydrophobise the surface (Hoefnagels et al., 2007). Gao et al. coated cotton with a series of polyhedral oligomeric silsesquioxanes to develop excellent water repellency (Gao et al., 2011). Fig. 14.7 shows an example of superhydrophobic self-cleaning structures found in nature.

Extensive work on the development of self-cleaning surfaces based on $TiO_2$ has been carried out by several research groups across the world. A photocatalyst is a substance which is photosensitive in nature and exhibits a strong oxidation effect in the

**Figure 14.7** Superhydrophobic self-cleaning structures found in nature.

presence of light. A self-cleaning property can be imparted to a textile surface by coating it with a photocatalytic oxide of a transition metal (Fujishima et al., 2000). The decomposition mechanism of organic contaminants by excited anatase $TiO_2$ radicals is shown in Fig. 14.8.

Anatase $TiO_2$ produces self-cleaning through two routes: photocatalytic oxidation and superhydrophilicity. Titania is photocatalytic because it is a semiconductor, meaning that a moderate amount of energy is required to lift an electron from its valence band to the conduction band. In the non-excited state, the photocatalyst is at the ground state with the electrons being localised in the valence band. When the surface is illuminated, electrons in $TiO_2$ absorb energy and move from the band gap to the conduction band, provided the photon energy is equal to or greater than the band gap energy (>3.2 eV) of the photocatalyst. In the activated state, the photocatalyst generates electron−hole pairs as highly active electrons are formed in the conduction band while positive holes are created in the valence band. The generation of these electron−hole pairs is the cause of the light-induced semiconductor properties of $TiO_2$ (Tung and Daoud, 2011; Veronovski et al., 2009). Apart from $TiO_2$, bismuth vanadate and benzophenone also have been tested for their self-cleaning effects.

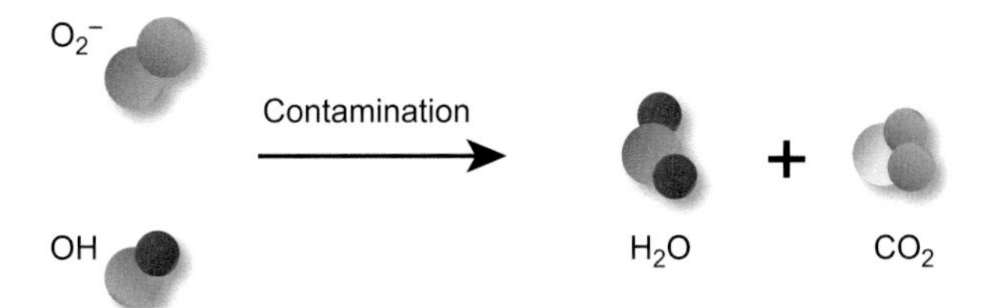


**Figure 14.8** Decomposition of contaminants by excited radicals (Gupta and Gulrajani, 2014).

### 14.3.11 Easy-care finishes

The most important formaldehyde-free sustainable easy-care finishes are polycarboxylic acids such as butanetetracarboxylic acid (BTCA), citric acid and maleic acid. BTCA is one of the best performing polycarboxylic acids and has been investigated extensively. Its di-anhydride derivative reacts with the hydroxyl groups of cellulose (Yang, 1993). Sodium hypophosphite ($NaH_2PO_2$) is the most effective catalyst, having good easy care properties and shrinkage control, comparable to formaldehyde containing finishes such as DMDHEU. The disadvantage of sodium hypophosphite is that it is expensive, higher amounts are needed than normal, and it leads to discolouration as it is a reducing agent for textiles dyed with sulphur or reactive dyes (Yang, 2013).

Non-phosphorus sodium propionate was examined for its catalyst performance for BTCA and was found to have similar easy care effect as sodium hypophosphite (Lee and Kim, 2001). The main advantage is the superior behaviour with respect to the strength retention. Tear and abrasion retention were improved by adding a softener (Vanneste, 2014). A high crease recovery and acceptable strength loss were observed when multi-finishing cotton with BTCA, in combination with sodium acetate trihydrate as catalyst, and chitosan. In this finishing system, BTCA plays a dual role: it cross-links with cellulose and reacts with chitosan, enabling the chitosan to be bound to the cotton resulting in wash resistance. An antimicrobial effect against Gram-positive and Gram-negative bacteria was still observed after 15 washes Hebeish et al. (2011).

Citric acid (CA) is also recognised as a successful non-formaldehyde easy care cross-linking agent. Yellowing of fabric caused by CA must be addressed. Combining CA with the catalyst $NaH_2PO_2$ led to improvement of the easy care behaviour, whiteness and breaking strength. The easy care finishing process of cotton with CA was optimised by varying the concentration of CA and trisodium citrate catalyst as well as the curing temperature. The higher the concentration of CA, the better the easy care properties. However, at the same time a reduction of tensile strength is observed. Decreasing the CA concentration and the curing temperature improves the whiteness. No effect of the catalyst was observed on this parameter. The loss of tensile strength is limited to 10% at a curing temperature of 180°C (Ramachandran et al., 2009).

Maleic acid (MA) is able to establish an esterification reaction with cotton but is not able to cross-link two cellulose molecules. However, at high curing temperatures, 175°C instead of 160°C, when hypophosphite is used as a catalyst along with maleic acid, both esterification and cross-linking reactions can occur with cotton (Peng et al., 2012). Combining maleic acid with sodium hypophosphite imparts good easy care properties, due to the formation of multiple carboxylic groups per molecule. To have this excellent effect, a sufficiently high application level is needed (Kim et al., 2000). A cyclodextrin-based cross-linker was developed to enhance the wrinkle recovery angle of fabrics. Cyclodextrin was oxidised using hydrogen peroxide and carboxylated with malic acid to introduce multiple carboxylic acid groups. The wrinkle resistance of the treated fabrics was found to be durable for up to 25 laundry washings (Patil and Netravali, 2019).

Polyamino carboxylic acids (PACAs) have been used to impart easy care properties in cotton fabrics. PACA can be obtained by reaction of polyvinylamine and bromoacetic acid. A study showed that PACAs are effective cross-linking agents for cotton using sodium hypophosphite as catalyst (Dehabadi et al., 2013). Another non-formaldehyde easy care finish that has been evaluated is glutaraldehyde. By reaction of its aldehyde groups with the OH-groups of cotton, hemiacetal is formed, which in a further stage reacts into acetal. A clear temperature dependence of the conversion of the carbonyl in the aldehyde groups is observed. The higher the temperature, the higher the conversion and the better the easy care properties (Yang et al., 2000).

### *14.3.12 Cosmetic and odour-resistant finishes*

According to the definition of the European Cosmetic Directive, cosmetotextiles are any textile product containing a substance or preparation that is released over time on different superficial parts of the human body, notably on human skin, and containing special functionalities such as cleansing, perfuming, changing appearance, protection, keeping in good condition or the correction of body odours. These textiles provide cosmetic and biological performances such as pleasant feeling, energising, slimming, refreshing, vitalising, skin glowing, anti-ageing, body care, fitness and healthiness. Cosmetotextiles are able to impart skincare properties, reduce ageing and improve feelings of wellness or wellbeing (Singh et al., 2011). Textiles can possess skincare properties through the release of cosmetic substances from the textile on the skin. It can be realised by a permanent and continuous delivery of active substances onto the skin. Furthermore, the active substances have to be protected against oxidation and other possible reactions.

The active substances have to be transferred to the skin in a controlled manner to act as cosmetic or pharmaceutical compounds. The bioactive compounds are released when the textile comes into contact with the skin (Cravotto et al., 2011). The incorporation of active substances in cosmetotextiles can improve application strategies because such systems can act as reservoirs and enable progressive delivery (Alonso et al., 2013). Substances may be antioxidants, fragrances, skin softeners, insect repellents, vitamins or UV blockers.

The odour-resistant textiles have the function of containing unpleasant odours. In regard to the end-use of textiles, odour resistance can be divided into two categories: odour-resistant textiles and fragrant apparels. The rise of an unpleasant odour from the surfaces after long use is always a problem. Thus, anti-odour finishing is necessary for these articles (Fung and Hardcastle, 2001). Among other possibilities, cyclodextrins incorporated into textiles can absorb or remove odour. Cyclodextrins are polysaccharides built from six to eight ($\alpha = 6$, $\beta = 7$, $\gamma = 8$) D-glucose units. The D-glucose units are covalently linked at the carbon atoms $C_1$ and $C_4$. The radii of the rigid cavities vary from 0.50 to 0.85 nm. In these cavities, guest molecules can be enclosed. The odour molecules being hydrophobic become trapped in the cavities of the cyclodextrins and are removed during laundering.

The cyclodextrin molecule does not show any affinity to any fibre material. For a permanent or nearly permanent fixation, the cyclodextrin molecule has to be modified.

**Figure 14.9** Representations of cyclodextrin and the cyclodextrin coating on textiles.

A cyclodextrin substituted with an alkyl chain can be fixed on polyester (Knittel et al., 1991). The most promising candidates for fixation of cyclodextrins are calixarenes and dendrons. The substitution of one tert-butyl group in a calixarene molecule leads to a derivative which can easily be immobilised on polyester fibres (Jansen et al., 2003). Fig. 14.9 shows two representations of cyclodextrin and the cyclodextrin coating on textiles.

Another very interesting class of container molecules is dendrimers. Dendrimers are spherical and high symmetric. They are formed from repetitively branched units. The chemical nature of these branches is responsible for the formation of hydrophilic or hydrophobic cavities. The fixation of a non-polar dendron on cotton results in hydrophobisation and the fixation of a polar dendron on polyester in a hydrophilisation of the textile material. Both dendrons are able to store and release molecules within their cavities (Buschmann and Schollmeyer, 2007).

## *14.3.13 Softening finishes*

Chemical softeners are widely used on textiles to develop a softer feel on the finished textiles. Softness is considered to be a dominant factor in offering comfort property. Since the fabric softness is felt by skin touch, they should be dermatologically safe to use and non-toxic to humans and the environment and also biodegradable. Softener contains both a hydrophobic or fatty part which does not mix with water and a hydrophilic part enabling it to disperse in water. The surfactant normally lowers the surface tension and ionises giving hydrophobic groups as cations and the hydrophilic part is anionic in nature (Guo, 2003). Almost all the fibres give rise to negative zeta potential in water and the extent of it depends on the relative hydrophilicity/hydrophobicity of the fibre, the presence of ionisable functional groups, etc. (Wahle and Falkowski, 2002). It is this zeta potential which also governs the extent of adsorption of surfactant by the fibre and thus consequently has an effect on softening performance (Stana-Kleinschek et al., 2002).

Innovations in silicone-based softening technology are leading to sustainability, high efficiency and reduced effluents. Silicone-based softeners are less temperature sensitive than most organic surfactants. They are supplied as aqueous emulsions by dispersing silicone oil in water using a proper emulsifier (Somasundaran et al., 2010). A number of factors influence the emulsifying ability of functional silicones such as hydrophilicity of the polymer, nature of the functional groups, the extent of modification and the method of emulsification (Mehta et al., 2009).

Reactive silicones are modified dimethylpolysiloxane polymers, containing functional amino mercapto and epoxy groups capable of reacting with fibres. Softeners of this type containing epoxy, hydroxyl or amino groups are used on cellulosic fabrics due to the ability of silananol or reactive epoxy groups to react with such fibres. Though very effective and water soluble, these softeners tend to yellow and thus are used in blends with cationic softeners (Lacasse and Baumann, 2004; Teli, 2014).

There have been several sustainable advances in the field of softening finishes and the finishing formulations are modified taking into account their application on a variety of textile blends and the demand for multiple performance properties. Similar to other finishes, the softening finishes also face various challenges during their application and there have been a lot of research efforts aiming to enhance their sustainability and environmental performance. Wetsoft eco and Wacker Finish eco are ecologically realigned products using the same processes and formulations. Instead of using fossil methanol during the manufacturing process, biomethanol produced from hay, grass cuttings or other plant residues is used (Wacker, 2023). Tubingal Rise from CHT Group, Germany, is a textile softener made of recycled silicone and renewable bio-based emulsifiers. It is the first softener developed for the circular economy with the make—use—reuse concept. Post-consumed silicone products are collected, recycled back to their purity and subsequently further processed to Tubingal Rise. Besides recycled silicone, renewable bio-based emulsifiers are used in the manufacture and it is suitable for all kind of fibres (CHT Group, 2023).

## 14.4 Smart coating based on anchor peptides

Anchor peptides, also called adhesion-promoting peptides or sticky peptides, are natural, amphiphilic molecules consisting of approximately 20—100 amino acids that are able to bind to synthetic materials including polyethylene, polypropylene and silicones as well as to natural surfaces like plant leaves (Meurer et al., 2017; Rübsam et al., 2017). Surface-binding anchor peptides promise huge potential to become a widely applied smart surface functionalisation technique, which could revolutionise the toolbox of currently used conventional chemical and mechanical modifications of textiles. Even at room temperature, surface-binding peptides are capable of binding to numerous materials and surfaces (Rübsam et al., 2017). The mechanism of smart coating by bioconjugation and bifunctional peptides is shown in Fig. 14.10.

Through the use of anchor peptides, surfaces can be immobilised with different compounds, such as enzymes, bioactive peptides, antigens, functional finishing agents, etc. As an example, a versatile toolbox for surface functionalisation of polymers, metals and silicon-based materials was established, which enabled immobilisation by click-chemistry of biotin, fluorescent molecules or antibiotics. For the generation of biohybrid coatings, a toolbox for click chemistry on surfaces was developed connecting chemical molecules or polymers with anchor peptides (Nöth et al., 2021). In

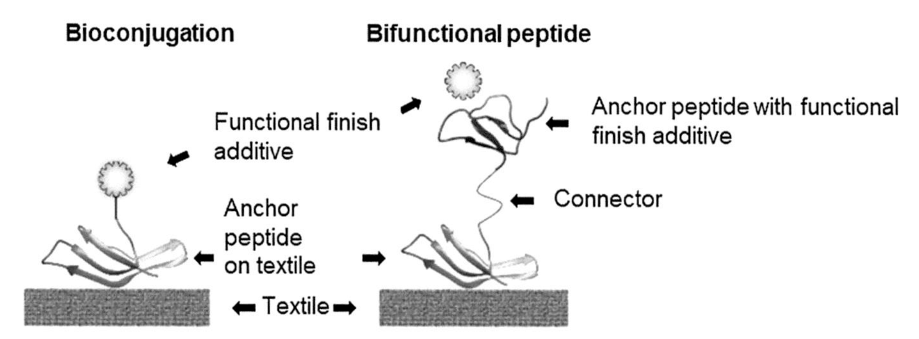


**Figure 14.10** Smart coating by bioconjugation and bifunctional peptides.

the same direction, a universal immobilisation platform (Matter-tag) for enzymes on polymers, metals and silicon-based materials was developed (Dedisch et al., 2020).

Other than the aforementioned methods for smart surface functionalisation, bifunctional peptides are an alternative powerful way to immobilise functional molecules on a material surface (Apitius et al., 2019a; Schwinges et al., 2019). Bifunctional peptides consist of two anchor peptides separated by a linker to prevent intermolecular interactions between the anchor peptides. One anchor peptide binds specifically to a material surface such as textiles and the other anchor peptide interacts selectively with the desired functional finish molecule (Apitius et al., 2019b; Krause et al., 2023). The finishing of textiles with anchor peptides can be carried out in a cost- and resource-efficient way in an aqueous solution, and in the form of a dense monolayer (occupancy densities usually >90%). In this way, material surfaces can be equipped with specific functionalities with high efficiency. Fig. 14.11 shows the different steps involved in the smart biohybrid coating of textiles.

Anchor peptides are capable of binding textiles with high selectivity and binding strength and can be used for developing sustainable smart functional coatings. Therefore, anchor peptides can be considered as good candidates for the replacement of PFAS in textile finishing. In the SmartBioFinish project, funded in the Innovationsraum: BioTexFuture by BMBF, Germany, RWTH-ITA and partners have developed an anchor peptide based smart coating technology. The peptides are screened and are functionalised either by covalent functionalisation (bioconjugation) or by bifunctional peptides with hydrophobic and oleophobic additives. This technology could

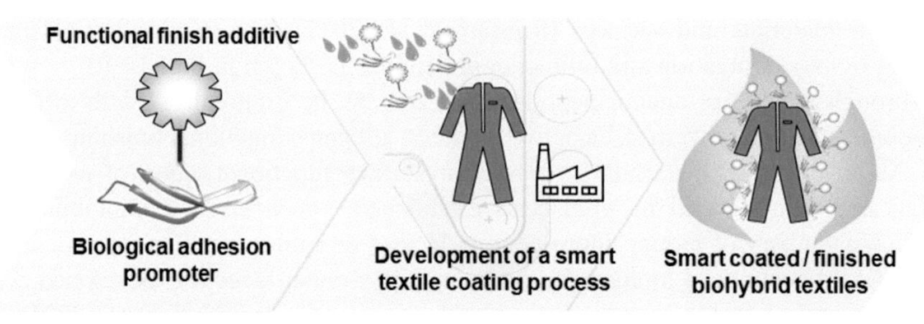


**Figure 14.11** Development of a smart biohybrid coating process.

contribute to the biotransformation of textile industry, as it is capable of replacing PFAS for developing water and oil repellent textiles. It is also possible to create bio-hybrid flame-retardant finishes on technical textiles for developing smart flame-retardant personal protective equipment (Heesemann et al., 2023; Krause et al., 2023).

## 14.5 Future trends

High-performance functional finishes have been successfully applied to a wide variety of textiles for imparting various technical properties. There is an upcoming trend in functional finishing to develop sustainable multifunctional textiles, which are highly durable and safe to the user as well as for the environment. This will require a careful balance between the compatibility of different finishing chemicals and treatments. Even though different functional properties can be developed on textiles, there is real concern over the health and safety of workers and the sustainability of the final product. In the case of nanomaterials, their extremely small size also means that they are much more readily taken up by the human body than larger sized particles.

Taking into account the sustainability aspects, various environmental and safety legislations are being imposed by different governments, which include the regulations from the European Commission such as REACH. Due to the increased consumer awareness, several ecological labels assuring the safety of end users are gaining importance. There has been a great improvement in the environmental profile of the functional finishes over the last few decades, but there remains a need to further improve the ecological aspects of functional finishes.

Modern research is now focused on developing novel and more sustainable techniques. Research related to functional finishing is being influenced not only in developing innovative and improved functionalities but also in fulfilling the demands in terms of health, safety and environmental protection. Apart from providing excellent protection, the developed functional textiles should offer good thermo-physiological properties and excellent skin sensory comfort for the wearer. Novel finishing possibilities such as nanofinishes, biotechnological finishes based on enzymes, dry finishing techniques like plasma or laser, layer-by-layer assembly techniques and anchor peptide-based smart coatings will play a key role in the future.

## 14.6 Conclusion

Most textiles pass through a finishing process during or at the final stage of their production, where their visual and functional characteristics are modified and influenced positively. Due to globalisation and free trade, the textile and apparel market has become consumer driven, and demands for multiple functionalities and sustainability are ever increasing. In this highly competitive scenario, manufacturers are attempting to offer high-quality functional textile materials at a cost-effective price. The functional textiles have opened up numerous applications in diverse fields and the utilisation of these textiles with enhanced properties is expanding rapidly. The future trend in functional finishing is to develop multifunctional textiles, which are highly efficient, durable, cost effective and manufactured in a sustainable manner. Research on new materials and sustainable processes has much to do in this respect and, there remains a long way to go. Anchor peptides, which are capable of binding substrates such as textiles with high selectivity and binding strength, can be used for developing sustainable smart functional coatings.

# References

Alonso, C., Martí, M., Martínez, V., Rubio, L., Parra, J.L., Coderch, L., 2013. Antioxidant cosmeto-textiles: skin assessment. European Journal of Pharmaceutics and Biopharmaceutics 84 (1), 192–199.

Ammayappan, L., Moses, J.J., 2009. Study of antimicrobial activity of aloevera, chitosan, and curcumin on cotton, wool, and rabbit hair. Fibers and Polymers 10 (2), 161–166.

Apitius, L., Buschmann, S., Bergs, C., Schönauer, D., Jakob, F., Pich, A., Schwaneberg, U., 2019a. Biadhesive peptides for assembling stainless steel and compound loaded microcontainers. Macromolecular Bioscience 19, 1900125.

Apitius, L., Rübsam, K., Jakesch, C., Jakob, F., Schwaneberg, U., 2019b. Ultrahigh-throughput screening system for directed polymer binding peptide evolution. Biotechnology and Bioengineering 116 (8), 1856–1867.

Baba, M., Snoeck, R., Pauwels, R., de Clercq, E., 1988. Sulfated polysaccharides are potent and selective inhibitors of various enveloped viruses, including herpes simplex virus, cytomegalovirus, vesicular stomatitis virus, and human immunodeficiency virus. Antimicrobial Agents and Chemotherapy 32 (11), 1742–1745.

Bajya, M., Majumdar, A., Butola, B.S., Verma, S.K., Bhattacharjee, D., 2020. Design strategy for optimising weight and ballistic performance of soft body armour reinforced with shear thickening fluid. Composites Part B: Engineering 183, 107721.

Bautista, L., 2014. Ballistic and impact protection finishes for textiles. In: Paul, R. (Ed.), Functional Finishes for Textiles. Woodhead Publishing, pp. 579–606.

Bernhard, A., Caven, B., Wright, T., Burtscher, E., Bechtold, T., 2022. Improving the ultraviolet protection factor of textiles through mechanical surface modification using calendering. Textile Research Journal 92 (9–10), 1405–1414.

Bischof, S., Flinčec, S., Kovačević, Z., Moron, M., Paul, R., 2012. Non halogen phosphorus containing flame retardants for cotton. In: Paper Presented in 6th International Textile Clothing & Design Conference (ITC&DC 2012) entitled Magic World of Textile. Dubrovnik, Croatia, 7-10 October.

Buschmann, H.-J., Schollmeyer, E., 2007. Dendrons for surface modification of polymeric materials. In: Mittal, K.L. (Ed.), Polymer Surface Modification: Relevance to Adhesion, 4. VSP, Leiden, Boston, pp. 209–218.

Carrington Textiles. 2023. https://www.carrington.co.uk/en/antiviral-fabric-finish/.

Chang, C.-P., Shih, C.-H., You, J.-L., Youh, M.-J., Liu, Y.-M., Ger, M.-D., 2021. Preparation and ballistic performance of a multi-layer armor system composed of kevlar/polyurea composites and shear thickening fluid (STF)-Filled paper honeycomb panels. Polymers 13, 3080.

CHT Group. 2023. https://solutions.cht.com/cht/web.nsf/id/pa_tubingal-rise-softener.html .

Cravotto, G., et al., 2011. A new cyclodextrin-grafted viscose loaded with aescin formulations for a cosmeto-textile approach to chronic venous insufficiency. Journal of Materials Science: Materials in Medicine 22, 2387–2395.

Curtis, C.F., Lines, J.D., Ijumba, J., Callaghan, A., Hill, N., Karimzad, M.A., 1987. The relative efficacy of repellents against mosquito vectors of disease. Medical and Veterinary Entomology 1, 109–119.

de Oliveira Braga, F., Bolzan, L.T., Lima Jr., E.P., Monteiro, S.N., 2017. Performance of natural curaua fiber-reinforced polyester composites under 7.62mm bullet impact as a stand-alone ballistic armor. Journal of Materials Research and Technology 6 (4), 323–328.

Dedisch, S., Wiens, A., Davari, M.D., Söder, D., Rodriguez-Emmenegger, C., Jakob, F., Schwaneberg, U., 2020. Matter-tag: a universal immobilization platform for enzymes on polymers, metals, and silicon-based materials. Biotechnology and Bioengineering 117 (1), 49–61.

Dehabadi, V.A., Buschmann, H.-J., Gutmann, J.S., 2013. Study of easy care and biostatic properties of finished cotton fabric with polyamino carboxylic acids. Journal of the Textile Institute 104, 414–418.

Dettenkofer, M., Block, C., 2005. Hospital disinfection: efficacy and safety issues. Current Opinion in Infectious Diseases 18 (4), 320–325.

Draney, K., Bates, J., 2022. Biodegradable superabsorbent polymers. In: TMS 2022 151st Annual Meeting & Exhibition Supplemental Proceedings. Springer International Publishing.

Drapeau, J., Rossano, M., Touraud, D., Obermayr, U., Geier, M., Rose, A., Kunz, W., 2011. Green synthesis of para-menthane-3,8-diol from Eucalyptus citriodora: application for repellent products. Comptes Rendus Chimie 14 (No. 7–8), 629–635.

ECHA. 2023. https://echa.europa.eu/hot-topics/perfluoroalkyl-chemicals-pfas. .

Egres Jr., R.G., Lee, Y.S., Kirkwood, J.E., Kirkwood, K.M., Wetzel, E.D., Wagner, N.J., 2004. Liquid armour: protective fabrics utilizing shear thickening fluids (STFs). In: 4th International Conference on Safety and Protective Fabrics. Pittsburgh, PA.

El-Tahlawy, K.F., et al., 2005. The antimicrobial activity of cotton fabrics treated with different crosslinking agents and chitosan. Carbohydrate Polymers 60 (4), 421–430.

Elshafei, A., El-Zanfaly, H., 2011. Application of antimicrobials in the development of textiles. Asian Journal of Applied Sciences 4, 585–595.

Fischer M. and Vögtle F. (1999), 'Dendrimers: from design to application', Angewandte Chemie International Edition, 38, 884–905.

Fujishima, A., Rao, T.N., Tryk, D.A., 2000. Titanium dioxide photocatalysis. Journal of Photochemistry and Photobiology, C: Photochemistry Reviews 1 (1), 1–21.

Fung, W., Hardcastle, M., 2001. Textiles in Automotive Engineering. Woodhead Publishing Ltd, Cambridge.

Gao, Y., He, C., Qing, F.L., 2011. Polyhedral oligomeric silsesquioxane-based fluoroether-containing terpolymers: synthesis, characterization and their water and oil repellency evaluation for cotton fabric. Journal of Polymer Science Part A: Polymer Chemistry 49 (24), 5152–5161.

Glampedaki, P., Petzold, G., Dutschk, V., Miller, R., Warmoeskerken, M.M.C.G., 2012a. Physicochemical properties of biopolymer-based polyelectrolyte complexes with controlled pH/thermo-responsiveness. Reactive and Functional Polymers 72, 458–468.

Glampedaki, P., Krägel, J., Petzold, G., Dutschk, V., Miller, R., Warmoeskerken, M.M.C.G., 2012b. Polyester textile functionalization through incorporation of pH/thermo-responsive microgels. Part I: microgel preparation and characterization. Colloids and Surfaces A: Physicochemical and Engineering Aspects 413, 334–341.

Gugliuzza, A., Drioli, E., 2013. A review on membrane engineering for innovation in wearable fabrics and protective textiles. Journal of Membrane Science 446, 350–375.

Guo, J., 2003. The Effects of Household Fabric Softeners on the Thermal Comfort and Flammability of Cotton and Polyester Fabrics. M.Sc. Thesis. Virginia Polytechnic Institute and State University.

Gupta, D., Gulrajani, M.L., 2014. Self cleaning finishes for textiles, editor: Paul R. In: Functional Finishes for Textiles. Woodhead Publishing, pp. 257–281.

Gupta, D., Khare, S.K., Laha, A., 2004. Antimicrobial properties of natural dyes against gram-negative bacteria. Coloration Technology 120 (4), 167–171.

Gyawali, R., Ibrahim, S.A., 2014. Natural products as antimicrobial agents. Food Control 46, 412–429.

Hebeish, A., Fouda, M.M.G., Aelsaid, Z., Essam, S., Tammam, G.H., Drees, E.A., 2011. Green synthesis of easy care and antimicrobial cotton fabrics. Carbohydrate Polymers 86, 1648–1691.

Heesemann, R., Bettermann, I., Paul, R., Rey, M., Gries, T., Feng, L., Schwaneberg, U., Hummelsheim, C., 2023. Development of a process for flame retardant coating of textiles with bio-hybrid anchor peptides. Journal of Vacuum Science and Technolology A 41, 053110.

HeiQ. 2023.https://www.heiq.com/products/textile-technologies/heiq-viroblock-antiviral. .

Hernandez, S.C., Bennett, C.J.C., Junkermeier, C.E., Tsoi, S.D., Bezares, F.J., Stine, R., Robinson, J.T., Lock, E.H., Boris, D.R., Pate, B.D., Caldwell, J.D., Reinecke, T.L., Sheehan, P.E., Walton, S.G., 2013. Chemical gradients on graphene to drive droplet motion. ACS Nano 7, 4746–4755.

Herrero, M., Mendiola, J.A., Plaza, M., Ibañez, E., 2013. Screening for bioactive compounds from algae. In: Advanced Biofuels and Bioproducts. Springer, pp. 833–872.

Hoefnagels, H.F., Wu, D., de With, G., Ming, W., 2007. Biomimetic superhydrophobic and highly oleophobic cotton textiles. Langmuir 23 (26), 13158–13163.

Huang, L.K., Gang, S., 2003. Durable and regenerable antimicrobial cellulose with oxygen bleach: concept proofing. AATCC Review 3 (10), 17–21.

Hubbard, A.M., Ren, Y., Papaioannou, P., Sarvestani, A., Picu, C.R., Konkolewicz, D., Roy, A.K., Varshney, V., Nepal, D., 2022. Vitrimer composites: understanding the role of filler in vitrimer applicability. ACS Applied Polymer Materials 4 (9), 6374–6385.

Huleihel, M., Ishano, V., 2001. Effect of propolis extract on malignant cell transformation by moloney murine sarcoma virus. Archives of Virology 146 (8), 1517–1526.

Hohenstein Institute. 2023. https://textile-network.com/en/Technical-Textiles/Hohenstein-Graphene-goes-textile. .

Jansen, K., et al., 2003. Mit Calixarenen Ausgerüstete Textile Materialien, Verfahren zu ihrer Herstellung und ihre Verwendung. German patent no. DE 10210115.

Joshi, M., Ali, S.W., Purwa, R., 2009. Ecofriendly antimicrobial finishing of textiles using bioactive agents based on natural products. Indian Journal of Fibre and Textile Research 34 (3), 295–304.

Junhao, F., Tao, T., Likang, Z., Huiwen, H., Meng, M., Yanqin, S., Si, C., Xu, W., 2023. High-toughness and biodegradable superabsorbent hydrogels based on dual functional cross-linkers. ACS Applied Polymer Materials 5 (5), 3686–3697.

Khandual, A., Pattanaik, P., Tripathy, H.P., Rout, J.N., Hu, J., Holderbaum, W., Paul, R., 2021. Prospective of germicidal far UV-C in face masks and PPE. World Journal of Textile Engineering and Technology 7, 36–39.

Kim, Y.K., 2014. Ultraviolet protection finishes for textiles. In: Paul, R. (Ed.), Functional Finishes for Textiles. Woodhead Publishing, pp. 463–485.

Kim, B.-H., Jang, J., Ko, S.-W., 2000. Durable Press finish of cotton fabric using malic acid as a crosslinker. Fibers and Polymers 1, 116–121.

Klopman. 2023. https://www.klopman.com/wp-content/uploads/2020/12/VIROFF-TEX_Web-1.pdf. .

Knittel, D., Buschmann, H.-J., Schollmeyer, E., 1991. Veredlung von natur- und synthesefasern durch fixierung von cyclodextrinderivaten. Textilveredlung 26, 92–95.

Krause, R., Bettermann, I., Paul, R., Rey, M., Gries, T., Feng, L., Schwaneberg, U., Hummelsheim, C., Kampas, L., 2023. Development of an alternative flame retardant finish for fire protection textiles and an adapted finishing process. In: The 10th European Conference on Protective Clothing: Protection Challenges in a Changing World. Arnhem, the Netherlands.

Lacasse, K., Baumann, W., 2004. Textile Chemicals–Environmental Data and Facts. Springer Publications Germany, pp. 381–393.

Lee, E.S., Kim, H.J., 2001. Durable press finish of cotton/polyester fabrics with 1,2,3,4-butanetetracarboxylic acid and sodium propionate. Journal of Applied Polymer Science 81, 654–661.

Liu, W., Chen, L., Wang, Y.-Z., 2012. A novel phosphorus-containing flame retardant for the formaldehyde-free treatment of cotton fabrics. Polymer Degradation and Stability 97, 5.

Lu, S.-Y., Hamerton, I., 2002. Recent developments in the chemistry of halogen-free flame retardant polymers. Progress in Polymer Science 27, 12.

Lv, F., Peng, Z., Zhang, L., Yao, L., Liu, Y., Xuan, L., 2009. Photoalignment of liquid crystals in a hydrogen-bonding-directed layer-by-layer ultrathin film. Liquid Crystals 36, 43–51.

Ma, M., Hill, R.M., 2006. Superhydrophobic surfaces. Current Opinion in Colloid & Interface 11 (4), 193–202.

Ma, J., Porath, L.E., Haque, M.F., Sett, S., Rabbi, K.F., Nam, S.W., Miljkovic, N., Evans, C.M., 2021. Ultra-thin self-healing vitrimer coatings for durable hydrophobicity. Nature Communications 12, 5210.

Ma, K., Cheung, Y.H., Kirlikovali, K.O., Wang, X., Islamoglu, T., Xin, J.H., Farha, O.K., 2023. Protection against chemical warfare agents and biological threats using metal-organic frameworks as active layers. Accounts of Materials Research 4 (2), 168–179.

Magicsrl. 2023. https://magicsrl.com/en.

Mahltig, B., 2014. Hydrophobic and oleophobic finishes for textiles, editor: Roshan Paul. In: Functional Finishes for Textiles. Woodhead Publishing, pp. 387–428.

Marsyahyo, E., Rochardjo, H.S.B., Soekrisno, 2009. Preliminary investigation on bulletproof panels made from ramie fiber reinforced composites for NIJ level II, IIA, and IV. Journal of Industrial Textiles 39 (1), 13–26.

Mehta, S.C., Somasundaran, P., Kulkarni, R., 2009. Variation in emulsion stabilization behavior of hybrid silicone polymers with change in molecular structure: phase diagram study. Journal of Colloid and Interface Science 333, 635–640.

Meurer, R.A., Kemper, S., Knopp, S., Eichert, T., Jakob, F., Goldbach, H.E., Schwaneberg, U., Pich, A., 2017. Biofunctional microgel-based fertilizers for controlled foliar delivery of nutrients to plants. Angewandte Chemie International Edition 56 (26), 7380–7386.

Moore, S.J., Hill, N., Ruiz, C., Cameron, M.M., 2007. Field evaluation of traditionally used plant-based insect repellents and fumigants against the malaria vector Anopheles darlingi in Riberalta, Bolivian Amazon. Journal of Medical Entomology 44 (4), 624–630.

Nasci, R.S., Zielinski-Gutierrez, E., Wirtz, R.A., Brogdon, W.G., 2012. Protection against mosquitoes, ticks and other insects and arthropods. In: CDC Health Information for International Travel ('The Yellow Book'). Centers for Disease Control and Prevention, Atlanta, GA.

Nöth, M., Zou, Z., El-Awaad, I., de Lencastre Novaes, L.C., Dilarri, G., Davari, M.D., Ferreira, H., Jakob, F., Schwaneberg, U., 2021. A peptide-based coating toolbox to enable click chemistry on polymers, metals, and silicon through sortagging. Biotechnology and Bioengineering 118 (4), 1520–1530.

Opwis, K., Gutmann, J.S., 2011. Surface modification of textile materials with hydrophobins. Textile Research Journal 81, 1594–1602.

Öztürk, M., Duru, M.E., Kivrak, S., Mercan-Doğan, N., Türkoglu, A., Ali Özler, M., 2011. In vitro antioxidant, anticholinesterase and antimicrobial activity studies on three Agaricus species with fatty acid compositions and iron contents: a comparative study on the three most edible mushrooms. Food and Cosmetics Toxicology 49, 1353–1360.

Patil, N.V., Netravali, A.N., 2019. Cyclodextrin-based "green" wrinkle-free finishing of cotton fabrics. Industrial & Engineering Chemistry Research 58 (45), 20496–20504.

Paul, R., 2016. Graphene Functionalisation of Protective Textiles. Poster Presented in the 8th International R&D Event in Turkish Textile and Clothing Sector. Organized by Uludag Textile Exporters Association (UTIB), Bursa, Turkey, 12-13 May.

Paul, R., Botet, J.M., Marsal, F., Casals, E., Puntes, V., 2010. Nano-cotton fabrics with high ultraviolet protection. Textile Research Journal 80 (5), 454–462.

Paul, R., Brouta-Agnésa, M., Esteve, H., 2012. Chapter 10: protective insulation materials in functional protective textiles. In: Bischof, S. (Ed.), Faculty of Textile Technology. University of Zagreb, Croatia, pp. 281–303.

Peng, H., Yang, C.Q., Wang, S., 2012. Nonformaldehyde durable press finishing of cotton fabrics using the combination of maleic acid and sodium hypophosphite. Carbohydrate Polymers 87, 491–499.

Pinheiro, C., Belino, N., Paul, R., 2017. Biodegradable Antimalarial Clothing. Poster Presented in Autex 2017 - Shaping the Future of Textiles. Corfu, Greece, 29-31 May 2017.

Pinheiro, C., Belino, N., Paul, R., 2020. Development of biodegradable and antimalarial textile structures, book: textiles, identity and innovation. In: Montagna, Carvalho (Eds.), Proceedings of the 2nd International Textile Design Conference D_TEX 2019. Taylor & Francis Group, London. June 19-21, 2019.

Quinto, E.J., Caro, I., Villalobos-Delgado, L.H., Mateo, J., De-Mateo-Silleras, B., Redondo-Del-Río, M.P., 2019. Food safety through natural antimicrobials. Antibiotics 8, 208.

Ramachandran, T., Gobi, N., Lakshmikantha, C.B., 2009. Optimization of process parameters for crease resistant finishing of cotton fabric using citric acid. Indian Journal of Fibre & Textile Research 34, 359–367.

Ramesh, C., Pattar, M.G., 2010. Antimicrobial properties, antioxidant activity and bioactive compounds from six wild edible mushrooms of western ghats of Karnataka, India. Pharmacognosy Research 2, 107.

Rübsam, K., Stomps, B., Böker, A., Jakob, F., Schwaneberg, U., 2017. Anchor peptides: a green and versatile method., 2017. Anchor peptides: a green and versatile method for polypropylene functionalization. Polymer 116, 124–132.

Rudolf GmbH. 2023. https://rudolf.de/technologies/bionic-finish-eco. .

Schmutterer, H., 1990. Properties of natural pesticides from the neem tree, Azadirachta indica. Annual Review of Entomology 35, 271–297.

Schreuder-Gibson, H.L., Truong, Q., Walker, J.E., Owens, J.R., Wander, J.D., Jones, W.E., 2003. Chemical and biological protection and detection in fabrics for protective clothing. MRS Bulletin 28, 574–578.

Schwinges, P., Pariyar, S., Jakob, F., Rahimi, M., Apitius, L., Hunsche, M., Schmitt, L., Noga, G., Langenbach, C., Schwaneberg, U., Conrath, U., 2019. A bifunctional

dermaseptin–thanatin dipeptide functionalizes the crop surface for sustainable pest management. Green Chemistry 21, 2316–2325.
Sforcin, J.M., Bankova, V., 2011. Propolis: is there a potential for the development of new drugs? Journal of Ethnopharmacology 133 (2), 253–260.
Singh, R., Jain, A., Panwar, S., Gupta, D., Khare, S.K., 2005. Antimicrobial activity of some natural dyes. Dyes and Pigments 66 (2), 99–102.
Singh, M.K., Varun, V.K., Behera, B.K., 2011. Cosmetotextiles: state of art. Fibres and Textiles in Eastern Europe 4 (87), 27–33.
Smith, W.C., 1999. An Overview of Protective Clothing – Markets, Materials, Needs, VII. Ballistic/Mechanical. Industrial Textile Associates, USA, pp. 4–5.
Somasundaran, P., Purohit, P., Gokarn, N., Kulkarni, R.D., 2010. Silicone emulsions-interfacial aspects and applications. Household and Personal Care Today 3, 35–42.
Srivastava, A., Majumdar, A., Butola, B.S., 2012. Improving the impact resistance of textile structures by using shear thickening fluids: a review. Critical Reviews in Solid State and Materials Sciences 37, 115–129.
Stana-Kleinschek, K., Ribitsch, V., Kreze, T., Fras, L., 2002. Determination of the adsorption character of cellulose fibres using surface tension and surface charge. Materials Research Innovations 6 (1), 13–18.
Sun, J., Wu, T., Liu, F., Wang, Z., Zhang, X., Shen, J., 2000. Covalently attached multilayer assemblies by sequential adsorption of polycationic diazo-resins and polyanionic poly(-acrylic acid). Langmuir 16, 4620–4624.
Tawatsin, A., Wratten, S.D., Scott, R.R., Thavara, U., Techadamrongsin, Y., 2001. Repellency of volatile oils from plants against three mosquito vectors. Journal of Vector Ecology 26, 76–82.
Teli, M.D., 2014. Softening finishes for textiles and clothing. In: Paul, R. (Ed.), Functional Finishes for Textiles. Woodhead Publishing, pp. 123–152.
Tokarev, I., Minko, S., 2009. Stimuli-responsive hydrogel thin films. Soft Matter 5, 511–524.
Tsafack, M.J., Levalois-Grützmacher, J., 2006a. Flame retardancy of cotton textiles by plasma-induced graft-polymerization (PIGP). Surface and Coatings Technology 201, 2599–2610.
Tsafack, M.J., Levalois-Grützmacher, J., 2006b. Plasma-induced graft-polymerization of flame retardant monomers onto PAN fabrics. Surface and Coatings Technology 200, 3503–3510.
Tung, W.S., Daoud, W.A., 2011. Self-cleaning fibers via nanotechnology: a virtual reality. Journal of Materials Chemistry 21, 7858–7869.
Turaga, U., Singh, V., Ramkumar, S., 2014. Biological and chemical protective finishes for textiles. In: Paul, R. (Ed.), Functional Finishes for Textiles. Woodhead Publishing, pp. 555–578.
Vanneste, M., 2014. Easy care finishes for textiles. In: Paul, R. (Ed.), Functional Finishes for Textiles. Woodhead Publishing, pp. 227–256.
Veronovski, N., Rudolf, A., Smole, S.M., Kreže, T., Geršak, J., 2009. Self-cleaning and handle properties of TiO2-modified textiles. Fibers and Polymers 10 (4), 551–556.
Wacker Chemie AG. https://www.wacker.com/cms/en-us/insights/wetsoft-eco-wacker-finish-eco.html. .
Wagner, N.J., Brady, F.J., 2009. Shear thickening in colloidal dispersions. Physics Today 62 (10), 27.
Wahle, B., Falkowski, J., 2002. Softeners in textile processing. Part 1: an overview. Review of Progress in Coloration 32, 118–124.
Woelfling, B.-W., Classen, E., Paul, R., Schmidt, A., 2016. Graphene surface modification of textiles for PPE applications. In: Paper Presented in The 90th Textile Institute World Conference. Poznan, Poland, 25-28 April.

Wohlleben, W., Subkowski, T., Bollschweiler, C., von Vacano, B., Liu, Y., Schrepp, W., Baus, U., 2010. Recombinantly produced hydrophobins from fungal analogous as highly surface-active performance proteins. European Biophysics Journal 39, 457–468.

Yang, C.Q., 1993. Effect of pH on nonformaldehyde durable press finishing of cotton fabric: FT-IR spectroscopy study – Part II: formation of the anhydride intermediate. Textile Research Journal 63, 706–711.

Yang, C.Q., 2013. Crosslinking: a route to improve cotton performance. AATCC Review 13, 43–52.

Yang, C.Q., Wei, W., Mcllwaine, D.B., 2000. Evaluating glutaraldehyde as nonformaldehyde durable press finishing agent for cotton fabrics. Textile Research Journal 70, 320–323.

Zhang, Z., et al., 2003. Antibacterial properties of cotton fabrics treated with chitosan. Textile Research Journal 73 (12), 1103–1106.

Zheng, J., Png, Z.M., Ng, S.H., Tham, G.X., Ye, E., Goh, S.S., Loh, X.J., Li, Z., 2021. Vitrimers: current research trends and their emerging applications. Materials Today 51, 586–625.

# Garment machinery for regenerative manufacturing

*Jenny Underwood and Saniyat Islam*
School of Fashion and Textiles, Royal Melbourne Institute of Technology University, Melbourne, VIC, Australia

## 15.1 Introduction

Technology innovations and improved environmental, social, and ethical practices are radically transforming the global textile and fashion industry, fundamentally changing how clothes are made, consumed, and experienced. From the integration of 2-D and 3-D virtual design and on-demand digital production, material science innovations, automation, and robotics, to new business models and a circular fashion system, the industry is transforming in multiple ways. Smart functional clothing solutions will support health care and well-being, provide protection in increasingly variable and extreme weather conditions, and augment and enhance performance. Simultaneously, society and consumer values, behaviour and expectations are changing. Conscious consumers are buying less but better and are wanting more meaningful experiences (Aakko and Koskennurmi-Sivonen, 2013).

Yet underlining these opportunities for new markets and products is the urgent need for the textile industry to transition to sustainable development practices as framed by the Unite Nation's (UN's) Sustainable Development Goals (SDGs) (United Nations, 2015), and to implement mitigation and adaptation strategies for climate change. The textile industry is asset and labour-intensive and as an industry, it is still largely working on industrial processes from the 19th century. It is the second highest employer globally, and one of the most polluting and wasteful industries. Even under optimistic assumptions, the industry's existing solutions and business models will not deliver the impact needed to transform the industry (Global Fashion Agenda, 2019; Agenda and Group, 2017).

The scale of change needed to limit global temperature rise to 1.5°C above the pre-industrial level, as called for by the UN's Intergovernmental Panel on Climate Change (Allan et al., 2021), and to rebalance the current unsustainable patterns of production and consumption are immense and time-critical. It will require a multi-stakeholder approach to connect all parts of the textile and fashion value chain. Collaboration, transparency, joint commitments, investment, and bold actions are required to fundamentally redefine business models and current imperatives of economic growth and rising consumerism (Global Fashion Agenda, 2019; Agenda and Group, 2017). However, as Kate Fletcher (Fletcher and Tham, 2019) pointed out, there 'is no evidence to support the idea that fashion is in a meaningful phase of sustainable transformation'.

**Sustainable Innovations in the Textile Industry. https://doi.org/10.1016/B978-0-323-90392-9.00017-3**

For the textile industry to address these challenges, technology will have a significant role to play. However, fundamental changes will not be easy for such an asset-intensive and fragmented industry, where production and consumption transcend. Such changes suggest new ways of working are needed that integrate the practices associated with design, manufacturing processes and distribution networks, and in doing so reconceptualise the relationship of producer and consumer, from what has been a largely linear system (take-make-waste) driven by growth to a circular system nourished by earth's logic. Innovations in garment machinery and manufacturing, be they incremental or radical, offer the potential to support the transition to a circular system of production and consumption, provided that a holistic systems approach is adopted.

Garment technology (Lina, 2015; Peterson, 2012) relates to the technology to realise the production of clothing, from design to manufacture. It includes a broad range of machinery for patternmaking, garment construction, apparel production, and knitwear, as well as fabric/material production of weaving, knitting and nonwovens.

This chapter outlines sustainable innovations in garment machinery as an integrated component for improved sustainable development based on regenerative practices to support the transition towards a circular system of production and consumption. A regenerative garment production framework is outlined and contextualised in relation to prior great societal transitional moments. Seamless knitting technology and nonwovens and sustainable colouration processes are then discussed as examples of regenerative garment production.

## 15.2 Historical pivot points for the fashion industry

Innovations and technological developments in garment machining and manufacturing are intricately linked to cultural shifts and societal change. In turn, cultural shifts transform garment machinery and production (Harris, 1993). Innovations have led to efficiencies, in speed of production and cheaper goods, but often at the expense of environmental and labour working conditions. Today, the textile and fashion industries are linear and fragmented systems. However, it was not always like this, nor does it need to be in the future. As the COVID-19 global pandemic has shown, changes in societal values and human behaviours that were already underway are accelerating (Castañeda-Navarrete et al., 2021).

It is useful to contextualise the transformation happening currently relative to prior significant societal transitional moments. Broadly, the textile industry has had four transitional phases:

- Transition to early settlements and early forms of clothing production
- Transition to pre-industrial clothing production – hand-powered
- Transition to industrial clothing production – machine-powered
- Transition to post-industrial clothing production – digitalisation

These four transitional phases discussed, highlight the interconnected relationship among raw materials, textile production, and garment manufacturing relative to shifts in consumption patterns. The social, political, and economic dimensions are considered alongside the technical. These broad transitional phases feature the dynamic and changing nature of garment technology and lead us to what might be considered the next transition for clothing production.

### 15.2.1 Transition 1: Early forms of clothing production

The transition of family units from being nomadic to forming early settlements aligned with the shift from animal fur and skins as the dominating clothing to making textile cloth using simple tools (loom and needle) and clothing (sewing needle) (Harris, 1993). Materials were limited to natural fibres locally sourced. Cloths were made into garments with little to no stitching of seams, no cutting, and no material waste. Nonwoven cloths were made for cooler climates, and woven cloths for warmer climates, made primarily from linen (Elmogahzy, 2019). Over time lengths of cloth were sewn together as simple tunics, and sashes used to define the waist, and pleats to give movement. Important developments of this period were the sewing needle, spinning yarn, looms for band weaving, simple looms with the shuttle and heddle, an early form of knitting, 'nålebinding' (binding with a needle) (Black, 2012). The processes of making cloth (textiles) and clothing (garments) were intricately connected. Centred on local knowledge and artisanship, textile fabrics and garments were made at home and in small workshops, as cottage industries emerged (Weibel, 1952). As materials were valuable, garments were intended to last a lifetime, and nothing was wasted. Production and consumption were a collective process within the family unit providing only what was needed.

### 15.2.2 Transition 2: Pre-industrial clothing production

The development of early city centres, trade routes and the early mechanisation of textile cloth saw textile and clothing production begin to separate. Regional areas began to specialise and produce excess quality textiles, and through trade exchanges of textiles and goods were made. This brought about a greater choice, with textiles made from natural fibres sourced from afar. The wearing of certain textile cloth reflected status in society and 'fashion' as a concept began to emerge.

As the textile industry grew, craft and merchant guilds emerged to promote economic and social interests providing training, employment, and protection of trade. Key inventions aided hand-powered production, including the spinning wheel to support the knitting industry, and the knitting needle as a more efficient process to make socks. Over time knitters began to use five needles to knit circular tubes instead of flat pieces (Black, 2012). While cloth was beginning to be mechanised, garments were still hand-sewn together. Tailors developed more sophisticated garment patterns and cloth-cutting emerged. Initially, cutting cloth was straight, and over time, curves were introduced. With more elaborate stitching and cutting, more cloth was used.

While textile cloths began to be sourced globally, garment making by necessity remained close to wear in the city. Clothing was made-to-measure for each individual wearing a garment, either tailor-made or homemade. The ready-to-wear industry was nascent, with accessories such as detachable sleeves and collars, gloves, and hats based on a standard wearer.

### 15.2.3 Transition 3: Industrial clothing production

Urbanisation, scientific discoveries, and the Industrial Revolution saw the textile industry transform from hand production methods to machines. Stream power and waterpower led to the mechanisation of spinning, weaving looms with the flying shuttle, knitting machines, and the invention of the sewing machine, all of which led to quicker methods of garment assembly.

With the scaling up of production, the organisation of work shifted to the factory system and the assembly line, and this further drove urbanisation as people sought work. Production of textiles was linked to an economic system that valued profits, reduced costs to increase production, and led to the widening of the distribution of wealth. Environmental degradation and pollution (air, water, and land) and increasingly dangerous working conditions were not accounted for (Deng et al., 2018).

Textile manufacturing was further hidden away from garment making, as clothing was still predominantly made to measure according to the measurements of the wearer. It was not until the mid-to-late 19th century that the retail department store emerged, and the consolidation of ready-to-wear garments and products began to take hold. As clothes began to be standardised, clothes were now either tailor-made for the wealthy who could afford homemade or ready-made (Harris, 1993).

### 15.2.4 Transition 4: Post-industrial clothing production

With globalisation and the emergence of service-oriented economies and digital information, workers further elongated and fragmented the supply chain of textiles and fashion. In regions where service industries overtook manufacturing, patterns of consumption and its associated by-products of 'waste' grew unsustainably, and greenhouse gas emissions emerged as the existential threat humanity faces (Tonn, 2009).

Globalisation in the late 20th century saw the linear supply chain become more fragmented, volatile, and elongated as brands and retailers dominated. Production and consumers are now split across international borders aligned to the Global North and Global South (Payne, 2020). Scientific developments led to synthetic fibres and synthetic blends as alternatives to natural materials, production became faster and cheaper, with garments being ready-made to a standard size. Individuals need to fit the clothes, not the clothes of the individual. With garments becoming cheaper, more could be purchased and worn less, and discarded rather than mended or handed on. This pattern of production and consumption draws on increased resources, generating more waste in the system at all stages from pre-consumer waste, use phase (fewer wears per garment) and disposal. In improving production efficiencies and quickening supply chain lead times, digitalisation, real-time data harvesting, and automation have only amplified these issues (Moretto et al., 2018).

While much of garment production was being 'faster and cheaper,' another market was also emerging by the early 21st century related to wearable technologies. This new wearable sector has seen the need for interdisciplinary approaches to garment design and production with increased complexity with the use of advanced conductive and smart materials, and the development of wearable systems, to provide a means to monitor health and personal information as well as communicate with other devices (Elmogahzy, 2019).

## 15.3 Circular economy and regenerative clothing production

Regenerative fashion refers to the clothing produced in ways that support circularity. As the world emerges from the global COVID-19 pandemic, changes in societal values and human behaviours that were already underway are accelerating. With the urgent need to address climate change and consumption patterns, the circular economy is seen as a way forward. The circular economy (MacArthur, 2014) is based on shifting from the linear economy (make, use, dispose) to one that seeks to keep resources in use for as long as possible, extract the maximum value from them while in use, and then recover and regenerate products and materials (MacArthur, 2017). Two interconnected networks are emerging in the textile and fashion industry:

- One network focused on local small-scale prosumption (involving both production and consumption) supported by technology innovations to keep garments in circulation through new fashion design models of mending and additional customisation in the local region
- Another network focused on being about global 'smart' production requiring multidisciplinary teams to work in interdisciplinary ways to reclaim garment materials for new products

Both networks are nurtured through regenerative approaches, being driven by processes rather than products. An example of how this might work is already beginning to be seen in the weft knitting sector's seamless knitting technology and the garment sewing technology sector.

Climate change is a dynamic process affecting global temperatures and carbon dioxide levels in the atmosphere. The solar radiation quality and heat stress have become a significant concern and climate change should be acknowledged as a global priority (Gilligan, 2007; Matheny, 2007). Within the textile and fashion industry, a step back to reconcile the concept of fast and luxury fashion driven by capitalism and the asynchronous use of material together with the failure to introduce design choices for consumers accommodating recycling or reuse options, have created enormous problems of ethics, morality, environmental responsibility, and responsible consumption of the mass population. It is evident from the state of global textile waste issues mentioned in the discussion in this chapter, that the 'triple p' approach to sustainability has aimed mostly at 'profit', sacrificing the needs of people and the planet (Islam, 2021). There is a consensus in the developed economies on slow and responsible fashion and this needs to be treated as a mainstream approach from all levels of society and socio-economic backgrounds. The global textiles and fashion industry need to address their share of climate change (Rockström et al., 2009) caused by human production and consumption practices from this point forward and take the responsibility to mitigate the threat of extinction of the human race (Matheny, 2007) from the face of the planet (Dietz et al., 2018).

In context to the current climate crisis (Rockström et al., 2009), the goal of sustainable development is an important one, but as an all-encompassing worldwide framework, it can be challenging to know how to implement practical, localised solutions. Similarly, framework-based approaches that focus on the technical dimensions of an individual garment are inadequate to deliver meaningful social and environmental change. There is a need to connect and view macro and micro-level frameworks as interdependent. To do this, a viable way forward for the textiles and fashion industry is to incorporate the concept of regenerative practices (Pawlyn and Ichioka, 2022; Mang and Haggard, 2016).

At its core, sustainable development seeks to meet peoples' needs of today without compromising the needs of future generations to meet their needs and its three pillars – environmental, economic, and social – are well established (Tonn, 2009; Brundtland et al., 1987). But the pursuit of economic growth has often resulted in environmental degradation and social disparities (Islam, 2020). Criticism of this approach suggests that the three pillars are trade-off each other rather than being integrated and can function as a compliance mechanism and measurement approach, rather than seeking systemic change to benefit all (Goel, 2010). To overcome this, drawing on regenerative practices may be useful.

Regenerative practices seek to develop restorative systems through feedback loops to benefit all living species and the planet, from the individual, community, and society (Mang and Haggard, 2016). Regenerative refers to processes that restore, renew, and revitalise and are linked to the concept of the circular economy (Morseletto, 2020). It highlights how current economic systems are inadequate and need to be restructured. To do this, a process-orientated whole systems approach needs to be adopted from a wider perspective to address biological and technical cycles.

For the textile industry taking a holistic systems approach is challenging given the global and complex nature of its value chain. As a result, the industry has often considered sustainability from the product level and the technical dimensions. For example, there are many sustainability-related approaches focused on technical dimensions such as life cycle assessment (LCA) (Zhang et al., 2018) (refer to Fig. 15.1), the 3R's (reduce, reuse, and recycle), cradle to cradle (Braungart et al., 2007), zero liquid discharge (Jenny and Anna, 2017), and zero discharge of hazardous chemicals (ZDHC) (Nimkar, 2018), zero waste in manufacturing (McQuillan, 2020), green manufacturing (Rusinko, 2007), low impact products, waterless products and carbon neutral products are becoming widely practised in the textiles and clothing sector. Through LCA the environmental impacts in a product's five life cycle stages (Fig. 15.1) of raw material acquisition stage, manufacturing stage, distribution and transportation stage, consumption and usage stage, and disposal stage can be assessed.

There are also various industrial attempts such as coalitions and collaborative efforts in harmonising the assessments of the environmental and social impacts of the textiles. Presiding over all these approaches is the United Nation's Sustainable

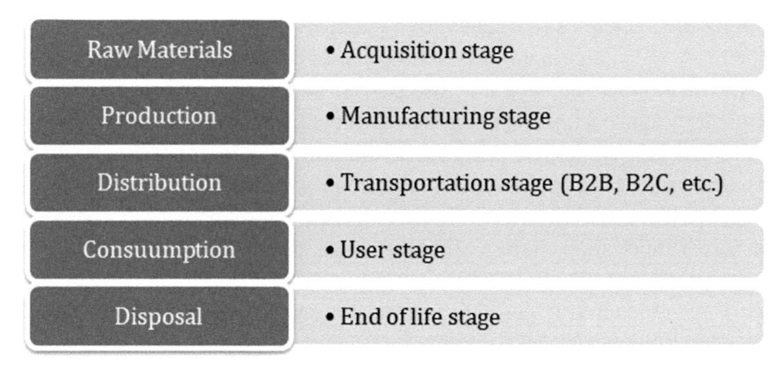


**Figure 15.1** Life cycle stages of textile and fashion products.

Development Goals (SDGs) framework, which seeks to unify and build a better world by the year 2030 as shown in Table 15.1. The 17 SDGs work as drivers to improve world issues under the themes of water, energy, climate, oceans, urbanisation, transport, science, and technology (United Nations, 2015, 2019). These approaches all play an important part in the textile industry's strategies to reduce its environmental impacts, but to be truly effective, they require a more transparent and interconnected pathway (Gardetti and Muthu, 2020).

**Table 15.1** Sustainable development goals of the United Nations (United Nations, 2019).

| The sustainable development goals | |
|---|---|
| **Goal 1** | No poverty |
| **Goal 2** | Zero hunger |
| **Goal 3** | Good health and well-being |
| **Goal 4** | Quality education |
| **Goal 5** | Gender equality |
| **Goal 6** | Clean water and sanitation |
| **Goal 7** | Affordable and clean energy |
| **Goal 8** | Decent work and economic growth |
| **Goal 9** | Industry, innovation and infrastructure |
| **Goal 10** | Reduced inequalities |
| **Goal 11** | Sustainable cities and communities |
| **Goal 12** | Responsible consumption and production |
| **Goal 13** | Climate action |
| **Goal 14** | Life below water |
| **Goal 15** | Life on land |
| **Goal 16** | Peace, justice and strong institutions |
| **Goal 17** | Partnerships to achieve the goals |

Adopting the principles of regenerative practices could provide the textile industry with a better way to integrate more technically driven approaches such as LCA with the holistic worldwide approach of the SDGs. Bringing together these distinct but complementary approaches could deliver production processes that are regenerative and able to integrate ecological and societal dimensions with the technical dimensions. But such approaches must be a shared responsibility between key stakeholders of manufacturers, brands, and consumers. There is no one solution here, but rather a myriad of interconnected innovations across all facets of garment production.

To understand sustainable innovation for garment machinery and manufacturing, a framework for regenerative garment production was developed, as illustrated in Table 15.2. This framework seeks to align and integrate the SDGs and LCA to support the longer-term goal of sustainable development and consists of eight elements:

- Material choice
- Material optimisation
- Energy emissions
- Chemical usage
- Place-making
- People
- Consumer practices
- Collaborative partnerships

Each of the eight elements together form part of a holistic system of intersecting interdependent network to facilitate lasting sustainable innovation. Interdependency is the key here as this approach enables to consider how machine innovations, individual products, and human decisions 'combine in cumulative, layered, holistic effects that influence entire systems' (Fletcher and Tham, 2019).

## 15.4 Sustainable innovations in garment manufacturing

### *15.4.1 Seamless 3-D knitting technology*

Weft knitting sectors demonstrate how taking a systems approach supports regenerative garment production. For this sector, the most significant innovation has been the development of seamless knitting technology and its associated software. Seamless technology offers enormous potential to provide a 'technology-enabled option to the sustainable fashion system' (Hethorn and Ulasewicz, 2008).

The leading flatbed knitting machine manufacturers are Shima Seiki and Stoll (Sieki, 2022; Stoll, 2022), which have both developed machine capabilities to produce 3-D seamless garments, as shown in Fig. 15.2. Seamless knitting technology allows for a fully automated computer-controlled method of production, by simultaneous knitting of three tubes (sleeve, body, sleeve) to create a garment. At the underarm connection point the two sleeves effectively 'slide' into the body of the garment to form one tube (Underwood, 2009) Through a combination of stitch transfer and holding stitches, the

**Table 15.2** Regenerative garment production framework.

| Elements | Garment machinery and manufacturing to facilitate | LCA stages | SDGs |
|---|---|---|---|
| Material choice | Material selection and production to transition away from emissions intensity of raw materials to enable use of alternative and new recyclable materials in manufacturing. | Raw material acquisition | 15, 9, 12 |
| Material optimisation | Material optimisation, and waste minimisation/ elimination through all stages of garment production | Manufacturing stage | 9 |
| Energy emissions | Decarbonising manufacturing and reducing emissions through improved energy efficiencies and transition to clean renewable energy sources | All five stages | 7, 13 |
| Chemical Usage | Minimising and converting to less harmful chemicals through all stages of garment production | Raw material Acquisition manufacturing | 6, 9 |
| Place-making | Improving the local environment, waterways, and air quality, reducing pollutants through cleaner technology, | All five stages | 14, 15 |
| People | Improving garment employee conditions through health, and gender quality and providing employment, opportunities for reskilling, and decent work | Raw material acquisition manufacturing transportation and distribution | 4, 5, 8 |
| Consumer practices | Changing consumer behaviour from consuming to being custodians of materials for a circular economy. Garments are valued as an ongoing resource that circulates in a local region through reuse and repair or is returned to the producer for reprocessing. | Consumption and usage stage | 11, 12 |

*Continued*

Table 15.2 Regenerative garment production framework.—cont'd

| Elements | Garment machinery and manufacturing to facilitate | LCA stages | SDGs |
|---|---|---|---|
| Collaborative partnerships | Systems thinking and collaboration to support regenerative practices. Connecting production, brand operations and consumers to support closed-loop recycling processes. | All five stages | 12, 17 |

*LCA*, life cycle assessment; *SDGs*, Sustainable Development Goals.

**Figure 15.2** Shima Seiki Wholegarment flatbed knitting machine.

sweater can be shaped by increasing and decreasing the number of stitches in each tube. In addition, different stitch patterns such as ribs, lace, and links-links patterns, as well as different yarn textures and colours can be designed into the garment. By either knitting one, two or three tubes an ever-increasing range of garment types, sleeve types and necklines can be achieved, such as cardigans, sleeveless sweaters, skirts, and pants.

Coupled with the machinery capabilities, there has been the development of sophisticated software that integrates design and technical programming via online libraries of yarns, stitch structures and patterns, and garment types. A garment can effectively be designed by selecting, mixing, and matching these together as a digital visualisation, while the technical programme is simultaneously built. The software for Shima Seiki's Wholegarment is the SDS-ONE Apex series (Sieki, 2022) and for Stoll's Stoll-Knit (Stoll, 2022) and Wear, is the M1plus and the Eneas software packages (Underwood, 2009).

The key advantage of seamless knitting technology is that it brings together and streamlines several garment production processes. Conventional knitwear be it 'cut and sew' or 'full fashioned' requires the fabric to be knitted as a length to be cut or as partial or fully fashioned panels, then sewed together, and then finished with details added and trims linked on. With seamless knitting technology, all these steps are clubbed together. Another advantage is that the garment, once designed, can be mass-manufactured at a lower cost and be readily modified to provide mass-customisation. Consequently, increased preparation time in the design and technical programming of the garment can be offset by the significant savings made in reducing the need for post-knitting labour, resulting in cost savings, greater efficiencies, and waste minimalisation (Choi and Powell, 2005)

By taking advantage of fibre and yarn developments, with material optimisation and waste minimisation through a streamlined garment production process, seamless knitting technology, has the potential to support low-emissions manufacturing through improved energy efficiencies. As knitting machines become more efficient, achieving faster knitting speeds and more reliability, the technology needs to transition to renewable energy sources.

Considering the framework for regenerative garment production, seamless technology can demonstrate how small innovations can affect the whole system of production. For example, seamless knitting machines increasingly support the use of more diverse yarns and material selection. For example, the development of the bi-partile compound needle (which provides a short, smooth, and simple action, compared to the conventional latch needle), the control of individual needle selection and stitch transfer, the sinker device (a thin metal plate, that is positioned right between the adjoining needle bed), and variable take-down rollers, and split carriage systems for increase speed and energy efficiencies. These have made it possible to control individual knit stitches on active and inactive needles so that only selected needles are knitted, while the stitches on non-selected needles are held. The resultant fabric provided a way to shape knitting, known as partial knitting and formed the bases of the seamless knitting of a tube where stitches are stored and transferred between beds. These machine innovations have resulted in greater knitting reliability and versatility, enabling a wide range of yarns and are a small but important part of the holistic systems approach needed towards achieving closed-loop recycling solutions.

Seamless technology also supports the potential for closed-loop garment systems. International brands and retailers, through their established distribution channels, have the means to collect and return seamless knitted garments to their factories for reprocessing. There is the potential to unknit a garment and then reknit the yarn into a new garment (Sieki, 2022). This approach of reknitting is already deployed within the pre-consumer stage of production as a waste minimisation strategy and could be extended to include post-consumer garments. How it might work for the post-consumer stage and at scale would need careful consideration, particularly if a garment is discarded because of damage through wear. This could lead to an expanded role for the retailer to include a collection and quality assessment process, so that a seamless knitted garment, when returned, is assessed for garment damage. The garment can then be either returned to the factory for repurposing into a new product or sent into a local network for mending and customisation and then resold. This approach opens the potential for small-scale manufacturing that is about service-based practices supporting ongoing bespoke garment customisation, visible mending, repurposing garments, and alternation services. Similarly, a knitted garment made of luxury materials could be shared and rented via an online retail platform. These approaches of knitwear garment circularity highlight the need to develop strong partnerships and shared investment across the value chain, integrating yarn producers, knitwear producers, brands and retailers, and the consumer.

### 15.4.2 Nonwoven products system

Nonwovens are fabric-like materials made by bonding fibres together through chemical, mechanical, and heat processes (Senthil and Punitha, 2017). The key difference in nonwoven production is the omission of the yarn manufacturing stage which governs fabrication processes of both knit and woven materials. Nonwoven sector, as part of the textile industry shows great potential to be a leader in how regenerative garment production might look in the future. Nonwovens, like all textile sectors, have been steadily moving towards more sustainable production processes (Asabuwa Ngwabebhoh et al., 2021).

Nonwoven materials play an increasingly influential role in the global manufacturing of many consumer products. Existing market sectors such as health, hygiene and automotive applications are expanding, and others are being added, with much of the growth driven by advances in technology and the need for more sustainable products (Consulting, 2022). The use of specialised fibres coupled with new construction techniques for novel and more sustainable structures has become one of the trends in the nonwoven industry. The nonwoven sector can be considered an indispensable sector of textiles which produces a broad spectrum of important and inevitable products for daily life (Chaudhari et al., 2008).

The term sustainability can be discussed at any level and one of them is product level. Product sustainability as a term must be balanced in the entire life cycle stages of a product. Ideally, a product must be sustainable through its entire life cycle as shown in Fig. 15.1. Innovation in the nonwoven industry is occurring much more rapidly today than it did within the last 2 decades. The main drivers for this include increasing demand for nonwovens in existing markets and the new applications and opportunities that are opening every day. The environmental impact of products has re-emerged as an important challenge for the nonwoven industry as the global economy has recovered. The concern over single-use and microplastics is a strong consumer trend which is leading to nonwoven businesses re-evaluating their products (Goswami and O'Haire, 2016).

Fast innovation in raw material research (Santos et al., 2021) and their availability are driving the market away from oil-based synthetics and into biopolymers and natural products (Consulting, 2022). Nonwoven product designs are also changing with the manufacturers now considering what happens at the end of the service life and how products can be designed for a second or even third end use (Rubino et al., 2021).

The stratification and customisation of nonwoven garments are becoming an eminent challenge for nonwoven manufacturers worldwide. In developed economies, where the consumers' fundamental needs have been provided for by mass-produced products, business opportunities will evolve from increasingly customised products and services. This trend towards increasingly tailored garments and services to satisfy individual customer needs translates to the offerings from nonwoven companies need to be designed around flexibility and should include the end-user in the development process. The tailored late-phase functionalisation for materials produced through nonwoven fabrication could be one of the tools that can help nonwoven manufacturers meet this challenge of sustainable production.

### 15.4.2.1 Broad spectrum of solutions

Garments are designed for different shelf life in context to their use and disposal like single-use or multiple uses. There are many industry-wide debacles in consolidating which garment is more sustainable - be it single-use or multiple uses. There is no simple answer that can be provided for these debates except for conducting a research review of a systematic LCA study which encompasses all the life cycle stages of a product and there is a vast research gap between the industry and a framework approach (Table 15.1) would provide a pathway to resolve these issues. Disposable products after single-use should be carefully considered in the current sustainability-conscious society. At this point comes the important aspect of sustainability in the nonwoven sector as the nonwoven segment of textiles is catering to the needs of a throwaway society. This, in turn, could be utilised for sustainable fast turnaround solutions, as shown in Fig. 15.3.

Fabrication of sustainable products in various aspects such as sourcing sustainable or green raw materials, renewable energy inputs, sustainable auxiliaries, green chemicals, and other production inputs, consuming less water and recovery, mitigating the amount of waste output and an overarching design for regenerative production processes, etc., are the need of the hour for the nonwoven sector as far as the production phase is concerned.

As a product, a nonwoven has an added edge in sustainable production in terms of its low manufacturing time with a fast production cycle and could be the answer to recycle/downcycle or upcycle materials from other manufacturing streams. A comparatively lower number of production processes to produce a product compared to

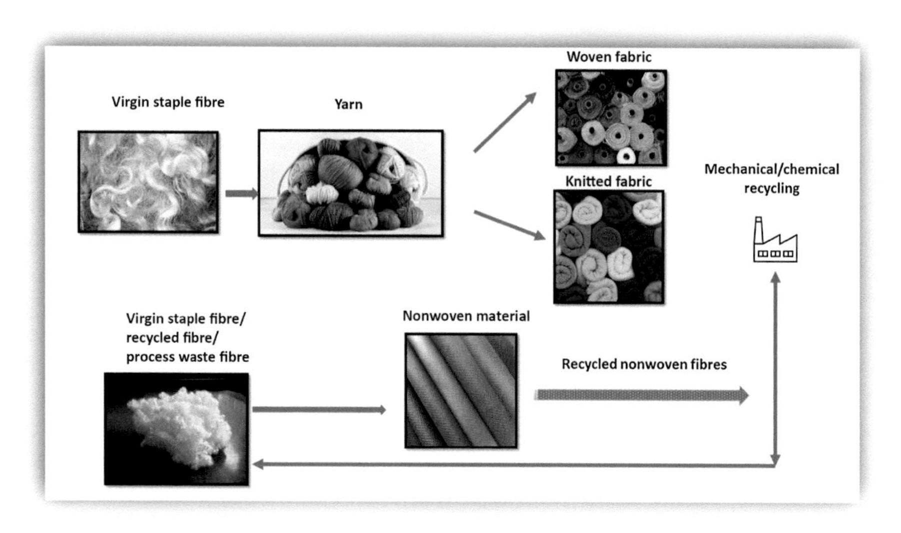


**Figure 15.3** Nonwoven technology as a regenerative practice.

competing production processes in the woven and knit categories certainly paves an opportunity window for any nonwoven manufacturing sector.

There is a plethora of options to utilise green or sustainable raw materials (Islam, 2020) such as recycled fibres and the use of green or sustainable chemicals (Nimkar, 2018). Also, many ways to make a nonwoven manufacturing line entirely green and sustainable exist. These include being less energy-intensive, less water-intensive, using energy and water from renewable resources, following 3R's in the production cycle, and more (Senthil and Punitha, 2017). Currently, researchers working on these specific areas to attain sustainability in the textile sector, which no doubt could be adopted and incorporated in the nonwoven sector simultaneously such as automation of design to manufacturing (Davies, 2018).

### *15.4.2.2 Automation of design to manufacturing*

This is one side of the story in terms of sustainable nonwovens; however, automation (Cloppenburg, 2019), artificial intelligence (Gültekin et al., 2020; Yekrang et al., 2016), and 3-D and virtual platforms (Wiegmann and Becker, 2007) will play pivotal roles in designing nonwovens in the extremely near future. These strategies may also contribute to alleviating the rising concern of developed countries on imported resources and over-securing access to critical raw materials, some of which may play a fundamental role in deploying low-carbon, renewable energy technologies. There needs to be a direct policy intervention on material resource efficiency (Islam, 2020). At the other end of the textile materials chain, still now globally a large amount of waste is generated but is increasingly moving towards more recycling and nonwoven manufacturing could certainly provide an avenue to mitigate a large amount of waste ending up in landfills. Waste-related targets, specifically SDG 12, could be a good starting point for the textile and fashion industry and requirements will help governments to increase recycling, although the prospects for reducing waste generation are less certain (Islam, 2021).

### 15.4.3 Garment sewing technology

Recent developments in automation and digitalisation are transforming traditional cut-and-sew garment production. This is bringing the garment design, sampling, and production phases closer together, by integrating virtual design with physical fabric and garment manufacturing into a frictionless streamlined workflow. Central to this is the need for collaborative partnerships across the supply chain to ensure compatibility and efficiencies.

An example of this approach is the partnership between fashion design software company CLO 3-D (Wang and Liu, 2020) with SwatchOn the wholesale online fabric distributor, Lectra, who specialise in automated cutting equipment, and Alvanon, who focus on fit tools (physical and digital mannequins). While each company is focused on a highly specialised aspect of the industry, together they can provide the user of CLO 3-D software with an integrated workflow. A designer using CLO 3-D to create and modify a virtual garment can access SwitchOn's extensive digitalised 3-D fabric library, knowing that the physical fabric can be ordered for production and that the garment can be fit-tested on a range of digital bodies in real time and later fitted to a physical mannequin. As this process is occurring, the digital garment pattern is also being created and updated with any modifications. The result is the finalised digital garment pattern can then be sent directly to a digital plotter and cutting tool. This process has the potential to reduce the time needed for sampling while bringing efficiencies through more informed material choices while minimising material waste.

Further to this, developments in automated garment assembly and sewing technologies are emerging as viable alternatives to labour-intensive manual processes. In garment manufacturing, the complex nature of 2-D pattern-making to 3-D garment construction and assembly, there is significant time and labour in material handling, such as lifting, moving, and positioning the cut or semi-finished fabric components, and in joining of fabric components (Salahuddin and Lee, 2022). Breaking down the tasks across a garment's assembly has allowed for innovative developments in various automation systems and advanced tools to be used such as automated material handling devices, inspection processes, pick/place robots, and high-speed sewing machines (Nayak and Padhye, 2018).

While automation in sewing is still in its infancy there is growing focus on this area due to increased labour costs in many developing countries, as well as a global shortage of skilled labour. Sewing is the most important textile joining technology, representing 85% of all joining methods (Jung et al., 2020). The type of fully- and semi-automated process depends on the complexity of a garment's construction, be it as a straight, contoured, or curved 2-D or 3-D seam and the greater the precision and handling required.

There are now several semi-automated sewing machinery commercially available such as Japan-based Juki Corp., Italy-based RI.MA.C. S.r.l., and Germany-based

Durkopp Adler AG (Jung et al., 2020). Machinery can also cater for specialised functions such as buttons, buttonholes, and binding. Within the machine, developments such as an automatic bobbin changing system. Fully automated sewing at a large scale is possible with Sewbots by SoftWear Automation Inc., Atlanta. It can track individual threads at the needle and coordinate the precise movement of the fabric using a four-axis robotic arm (Menghi et al., 2018).

### 15.4.4 Garment dyeing and washing

The garment industry is critical for meeting the rapidly changing demands of fashion. Today's market has an increasing need for short-run repeat orders, with low skill input, low reject rates, complex designs, unique washes, and value additions (Deraniyagala, 2017). Garment dyeing is an important process that enables these demands to be met using simple, inexpensive equipment with modest space requirements and without the production of effluent (Nayak et al., 2020).

The Society of Dyers and Colourists' recent white paper (SDC, 2022), highlighted how technology and innovation can reduce the carbon footprint of textile dyeing. The paper focuses on how nature can influence and provide solutions for clean colouration. Archroma and Colorifix are two European companies that have been featured for their use of natural hues in textile dyeing. Archroma has created a limited palette of dyes based on non-edible agricultural and herbal waste, while Colorifix uses microorganisms to grow colour based on natural DNA codes. Both companies have opened opportunities for dyehouses to incorporate more natural dyeing into their existing operations (SDC, 2022). Archroma's EarthColors dye range is used by various brands, including Primark, Patagonia, and Esprit. Colorifix looks for colours generated by organisms in nature and uses DNA sequencing to pinpoint the genes, leading to the production of the pigment. The company's products have received interest from mills and brands (SDC, 2022). The two companies have taken a comprehensive approach to dyeing processes, focussing on replacing chemistry with biology, in a bid to minimise the environmental impact of industrial dyeing (Teo et al., 2022).

In the fashion industry, predicting which colours will be in demand can be challenging. To address this, the general trend is to defer the finishing process until after the garment is made. Garment dyeing of cotton in its finished form is increasing in importance, as it allows for the use of the colours in demand and subsequent finishing for the required resistance to wrinkles and dimensional stability. However, the industry is always exploring new garment finishing systems that can be used in conjunction with garment dyeing (Natarajan et al., 2022). There is a wide variety of dyes available that can be used to create unique looks on garments, including direct, reactive, sulphur, and pigments (Raj et al., 2022). Sulphur and pigments are used to achieve worn-out looks.

Indigo dyeing of cotton denim yarns creates highly coloured effluents due to the low affinity of the dye to cotton. However, in the garment dyeing of denim, there is a possibility to achieve sustainability by reusing the dyebath continuously without replenishment. This can result in a series of lighter shades, without releasing any coloured effluent (Deo and Paul, 2004a). On the other hand, a standing bath technique with continuous replenishment can result in similar shades repeatedly (Deo and Paul, 2004b). Garment washing processes on denim can produce special effects, such as distressed denim, quick wash denim, tinted denim, grainy look denim, sandblast denim, etc. The new developments include laser technology, stoneless stone wash effect, ion wash, mud wash, ozone fading, and water jet fading (Paul, 2015).

Smart colourants are a type of dye that responds to different environmental conditions, such as temperature or light (Thadepalli and Roy, 2022). These colourants for garments are becoming increasingly popular (Shaban et al., 2022). They are called smart because they sense conditions in their environment and respond to those conditions (Karpagam et al., 2017). Smart colourants can be classified into three categories: photochromic colour, thermochromic colour, and glow-in-the-dark colour. Photochromic colour changes from clear when indoors to colour when taken outdoors, responding to UV radiation. Thermochromic colour changes with temperature, and glow-in-the-dark colour emits light after excitation has ceased. Photochromic colours can be applied to garments using three methods: spray, brush, and screen print. Similarly, thermochromic colours can be applied using the same three methods. In all cases, the colours are directly applied to the garment and cured at 150°C for a set amount of time. Glow-in-the-dark colours hold their afterglow for up to 2 h and can also be applied to garments (Çay et al., 2017).

The garment colouration industry is constantly evolving, with new garment finishing systems and dyes such as biogenic dyes (Fried et al., 2022) being developed to meet the rapidly changing demands of the fashion market (Costa et al., 2020). With new developments in technology, such as laser technology and ion wash, the garment industry is sure to continue to grow and adapt to meet the ever-changing demands of the fashion market. In addition, wastewater treatment and separation of colour have become an important aspect of the industry.

## 15.5 Future perspectives

Looking ahead to the future of sustainable clothing production, the transition towards a circular economy will be critical in meeting the UN's SDGs soon (UNFCCC, 2018). To achieve this, the fashion industry will have to adopt innovative approaches to garment production that prioritise reducing waste and minimising the use of natural resources. Some promising approaches to achieving this are using advanced technologies such as artificial intelligence (AI), augmented reality, robotics, and blockchain platforms. These technologies can be used throughout the garment production process, from manufacturing to post-consumer resource sorting, to minimise inefficiencies and waste generation.

In smart garment manufacturing, AI can be used to optimise decision-making processes, creating a 'decision-making tree' approach that considers a range of factors, including sustainability, resource efficiency, and cost-effectiveness. This will enable brands and manufacturers to make informed decisions about the materials, processes, and technologies used in garment production, helping to minimise waste and reduce the industry's impact on the environment.

Augmented reality can also be used to improve the efficiency of clothing production. By allowing designers and manufacturers to visualise garments in 3-D, it is possible to identify potential design flaws and adjust before garments are manufactured, reducing the need for multiple iterations, and minimising waste (Wang and Liu, 2020). Similarly, robotics will increasingly be used to automate processes such as synchronous sorting and streamlined recycling of post-consumer materials, further reducing waste, and improving resource efficiency.

Blockchain platforms can play a crucial role in achieving greater traceability and transparency in the fashion industry (Yadav et al., 2023). By creating a secure and decentralised system for tracking the entire supply chain, blockchain technology can help to identify and address issues such as labour exploitation, environmental damage, and counterfeit goods (Badhwar et al., 2023).

By blockchain, consumers can gain greater transparency into the origins and production methods of the garments they purchase. This can help to increase consumer awareness and demand for sustainable and ethical clothing production practices, thereby driving positive change within the industry. Moreover, blockchain can also facilitate the sharing of information between different players in the supply chain, enabling greater collaboration and accountability. By creating a secure and tamper-proof record of transactions, blockchain technology can help to reduce the risk of fraud and increase trust between different parties (Badhwar et al., 2023). As the fashion industry continues to embrace sustainable practices and advanced technologies, one can look forward to a future in which clothing production is more efficient, less wasteful, and more environmentally friendly.

## 15.6 Conclusion

Within the textile and fashion value chain, increasing resource efficiency, preventing material waste, and shifting to post-consumer 'waste' are being revalued as new resources at the core of the circular economy. For such a transition to happen, it will require a different kind of mindset, one that can creatively navigate multiple overlapping systems that consider and integrate economic, social, ecological, and political dimensions. Through the integration of the SDGs and LCA, a framework of eight elements for regenerative garment manufacturing is proposed in this chapter.

This framework demonstrates how future garment production practices could transform the industry to address raw material scarcity, deploy low-carbon, renewable energy technologies, and prevent environmental degradation, as well as provide a means for improved working and worker conditions. While each element of the framework offers significant ways to address sustainability, it is only when all elements are linked together and considered a holistic regenerative system that the smaller individual innovations can have a disproportionately large positive or negative impact across the entire system. A shift towards a multi-faceted approach, based on collaborative partnerships, shared responsibilities, and a system-oriented mindset to identify and address small interventions within the garment manufacturing sector, could provide the greatest potential to achieve longer-term sustainable development.

# References

Aakko, M., Koskennurmi-Sivonen, R., 2013. Designing sustainable fashion: possibilities and challenges. Research Journal of Textile and Apparel 17, 13–22.

Agenda and Group, 2017. Pulse of the fashion industry. Global Fashion Agenda & The Boston Consulting Group.

Allan, R.P., Hawkins, E., Bellouin, N., Collins, B., 2021. IPCC, 2021: Summary for Policymakers.

Asabuwa Ngwabebhoh, F., Saha, N., Saha, T., Saha, P., 2021. Bio-innovation of new-generation nonwoven natural fibrous materials for the footwear industry: current state-of-the-art and sustainability panorama. Journal of Natural Fibers 1–11.

Badhwar, A., Islam, S., Tan, C.S.L., 2023. Exploring the potential of blockchain technology within the fashion and textile supply chain with a focus on traceability, transparency, and product authenticity: a systematic review. Journal of Fisheries International B 6, 1–19.

Black, S., 2012. Knitting: Fashion, Industry, Craft. V&A Publishing.

Braungart, M., Mcdonough, W., Bollinger, A., 2007. Cradle-to-Cradle Design: Creating Healthy Emissions Ea Strategy for Eco-Effective Product and System Design.

Brundtland, G.H., Khalid, M., Agnelli, S., Al-Athel, S., Chidzero, B., 1987. Our common future. Journal of Northwestuniversity (Philosophy and Social Sciences Edition) 8.

Castañeda-Navarrete, J., Hauge, J., López-Gómez, C., 2021. COVID-19's impacts on global value chains, as seen in the apparel industry. Journal of Development Planning 39, 953–970.

Çay, A., Kumbasar, E.A., Morsunbul, S., 2017. Exergy analysis of encapsulation of photochromic dye by spray drying. In: IOP Conference Series: Materials Science and Engineering. IOP Publishing, 022003.

Chaudhari, S., Mandot, A., Milin, P., Karansingh, M., 2008. A Review on Nonwoven Fabrics Used in Apparel.

Choi, W., Powell, N.B., 2005. Three dimensional seamless garment knitting on v-bed flat knitting machines. Journal of Textile Apparel, Technology Management 4, 1–33.

Cloppenburg, F., 2019. Digitalization of Nonwoven Cards.

Consulting, S. M. R, 2022. Nonwoven Fabrics - Global Market Outlook (2020–2028. MarketResearch.com. Available: https://www.marketresearch.com/Stratistics-Market-Research-Consulting-v4058/Nonwoven-Fabrics-Global-Outlook-14737380/.

Costa, C., Azoia, N., Silva, C., Marques, E., 2020. Textile industry in a changing world: challenges of sustainable development. University of Porto Journal of Engineering 6, 86–97.

Davies, N., 2018. A major transformation in non woven textiles? AATCC Review 18, 30–35.

Deng, D., Aryal, N., Ofori-Boadu, A., Jha, M.K., 2018. Textiles wastewater treatment. Water Environment Research 90, 1648–1662.

Deo, H.T., Paul, R., 2004a. Reuse of indigo vats for shade development. International Dyer 189 (2), 14.

Deo, H.T., Paul, R., 2004b. Standing bath technique for indigo denim dyeing. International Textile Bulletin 50 (2), 66.

Deraniyagala, H., 2017. Textile colour waste and sustainability. In: Colour Design: Theories and Applications, second ed.
Dietz, S., Bowen, A., Doda, B., Gambhir, A., Warren, R., 2018. The economics of 1.5 C climate change. Annual Review of Environment and Resources 43, 455−480.
Elmogahzy, Y.E., 2019. Engineering Textiles: Integrating the Design and Manufacture of Textile Products. Woodhead Publishing.
Fletcher, K., Tham, M., 2019. Earth Logic: Fashion Action Research Plan. JJ Charitable Trust.
Fried, R., Oprea, I., Fleck, K., Rudroff, F., 2022. Biogenic colourants in the textile industry−a promising and sustainable alternative to synthetic dyes. Journal of Geosciences of China 24, 13−35.
Gardetti, M.A., Muthu, S.S., 2020. The UN Sustainable Development Goals for the Textile and Fashion Industry. Springer.
Gilligan, I., 2007. Neanderthal extinction and modern human behaviour: the role of climate change and clothing. Journal of Women and Aging 39, 499−514.
Global Fashion Agenda, 2019. Pulse of the Fashion Industry, 2019 Update.
Goel, P., 2010. Triple bottom line reporting: an analytical approach for corporate sustainability. Journal of Finance, Accounting and Management 1.
Goswami, P., O'haire, T., 2016. Developments in the use of green (biodegradable), recycled and biopolymer materials in technical nonwovens. Advances in Technical Nonwovens 97−114.
Gültekin, E., Çelik, H.I., Nohut, S., Elma, S.K., 2020. Predicting air permeability and porosity of nonwovens with image processing and artificial intelligence methods. The Journal of the Textile Institute 111, 1641−1651.
Harris, J., 1993. 5000 Years of Textiles. British Museum Press.
Hethorn, J., Ulasewicz, C., 2008. Sustainable Fashion. Fairchild Books.
Islam, S., 2020. Sustainable raw materials: 50 shades of sustainability. In: Sustainable Technologies for Fashion and Textiles. Elsevier.
Islam, S., 2021. 15 - Waste management strategies in fashion and textiles industry: Challenges are in governance, materials culture and design-centric. In: Nayak, R., Patnaik, A. (Eds.), Waste Management in the Fashion and Textile Industries. Woodhead Publishing.
Jenny, G., Anna, C.J., 2017. The impact of 'zero' coming into fashion: zero liquid discharge uptake and socio-technical transitions in Tirupur. Water Alternatives 10, 602−624.
Jung, W.-K., Park, Y.-C., Lee, J.-W., Suh, E.S., 2020. Remote sensing of sewing work levels using a power monitoring system. Applied Sciences 10, 3104.
Karpagam, K., Saranya, K., Gopinathan, J., Bhattacharyya, A., 2017. Development of smart clothing for military applications using thermochromic colorants. The Journal of the Textile Institute 108, 1122−1127.
Lina, W., 2015. Material Science and Garment Technology towards Circular Economies within the Fashion Industry.
MacArthur, E., 2014. Towards the circular economy: accelerating the scale-up across global supply chains. In: World Economic Forum Reports. Ellen MacArthur Foundation.
Macarthur, F.E., 2017. A new textiles economy: redesigning fashion's future. Journal of Ellen Macarthur Foundation 1−150.
Mang, P., Haggard, B., 2016. Regenerative Development and Design: A Framework for Evolving Sustainability. Wiley.
Matheny, J.G., 2007. Reducing the risk of human extinction. An International Journal 27, 1335−1344.
Mcquillan, H., 2020. Zero Waste Systems Thinking: Multimorphic Textile-Forms. Högskolan i Borås.

Menghi, C., García, S., Pelliccione, P., Tumova, J., 2018. Towards multi-robot applications planning under uncertainty. In: Proceedings of the 40th International Conference on Software Engineering: Companion Proceeedings, pp. 438–439.
Moretto, A., Macchion, L., Lion, A., Caniato, F., Danese, P., Vinelli, A., 2018. Designing a roadmap towards a sustainable supply chain: a focus on the fashion industry. Journal of Cleaner Production 193, 169–184.
Morseletto, P., 2020. Restorative and regenerative: exploring the concepts in the circular economy. Journal of Industrial Ecology 24, 763–773.
Natarajan, G., Rajan, T.P., Das, S., 2022. Application of sustainable textile finishing using natural biomolecules. Journal of Natural Fibers 19, 4350–4367.
Nayak, R., Nguyen, L.V.T., Panwar, T., Jajpura, L., 2020. Sustainable technologies and processes adapted by fashion brands. In: Sustainable Technologies for Fashion and Textiles. Elsevier.
Nayak, R., Padhye, R., 2018. Introduction to automation in garment manufacturing. In: Automation in Garment Manufacturing. Elsevier.
Nimkar, U., 2018. Sustainable chemistry: a solution to the textile industry in a developing world. Current Opinion in Green and Sustainable Chemistry 9, 13–17.
Paul, R. (Ed.), 2015. Denim: Manufacture, Finishing and Applications. Woodhead Publishing Ltd (Elsevier), Cambridge, UK. ISBN: 9780857098436.
Pawlyn, M., Ichioka, S., 2022. Flourish: Design Paradigms for Our Planetary Emergency. Triarchy Press.
Payne, A., 2020. Designing Fashion's Future: Present Practice and Tactics for Sustainable Change. Bloomsbury Publishing.
Peterson, J., 2012. Customisation of Fashion Products Using Complete Garment Technology. Tampere University of Technology.
Raj, A., Chowdhury, A., Ali, S.W., 2022. Green chemistry: its opportunities and challenges in colouration and chemical finishing of textiles. Sustainable Chemistry and Pharmacy 27, 100689.
Rockström, J., Steffen, W., Noone, K., Persson, Å., Chapin III, F.S., Lambin, E., Lenton, T.M., Scheffer, M., Folke, C., Schellnhuber, H., 2009. Planetary boundaries: exploring the safe operating space for humanity. Ecology and Society 14.
Rubino, C., Aracil, M.B., Liuzzi, S., Stefanizzi, P., Martellotta, F., 2021. Wool waste used as sustainable nonwoven for building applications. Journal of Cleaner Production 278, 123905.
Rusinko, C., 2007. Green manufacturing: an evaluation of environmentally sustainable manufacturing practices and their impact on competitive outcomes. IEEE Transactions on Engineering Management 54, 445–454.
Salahuddin, M., Lee, Y.-A., 2022. Automation with robotics in garment manufacturing. In: Leading Edge Technologies in Fashion Innovation: Product Design and Development Process from Materials to the End Products to Consumers. Springer.
Santos, A., Ferreira, P., Maloney, T., 2021. Bio-based materials for nonwovens. Cellulose 28, 8939–8969.
SDC, 2022. Destination Low Carbon: Global Technology and Innovation Reducing the Environmental Footprint of Textile Coloration. The Society of Dyers and Colourists (SDC).
Senthil, K., Punitha, V., 2017. An overview of nonwoven product development and modelling of their properties. Journal of Textile Science and Engineering 7, 1–5.
Shaban, H., El-Thalouth, A., Zaher, A., Hebeish, A., Shahin, A., 2022. Green and eco-friendly indigo printing using new smart nano-colorants and their application in combined printing and antibacterial finishing. Egyptian Journal of Chemistry 65, 1–2.

Sieki, S., 2022. Whole Garment. Available: https://www.shimaseiki.com/sustainability/.
Stoll, 2022. K.Innovation create design. Karl Mayer Stoll Textilmaschinenfabrik GmbH.
Teo, S.H., Ng, C.H., Islam, A., Abdulkareem-Alsultan, G., Joseph, C.G., Janaun, J., Taufiq-Yap, Y.H., Khandaker, S., Islam, G.J., Znad, H., 2022. Sustainable toxic dyes removal with advanced materials for clean water production: a comprehensive review. Journal of Clear Production 332, 130039.
Thadepalli, S., Roy, S., 2022. Implications of sustainability on textile fibres and wet processing, barriers in implementation. In: Sustainable Approaches in Textiles and Fashion: Manufacturing Processes and Chemicals. Springer.
Tonn, B.E., 2009. Obligations to future generations and acceptable risks of human extinction. Journal of Francais 41, 427–435.
Underwood, J., 2009. The Design of 3D Shape Knitted Preforms. RMIT University.
UNFCCC, 2018. Fashion Industry Charter for Climate Action. Germany Available: https://unfccc.int/climate-action/sectoral-engagement-for-climate-action/fashion-charter.
United Nations, 2015. Transforming Our World: The 2030 Agenda for Sustainable Development. United Nations, Newyork.
United Nations, 2019. In: The Paris Agreement. United Nations, Newyork.
Wang, Y.-X., Liu, Z.-D., 2020. Virtual clothing display platform based on CLO3D and evaluation of fit. Journal of Fiber Bioengineering and Informatics 13, 37–49.
Weibel, A.C., 1952. Two Thousand Years of Textiles: The Figured Textiles of Europe and the Near East.
Wiegmann, A., Becker, J., 2007. Virtual characterization of the pore structure of nonwoven. Proceedings of the International Nonwoven Technical Conference 670–677.
Yadav, N., Luthra, S., Garg, D., 2023. Blockchain technology for sustainable supply chains: a network cluster analysis and future research propositions. Environmental Science and Pollution Research 1–21.
Yekrang, J., Sarijeh, B., Semnani, D., Zarrebini, M., 2016. Prediction of heat transfer and air permeability properties of light weight nonwovens using artificial intelligence. Indian Journal of Fibre and Textile Research Journal 40, 373–379.
Zhang, Y., Kang, H., Hou, H., Shao, S., Sun, X., Qin, C., Zhang, S., 2018. Improved design for textile production process based on life cycle assessment. Clean Technologies and Environmental Policy 20, 1355–1365.

# Sustainable textile care and maintenance

16

*Edith Classen*
Hohenstein Institute, Boennigheim, Germany

## 16.1 Introduction

Textiles and clothing are used for a long time and should be clean and odour-free over the whole time of use. Therefore, cleaning textile products is important for the long use and maintenance of clothing. The cleaning process uses washing machines, water or organic solvents, energy and detergents, and produces wastewater. Since the 1960s, much effort has been made to optimise the washing process and cleaning results and to reduce energy and material consumption as well as the load on wastewater to minimise the environmental impact of laundry care. Washing textiles is routine work in homes and in textile service companies. Laundries and textile service companies are important for the health care sector, retirement homes, the hotel and hospitality sector, workwear and personal protective clothing. In households, washing is carried out in water and most textiles are washed in washing machines. For fabrics that cannot be washed in water, commercial dry cleaning is used. In commercial and industrial laundries, liquid solvents such as water or organic solvents are used. Advantages of water as a solvent are that water has high solvency for many substances and that it has good environmental and occupational safety. Advantages of organic solvents in dry cleaning are the gentle cleaning of fabrics and the easy removal of stains.

Cleaning removes dirt, especially from clothing. Fabrics can be stained from the atmosphere, bodily excretion and impurities derived from domestic, commercial or industrial activity. Stains can be divided in different classes based on solubility: water-soluble materials such as inorganic salts, sugar, urea and sweat; pigments such as metal oxides, carbonates, silicates, humus, and carbon black (soot); fats such as animal fat, vegetable fat, sebum, mineral oil and wax; proteins such as blood, egg, milk and skin residues; carbohydrates such as starch; and bleachable dyes from fruit, vegetables, wine, coffee and tea. Washing is a complex process involving numerous physical and chemical influences. The first step is to remove dirt by the water or aqueous surfactant solution of poorly soluble residues and dissolution. The second step is to stabilise dissolved soil in the wash liquor to avoid redeposition onto the fibres. Cleaning is influenced by various components such as the solvent (water or organic solvent), soil, textiles, washing equipment and detergent. The wash performance is influenced by textile properties, the type of stain, the water quality, the washing technique and the detergent composition.

To ensure that textiles can be washed under the right conditions, care symbols attached to a ribbon on the textile were introduced and indicate the manufacturer's

Sustainable Innovations in the Textile Industry. https://doi.org/10.1016/B978-0-323-90392-9.00015-X

suggestions for washing, dry cleaning and ironing clothing. In the early 1970s, Groupement International d'Etiquetage pour l'Entretien des Textiles, the European Association of Textile Care Labelling, and the International Organization for Standardisation (ISO) developed the international standard for textile labelling symbols. The care symbols are internationally recognized standards for care labels and pictograms and their correct use (GINETEX, 2022; ISO 3758 2012).

The washing of textiles has an environmental impact and releases detergents, dirt and microplastics into household wastewater. During washing, textiles are exposed to different strains that may cause fibres to be released into the washing effluent. In Europe, most households are connected to a sewage system and wastewater treatment. It is estimated that 13,000 tons of textile microfibres are released to surface water every year, accounting for 8% of total primary microplastic releases into water (Manshoven et al., 2022).

Regarding climate change, sustainability in the textile sector is becoming increasingly important. In March 2022, the European Union (EU) published the European Strategies for Textiles. In the EU, the consumption of textiles has, on average, the fourth highest impact on the environment and climate change, after food, housing and mobility. It is also the third highest area of consumption for water and land use, and the fifth highest for the use of primary raw materials and for resulting in greenhouse gas emissions (EU, 2022). To achieve more sustainability in the textiles ecosystem, a change from the current linear path to a circular textile industry is necessary.

In a circular economy, textile products are designed, produced and discarded, and the impact on climate change, unsustainable resource use and environmental pollution should be reduced in the textile value chain (Hemkhaus et al., 2019). The maintenance of textiles is important from the viewpoint of reducing the wastewater load, fibre debris and energy during manufacturing, as well as achieving a long, healthy and comfortable clothing life. Washing textiles is an important aspect for longevity in the circular economy of textiles. The outsourcing of services is an increased trend for textiles and influences the environmental impact of washing. Rental textile services include all services to clients that relate to the selection of textiles, garment manufacturing, stock management, logistics and delivery, care and maintenance, and, when necessary, repair or replacement of damaged textiles.

## 16.2 Historical aspects and current scenario

Since people have worn clothes, these clothes have been washed. Washing and cleansing is a complex process involving numerous physical and chemical parameters and influenced by detergents and the mechanical influences. In ancient Egypt and Rome, washing was done in watercourses and consisted mainly of mechanically treating clothing by heating, treading, rubbing and similar procedures. Water carried materials that caused stains and odours out of the textiles.

Washing additives such as soda and soap were also used by the Egyptians. It was found early on that the quality of water has an influence on the wash performance. Washing with rainwater resulted in a better performance than did normal water.

Cold water gave a poorer wash result than did hot water. For the past 1000 years, soap has been used as an all-purpose washing and laundering agent (Jakobi and Löhr, 1987). In the 19th century, a major innovation in washing started with the development of new detergents and washing machines. Stain removal during washing process is improved by increasing the mechanical input, chemical input, wash time and temperature. Moreover, interactions among washing components and different classes of stains are important.

### 16.2.1 Detergents

The development of synthetic detergents started in the 19th century. Sodium silicate (water glass) was used in the United States in the 1860s (Aftalion, 2001). In 1876, Henkel sold a sodium silicate-based product as a *Universalwaschmittel* in Germany, which could be used with soap. In 1878, soda and sodium silicate were mixed to produce Germany's first brand detergent, *Bleichsoda* (Jakobi and Löhr, 1987). In 1907, soap became an ingredient in multicomponent systems for the routine washing of textiles, and the bleaching agent sodium perborate was added to launch the first self-acting laundry detergent (Persil) (Jakobi and Löhr, 1987). The development of washing machines was another important milestone for the development of new detergents. The formulation of washing agents was changed, and soap was replaced with synthetic surfactants (Bertsch, 1968). After its use, detergent enters the wastewater with the loosened dirt, which can have an impact on the environment.

In the 1950s, the widespread use of household washing machines, the increasing use of textiles and greater awareness of hygiene led to a sharp increase in the consumption of detergents. Numerous new ingredients (e.g. dirt antiredeposition agents, enzymes, fluorescent whitening agents, foam regulators and bleaching activators) were introduced to improve cleaning performance. These are the state of the art for modern detergents (Jakobi and Löhr, 1987).

In the washing process, water is a solvent for detergents and soluble salts within the stain and a transport medium for dispersed and colloidal stain components. Washing starts by wetting the fabric and penetrating stained fabrics. For fast and effective wetting, the high surface tension of water must be drastically reduced by surfactants (Jakobi and Löhr, 1987). Water hardness also has an important influence on the washing results. Water hardness is defined in terms of the amount of calcium and magnesium salts. It can vary considerably from one country to another and from season to season. The best degree of hardness for washing is soft water. In Europe, soft water is relatively rare compared with the United States and Japan (Werdelmann, 1974).

Washing components in modern detergents are surfactants, water-soluble complexing agents and water-insoluble ion exchangers apart from bleaches. The main components are surfactants, builders and bleaches. Secondary components are auxiliary agents present only in small quantities; each has a specific purpose. Surfactants are water-soluble, surface-active agents consisting of a hydrophobic portion attached to hydrophilic or solubility-enhancing functional groups. Surfactants can be divided in four classes: anionic, nonionic, cationic and amphoteric. With an increase in the chain length of surfactants, the effectiveness of washing increases. Surfactants have numerous properties, such as specific adsorption, stain removal, low sensitivity to water hardness, dispersion properties, the antireposition capability of dirt, high solubility, wetting power, desirable foam characteristics, neutral odour, low intrinsic colour, storage stability and good handling characteristics (Jakobi, 1981; Cox et al., 1984).

Builders soften the water through ion exchange and complexing action with calcium and magnesium ions. Thus, they support the washing effect effectively and efficiently. Historically, alkalis used in detergent manufacturing were derived from the ash of plants. The term alkali denotes a substance that is chemically a base. It is neutralised with an acid and is composed of several types of materials. The most commonly used alkalis manufactured commercially are alkaline substances such as sodium carbonate and sodium silicate, complexing agents such as sodium diphosphate sodium triphosphate or nitrilotriacetic acid (NTA) and ion exchangers such as water-soluble polycarboxylic acid and insoluble zeolites. In industrial laundry, soft water is produced by a softening water system, such as ion exchanger systems. Therefore, builders are not used in industrial laundry (Chopade and Nagarajan, 2000).

Colour stains and yellowing can be removed from laundry with the help of bleach. Bleaches are used to clean, whiten and brighten fabrics and remove stubborn stains. They convert dirt into colourless, soluble particles that can be removed and carried away by detergents and water. Bleaching agents also have a disinfecting effect on fungi and bacteria but can attack textiles. Oxygen-based bleaches are used in modern laundry detergents. They release oxygen from hydrogen peroxide or ozone. Bleaching agents of this type are adjusted to specific bleaching conditions by means of stabilizers. Hydrogen peroxide is insufficiently active as a bleach below 60°C and must be used with hot water. In the 1970s and 1980s, bleach activators such as tetraacetylethylenediamine (TAED) were developed to enable bleaching using cooler washing temperatures. These compounds react with hydrogen peroxide to produce peracetic acid, which can bleach at lower temperatures (Smulders et al., 2002). In heavy-duty detergents, bleaches become effective only at higher temperatures. Detergents for delicate and coloured fabrics, on the other hand, contain no bleach. For household textile cleaning in particular, oxygen-based bleach is widely used as hydrogen peroxide in liquid bleach and sodium perborate in powder bleach. Chlorine-based bleach is used in industrial textile cleaning and textile processing. The disadvantage of chlorine-based bleach is that the colour can change because bleach does not differentiate between dirt and colour molecules and is not environmentally friendly.

Auxiliary agents are enzymes, dirt antiredeposition agents, foam regulators, corrosion inhibitors, fluorescent whitening agents, fragrance, dyes, fillers and formulation aids. One important auxiliary agent group is enzymes. Enzymes such as proteases, amylases, and lipases catalyze the reaction between stains and the water solution, making dirt removal more efficient. Table 16.1 lists enzymes and their functionality in detergents.

Fat-based stains can leave unsightly marks after washing. They can also appear after repeated washes and become more pronounced. Detergents contain enzyme mixtures to removing stains. Enzymes in detergents should have activity optimum at an alkaline pH, good efficacy at a low wash temperature between 20°C and 40°C,

**Table 16.1** Enzymes for detergents and the degradation of stains.

| Enzyme | Degradation | Type of soil |
|---|---|---|
| Proteases | Protein-based stains | Grass, blood, egg |
| Amylases | Starch-based stains | Pasta, potatoes, baby food |
| Lipases | Fat-based stains | Butter, oil, human sebum |
| Cellulases | Cellulose-based stains | Soot, clay, rust |
| Mannanases | Mannans-containing stains | Soils from substances such as chocolate, ice cream, barbecue sauce |
| Pectase lyases | Pectin-based soils | Fruits, vegetables, jams |

stability at wash temperatures up to 60°C and high stability in the presence of other detergent ingredients such as surfactants, builders and activated bleach during both storage and use. Enzymes should show low specificity to soils (i.e. a specificity broad enough to enable the degradation of a large variety of proteins, starches and triglycerides) (Kissa, 1986). Low-temperature washes with detergent enzymes can prevent damage to fabric, which is important to improve the longtime use of fabrics.

In the 1950s, fragrance was first added to detergents. It should provide detergents with a pleasant odour and mask certain odours arising from the wash liquor during washing. Fragrances are intended to confer a fresh, pleasant odour to the laundry itself. Therefore, long-lasting fragrances resulting from detergents or fabric softeners have been used since the 1990s. Detergent fragrance is present at low concentrations. These are complex mixtures of many individual ingredients, depending on the detergent formulation and the material of the fabrics. Chemical stability relative to other detergent ingredients is important, as is the limited volatility of the individual fragrances. The fragrance industry has a large selection of synthetic compounds and a wide range of natural fragrances. The selection is considerably restricted owing to the instability of many fragrances and because some tend to discolour textiles (Smulders et al., 2002).

Detergents, which mainly contain surfactants and softener, end up in sewage treatment plants after use. In the past, phosphates were used as softeners, which were difficult to degrade in sewage treatment plants and led to the overfertilization of water bodies. Today, phosphates have been replaced by other substances such as citric acid or silicates such as zeolite A, which are considered to be relatively harmless to the environment. However, ethylenediaminetetraacetic acid is poorly biodegradable (Nörtemann, 1999) and NTA is possibly carcinogenic (Epstein, 1971). Other ingredients such as dyes and fragrances can be problematic for the environment. For example, musk fragrances such as galaxolide and tonalide are biodegradable only slowly or not at all. Also, not all optical brighteners used are biodegradable. Detergents that are labelled as antibacterial also contain biocides to kill pathogens and pests. In waterbodies, biocides can harm aquatic life; they also impair the effectiveness of biological sewage treatment plants (Pöppelmann and Goldmann, 2008). To keep the chemical input into the environment as low as possible when washing, environmental and consumer advisors recommend consumers to follow information on packaging during use so as not to overuse detergent (Pöppelmann and Goldmann, 2008).

### 16.2.2 Regulations concerning detergents

Before 1960, tetrapropylene benzene sulphonate was mainly used as a synthetic surfactant because of its simple synthesis. However, owing to its branched molecular structure, it is difficult for microorganisms to decompose detergents, which could emerge unchanged from sewage treatment plants. In the 1960s, this led to a dangerous increase in surfactant concentrations in water. Mountains of foam in rivers and lakes made it clear that detergents had a considerable impact on the environment. To protect the environment, the first German Detergent Law was established in 1961 because of increasing surfactant concentrations in water and the considerable impact on the environment (Jakobi and Löhr 1987; Gesetz über Detergentien, 1961). This started the development of biodegradable surfactants (Verordnung über die Abbaubarkeit von Detergentien, 1962).

Surfactants with unbranched side chains have since been used, which have better biodegradability. In the 1970s, high levels of phosphate in water were caused by detergents, leading to eutrophication. To lower the phosphate content in water, in 1980, manufacturers in Germany were obliged to reduce the maximum permitted amount of phosphate in detergents. In 1986, about half of all laundry detergents were phosphate-free. Today there are almost only phosphate-free detergents on the market (Rudolph and Block, 2001). In Germany, the Washing and Cleaning Agents Law of 1975 (amended in 1994 and 2020) sets requirements for the environmental compatibility of washing and cleaning agents. In addition, in 1993, after foundations were laid with strict criteria concerning complete biological degradability and regarding toxicity to water organisms, the Blue Angel (a voluntary product symbol) was first awarded to a cleaning agent in the modular construction system to support product consumers in making households environmentally friendly.

In 1995, criteria that had been worked out under German leadership were passed for a European Environmental Label for cleaning agents (Rudolph and Block, 2001). Blue Angel eco-labelled detergents should contribute to sustainable development by preferentially using renewable raw materials in their manufacture, which are produced under sustainable conditions. European Detergents Regulation Number 648/2004 establishes common rules to enable detergents and surfactants to be sold and used across the EU while protecting the environment and human health. The European Detergents Regulation replaced five existing directives on the biodegradability of surfactants in detergents and regulates the definition of terms, requirements for the final aerobic biodegradability of surfactants in detergents, exemptions for placing surfactants in detergents on the market if requirements regarding final aerobic biodegradability are not met, labelling regulations for detergents and information requirements for manufacturers. The regulation enables national authorities to ban a specific detergent if it poses risks to human health or the environment. After 2011, phosphates in consumer laundry detergent were restricted and banned by June 2013 (EU, 2022).

### 16.2.3 Washing technology

Another important aspect of washing is the mechanical impact on the washing process and the development of washing machines. In 1691, the first machine was patented in England (Stanley, 1995). Washing machines were built as containers or basins with grooves, fingers or paddles to support scrubbing and rubbing the textiles. The first machines were hand-powered, and the user employed a stick to press and rotate textiles to agitate them to remote dirt. At the beginning of the 20th century, the electric drive of mechanical washing machines appeared. In the following years, new detergents were developed that were optimised for use in these washing machines (Jakobi and Löhr, 1987). The machines were not sold to private households until the 1920s. The first fully automatic washing machines came on the market in United States in 1946, and in Germany in 1951. This was a breakthrough for the washing machine compared with the industrial laundry. Human intervention was no longer necessary between process steps. Household washing machines use water as the solvent whereas industrial washing machine use water or solvent, and special technologies were developed.

#### 16.2.3.1 Household washing machine

Different types of washing machines are on the market. In horizontal axis machines, the bottom of the washtub is filled with water. In vertical axis machines, the whole tub is filled with water. Horizontal axis machines consume much less water per wash cycle than do vertical axis machines. The main tasks of all types of washing machines are to provide hygienically clean laundry and to preserve its value. Vertical axis machines are most widespread in North America, Australia and Asia. Often, horizontal axis machines have a minimum programme temperature of 30°C, so these machines use electricity to heat water even at the coldest programme selectable (Pakula and Stamminger, 2010).

Major components include the cabinet, wash drum, tub, motor, water pump, thermostat and pressure switch, detergent compartment, and control panel/circuit board. In the past, weight was necessary for sufficient stability and smooth running. Without a certain weight, imbalance developed and the washing machine wobbled. The harder the machine spins, the greater the risk is of imbalance and wet laundry in the drum collects in a heap. Force generated by the speed of the drum and the weight of the laundry is concentrated in one place. New washing machines are therefore equipped with imbalance control, which detects imbalance with the help of sensors and counteracts it. Therefore, new washing machines have less weight.

There is a trend towards increasing the loading capacity in washing machines. The average capacity (i.e. load size) of the appliance models offered for sale, for example, increased from 4.9 kg in 2000 to 7.04 kg in 2013 (Boyano et al., 2017). The range of washing machines has grown considerably: in 1997, primarily appliances smaller than 6 kg were sold. Currently, most appliances have a capacity of 5–9 kg, and appliances with up to 15 kg are available (Boyano et al., 2017). Moreover, there was a decline in the average size of households, as a result of which the volume of laundry per household tended to be smaller (Rüdenauer and Gensch, 2016). There is a trend towards differentiation in the programme design of washing machines towards more specific programs, and in special washing agents (Rüdenauer and Gensch, 2016).

The proliferation of washing machines in households led to significant time savings in textile care, but demands regarding hygienic and aesthetics aspects of the cleanliness of the resulting wash increased (König, 2000). In the 1960s and 1970s, washing machines became inexpensive standard items in industrialised countries. In 2010, most of the population of seven billion still handwashed their clothes. It is estimated that only two (of seven) billion people have access to washing machines (Mingels, 2016). Most laundry washing today is conducted with machines (80%), even in less-developed countries (Laitala et al., 2017). Handwashing is still practiced in developed countries, especially with delicate or nonmachine-washable fabrics.

### *16.2.3.2 Regulations for washing machines*

In 1992, the EU established an energy consumption labelling scheme for white goods and others (EU, 1992). The energy efficiency of washing machines is rated in terms of a set of energy efficiency classes from A to G on the label, in which A is the most energy efficient and G is the least efficient. The labels also provide other useful information to the customer as they choose among various models. Because of technical progress in energy efficiency, more devices received a good label and the labelling scheme has been further developed (EU, 2010), replaced by Directive 2010/30/EU. The scale was extended to class A+++ and a new type of label exists with pictograms rather than words, to allow manufacturers to use a single label for products sold in different countries. In 2017, updated labelling requirements were developed. The classification was simplified, using only letters from A to G. Rescaling should lead to better differentiation among products that, under the current label classification, all appear in the same top categories (EU, 2017).

In the EU, ecodesign is important in connection with the circular economy. On March 1, 2021, ecodesign regulations for washing machines became mandatory for all manufacturers and importers who want to sell products in the EU and established ecodesign requirements for washing machines (EU, 2019). This regulation better meets the needs of users and contains elements to improve the repairability and recyclability of devices. Spare parts (e.g. door hinges and door seals, door latches and plastic accessories such as detergent containers) whenever generally available for 10 years, professionally competent repairers also have access to spare parts and information that are required for the repair and professional maintenance of washing machines and washer-dryers.

### *16.2.3.3 Industrial washing*

Textile services and commercial laundries uses industrial or commercial washing machines with water, and dry cleaning machines with organic solvents or wet cleaning machines based on dry cleaning but with the use of water.

## Industrial washing machines

Industrial or commercial washing machines differ from household machines in that they have a more powerful motor and higher loading capacity. In commercial laundries, washing machines and continuously operating washing lines (continuous systems) are used. Industrial washing machines can be mounted on heavy-duty shock absorbers and fixed to a concrete floor, enabling them to siphon water from heavy loads. Noise and vibrations, which are undesirable in household washing machines, do not interfere with industrial washing. The machines can be mounted on hydraulic cylinders, enabling the entire washing machine to be lifted and tilted and textiles to be tipped automatically from the washing drum onto a conveyor once the wash cycle is complete.

Many industrial washing machines have an automatic dosing system for five or more chemical types. With the dosing system, detergent and additives can be directly drawn from large liquid chemical storage drums and injected into various wash and rinse cycles. It is possible to manage various washing and rinsing cycles using computer control.

The tunnel washer, also called a continuous batch washer, is a special type of continuous-processing washer. In a tunnel washer, textiles move slowly and continuously through a long, large-diameter, horizontal-axis rotating tube in the manner of a conveyer belt with different processes at different positions, while water and washing chemicals move in the opposite direction. Thus, textiles move through pockets of progressively cleaner water and fresh chemicals. Soiled textiles can be continuously fed into one end of the tunnel while clean textiles emerge from the other. Continuous systems are continuously or intermittently working washing lines in which items are washed fully automatically and continuously, without machine downtime owing to loading and unloading. Modern computer-controlled tunnel washing systems can monitor and adjust chemical concentrations in individual pockets (Matt, 2018). In the case of special hygienic requirements (e.g. hospital laundry), the machines are designed so they can be accessed from two separate rooms. Loading takes place on the impure side, and unloading on the clean side. The machines are heated with superheated steam, thermal oil or electrically. Heating is the most energy-consuming process. Wash cycle times are often shorter than in household washing machines, typically by 40 min per wash load, to increase throughput.

For years, laundries have been using various technologies to reduce resource consumption such as energy, water and detergents to reduce the cost of the laundry process. This also leads to a reduction in the environmental impact and is essential for sustainable development. The optimization of water consumption can be done by recirculating and reusing water and by treating processed water for recycling. Energy consumption can be reduced by implementing energy-efficient technologies such as heat recovery systems and low-temperature washing and by paying more attention to maintaining the machines.

In Germany, wastewater from industrial laundries must meet requirements for discharge into public waters. Industrial laundries use various technologies to clean wastewater (GE, 2004). To minimise the impact on the environment, wastewater should be cleaned and only environmentally friendly washing detergents should be used. The main waste from textile services is discarded textiles. Textiles should be reused as long as possible. If they are no longer suitable, textiles should be recycled.

## Dry cleaning

Dry cleaning is a cleaning process for fabrics using solvents other than water. Several types of fabrics are sensitive to water, depending on the fibre type, such as wool and silk. After washing in the washing machine, damage such as wrinkles or shrinkage, changes in dimension and form, and changes in colour and quality can occur. These fabrics can be cleaned by dry cleaning, a process that includes cleaning, drying, spotting and finishing such as pressing or ironing and is widely used in professional textile care. The process removes stains from textiles using a nonaqueous, apolar solvent instead of water. The advantages of organic solvents are that hydrophilic textiles do not swell in liquid solvents, they show no dimensional changes, and no fabric shrinkage, wrinkles, or colour bleeding occur. Oily stains are removed at low temperatures, whereas when washing with water high temperatures are needed. A typical wash cycle takes 8−15 min, depending on the type of garment and the degree of pollution. During the first 3 min, solvent soluble stains dissolve in the organic solvent and loose, insoluble stains dissolve.

With the solvent perchloroethylene (PCE), it takes 10−12 min after the stain is loosened to remove ground-in insoluble dirt from the garments. With hydrocarbon (HCS) solvents, it takes a minimum of 25 min per cycle because of the much slower rate of solvation of soluble stains. At the end of the wash cycle, the machine starts a rinse cycle in which the textile load is rinsed with a freshly distilled solvent from the solvent tank. This pure solvent rinse prevents discolouration caused by dirt particles being reabsorbed onto the garment surface by the dirty working solvent. After the rinse cycle, the machine begins extraction, which recovers the solvent for reuse. The recovering rate depends on the solvent and the machine technology. Cleaning is done at a normal temperature because the solvent is never heated in dry cleaning. When no more solvent can be spun out, the machine starts the drying cycle.

The original dry solvent technique removed only oil-soluble stains, but currently used methods remove oil and many water-soluble and some insoluble materials with the help of detergents and various agents. Dry cleaning detergents consist of surfactants, cosolvents, lubricants, antistatic compounds and water. The components increase the cleaning efficiency, improve the handling of dry-cleaned textiles, and prevent an electrostatic charge on textiles. Dry cleaning solvents are formulated as liquids and can be added by automatic dosing equipment in the cleaning machine. Because of the high costs of solvents, it is important to recycle them to avoid a single use. Three commonly used recycling processes are filtration, adsorption and distillation. Dry cleaning machines work in closed circuits. The solvent is permanently recycled and produces no wastewater. Dirt is removed is concentrated in the residue, which is collected. Problematic soil from workwear, oil and heavy metals cannot enter wastewater streams by this system.

Spotting is the removal of stains from textiles during professional textile cleaning. Most dry cleaners have their own type of spotting, commonly known as prespotting.

Prespotting is a treatment that removes stains before dry cleaning or wet cleaning a garment. It is chemically soluble or water-soluble. Oil-based stains may require prespotting for effective removal. For finishing, dry-cleaned garments are steamed, pressed and/or ironed with special technical equipment.

In the 19th century, dry cleaning started with the use of petroleum-based solvents that were highly flammable. This problem was solved by the development and use of chlorinated solvents. Since 1930, the most common solvent in conventional dry cleaning has been PCE, with good cleaning performance and nonflammability. The PCE recovery rate in the cleaning process is greater than 98% (Cinet, 2013). PCE has negative effects on health and the environment and is classified into group 2A by the International Agency for Research on Cancer (Guha et al., 2012) and as a toxic air contaminant by the Environment Protection Agency (US EPA, 2006). PCE is regulated in many countries and states in Europe and the United States and will be phased out (Ceballos et al., 2021). Alternative, more environmentally friendly solvents for dry cleaning are HCS and liquid silicone. HCS such as iso-paraffins are petroleum-based solvents that are less aggressive but also less effective than PCE. HCS are also combustible. Because of the high flashpoint of HCS, they are regarded as volatile organic compounds (VOCs) by state and federal agencies, HCSs can have adverse impacts on ambient air quality (Whittaker and Shaffer, 2018).

In the 2000s, liquid silicone or decamethylcyclopentasioxane (D5) was developed as a more environmentally friendly alternative to PCE. In the cosmetics industry, D5 has been used for a long time. D5 is a clear and odourless silicone (siloxane)-based solvent considered to be nontoxic, nonhazardous and biodegradable. This silicone-based solvent degrades into sand ($SiO_2$), water and carbon dioxide, leaving no toxic residue when it is released into the atmosphere (Troynikov et al., 2016) With low surface tension, silicone-based solvents penetrate into textile fibres more effectively to loosen dirt. They are chemically inert and do not react chemically with fabrics or dyes during cleaning. This reduces the problems of dye removal and bleeding, minimises the abrasion and/or swelling of fabric fibres, results in less damage to fabrics and prints and reduces shrinkage (Troynikov et al., 2016). Another alternative solvent is dibutoxymethane (also known as butylal,1-(butoxymethoxy)butane and formaldehyde dibutyl acetal), which is biodegradable and halogen-free. Available data concerning the effects on human health (skin irritation, dermal and oral exposure and inhalation) are in sufficient, and more research is necessary (Ceballos et al., 2021).

## Wet cleaning process

In contrast to traditional dry cleaning, wet cleaning is a sustainable process of professional cleaning that avoids using chemical solvents; instead, it uses water as a solvent. Environmental and safety regulations implemented in many countries why dry cleaners switched to dry cleaning alternatives or wet cleaning. Wet cleaning uses water as a solvent combined with machines used in dry cleaning (Ono, 2002). The highly programmable machines can carry out special drum movements with pauses and speeds of rotation and have drums with three-dimensional draws to increase the detergent and solvent penetration into fabric. To use water as a solvent, the machines have wash programs for the better use of water and wet cleaning products and detergents are adapted to the wet cleaning process. Because the wet cleaning process uses water, this process produces no special waste compared with traditional dry cleaning process. Wet cleaning is an environmentally friendly and sustainable process. Detergents used in wet cleaning differ from those in washing machines. The wet cleaning detergents are highly concentrated products with enzymes for a high performance level and to protect fabrics (Whittaker et al., 2020).

After wet cleaning, garments should be thoroughly rinsed for detergent residue, which is not desirable. Wet cleaning consumes about 50% less energy than dry cleaning and is an energy-efficient process. Wet cleaning is less expensive than dry cleaning because water is cheaper than solvents. A wide range of fibre types such as wool, silk, linen, cotton and leather/suede, as well as wedding gowns and garments decorated with beads can be wet cleaned. Wet cleaned clothes need sometimes special care because of shrinkage, wrinkling or surface changes, but they are free from chemical odour, unlike dry cleaned clothes, which can retain a strong odour after dry cleaning. However, wet cleaning produces contaminated wastewater (Gottlieb, 2001). Wet cleaning is classified as an environmental cleaning process and with the environmental label Blue Angel (DE-UZ 104) (Blue Angel, 2018a).

#### 16.2.3.4 Tumble dryers

After washing, the textile must be dried before the next use. Traditionally, drying is defined as a process in which the liquid portion of the solution (moisture content) is evaporated from the fabric. Fabrics can dry in air by hanging or lying on a rack. The first hand-operated dryer was invented in France in 1800. At the beginning of the 20th century, the first electric tumble dryer came on the market. In the 1940s, industrial designer Brooks Stevens constructed the first clothes dryer with a glass window (Stahl, 2023). In 1958, the first European tumble dryer was developed by Miele (2023). However, these first tumble dryers were initially hardly used in private households, because laundry was usually hung out to dry in the traditional way in drying rooms (usually in the attic) or on clotheslines outside. After 1980, tumble dryers were increasingly used in private households in Europe owing to changes in lifestyle. To date, only the design and other features of the tumble dryer have changed, but the basic principle has remained the same.

Most dryers consist of a rotating drum (a tumbler) through which heated air is circulated to evaporate moisture while the tumbler is rotated to maintain air space between articles. There are spin dryers, vented dryers, condenser dryers and heat pump dryers on the market. The combination of a dryer with a fully automatic washing machine is termed a washer-dryer on the market. The main dryer product components are the drum, motor, fan, electric or gas heater and optional additional features such as automatic dry controls, self-cleaning lint screens, self-diagnostic microcontrollers and full-body insulation. For dryers, increased energy efficiency as the cost of energy is more significant. The capacity of dryers in households is 4–10 kg (IEA ETSAP, 2012). In 2013, Miele developed the first solar dryer directly using the sun's energy, without converting it into electricity. The basis for this is a solar thermal system that supplies the heating system with warm water via a stratified storage tank (Miele, 2013).

Commercial clothes dryers are segmented into single-load dryers with a capacity of around 9.1 $kg^2$ of laundry per load. This is slightly larger than household dryers with a capacity of 6 kg, larger capacity tumble dryers with a capacity of 9–34 $kg^2$ of laundry per load and industrial-sized dryers with a capacity of 54–318 $kg^2$ of laundry per load (IEA ETSAP, 2012). In industrial laundries, a finishing tunnel is also used to dry different kinds of garments (e.g., blouses, skirts, coats, trousers, T-shirts, sweatshirts) and is a continuous process. The wet garment passes through a steam chamber. In tunnel finishing, the garment passes through various stages and across different modules before it is ready to be used. Wrinkles in the garment are removed by strong hot air flow alongside the garments, and the garment is dried by cool air. Th output of tunnel finishers varies from 450 to 5000 pieces per hour (APPAREL, 2009). Since 2021, ecodesign regulations have been mandatory for dryers, as mentioned for washing machines (EU, 2019).

## 16.3 Impacts on health and hygiene

Washing should remove visible stains and dist and should lead to a hygienically clean textile surface to prevent disease transmission through textiles. Textiles in hospitals, retirement homes and so one need to be hygienically cleaned. In a normal household, the risk of infection transmission via laundry is low. Washing with hot water (>60°C) and detergent containing bleach or at lower temperatures (40–60°C) with disinfecting washing additives are listed applied hygiene (e.g. list of the Association for Applied Hygiene (VAH list in Germany) (German: VAH: Verbund für angewandte Hygiene e.V.)) (Exner and Simon 2011; VAH, 2015) removes heavily soiled contaminants from textiles, underwear and towels or cleaning rags in the household, and the washed textiles are decontaminated. The rest of the clothes can be washed at lower temperatures. If textiles are contaminated with pathogens, or if family members have a weakened immune system, textiles should preferably be washed at 90°C or 60°C with the addition of disinfecting washing additives (Bloomfield et al., 2011). Microbial species on human skin and mucosal biota can be transferred to fabrics after direct body contact (Bloomfield et al., 2011). Microbial contamination on textiles is linked to their use in different environments and includes dust, soil and food, which might point to specific areas such as health care facilities.

A hygienically clean textile can be achieved by reducing microorganisms and other microbial effects such as the formation of bad odours during process. The reduction in microorganisms and other microbial effects is traditionally solved by combining time, temperature, mechanics and chemistry, all of which either remove or kill microbial cells. To obtain a hygienically clean textile, pathogens must be removed or inactivated to avoid the cross-contamination of textiles or from washing machines to textiles, biofilm formation inside the washing machine and the recontamination of already washed textiles during rinsing as well as the formation of bad odours (Bockmühl et al., 2019).

Because most microorganisms found on textiles are also part of the human microbiome or the environment, they should not generally pose a health risk to humans.

Temperature, detergents or mechanical action in the washing process can inactivate or remove microorganisms from textiles. A sufficient level of hygiene must be guaranteed, and laundering is recommended at temperatures above 60°C with the use of bleach. Oxidizing compounds such as chlorine or active oxygen bleach and temperatures of 60°C or higher are important to ensure the efficient antimicrobial action of laundering (Bockmühl et al., 2019). The trend towards reducing energy costs in washing by lowering washing temperatures can impede microbial reduction during washing. Using activated oxygen bleach (AOB) containing detergent or other antimicrobial compounds can compensate for the lower temperature and short durations of wash cycles (Bockmühl et al., 2019). Bleaching agents are presumably the most important component determining the antimicrobial activity of laundering. In the United States and southern Europe, chlorine bleach has traditionally been used, and in western and northern Europe, AOB is used, which is based on perborate or percarbonate. AOB needs bleach activators such as TAED and induces the formation of peracetic acid, which occurs even below 60°C. Peracids can be formulated into solid detergent, providing high microbial reduction during laundering. Studies show that using AOB significantly increases antimicrobial efficacy, but microbial reduction varies depending on the microorganisms tested and the conditions (Honisch et al., 2014; Lichtenberg et al., 2006; Linke et al., 2011).

Besides AOB or chlorine bleach, quaternary ammonium compounds (QACs) are used. QACs have a long alkyl chain and are used as antimicrobials and disinfectants such as benzalkonium chloride, cetylpyridium chloride and tetraethylammonium bromide (Zhishen et al., 2001). QACs are cationic surfactants that are used particularly during rinsing after the main wash cycle, because they are deactivated by anionic surfactant detergents, which are usually part of detergents. This can lead to substances that may stay on textiles after laundering. QACs act against fungi, amoebas and enveloped viruses such as SARS-CoV-2 (Hora et al., 2020) by destroying the cell membrane or virus envelope (Hora et al., 2020; Cocco et al., 2015), except for endospores and non-enveloped viruses. The microbial contamination of textiles might be mediated by biofilms inside washing machines, and QAC-containing products might be able to compensate for this effect. To avoid biofilms in washing machines, the machine should be cleaned without a textile load by detergents with bleach at 90°C.

Industrial laundries that have implemented a hygiene system based on the Robert Koch Institute (RKI) requirements or a risk analysis and biocontamination control (RABC) quality management system can guarantee the microbiological quality of the expedition of textiles. In its Guideline on Hospital Hygiene and Infection Prevention, RKI set out 'Hygiene requirements for textiles from healthcare establishments, the laundry and washing process' (RKI, 1995). The RABC system, based on EN 14065, is used in many textile processing companies to guarantee the microbiological quality of end products (EN 14065, 2003; Heintz et al., 2007; Heintz and Bohnen, 2011). This is a control system in which the risk of biocontamination is stipulated for each process step in line with potential hazards, and corresponding measures are taken in advance to check and eliminate problems.

Another control system was established by the German Certification Association for Professional Textile Services (*Gütegemeinschaft Wäschepflege*) founded in 1953, an organization for commercial laundries that are particularly committed to maintaining high-quality standards in services (RAL Gütezeichen Wäschepflege, 2023). A member of this association can label the professional textile service if it participates in the monitoring systems. If the requirements of the monitoring system are met, the business can use the RAL-GZ 992 quality marks. Compliance with the strict RAL quality certification requirements is regularly monitored by the independent testing house, Hohenstein Institute. All certified laundries have a highly efficient self-monitoring system regularly checked on-site to ensure that it is up-to-date and applied correctly. Obligatory further training for employees ensures that technological advances are properly implemented.

The first certification mark was the RAL Quality Certification Mark for commercial linen (RAL-GZ 992/1). It was aimed particularly at hotels and catering businesses. In 1986, the quality mark was extended to include a hygiene certificate to offer hospitals a hygienic alternative to processing linen in-house (RAL-GZ 992/2). This mark indicates compliance with certain hygiene specifications, covering all standard regulations. The next mark is the RAL Quality Certification Mark for Linen from Food Processing Business (RAL-GZ 992/3). This RAL mark includes both process control under RAL-GZ 992/1 and a hygiene certificate. To support care facilities to ensure the hygiene security of the residents' own textiles, quality certification mark RAL-GZ 992/4 for resident's textiles from care homes was created in 2012. In contrast to the processing of general textiles in care homes, which according to the current recommendation by the RKI, 'Prevention of infections in care homes,' have to undergo a disinfectant wash cycle, the responsibility for the hygiene status of residents' textiles lies with the care facilities or appointed commercial laundries. The processing of residents' textiles is not subject to specific legal regulations and mainly includes garments (e.g. trousers, blouses or jumpers), undergarments (e.g. pajamas, underwear or socks) and nonwashable outer garments such as silk blouses. Hygienic handling and proper processing of all types of textiles in care homes is an important contribution to preventing infection (RAL Gütezeichen Wäschepflege, 2023).

## 16.4 Environmental impacts of washing

Washing is a material- and energy-consuming process and releases chemicals and microplastics into wastewater. The composition of wastewater depends on dirt on the fabric, the degree of dirt and the detergents. Wastewater from a laundry in which dirty items are washed contains mineral oils, heavy metals and dangerous substances that have chemical oxygen demand (COD) values of 1200−20,000 mg $O_2$/L. Wastewater from hospitals contains fat, the remains of food, blood and urine, with COD values of 400−1200 mg $O_2$/L. Laundries that wash items from households and hotels pollute water with COD values of 600−2500 mg $O_2$/L (Gosolits et al., 1999; Turk et al., 2005).

The most widely used systems for laundry wastewater treatment are conventional methods such as precipitation/coagulation and flocculation, sedimentation and filtration or a combination of these. Coagulation and flocculation aids are usually added to facilitate the formation of larger agglomerated particles. Adsorption on granular activated carbon after flocculation can improve treatment, and a wide range of compounds absorb on the large surface area of the carbon. Membrane processes and ultrafiltration are also used (Turk et al., 2005). Conventional methods are especially effective for minimising organic pollutants when wastewater is drained into communal sewage or directly into water. Because of the characteristics of separation, the consumption of chemicals can be reduced by membrane technology compared with coagulation and adsorption methods. No coagulators or active carbons are used for membrane filtration, and sludge production is lower.

Textiles have been identified as a major source of microplastics because they have the potential to release fibres during production, their use phase (wearing, washing and drying) and disposal (Haap et al., 2019; Rochman et al., 2015; Bahners et al., 1994; Wang, 2010). During washing, textiles are exposed to chemicals, temperature, long washing and mechanical stress, which may cause fibres to be released into the washing effluent and may contribute to fibre emissions (Bahners et al., 1994; Wang, 2010). The type of washing machine (front or top loader, machine capacity, rotation and spin dry rate and amount of water) can affect the mass loss of textiles (Wang, 2010; Carr, 2017; Browne et al., 2011). In addition, the apparel design including textile characteristics (e.g. fibre type, yarn construction and surface treatment) have considerable potential to affect fibre release (Bahners et al., 1994; Browne et al., 2011).

Different analytical methods were established to detect fibre release. The released fibres are filtered from wastewater after laundering, with subsequent manual counting of fibres recovered on a filter material by scanning microscopy or light microscopy (Bahners et al., 1994; Carr, 2017, Hernandez et al., 2017). To count debris, the homogeneous distribution of fibres on the filter surface has to be ensured (Bahners et al., 1994). After the liquid is separated through a filter, the amount of debris can be determined by gravimetric analysis of the filtered material (Wang, 2010; Carr, 2017; Browne et al., 2011; Napper and Thompson, 2016). For both counting and weight analysis, the selection of the filter pore size significantly affects the results. Small pore sizes increase the retention capacity of fibres, but clogging effects are more likely to occur (Napper and Thompson, 2016).

By dynamic image analysis, an optical detection system for particle characterisation used in pharmacy, food industry and geology, fibre characterisation and quantification can be done in one step within a short measuring time (Haap et al., 2019). The comparability of different studies is hindered by the diversity of analytical technologies and pretreatment procedures as well as the diversity of selected textiles and applied washing processes (Haap et al., 2019). Establishing a harmonised protocol to analyse fibres in wastewater from a textile laundry is a growing need to generate a comprehensive understanding of fibre abrasion during washing and develop feasible solutions. Standardized methods and practices in sampling, identification technologies and the quantification of microfibres are under development, as are three ISO standards (ISO 4484-1, 2023; ISO/DIS 4484-2, 2022; ISO/FDIS 4484-3, 2022). Microplastics have potentially hazardous effects through particle and chemical toxicity, and as conduits of harmful chemicals and microorganisms (Zhang and Liu, 2018).

## 16.5 Innovations leading to sustainability and energy efficiency

Sustainability is becoming increasingly important. Consumers want eco-friendly products and services. The trend of sustainable or green practices is strong and will continue for years to come. The cleaning process requires water, detergents and energy. Thus, many in the laundry industry are working to combat the exploitation of these resources by creating machines that use less energy and water.

### 16.5.1 Washing machine innovations

Energy consumption in laundries is high, and owing to rising energy costs, energy-saving technologies are in demand. Heating in the laundry is done by damp and electric energy. Water heating consumes about 90% of energy in the industrial laundry process. Oily stains need high temperature for cleaning. However, other types of stains can be removed by lower temperatures using enzyme-containing detergents. Switching the temperature setting from hot to warm can reduce energy use by half. Using the cold cycle reduces energy use even more. However, the use of enzymatic detergents is necessary. Table 16.2 lists innovative developments and improvements.

In the mid-1980s, a prototype ultrasonic washing machine was developed in Japan and is currently on the market in Asia. However, even after several decades, there is no ultrasonic washing machine in Europe. The ultrasonic washing machine is a Pulsator washing machine, in which textiles and washing water are moved by a shaft wheel on the floor. It is not the drum washing machine that is common in Europe. Textiles do not have a hard surface, and therefore dirt particles can be blown away only if the ultrasonic transducer approaches the textile directly. A comparison of the ultrasonic washing machine with a common European washing machine shows that the European washing machine removes dirt as well as the ultrasonic device, even with cold water and without a detergent (Wissenschaft, 2008).

Ultrasonic waves were also used to have a mechanical impact on the washing process. Ultrasonic waves are applied as a mechanical action to obtain further increases in the detergency with the aqueous solutions. Although detergency can be depend on the rotation speed of stirring and the output level of ultrasonic waves, ultrasonic waves are not as effective as mechanical action for washing textiles in aqueous detergent systems (Kazama, 2001).

In the 1990s, washing machines with a cold and hot water connection were developed to save energy. However, the disadvantage of warm water is that dirt-containing protein can be fixed in the fabric through denaturation if water runs at temperatures above 40°C at the beginning of the washing process. In addition, enzymes of detergents can be destroyed. Because washing tends to be carried out at lower temperatures and water levels, possible energy savings through a hot water connection are significantly lower. The concept of a washing machine with a cold and a warm connection was unsuccessful because of the high cost of house installations for a separate warm water connection for the washing machine.

**Table 16.2** Innovations leading to improved textile cleaning.

| Innovation and development | Saving potential | | | Modification | | |
|---|---|---|---|---|---|---|
| | **Energy** | **Detergent** | **Water** | **Machine** | **Detergent** | **Others** |
| Low-temperature washing | x | | | x | | |
| Ultrasonic washing machine | x | | | x | | |
| Hot and cold water connection | | x | | x | | |
| Minimalist design | | | | x | | |
| Automatic dosing | | x | | x | | |
| Dimension of machine | x | x | x | x | | |
| Disinfection and softening with ozone | | x | | x | | |
| Washing with magic balls | | x | | x | | x |
| Washing with pellets | | x | x | x | | x |
| Heat pump for heating | x | | | x | | |
| UV-C for disinfection | | x | x | x | | |
| Home connection | x | x | x | | | x |
| Washing machine of future | x | x | x | x | | |
| Dry cleaning with $CO_2$ | | | x | x | | |
| Solar energy (industrial laundry) | x | | | x | | |
| Biotechnological detergents | | x | | | x | |

Further developments of improvements for washing machines include a minimalist design, automatic dosing by doing away with the washing-up compartments, large displays, and lighting on the washing machine door. The dimensions of washing machines are also under development. Small machines that can be fixed to the wall for single households, machines with two drums (one above for large amounts of textiles and one below for small batches of textiles) are being developed for families. Another approach is to change the size and shape of the washing drum and the movement of the drums. Some machines also add tiny beads that absorb stains and provide a gentle mechanical wash action to lessen the amount and temperature of water needed, and therefore lessen the amount of detergent.

Washing machines with the in situ generation of ozone for disinfection and water softening to save detergent are also available. In 2006, a washing machine was introduced in Japan that washed clothes using ozone instead of water. It was produced in small numbers but did not catch on (Sanyo, 2006).

Washing nearly without detergent is possible with various washing balls that have different mechanisms (e.g. Magic Washing Ball, a plastic ball filled with special ceramic beads of 82 different minerals). The plastic ball is placed with the textile in the washing drum of the machine. The ceramic beads convert water into washing liquid. The surface tension is decreased, and dirt particles, fat and proteins are easily detached from the fabric. Only water and energy are needed to wash laundry, and for particularly heavily soiled laundry, only a small amount of detergent is needed.

In the United Kingdom, researchers from the University of Leeds and the company Xeros are developing a washing machine that uses nylon plastic pellets and just 1 cup of water per wash. Because of its polarised molecular structure, nylon attracts dirt. A slightly damp environment ensures that the dirt is absorbed by the nylon pellets, which can be used for several hundred washes. The nylon pellets are located in a second drum that revolves around the laundry drum, and the rotation brings pellets to the textiles and picks up dirt. After the washing process, the outer drum stops and the pellets are transported back into the outer drum by the rotation of the laundry drum. The machines are being tested in British and American laundries (Leeds, 2013).

The consumption values of washing machines (water and electricity) have decreased in recent decades, but the minimum has not yet been reached. Around 70% of the energy consumption of a wash cycle for modern washing machines is required to heat water. Another concept is to reduce the free liquor from 8 to 5.6 L of water. Therefore, a load detection system via mass inertia, a wetting process (spin and spray) and a heating system via steam heating was implemented. Owing to the higher concentration of detergent and the lower water content, a better washing effect and improved hygiene at low washing temperatures was achieved. On the one hand, the process enables higher washing temperatures with the same energy input. On the other hand, lower energy consumption for heating decreased the amount of water (Schambil, 2017).

In 2013, the first energy-efficient washing machines were built, which had a heat pump for hot water preparation instead of electrical resistance heating, reducing the need for electrical energy. They are available from different washing machine manufactures (e.g. VZug, Miele). During heating, cleaning water mixed with detergent is pumped through the condenser of the heat pump (warm side). Washing machines with heat pump technology reduce the energy requirement compared with conventional washing machines for a four-family household by 17%–60% (as of 2015). A major key to increasing the efficiency and operation of the heat pump is to recover thermal energy normally lost in the pumped wastewater with the heat pump.

An innovation uses UV-C technology to reduce the wastewater load. This leads to a reduction in chemicals needed in the washing process, high stain and dirt removal, gentle action on fibres and high degrees of whiteness (CHT, 2023). In the washing machine market, important topics are the home connection and the use of apps to make the use of the machine independent from the user. The connection to the Internet and voice control make washing machines smarter (Schambil, 2017).

The Swedish company Electrolux and Korean manufacturer LG go a step further and are developing washing machines of the future, which will wash without the use of detergents or water. At the Washing Machine of the Future 2050 competition, they presented a concept for a washing machine called Orbit, which washes laundry sparkling clean in just 5 min without a drop of water. Dry ice replaces the water. The Orbit machine consists of a transportable drum made of a superconducting material levitating in a ring consisting of an electrically conductive battery. The drum moves inside the ring when electrical resistance approaches zero. Textiles and dry ice filled the washing drum. Frozen carbon dioxide changes from a solid to a gaseous state and thus releases high pressure on clothing when the drum rotates. Carbonic acid reaches dirt to be separated through a pipe. Carbon dioxide should be usable indefinitely without replacement, and the batteries charge themselves. The machine should continue to have near-field communication and be automatically networked with other household appliances without the user having to do anything. The machine may be operated via a smartphone app, which also shows where a wash cycle is and should report the end of the wash cycle via a pop-up message. However, this is only a design study; such a machine is not coming before 2050 (Czycholl, 2013).

### 16.5.2 Dry cleaning with $CO_2$

Another alternative dry cleaning solvent is liquid carbon dioxide ($CO_2$), in which carbon dioxide is pressurised into a liquid solvent that safely cleans clothing (Sutando, 2014). Research on dry cleaning with liquid $CO_2$ started in the early 1970s with a patent by Maffei (1977). Since the invention in 1970s, $CO_2$ dry cleaning technology has undergone significant technical developments. Its commercialisation was difficult because of several barriers, such as the poor solubility of many chemical compounds in $CO_2$ and relatively high initial investment costs because of high-pressure equipment. The potential of $CO_2$ to replace organic solvents in dry cleaning is particularly evident from a sustainability perspective. However, further developments (e.g. surfactants) and a better understanding of the cleaning mechanism and textile movement are necessary. In 2014, around 20 commercial $CO_2$ machines in the United States and 10 in Europe (Sweden and Denmark) were in operation (Sutando, 2014).

Cleaning textiles with $CO_2$ is a method to wash without water. $CO_2$ is a sustainable alternative for PCE because it fulfils the basic properties of a green solvent, which is low or nontoxic, chemically stable, readily available and easily recyclable (EPA, 2006). $CO_2$ dry cleaning uses natural liquid $CO_2$ or that recycled from industry combined with recyclable cleaning agents to clean soiled textiles. Then, it flushes and dries using $CO_2$'s state changing cycles from a gas to a liquid, which absorbs all dirt in this form and then back to a gas. At the transition from liquid to the gaseous state, dirt particles are released again. The released dirt particles can be disposed of while $CO_2$ remains in the cycle and continues to be used. The advantages of this process are that $CO_2$ is more environmentally friendly.

In 1995, the Hughes Aircraft Company built the first commercial machine (DryWash) and process that uses $CO_2$ as a cleaning solvent for fabrics. In the DryWash system, garments are held in a perforated basket inside the cleaning vessel, and a premixture of liquid $CO_2$ and additives is added Beckman (2004). A commercial development in $CO_2$ dry cleaning equipment is the launching of $CO_2$Nexus machines in the United States (Sutando, 2014). The Tersus series was designed for regular dry cleaning operations.

Machines based on Micell Technologies were sold in the United States in the beginning of 2000. In Europe, Electrolux was the main supplier of these $CO_2$ dry cleaning machines. The first $CO_2$ dry cleaner in Europe was founded in Stockholm, Sweden, in 2004, and the next in Amersfoort, Netherlands. The stores in Netherlands were bought by Linde Gas, who collaborated with Electrolux to sell the technology in Europe under the name Fred Butler. In 2011, Linde decided to stop this project. In 2013, the only dry cleaners using $CO_2$ in Europe were Kymi Rens (Aalborg, Denmark) and Fred Butler (Copenhagen, Denmark) (Sutando, 2014).

An advantage of this process is that $CO_2$ is environmentally friendly. However, stain removal in $CO_2$ dry cleaning is relatively low because of high interaction forces between particles and textiles, and the low density difference between the liquid and gas phases of $CO_2$ and low viscosity. The low density difference causes a low level of mechanical action. The low viscosity causes low momentum transfer. The redeposition of soil occurs when the released dirt is not properly stabilised in or removed from the cleaning medium. Redeposition usually cannot be reversed, and that leads to greying of the fabric and unsatisfactory cleaning results. The reposition cannot be reduced by adding a rinsing step (Sutando, 2014). A rinsing step distributes redeposited particles more evenly. Redeposition was more severe using a longer washing time. $CO_2$ dry cleaning needs substantially higher pressures compared with dry cleaning with other solvents, and it requires equipment with higher investment costs. Dry cleaning with $CO_2$ is labelled Blue Angel DE-UZ-126 (Blue Angel, 2018b). The label is given to cleaning processes that do not emit pollutants into the air or wastewater even though textiles are cleaned.

### 16.5.3 Innovations in industrial laundry

In the research project SoProW, the Fraunhofer Institute for Solar Energy Systems (Freiburg, Germany), the Hohenstein Institute für Textilinnovation (Boennigheim, Germany) and laundry companies investigated introducing solar energy to industrial laundries (Hecht, 2016; Hecht et al., 2016). After an analysis of the processes in 10 laundries, various integrations of solar energy systems were simulated and evaluated. Wash water heating, in which steam is injected into water, is well-suited. Another process that can be supported by solar energy is the generation of steam sprayed onto the laundry. During this process, the system loses steam and water, and these must be processed and returned to the boiler. Heating of this boiler water is another process into which solar heat can be integrated. In the SoProW demo project, the technology was installed in three German textile service companies and served as lighthouse projects for similar companies. The partners developed a concept and guidelines for the laundry industry (Zischke, 2020).

The introduction of smart laundries go hand in hand with the installation of sensors inside the wash process that measure the most important parameters of the washing process in a timely fashion and can directly regulate the procedure. The outcome is that one wash cycle needs only the time necessary for good cleaning, which saves time and energy. The machines run more efficiently, and wear and tear on the machines are reduced. Because the machines are more efficient, the need for launderers to rewash garments are reduced and time is saved. Smart technologies have the ability to track and analyse utility use and other key metrics of laundry operations. That makes it easy to measure the impact of the process (Robinson, 2019). The integration of smart technology in laundries improves cleaning results. With smart technology, the machine can give feedback about the status of washing, adjust cleaning criteria and optimise washing conditions during the real process. This can lead to a more sustainable cleaning of fabrics by saving water, detergents and energy and can lead to a longer use time for the fabrics.

### 16.5.4 Innovative detergents

The basic principle of washing machines has hardly changed over many decades. Washing textiles is still a water-, energy- and detergent-intensive process. Surfactants in detergents are washing-active substances that reduce the surface tension of water and enable the dispersion of non-water soluble fats. Most surfactants are based on petroleum, and others on palm kernel oil or coconut oil. A combination of different surfactants is usually employed. Around 20 million tons of surfactants are produced worldwide. In western Europe, about half is accounted for by detergents and cleaning agents. An annual growth of up to 4% is forecast by 2024. Detergents can harm the environment and are not environmentally friendly because they always contain substances that are not easily or completely biodegradable, are harmful to aquatic organisms or accumulate in the environment or organisms.

Almost all surfactants can be obtained based on animal or vegetable oils and fats. Owing to technical properties, these are mostly oils from tropical crops with a high proportion of lauric acid (or example, coconut or palm kernel oil). To achieve a better ecological balance, bio-based surfactants based on European oil plants such as rapeseed, olives, flax and sunflower are increasingly being used. These usually have a low lauric acid content, and therefore the chemical-synthetic steps from vegetable oil to surfactant are complex. Many synthetic surfactants or their degradation products are not fully biodegradable, which can pollute water bodies.

According to the European Detergents Regulation, a surfactant must be biodegradable when exposed to oxygen, and thus guarantee the reduction of surfactants in sewage treatment plants. According to the Detergents and Cleaning Agents Act, the primary degradability of surfactants must be at least 80%. However, the law does not establish the final degradation of the resulting degradation products into water, minerals and $CO_2$. This leads to major differences in practice among different surfactants. One example is alkylphenol ethoxylates. In sewage treatment plants, alkylphenol ethoxylates are broken down into nonylphenol, which is considered an endocrine disruptor and can damage the hormone system. In principle, alkylphenol ethoxylates can based on vegetable oils and are biologically degradable; however, the product of the degradation process is harmful.

Natural products such as soap nuts or chestnuts are also used as natural detergents. Soap nuts or Indian soap berries have been known and used for centuries by many Indian households and have become increasingly popular. They are the fruit of the *Sapindus mukorossi* tree, a native of the Himalayas and the mountainous region between India and Nepal. The shells contain a natural cleaning agent called saponin, which produces an effect comparable to soap (Soapnut, 2023).

Surfactants can be produced biotechnologically by microorganisms, based on renewable raw materials. These biosurfactants are biodegradable and insensitive to water hardness, and despite their strong cleaning power, they are kind to the skin. The potential of biosurfactants was discovered in the late 1980s. After 10 years of development, they were manufactured commercially. The greatest technical challenge is controlling foam formation in large-production volumes. In nature, some microorganisms naturally produce surfactants to dissolve fats on which they feed. An example of this is the pathogen *Pseudomonas aeruginosa*, which produces rhamnolipids. Rhamnolipids are produced by bacteria of the *Pseudomonas* genus, which require sugar or glycerol as a raw material. They have a good cleaning effect and are completely biodegradable with and without oxygen. Sophorolipids are produced by yeast fungi, which can feed on vegetable oils in addition to sugars. With more than 400 g/L, they have achieved the best product yield among biosurfactants and are processed commercially in skin creams. Surfactin is produced by the bacterium *Bacillus subtilis* and is the best-known biosurfactant that is researched for commercialisation. Cellobiose lipids and mannosylerythritol lipids are produced from the family of nonpathogenic smut fungi Ustilaginaceae, including *Ustilago maydis* and *Moesziomyces aphidis*, but their use is still at the research stage. Other biosurfactants are Emulsan (produced by *Acinetobacter calcoaceticus*) and Liposan (by *Candida lipolytica*).

In Germany, the innovation alliance for functionally optimised biosurfactants was founded by several research institutions and companies to promote the development and biotechnological production of biosurfactants from domestic, renewable raw materials and residues, The aims of the alliance are to discover different biosurfactants for as many areas of application as possible and to scale them up to larger sample quantities. Researchers are investigating the biotechnological production of surfactants by fungi or bacteria. In this way, biosurfactants based on renewable raw materials such as sugar and vegetable oil can be produced microbially. These are more easily degradable and less toxic. Fats are dissolved in the same way as conventional surfactants. However, production is complex, and the fermentation yield is still too low. For the mass market, the fermentation yield needs to be improved.

In bioreactors, microorganisms are supposed to mature faster in different nutrient media. Various projects have been initiated by the innovation alliance. The BestBioSurf project combines bioinformatics, synthetic biology and a metabolic pathway design to optimise established biosurfactants and their fermentation processes. The EU project CARBOSURF investigated the development of novel nature-like biosurfactants that can be produced from agricultural waste products. The aim of the BioProMare initiative is to research and develop the biotechnological potential of life forms in the sea. In the GlycoX project, the development of the glycolipid biosynthesis pathway in a bacterium discovered in the North Sea is meant to produce novel biosurfactants. LIPOMAR is a subproject with the aims of identifying high-quality surfactants from macroalgae and transferring them to microbial fermentation (Bioallianz, 2022).

## 16.6 Future perspectives

The EU Ecodesign regulations for washing machines and dryers, and the transformation of the economy from a linear throwaway society to a circular model will have an impact on further developments in textile care and maintenance. A circular economy means that raw materials are used as long and as frequently as possible, and natural resources are used in a closed circle, without the need to use new resources. The longevity, repairability and recycling of washing machines and dryers are important aspects for reducing the use of raw materials. These aspects should be integrated more in further development. The energy consumption of washing machines und dryers in production, during use and at the end of life should be reduced.

The washing process needs water and generates wastewater. Wastewater from washing machines is not filtered, and residues of detergents, textile debris and dirt pollute water. The development of filters can reduce contamination in wastewater. Research is also focused on developing innovative detergents, and the industrial production of sustainable detergents in sufficient quantities is necessary. The transition to a sustainable production needs a lot of effort in research and production. Preliminary steps towards sustainability are already achieved. Further steps to reduce material and energy consumption with innovative ideas have to follow in coming years.

## 16.7 Conclusion

Washing clothing is important for the long-term use of clothes. For years, washing has been influenced by the consumption of a solvent (water or organic solvent), detergents, the mechanical impact of the washing machine, the washing time and temperatures. To reduce the environmental impact of washing, much effort was made in the past to develop more environmentally friendly detergents and reduce the amount of detergents and energy without reducing cleanness.

Increasingly, cooperation between appliance manufacturers and detergent producers can be observed to find new solutions. There are trends toward washing without water, without detergent, at low temperatures for energy savings, as well as changes in the technique of the drum of the washing machine. Smart washing machines are coming to the market that can control and improve washing to save detergents, water and energy. The ecological production of the components of detergents is being investigated and is based on biotechnology methods. One important aspect is the technology transfer from the laboratory stage to commercial production.

More changes will come with the introduction of the circular economy of textiles. This will lead to changes in the production, use, repair and reuse of products of daily life, and this will include textiles and washing machines. Resources such as materials and energy should be used in a circle, and this can lead to innovations in washing. Washing without water does not produce wastewater. The technology for wastewater treatment is unnecessary, and energy and materials might be saved. Washing without detergents using low temperatures saves material resources and the wastewater treatment becomes unnecessary. Further developments are needed to find manageable and usable solutions in future washing. Until new, innovative, resource-saving solutions for washing textiles are found and established on the market, traditional washing in water or solvents will continue to take place with material- and energy-saving processes.

## References

Aftalion, F., 2001. A History of the International Chemical Industry. Chemical Heritage Press, p. 82.

APPAREL, 2009. https://apparelresources.com/technology-news/manufacturing-tech/tunnel-finishers-making-garments-wrinkle-free/.

Bahners, T., Ehrler, P., Hengstberger, M., 1994. Erste Untersuchungen zur Erfassung und Charakterisierung textiler Feinstäube. Melliand Textilberichte 1, 24–30.

Beckman, E.J., 2004. Supercritical and near-critical $CO_2$ in green chemical synthesis and processing. The Journal of Supercritical Fluids 28, 121–191.

Bertsch, H., 1968. Tenside 5, 185–188.

Bioallianz, 2022. https://www.allianz-biotenside.de/#:~:text=Mit%20der%20Innovationsallianz%20Biotenside%20haben,die%20bislang%20aus%20fossilen%20Rohstoffen,

Bloomfield, S.F., Exner, M., Signorelli, C., Nath, K.J., Scott, E.A., April 2011. The infection risks associated with clothing and household linens in home and everyday life settings, and the role of laundry. International Scientific Forum on Home Hygiene, pp. 1–43.

Blue Angel, 2018a. DE.UZ 104, Blauer Engel Wet Cleaning Services DE-UZ 104-200904-en Criteria-2018-10-11.pdf. https://www.blauer-engel.de/en/certification/basic-award-criteria#UZ104-2021.

Blue Angel, 2018b. DE-ZU 126, Blauer Engel Kohlendioxidreinigungsdienstleistung DE-ZU 126-200904-die Kriterien-2018-07-24.pdf. https://produktinfo.blauer-engel.de/uploads/criteriafile/en/DE-UZ%20126-200904-en%20Criteria-2018-10-15.pdf.

Bockmühl, D.P., Schages, J., Rehberg, L., July 1, 2019. Laundry and textile hygiene in healthcare and beyond. Microbial Cell 6 (7), 299–306.

Boyano, A., Cordella, M., Espinosa, N., Villanueva, A., Graulich, K., Rüdenauer, I., Alborzi, F., Hook, I., Stamminger, R., 2017. Preparatory study for ecodesign and energy label for household washing machines and washer dryers. https://doi.org/10.2760/029939. Report number: ISBN 978-92-79-74183-8. Affiliation: European Commission.

Browne, M.A., Crump, P., Niven, S.J., Teuten, E., Tonkin, A., Galloway, T., Thompson, R., 2011. Accumulation of microplastic on shorelines worldwide: sources and sinks. Environmental Science & Technology 45, 9175–9179.

Carr, S.A., 2017. Sources and dispersive modes of micro-fibers in the environment. Integrated Environmental Assessment and Management 13, 466–469.

Ceballos, D.M., Fellows, K.M., Evans, A.E., Janulewicz, P.A., Lee, E.G., Whittaker, S.G., 2021. Perchlorethylene and dry cleaning: it's time to move the industry to safer alternatives. Frontiers in Public Research 9, 638082.

Chopade, S.P., Nagarajan, K., 2000. Detergent Formulations: Ion Exchange. Michigan State University, East Lansing, MI, USA Copyright 2000 Academic Press.

CHT, 2023. Documents/Literatur%20Waschen/waschen/CHT-Smart-Technology-for-Textile-Care-EN.pdf. https://solutions.cht.com/cht/medien.nsf/gfx/med_MJOS-AUNERJ_3BA4AA/$file/SMART-UV-POWER-EN.pdf?OpenElement&v=0.233,671229114214.

Cinet, 2013. Safe and Sustainable Use of Tetrachloroethylene in Professional Textile Cleaning by Best Practice Approach, https://www.cinet-online.com/uploads/files/CINET_Safe%20and%20sustainable%20processing%20with%20PERC.pdf, 27.04.2022.

Cocco, A., de Oliveira da Rosa, W.L., Fernandes da Silva, A., Guerra Lund, R., Piva, E., 2015. A systematic review about antibacterial monomers used in dental adhesive systems, current status and further prospects. Journal of Dental Material 31 (11), 1345–1362.

Cox, M.F., Matson, T.P., Berna, J.L., Moreno, A., Kawakami, S., Suzuki, M., 1984. Journal of the American Oil Chemists' Society 2, 330.

Czycholl, H., 2013. https://www.welt.de/wissenschaft/article119094038/Die-Waschmaschine-der-Zukunft-waescht-ohne-Wasser.html.

Desinfektionsmittel-Kommission im VAH, April 2, 2015. In: Requirements and Methods for VAH Certification of Chemical Disinfection Procedures. mhp Verlag, Wiesbaden. Available from: https://vahonline.de/files/download/ebooks/eBook_VAH_RequirementsandMethods.pdf. Verbund für Angewandte Hygiene (VAH): List of Disinfectants.

EN 14065, 2003. Textiles – Laundry Processed Textiles – Biocontamination Control System.

EPA, 2006. The 12 Principles of Green Chemistry, United States Environmental Protection Agency. (Retrieved 31 July 2006).

Epstein, S.S., 1971. Toxicological and environmental implications on the use of nitrilotriacetic acid as a detergent builder—1. International Journal of Environmental Studies 2, 291–300.

EU, 1992. Council Directive 92/75/EEC of 22 September 1992 on the Indication by Labelling and Standard Product Information of the Consumption of Energy and Other Resources by Household Appliances. Europe. http://data.europa.eu/eli/dir/1992/75/oj.

EU, 2010. Directive 2010/30/EU of the European Parliament and the Council of 19 May 2010 on the Indication by Labelling and Standard Product Information of the Consumption of Energy and Other Resources by Energy-Related Products. https://eur-lex.europa.eu/eli/dir/2010/30/oj.

EU, 2017. Regulation (EU) 2017/1369 of the European Parliament and of the Council of 4 July 2017 Setting a Framework for Energy Labelling and Repealing Directive 2010/30/EU. https://eur-lex.europa.eu/legal-content/EN/TXT/PDF/?uri=CELEX:32017R1369.

EU, 2019. Regulation (EU) 2019/2023 of 1 October 2019 Laying Down Ecodesign Requirements for Household Washing Machines and Household Washer-Dryers Pursuant to Directive 2009/125/EC of the European Parliament and of the Council, Amending Commission Regulation (EC) No 1275/2008 and Repealing Commission Regulation (EU) No 1015/2010. http://data.europa.eu/eli/reg/2019/2023/oj.

EU Strategy for Sustainable and Circular Textiles, 2022. Brussels, COM 141 (final).

Exner, M., Simon, A., 2011. Infektionen? Nein danke! Wir tun was dagegen! DLH e.V.-Broschüre.
GE, 1962. Verordnung über die Abbaubarkeit von Detergentien in Waschmitteln und Reinigungsmitteln, vol. 1. BGBl, pp. 698–706.
GE, June 17, 2004. Verordnung über Anforderungen an das Einleiten von Abwasser in Gewässer (Abwasserverordnung -AbwV)m Anhang 55 Wäschereien, Anhang 55 - Abwasserverordnung (AbwV). BGBl, pp. 1180–1181.
Gesetz über Detergentien in Wasch- und Reinigungsmitteln, 5.9.1961. BGBl, p. 1653.
GINETEX, Who We Are. ginetex.net. (Retrieved 27 April 2022).
Gosolits, J., et al., 1999. Waschmittel und Wasserrecycling in Gewerblicher und Krankenhauswaeschereien (Phase 1), Project BMBF-FKZ:01 RK 9622. Bekleidungsphysiologisches Institut Hohenstein, Hohenstein, Germany.
Gottlieb, R., 2001. Dry cleaning's dilemma and opportunity: overcoming chemical dependencies and creating a community of interests. In: Gottlieb, R. (Ed.), Environmentalism Unbound. Massachusetts Institute of Technology, Cambridge, MA, pp. 101–339.
Guha, N., Loomis, D., Grosse, Y., Lauby-Secretan, B., Ghissassi, F.E., Bouvard, V., et al., 2012. Carcinogenicity of trichloroethylene, tetrachloroethylene, some other chlorinated solvents, and their metabolites. The Lancet Oncology 13 (12), 1192–1193.
Haap, J., Classen, E., Beringer, J., Mecheels, S., Gutmann, J.S., 2019. Microplastic fibers released by textile laundry: a new analytical approach for the determination of fibers in effluents. Water 11, 2088. https://doi.org/10.3390/w11102088.
Hecht, K., 2016. Möglichkeiten für die Integration solarer Prozesswärme in Wäschereien. In: Vortrag. Jahrestagung Gütegemeinschaft sachgemäße Wäschepflege e. V., Kassel, October 20–22, 2016.
Hecht, K., Haap, J., Beeh, M., Classen, E., 2016. Concepts for integrating solar process heat in industrial laundries. In: Poster. SEPAWA-Kongress, Fulda, October 12–14, 2016.
Heintz, M., Bohnen, J., 2011. Hygiene in commercial laundries. Hygiene & Medizin 36 (7/8), 292–298.
Heintz, M., Krämer, J., Vossebein, L., 2007. Risk analysis and biocontamination control – hygiene measures in commercial laundries. Tenside Surfactants Detergents 44 (5), 274–280.
Hemkhaus, M., Hannak, J., Malodobry, P., Janßen, T., Griefahn, N.S., Linke, C., 2019. Kreislaufwirtschaft-Textilsektor-2019, GIZ-Studie.
Hernandez, E., Nowack, B., Mitrano, D.M., 2017. Polyester textiles as a source of microplastics from households: a mechanistic study to understand microfiber release during washing. Environmental Science & Technology 51, 7036–7046.
Honisch, M., Stamminger, R., Bockmühl, D.P., 2014. Impact of wash cycle time, temperature and detergent formulation on the hygiene effectiveness of domestic laundering. Journal of Applied Microbiology 117 (6), 1787–1797.
Hora, P.I., Pati, S.G., McNamara, P.J., Arnold, W.A., 2020-06-26. Increased use of quaternary ammonium compounds during the SARS-CoV-2 pandemic and beyond: consideration of environmental implications. Environmental Science and Technology Letters 7 (9), 622–631.
IEA Energy Technology Systems Ananlysis Programme (ETSAP), June 2012. Technology Brief R09 "Dryers".
ISO 4484-1, 2023. Textiles and Textile Products — Microplastics from Textile Sources — Part 1: Determination of Material Loss from Fabrics During Washing.
ISO/DIS 4484-2, 2022. Textiles and Textile Products — Microplastics from Textile Sources — Part 2: Qualitative and Quantitative Evaluation of Microplastics.

ISO/FDIS 4484-3, 2022. Textiles and Textile Products — Microplastics from Textile Sources — Part 3: Measurement of Collected Material Mass Released from Textile End Products by Domestic Washing Method.

Jakobi, G., 1981. In: Stache, H. (Ed.), Tensid-Taschenbuch, second ed. Hanser Verlag, München-Wien, pp. 253–337.

Jakobi, G., Löhr, A., 1987. Detergents and Textile Washing: Principles and Practice. VCH, Weinheim.

Kazama, K., 2001. Comparison of wash with water and without water. Journal of Textile Engineering 54, 445–448.

Kissa, E., 1986. Detergency: Theory and Technology, Surfactant Science Series, vol 20, p. 333.

König, W., 2000. Geschichte der Konsumgesellschaft, VSWG-Beihefte. Franz Steiner Verlag Stuttgart. ISBN 3-515-07650-6.

Laitala, K., Klepp, I.G., Hennty, B., 2017. Global laundering practices—alternatives to machine washing. Household and Personal Care Today 12, 10–16.

Leeds, 2013. Virtually Waterless Washing Machine. Xeros Ltd, University of Leeds, leeds.ac.uk.

Lichtenberg, W., Girmond, F., Niedner, R., Schulze, I., 2006. Hygieneaspekte beim Niedrigtemperaturwaschen. SÖFW-Journal 132, 28–34.

Linke, S., Gemein, S., Koch, S., Gebel, J., Exner, M., 2011. Orientating investigation of the inactivation of Staphylococcus aureus in the laundry process. Hygiene + Medizin 36, 8–12.

Maffei, R.L., 1977. Extraction and Cleaning Processes. US Patent 4012194.

Manshoven, S., Smeets, A., Malarciuc, C., Tenhunen, A., Mortensen, L.F., 2022. Microplastic Pollution from Textile Consumption in Europe, Eionet Report - ETC/CE 2022/1.

Matt, P., 2018. Efficient, Flexible Tunnel Washers. American Laundry News. American Trade Magazines (Retrieved 24 April 2022).

Miele Information, 2023. Wäschetrockner. www.waeschetrockner.net/miele-waeschetrockner.

Miele Pressemitteilung, 2013. Miele führt den ersten Solartrockner ein, Presserelease, Nr. 054.

Mingels, G., 2016. Haushaltsgeräte in Der Spiegel Nr. 15.

Napper, I.E., Thompson, R.C., 2016. Release of synthetic microplastic plastic fibres from domestic washing machines: effects of fabric type and washing conditions. Marine Pollution Bulletin 112, 39–45.

Nörtemann, B., 1999. Biodegradation of EDTA. Applied Microbiology and Biotechnilogy 51, 751–759.

Ono, M., 2002. Survey on the professional wet cleaning. Journal of Textile Engineering 55, 202–206.

Pakula, C., Stamminger, R., November 2010. Energy Efficiency. https://doi.org/10.1007/s12053-009-9072-8.

Pöppelmann, C., Goldmann, M., 2008. Umweltgerecht einkaufen: Worauf Verbraucher achten sollten. DIN-Ratgeber. Beuth Verlag, Berlin/Wien/Zürich, pp. 86–88.

RAL Gütezeichen Wäschepflege, 2023. https://www.waeschereien.de/en/home.

RKI Guideline, 1995. Richtlinie für Krankenhaushygiene und Infektionsprävention Ziffer 4.4.3 und 6.4. Robert Koch-Institut (RKI) (Hrsg.), Bundesgesundheitsbl Gesundheitsforsch Gesundheitsschutz 7/95.

Robinson O., 2019. https://www.future-of-laundries.com/fileadmin/user_upload/Hohenstein_FoL/Schulungen/downloads/45_project-intwash-developing-digitalisation-platform-for-laundry-industry.pdf.

Rochman, C.M., Tahir, A., Williams, S.L., Baxa, D.V., Lam, R., Miller, J.T., Teh, F.-C., Werorilangi, S., Teh, S.J., 2015. Anthropogenic debris in seafood: plastic debris and fibers from textiles in fish and bivalves sold for human consumption. Scientific Reports 5, 14340.
Rudolph, K.-U., Block, T., October 2001. The German Water Sector – Policues and Experiences. BMU and UBA, Berlin-Bonn-Witten.
Rüdenauer, I., Gensch, C.-O., 2016. Saving Potential of Miele Washing Machines with the PowerWash 2.0 Technology. Öko-Institut e.V.
Sanyo, February 2, 2006. SANYO Announces the World-First Drum Type Washing Machine with 'Air Wash' Function, Japan.
Schambil, F., 2017. Innovationen rund ums Waschen & Reinigen. 8. SÖFW Journal 52–56.
Smulders, E., Röhse, W., von Rybinski, W., Steber, J., Sung, E., Wiebel, F., 2002. In: Smulders, E. (Ed.), Laundry Detergents. Wiley-VCH Verlag GmbH, Weinheim.
Soapnut, 2023. https://earthbits.com/blogs/earthbits/a-guide-to-soapnuts-what-are-soap-nuts-and-how-to-use-them.
Stahl, 2023. https://stahl-waeschereimaschinen.de/en/useful-information/how-long-has-the-tumble-dryer-been-around/.
Stanley, A., 1995. Mothers and Daughters of Invention: Notes for a Revised History of Technology, Autumn Stanley. Rutgers University Press, p. 301.
Sutando, S., 2014. Textile Dry Cleaning Using Carbon Dioxide: Process, Apparatus and Mechanical Action (PhD-thesis). Uni Delft.
Troynikov, O., Watson, C., Jadhow, A., Nawaz, N., Kettlewell, R., 2016. Towards sustainable and safe apparel cleaning methods: a review. Journal of Environmental Management 182, 252–264.
Turk, S.S., Petrinić, I., Simonič, M., May 2005. Laundry wastewater treatment using coagulation and membrane filtration. Resources, Conservation and Recycling 44 (2), 185–196.
US EPA, 2006. Final Amendments to Air Toxics Standards for Polyethylene Dry Cleaners- Fact Sheet. https://www.epa.gov/sites/production/files/2015-06/documents/fact_sheet_dry_cleaning_july2006.pdf.
Werdelmann, B., 1974. Soap, cosmetics, chemical specialities, 50, p. 36.
Wang, Y., 2010. Fiber and textile waste utilization. Waste and Biomass Valorization 1, 135–143.
Wissenschaft.de, January 15, 2008. https://www.wissenschaft.de/allgemein/wundersame-waschkraft/.
Whittaker, S.G., Fellows, K., Pedersen, A.E., 2020. Converting PERC Dry Cleaners to Professional Wet Cleaning: A Pilot Program. Hazardous Waste Management Program in King County, Seattle, WA.
Whittaker, S.G., Shaffer, R.M., 2018. Dry Cleaning with High Flashpoint Hydrocarbon Solvents, Report No. IHWMP 0275. Hazardous Waste management program in King County, Seattle WA.
Zhang, G.S., Liu, Y.F., 2018. The distribution of microplastics in soil aggregate fractions in southwestern China. Science of the Total Environment 642, 12–20.
Zhishen, J., Dongfeng, S., Weiliang, X., 2001. Synthesis and antibacterial activities of quaternary ammonium salt of chitosan. Carbohydrate Research 333 (1), 1–6.
Zischke, T., 2020. Report SoProW-Demo-Demonstartion of Energy Efficiency and Solar Process Heat in Indsutrial Laundries, (FKZ 0325999A, BMWi).

## Sources of further information

ACI, November 8, 2021. The Role of Enzymes in Detergent Products. American Cleaning Institute (ACI).

EU 2022, European Commission, December 14, 2011. EP Supports Ban of Phosphates in Consumer Detergents. europa.eu (Retrieved 10 April 2022).

# Sustainable supply chain and logistics of fashion business

*Manoj Kumar Paras*[1] *and Rudrajeet Pal*[2,3]
[1]National Institute of Fashion Technology, Kangra, Himachal Pradesh, India; [2]The Swedish School of Textiles, University of Borås, Borås, Sweden; [3]Department of Industrial Engineering and Management, University of Gävle, Gävle, Sweden

## 17.1 Introduction

The business of fashion broadly includes design, manufacturing and retailing. Initially, fashion products such as clothes were tailored for individual customers according to their measurements. The fashion business became an industry when tailored manufacturing transformed into large manufacturing during mid-19th century. The fashion businesses have witnessed significant progress with increased demand and diversification in clothing and other fashion products. The business further developed due to an increase in design focus, the establishment of fashion business houses, and technological and retail development during the 20th century (Lowson et al., 1999). In the current scenario, the fashion business is significantly large; therefore, the supply chain and logistics need to be managed efficiently to remain competitive. Suppliers, manufacturers, and retailers need to work in a collaborative model to meet market demand. Optimal supply chain and logistics of the fashion business could be achieved by balancing demand and supply, increasing profits, and reducing cost (Alikhani et al., 2019). This chapter aims to provide insights into the fashion business's logistics and supply chain.

The fashion industry is one of the major sectors for developed and developing economies. The sector is exceptionally diverse and heterogeneous (Bruce et al., 2004). The fashion industry (textile and apparel) falls under small- and medium-scale industries. Textile industries comprise spinning units, dyeing units, fabric manufacturing units, and apparel manufacturing units. Apparel manufacturing includes design, sampling, material procurement, manufacturing, distribution, and retailing (Keiser and Garner, 2008). Apparel manufacturing consists of three steps, i.e., cutting, sewing, and finishing. All three processes of apparel manufacturing are labour intensive. Hence, to minimise the cost, apparel manufacturing can be offshored to developing economies (Bruce et al., 2004). In pursuit of lower manufacturing costs, the apparel manufacturing firms in developing countries took advantage of lower environmental and social awareness and regulation (Shen, 2014). The global apparel manufacturing industry is the second most polluting (after the oil industry) and socially challenged industry. Adaptation of sustainability in the apparel industry provides strong motivations to reduce raw material and hazardous chemical consumption towards the minimisation of environmental

**Sustainable Innovations in the Textile Industry. https://doi.org/10.1016/B978-0-323-90392-9.00007-0**

pollution. The current value chain of apparel has become more complex with the rise in product complexity. The season-dependent and fast fashion concepts in the apparel industry reduce the life of apparel. As a result, consumers discard clothes after the season, not after the end-of-life (Dervojeda et al., 2014).

## 17.2 Historical aspects and current scenario

The textile and apparel industries have a long history of being influenced by a complex mixture of market and non-market forces. The major fields of logistics are procurement, production, distribution, after-sales, and disposal. The term logistics originates from the French word *logistique* or *loger*, which first appeared in the book *The Art of War* by Baron Henri, who was a general in the French army during Napoleon's time. Its original use was to describe the science of movement, supplying, and maintenance of military forces in the field. Later, it was used to describe the management of materials flow through an organisation, from raw materials through to finished goods.

During the late 20th century, textile and apparel production migrated to the developing world economies with a supply chain structure not significantly different from that used by the industry of post-World War II. After 1946, the US textile and apparel industry was composed of hundreds of small firms. Using low-skill, labour-intensive processes, the yarn, fabric, fabric processing, and product forming firms produced output that was used by other firms to supply final products for apparel, home furnishings, and/or industrial uses (Cooper, 2010). The term supply chain management entered the public domain when Keith Oliver, a consultant at Booz Allen Hamilton used it in an interview for the *Financial Times* in 1982. The take up of the term was slow, but it gained importance in the mid-1990s. In the late 1990s, it rose to prominence as a management buzzword, and operations managers began to use it in their titles with increasing regularity.

In the current globalised economy, the best customer value does not come from individually competing firms, but from the harmony of individual firms acting together as a supply chain. Supply chains can optimise customer value through the integration of customer perceptions, product design, process design, and an overlaying supply chain network design supported by transportation and information channels (Cooper, 2010). In logistics, energy is a resource that factors significantly into not only economic performance but also environmental sustainability. In the process, however, logistics operations consume huge energy resources and, in turn, generate a major share of greenhouse gas emissions (Browne, 2005).

## 17.3 Supply chain of fashion business

The supply chain comprises a chain of different partners that perform functions such as material procurement, production, distribution, etc. This network exists in manufacturing and in-service organisations (Handfield and Nichols, 1999). The value chain can be conceptualised as a chain of activities through which a product passes from the conception stage to the disposal stage (Kaplinsky and Morris, 2001). A typical textile supply chain starts from cotton farming and ends when the consumer discards the apparel. The whole value chain of apparel is multi-echelon and distinct to the product, supplier, and consumer (Keiser and Garner, 2008). Each echelon of the apparel supply chain can work independently or vertically integrated to have all value creation activities under one roof.

Value creation at each step is intended to increase the worth of products to earn higher revenue. A designer develops an apparel design based on the fashion forecast. The manufacturing companies receive orders and design apparel. The product development process, which includes sampling, is done before mass production. The bill of material for apparel is generated to start the procurement of fabrics and trims. The industrial engineering department develops an operation bulletin to start manufacturing apparel based on samples. The manufacturing process of apparel starts with the cutting of fabrics. The cutting panels are fed to sewing lines, and stitched apparel comes out of line. The stitched apparel is finished and packed according to the order. The packed order is shipped and distributed to the retailers. End consumers purchase the apparel, and use and discard it after the end of use or end of life. Fig. 17.1 depicts the fashion supply chain.

Earlier, the focus was given to a unidirectional flow of material across the supply chain, but the focus has been shifted to bi-directional flow. Due to enhanced complexity, there is a need to ensure efficient recovery at each stage. This is mainly based on optimisation and cost minimisation (Yang et al., 2013). Various efforts have been made to bring back the disposed-off product by closing the loop of the forward chain. This can be done with the help of different reverse chain activities, i.e., reuse, refurbishing, repair, remanufacturing, recycling, or redesign (Jayaraman and Luo, 2007). A value chain in which downstream and upstream activities are combined is called a closed-loop value chain. Gradually, the closed-loop value chain emerged as an independent subject and research area (Paksoy et al., 2011). Many government and non-government organisations have also extended their support and contributed significantly to maintaining closed-loop supply chain operations to save natural resources for future generations (Dururu et al., 2015).

Natural resource depletion and environmental degradation can be controlled by avoiding the entry of virgin (new) raw materials (Chen and Burns, 2006; Kuo et al., 2014). The concept of a closed-loop value chain is based on the principle of 3R (i.e., reduce, reuse, and recycle) or 6R (reduce, reuse, recycle, recover, redesign, and remanufacturing) (Govindan et al., 2015; Wang and Hsu, 2010). This controls the entry of new materials into the value chain (Ji, 2007).

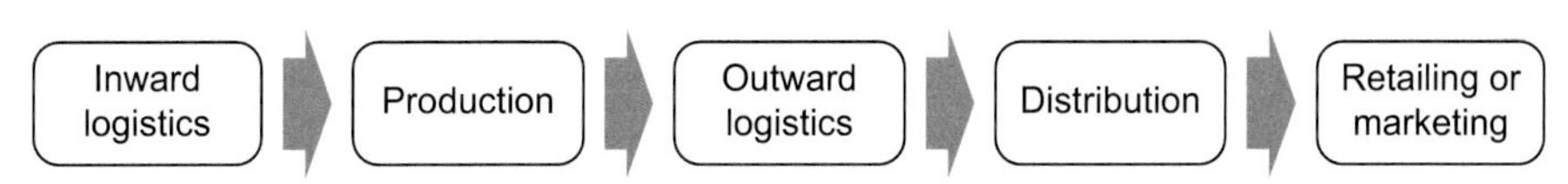


**Figure 17.1** Fashion supply chain.

## 17.4 Reverse supply chain

Fleischmann et al. (2000) identified the main difference between traditional and reverse supply chains. The traditional supply chain involves endogenous factors, so the quality, quantity, and timing can be controlled according to capacity and demand from the customer. At the same time, the reverse supply chain involves exogenous factors, highly unpredictable and influenced by a high level of uncertainty. The nature of the reverse supply chain is also uncertain as the recovery of the returned product is not triggered by market demand. Furthermore, the volume of reverse supply is low through a large number of collection or return points. Fleischmann (2001) has highlighted the characteristics of the recovery network in the reverse supply chain, i.e., coordination (between disposer and reuse market), supply uncertainty and disposition activities. Coordination is required between the disposer market (the place where the user releases the used product) and the reuse market (the place where the demand for the used product occurs). However, the reuse market's availability of used products is difficult to achieve due to supply uncertainty in the recovered network. Disposition is the main activity, and inspection of the product should be performed as early as possible to determine recovery options. Fleischmann et al. (2004) highlighted the mechanism of traditional and recovery networks; the pull strategy is adopted in the traditional network, while the reverse network push mechanism is generally accelerated by government legislation or consumer awareness. In the traditional supply chain, product differentiation is delayed to achieve higher value. In contrast, in the reverse supply chain, the condition of value is assessed as soon as recovered to maximise value.

The main components of the reverse supply chain are drivers, recovery options, product characteristics, actors, and phases. Economic and environmental factors are the main drivers of the recovery network. Supply chain factors (internal) or factors affecting the surrounding (external) factors determine driving factors. External factors can be summarised as economic, marketing, legislative, and aftermarket protection. An internal driver depends upon manufacturing and distribution facilities (Fleischmann, 2001; Rogers and Tibben-Lembke, 2001). The recovery option depends upon the product's condition, i.e., direct reuse, resale, repair, refurbishing, remanufacturing, cannibalisation, recycling, or landfill (Fernández, 2004). Status of product returns could be production returns, commercial returns, product recalls, warranty and service returns, and end-of-use and end-of-life returns (Fleischmann, 2001; Krikke et al., 2013). The number of reverse supply chain actors varies according to the recovery network.

The disposed of product can be returned to the forward chain, which can be called a closed-loop chain. The actors in a closed-loop chain could be suppliers, manufacturers, wholesalers, or retailers. The product can be returned through the alternative chain that could be specialised or common for all products. The actors for alternative recovery networks can be recycling and sorting companies, dedicated logistic or transport providers, municipalities or government agencies, brokers, or other profit-making companies. Several recovery phases vary according to the recovery network. In general, recovery phases appear as acquisition or collection, transportation (centralised or decentralised), inspection/selection or separation/sorting, reprocessing or reconditioning and disposal (Fleischmann, 2001; Guide et al., 2003; Krikke et al., 2004).

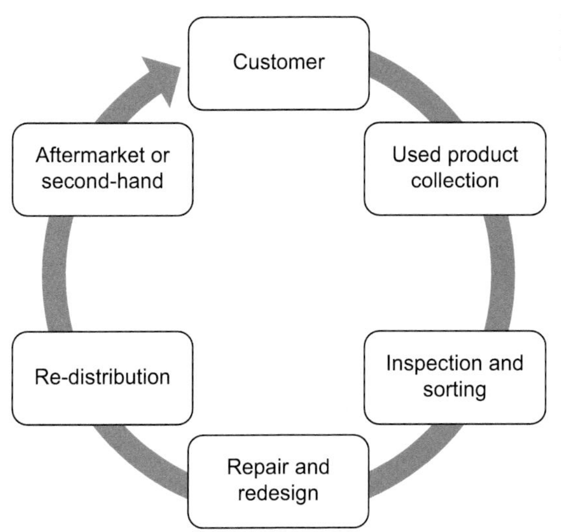


**Figure 17.2** Reverse supply chain of fashion.

There are several processes/stages in the fashion supply chain, starting from spinning cotton to apparel stitching. Every process adds some value to the product. The textile industry is considered to be the second largest industry globally, consuming nearly 10% of global energy, after the food industry. Many hazardous chemicals are used and emitted in the apparel manufacturing processes. To overcome this, some firms have taken initiatives to reduce the harmful effects of toxic chemicals. However, a significant minimisation seems difficult to achieve (Kuo et al., 2014). The situation is alarming as the consumption and disposal rates have also increased. Textile waste could threaten the environment because discarded apparel ultimately goes to landfilling or incineration. Fig. 17.2 represents the reverse supply chain of the fashion.

Fashion and the associated apparel industry are widely blamed for their environmental impacts. Even in developed countries, up to two-thirds of textile waste is condemned to landfill. At the landfill, fibres decay slowly and emit toxic and harmful gases (Cumming, 2017). A closed-loop system is the right alternative to reduce the volume of textile waste. The majority of the materials used in the production of clothes can be recovered from discarded products and can be returned to the consumption cycle. Although most textile materials are technically recyclable, it is difficult and costly to execute in practice. The recycling of fibre blends can generate lower quality fibre. Mono-fibre clothes have been proposed for better recyclability, however mono-fibre clothes cannot provide a solution to the fast-changing apparel industry fashion trends.

## 17.5 Closed-loop supply chain

A classical supply chain consists of suppliers, manufacturers, distributors, retailers, and end consumers. In contrast, a reverse supply chain is a process of recovering value from disposal from the point of consumption. Reverse logistics can be defined as the process of planning, implementing, and controlling the efficient, cost-effective flow of raw materials, in-process inventory, finished goods and related information from the point of consumption to the point of origin to recapture value or proper disposal (Rogers and Tibben-Lembke, 2001). The forward and reverse value chains are considered simultaneously in a closed-loop value chain. A closed-loop supply chain (CLSC) is management to design, control, and operate a system to maximise value creation over the entire life cycle of a product with the dynamic recovery of value from different types and volumes of returns over time (Govindan et al., 2015). Fig. 17.3 represents forward and reverse logistics.

The closed-loop supply chain is established to create revenue by acquiring products from consumers and recapturing the residual or unused value (Guide and Wassenhove, 2009). CLSC management is aligned with environmental and natural resource protection legislation to minimise waste creation. The current social and environmental legislations focus on natural resource conservation, sustainable product design, waste management, and pollution control (Wang and Hsu, 2010). Environmental legislation and economic benefits have amplified the significance of the closed-loop supply chain (Sasikumar and Haq, 2011). CLSC has recently raised interest among researchers and practitioners in addition to the legal and commercial benefits. The acquisition of products on the reverse side of the supply chain is complex compared to raw material sourcing. The operation of CLSC is largely dependent on the acquisition of products from consumers (Morana and Seuring, 2007). Product acquisition is the process of getting back products from the user. Reverse logistics (RL) moves collected products back to different supply chain partners (Morana and Seuring, 2007). Krikke et al. (2004) have identified four main kinds of product acquisition, i.e., end-of-life, end-of-use, commercial, and re-useable. Collection refers to the activities of product acquisition. The collected products are inspected, re-processed, transformed, re-marketed, and re-distributed (Fleischmann et al., 2000).

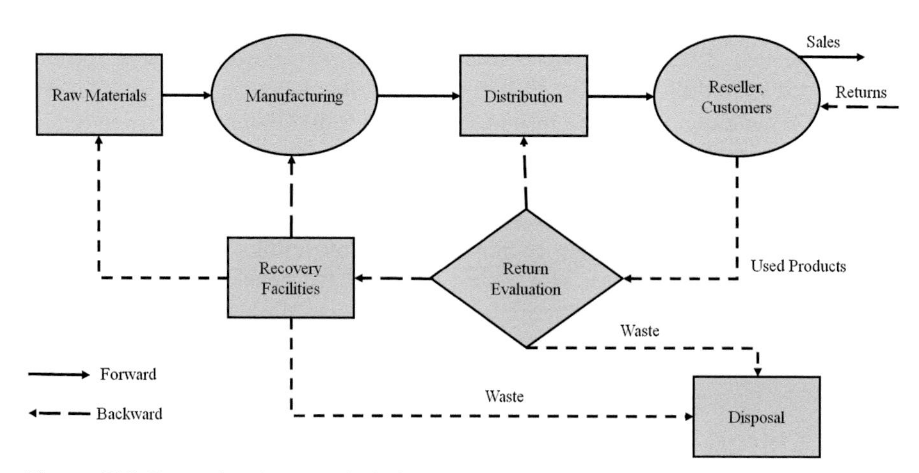


**Figure 17.3** Forward and reverse logistics.
Adapted from Govindan et al. (2015).

CLSC can be observed from the perspectives of processes, customers, and products. The process perspective looks into product return procedures. CLSC constitutes recovery processes and the forward supply chain processes such as sourcing, manufacturing, assembly, and distribution. Returned products are categorised according to quality and economic evaluation. Testing and sorting are carried out to decide the state of products and the most optimal reuse options such as direct reuse, repair, refurbishing, remanufacturing cannibalisation, and recycling. Re-marketing is done to create and develop a market for the recovered products (Morana and Seuring, 2007). Products in good condition are used in the existing state after being cleaned and repackaged. The products that require little rework are repaired to bring parts and products back to working condition. The collected products are completely disassembled during the remanufacturing process, and defected parts are replaced with new ones. The remanufactured products can be sold in the market along with new products. In the refurbishing process, the products are improved to achieve better functional and aesthetic quality but which is lower than for the original products. Cannibalisation is a process of parts extraction from a product that cannot be restored for use. If returned products or their parts are found not to be suitable for use, then recycling can be done to extract materials. The products could be incinerated if neither parts nor material recovery is possible. Energy can be recovered by burning products through an incineration process. If none of the recovery processes is suitable, physical products can be disposed of at a waste landfill (Nunen and Zuidwijk, 2004).

The customer perspective of CLSC relates to customer behaviour. Customers' behaviour can be related to purchasing, re-use, repair, discarding, and disposal of information. The analysis of consumer data provides useful perspectives for product use and return. The product data are directly or indirectly collected from products to analyse CLSC from product perspectives. The analysis of product data can be used to redesign the product for efficient recovery. These data will also help to improve forward supply chain operation and the recovery stage (Nunen and Zuidwijk, 2004). There is a need for industry-specific research on the closed-loop supply chain or reverse value chain to study each process of reverse logistics (Kumar and Putnam, 2008). Fig. 17.4 shows the closed-loop supply chain of the fashion.

Paras et al. (2017) have developed an empirical model of a closed-loop supply chain for the fashion industry. This model has been developed and analysed to enhance the understanding of the closed-loop supply chain for the apparel industry. Dervojeda et al. (2014) have highlighted the importance of smaller loops to save energy and effort to extend the product's life. The reverse value chain in the apparel industry can be incorporated in three different ways or loops. Firstly, retailers and distributors can redesign the leftover stock of products to make them saleable. Secondly, consumers can extend the life of self-used clothes by getting them redesigned. Thirdly, formal reuse can be adopted by selling and purchasing used apparel from the second-hand market. Based on the qualitative analysis and above discussion from the case study findings, a model for a closed-loop supply chain is proposed as follows.

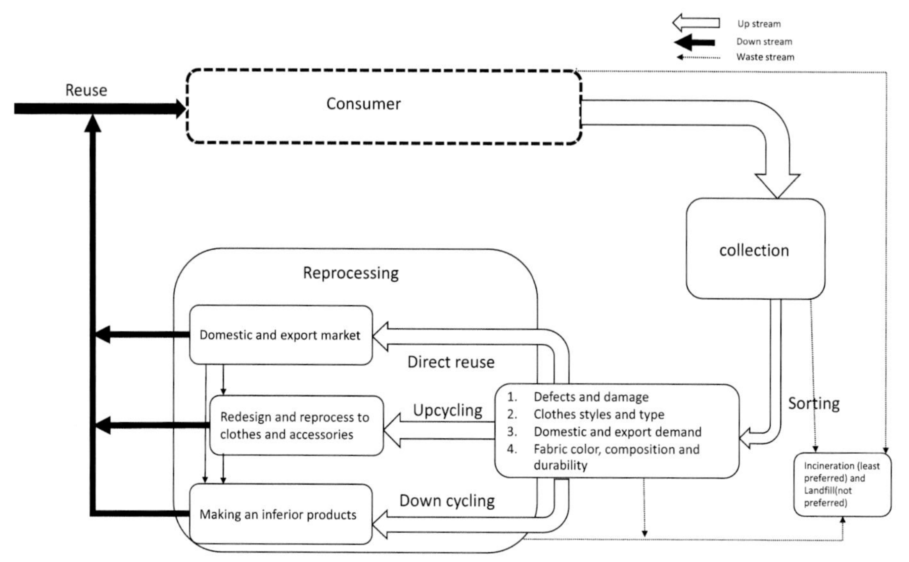


**Figure 17.4** Model of the closed-loop supply chain of fashion.
Adapted from Paras et al. (2017).

## 17.5.1 Factors affecting closed-loop supply chain

Several parameters influence the process of reverse logistics in the closed-loop supply chain. Important factors are listed and discussed below. Deriving from the literature, the major determinants of a closed-loop supply chain may include a business system, product designs, product price, product reuse information, legislation, and consumers' environmentally conscious attitude. The most important factors have been identified that could positively impact the reverse value chain business of apparel. This has been illustrated in Fig. 17.5.

### 17.5.1.1 System

In the majority of cases, clothes reuse activities are carried out by charity organisations. These organisations do not have the proper infrastructure to perform reverse value chain activities. The volunteers working with charity organisations are mostly semi-skilled or retired persons. They tend to carry out activities in a traditional way. Therefore, there is a need to develop the business model in collaboration with charity organisations. The main source of the collection of old clothes is through donation; hence the business model cannot be developed without incorporating social factors (Parsons, 2002). The collection, sorting, and reprocessing are supposed to be studied by considering social factors. Moise (2008) also

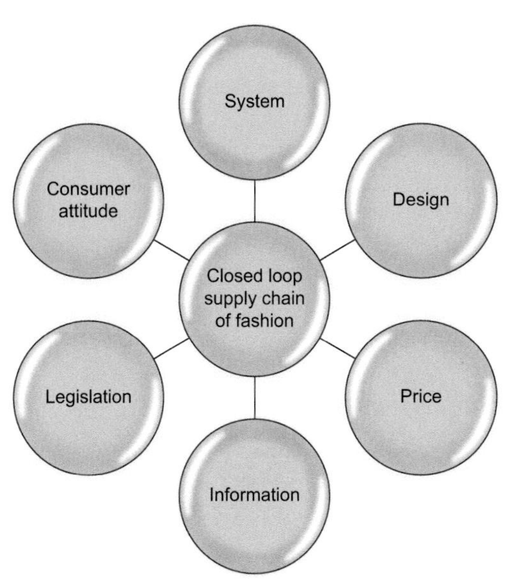


**Figure 17.5** Factors affecting the closed-loop supply chain of fashion.
Adapted from Paras et al. (2017).

highlighted the need for separate mapping of the activities of the reverse side of the value chain. This will help to understand the business of collection, sorting, and reprocessing at a minute level. In the current situation, there are different players for each of the processes involved in the reverse value chain of the apparel industry. Fragmentation and lack of collaboration are identified as major factors responsible for inefficiencies. This highlights the need for the complete network of the system, which might comprise different subsystems for the whole value chain (Abraham, 2011).

#### *17.5.1.2 Fashion design*

Fashion products are highly dependent on the style and trends prevailing at a certain time. This creates interest in styles which were in great demand during previous decades. This has created a unique market for fashion items popularly known as vintage products (DeLong et al., 2005). Different kinds of value addition, such as print and embroidery, can further help in increasing the aesthetic value of old products (Hiller Connell, 2011). However, the repair and redesign of old products are labour-intensive processes. These process can be simplified by considering the future reuse of products or components during the initial design stage (Das and Chowdhury, 2012).

#### 17.5.1.3 Cost factor

In the past, worn clothes were exchanged between friends and family (Sanderson, 1997). However, now, this informal exchange has taken the shape of formal exchange via stores, where old or used clothes are sold at affordable prices. Consumers go to second-hand stores in search of quality products at a lower price (Guiot and Roux, 2010). This highlights that the price of the used product is another important factor which decides its saleability (Horvath et al., 2005). The price of old products should be kept low so that they attract customers. There should be a significant price difference between old and new products so that consumers find older products to be more affordable (Atasu et al., 2008).

#### 17.5.1.4 Information system

In the current situation, not a single supply chain can function without the help of a management information system or enterprise resource planning. Information has become a key driver of any business. This also plays an important role in reverse logistics' operation and strategic decision-making process (Ferguson and Browne, 2001). Different technologies provide timely and accurate information in the forward supply chain. Similar technologies can be used for reverse logistics activities in the apparel industry (Das and Dutta, 2015). The trade in second-hand apparel is subjective due to several variable parameters. This leads to the problem of quality which can be reduced by proper communication among the different partners (Abimbola, 2012). Proper information between consumers and partners of reverse logistics will help in meeting the requirements of each other.

#### 17.5.1.5 Legislation

Successful waste management practices need intervention from local government authorities; the same applies to the recovery of textiles and apparel. For the first time, in 1970 in the UK, it was found that local authorities collaborated with grassroots people to collect waste (Oldenziel and Weber, 2013). This cooperation was quite successful. In a recent development, the Ulsan Metropolitan authority in Korea has signed a memorandum of understanding with LG Electronics to collect all electronic waste irrespective of brand and product (Das and Dutta, 2015). Hence, the success of non-profitable reverse logistics more or less depends on favourable government legislation and support. This highlights that there is a need for similar government intervention for clothing and textile recovery also.

#### 17.5.1.6 Consumer attitude

Disposal of clothes into a collection bin or charity shop is highly dependent on the individual's behaviour and attitude. Environment-conscious consumers are more inclined towards proper disposal (Bianchi and Birtwistle, 2012). It has also been found that consumers that consider themselves responsible for the environment tend to visit second-hand shops very frequently. In comparison, other types of consumers who do not like to shop for or wear used clothes tend to purchase new clothes. Hence, it can be stated that an environmentally conscious attitude of consumers influences purchase decisions and disposal (Lim et al., 2012).

## 17.6 Sustainable innovations in supply chain

Sustainable development meets the requirements of the present without compromising the needs of future generations (Seuring and Müller, 2008). The central concept of sustainability is a triple bottom line that discusses a minimum and balanced approach between environmental, economic, and social dimensions of sustainability (Elkington and Rowlands, 1999). A sustainable supply chain considers all three dimensions of sustainability, i.e., economic, environmental, and social, while managing material, information, and capital internally and externally. A sustainable supply chain encompasses the concept of a green and environmental supply chain at the broader level (Seuring and Müller, 2008). In the sustainable supply chain, fulfilling social and environmental criteria is important, and achieving customer needs and economic criteria are required to remain competitive. Fig. 17.6 depicts the three pillars of sustainability.

A sustainable supply chain is operated in a way that generates competitive returns on its capital assets without sacrificing the legitimate needs of internal and external stakeholders and with due regard for the impact of its operations on people and the environment (Kleindorfer et al., 2005, p. 489). Seuring and Müller (2008) have identified two strategies for a sustainable supply chain. First is supplier management for risks and performance, and advocates considering environmental and social criteria for supplier economic evaluation. The second strategy is supply chain management for sustainable products that demand the life cycle-based standards for the environmental and social performance of products.

A sustainable supply chain is concerned with managing raw materials, inventories, manufacturing, and finished goods from the points of origins of raw material to consumption by the end-user in such a way as to minimise the negative impacts on the triple bottom line (Zaarour et al., 2014). The sustainable supply chain also considers reverse value chain operations such as reduction, reuse, repair, recovery, disassembly, refurbishing, remanufacturing, and recycling, as this operation has huge impacts on the

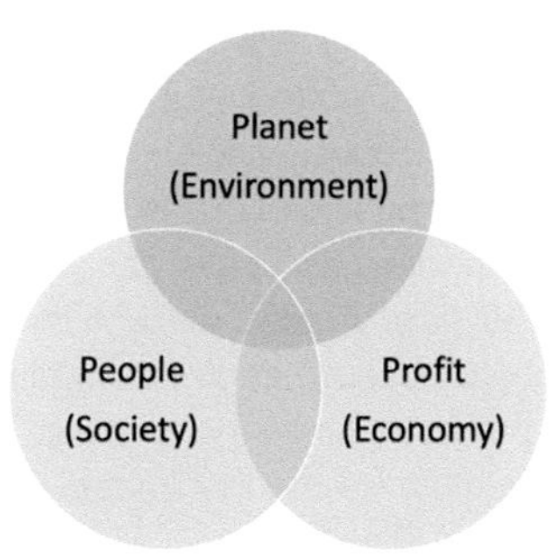


**Figure 17.6** Sustainability pillars.

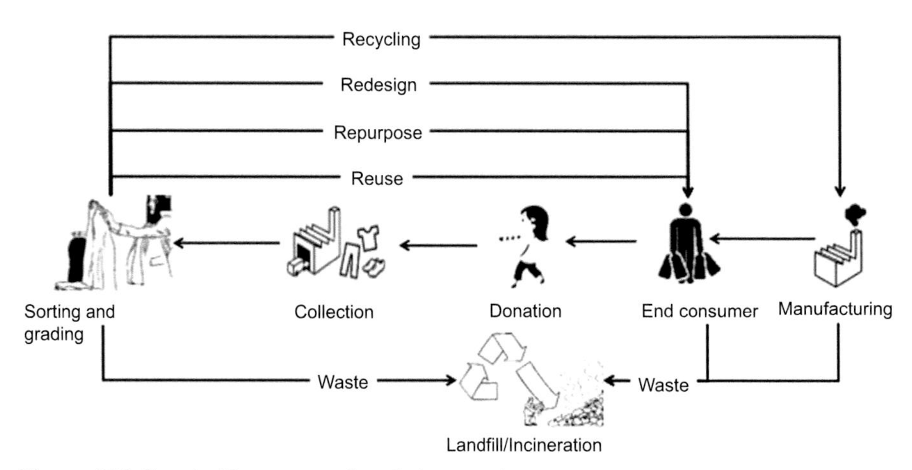


**Figure 17.7** Sustainable reverse value chain operations.

environmental, social, and economic dimensions of the triple bottom line of sustainability (Guide et al., 2003). This has been illustrated in Fig. 17.7.

Therefore, a sustainable supply chain has a multi-echelon business involving forward and reverse value chains from the cradle to the grave and vice versa (Zaarour et al., 2014). Due to lesser economic benefits, sustainability in the supply chain is mainly inspired externally by government, customers, and/or other stakeholders (Gold et al., 2010). Industries face pressure on environmental and social issues from government legislation, competitors, and other stakeholders such as consumers and brand owners. Therefore, organisations are forced to adopt sustainable practices to improve social and environmental measures (Diabat et al., 2014).

### 17.6.1 Sustainable fashion business model

Sustainability in the fashion industry is suggested through minimisation and alternative materials use and products. Unconventional ideas of leasing, renting, formal sharing/reusing, and upcycling have improved the products/material use of the product in the supply chain. Sustainability can be achieved through repairing and returning fashion products. Repairing or replacement services are even provided till the end-of-life of the fashion products. Higher product utilisation is also achieved through leasing, renting, and sharing, where the brand retains ownership of the product. Product longevity can also be achieved through the swapping of clothes. The emergence of fashion/clothing libraries in Scandinavian countries is a collaborative phenomenon of sustainable practice. Producers are liable for safe disposal as per country-specific legislation. The producer or sourcing partner either takes care of reverse logistics or pays a fee to manage the end-of-life product. Different supply chain partners and other actors such as charity organisations and service providers are taking initiatives such as product take-back, repairing, and upcycling to achieve sustainability (Paras, 2018). There exist several sustainable models in the fashion business, which are elaborated further below (Pal, 2016).

#### 17.6.1.1 Repair service

Free repairing of products is provided through repair stores, or the same is facilitated through the brand outlet. Damaged products are collected at the brand store and repair is performed in the company. In some other cases, do-it-yourself (DIY) kits are provided with awareness and know-how through company social media handlers and websites.

#### 17.6.1.2 Renting service

Traditionally, an offline rented service was available for stage performance costumes. This service has been gradually extended for party wear, childrenswear, etc. At present, almost all kinds of fashion products are available for rent. Laundry and washing are also arranged with the pick-up and delivery through a third party.

#### 17.6.1.3 Fashion libraries

This highly innovative model works on the line of university (book) libraries. Clothes can be borrowed against the membership fees paid for a 3- or 6-month period. Like books, clothes can be borrowed for a particular period (possibly 1–2 weeks). Borrowers are responsible for returning the fashion product without damage after properly washing it.

#### 17.6.1.4 Collection, sorting, and redesign service

Mostly, charities and non-profit organisations are involved in these services. Various collection methods such as collection boxes, door-to-door, and shop collections are used. Collected clothes are segregated into different groups before selling or exporting. Clothes or other fashion products are redesigned through co-creation or workshops based upon the availability of resources.

#### 17.6.1.5 Swapping service

Collaborative consumption can be promoted in communities through social networking. Sale and purchase can also be provided through an online platform by receiving fixed fees and courier charges. Even offline swapping is performed among members of a particular social group (such as a university, society, etc.). Each individual brings some clothes, and in return, they receive other clothes of their choice and needs.

#### 17.6.1.6 Digital services

Digital platforms provide a decision support system. These platforms also provide technical knowledge and skill to repair and redesign. Nearest clothes drop-off locations for used clothes are identified through GPS technology. It also offers discount vouchers in exchange for the disposal of clothes.

New sustainable business models are emerging in the fashion industry to reduce environmental impacts. Most fashion manufacturers operating in a linear and unsustainable business model should incorporate environmental value and explore more sustainable resource usage methods (Pal and Gander, 2018).

## 17.7 Future perspectives

There exist several future directions towards achieving sustainability in the supply chain and logistics. The influence of the closed-loop supply chain's structure on the product's design could be a worthy area of study in accordance with upcoming textile regulations. Various governments around the world are planning to achieve sustainable production and consumption through a circular economy. The fashion products sold and imported should be circular and environment friendly, and these products must contain recycled material as much possible. The new strategy is promoting innovation in the textile sector to avoid waste and convert waste into secondary materials. The strategic focus is on reusable, repairable, recyclable, durable, and energy-efficient textile and fashion products by 2030. This could be achieved through improved product design, increased reuse and recycle initiatives, waste collection, green procurement and production, and a sustainable lifestyle.

The digital product passport is also a part of the upcoming textile sustainability to achieve transparency. The digital system will enable access to product-specific characteristics for the manufacturer, consumer, and other stakeholders of the supply chain. Such information includes composition, location, and manufacturer details of the products. Hence, the digital system is futuristic and capable of facilitating a circular textile system (Stretton, 2022).

A comparative study on different types of supply chains, based on empirical findings, holds potential for future research. A study with quantitative data from different supply chains could be useful to identify the merits and demerits. An empirical investigation of different types of sustainable models in the fashion supply chain could be a focus area in the near future. An in-depth analysis could also be performed to determine the influence of one factor on other factors. Similarly, the relationship between the business know-how information, government regulations, legislation, and consumer awareness could be another future research area.

## 17.8 Conclusion

The supply chain and logistics of the fashion business have received increased attention due to the high consumption and disposal of fashion products. This chapter discusses various supply chain and logistics forms that exist in the fashion industry, such as forward supply chain, reverse supply chain, closed-loop supply chain, and sustainable supply chain. The fashion supply chain is discussed in which products move from the design conceptualisation stage to the end consumer. Product passes through different downstream stages such as manufacturing, distribution, and retail. The product can move upstream once it reaches end-of-use or end-of-life through reverse logistics. Various steps are involved in the reverse supply chain, such as collection, sorting, resale, or redesign. Depending on the condition, products can be moved to the second-hand market for direct reuse. If the condition of collected fashion products is not good, they can undergo redesign or recycling. This reverse logistics option closes the loop to form a closed-loop supply chain.

Five factors have been identified to successfully implement a sustainable closed-loop supply chain for fashion products. The existence of a proper supply chain is an important factor for the execution or success of a closed-loop supply chain. In the absence of vital steps/stages, implementing a closed-loop supply chain may not be efficient. The eco-design or construction of fashion products is another important aspect of ensuring the smooth flow of fashion products. The approach of correct information or know-how of fashion business/products is critical among the partners of the reverse supply chain. One of the major success factors of the closed-loop supply chain is consciousness among consumers. A conscious consumer can play a vital role by purchasing good-quality fashion products in the downstream channel and discarding the same in the upstream supply chain. Renting, swapping, fashion libraries, and GPS-based digital services for locating clothes donation centres are among the methods identified for achieving a sustainable supply chain.

## References

Abimbola, O., 2012. The international trade in secondhand clothing: managing information asymmetry between west African and British traders. Textile: The Journal of Cloth and Culture 10 (2), 184–199.

Abraham, N., 2011. The apparel aftermarket in India - a case study focusing on reverse logistics. Journal of Fashion Marketing and Management 15 (2), 211–227.

Alikhani, R., Torabi Ali, S., Altay, N., 2019. Strategic supplier selection under sustainability and risk criteria. International Journal of Production Economics 208, 69–82.

Atasu, A., Guide Jr., V.D.R., Van Wassenhove, L.N., 2008. Product reuse economics in closed-loop supply chain research. Production and Operations Management 17 (5), 483–496. https://doi.org/10.3401/poms.1080.0051.

Bianchi, C., Birtwistle, G., 2012. Consumer clothing disposal behaviour: a comparative study. International Journal of Consumer Studies 36, 335–341. https://doi.org/10.1111/j.1470-6431.2011.01011.x.

Browne, M., 2005. Special issue introduction: transport energy use and sustainability. Transport Reviews 25 (6), 643–645. https://doi.org/10.1080/01441640500422140.

Bruce, M., Daly, L., Towers, N., 2004. Lean or agile: a solution for supply chain management in the textiles and clothing industry? International Journal of Operations & Production Management 24 (2), 151–170. https://doi.org/10.1108/01443570410514867.

Chen, H.-L., Burns, L.D., 2006. Environmental analysis of textile products. Clothing and Textiles Research Journal 24 (3), 248–261.

Cooper, W.D., 2010. Textile and apparel supply chains for the 21st century. Journal of Textile and Apparel Technology and Management 6 (4), 1–10.

Cumming, D., 2017. A case study engaging design for textile upcycling. Journal of Textile Design Research and Practice 4 (2), 113–128. https://doi.org/10.1080/20511787.2016.1272797.

Das, D., Dutta, P., 2015. Design and analysis of a closed-loop supply chain in presence of promotional offer. International Journal of Production Research 53 (1), 141–165. https://doi.org/10.1080/00207543.2014.942007.

Das, K., Chowdhury, A.H., 2012. Designing a reverse logistics network for optimal collection, recovery and quality-based product-mix planning. International Journal of Production Economics 135 (1), 209–221. https://doi.org/10.1016/j.ijpe.2011.07.010.

DeLong, M., Heinemann, B., Reiley, K., 2005. Hooked on vintage! Fashion Theory - Journal of Dress Body and Culture 9 (1), 23–42.

Dervojeda, K., Verzijl, D., Rouwmaat, E., 2014. EU-Circular-Supply-Chains. Business Innovation Observatory.

Diabat, A., Kannan, D., Mathiyazhagan, K., 2014. Analysis of enablers for implementation of sustainable supply chain management–a textile case. Journal of Cleaner Production 83, 391–403.

Dururu, J., Anderson, C., Bates, M., Montasser, W., Tudor, T., 2015. Enhancing engagement with community sector organisations working in sustainable waste management: a case study. Waste Management & Research: The Journal of the International Solid Wastes and Public Cleansing Association, ISWA 33 (3), 284–290. https://doi.org/10.1177/0734242x14567504.

Elkington, J., Rowlands, I.H., 1999. Cannibals with forks: the triple bottom line of 21st century business. Alternatives Journal 25 (4), 42.

Ferguson, N., Browne, J., 2001. Issues in end-of-life product recovery and reverse logistics. Production Planning & Control 12 (5), 534–547. https://doi.org/10.1080/09537280110042882.

Fernández, I., 2004. Reverse Logistics Implementation in Manufacturing Companies. University of Vaasa.

Fleischmann, M., 2001. Reverse logistics network structures and design. In: Guide Jr., V.D.R., Van Wassenhove, L.N. (Eds.), Business Aspects of Closed-Loop Supply Chains: Exploring the Issues. Carnegie Mellon University Press, 2003, Pittsburgh, PA.

Fleischmann, M., Krikke, H.R., Dekker, R., Flapper, S.D.P., 2000. A characterisation of logistics networks for product recovery. Omega 28 (6), 653–666.

Fleischmann, M., van Nunun, J.A., Grave, B., Gapp, R., 2004. Reverse Logistics – Capturing Value in the Extended Supply Chain. ERIM Report No. ERS-2004-091-LIS, Rotterdam.

Gold, S., Seuring, S., Beske, P., 2010. The constructs of sustainable supply chain management–a content analysis based on published case studies. Progress in Industrial Ecology, an International Journal 7 (2), 114–137.

Govindan, K., Soleimani, H., Kannan, D., 2015. Reverse logistics and closed-loop supply chain: a comprehensive review to explore the future. European Journal of Operational Research 240 (3), 603–626. https://doi.org/10.1016/j.ejor.2014.07.012.

Guide, V.D.R., Jayaraman, V., Linton, J.D., 2003. Building contingency planning for closed-loop supply chains with product recovery. Journal of Operations Management 21 (3), 259–279. https://doi.org/10.1016/s0272-6963(02)00110-9.

Guide Jr., V.D.R., Van Wassenhove, L.N., 2009. The evolution of closed-loop supply chain research. Operations Research 57 (1), 10–18. https://doi.org/10.1287/opre.1080.0628.

Guiot, D., Roux, D., 2010. A second-hand shoppers' motivation scale: antecedents, consequences, and implications for retailers. Journal of Retailing 86 (4), 383–399.

Handfield, R.B., Nichols, E.L., 1999. Introduction to Supply Chain Management, vol. 999. Prentice Hall, Upper Saddle River, NJ.

Hiller Connell, K.Y., 2011. Exploring consumers' perceptions of eco-conscious apparel acquisition behaviors. Social Responsibility Journal 7 (1), 61–73.

Horvath, P.A., Autry, C.W., Wilcox, W.E., 2005. Liquidity implications of reverse logistics for retailers: a Markov chain approach. Journal of Retailing 81 (3), 191–203. https://doi.org/10.1016/j.jretai.2005.07.003.

Jayaraman, V., Luo, Y., 2007. Creating competitive advantages through new value creation: a reverse logistics perspective. Academy of Management Perspectives 21 (2), 56–73.

Ji, G., 2007. Virtual enterprise and performance evaluation by using exergoeconomics in closed-loop supply chain. In: Xu, L.D., Tjoa, A.M., Chaudhry, S.S. (Eds.), Research and Practical Issues of Enterprise Information Systems II, IFIP — The International Federation for Information Processing, vol. 254. Springer, Boston, MA.

Kaplinsky, R., Morris, M., 2001. A Handbook for Value Chain Research, vol. 113. IDRC Ottawa.

Keiser, S.J., Garner, M.B., 2008. Beyond Design. Fairchild Publications, New York.

Kleindorfer, P.R., Singhal, K., Wassenhove, L.N., 2005. Sustainable operations management. Production and Operations Management 14 (4), 482–492.

Krikke, H., Hofenk, D., Wang, Y., 2013. Revealing an invisible giant: a comprehensive survey into return practices within original (closed-loop) supply chains. Resources, Conservation and Recycling 73, 239–250. https://doi.org/10.1016/j.resconrec.2013.02.009.

Krikke, H., le Blanc, I., van de Velde, S., 2004. Product modularity and the design of closed-loop supply chain. California Management Review 46 (2), 23–39.

Kumar, S., Putnam, V., 2008. Cradle to cradle: reverse logistics strategies and opportunities across three industry sectors. International Journal of Production Economics 115 (2), 305–315. https://doi.org/10.1016/j.ijpe.2007.11.015.

Kuo, T.-C., Hsu, C.-W., Huang, S.H., Gong, D.-C., 2014. Data sharing: a collaborative model for a green textile/clothing supply chain. International Journal of Computer Integrated Manufacturing 27 (December 2014), 266–280. https://doi.org/10.1080/0951192X.2013.814157.

Lim, W.M., Ting, D.H., Wong, W.Y., Khoo, P.T., 2012. Apparel acquisition: why more is less? Management & Marketing 7 (3), 437–448.

Lowson, B., King, R., Hunter, A., 1999. Quick Response: Managing the Supply Chain to Meet Consumer Demand. Wiley.

Moise, M., 2008. The importance of reverse logistics for retail activity. Amfiteatru Economic 10 (24), 192–209.

Morana, R., Seuring, S., 2007. End-of-life returns of long-lived products from end customer - insights from an ideally set up closed-loop supply chain. International Journal of Production Research 45 (18–19), 4423–4437. https://doi.org/10.1080/00207540701472736.

Nunen, J., Zuidwijk, R.A., 2004. E-enabled closed-loop supply chains. California Management Review 46 (2), 40–+.

Oldenziel, R., Weber, H., 2013. Introduction: reconsidering recycling. Contemporary European History 22 (3), 347–370.

Paksoy, T., Bektas, T., Ozceylan, E., 2011. Operational and environmental performance measures in a multi-product closed-loop supply chain. Transportation Research Part E: Logistics and Transportation Review 47 (4), 532–546. https://doi.org/10.1016/j.tre.2010.12.001.

Pal, R., 2016. Extended responsibility through servitization in PSS: an exploratory study of used-clothing sector. Journal of Fashion Marketing and Management: International Journal 20 (4), 453–470.

Pal, R., Gander, J., 2018. Modelling environmental value: an examination of sustainable business models within the fashion industry. Journal of Cleaner Production 184. https://doi.org/10.1016/j.jclepro.2018.02.001.

Paras, M.K., 2018. Reuse-Based Reverse Value Chain for Sustainable Apparel Industry. Högskolan I Borås. Doctoral Dissertation.

Paras, M.K., Pal, R., Ekwall, D., 2017. Systematic literature review to develop a conceptual framework for a reuse-based clothing value chain. International Review of Retail Distribution & Consumer Research 28 (3), 1–28. https://doi.org/10.1080/09593969.2017.1380066.

Parsons, E., 2002. Charity retail: past, present and future. International Journal of Retail & Distribution Management 30 (11/12), 586–594.

Rogers, D.S., Tibben-Lembke, R., 2001. An examination of reverse logistics practices. Journal of Business Logistics 22 (2), 129.

Sanderson, E.C., 1997. Nearly new: the second-hand clothing trade in eighteenth-century Edinburgh. Costume 31, 38–48.

Sasikumar, P., Haq, A.N., 2011. Integration of closed loop distribution supply chain network and 3PRLP selection for the case of battery recycling. International Journal of Production Research 49 (11), 3363–3385. https://doi.org/10.1080/00207541003794876.

Seuring, S., Müller, M., 2008. From a literature review to a conceptual framework for sustainable supply chain management. Journal of Cleaner Production 16 (15), 1699–1710. https://doi.org/10.1016/j.jclepro.2008.04.020.

Shen, B., 2014. Fashion supply chain: lessons from H&M. Sustainability 6 (9), 6236–6249. https://doi.org/10.3390/su6096236.

Stretton, C., 2022. Digital Product Passports (DPP): What, How, and Why? at: https://www.circularise.com/blogs/digital-product-passports-dpp-what-how-and-why. (Accessed 20 April 2023).

Wang, H.-F., Hsu, H.-W., 2010. A closed-loop logistic model with a spanning-tree based genetic algorithm. Computers & Operations Research 37 (2), 376–389. https://doi.org/10.1016/j.cor.2009.06.001.

Yang, P.C., Chung, S.L., Wee, H.M., Zahara, E., Peng, C.Y., 2013. Collaboration for a closed-loop deteriorating inventory supply chain with multi-retailer and price-sensitive demand. International Journal of Production Economics 143 (2), 557–566. https://doi.org/10.1016/j.ijpe.2012.07.020.

Zaarour, N., Melachrinoudis, E., Solomon, M.M., Min, H., 2014. The optimal determination of the collection period for returned products in the sustainable supply chain. International Journal of Logistics Research and Applications 17 (1), 35–45. https://doi.org/10.1080/13675567.2013.836160.

# Closed-loop postconsumer textile recycling

# 18

*Jens Oelerich and Jan W.G. Mahy*
Saxion University of Applied Sciences, School of Creative Technology, Enschede, The Netherlands

## 18.1 Introduction

Because the production of textiles needs raw materials such as fibres, process chemicals, energy and water, this has an impact on the environment. However, the trend of overconsumption that has outgrown the logical use of textiles has led to a tremendous increase in production and an associated impact. The doubling of production in only 15 years, between 2000 and 2015 (Ellen MacArthur Foundation, 2017), was partly initiated by the growth in ambition of fashion retailers and brands to produce inexpensive and lower-quality textiles and fulfil the wishes of the market for fast and fashionable clothing. This is called the 'fast fashion' trend, and it catalysed the unsustainable behaviour of consumers to buy many more textiles than were needed for protection and comfort, such as those that were used only a few times before being discarded. This trend causes an enormous amount of textile waste. Approximately 11 kg of textiles are discarded per person every year in the European Union (EU), which adds up to 5.8 million tons of annual textile waste (EEA, 2019).

Looking at the increasing frequency of new seasonal collections launched by the fashion industry and the decreasing quality of textile materials used and presented, clothing is worn and washed less often before it is discarded. This accelerates the consumption of clothing and creates unsustainable systems of overproduction and overconsumption, which directly conflict with Sustainable Development Goal 12 in ensuring sustainable consumption and production patterns (European Commission, 2022a). Fast fashion trends, the growing world population and the low rates of use, reuse, repair, and closed-loop recycling of textiles lead to the ever-increasing demand for primary raw materials from natural and fossil resources, especially from outside Europe (European Environmental Agency, 2022).

The dumping of postconsumer textile waste is a huge urban waste problem. There is a common practice of commercial and charity organisations collecting discarded textiles. Thus, a small portion is recovered, but the rest is normally discarded as solid urban waste. There is a huge untapped potential for discarded postconsumer textile waste, and the recycling possibilities are unlimited. To recycle on an industrial scale, collected clothes should first be sorted. Several systems are available for sorting textile waste according to colour, fibre type and chemical composition. The sorted textile waste can be collected separately to develop different high-added value products (Paul, 2012).

**Sustainable Innovations in the Textile Industry. https://doi.org/10.1016/B978-0-323-90392-9.00005-7**

Textiles mainly consist of fibres that are long (2−50 cm) and fine (10−40 μm in cross-section). To convert them into yarns, they need to have a degree of flexibility and surface roughness (Mather and Wardman, 2015). Fibres can come from virgin sources (both natural and synthetic) or recycled textiles. In both cases, the quality of the fibres is often determined by their length and fineness, which likewise determines their possible applications. Finer and longer fibres such as merino wool, silk and synthetic microfibres can be processed into fine yarns, which can be woven or knitted in lightweight and high-quality fabrics. Coarser or shorter fibres, such as mechanically recycled cotton fibres, can often be processed only into coarser yarns and coarse textile structures such as felt and other nonwovens.

## 18.2 Historical aspects and current scenario

Textile recycling has a long tradition. In the early 20th century, clothing was less affordable and fibre production was more demanding. Reuse and recycling were necessary because natural fibre sources were limited and synthetic fibres were still in development. In particular, wool recycling from textiles to fibres has a long tradition (Cardato, n.d.). In the 1980s and 1990s, polyester fibre production increases with the constant amount of cotton fibres available (Niinimäki et al., 2020). With synthetic fibres at lower prices, clothing became more affordable. At the start of the 21st century, the excessive availability of inexpensive clothing decreased the need to repair and reuse textiles. In addition, more fibre blends were introduced to the market. Open-loop recycling into nonwoven material for automotive or other insulation applications became the state of the art and is still the most commonly used technology for textile recycling (Ellen MacArthur Foundation, 2017)

Recycled polyethylene terephthalate (PET) fibres from PET bottles have been introduced to the market and have gained a larger share. However, these fibres are seen as an open-loop solution because the recycled materials come from outside the clothing sector. In addition, recycled PET fibres from bottles are under debate because food-grade PET leaves food-grade applications when it is recycled into textile fibres. This is considered unwanted downcycling because there is a high demand for recycled PET in the food industry as well.

There is increased consensus that the textile industry should solve its circularity challenges with materials of its own industry. According to this trend and economic considerations, the textile industry started to recycle postindustrial textile waste back into fibres for textile products (closed-loop recycling). Spinning companies installed machinery to open yarn leftovers and fabric cutting waste and thus decrease the use of virgin material. Advantages of postindustrial textile recycling are short logistics and well-defined waste streams.

Few economic incentives exist for the recycling of large amounts of postconsumer waste textiles. Extended producer responsibility (EPR) regulation can give such incentives (Staatsblad, 2023). With an EPR in place in an increasing number of EU member states, producers can be made responsible for treating their waste streams, including the end-of-life options of their products (OECD, 2016). Making producers financially responsible for collecting and recycling waste streams, together with proper guidelines, is critical for successful EPR systems (Gupt and Sahay, 2015).

New regulations should raise the demand for postconsumer recycled (PCR) materials. According to the European Waste Framework Directive (2008/98/EC), all EU member states should implement a waste management hierarchy dealing with prevention, reuse, recycling and disposal in their waste management plans (European Union, 2008) and a separate collection of textile waste streams by 2025 (European Union, 2018). The publication of the EU strategy on sustainable textiles (European Commission, 2022a) as well as regulations on a national level (i.e., EPR regulations in the Netherlands) (MI&W, 2020; Staatsblad, 2023) will further increase the demand for recycling solutions for postconsumer waste streams. Coming from a situation in which textile recycling was driven first by a material shortage and later by economic considerations, ecologic concerns and legislation now are the driving forces to foster textile recycling.

## 18.3 Circular textile economy

All natural and synthetic fibres consist of polymers (i.e. long-chain molecules). The chemical structure of these polymers contributes to the properties of the fibres. Moreover, the polymer chain length, crystallinity and morphology of the fibres are important characteristics. Because many (chemical) recycling methods are based on the intrinsic chemical structures of textile materials, it is important to understand the differences and characteristics of textile polymers. Harmsen et al. provided an overview of the most important chemical structures in textile fibres and their relation to recycling technologies (Harmsen et al., 2021).

In the past, textiles were produced to last one or more generations, but currently many textiles are even deliberately and fashionably damaged, weakening their textile structure before they reach the customer. This can be achieved by minor washings, laser structuring to preripped garments, and so forth (Fig. 18.1a). In addition, with the need to create new collections every season, designers and product developers

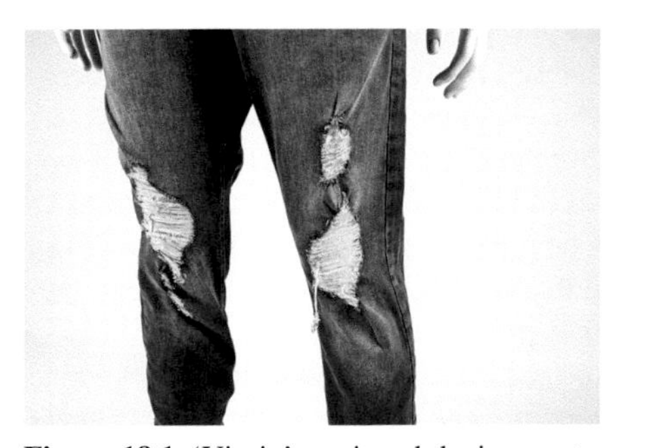

**Figure 18.1** 'Virgin' preripped denim; postconsumer clothing with sequins and prints.

were forced to create new and unusual looks with complex material mixtures and accessories that tend to be challenging for later circularity (Fig. 18.1b).

In 2022, the European commission published the EU strategy for sustainable and circular textiles. In that position paper, the Commission emphasised that 'the textiles ecosystem requires deep changes in the currently prevailing linear way in which textile products are designed, produced, used and discarded' (European Commission, 2022a).

In the first instance, recycling is not the preferred option in the circular economy. According to the R strategy introduced as Lansink's ladder by Dutch politician Ad Lansink and further specified by Potting et al. (2016) (Fig. 18.2), strategies to prevent material use in the first place (R0–2) and the elongation of textile life cycles (R3–7) are preferred options to decrease the impact of textile materials. However, at some point every material reaches a physical end of life. At that stage, textile materials should be recycled (R8) to generate new fibres for textiles (closed-loop recycling) or materials for other kinds of applications, such as thermal or acoustical isolation materials (open-loop recycling).

### 18.3.1 Textile recycling

When textile materials reach the physical end of life and value retention option from the R strategy are no longer feasible, they should be recycled. In principle, every textile item can be recycled in one way or another, depending on the effort to process it. Hawley described this thus: Almost everything is recyclable; the question is how much energy, chemistry and processing has to be performed to recycle something, and whether it is worth it, considering the costs and the environmental impact of recycling (Hawley, 2009).

Currently, few recycling options are ecological and economically viable to be implemented on a larger scale in industry. Economic viability strongly depends on the related costs of recycling compared with the costs of virgin materials on the market. Imbalance between costs of recycling processes and the cost of virgin materials will likely shift in the near future owing to upcoming legislation and increasing prices for both natural (e.g. cotton, wool) and fossil-based (e.g. synthetic) materials (European Commission, 2022a,b,c). However, the ecological consequences should also be thoroughly investigated for each new recycling technology.

EU Waste Framework Directive Art 3(17) describes recycling as any recovery operation by which waste materials are reprocessed into products, materials or substances, whether for the original or other purposes. Energy recovery operations that process waste in primary resources as energy or fuels are excluded from this definition (Waste Framework Directive Art 3(17)). The draft ISO Standard DIS 5157 Textiles – Environmental Aspects – Vocabulary further clarifies the term of recycling for textiles as the action of reprocessing a material or component that has previously been processed for inclusion in a product, with the side note that the process may be chemical, mechanical or thermal/thermomechanical (Duhoux et al., 2021).

This chapter uses definitions that are in line with the Waste Framework Directive and the study on the technical, regulatory, economic and environmental effectiveness of textile fibre recycling (Duhoux et al., 2021), in which the categories chemical, mechanical and thermal/thermomechanical are used. Chemical recycling has been further

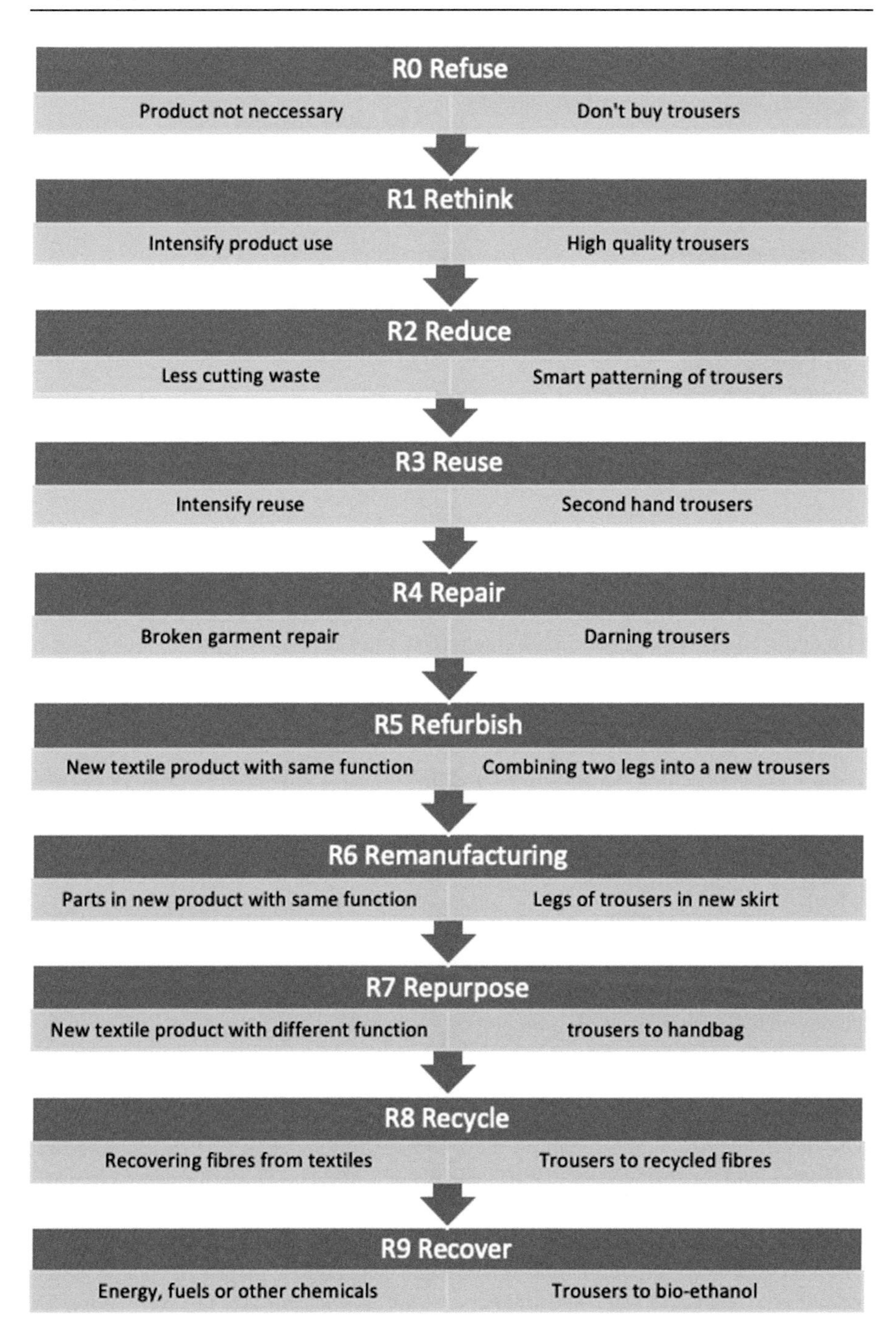


**Figure 18.2** R strategy for textile economy, with cotton trousers as example.
Adapted from Potting et al. (2016).

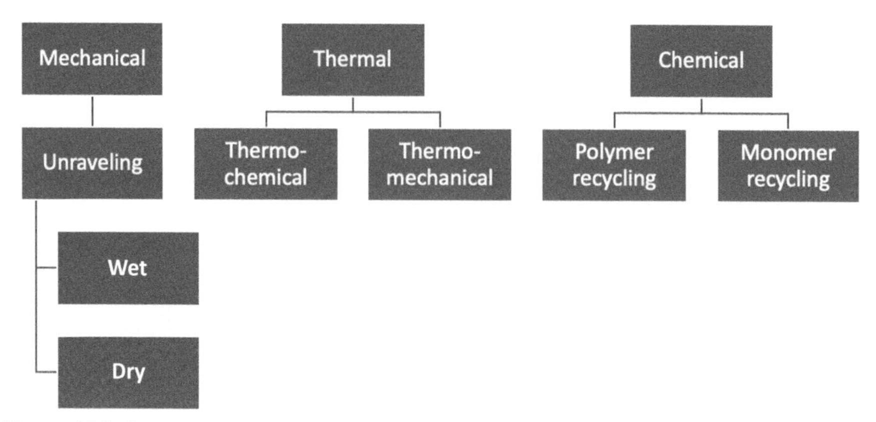


**Figure 18.3** Categorisation of closed-loop textile recycling technologies.
Adapted from Duhoux et al. (2021).

specified for recycling methods that use biochemical and thermochemical processes to clarify the nature of the reaction medium. Because only closed-loop recycling technologies are included in this chapter, processes that lead to materials useful only for open-loop applications are excluded. The categorisation of textile recycling technologies is presented in Fig. 18.3.

#### *18.3.1.1 Mechanical recycling*

In mechanical recycling, physical forces such as cutting, shredding, tearing or carding are used to generate individual fibres from textile structures such as yarns, fabrics or garments (Lindström et al., 2020). The mechanical recycling of textiles is a technology on an industrial scale in which all kinds of material types and structures can be processed. The opened textile materials are used to a large extent in nonwoven applications such as isolation panels for homes, washing machines or cars, drainage systems and geotextiles. For these applications, it is important to guarantee the material mix, to achieve, for example, flame retardant properties. The degree of openness and the fibre length are typically less important.

To respin textile fibres into yarns for textiles (i.e. closed-loop recycling), the yarns should be well-opened and the fibres should remain as long as possible. Optimisation of these processes is the subject of many academic and industrial investigations (Duhoux et al., 2021). In mechanical recycling, textiles are exposed to heavy mechanical forces, and thus fibres tend to break and become shorter. This affects the spinnability of fibres into new yarns (Ütebay et al., 2019).

Mechanical wool recycling has a long tradition, especially in the Prato region in Italy. Wool fibres, which are of animal origin and consist of polypeptide polymer chains, are typically longer and thinner as well as morphologically more complex than other natural fibres, such as those of vegetal origin. The fibres have high mechanical interaction leading to strong, thin yarns. This also requires that products from wool be made (during production) and handled (during use) with care, such as during washing. In addition, wool is predominantly processed into knitted fabrics that are easier to open in mechanical recycling processes than woven fabrics. This leads to longer recycled fibres from mechanical recycling processes, and the respinning of yarns with high amounts of recycled content is achieved (Cardato, n.d.).

For economic reasons, postindustrial residue such as yarn leftovers from spinning and fabric cutting waste is typically recycled into open fibres and added to virgin material (Russell et al., 2016). To obtain recycled fibres from postconsumer sources with uniform quality and colour, fabrics need to be sorted and cleaned before they are turned into fibres by cutting and tearing. For postconsumer waste streams, it is important to minimise fibre breakage during the process, to achieve recycled fibres with a sufficient length for yarn spinning (Ütebay et al., 2019).

Mechanical recycling of cotton and polyester/cotton blends is the focus of academic and industrial investigations. Mechanical forces are also used as preprocessing steps for thermal or chemical recycling processes. Here, cutting and milling activities are often performed to decrease the size, which increases the surface of the material for the better accessibility of biochemicals.

#### *18.3.1.2 Thermal recycling*

In thermal textile recycling, heat is used to recover polymers or building blocks from textile materials (not to be confused with incineration). Thermoplastic polymers from polycondensates (e.g. polyester, polyamide) are molten, and granulates are melt-spun into highly uniform filament fibres. These processes are comparable to film processes used for solid plastic waste from packaging. Before a new textile fibre can be produced, shredding, cleaning, drying and degassing are typically performed as pretreatment steps in thermal textile recycling.

To spin new textile fibres, virgin material or melt-filtration often needs to be added to achieve the necessary quality of the spinning dopes. Owing to similarities with virgin melt-spinning processes, thermal textile recycling can be economically viable and have appropriate melt filtration implemented in existing processes. However, pure waste streams, predominantly production waste and specific consumer waste, are often currently processed (Duhoux et al., 2021). Thermochemical recycling produces low-molar mass components or heat as feedstock for the chemical industry. Because the scope of this chapter is closed-loop postconsumer recycling, thermal recycling and thermochemical recycling will not be discussed in more detail here.

#### *18.3.1.3 Chemical recycling*

In chemical recycling, textile materials are treated with chemicals to recover polymers or monomers (the building blocks of polymers) from materials. In polymer recycling, polymers in textiles are purified, adjusted to the necessary quality and respun into fibres. This can be done by disassembling the textiles and purifying the polymer before it is dissolved, or vice versa. In monomer recycling, polymers in textiles are broken down into their constituent building blocks (monomers) and rebuilt to polymers after purification to generate new textile fibres. Usually, the smaller the molecules are in chemical recycling, the easier the processing and purification are of the material. However, breaking down polymers into monomers requires energy and likely increases the environmental impact of the process (Duhoux et al., 2021).

For cotton, polymer recycling is the only closed-loop chemical recycling option, because cotton monomers (sugar molecules) cannot be rebuilt synthetically into polymers. Chemical recycling of cotton produces cellulose dissolving pulp, which is also the precursor for spinning regenerated cellulose fibres such as viscose and lyocell. Chemical recycling of cotton includes mechanical preprocessing to remove nontextile materials, purification of cellulose and sometimes adjustments to the degree of polymerisation (DP), which resembles the polymer chain length of the cellulose (Oelerich et al., 2017). For virgin viscose and lyocell, cellulose pulp from wood is used. Chemically recycled cotton fibres are also produced from cotton cellulose pulp via different fibre spinning processes:

- Viscose process, a large-scale industrial process using carbon disulphite
- Lyocell process, a large-scale industrial process using *N*-methylmorpholine *N*-oxide (NMMO)
- Carbamate process, a smaller-scale industrial process using urea
- Ionic liquid process, a pilot-scale process using different ionic liquids

Polymer recycling can also be achieved with solvents that dissolve textile polymers. When these dissolution processes are selected for one polymer type, blends of textile materials can be separated from each other and purified from dyestuffs and other contaminates (Sherwood, 2020).

In monomer recycling, synthetic polycondensation polymers such as polyester and polyamides are broken down into their building blocks. This depolymerisation can be performed in different reaction media, resulting in different kinds of monomers. These processes have been thoroughly investigated in academia (Payne and Jones, 2021), and several industrial players are looking for possibilities to upscale developed technologies (Duhoux et al., 2021). The advantage of this technology is that depolymerisation leads to the dissolution of monomers in the reaction medium, which can subsequently be purified from dyestuffs and undissolved impurities by filtration. Reaction media typically applied are water (leading to hydrolysis reactions), methanol (leading to methanolysis reactions) and glycols (leading to glycolysis reactions) (Payne and Jones, 2021). For polyamides, monomer recycling is the most well-established for polyamide-6, which is depolymerised in water to form its monomer, caprolactam, which can be purified and used for the repolymerisation of polyamide-6. This process has been industrialised to recycle carpets, fishing nets and clothing waste by Aquafil and is called the Econyl process (Econyl, n.d.).

Hydrothermal and biochemical recycling technologies are typically mixed processes of monomer and polymer recycling. Hydrothermal technology consists of an approach to (partially) degrade cotton, PET or both. These processes rely on water, pressure, temperature and green (enzyme) chemistry, in which the final output depends on the specific process applied (Duhoux et al., 2021).

Some biochemical recycling technologies use enzymatic degradation reactions of cotton to transform cotton from fibre blends into sugars or smaller molecules such as bioethanol and recover PET and other polymers from the mixture (Ribul et al., 2021). Even if the separated polymers could be used to make fibres for the textile industry, these textile waste valorisation technologies should be seen only as closed-loop solutions if PET is a major part of the blends and most of the material is recovered to be respun into textile fibres (Subramanian et al., 2021).

A more detailed classification of textile fibre recycling was proposed by Harmsen et al., in which this classification was combined with the systematisation of recycling routes. This comprehensive overview shows the relationship between the chemical reactivity of textile polymer structures and their recycling options (Harmsen et al., 2021).

### 18.3.2 Open-loop versus closed-loop recycling

Depending on the status of the material to be recycled (i.e. mono-material or blends), clean or contaminated, and open-loop or closed-loop applications are possible (Fig. 18.4). In textile recycling, open-loop processes are often described as producing materials that are not used again for the same application. Mechanically recycled textile fibres for nonwoven applications in the automotive industry are common examples. However, recycling processes that transform materials from other applications into fibres for use in the textile industry are also examples of open-loop recycling. PET bottles that are transformed into PET fibres for the textile industry are important recycled materials in the textile industry.

In closed-loop recycling, textile recycling technologies are used in which textiles are recycled into material like that being delivered. Thus, if textiles for

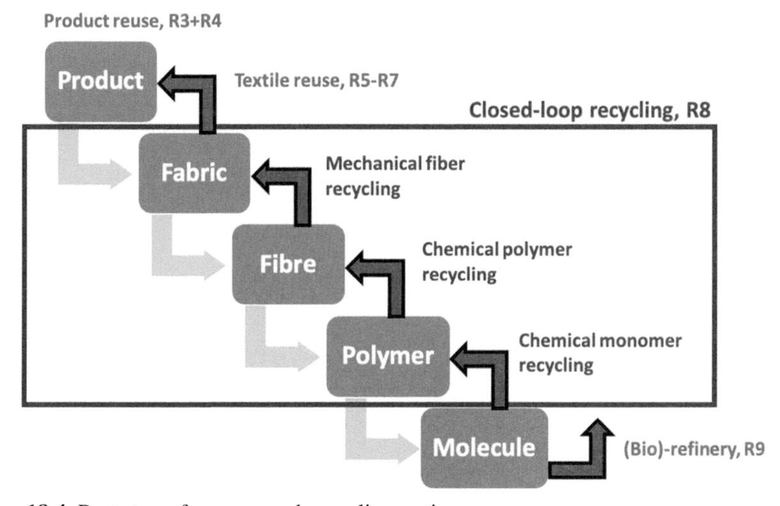


**Figure 18.4** R strategy for reuse and recycling options.

clothing are recycled into fibres, for closed-loop recycling these fibres should be used again in the clothing industry (Kumar and Yaashikaa, 2018). Often open-loop recycling is attributed as downcycling with lower-value products and closed-loop recycling as upcycling is seen to be the only viable solution for a circular economy. This comparison is not preferred because not all materials can be kept in closed-loop systems. Considering ecological and economic factors, the two recycling options need to coexist.

### 18.3.3 Postindustrial versus postconsumer recycling

The recycling of postindustrial textile waste (i.e. cuttings) is becoming advanced because challenges compared with recycling postconsumer waste streams are less complex (Vadicherla and Saravanan, 2014). The material, structure and colour of postindustrial waste stream are often uniform and known. Also, accessories are typically not added at that stage of production. In addition, supply chain partners that recycle postindustrial waste streams can use direct reversed logistics in which, for example, the leftovers of yarns and fabric cutting waste return to the originating spinning companies to open the structures again and recycle them directly in their own processes (Table 18.1). This saves virgin resources and is also a good business case for the

**Table 18.1** Differences between postindustrial and postconsumer textile waste streams.

| | Postindustrial textile waste | Postconsumer textile waste from laundries | Postconsumer textile waste from public collection |
|---|---|---|---|
| Appearance | Yarn leftovers, weaving selvages, cutting waste, etc. | Whole textiles such as bedsheets, towels or workwear | Whole garments |
| Colours | white or uniform | Often white or uniform | Mixed |
| Dyestuffs | Often uniform | Often uniform | Mixed |
| Structure | Uniform | Often uniform | Mixed |
| Sewing threads | None or few | Often uniform | All kinds |
| Accessories | None | Buttons, zippers, uniform within one batch | All kinds, buttons, zippers, sequins |
| Sorting | Easy when necessary | Easy | Difficult |
| Recycling options | All | All | Mechanical and chemical recycling if sorted |
| Supply chain from source to recycler | Often short and well-defined | Often well-defined | Longer and more complex |
| Commercial scale | Yes, for many waste streams | Yes, for several waste streams | Only for some waste streams |

companies. Hence, it has reached an industrial scale (Velener Textil, n.d.). Therefore, the recycling of postindustrial waste is not addressed in detail in this chapter.

With the regulations on the European level and the increasing demand for recycled fibres, postconsumer waste recycling will gain more attention in science and industry. To reach levels of >50% recycled fibres in new products by 2030, a dramatic change in the system needs to be accomplished. Postindustrial recycling should be included in the textile value chain, but the overall volumes of postindustrial waste are not intrinsically big enough, typically <10%, to feed the industry again, and these waste streams should also be reduced according to the principles of the R strategy (Fig. 18.2, R2).

As indicated in Table 18.1, textiles from postconsumer streams form the most challenging material stream for recycling. The quality of materials is reduced by contamination and a decrease in fibre quality during use (by washing and wear). In addition, the collection systems are generally organised on a municipal level and strongly influence the quality and uniformity of postconsumer waste streams.

### *18.3.4 Value retention by recycling*

In a circular economy, recycling is a tool used to retain the value of material already produced. Open-loop recycling is the standard recycling scheme for textile waste streams. It converts discarded textiles into raw materials to create new products, preventing them from being landfilled, and retaining some of their value. Although the quality of the recycled material is often less, open-loop recycling still reduces waste and conserves resources. Thus, it often benefits both the environment and the economy.

In the current economy, closed-loop recycling processes for postconsumer textiles are often not economically viable owing to the complex processes necessary for recycling and the low prices of virgin materials. In a large-scale circular system for textiles, all value chain stakeholders involved need to create value for their own businesses by retaining the value of the material. For postconsumer recycling, consumers need to have the incentive to hand in unusable cloth for collection systems, in which collectors profit from reselling second-hand cloth as well as from collecting recyclable fractions. Sorters then need to benefit from sorting operations that fit the needs of recycling companies. They should perform recycling processes on a fair playing field with virgin materials to provide recycled fibres to spinning mills, which also need to profit from using recycled instead of virgin fibres. The same holds for fabric manufacturing. The retailer should have an economic incentive to offer high-quality clothing to customers, which subsequently is maintained with all actions from the R strategy (Potting et al., 2016) until the materials reach the recycling level again.

This scenario may look unrealistic and is unlikely to be realized soon, but as a good example for wool fibres, a similar circular system has already been achieved in Italy. In the Prato region, many companies benefit from the good circularity of high-quality wool fibres, in which an equal playing field regarding virgin and recycled fibre prices exists (Cardato, n.d.).

## 18.4 Sustainable innovations leading to resource efficiency

Textile recycling was previously dominated by classical mechanical processes to isolate and reclaim fibres. For closed-loop applications, recycled fibres need to fulfil certain requirements such as a sufficient fibre length, purity and colour uniformity. Thus, chemical treatments attract more attention. They are used to purify, discolour and prepare textile fibres and polymers to enhance the possibility of respinning these fibres into high-quality yarns. The goal of each closed-loop textile recycling activity should be a material that can be processed into products with a good balance between quality and effort in terms of the economic and ecological impact.

### *18.4.1 Design for recycling and design with recycled materials*

The aspects of textile and garment design have a crucial role in recycling. Designers and product developers determine the choice of material and textile production process of a textile product that must be recycled at the end of the life cycle (design for recycling) and can also foster the use of recycled materials (recycling in design) by creating attractive textile products containing recycled materials.

The design of textile clothing strongly influences its overall ecological impact. It starts with the choice of the fabric material. Material blends usually hamper the recyclability of textiles because of the more complex separation of materials in chemical recycling processes and the difficult adjustment of tearing lines in mechanical recycling processes. Pure cotton goods such as T-shirts or denim are preferred waste streams for mechanical and chemical recycling. However, even with these streams, recycling technologies must deal with material impurities such as polyester sewing threads or metal buttons and zippers. Pure polyester-based textiles are the preferred waste stream for chemical recycling companies such as CuRe Technologies and Ioniqa Technologies. Aquafil developed the Econyl process, which can recycle pure polyamide-6 waste streams mainly from fishing nets or carpets (Econyl, n.d.). The EU Horizon 2020 research and innovation project, CISUFLO, aims to minimise the environmental impact of the EU flooring sector by setting up a systemic framework for circular and sustainable floor coverings, considering both technical feasibility and socioeconomic factors (Jordan et al., 2022).

Aside from these more or less pure (mono-material) waste streams, polycotton is a textile blend often used in the textile industry. Workwear as well as fashion items are typically made from varying compositions of polyester and cotton. Because of the large amounts available, polycotton waste streams have gained attention in science and industry. Polycotton textiles are a good waste stream for mechanical recycling, especially when the material was exposed to industrial laundry processes including bleaching, which damages the cotton content and makes polyester assessable for recycling (EuraMaterials et al., 2021). In chemical recycling, polycotton textiles can be used with several methods, all using a technology to separate the materials and recover fibres, polymers, monomers or other relevant chemicals (Damayanti et al., 2021; Harmsen et al., 2021).

The addition of small quantities of other materials such as elastomeric fibres, finishes, prints, sewing threads, labels and accessories that use materials other than fabric are seen as contamination from a recycling point of view. This adds complexity to recycling these textiles, often making additional recycling steps necessary or recycling impossible. This shows that to a large extent, recyclability is determined in the design of goods. Despite these considerations, the recyclability of textiles is rarely considered in the design of textiles (Watson et al., 2017), but it should be reflected in eco-design requirements for the textile industry to be established, as suggested by the European Commission (2022b).

Designers also have a crucial role in the transition to a circular economy. Designers and product developers can decide to use recycled materials in their collections and can stimulate consumers to buy products with recycled materials through attractive designs and well-balanced material choices. Different recycling technologies can offer materials such as coloured regenerated cellulose fibres from chemically recycled cotton, which will vary in colour with each batch of the waste stream but allow designers to make circular, sustainable and fashionable clothing (Ma et al., 2019).

### 18.4.2 Innovations in sorting technologies

Because the recycling of mixed fibres is complex, each recycling technology benefits from homogeneous and well-defined waste streams (European Commission, 2022a,b,c). Postindustrial waste streams and waste from laundries is typically easy to sort and homogeneous. For some waste streams such as knitted sweaters or shirts (Harmsen et al., 2021), especially from wool (Cardato, n.d.), sorting by hand into homogeneous structures and different colour batches can be economically feasible (Wolkat, n.d.). Likewise, postconsumer denim can be sorted into distinct waste streams owing to the recognisable structures of denim. This makes denim a favourite textile waste stream for mechanical recycling (RVO - Green Deal, 2020) and chemical recycling. However, even with denim, not all desired distinctions can be made. For most recycling technologies, it is important to know whether textiles contain elastomeric materials. This differentiation cannot be achieved solely by hand sorting. Unfortunately, most detection technologies fail to detect elastomers in textile waste streams. For this reason, and because sorting by hand is labour-intensive and unrealistic for complex waste streams, automatic identification and sorting technologies are required.

### 18.4.3 Automatic identification and sorting

The sorting of different textile materials by spectroscopic methods combined with mathematical algorithms (Zhou et al., 2019) has achieved high success rates of up to 100% of differentiation among cotton, linen, wool, silk, polyester, polyamide and viscose (Riba et al., 2020). For mechanical recycling, it is important to know the structure of the textile material. Knitted and more open structures are usually easier to fibreise and to make into longer fibres after recycling as closed and woven fabrics. The automatic classification of fabric structures could be used to predict the recyclability of textile waste streams via a mechanical recycling process. The differentiation of woven fabric patterns has been achieved by optical imaging techniques and mathematic algorithms (Kuo and Kao, 2007), but it has not been applied in automatic sorting systems.

Because most postconsumer textiles are fibre blends, it may be necessary to determine blended fibres and blend ratios to find the most suitable recycling options. For cotton−polyester blends, which represent a significant fraction of postconsumer textile waste streams, hyperspectral near-infrared (NIR) imaging was proposed by Mäkelä et al. to develop machine vision for textile characterisation and recycling (Mäkeläa et al., 2020). They characterised polyester−cotton blends within the range of 0%−100% for polyester with prediction errors of 2.2%−4.5%.

Because all spectroscopic technologies use reference samples, building a comprehensive library with spectrum information is important for high prediction accuracy (Liu et al., 2020). To facilitate efficient and accurate sorting, these libraries should be organised internationally, made publicly accessible, secure and transparent, and have independent governance, which would enable the standardisation of sorting activities and achieve reliable waste stream compositions for recycling companies.

Two large international projects investigated the development of automatic textile sorting lines. In the Swedish project Siptex, a sorting facility with 4 NIR/visible region of the spectrum (VIS) detection units (detection based on the wavelength of light in the NIR and VIS) and a maximum capacity of 4.5 tons/h was constructed in Malmö, Sweden. This facility can sort three fibre types or colours at the same time (Sysav, n.d.). The Fibresort sorting line developed in the EU project with the same name uses a similar optical sorting technology to scan and sort textile waste streams into uniform categories of materials with specific colours, structures and/or compositions (Wieland Textiles, n.d.). Köhler provided a more detailed review about manual and automatic sorting technologies (Köhler et al., 2021). However, the current effectivity (accuracy of detection) and efficiency (sorting speed) are insufficient to deal with increasing PCR textile waste streams.

### 18.4.4 Pretreatment of textile waste streams

After waste streams are sorted into specific material fractions, pretreatment steps can further purify textile waste by removing accessories or unwanted impurities such as elastomers. Accessories that are stitched onto the textiles may be removed if stitching yarns are broken. Even the complete disassembly of clothes might be possible if sewing threads are removed. This would also separate layers of different textile materials from each other (e.g. the interlining from the main fabric in suits).

Wear2 technologies is based on sewing threads that lose strength when treated with microwave irradiation. With this technology, clothing can be disassembled and stitched accessories can be removed, enabling each part of the clothes to be recycled separately, giving higher-quality output. Especially for workwear, this is a promising technology because these textiles are often labelled with tags and kept in closed circles within industrial laundries (Wear2, n.d.). Resortecs technology uses yarns with low melting temperatures to weaken seams when heated in industrial disassembling ovens. With this technology, all parts stitched with Resortecs yarns can be disassembled, enabling separate recycling (Resortecs, n.d.).

Clothing items often contain elastomeric fibres to increase the flexibility of garments and provide more comfort. In textile recycling, however, elastomeric fibres are seen as impurities that hamper or even impede closed-loop recycling processes. A proprietary pretreatment process that claims to remove over 99% of elastane present in mixed cotton textiles was developed on a laboratory scale by the Wageningen University and Research. After treatment with solvents, the elastane content in a cotton textile shirt was reduced from 5% to less than 0.05% (van den Oever et al., 2019). This technology was patented, and upon further development, it might improve the recyclability of cotton textiles in mechanical or chemical recycling processes (Van Dam et al., 2020).

Discolouration is a pretreatment step that can add value to recycled materials, because white fibres are expected to generate a higher market price as coloured fibres. To discolour textile materials, two approaches can be followed. The dye molecule can be chemically modified to lose colour, or the dye molecule can be separated from textile fibres. Bleaching processes that modify the dye molecule to lose colour can also harm the textile material, and bleaching agents are often chlorinated chemicals considered harmful to be to the environment. However, with the combination of a leaching process using nitric acid, a dissolution process using dimethyl sulphoxide and a bleaching process with sodium hypochlorite and diluted hydrochloric acid, textile dyes were removed from polycotton blends. After the dissolution of the polyester part, white cotton was recovered with a recovery rate of 93%. In addition, closed-loop systems for the acid, solvents and polyester were constructed to develop a sustainable recycling process (Yousef et al., 2019). An alternative to separate dyes and fibres by disrupting physical interactions between the materials. Mu et al. found this to be possible by reducing of the fibre density using solvents and temperature. With their process, disperse dyes, direct dyes and acid dyes can be removed from polyester, cotton and nylon fibres without changing the chemical structure of the dye (Mu and Yang, 2022).

## 18.4.5 Innovations in mechanical recycling

Innovations in mechanical textile recycling typically have the goal of increasing the average length of recycled fibres. This can be done by choosing of the raw material, constructing the tearing line and choosing process parameters such as the processing speed. In addition, pretreatment processes can influence the outcome of the tearing process. Lindström et al. developed a process in which polyethylene glycol (PEG)-based lubricants were used as pretreatment agents for mechanical recycling. The researchers found that the fabrics disassembled more easily when they were pretreated with PEG 4000. This was also visible in the recycled fibres. The researchers found that the lubricant reduced cohesion between fibres in mechanical recycling. Interfibre cohesion was tested and it was possible to predict a more efficient tearing process with lubricant-treated fabrics.

Pretreatment with PEG led to higher tearing efficiency and longer recycled fibres after mechanical recycling, especially with polyester waste streams. The treatment enabled rotor spun yarn to be spun from 100% recycled fibres. However, the reduced fibre cohesion influenced the strength of the rotor-spun yarns. Yarns with PEG 4000-treated fibres had lower strength than did yarns that were washed with water after spinning to remove the lubricant (Lindström et al., 2020).

Because the mechanical recycling of textiles is already performed on a large scale, the tearing of different textile waste streams can be investigated for production machinery. In the EU-Interreg project Retex, a broad industrial consortium investigated several value chains for closed-loop mechanical recycling. Researchers found that cotton and polycotton waste streams can be successfully recycled into yarns with good properties, but that economic viability was still challenging. Only if the costs for incineration increased or additional taxes were added to virgin materials would yarn alternatives with recycled materials become competitive (EuraMaterials et al., 2021). Nevertheless, one value chain has reached market introduction. The Belgian yarn spinning and weaving company Utexbel introduced Dr Green fabric produced from post-consumer recycled fibres, recycled polyester from polyester bottles and cotton (Utexbel, 2021).

### 18.4.6 Innovations in chemical recycling

Different innovative processes have been developed for chemically recycling textile waste. The general advantage of these processes compared with mechanical recycling is that the fibre length of chemically recycled fibres can be chosen, because the product of chemical recycling is an endless filament that can subsequently be cut into staple fibres of the desired length. This makes yarn spinning with these fibres much easier and enhances the strength of the yarns.

Spinning of regenerated cellulose fibres with ionic liquids is the process with the largest innovation potential. Ionic liquids have advantages in that they are less toxic than carbon disulphite, which is the main chemical used in the traditional viscose fibre process, and they are less potentially dangerous than NMMO used for the lyocell process. The disadvantages of most ionic liquids are that they are expensive and recovery processes are far less developed (Ma et al., 2019).

In 2016, Asaadi et al. demonstrated that high-quality chemically recycled fibres can be produced from postconsumer waste streams using 1,5-diazabicyclo[4.3.0]non-5-enium acetate ([DBNH]OAc) as an ionic liquid cellulose solvent for dry-jet wet spinning. The regenerated cellulose fibres showed high tenacities that exceeded the mechanical properties of most cellulose fibres (Asaadi et al., 2016).

This solvent system was also used to separate cotton and polyester from intimate fibre blends in postconsumer textiles. The different chemical structure of the polymers enabled efficient separation by the dissolution of cellulose and filtration of the undissolved polyester. The dissolved cellulose from cotton was subsequently regenerated by dry-jet wet spinning into textile-grade fibres. These fibres had properties similar to those of commercial lyocell fibres, with a breaking tenacity of 27–48 cN/tex. The cellulose was almost completely removed from the polyester residue (1.7–2.5 wt%), but the polyester tensile properties were significantly reduced. The authors stated that to commercialise this technology, it is important to identify and sort the waste streams, and the technology needs to be efficient and reliable. Degradation of the polyester might be limited by further optimisation of the dissolution and filtration steps. This could lead to a chemical recycling strategy that enables the closed-loop recycling of both the cotton and polyester waste streams from polycotton blends (Haslinger et al., 2019a,b).

The dyeing process greatly affects the lifecycle of textile products, and the discolouration of textile waste is not always achievable without the use of process chemicals such as bleaching agents and energy, leading to wastewater consequences. (Yousef et al., 2019). Hence, reusing dyestuffs in postconsumer waste streams would significantly reduce the impact of textile fibres (Ma et al., 2019; Haslinger et al., 2019a,b).

Ma et al. developed a process in which process conditions can be controlled so that the colour of postconsumer denim is retained in the chemically recycled fibres or is removed from cellulose before wet spinning. Ionic liquid 1-butyl-3-methylimidazolium acetate ([Bmim]OAc) combined with dimethylsulphoxide (DMSO) was used to dissolve cellulose from denim to create regenerated cellulose fibres. This process shows the possibility of producing white or coloured fibres, depending on the market needs. By retaining colour in recycled fibres, there is no need to redye fibres after production in traditional textile dyeing processes, which considerably reduces environmental impacts by avoiding water, energy and chemical use. With DMSO as a cosolvent combined with the ionic liquid (ratio of 1:4), the drawbacks of spinning using only ionic liquids were significantly reduced. Cellulose was dissolved faster by reducing the viscosity of the solution and process costs decreased by 77%. The properties of the chemically recycled fibres were comparable to those of standard viscose fibres used in the textile industry even when the dyestuff was retained in the fibres. Ma et al. also demonstrated that the process can be designed as a closed-loop system when solvents are recovered by distillation (Ma et al., 2019).

The pretreatment process for coloured postconsumer waste textiles was studied in detail by Nasri-Nasrabadi et al., who found that pretreating waste cotton fabrics had a significant influence on the colour strength of the regenerated cellulose fibres. When sodium hydroxide was used to reduce the average polymer chain length (DP) of cellulose, the colour strength was significantly reduced. This was not the case when sulphuric acid was used to reduce the DP. The colour fastness of the chemically recycled fibres was excellent in both cases. This demonstrates the benefits of coloured chemically recycled fibres. The dyestuff is fully incorporated into the fibre structure comparable to traditional dope dyed fibres, which leads to excellent colour fastness properties (Nasri-Nasrabadi et al., 2020).

The similarities to dope dyeing can also be observed when two complementary coloured waste textiles are mixed. New colours can be created by blending two waste streams. This was demonstrated by the creation of purple and green coloured chemically recycled fibres from red/blue- and blue/yellow-coloured waste textiles, respectively (Nasri-Nasrabadi et al., 2020).

The same process was used to create bicomponent regenerated cellulose fibres from waste textiles. For the outer shell of the fibres, blue denim waste was dissolved in a [Bmim]OAc−DMSO mixture at a ratio of 1:4, and white cotton waste was used for the core. By extruding the two different dopes through a special axial needle spinneret, bicomponent regenerated fibres with a white core and a blue shell were created. Because a significant part of the environmental impact of the textile industry comes from dyeing, retaining the colour of textile waste can lead to considerable savings of this impact. Optimised bicomponent fibres were constructed from almost equal amounts of blue and white cotton waste to achieve the same colour strength as similar fibres consisting only of blue denim cotton waste. Moreover, the mechanical properties were similar to those of regenerated cellulose fibres from one waste stream, whereas a discrete colour separation was observed between the core and the shell. This shows the potential of bicomponent fibres with the coloured material outside and white material inside (Rosson and Byrne, 2022).

Haslinger et al. demonstrated for the [DBNH]OAc ionic liquid process that preconsumer and postconsumer waste cotton dyed with several vat and reactive dyes can be processed into chemically recycled fibres with excellent physical properties. Recycled fibres from dyed preconsumer waste had lower tenacities and elongations of break of 40.3−48.5 cN/tex and 7.4% to 12.8%, respectively, compared with recycled fibres from dyed postconsumer waste, for which tenacities of 40.3−48.5 cN and elongations to break of 7.4% to 12.8% were found. These physical properties are superior to commercial viscose or lyocell fibres used in the textile industry, which shows the high potential of chemical recycling of postconsumer waste streams with ionic liquids. The colours of the dyed waste streams changed slightly during spinning, but most recycled fibres showed bright colours suitable for producing textile demonstrators (Haslinger et al., 2019a,b). The colour fastness properties for the chemically recycled fibres were good, as in the process of Ma et al., and wet rubbing fastness increased significantly after recycling (Ma et al., 2019; Haslinger et al., 2019a,b).

Research at the Saxion University of Applied Science showed that blue postconsumer cotton overalls can be chemically recycled into blue lyocell fibres with properties similar to those of commercially available lyocell fibres. After discarded overalls were collected at an industrial laundry and sorted into only blue overalls, the noncellulosic parts were removed by hand and cotton was transferred into cellulose pulp that could subsequently be converted into blue lyocell-type chemically recycled fibres (Fig. 18.5) (Saxion, 2019).

Many promising technologies investigated in academia for chemical recycling cotton based textiles show the high potential of chemical recycling technologies. Some general conclusions that can be drawn:

- Chemically recycled fibres have high potential for innovations in textile recycling.
- The fibre length of chemically recycled fibres can be chosen to meet the requirements of yarn spinning companies.

**Figure 18.5** Recycled blue overalls and recycling steps of chemically recycled blue fibres.

- The mechanical properties of chemically recycled cotton fibres are often similar to or better than those of commercially available regenerated cellulose fibres.
- The colour of postconsumer waste textiles can be retained in chemically recycled fibre, which reduces the environmental impact from discolouration and the dyeing of recycled fibres.
- Closed-loop chemical recovery processes need to be developed to meet legislative and environmental needs.
- Chemical recycling processes need to prove their potential environmental benefits by a thorough impact analysis.

### 18.4.7 Innovations in production with recycled fibres

Chemical processes change the properties of waste cotton towards properties of regenerated cellulose fibres. This is not always desirable in the textile industry and even if the fibre properties of chemical recycled fibres are promising, the comfort, feel and look of these fibres often do not resemble those of cotton. Recycled fibres need to fit into upstream textile processes such as spinning, weaving or knitting and finishing. Thus, the closed-loop recycling of textiles does not stop at the fibre level. The smart fibre blending of cellulose-based chemically recycled cotton fibres with virgin cotton or better mechanically recycled cotton could be beneficial. The spinning of yarns has a crucial role in this.

Mechanically recycled fibres are usually shorter than virgin fibres (Ütebay et al., 2019). Spinning companies must adjust processes and machinery to be able to process these fibres on regular machinery. Spinning machine manufacturers such as Rieter AG (Schwippl, 2020; Spatafora and Schwippl, 2020) and Trützscher (Trützschler, n.d.) investigated how mechanically recycled fibres can be processed into high-quality yarns with the highest efficiency. They found that the quality of the recycled material strongly determines the optimal end spinning process. For cotton, the short fibre content determines whether the material can be used for rotor or ring spinning processes.

**Table 18.2** Recycling classification by Rieter.

| Classification | Short fibre content (%) | Mean fibre length (mm) | 5% fibre length (mm) |
|---|---|---|---|
| Very good | 45 | 17 | 31 |
| Good | 55 | 14 | 29 |
| Medium | 60 | 13 | 28 |
| Poor | 78 | 10 | 22 |
| Cotton (reference) | 24 | 21 | 34 |

Adapted from Schwippl (2020).

To grade the quality of the mechanically recycled fibres they received, Rieter developed a classification system. Mechanically recycled fibres are graded on the short fibre content, mean fibre length and average length of the longest 5% of fibres (Table 18.2). This enables a good estimation of the processability of the fibres and the resulting yarn quality.

The quality of the recycled fibres is important for processing them into yarns, but so are some machine modifications combined with adjustments in the process sequence and the setting in almost all process steps. For rotor spinning with mechanical recycled fibre, shortened processes and the design of special modules on the carding machine are beneficial and strongly increase the quality of the yarns (Spatafora and Schwippl, 2020). With the correct setup, rotor spinning leads to the lowest unevenness compared with other spinning processes. In comparison, ring spinning provides the highest yarn tenacity. This is caused by the improved fibre integration in ring yarns. With additional twining on the compact spinning machine, the strength and unevenness of yarns are further increased. Ring spinning is possible with 75% of recycled fibres when the quality is equal of a 'medium' grade or higher in the classification. The improved processes and the possibility of using mechanically recycled fibres in several spinning processes enables yarns to be produced for several applications in woven fabrics as well. Rieter also investigated the economic viability of spinning with mechanically recycled fibres and found that up to 75% recycled content for rotor spinning and up to 60% for ring spinning are commercially interesting (Schwippl, 2020).

## 18.5 Ecological and socioeconomic impacts of recycling

A transition to circularity in the textile industry could lead to a large reduction in the environmental impact in the EU and worldwide. The European Commission summarised the current impact of textiles in the EU as fourth highest impact on the environment and climate change after food, housing and mobility, and as the third highest pressure on water and land use (European Commission, 2022a).

### 18.5.1 End-of-life scenarios for textiles

In 2015, 73% of the textile waste material was still incinerated or landfilled (Ellen MacArthur Foundation, 2017), which had a tremendous negative impact on the environment. The impact of end-of-life textiles adds up to a large impact for these textiles during production. The data show that waste management systems to recover textile materials from other waste streams can reduce the end-of-life impact of textiles.

The European Commission aims to reduce this impact by reaching 100% circularity in 2050. The Netherlands has set the goal to achieve 50% of sustainable fibres in the textile industry by 2030 and 100% circular fibres by 2050. These goals are highly ambitious, realizing that textile consumption is still growing and the current percentage of closed-loop recycled fibres in new textiles is $<1\%$ (Ellen MacArthur Foundation, 2017).

### 18.5.2 Increasing sustainability by recycling

Following the R strategies (Fig. 18.2) can help to reach these goals. Using less materials and keeping them longer in the life cycle will diminish the need for new fibre. Recycling materials at the end of the physical life cycle can provide raw materials that enter a new life cycle and achieve circularity. Realizing a circular economy does not automatically mean a sustainable economy. Processes in circular value chains need energy to transport materials for reselling, repair or refurbishing and energy, water and chemicals for many recycling activities. These impacts need to be considered, and circular activities need to prove their sustainability by life cycle assessments (LCAs).

Sandin et al. provided strong arguments for a more sustainable industry through reuse and recycling. In that review, the authors studied 41 relevant publications on the reuse and recycling of textiles and found strong support for the reduced environmental impact of these activities compared with incineration and landfilling. The study revealed that most publications about the environmental impact of recycling activities focused on climate change, whereas other impact categories such as water and land use were poorly investigated (Sandin and Peters, 2018).

Especially for these two indicators, recycling processes are expected to outperform virgin material production, at least for cotton. Cotton usually needs large amounts of water to grow (La Rosa and Grammatikos, 2019) and occupies agricultural land that could otherwise be used for food production. ‘Organic’ or ‘bio’ cotton may need less chemicals, but it needs more water because less cotton is harvested per surface area.

### 18.5.3 Environmental aspects of mechanical recycling

Mechanically recycled cotton, for instance, needs no water or insignificant amounts of it (Esteve-Turrillas and de la Guardia, 2017) and no agricultural land. However, postconsumer waste streams most likely lead to low yields of fibre with a length useful for yarn spinning (European Commission, 2022a,b,c).

Because the impact of the mechanical recycling itself is limited to the energy use of tearing machines, it seems to be reasonable to conclude that environmental savings compared with virgin materials is growing with an increasing environmental burden on the virgin material replaced. For instance, if a mechanically recycled aramid fibre could replace virgin aramid fibres, it would prevent a significant environmental impact. The benefit for the environment is largest if recycled fibres can replace virgin fibres in the same application. Fidan et al. showed that mechanically recycled cotton fibres can replace 20% of virgin fibres in denim production without significantly reducing the quality of the textile. They also demonstrated that using recycled fibres considerably decreases the environmental impact of the products. In particular, the eutrophication potential and water use were significantly reduced by replacing virgin cotton fibres with mechanically recycled cotton (Fidan et al., 2021).

From an environmental point of view, the use of coloured postconsumer waste streams has the benefit of potentially avoiding downstream dyeing processes. Thus, Esteve-Turrillas and Guardia investigated the environmental impact of mechanically recycled cotton compared with virgin conventional and organic cotton production. According to all production processes of the dyed cotton yarn, yarn with coloured mechanically recycled fibres clearly outperformed virgin cotton yarn, even from organic farming, in all assessed environmental impact categories. This is mostly due to the avoided dyeing process, which was calculated to contribute 2.10–5.19 kg $CO_2e$ of global warming potential to the production of dyed cotton textiles (Esteve-Turrillas and de la Guardia, 2017).

### 18.5.4 Environmental aspects of chemical recycling

Sandin et al. identified some scenarios in which reuse and recycling might not be beneficial to the environment. Low replacement rates of virgin materials or relatively clean production processes can change that general positive picture of recycling. Therefore, the quality of recycled materials is an important factor for the sustainability of recycling processes. Whereas mechanically recycled fibres are shorter than virgin fibres in recycled textiles, chemically recycled fibres can reach virgin quality or even outperform the mechanical properties of recycled fibres (Asaadi et al., 2016).

When sorted on specific colours, chemical recycling processes can retain colour from waste textiles as well. For cotton, chemical recycling processes that do not alter the chemical structure of cellulose or the dyestuff can lead to coloured recycled fibres that might avoid upstream dyeing processes. The same is true for mechanical recycling processes (Haslinger et al., 2019a,b; Ma et al., 2019).

The environmental effects of chemical recycling technologies are as diverse as the processes themselves. Many processes are not yet assessed at the industrial scale yet and it is difficult to make estimations owing to varying scales and technologies. Thus, studies often show positive and negative scenarios. The chemical recycling of textile waste via carbamate technology, for instance, shows a comparable carbon footprint for viscose fibres from virgin wood pulp produced in Asia, which can be significantly reduced if fibre production is integrated in the pulping factory, enabling the efficient circulation of process chemicals and energy (Paunonen et al., 2019). Pulping processes from textile waste streams themselves can vary significantly in the environmental impact, depending on the purity of the waste stream. A 1-kg waste stream containing 10% PET increases the climate change impact by 0.38 kg $CO_2e$ compared with a 100% cotton waste stream because of the increased use of process chemicals and energy (Oelerich et al., 2017). Rosson et al. performed an LCA for the pretreatment of cotton textile waste to lower the polymer chain length of cellulose polymers by sulphuric acid or sodium hydroxide pretreatment. They found that acid treatment had a significantly lower environmental impact than treatment with hydroxide (Rosson and Byrne, 2020).

### 18.5.5 Environmental aspects of biochemical recycling

Biochemical recycling processes could also contribute to a sustainable textile industry. Ribul et al. analysed several biochemical processes, which mostly used the biochemical degradation of cotton into other valuable products and polyester polymers that could be used for respinning into polyester yarns. Because of the selectivity of biocatalytic processes, these recycling technologies could have multiple advantages over other recycling technologies, but Ribul et al. concluded that enzymatic and biological processes must be monitored for their energy and water demand, because they use larger amounts of water (Ribul et al., 2021). Often larger-scale factories are needed to reach the necessary circularity of chemicals and energy.

Leal Filho et al. investigated the socioeconomic benefits of textile recycling. They concluded that for all fibres and recycling methods currently in use, recycling is still a better environmental and socioeconomic option whose benefits far outweigh mere incineration. In addition, they found that the socioeconomic benefits were highest for wool recycling, but that polyester fibre recycling is also beneficial because the raw materials are derived from fossil fuels, which adds a tremendous burden on the environment (Leal Filho et al., 2019). Independence from fossil raw materials is being seriously discussed, and recycling polyester materials can contribute to this goal. Some general points can be outlined:

- Recycling textile fibres often is beneficial to the environment compared with virgin material when the quality of recycled fibres is good enough to reach the high replacement rates of virgin fibres
- Mechanical recycling processes generally have low carbon emissions, use less water than most virgin natural fibres and need no agricultural land, but the mechanical properties of recycled fibres are generally lower
- Chemical recycling processes for cotton need larger-scale factories to recycle and circulate chemicals efficiently, and they use low levels of sustainable energy, such as by combining pulping and fibre spinning processes. The mechanical properties of chemically recycled fibres can be engineered to at least those of virgin fibres
- The use of biochemical processes such as employing enzymes can lower the impact of recycling activities because of the specific activity of enzymes at lower temperatures (i.e. energy input levels)
- Recycling processes have ecological as well as socioeconomic benefits, especially when the lower dependency of raw fossil materials is intended

## 18.6 Future perspectives

The current situation of overconsumption and consequent unsustainable production was initiated by the availability of inexpensive raw materials and labour, fuelling fast-changing fashion trends. By mandatory EU regulation, to be implemented progressively towards 2050, the fashion industry is required to stop and reverse this trend by the transparent and unconstrained use of more sustainable and recycled materials, implementing the legislation and the required eco-design considerations mentioned earlier.

Only with incentives and regulations to lower the consumption of virgin materials, including EPR-based tariffs and import taxes for unsustainably produced goods, as well as the use of novel recycled materials from emerging technologies and smart material cascading, can the goal of a high degree of circularity in Europe by 2050 be reached. An optimised circular economy of textiles, based on innovations in recycling could look like this:

- Textile designers and product developers use upcoming eco-design regulations to develop textile products with high quality and a clear recycling process with the end-of-life phase in mind.
- The consumer uses the textile items with care and for a prolonged time to minimise the impact of the item for each use, and hands over unused items for the best possible follow-up application.
- At the end-of-life, a material passport should indicate the intended recycling process and show collectors and sorters how to handle the garment in the right way.
- Recycling companies receive optimised feedstock streams for the recycling technology, from which textile fibres with sufficient properties such as length and desired colour can be achieved.
- Designers use the highest amounts of recycled fibres possible for their products.
- The production of textiles explores new business models, including, for example, 'product as a service' or 'made-to-order' production processes, avoiding overstock and the fast succession of new collections each season.
- Where possible, logistic chains are shortened as PCR textile material streams become 'local sources' to be used regionally for new products.

Mechanical and chemical recycling technologies have both advantages and disadvantages that need to be approached in downstream processing. A promising technological innovation can come with the chemical recycling of cotton and the thermomechanical recycling of polyester, in which dyestuff from textile waste streams stays in the material. This would save on the environmental impact by avoiding the need to discolour the waste material, as well as the dyeing processes of the resulting fibres.

However, working with these technologies would mean that designers and product developers need to work with predefined colours that might vary with every batch. Designers have shown that they can deal with these constraints, because mechanically recycled fibres from postconsumer textiles also have predefined colour that need to be integrated into the product.

These new material streams will also require adjustments in downstream processing, such as spinning or weaving and knitting. By the controlled blending of recycled fibres with varying colours, yarns and fabrics with constant colours can be produced. All of these adjustments in the processes and markets need to come with a change in legislation and consumer behaviour. Consumers need to be aware of their role in the system, and waste management schemes need to be in place (Hole and Hole, 2019). Innovations in closed-loop postconsumer recycling can lead to significant reductions in the use of virgin fibres, but they need to be embedded in circular systems in which all stakeholders are aligned.

## 18.7 Conclusion

The textile industry is facing a critical need to minimise its impact on the environment. The circular economy is one approach to achieve this goal. With regulations already in place, or expected to be in the near future, there is a growing emphasis on increasing the use of circular materials in textile production. To achieve this, designers and product developers must take a leading role in creating long-lasting textiles with recycled materials.

In addition, sustainable innovations in recycling and processing textiles need to be facilitated to scale up and make textile circularity possible. These innovations must be assessed on their economic and ecological impact on the entire supply chain. Finally, stakeholders in the circular economy for textiles must work together to build transparent and fair regional supply chains that promote circularity and minimise waste.

## References

Asaadi, S., et al., 2016. Renewable high-performance fibres from the chemical recycling of cotton waste utilizing an ionic liquid. ChemSusChem 22 (9), 3250–3258.

Cardato, n.d. Cardato - Made in Prato. Available at: http://www.cardato.it/en/what-is-the-carded-wool/. (Accessed 29 April 2022).

Damayanti, D., et al., 2021. Possibility routes for textile recycling technology. Polymers 21 (13), 3834.

Duhoux, T., et al., 2021. Tudy on the Technical, Regulatory, Economic and Environmental Effectiveness of Textile Fibres Recycling Final Report. Publications Office of the European Union, Lucembourg.

Econyl, n.d. Discover Econyl. Available at: https://www.econyl.com/the-process. (Accessed 27 April 2022).

EEA, 2019. Textiles and the Environment in a Circular Economy (s.l.: s.n).

Ellen MacArthur Foundation, 2017. A New Textiles Economy: Redesigning Fashion's Future s.l. http://www.ellenmacarthurfoundation.org/publications.

Esteve-Turrillas, F., de la Guardia, M., 2017. Environmental impact of recover cotton in textile industry. Resources, Conservation and Recycling 116, 107–115.

EuraMaterials, Fedustria, Centexbel, c., 2021. Retex. Available at: https://www.dotheretex.eu/resultaten. (Accessed 29 April 2022).

European Commission, 2022a. EU Strategy for Sustainable and Circular Textiles. European Commission, Brussel.
European commission, 2022b. Proposal for a Regulation of the European Parliament and of the Council Establishing a Framework for Setting Ecodesign Requirements for Sustainable Products and Repealing Directive 2009/125/EC. European commission, Brussel.
European Commission, 2022c. Sustainable and Circular Textiles by 2030. Publications Office of the European Union, Luxembourg.
European Environmental Agency, 2022. Textiles and the Environment: The Role of Design in Europe's Circular Economy. Available at: https://www.eea.europa.eu/publications/textiles-and-the-environment-the. (Accessed 29 April 2022).
European Union, 2008. Directive 2008/98/EC Of the European Parliament and of the Council. Available at: https://eur-lex.europa.eu/legal-content/EN/TXT/?uri=CELEX:32008L0098. (Accessed 29 April 2022).
European Union, 2018. Directive (EU) 2018/851 of the European Parliament and of the Council. Available at: https://eur-lex.europa.eu/legal-content/EN/TXT/?uri=uriserv:OJ.L_.2018.150.01.0109.01.ENG. (Accessed 29 April 2022).
Fidan, F., Aydoğan, E., Uzal, N., 2021. An integrated life cycle assessment approach for denim fabricproduction using recycled cottonfibres and combined heat and powerplant. Journal of Cleaner Production 287, 12439.
Gupt, Y., Sahay, S., 2015. Review of extended producer responsibility: a case study approach. Waste Management and Research 7 (33), 595−611.
Harmsen, P., Scheffer, M., Bos, H., 2021. Textiles for circular fashion: the logic behind recycling options. Sustainability 13, 971.
Haslinger, S., et al., 2019a. Upcycling of cotton polyester blended textile waste to new man-made cellulose fibres. Waste Management 97, 88−96.
Haslinger, S., et al., 2019b. Recycling of vat and reactive dyed textile waste to new coloured man-made cellulosefibres. Green Chemistry 20 (21), 2298−5610.
Hawley, J., 2009. Understanding and improving textile recycling: a systems perspective. In: Blackburn, R. (Ed.), Sustainable Textiles, Life Cycle and Environmental Impact. Woodhead Publishing, Philadelphia, pp. 179−199.
Hole, G., Hole, A.S., 2019. Recycling as the way to greener production: a mini review. Journal of Cleaner Production 212, 910−915.
Jordan, J.V., Krause, R., Paul, R., Härpfer, C., Finetti, C., Paschen, A., Rüdiger, T., Chiarotti, U., de Vilder, I., Buyle, G., Pizza, A., 2022. Floor coverings of tomorrow and today - in search of a design for recyclers. In: Paper presented in Carpet Recycling UK Annual Conference, Solihull, UK, 29-30th June.
Köhler, A., et al., 2021. Circular Economy Perspectives in the EU Textile Sector, EUR 30734 EN. Publications Office of the European Union, Luxembourg.
Kumar, P., Yaashikaa, P., 2018. Recycled fibres. In: Muthu, S. (Ed.), Sustainable Innovations in Recycled Textiles. Springer, Singapore, pp. 1−17.
Kuo, C.-F.J., Kao, C.-Y., 2007. Self-organizing map network for automatically recogniz ing colour texture fabric nature. Fibres and Polymers 2 (8), 174−180.
La Rosa, A.D., Grammatikos, S.A., 2019. Comparative life cycle assessment of cotton and other natural fibres for textile applications. Fibres 7 (101).
Leal Filho, W., et al., 2019. A review of the socio-economic advantages of textile recycling. Journal of Cleaner Production 218, 10−20.
Lindström, K., Sjöblom, T., Persson, A., Kadi, N., 2020. Improving mechanical textile recycling by lubricantpre-treatment to mitigate length loss of fibres. Sustainability 12 (20), 8706.

Liu, Z., Li, W., We, Z., 2020. Qualitative classification of waste textiles based on near infrared spectroscopy and the convolutional network. Textile Research Journal 90 (9–10), 1057–1066.

Ma, Y., Zeng, B., Wang, X., Byrne, N., 2019. Circular textiles: closed-loop fibre to fibre wet spun process for recycling cotton from denim. ACS Sustainable Chemistry & Engineering 14 (7), 11937–11943.

Mäkeläa, M., Rissanena, M., Sixta, H., 2020. Machine vision estimates the polyester content in recyclable waste textiles. Resources, Conservation and Recycling 161, 105007.

Mather, R.R., Wardman, R.H., 2015. The chemistry of textile fibres. In: The Chemistry of Textile Fibres. Royal Society of Chemistry, Cambridge, pp. 7–12.

MI & W, 2020. Policy Programme for Circular Textile 2020-2025. Ministry of Infrastructure and Water Management, The Hague.

Mu, B., Yang, Y., 2022. Complete separation of colourants from polymeric materials for cost-effective recycling of waste textiles. Chemical Engineering Journal 427, 131570.

Nasri-Nasrabadi, B., Wang, X., Byrne, N., 2020. Perpetual colour: accessing the colourfastness of regenerated cellulose fibres from coloured cotton waste. Journal of the Textile Institute 12 (111), 1745–1754.

Niinimäki, K., et al., 2020. The environmental price of fast fashion. Nature Reviews Earth and Environment 4 (1), 189–200.

OECD, 2016. Extended Producer Responsibility, Updated Guidance for Efficient Waste Management. Available at: https://www.oecd-ilibrary.org/sites/9789264256385-en/index.html?itemId=/content/publication/9789264256385-en. (Accessed 29 April 2022).

Oelerich, J., Bijleveld, M., Bouwhuis, G.H., Brinks, G.J., 2017. The life cycle assessment of cellulose pulp from waste cottonvia the SaXcell™process. IOP Conference Series: Materials Science and Engineering 254, 192012.

Paul, R., 2012. Technology to the rescue. In: Trash Talking: Textile Recycling. Waste Management World, pp. 32–37. November-December.

Paunonen, S., et al., 2019. Environmental impact of cellulose carbamate fibres from chemically recycled cotton. Journal of Cleaner Production 222, 871–881.

Payne, J., Jones, M., 2021. The chemical recycling of polyesters for a circular plastics economy: challenges and emerging opportunities. ChemSusChem 19 (14), 4041–4070.

Potting, J., Hekkert, M., Worrell, E., Hanemaaijer, A., 2016. Circular Economy: Measuring Innovation in Product Chains. PBL Netherlands Environmental Assessment Agency, The Hague.

Resortecs, n.d. Resortecs, Recycling Made Easy. Available at: https://resortecs.com/technology. (Accessed 29 April 2022).

Riba, J.-R., Canterob, R., Canalsb, T., Puig, R., 2020. Circular economy of post-consumer textile waste: classificationthrough infrared spectroscopy. Journal of Cleaner Production 272, 123011.

Ribul, M., et al., 2021. Mechanical, chemical, biological: moving towards closed-loop bio-based recycling in a circular economy of sustainable textiles. Journal of Cleaner Production 326, 129325.

Rosson, L., Byrne, N., 2020. Comparative gate-to-gate life cycle assessment for the alkali and acid pre-treatment step in the chemical recycling of waste cotton. Sustainability 12, 8613.

Rosson, L., Byrne, N., 2022. Bicomponent regenerated cellulose fibres: retaining the colour from waste cotton textiles. Cellulose 7 (29), 4255–4267.

Russell, S., Swan, P., Trebowicz, M., Ireland, A., 2016. Natural fibres: advances in science and technology towards industrial applications. In: Fangueiro, R., Rana, S. (Eds.), Review of Wool Recycling and Reuse. Springer, Dordrecht, pp. 415–428.

RVO - Green Deal, 2020. C-233 Green Deal on Circular Denim, "Denim Deal". Available at: https://www.greendeals.nl/green-deals/green-deal-circulaire-denim. (Accessed 29 April 2022).

Sandin, G., Peters, G.M., 2018. Environmental impact of textile reuse and recyclinge a review. Journal of Cleaner Production 184, 353–365.

Saxion, 2019. From Workwear to Workwear. Available at: https://www.saxion.nl/onderzoek/smart-industry/sustainable-functional-textiles/saxcell. (Accessed 29 April 2022).

Schwippl, H., 2020. The Increasing Importance of Recyclingin the Staple-Fibre Spinning Process. Rieter Machine Works Ltd, Winterthur.

Sherwood, J., 2020. Closed-loop recycling of polymers using solvents. Johnson Matthey Technology Review 64, 4–15.

Spatafora, J., Schwippl, H., 2020. The Ideal Rotor Spinning Process for a High Short-Fibre Content. Rieter Machine Works Ltd, Winterthur.

Staatsblad, 2023. https://zoek.officielebekendmakingen.nl/stb-2023-132.html.

Subramanian, K., et al., 2021. An overview of cotton and polyester, and their blended waste textile valorisation to value-added products: a circular economy approach – research trends, opportunities and challenges. Critical Reviews in Environmental Science and Technology.

Sysav, n.d. Siptex - Textile Sorting. Available at: https://www.sysav.se/en/siptex#block4. (Accessed 27 April 2022).

Trützschler, n.d. The True Way to Your Yarn from Recycled Material, Trützschler Group SE, Mönchengladbach.

Ütebay, B., Çelik, P., Çay, A., 2019. Effects of cotton textile waste properties on recycled fibre quality. Journal of Cleaner Production 222, 29–35.

Utexbel, 2021. Persbericht Dr. Green. Available at: https://utexbel.com/nl/persbericht-dr-green/. (Accessed 29 April 2022).

Vadicherla, T., Saravanan, D., 2014. Textiles and apparel development using recycled and reclaimed fibres. In: Muthu, S. (Ed.), Roadmap to Sustainable Textiles and Clothing. Springer, Singapore, pp. 139–160.

Van Dam, J., Knoop, J., Van den Oever, M., 2020. Method for Removal of Polyurethane Fibres from a Fabric or Yarn Comprising Polyurethane Fibres and Cellulose-Based Fibres. The Netherlands, Patent No. WO 2020/130825 A1.

van den Oever, M., van Dam, J., Paulien, H., Bolck, C., 2019. Bio2HighTex/Wastexcel: New Solutions for the Textile Chain. Available at: https://www.wur.nl/en/project/Bio2HighTexWastexcel-new-solutions-for-the-textile-chain.htm. (Accessed 29 April 2022).

Velener Textil, n.d. Our No-Waste Model for Cotton Processing. Available at: https://www.velener.de/en/wecycled.html. (Accessed 27 April 2022).

Watson, D., et al., 2017. Stimulating Textile-To-Textile Recycling. The Nordic Council of Ministers, Copenhagen.

Wear2, n.d. Wear2 ® Microwave Technology. Available at: https://wear2.com/en/wear2-microwave-technology. (Accessed 29 April 2022).

Wieland Textiles, n.d. Innovation: Fibresort. Available at: https://www.wieland.nl/en/innovation-fibresort/. (Accessed 27 April 2022).

Wolkat, n.d. International Textile Innovators. Available at: https://www.wolkat.com/en/steps. (Accessed 29 April 2022).

Yousef, S., et al., 2019. A new strategy for using textile waste as a sustainable source of recovered cotton. Resources, Conservation and Recycling 145, 359–369.

Zhou, C., et al., 2019. Rapid identification of fibres from different waste fabrics using the near-infrared spectroscopy technique. Textile Research Journal 17 (89), 3610–3616.

# Sources of further information

Ma, Y., Rosson, L., Wang, X., Byrne, N., 2020. Upcycling of waste textiles into regenerated cellulose fibres: impact of pretreatments. The Journal of the Textile Institute 5 (111), 630–638.

Ma, Y., et al., 2021. Regenerated cellulose fibres wetspun from different waste cellulose types. Journal of Natural Fibres 12 (18), 2338–2350.

# Textile effluent treatment and recycling

# 19

*Sherif A. Younis, Mohamed Elshafie and Yasser M. Moustafa*
Analysis and Evaluation Department, Egyptian Petroleum Research Institute (EPRI), Nasr City, Cairo, Egypt

## 19.1 Introduction

Textiles are one of the most important and leading sectors in the global manufacturing industry. They play a crucial role in modern society, fulfilling our clothing requirements (Deng et al., 2020). The textile sector in the European Union (EU) accounts for 30% of the global textile market, while the EU single market is the most important worldwide in terms of size, quality, and design. In addition, as a labour-intensive area, human resources are crucial for the development and survival of the textile sector worldwide (Zhezhova et al., 2021). Worldwide, the textile industry is the second-largest employer (i.e., after agriculture) in developing and developed countries, providing more than 100 million jobs (direct or indirect) and contributing to 6% of production among all manufacturing sectors globally (Kumar et al., 2017).

Based on the Statista database (www.statista.com), China and the European Union (EU) are the top global textile exporters with approximate values of 154 and 64 US billion dollars, respectively, in 2020. However, according to the European Commission, the EU textile market is a leader globally (considering size, quality, and design) and represents more than 30% of the world market (Hossain and Khan, 2020). The Statista database also shows that the current EU textile sector employs around 1.7 million people, with a turnover of EUR 166 billion. Specifically, among EU countries, Germany, France, Spain, Portugal, and Italy are at the forefront, with more than one-fifth of the global textile industry at a turnover value of more than 160 US billion. As a global economic opportunity, the textile market was expected to generate a turnover of approximately \$1032.1 billion in 2022, then rise to \$1420 billion in 2030 with a compound annual growth rate (CAGR) of 4.0%. Hence, the textile industry's growth undoubtedly influences the sustainable development of society, the environment, and the economy. Driven by the 2030 Sustainable Development Goals (SDGs), the increasing consumption of textile products has raised numerous concerns about the challenges of sustainable management of solid and liquid wastes generated in the textile sector and their negative implications on the environment, economy, and society (Cai and Choi, 2020).

Due to the complex manufacturing processes, textiles are one of the most polluting sectors for the environment due to the extensive use of chemicals and vast amounts of water, resources (e.g., cotton, wool, linen, chemicals, and dyes),

**Sustainable Innovations in the Textile Industry. https://doi.org/10.1016/B978-0-323-90392-9.00001-X**

and energy (i.e., electricity) (Cai and Choi, 2020; Keßler et al., 2021). It is one of the most polluting, resource-consuming, and labour-intensive industries (Paul et al., 1995; Kumar et al., 2017). In particular, the textile industry consumes massive amounts of water (approximately 93 billion cubic metres [BCM] of water globally), around 4% of the annual freshwater withdrawal, especially in the dyeing process (Hossain and Khan, 2020). Beyond dyeing/textile production, the global consumption of water during maintenance and washing processes may also reach 20 BCM annually. However, the generated textile wastewater contains multiple toxic chemicals used during manufacturing. According to the statistics of the World Bank, around 72 toxic chemicals can be discharged into textile wastewater effluent, especially after the dyeing process.

Since the textile sector consumes an intensive amount of freshwater, this directly yields tremendous wastewater effluent generation that imposes high pressure on the natural ecosystem. For instance, dyeing is the main step in consuming various dyes and chemicals for colouring fibres in textile manufacturing. Some of these dyes and chemicals become part of the textile effluents after the finishing process (Paul et al., 1995; Yaseen and Scholz, 2019). Approximately 2%–9% of the consumed dyestuffs in textiles are projected to be discharged into the wastewater effluent. Approximately 96% of dyes released had an $LC_{50}$ to fish above 10 $mg/dm^3$ (only 2% with an $LC_{50} < 1.0$ $mg/dm^3$) (O'Neill et al., 1999).

If textile effluent is not adequately treated or handled, the receiving water bodies (rivers and streams) can be seriously polluted (i.e., the textile industry is responsible for around 17%–20% of worldwide water pollution) (Cai and Choi, 2020; Hossain and Khan, 2020). As a result, colourful water bodies can be seen frequently in developing countries due to textile factories' operations without adequate investment in wastewater treatment technologies (El-Fawal et al., 2020; Younis et al., 2020). In addition, the improper disposal of textile effluents without effective treatment has adverse environmental, social, economic, and health impacts (Cai and Choi, 2020; Provin et al., 2021). As such, it is vital to find innovative technological solutions to reduce the negative consequences of textile effluents through upgrading industrial wastewater treatment plants and consumers'/stakeholders' knowledge about effective waste management to achieve a more sustainable consumption scenario (Rajput et al., 2021).

Therefore, it is necessary to adopt cleaner production practices (CPPs) to increase the efficiency of raw materials, clean water, and energy, not generate, reduce, or recycle waste from manufacturing processes (de Oliveira Neto et al., 2019; Provin et al., 2021). Herein, this chapter aims to provide a comprehensive survey about textile effluents, taking into account the historical aspects of the textile industry and effluent characteristics. This chapter also highlights the advantages and limitations of traditional vs. advanced technologies used to control effluent pollution for safe reuse. Also, it discusses the potential reuse cycles of textile effluents and the socioeconomic awareness related to textile effluents for reuse purposes to achieve sustainability within textile plants.

## 19.2 Historical aspects and current scenario

The history of dye use has been traced back to the Middle Stone Age/Middle Palaeolithic in Africa and Europe (150,000–30,000 BCE) when used in the caves of Australia (Arnhem Plateau), Indonesia (Petakiri), Spain (El Castillo), France (Chauvi), and Romania (Calipuaya) (Ardila-Leal et al., 2021). Among dyes/pigments used, red and black pigments derived from natural rocks (or other geological components) were widely used in prehistoric times as they are present in almost all settlements and quarries from the Palaeolithic to the Upper Palaeolithic (35,000–10,000 BCE) periods (Hovers et al., 2003). It has been reported that iron oxide or ochre was the major source for the red, orange, and yellow pigments, while the black colour comes from carbon stones (Behrmann and Gonzales, 2009). With the development of human civilisation in the late Neolithic (6000–3500 BCE) and Bronze Ages (3000–1200 BCE), which ended the nomadic unstable lifestyle, dyes began to be used in paints (the walls of temples, tombs, or houses) and for dyeing fibre tissues, skin, and hair (Abel, 2012; Pugh and Cecil, 2012; Chekalin et al., 2019). However, it should be noted that different dyes with varying compositions/structures were used, depending on the purpose.

From a historical point of view, the decoration of ceramic in the Maya civilisation (2000 BCE to CE 900) was done using red pigments rich in iron and chromium, while red pigment rich in cerium was used for the red and black decorations (Pugh and Cecil, 2012). Also, blue dye was obtained from deposits of lapis lazuli (a precious and expensive source like gold) due to the difficulty in obtaining non-ochre and black colours in prehistoric times (Abel, 2012). The ancient Egyptians developed the first artificially produced blue dye, known as Egyptian blue, between 2900–2750 BCE during the Fourth Dynasty (2630 BCE, Sneferu [Pharaoh]) (Berke and Wiedemann, 2000; Berke, 2002). After that, the Chinese developed the Han Blue pigment at around 500 BCE (Berke and Wiedemann, 2000), followed by the Maya Blue pigment by the Mayan civilisation in the Upper Preclassical period (350–150 BCE) (Fernández-Sabido et al., 2012). As a matter of fact, the Middle Ages – the fifth to fifteenth centuries – were characterised by the use of bright, clear, and specific colours, along with progress in dyeing and the transfer of colourants across Europe (Abel, 2012).

During the Renaissance (15–16th centuries CE), cultural and artistic progress encouraged advances in developing alternative pigments for gold at low cost (Barnett et al., 2006). In 1630, Cornelius Drebbel produced the first dye, erroneously classified as synthetic, by mixing cochineal red (obtained from insects) with tin to improve the stability of natural pigments (Becker, 2016). In 1704, Diesbach made the first dye known as Red Lake, while the highly toxic Emerald Green dye, or Scheele Green, composed of copper aceto-arsenite, was developed by Carl Scheele in 1788 and used until 1960 (Barnett et al., 2006). In 1854, the first synthetic dye called 'mauveine', derived from coal tar, was developed by Henry Perkin (Robinson et al., 2001; Abel, 2012). Accordingly, Perkin solved the problem of industrialisation and announced the initiation of various procedures to produce new, low-cost synthetic dyes using coal tar distillates (Johnston, 2008). Since then, these low-cost synthetic dyes began to be used in place of some natural dyes (e.g., Alizarin) (Abel, 2012). In 1869, the production of dyes diversified, with reports of over 100,000 synthetic products today (Paz et al., 2017).

Most artificial dyes are dissolved in water and have a particle size ranging from 0.025 to 1.0 μm. Artificial dyes have many benefits in the textile and tanning industries in terms of their low cost, high chemical stability under various physicochemical and biological processes, and the maintenance of colour intensity (Ardila-Leal et al., 2021). Synthetic dyes have a considerable structural diversity due to the presence of extended conjugation with other chromogenic (e.g., −N=N- [azo], =C = O [carbonyl], NO or $-NO_2$ [nitrous or nitro], and C=S [sulphur]) or auxochrome (e.g., COOH [carboxyl], $-NH_3$ [amine], −OH [hydroxyl], and $-HSO_3$ [sulphonate]) moieties. The chromophores make dyes absorb the visible light spectrum (400−700 nm). Also, the auxochrome moieties are responsible for increasing dye interaction with fibre (intensity [tone] and affinity).

It is to be noted that the dyestuffs can be categorised based on (i) the type of chromogenic group into 12 classes and/or (ii) the mode of action/application into nine categories (such as acid, basic, reactive, disperse, mordant, vat, direct, metal-complex, and sulphur dyes). Among these groups, the azo dye is the most used textile dyestuffs (60%−70%), followed by reactive dyes (about 20%−30% market share), and anthraquinone type (Vandevivere et al., 1998). The high market share of these types is attributable to their use in dyeing cotton, which represents half of global fibre consumption. Also, Europe was the largest producer of synthetic dyes at the beginning of the 20th century. At the same time, China and India have become the largest producers and suppliers of textile/dyed fabrics worldwide today (Tkaczyk et al., 2020).

The increasing need for textile products has accelerated the industrialisation of the textile industry and its discharged effluent (around 0.069−0.317 $m^3$/kg of fabric produced/day) to natural water bodies (Sarayu and Sandhya, 2012). Due to the high contents of organic (e.g., dyes) and inorganic (heavy metals) constituents in textile effluents, this discharge may exceed natural purification capacities in developing countries, making it one of the major sources of water pollution globally (Sarayu and Sandhya, 2012; Yaseen and Scholz, 2019). Discharging coloured effluents into natural water bodies also increases costs for downstream industries, resulting in higher costs for consumers. For industries located in coastal areas, the intrusion of untreated effluents into the sea also threatens the desalination process. In particular, effluent characteristics mainly depend on (1) the type of fabrics (e.g., cellulose, protein, or artificial) produced, (2) the textile processing steps (e.g., sizing/de-sizing, scouring, bleaching, dyeing, printing, or finishing) (Fig. 19.1a), and (3) the effectiveness of wastewater treatment plants used in the textile industry (Fig. 19.1b) (Pang and Abdullah, 2013).

## 19.3 Environmental impacts of textile effluents

### *19.3.1 Characteristics of textile effluents*

The textile sector is the largest source of industrial wastewater due to the high water consumption of various wet treatment processes (e.g., about 200 L of water consumed

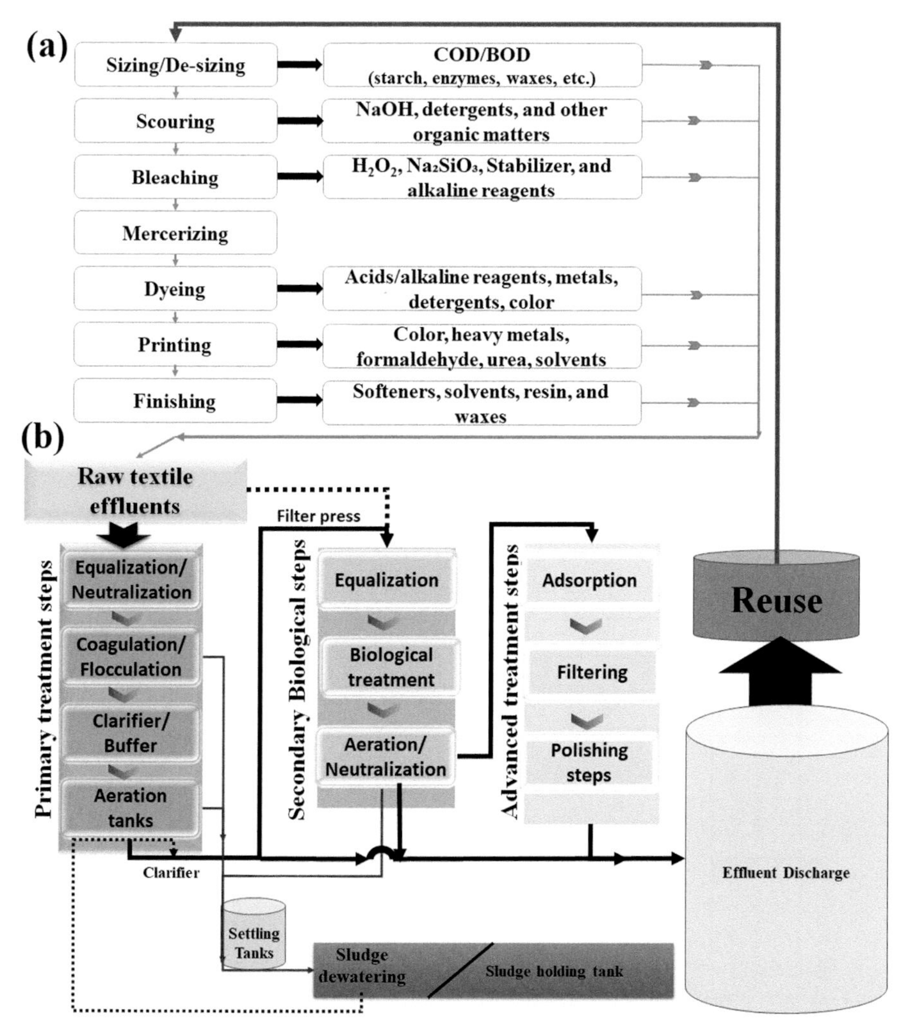


**Figure 19.1** Textile effluents: (a) pollution sources after each step in wet processing and (b) typical construction of textile wastewater treatment plants.

per kg of fabric produced per day, and 80% of this volume is discharged as wastewater effluents) (Holkar et al., 2016). Also, it has been noted that nearly all textile manufacturing effluents discharged into the environment are highly polluted with a variety of chemicals such as alkalis, acids, $H_2O_2$, dyes (colouring agents), surfactants, solvents, starch, dispersing agents, and mineral soaps, dissolved solids/salts, and suspended solids (SS). In dyeing and printing steps, around 1%–50% of colouring materials employed in fabrics are unfixed and lost in wastewater, depending on the fabric type and dyeing chemicals and technique. Table 19.1 presents the concentration range of major constituents in textile effluents collected from various manufacturing plants in different countries based on the reported data in the literature (Bisschops and Spanjers, 2003; Yaseen and Scholz, 2019).

**Table 19.1** Characteristics of effluents generated from the textile industry.

| Textile process/Location** | Fabric type | pH (−) | *Colour (ADMI) | Total solids (mg/L) | Dissolved solid (mg/L) | *BOD (mg/L) | *COD (mg/L) | $NH_4$ (mg/L) | $NO_3$–N (mg/L) | *TOC (mg/L) | Turbidity (NTU) | $PO_4$ (mg/L) | $H_2O$ usage (L/kg fibre) |
|---|---|---|---|---|---|---|---|---|---|---|---|---|---|
| **Major constituents of textile effluents from different processes** | | | | | | | | | | | | | |
| Desizing | Not specified | 6–8 | – | 7600–42,900 | – | 200–5200 | 4600–12,000 | – | – | – | 930 | – | 12.5–35 |
| | Wool | – | – | – | – | – | – | – | – | – | – | – | – |
| | Synthetic | – | – | – | – | – | – | – | – | – | – | – | – |
| | Cotton | 8.8–9.2 | 64–1900 | – | 530–6900 | – | 950–20,000 | 9–19 | | 250–2750 | | 4–10 | – |
| Scouring | Not specified | 7.3–13 | 21–694 | 7600–17,400 | – | 100–2900 | Up to 8000 | – | – | – | – | – | – |
| | Wool | 7.6–10.4 | 2000 | 28,900–49,3000 | – | 2270–60,000 | 5000–9000 | 604 | – | 5800 | – | 89.3 | 4–77.5 |
| | Synthetic | 8–10 | – | – | – | 500–2800 | – | – | – | – | – | – | 17–67 |
| | Cotton | 7.2–13 | 694 | – | – | 100–2900 | 8000 | – | – | – | – | – | 2.5–43 |
| Bleaching | Not specified | 7–12.4 | 153 | 2300–14,400 | 4800–19,500 | 80–1700 | 1060–13,500 | 2.15–18.6 | 3.3–9.4 | – | – | – | – |
| | Wool | 6 | – | 910 | – | 400 | – | – | – | – | – | – | – |
| | Synthetic | – | – | – | – | – | – | – | – | – | – | – | – |
| | Cotton | 6.5–13.5 | 150 | 2300–14,400 | 4760–19,500 | 90–1700 | 288–13,500 | 8–29 | – | 320 | – | 6–60 | 30–50 |
| Mercerising | Not specified | 5.5–9.5 | – | 600–1900 | 4300–4600 | 50–100 | 1600 | – | – | – | – | – | – |
| Soaping | Not specified | 12 | 38 | – | – | – | 578 ± 23.5 | – | – | – | – | – | – |
| Dyeing/printing | Not specified | 5–10 | 1450–4750 | 500–50,000 | 50 | 10–1800 | 258–49,170 | 0.48–370 | 3.7–8.3 | – | – | – | 20–300 |
| | Wool | 4.6–8 | 2225 | – | – | 400–2000 | 7920 | – | – | – | – | – | 280–520 |
| | Synthetic | 11.7 | 1750 | 150–250 | – | 530–590 | 620–1515 | 129 | – | – | – | 21 | – |
| | Cotton | 9.2–10.1 | 1450–4750 | – | – | 970–1460 | 1115–4585 | – | – | – | – | – | 150 (dyeing) |

**Typical physicochemical properties of effluents discharged in various countries**

| | | | | | | | | | | | | | |
|---|---|---|---|---|---|---|---|---|---|---|---|---|---|
| Mexico (Rinsing step of the denim textile industry) | – | 6.84 | 330 | 6000–7000 | 1500–12,000 | 91.9 | 344 | – | 1.9 | 84.92 | 104.66 | 287.1 | – |
| India (Textile effluents) | – | 4.3–12.7 | 240–290 | 450–6510 | 430–49,440 | 108–1750 | 195–4400 | – | 73–85.6 | 101–7784 | 12.5–16.6 | 72.8–86.8 | – |
| Malaysia (Textile effluents) | – | 3.85–11.4 | 76–1777 | 39.33–11,689 | 14.00–11,564 | – | 231.7–990 | 0.47–50.83 | 1.23–5.60 | – | 63–74 | 0.07–4.01 | – |
| Spain (Dyeing and rinsing processes) | – | 6.9–7.8 | 0.39–300 | – | 1456–1568 | – | 200–806 | – | – | – | 4.02–12.6 | | – |
| Iraq (Textile effluents) | – | 5–12.9 | 50–85 | – | 600–1350 | 15–149.3 | 120–225.08 | – | – | – | 26–30 | 0.64 | – |

**ADMI*, American Dye Manufacturers Institute unit; *COD*, Chemical oxygen Demand; *TOC*, Total Organic Carbon. Dissolved solids refer to all metals in discharged effluents (e.g., alkaline earth, transition, heavy metals, etc.).

**Note the sources of pollution in each textile process are as follows: (i) Desizing [e.g., starch, acetic acid, water-soluble sizes/polyvinyl alcohol, enzymes, ammonia lubricants, wax, NaOH, desizol THL, wettanol DBL (sodium ethyl hexyl sulphate), and seraquest], (ii) Scouring [e.g., NaOH, $NaHSO_4$, acetic acid, soda ash, disinfectant, pectinase, fats, surfactant, insecticide residue, oils, lubricants, wax, wetting agent (Wettanol), stabiliser, serafile, and hydrose], (iii) Bleaching [e.g., NaOCl, $H_2O_2$, sodium silicate, seraquest, and organic stabilizer], (iv) Mercerising [e.g., NaOH, liquid ammonia, and Wettanol DBL (sodium ethyl hexyl sulphate)], (v) Soaping [e.g., Salt, alkali agents, and soda ash], and (vi) Dyeing/printing [e.g., $Na_2CO_3$, NaOH, NaCl, $Na_2SO_4$, ammonia, sodium alginate, Resist salts, urea, sodium bicarbonate, Seraquest M-USP, OT paste Thickening agent (kerosene and water-based polyacrylate copolymer), Wetting agent (alkyl phenol ethoxylates and fatty alcohol phenol ethoxylates), reducing agent ($Na_2S$ and glucose), organic acid carriers, oxidizing agents, dyestuffs, formaldehyde (binder), oils, thickeners, and surfactants].

As can be seen in Table 19.1, the composition of textile effluents varies with respect to the operational textile step and the origin of service activities (i.e., location). Also, the composition of textile effluents changed significantly from one country to another, possibly due to variations in the manufacturing steps and/or treatment stages in textile industries (Yaseen and Scholz, 2019). Given the wide variety of chemicals used in textile processing (Table 19.1), defining and monitoring textile effluents' chemical composition is challenging due to the massive amount of chemicals used after each step in textile mills. Hence, the effective treatment of textile effluents has become a critical problem since it has a wide variety of pollutants, which are characterised by high BOD/COD levels from 210/340 to 5500/17,900 mg/L, colouring from 300 to 3500 colour units, and total suspended solids (TSS) from 50 to 24,000 mg/L(Valdez-Vazquez et al., 2020). It has also been found that conventional wastewater treatment plants (WWTP) cannot adequately treat a massive load of chemicals disposed of in the discharged effluents. Accordingly, the major effluent parameters (e.g., colour, turbidity, $NO_3{-}N$, $NH_4{-}N$, dissolved solids/metals, TOC, COD, and BOD) are monitored to compare their levels with the standard regulation limits before final disposal into the environment.

For example, the mercerisation process uses 18 wt.%−24 wt.% of NaOH to treat cotton fabric to shine and accelerate the dye uptake (dye stabilisation onto fabric) (Holkar et al., 2016). After the mercerisation process, washing the fabrics can result in a NaOH (alkaline reagent) load in the wastewater. The frequent use of bleaching agents (e.g., peroxide, hypochlorite, and peracetic acid) to remove natural fabric colour is also considered one of the textile industry's pollution sources, as these bleaching agents become part of the textile effluent constituents (Dey and Islam, 2015). Several auxiliary chemicals are used during the dyeing/printing process to increase the dye's absorption onto the fabrics (Yaseen and Scholz, 2019).

All textile dyes (e.g., azo, xanthene, indigo, anthraquinone, and triaryl methane) and auxiliary chemicals used in the textile sectors (Table 19.1) may become part of textile effluents, which, in turn, cause an unacceptable appearance and harmful effect on water bodies. Among all textile steps, the dyeing and finishing processes produce 17%−20% of total industrial effluents based on the World Bank estimation (Holkar et al., 2016). Considering such pollution severity, the effluent generated from the de-sizing process is classified as strong wastewater with a high BOD (1700−5200 mg/L) and COD (4600−5900 mg/L) values due to the massive use of various sizing agents (i.e., starch materials) (Bidu et al., 2021).

After the dyeing process, the effluents are characterised by high pH, colour index, solid contents, metals, COD, BOD, and amines/nitrates values (Table 19.1). For instance, after cotton dyeing, the effluent generated is characterised by high pollution (i.e., colour index up to 4750 based on the American Dye Manufacturers Institute [ADMI] unit) due to the leaching of reactive dyes which are chemically resistant to biological degradation (Holkar et al., 2016). Specifically, dyehouse effluents could contain around 30% of the consumed reactive dyes in textiles due to their hydrolysis in the alkaline conditions (i.e., the typical concentration of reactive dyes in dyehouse effluent is 0.6−0.8 g dye/$dm^3$) (Vandevivere et al., 1998). Based on the ADMI unit, the estimated dye concentration (colouring agent) in the dyeing effluents varies from 10 to 800 mg/L (Gähr et al., 1994; Shelley, 1994; Vandevivere et al., 1998).

In Iraq, the effluent dye concentration discharged from some textile mills ranged between 20 and 50 mg/L (Abid et al., 2012), while the outflow concentration of acid orange dye from the clarifier step of the textile industry in India was determined to be in the range of 45 mg/L (Sivakumar et al., 2013). In India, around 80% of total dyestuffs produced (1.3 million tons) is consumed by the textile sector due to the high demand for cotton and polyester globally (Holkar et al., 2016).

Chemically speaking, as some textile dyes are metal complexes (i.e., 1:1 or 1:2 metal-organic complexes), the concentrations of heavy metals (e.g., Co, As, Cd, Pb, Ni, Cr, and Zn ions) in dyeing effluents were in the range between 0.001 and 110 mg/L higher than the values reported in Industrial Effluent Discharge Standards (Standard-B) for discharge downstream of any raw water intake (Bakar et al., 2020; Panigrahi and Santhoskumar, 2020; Reddy and Osborne, 2020; Velusamy et al., 2021). As a result, the decolourisation of coloured effluent and removal of heavy metals should be made on-site prior to the final discharge. However, it should be noted that almost all traditional wastewater treatment plants (WWTPs) in the textile sector involve chemical (equalisation, coagulation/flocculation, and clarifier units), biological (bioreactor sludge), and physical/polishing (e.g., sand filtration and carbon/biochar adsorption) methods (Fig. 19.1b). In the traditional WWTP, most dyes are partially decomposed due to the high stability and structural complexity, resulting in the decolourisation of coloured effluents while producing other toxic aromatic substances (as by-products) in the effluent. Accordingly, it is important to upgrade the traditional WWTP in the textile sector (Fig. 19.1b) with advanced treatment technology able to treat coloured effluents effectively without generating other hazardous intermediate by-products in the effluents. Hence, many efforts have been made to find sustainable and innovative technologies for managing textile effluents to comply with the permissible standards of the relevant environmental quality act before final disposal into surface water bodies.

### 19.3.2 Impacts and regulations for textile effluents

According to David Malpass, the 13th president of the World Bank Group, water quality deterioration is stalling economic growth, worsening health conditions, reducing food production, and exacerbating poverty in many countries. Considering the composition and volume of the discharged effluents, the textile sector has been classified as the most polluting source of the natural ecosystem. Hence, the rapid growth of the textile sector in the future may accelerate pollution of the aquatic environment, making water a limited and scarce resource. Such water pollution and scarcity may impose a costly burden on productivity. As a case study, it was reported that the annual loss of some cities in China due to water shortage and pollution is estimated at $11.2 billion/year (industrially) and $3.9 billion/year on human health, which is definitely an underestimate (Lu et al., 2010).

To overcome water pollution and scarcity, a number of stricter regulations have been set in many countries regarding the discharge of industrial effluents into natural water bodies in the past few years (Kao et al., 2001; El-Gohary et al., 2013). For instance, in 1974, ETAD (Environmental and Toxicological Association of the Dye Manufacturing Industry) cooperated with the government and public officials to reduce the harmful effects of their products on the environment and protect human health. These aspects are also regulated in Great Britain by the Environment Agency (EA) in England and Wales and SEPA (Scottish Environmental Protection Agency) in Scotland. In developed countries, various legislations have also been applied to strictly regulate the discharge of dyeing wastewater. The United Kingdom, for example, has enforced a law to restrict the release of synthetic dyes into the marine environment.

The European Commission (EC) has also imposed several strict regulations regarding the presence of dyes in wastewater (Samsami et al., 2020). With regard to legislation, it must also be noted that there is still no international consensus on the acceptable standards for discharging textile effluents into natural ecosystems. Environmental legislation in many developed countries, such as Canada, the United States, and EC countries, forces the textile sector to determine the concentration levels of dyes, heavy metals, and other harmful additive chemicals in treated wastewater before its discharge into the environment. In other regions, such as Pakistan, Malaysia, and India, the Central Pollution Control Board regulates the limits on releasing effluents into the environment.

In response to these constraints, the textile sector has embraced drastic changes and rapid innovations in the generation, treatment, and reuse of textile wastewater (Fig. 19.2) (Vandevivere et al., 1998; Adane et al., 2021). In addition, dyestuff manufacturers have developed new types of dyeing aids to comply with benign environmental requirements, such as (i) non-alkyl phenol-derived emulsions, (ii) non-alkyl phenol-derived dye alternatives to halogens or heavy metals, (iii) chlorine-free bleaching agents, and (iv) synthetic thickening agents to reduce the amount of wasted dye in

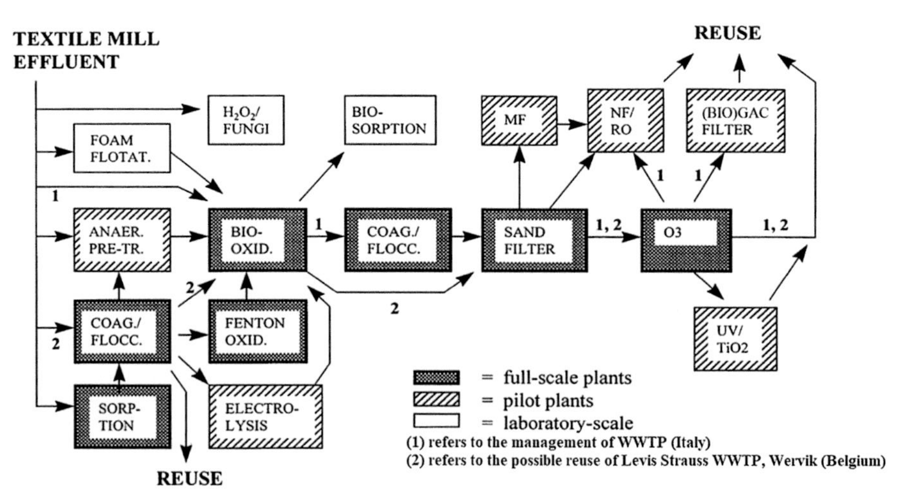


**Figure 19.2** Possible commercial and lab-scale treatment technologies for the reuse of textile effluents (Vandevivere et al., 1998).

the textile processing (Vandevivere et al., 1998). Also, reclamation and reuse of industrial wastewater (such as textile effluents) have become an attractive option of great importance to expand the available water resources (non-conventional resources) sustainably and to protect the environment. In this regard, a number of practices have been proposed for the safe reuse of treated wastewater, over the past decade, such as (i) recreational holding, (ii) urban uses in toilet flushing, (iii) green spaces or nonedible crop irrigation, (iv) industrial applications, and (v) augmenting water supply through reservoir/groundwater recharge. In fact, the reuse practice of reclaimed wastewater is often a reliable option to meet the hydraulic deficit in many regions worldwide.

## 19.4 Innovations in wastewater treatment plants

This section attempts to summarise and compare the various innovative processing technologies used to treat and reuse textile effluents, emphasising industrial plants to accurately assess the applicability of these new technologies in real situations. Different technologies are also used sequentially (i.e., hybrid process) in the textile WWTPs (see Fig. 19.2) to effectively treat wastewater for possible reuse and/or safe discharge. These treatment methods include physicochemical (e.g., sand/membrane filtration, coagulation/flocculation, adsorption, distillation, membrane separation), chemical oxidation (e.g., $H_2O_2$, ozonation, and electro/photo-chemical), and biological (e.g., anaerobic/aerobic degradation and enzyme-assisted biochemical degradation) (Upadhye and Joshi, 2012; Holkar et al., 2016; Samsami et al., 2020). In general, the combined technologies complement each other's strengths and overcome individual technologies' weaknesses. Table 19.2 shows that each treatment method could have distinctive features in one way but be limited in another based on the type and composition of wastewater effluents. Accordingly, the sequential use of physical, chemical, and biological technologies in conventional textile WWTPs can meet the permitted specifications under the Environmental Quality Act for the discharge of textile effluents into the environment but does not have the potential to enable the reuse of treated effluents in textile operations.

As a primary treatment step, filtration and coagulation/flocculation techniques have been extensively used to treat coloured effluent contaminated with particulate matter and organic compounds, including dyes (e.g., disperse, reactive, and/or vat dyes) (Madhav et al., 2018). In the coagulation/flocculation process, the treatment efficiency depends on many variables such as the type and dose of chemical reagents, pH, alkalinity, temperature, ionic strength, mixing conditions, and contact time. Although all WWTPs use a coagulation/flocculation unit as a primary step in treating wastewater, this process generates a large amount of solid sludge (as a secondary waste disposal problem) contaminated with heavy metals. Hence, filtration methods can be preferably applied for the enhanced removal of colour and heavy metals from dye effluents, and, subsequently, this step dramatically reduces the BOD/COD values of treated wastewater. From an application standpoint in textile WWTPs, most nanofiltration (NF) systems had significant rejection capabilities against inorganic salts and water-soluble

**Table 19.2** Characteristics of textile WWTP technologies.

| Step | Treatment method | Examples | Target pollutants | Advantages | Limitations | Potential applicability of treated effluent |
|---|---|---|---|---|---|---|
| Physical/chemical methods | Gravity separation | Sedimentation | Suspended matter, and oil and grease | Simple, low energy consumption, no chemical uses | Large spacing and long retention time. It only separates undissolved particulate matter | No reuse recommended |
| | Chemical precipitation | Coagulant/ flocculent | High organics, metals, COD/ BOD, turbidity contents | A simple method, efficient, and economically advantageous. Adaptable to high-pollutant loads. Extremely useful for metals and fluoride removal with a significant reduction in COD and BOD. Remarkably efficient for insoluble contaminants removal (e.g., SS and colloidal particles). Dewatering qualities and produced sludge with good settling | Requires addition of non-reusable chemicals. High chemical consumption (lime, oxidants, and $H_2S$). Ineffective in the removal of metal ions at low concentrations. Sludge production is high and disposal problems | Landscape, golf course irrigation, toilet flushing, vehicle washing, industrial reuse |
| | Filtration | Sand and activated carbon filters | Suspended matter, dyestuffs, oil and grease, chemicals, and heavy metals | A simple physical filtration, cost-effective, energy-saving, and regenerable | Not effective against dyes, colour, heavy metal ions removal | |
| | Membrane separation | Micro/ultra/nano-membrane filtration | Suspended matter, oil and grease, dyes, dissolved metals, and heavy metals | High removal of heavy metals, lower space requirement, simple and easy to fabricate. Rapid and efficient, even at high concentrations. Produces a high-quality treated effluent. No chemicals required. Low solid waste generation. Eliminates all types of dyes, salts, and mineral derivatives | Very expensive and the energy requirement is high due to frequent membrane fouling, high-pressure requirements, low permeate flux, and frequent maintenance and operation costs | The treated effluent can be disposed of to received water bodies and/or reused in landscape, afforestation, or irrigation of non-edible plants |
| | Adsorption | Traditional/ functionalised adsorbents | Dissolved organic and metal pollutants | Easy operation, less sludge production, utilisation of low-cost adsorbents. High adsorption capacity, regenerable, wide pH range, and low operation cost. Excellent ability to separate a wide range of pollutants. A highly effective process with fast kinetics. Excellent | Desorption and nanosized adsorbents require moderate materials, while the regeneration could result in secondary pollution. Relatively high investment and regeneration are expensive | |

| | | | | | | |
|---|---|---|---|---|---|---|
| Biological methods | Aerobic degradation<br>Anaerobic degradation<br>Enzymatic biochemical reaction | Active sludge<br>Microbial consortium<br>Enzymes | Soluble organics and dyes | This technology is beneficial in removing heavy metals. The application of microorganisms for the biodegradability of organic contaminants is simple. Economically attractive method but time-consuming | Requires maintenance of microorganisms and pre-treatment like physicochemical treatment is inefficient on non-biodegradable compounds or toxic compounds. Generation of biological sludge and uncontrollable degradation products. Needs to be developed | Irrigation of green area non-edible crops, landscape impoundments, groundwater recharge, non-potable aquifer, wetlands, and industrial cooling processes |
| Advanced treatment methods | Ion exchanger | Ion-exchanger | Heavy metals and other dissolved metal ions, especially high valence cations | Speedy reaction, a high transformation of components and high treatment efficiency. Straightforward method and well-established process. Easy to use with other techniques, for example, precipitation and filtration in an integrated wastewater process. Rapid and efficient process and produces a high-quality treated effluent. Relatively inexpensive and efficient for metal removal cleans up to ppb levels. Resins can be selective for certain metals. Attractive and efficient technology for the recovery of valuable metals | Removes only limited metal ions, expensive due to synthetic resin/ beads, low selectivity regeneration could result in secondary pollution. Large volume requires large columns. Performance sensitive to pH of effluent, conventional resins not demanding. Selective resins have limited commercial use which is not adequate for certain target pollutants such as disperse dyes, drugs, etc. | Food crop irrigation, indirect potable reuse, water reservoir augmentation, agriculture |
| | Electrochemical | Oxidation process | Organic pollution | Efficient method for the removal of metal ions and low chemical usage. Adaptation to different pollutant loads and different flow rates. More effective and rapid organic matter separation than in traditional coagulation | The initial investment is high and also needs a high electrical supply. Requires the addition of chemicals such as coagulants, flocculants, and salts | |

dyes (a molecular weight cut-off of 200 up to 4500 Da), employing size exclusion (pore size ranges from ~0.5 to 2.0 nm) and electrostatic repulsion methods. Despite these benefits, they are characterised by a relatively low water permeability and thus require a high operating pressure (i.e., more energy consumption) to achieve the capabilities to increase water flux in real operational conditions (Han et al., 2018).

It should also be noted that it is not preferred to use pressure-driven membrane filtrations (like NF and reverse osmosis [RO]) in the primary treatment step of textile WWTPs because of the frequent fouling (i.e., membrane pore clogging by suspended matter and dye molecules) and high operational cost (in terms of energy consumption). Due to these limitations, great attention has been drawn to the adsorption technique as the most effective, eco-friendly procedure for decolourising textile effluent containing a variety of dyes. In this regard, several factors must be considered for the economic application of the adsorption process; the most important is designing a low-cost absorbent with a high affinity and capacity for colour removal and a good ability to regenerate multiple times without affinity loss. To comply with these requirements, a large number of adsorbent materials have been designed and reported for dye removal using biotic, natural, or synthetic resources such as agro-biowastes (Mishra et al., 2021), bentonite clay (Rashid et al., 2021), activated carbon (Bakar et al., 2021), and polymeric resins (Hassan and Carr, 2018; Kumar et al., 2019). However, the utilisation of these adsorbents has been restricted by high regeneration costs, low capacity/selectivity, and/or dumping (i.e., sludge production). As such, adsorption should be installed as a secondary treatment process to treat effluent with low concentrations of target pollutants.

In the biological treatment process, the azoreductase-catalysed reduction was commonly used to break azo bonds under anoxic conditions. However, such an anaerobic degradation process could generate stable aromatic amines with higher toxicity than parent azo dyes. Hence, both anaerobic and aerobic systems were consistently integrated industrially to ensure the effective degradation of intermediate amines into environment-friendly by-products. Accordingly, mixed aerobic and anaerobic (anoxic) microorganisms are frequently used to reclaim dyeing wastewater simultaneously or sequentially. The biological treatment process is cost-competitive, consumes less water, and produces less sludge (relative to other physicochemical methods). Nonetheless, it has been reported that the aerobic biodegradation route is insufficient to mineralise textile dyes (such as azo dyes), producing unknown intermediates into the effluents. It also needs sample space and a long hydraulic retention time for operation. The efficiency of biological processes also depends on the selected microbes' adaptability and enzymes' activity (Rashid et al., 2021).

Chemical technologies are rarely used in textile WWTPs due to their high operating cost, high energy consumption, and the need for large amounts of chemicals compared to physical and biological treatments. For example, the advanced oxidation processes (AOPs) using chlorine, $H_2O_2$, and/or $O_3$ (as key oxidising agents) are used to break down recalcitrant dye and organic pollutants dissolved in textile streams (i.e., decolourisation process). However, this chemical oxidation requires certain activation techniques for reaction initiation (such as UV light source [photochemical], electrical current [electrochemical], ultrasound, or other co-oxidant like iron sulphate [Fenton catalysis]) (Lin and Chen, 1997; Deng et al., 2020; El-Fawal et al., 2020). Following AOPs, these activation techniques induced the generation of radical oxidising species (ROS: $^{\bullet}OH$, $^{\bullet-}O_2$, and/or $HO_2^{\bullet}$) for oxidative destruction mechanisms of organic pollutants (e.g., dyes) (Samsami et al., 2020).

Although the chemical oxidation processes are beneficial in terms of no sludge or fouling formation, the oxidation breakdown of dyes could generate diverse hazard intermediates (such as aromatic amine by-products). In addition, these processes are unsuitable for removing dispersed/insoluble contaminants (such as suspended matter or dispersed dyes) and heavy metals from textile effluents. Also, the chlorine-based decolourisation process is not preferable environmentally because of the harmful effects of chlorine on receiving waters and aquatic animals (Adane et al., 2021).

With regard to environmental and economic feasibility, it is also undesirable to use treatment technologies with low efficiency, high operating cost, require a long processing time, and produce other harmful by-products after treatment. For instance, hydrolysis is one of the common approaches used to remove natural desizing agents (e.g., starch) by dilute acid, amylases enzyme, reaction with hot caustic soda, or detergent. However, this method is comparatively less efficient and requires a relatively tiny amount of water during the treatment process. In addition, starch hydrolysis by-products cause a significant increase in the BOD value of desizing effluent. Comparatively, using oxidative desizing approaches, such as the cold pad-batch method or the oxidative pad-steam process, can effectively remove natural desizing agents and cleanse the fabric (i.e., extracting the impurities). Also, the UF membrane can be applied to remove and recover synthetic/natural desizing chemicals (such as starch and PVA), which reduces the cost of desizing chemicals by promoting reuse and reducing water contamination (Saxena et al., 2017).

Currently, many promising technologies have also been suggested and applied for the enhanced treatment and reuse of textile effluents, such as membrane bioreactor (MBR) (Samsami et al., 2020; Othman et al., 2021), photocatalytic membrane reactor (PMR) (Sathya et al., 2021), and direct-contact membrane distillation (DCMD) (Ramlow et al., 2017). In particular, MBR and PMR are among the reported promising hybrid technologies for treating textile wastewater to produce high-quality textile effluents for agricultural and industrial reuse. These hybrid technologies are designed by combining the membrane filtration with biological processes or photocatalysis method to yield MBR or PMR systems.

Unlike the conventional treatment methods, these hybrid technologies require a smaller area for installation due to the direct setting up of the membrane inside the reactors. They also possess other advantages in terms of low maintenance, small footprint, continuity of high-quality treated water, high efficiency in removing contaminants, and low sludge production. Following the green chemistry principle, the hybridised PMR or MBR techniques can be effectively used to (i) abate the membrane fouling and (ii) minimise textile pollution using photo/bio-catalysis and membrane separation in a one-step continuous mode (Goswami et al., 2018; Bhattacharya and Ambika, 2022). Fig. 19.3 shows the suggested flowchart for the possible integration of PMR and MBR systems in textile WWTPs.

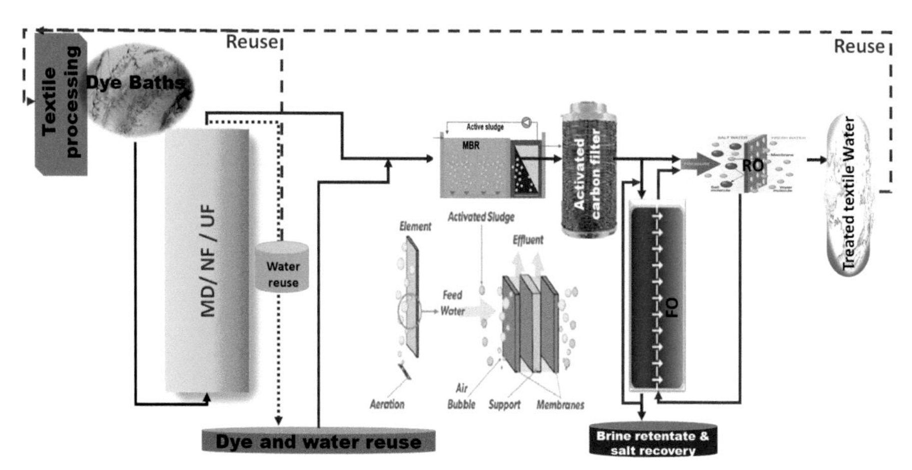


**Figure 19.3** Flow diagram for integrating PMR and/or MBR technologies in traditional WWTPs.

Compared with the traditional biological process, membrane filtration replaces sedimentation in the secondary clarifier to separate the treated textile effluent and the activated sludge in the MBR (Fig. 19.3). In the case of the PMR unit, there are two configurations based on the photocatalyst system design: (i) a suspended photocatalyst in the reaction medium and (ii) an immobilised photocatalyst on the membrane surface (Fig. 19.3). It is to be noted that the efficiency of the PMR reactor depends on the reactor design, the chemical structure of textile pollutants, the retention time of effluents in the reactors, and the properties/composition of the catalytic and membrane materials, which, in turn, control the behaviour of the pollutants in the reactor and reaction kinetics (Othman et al., 2021; Bhattacharya and Ambika, 2022). For instance, in these membrane reactors, the membrane should have a high mechanochemical/thermal stability, a large surface area, and a high surface roughness to control the membrane selectivity, resulting in reduced fouling and improved dye separation efficiency and permeability.

The higher surface roughness also increased the membrane's ability to hold photocatalytic molecules in the PMR technique. For example, most PMR/MBR technologies prefer inorganic ceramic membrane integration (e.g., $SiO_2$, $Al_2O_3$, $ZrO_2$, and $TiO_2$) compared to polymeric materials (e.g., polyvinylidene fluoride, polyamide, polyethylene terephthalate, and polyethersulphone) (Samsami et al., 2020; Sathya et al., 2021; Bhattacharya and Ambika, 2022). This is possibly attributable to the desired characteristics of the porous ceramic membranes in terms of higher mechanical/chemical/thermal stability, longer life span, and antifouling properties. However, the production cost of ceramic membranes still limits their marketability for application in textile wastewater treatment.

Hence, the feasibility of using PMR in the real application is still disputed as issues related to the photocatalytic process must be improved by: (1) developing a photocatalyst with high photo-efficiency under solar irradiation spectra to reduce energy

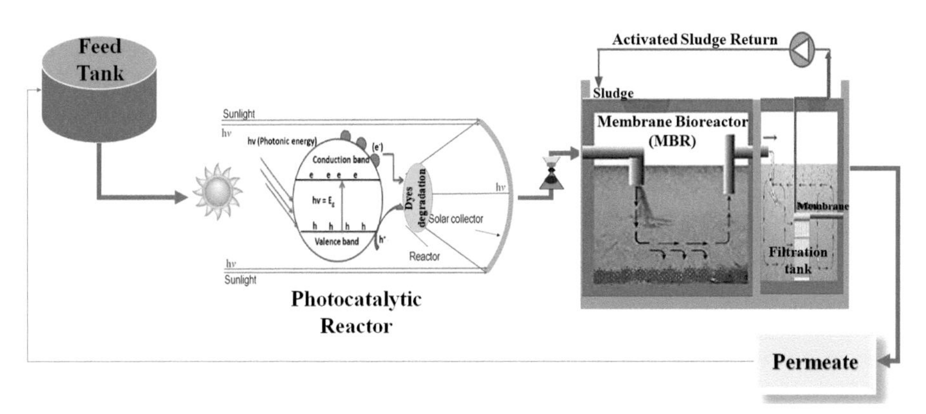


**Figure 19.4** Schematic representation of lab-scale PMBR technology for treating textile effluents.

consumption (i.e., operation costs), (2) addressing the immobilisation strategies of the photocatalyst in the membrane reactor to increase its effective separation from the reaction media, and (3) improving the photocatalytic reaction under broader pH conditions to reduce the addition of oxidising agents. To address these issues, a new hybrid system, namely photocatalytic membrane bioreactor (PMBR; Fig. 19.4), was created by combining PMR with MBR to improve the treatment efficiency in a shorter reaction time (faster reaction kinetics), reduce membrane fouling, and minimise energy consumption (operation cost) (Bhattacharya and Ambika, 2022).

As a case study, a PMBR system was developed of a mild steel rectangular photocatalytic reactor (PCR) connected with a polyethersulphone submerged hollow-fibre membrane bioreactor (20 L working volume) (Sathya et al., 2021). In the PCR, a composite bead of tungsten oxide ($WO_3$) photocatalyst incorporated with 1% graphene oxide (GO) was used to suppress the recombination rate of electron−hole pairs during photocatalysis. Upon using the integrated PMBR system for treating textile wastewater collected from the Indian industry, the COD/colour levels were significantly reduced by 48%−25% after 3 h PCR, then increased up to 76%−70% with the integration of PCR with the MBR unit (PMBR) for 10 h hydraulic retention time at 100 kPa of transmembrane pressure.

The DCMD is a promising technique for the advanced treatment of textile effluents at low pressure and mild temperature conditions relative to pressure-driven membrane processes (Ramlow et al., 2017; de Sousa Silva et al., 2020). Compared with the coagulation/flocculation method, the DCMD unit does not require a large area and/or chemical reagents to operate due to the high mass transfer area per unit volume during processing (i.e., low vapour area) (Fortunato et al., 2021). In the case that the DCMD process is applied directly near the dyeing machine, DCMD can also exploit the heat energy in the dye house (temperature above 100°C) and consume less energy during operation (Ramlow et al., 2017; Li et al., 2018). Although the technical part of DCMD application in treating textile effluents is underway, there remains a lack of economic analysis and economic feasibility (in terms of thermal/electricity consumption and membrane replacement costs) to compare potential advantages with conventional techniques (Laqbaqbi et al., 2019). In addition, since textile dyes are non-volatile substances, the application of the DCMD process mainly focuses on water recovery while separating dyes from coloured effluents (i.e., decolourisation) (Li et al., 2018; Ramlow et al., 2020). Therefore, the concentrated dyebath resulting from the residual feeding solution needs further processing.

Some sustainable alternatives are also suggested to treat the residual water solution, for example, recovery of dye (and other auxiliaries) from the water by adsorption, concentration, and evaporation techniques. However, some economic advantages are afforded in reusing textile effluent (treated wastewater) industrially with possible heat recovery (thermal energy) from hot fabric streams during DCMD operation (El-Abbassi et al., 2013). According to the literature, the DCMD process can reject 40%–100% colour dye, 72%–80% COD, and 80%–100% TOC from textile effluents with a feed velocity/flux rate range of 0.002–4 L/min over a temperature of 15–90°C (Ramlow et al., 2017; Li et al., 2018; de Sousa Silva et al., 2020). It is, therefore, necessary to carry out a scale-up and market feasibility studies to evaluate the economic planning upon applying the DCMD process in textile WWTP, as the return on investment cost is likely to be rapid due to its economic advantages.

Accordingly, with the development of the machine learning (ML) approach, many studies have applied ML techniques (especially artificial neural networks [ANNs] and artificial intelligence [AI]) to optimise and integrate the newly developed systems in wastewater treatment processes. The ML models can provide a quick and innovative management and monitoring tool to simulate the efficiency of the newly developed/suggested treatment techniques upon implementation in actual WWTPs using the correlations among data obtained on a bench-scale analysis (Alavi et al., 2022). Aside from these algorithms, the ANN could be utilised to forecast the effluent quality due to its rapid operation and good fitting effect for nonlinear concerns (Xie et al., 2022). At the same time, for pollution prevention, the textile dyeing and finishing sectors have made various strides towards waste minimisation by replacing or abandoning hazardous chemicals, implementing lower liquor-ratio dyeing machines, and/or using low/no-water textile processing (Vajnhandl and Valh, 2014).

## 19.5 Sustainable innovations leading to energy and resource efficiency

With the expected economic growth scenario, the global industrial demand for water is projected to increase up to 1500 billion $m^3$ (around 22% of total withdrawals) by 2030. Due to the ever-increasing demand for clean water production, the reuse cycle of industrial effluents (especially textile effluents) has become a fundamental challenge due to their massive load with many contaminants. In addition, the current situation reveals that the typical textile industry prefers to use high-quality freshwater in all production operations. Within the strategic plan of the Water Supply and Sanitation Technology Platform (WssTP) of the European Commission, the implementation of a sustainable reuse system has become a goal to (i) address the increasing water demand across the industrial sector globally, (ii) climate impacts on water availability (i.e., water scarcity) and environmental degradation, and (iii) protect public health (Schneider, 2006; Vajnhandl and Valh, 2014).

On this basis, recycling and reusing textile resources are becoming increasingly popular as cleaner production (CP) alternatives to achieve sustainability in the textile chain, resulting in the increased consumption of raw materials (water supply chain) that are becoming scarce. For instance, the advanced textile plants recently adopted the CP concept to mitigate the impact of effluent hazards on the environment for a cleaner and more sustainable ecological quality outcome. The CP concept is an economically feasible plan to reuse natural resources and energy throughout the life cycle of the textile chain, offering exemplary implementation for ecological protection (i.e., diminishing waste of production systems and risks for humans) (de Oliveira Neto et al., 2022). It is to be noted that CP adoption in textiles is also not limited to the innovation of process management but to developing alternative materials, improving technical operation, investing in new treatment tool acquisition, and adopting life cycle assessment (LCA) for sustainable effluent reuse purposes. From the standpoint of the Sustainable Development Goals (SDGs), one of the crucial challenges is aligning the economic elaboration matched with the objectives of contributing to cleaner and sustainable environmental conditions. Many obstacles still need to be overcome to effectively manage the effluents emitted within the textile sector.

In the textile dyeing and finishing sectors, on-site treatment and recycling of textile effluents is also not a common practice because of the limited resources for acquiring closed water loops. Therefore, several suggested practices have been tested and evaluated to achieve textile sustainability related to water consumption and the recovery of energy, chemicals, and unexhausted substances (Saxena et al., 2017). In this regard, textile effluent could be effectively used as a raw source for energy production (hydrogen energy), a valuable chemical resource, and clean water in wet processing (e.g., desizing a new grey fabric).

For instance, various EU projects (e.g., Fototex, Prowater, Adopbio, Battle, Purifast, and Aquafit4use) have been implemented in the last decade to tackle issues of textile effluent treatment for subsequent reuse purposes, achieving zero-liquid discharge (ZLD) (Vajnhandl and Valh, 2014). Among these EU projects, AquaFit4use funded by the EC was to reduce fresh, high-quality water use through innovative cross-sectoral, integrated technologies (such as UF/NF membranes with MBR anaerobic process, evapo-concentration, and advanced oxidation approaches), taking into account all textile production characteristics (Fig. 19.5) (Vajnhandl and Valh, 2014). Considering the LCA tool, these integrated technologies enabled a 98% colour reduction with complete removal of suspended solids (turbidity $\approx$100%), COD (above 90%), and salinity (>67%). These combined technologies also reduce the volume of effluents to be disposed of via the potential reuse of distillate (high-purity water) obtained during the evapo-concentration process.

The ZLD concept could also be achieved in textile plants via a sustainable reuse cycle of textile effluents in wet processing, following these steps: (a) traditional removal or oxidation of scouring and bleaching chemicals by hydrolysis or oxidative approaches, (b) recovering unexhausted chemicals by ultrafiltration, and (c) mixing

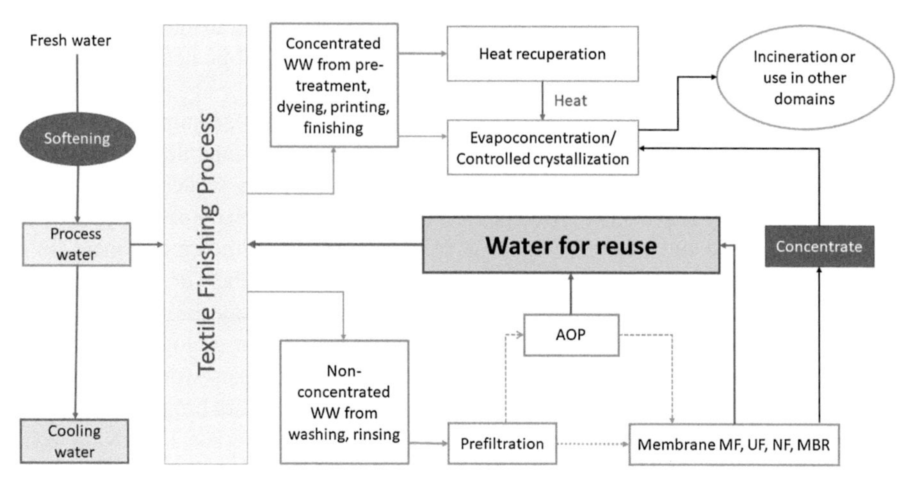


**Figure 19.5** Concept scheme of the AquaFit4use project for treating concentrated and non-concentrated textile effluents (Vajnhandl and Valh, 2014).

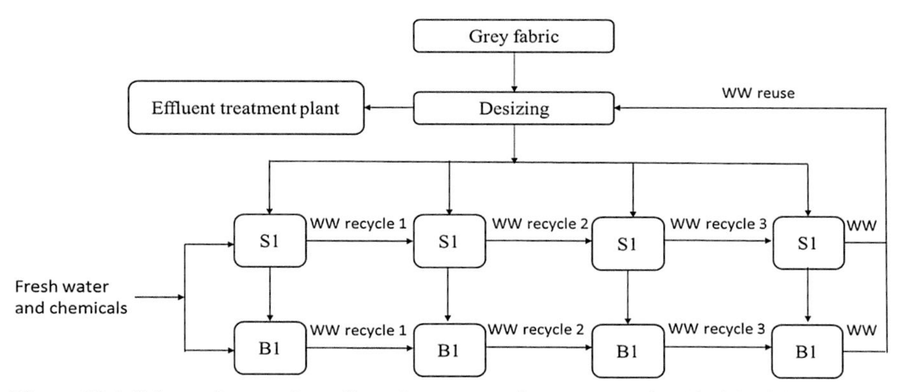


**Figure 19.6** Scheme for reuse/recycling of wastewater in wet processing (desizing) (Harane and Adivarekar, 2017).

treated effluents with influent water at 50:50, 70:30, and 30:70 proportions, respectively (Fig. 19.6) (Harane and Adivarekar, 2017). Bahadur and Bhargava (Bahadur and Bhargava, 2019) also reported a three-stage UV-$TiO_2$ photocatalytic pilot-scale PCR system (Fig. 19.7) to treat and reuse/cycle dyeing effluents, achieving ZLD goals. The proposed process accomplished BOD and COD reductions to 95% and 91%, respectively, with the appearance of a minor amount of non-toxic sludge (0.4 kg $m^{-3}$), with estimated operating expenses of around 1.55 USD/$m^3$ and a complete reduction of water hardness, total nitrogen, and decomposition of organic nitrogen along with a considerable reduction in chloride content, TSS, and TDS.

Lin and coworkers (Lin et al., 2019) also developed an integrated pilot-scale filtration composed of tight ultrafiltration (TUF)/bipolar membrane electrodialysis (BMED) systems (Fig. 19.8). Such a BMED system was suggested to recover the value

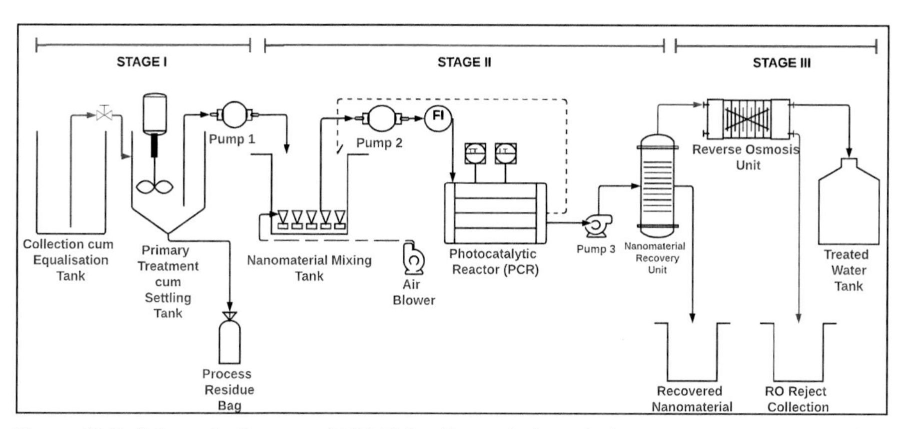


**Figure 19.7** Schematic diagram of UV-$TiO_2$ pilot scale for a dyeing wastewater treatment plant (Bahadur and Bhargava, 2019).

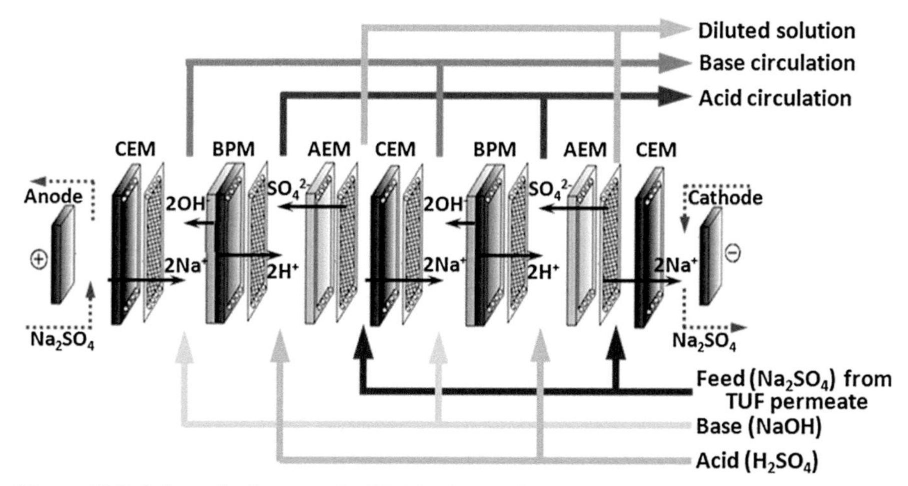


**Figure 19.8** Schematic diagram of a UF/bipolar membrane pilot-scale system for resource recovery from textile effluents (Lin et al., 2019).

resources (i.e., dye and acid/base chemicals) from the dyeing effluent matrix and achieve sustainable control of saline textile wastewater for clean water production (i.e., ZLD goals). The TUF system has been characterised by its capability for a diafiltration process with an efficiency of >99.42% salt permeation (NaCl and $Na_2SO_4$) and recovery of >99.6% dye (reactive and direct dyes) as a concentrated solution. However, the BMED enabled a subsequent treatment of acid, base, and salt-containing TUF permeate to reduce membrane fouling and generate pure water with an energy consumption of 4.23 kWh/kg. Hence, this hybrid TUF-BMED process indicates powerful technical applicability for resource sustainability and recovery from textile effluent, providing zero liquid discharge and closing the material loop (Lin et al., 2019).

Many researchers have also studied expanding the outcome of the NF process by optimising membrane composite for concentrating and recovering consumed dye and salt in the dyeing process. For example, Lin et al. (Lin et al., 2015) investigated the dye/salt fractionation capacity of two commercial NF membranes, Sepro NF 2A and Sepro NF (Ultra, USA). The outcome data revealed that both NF systems had a comparable rejection rate of 99.9% for direct red, with a salt permeation rate of 95.2%, which is promising for the dye/salt mixture recovery. In another study, the use of loose polyacrylonitrile (PAN) NF membrane coated with a gallic acid modified polyethyleneimine (PEI) layer was prepared and tested for the separation recovery of the dye/salt mixture (Zhao and Wang, 2017). The modified NF systems demonstrated (i) a 97% rejection rate for dyes and (ii) a 95% salt permeation rate over long-term stability, with a remarkable antifouling performance (i.e., 8.1% irreversible humic acid fouling). A self-standing forward osmosis (FO) membrane based on a poly(triazole-co-oxadiazole-co-hydrazine) (PTAODH) skeleton was also found to exhibit superior antifouling and antibacterial activity against *E. coli* and *S. aureus* with depressed salt flux (>98% salt and dye recovery) than commercial FO membranes (Li et al., 2018; Li et al., 2019). The outstanding efficiency of the PTAODH membrane was attributed to its smooth surface, high negative surface charge, and dense structure (Li et al., 2019).

Due to the rapid increase in the world's population, scientists have also begun to propose solutions to use treated textile effluents for irrigation and agriculture to meet the increasing demand for food and to address water scarcity. Reusing treated textile effluents as unconventional water resources in irrigation, especially for arid and semi-arid countries, offers a great opportunity to restrain freshwater depletion and groundwater over-abstraction (Lahlou et al., 2021). From an economic point of view, textile effluents provide a good resource for land fertilisation due to micronutrients (e.g., Mg, Ca, Cu, B, Zn, and S) required for plant growth and major nutrients (NPK) crucial for plant productivity problems. Alteration to utilise wastewater represents a stage towards sustainable crop production besides easing stress on the water and energy sectors (Lahlou et al., 2021; Ma et al., 2021).

However, the significant environmental threats confronted by textile effluent utilisation in irrigation include colour, dissolved solids, detergents, organic pollutants, and heavy metal ions, which reduce soil fertility (Abubacker and Kirthiga, 2017). Therefore, when closing the effluent reuse cycle in irrigation, the main challenge is attaining the 'right water quality' for safe use, and preserving process stability and agriculture product quality. Following this requirement, scientists and the WHO/FAO organisations have proposed different indicators to assess and evaluate the sustainable reuse cycle of treated effluents in irrigation and soil fertility. These indicators include the irrigation water quality (IWQ), water quality index (WQI), contamination factor (CF), heavy metal pollution index (HPI), pollution load index (PLI), and ecological risk index (RI).

The WQI is a standard calculation to consistently evaluate the quality of water bodies and their potential threat to various uses, such as recreation and aesthetics, habitat for aquatic life, irrigation/farming, livestock, and drinking water supplies (Agidi et al., 2022). This index was determined using the quality rating scale (qi) and the relative weight (Wi) for each boundary element under consideration in the treated effluent sample (relative to its standard [Si] permissible limit).

Following the WQI values, the Canadian Council of Ministers of Environment (CCME) classified the quality of treated effluents into five groups (Akter et al., 2016): (i) excellent with a virtual absence of threat or impairment (CCME WQI Value 95−100) – close to natural or pristine levels; (ii) good with a minor threat rarely depart from natural or desirable levels (CCME WQI Value 80−94); (iii) fair but occasionally threatened or impaired (CCME WQI Value 65−79); (iv) marginal, frequently threatened or impaired (CCME WQI Value 45−64); and (v) poor with highly threatened conditions (CCME WQI Value 0−44).

Compared with the WQI indicator, the CF, HPI, PLI, and RI are pollution indicators, measuring the relative ratios of target elements in the effluent/sediment relative to their upper permissible or background values in uncontaminated media (or national discharge standard limits) (Hakanson, 1980; Li et al., 2021). For example, the CF values are interpreted as $CF < 1$ = low contamination; $1 < CF < 3$ = moderate contamination; $3 < CF < 6$ = considerable contamination, and $CF > 6$ = very high contamination. The HPI and PLI indicators can also help to assess the overall toxicity of the treated effluent upon their discharge or reuse in target applications, as follows: $PLI < 1.0$ at $RI < 40$ (safe treated water, with no contamination); $1.0 < PLI < 2.0$ at $40 \leq RI < 80$ (moderately contaminated effluents with moderate potential ecological risk); $2.0 < PLI < 3.0$ at $80 \leq RI < 160$ (considerably contaminated effluent with potential ecological risks); and $PLI > 3$ at $160 \leq RI$ (strongly contaminated effluent with high to very high ecological risk upon reuse in agriculture). Based on the FAO, many other indicators (Table 19.3) are also used to assess the suitability of the reuse cycle of treated effluent in agriculture applications.

Considering the above indicators, there is an increasing concern about applying textile effluents for irrigation purposes due to the complex compositions that contain a mixture of chemicals, auxiliaries, heavy metals, chloride, and various dyestuffs of different classes (Zengınbal et al., 2018). As the IWQ metric depends on quantity and the type of salts in the used water, high heavy metal and chloride concentrations in textile effluents could deteriorate soil quality (e.g., soil salinisation, change soil pH, enrich microbial soil, and decrease soil enzymes [DHA, urease, and FDA], etc.) (Monika et al., 2021). Accordingly, using low-quality textile effluent in irrigation could reduce permeability, fertility, and exposure to particularly toxic metal ions (such as Cd, Cr, Pb, etc.).

As the textile industry is an energy-intensive sector, recycling textile effluents should increase this sector's energy consumption (i.e., 24.9% of the total thermal energy used is lost for wastewater treatment in a dyeing plant), accounting for 10% of $CO_2$ emissions (1.2 billion tons/year) (Okafor et al., 2021). Hence, reducing energy consumption and greenhouse gas (GHG) emissions should be considered to achieve sustainability in effluent recycling, considering the circular economy goals. In this regard, the production of green, renewable energy from textile effluent has attracted scholars concerned about energy recovery from waste resources. For example, hydrogen recovery during the catalytic degradation/conversion of organic dye pollutants is an integral component of sustainable development systems. The photo/thermal/

**Table 19.3** FAO indicators to assess the suitable reuse of treated effluents in agriculture and soil sustainability.

| Indicators | Interpretation for unsuitable treated effluents | Equation |
|---|---|---|
| Sodium absorption ratio (SAR) | High SAR values > 26 indicate a reduction of soil permeability upon irrigation with treated effluent | $SAR = \frac{Na^+}{\sqrt{\frac{Ca^{2+} + Mg^{2+}}{2}}}$ |
| Bicarbonate hazard (BH) | High bicarbonate levels increased Ca and Mg precipitation and SAR value (i.e., the treated effluent is not suitable for agricultural use) | $BH = HCO_3^- - (Ca^{2+} + Mg^{2+})$ |
| Magnesium hazard (MH) | High magnesium concentration negatively influences crop yields as the soil becomes more alkaline | $MH = \left(\frac{Mg^{2+}}{Ca^{2+} + Mg^{2+}}\right) \times 100$ |
| Percent sodium (%Na) | High sodium percent leads to aggregates of some soils, causing the sealing of soil pores and a reduction in permeability | $\% Na = \left[\frac{Na^+ + K^+}{Ca^{2+} + Mg^{2+} + Na^+ + K^+}\right] \times 100$ |
| Kelly's ratio (KR) | Water with KR > 1 suggests unsuitability for irrigation due to the hazardous alkalinity of treated effluent | $KR = \frac{Na^+}{Ca^{2+} + Mg^{2+}}$ |
| Chloro-alkaline indices (CAI) | Regulates the chemical and pollutant transport in the aquifer and soils | $CAI = \left[\frac{Cl^- - Na^+ + K^+}{Cl^-}\right]$ |

electro-catalytic conversion of organic contaminants to hydrogen is more thermodynamically favoured than generating hydrogen from water splitting (Rioja-Cabanillas et al., 2021).

Production of yarns with 80% recycled uniforms could reduce energy use and $CO_2$ emissions by 42%–33% relative to a non-recycled thread (Koszewska, 2018). Developments in the textile sector are also growing to increase energy recovery efficiency. For instance, cotton waste with high cellulose content (85%–95% dry wt.) could be used for biofuels. Using N-methylmorpholine-N-oxide (NMMO) to treat cotton/polyester blends has also been shown to generate biogas (methane) (Menon and Rao, 2012; Samsami et al., 2020). In addition, heat energy could be produced by thermal treatment of waste textile using the combustion technology fluidised bed such as a combined heat and power plant (CHP) in Sweden (Okafor et al., 2021).

In sum, considering effluent reuse, although recycling of textile effluents has recently attracted attention, its reuse is a daunting task in the textile sector due to the features of textile effluents as the heaviest water polluters. Hence, the degree of effluent reuse adopted differs between various industrial sectors and is strongly dependent on typical situations, such as the local circumstances, type of manufacturing plants, WWTP technologies, and the applications and scales where treated effluent is used. However, effluent reuse near the manufacturing processes has been favoured and applied for decades (cooling towers, galvanic/paper industry, etc.). Also, if textile effluents are thoroughly processed based on CP concepts, it is possible to reuse up to 90%.

## 19.6 Socioeconomic awareness and future perspectives

Textile effluents are the heaviest contaminated waters to be treated due to their features of high colour density, salinity, turbidity, COD/BOD values, and high concentration of recalcitrant organics (i.e., more than 3600 different dyes and 8000 chemicals used in this sector) (Hussain and Wahab, 2018). As a matter of fact, textile effluents directly threaten the environment and human health if left untreated. Hence, a sharp increase in awareness regarding the environmental aspects of the textile industry became noticeable over the last 2 decades of the 20th century. This is mainly due to the immense pressure put on the textile industry to achieve ZLD as the legislation regarding environmental protection has been tightened.

In this context, evaluating the complete ecological aspect of the textile sector is challenging due to variation in the manufacturing and post-use phases of effluents based on the implemented technologies and use location, which, in turn, change the carbon/water footprint (Kumar et al., 2017). For instance, as far as textile recycling is concerned, many obstacles prompt a low recycling efficiency rate. In terms of water footprint, water consumption can decline from 290 to 15 $m^3/t$ in the dyeing process if the manufacturer switches from the paddle dying to the cold pad-batch dyeing process (Ren, 2000). However, from the process perspective, cotton fabric is responsible for 2.6% of global water withdrawal. Also, it has been shown that the blue and grey water footprints for the textile sector can reach 135 and 843 L per kg fabric, respectively. In sum, global statistics indicate that 4% of global freshwater withdrawal can be attributed to textile production (including cotton farming), which uses around 93 BCM of water annually.

Considering the location aspect, energy consumption during the fabrication of 1 kg of polyester could reach 127 MJ in the United States to 104 MJ in Europe and 109 MJ in the UK (Collins and Aumônier, 2002; Cherrett et al., 2005). Moreover, approximately >70% of total energy could be consumed during the use-phase of cotton fibre-based apparel, which varies based on the country or user behaviour (Yasin et al., 2016). Based on a waste management strategy (LCA analysis), it is difficult to attain a realistic picture of the current situation for the already implemented textile (waste) water reuse purposes. The current situation shows that most textile companies prefer to use high-quality fresh water during all production processes. Therefore, wastewater is discharged into the sewage system without proper treatment. This suggests that the costs of developing technologies for WWTPs are unaffordable due to the often long payback periods. In Germany, for example, around 11% of the total textile industry plants (127) discharged their effluent directly into surface waters, while the remaining plants released their effluents into municipal wastewater streams (Vajnhandl and Valh, 2014; Koszewska, 2018). Hence, the efficiency of textile recycling is highly dependent on the strength of the management system.

Indeed, in the textile field, the major effluent management and reuse/recycling practices are the collection, physical mixing, neutralisation, and transportation to municipal WWTPs, while their penetration into the reuse cycle for production remains challenging. In fact, textile effluent/waste cannot become a valuable resource without a 'culture' of waste fractionation and sorting at source. In this context, only a few countries are currently able to reuse ≥50% of their potentially recyclable wastes. For instance, the rate of the recycling process in Brazil, China, Russia, and South Africa is less than 12%, 20%, 11%, and 10%, respectively. In developed economies (e.g., France, Canada, and Italy), the reuse/recycling rate could reach up to 32%–43%. Unfortunately, in recent times, the textile recycling rate in some EU countries has declined by 10%–20%, especially in the UK, Spain, France, Germany, and Belgium (Leal Filho et al., 2019). The significant barriers to textile recycling include: economic viability, textile waste composition, technological limitation, lack of and/or limited LCA information, and weak management policies.

In terms of economic viability, most virgin textile effluents are unsuitable for recirculation purposes due to the high operation cost required to generate good-quality water for target applications. In addition, effluent recycling is not always feasible due to difficulties in separating and removing the various types of textile pollutants, such as heavy metals, organic dyes, chemical additives, etc. Furthermore, increased textile pollution strength complicates the treatment process, and increases energy consumption and the environmental carbon footprint of the effluent. Moreover, the high concentrations of dissolved chemicals and dyes hinder recycling by lowering the water quality (i.e., the presence of toxic textile dyes and heavy metals).

Added to that, the lack of technology/methods for the recycling process is another major reason hindering the effective reuse/cycling process of textile waste effluents. It is to be noted that most existing WWTP technologies/practices have low efficacy in removing dyes and other dissolved contaminants from waste effluents, requiring significant investment in the RIA plan to overcome existing limitations and introduce more efficient and effective techniques. It should also be noted that public awareness of the merits of effluent recycling is limited, causing market inefficiency and contributing to a low reuse rate. It is also reported that waste recycling policies may be ineffective without a change in citizens' behaviour.

Similarly, inadequate knowledge of the recyclability of textile effluents and the composition of potentially dangerous components could negatively affect the process efficiency and recycling rate. There are also two fundamental consumer-related barriers to recycling: the attitude barrier (non-commitment to the ideals of recycling) and the knowledge barrier (ignorance of what to recycle), which negatively impact governmental management systems. The absence of integrated policies and frameworks or generally accepted standards to improve the overall textile efficiency could significantly influence the effluent recycling eco-system. These barriers discourage investment in textile recycling, making recirculation and reuse not economically viable and limited.

Therefore, there is an urgent need to formulate harmonised and regulation policies to address the technical standards related to the development of the textile recycling process. In addition, the re-orientation of public/government awareness to view textile effluents as a potential raw material should be considered to provide sustainable social, economic, and environmental benefits. Such benefits could be attained by developing the waste management system (especially in developing countries), updating effective legislation, changing consumer behaviour and environmental awareness, and increasing technical ability. This is due to the lack of standardised LCA studies for the textile chain due to reliance on the data used, assumptions, the considered impact categories, and the chosen LCA method (Leal Filho et al., 2019).

Also, the lack of data consistency and the availability of inventories on the consequence of textile chemicals continue to hamper the LCA of the textile industry's environmental/water footprint. On the above basis, to achieve 'textile water-efficiency', it is important to strengthen collaboration in research, innovation, and action (RIA) plans at each industrial plant based on locally available LCA data analysis. Based on the WssTP, it is also recommended to accelerate the innovation transfer to the market related to large-scale demonstrators for maximising the research impact on textiles for sustainability achievement. As the latest RIA action, the European Innovation Partnership (EIP) on water-related challenges is aimed at supporting and facilitating the development of innovative action solutions for water-reuse/recycling purposes. In this relationship, the RIA-EIP strategy aimed to bring all relevant problem-oriented approaches together to identify the priority areas for action, identify those barriers to innovation, and propose solutions for overcoming those barriers, following the existing European, national, and regional initiatives.

Based on the WssTP, it is also recommended to accelerate innovation transfer to the textile-related market for sustainability. Also, as the most recent action of the RIA, the European Innovation Partnership (EIP) on Water-related challenges aims to support and facilitate the development of innovative business solutions for water reuse/recycling purposes. In this relationship, the RIA-EIP Strategy aims to bring together relevant problem-oriented approaches together to: (i) identify priority areas for action, (ii) identify barriers to innovation based on the detailed physicochemical description of waste streams and their separation and segregation concepts, and (iii) propose solutions to overcome these barriers in accordance with existing national and regional initiatives.

## 19.7 Conclusion

The textile sector is one of the oldest and most technologically complex industries worldwide, and it is estimated that the textile sector consumes around 79–93 billion cubic metres of water per year, which is 4% of all freshwater extraction globally. By 2030, water consumption in the textile sector is expected to increase by 50%. Due to the rapid growth of the textile sector in the future, these expectations raise many concerns about the pollution of the aquatic environment, which subsequently makes water a limited and scarce resource and imposes a costly burden on productivity. It is estimated that approximately 0.069–0.317 $m^3$ of liquid waste/kg of fabric produced is generated daily and discharged into the aquatic environment. This textile effluent waste is one of the most polluting sources for the environment, containing a high load of contaminants such as dyes, heavy metals, dissolved solids, detergents, surfactants, biocides, suspended matter, and more.

However, it should be noted that the characteristic/composition of textile effluent varies with respect to the (1) type of fabrics produced, (2) textile processing steps, (3) origin of service activities (i.e., location), and (4) effectiveness of technologies implemented in textile wastewater treatment plants (WWTPs). In particular, based on the World Bank estimation, dyeing, and finishing processes produce 17%–20% of textile effluents and are characterised by high BOD (1700–5200 mg/L) and COD (4600–5900 mg/L) values. Due to its complex constituents and the high load of toxic contaminants, it is difficult to effectively treat textile effluent with traditional technologies implemented in textile WWTPs. Accordingly, there is a great challenge associated with the sustainable management of textile effluents due to their negative impacts on the environment, economy, and society, following the 2030 Sustainable Development Goals (SDGs).

To overcome the concerns and challenges associated with textile effluent management, it is vital to develop industrial WWTPs with innovative new technologies, such as advanced chemical oxidation processes, membrane bioreactor (MBR), photocatalytic membrane reactor (PMR), and direct contact membrane distillation (DCMD), to reduce the negative impacts of toxic pollutants in textile effluents. It is also important to develop consumer/stakeholder knowledge about effective waste management to achieve a more sustainable consumption scenario. Some sustainable alternatives suggested to treat the textile effluents include adsorption and evaporation techniques to improve the recovery of dye (and other auxiliaries) from the discharged wastewater solutions. Adding to that, the textile industry must evolve in such a way as to develop only environmentally friendly products to avoid any toxicity as efficiently as possible in the future.

In response to these requirements, dyestuff manufacturers have developed new types of dyeing aids to comply with SDGs, such as: (i) non-alkyl phenol-derived emulsions, (ii) non-alkyl phenol-derived dye alternatives to halogens or heavy metals, (iii) chlorine-free bleaching agents, and (iv) natural and synthetic thickening agents to reduce the amount of wasted dye in the textile processing.

Also, a number of stricter regulations have also been set in many countries to minimise the harmful effects of the textile sector on the environment, clean water availability, and human health, such as ETAD (Environmental and Toxicological Association of the Dye Manufacturing Industry) in Great Britain by the Environment Agency. Textile sector regulations in various countries, including Canada, the USA and European Union also require the assessment of dye, heavy metal and harmful additive chemical concentrations in treated wastewater prior to discharge into the environment. However, it should be noted that although the European Commission (EC) imposed several strict regulations regarding the discharge of textile effluents into the environment, there is still no international consensus on acceptable standards for discharging textile effluents into natural ecosystems.

Hence, over the last few years, the textile sector has embraced drastic changes and rapid innovations in the generation, treatment, and reuse of textile effluents. Also, reclamation and reuse of textile effluents have become an attractive option of great importance to expand the available water resources (non-conventional resources) sustainably and to protect the environment. In this regard, a number of practices have been proposed for the safe reuse of treated wastewater over the past decade, such as (i) recreational holding, (ii) urban uses in toilet flushing, (iii) green spaces or nonedible crop irrigation, (iv) industrial applications, and (v) augmenting the water supply through reservoir/groundwater recharge.

In addition, with the development of the machine learning (ML) approach, many studies have applied artificial neural networks (ANNs) and artificial intelligence (AI) to optimise and integrate the newly developed water treatment technologies in industrial WWTPs to minimise waste generation by implementing lower liquor-ratio dyeing machines, and/or using low/no-water textile processing. In fact, within the strategic plan of the Water Supply and Sanitation Technology Platform (WssTP) of the EC, the reuse practice of reclaimed textile effluent is often a reliable option to meet the hydraulic deficit in many regions worldwide. Also, it could be effectively reused as a raw source for energy production (hydrogen energy), a valuable chemical resource, and clean water in wet processing (e.g., desizing of new grey fabric), achieving the ZLD concept. Nonetheless, as the textile industry is an energy-intensive sector, recycling textile effluents should increase thermal energy consumption, which, in turn, increases the emission of greenhouse gas (GHG). To minimise this challenge, the production of green, renewable energy from textile effluent has attracted the attention of scholars concerned about energy recovery from waste resources to achieve the goal of sustainability in effluent recycling. In sum, this chapter reveals that the trend of recycling effluents in the textile industry is an effective way to achieve sustainability goals, while more efforts are needed by water technologists and textile industry experts to reduce water consumption in the industry, considering the circular economy concept.

# References

Abel, A., 2012. The history of dyes and pigments: from natural dyes to high performance pigments. In: Colour Design. Elsevier, pp. 557–587.

Abid, M.F., Zablouk, M.A., Abid-Alameer, A.M., 2012. Experimental study of dye removal from industrial wastewater by membrane technologies of reverse osmosis and nanofiltration. Iranian Journal of Environmental Health Science & Engineering 9 (1), 1–9.

Abubacker, M.N., Kirthiga, B., 2017. Long Term Impact of Irrigation with Textile Waste Water and an Ecofriendly Approach for Heavy Metal Degradation, pp. 133–161. https://doi.org/10.1007/978-3-319-48439-6_12.

Adane, T., Adugna, A.T., Alemayehu, E., 2021. Textile industry effluent treatment techniques. Journal of Chemistry 2021. https://doi.org/10.1155/2021/5314404.

Agidi, B.M., et al., 2022. Water quality index, hydrogeochemical facies and pollution index of groundwater around Middle Benue Trough, Nigeria. International Journal of Energy and Water Resources. https://doi.org/10.1007/s42108-022-00187-z.

Akter, T., et al., 2016. Water Quality Index for measuring drinking water quality in rural Bangladesh: a cross-sectional study. Journal of Health, Population and Nutrition 35 (1), 4. https://doi.org/10.1186/s41043-016-0041-5.

Alavi, J., et al., 2022. A new insight for real-time wastewater quality prediction using hybridized kernel-based extreme learning machines with advanced optimization algorithms. Environmental Science and Pollution Research 29 (14), 20496–20516. https://doi.org/10.1007/s11356-021-17190-2.

Ardila-Leal, L.D., et al., 2021. A brief history of colour, the environmental impact of synthetic dyes and removal by using laccases. Molecules 26 (13). https://doi.org/10.3390/molecules26133813.

Bahadur, N., Bhargava, N., 2019. Novel pilot scale photocatalytic treatment of textile & dyeing industry wastewater to achieve process water quality and enabling zero liquid discharge. Journal of Water Process Engineering 32. https://doi.org/10.1016/j.jwpe.2019.100934.

Bakar, N.A., et al., 2020. Physico-chemical water quality parameters analysis on textile. In: IOP Conference Series: Earth and Environmental Science. Institute of Physics Publishing. https://doi.org/10.1088/1755-1315/498/1/012077.

Bakar, N.A., et al., 2021. An insight review of lignocellulosic materials as activated carbon precursor for textile wastewater treatment. Environmental Technology & Innovation 22, 101445.

Barnett, J.R., Miller, S., Pearce, E., 2006. Colour and art: a brief history of pigments. Optics & Laser Technology 38 (4–6), 445–453.

Becker, D., 2016. Color Trends and Selection for Product Design: Every Color Sells a Story. William Andrew.

Behrmann, R., Gonzales, J., 2009. Les colorants de l'art paléolithique dans les grottes et en plein air. L'anthropologie 113, 559–601.

Berke, H., 2002. Chemistry in ancient times: the development of blue and purple pigments. Angewandte Chemie International Edition 41 (14), 2483–2487.

Berke, H., Wiedemann, H.G., 2000. The chemistry and fabrication of the anthropogenic pigments Chinese blue and purple in ancient China. East Asian Science, Technology, and Medicine 17, 94–120.

Bhattacharya, A., Ambika, S., 2022. Progression and Application of Photocatalytic Membrane Reactor for Dye Removal: An Overview. In: Membrane Based Methods for Dye Containing Wastewater, pp. 49–77.

Bidu, J.M., et al., 2021. Current status of textile wastewater management practices and effluent characteristics in Tanzania. Water Science and Technology 83 (10), 2363–2376. https://doi.org/10.2166/wst.2021.133.

Bisschops, I., Spanjers, H., 2003. Literature review on textile wastewater characterisation. Environmental Technology 24 (11), 1399–1411. https://doi.org/10.1080/09593330309385684.

Cai, Y.J., Choi, T.M., 2020. A United Nations' Sustainable Development Goals perspective for sustainable textile and apparel supply chain management. Transportation Research Part E: Logistics and Transportation Review 141 (June), 102010. https://doi.org/10.1016/j.tre.2020.102010.

Chekalin, E., et al., 2019. Changes in biological pathways during 6,000 years of civilization in Europe. Molecular Biology and Evolution 36 (1), 127–140.

Cherrett, N., et al., 2005. Ecological Footprint and Water Analysis of Cotton, Hemp and Polyester. Stockholm Environmental Institute.

Collins, M., Aumônier, S., 2002. Streamlined Life Cycle Assessment of Two Marks & Spencer Plc Apparel Products. Marks & Spencer plc.

de Oliveira Neto, G.C., et al., 2019. Cleaner production in the textile industry and its relationship to sustainable development goals. Journal of Cleaner Production 228, 1514–1525. https://doi.org/10.1016/j.jclepro.2019.04.334.

de Oliveira Neto, G.C., et al., 2022. Overcoming barriers to the implementation of cleaner production in small enterprises in the mechanics industry: exploring economic gains and contributions for sustainable development goals. Sustainability 14 (5), 2944. https://doi.org/10.3390/su14052944.

de Sousa Silva, R., et al., 2020. Steady state evaluation with different operating times in the direct contact membrane distillation process applied to water recovery from dyeing wastewater. Separation and Purification Technology 230, 115892.

Deng, D., et al., 2020. Textiles wastewater treatment technology: a review. Water Environment Research 92 (10), 1805–1810. https://doi.org/10.1002/wer.1437.

Dey, S., Islam, A., 2015. A review on textile wastewater characterization in Bangladesh. Resources and Environment 5 (1), 15–44. https://doi.org/10.5923/j.re.20150501.03.

El-Abbassi, A., et al., 2013. Integrated direct contact membrane distillation for olive mill wastewater treatment. Desalination 323, 31–38.

El-Fawal, E.M., et al., 2020. Preparation of solar-enhanced AlZnO@carbon nano-substrates for remediation of textile wastewaters. Journal of Environmental Sciences 92, 52–68. https://doi.org/10.1016/j.jes.2020.02.003.

El-Gohary, F., et al., 2013. A new approach to accomplish wastewater regulation in textile sector: an Egyptian case study. Cellulose Chemistry & Technology 47, 309–315.

Fernández-Sabido, S., et al., 2012. Comparative study of two blue pigments from the Maya region of Yucatan. MRS Online Proceedings Library 1374, 115–123.

Fortunato, L., et al., 2021. Textile dye wastewater treatment by direct contact membrane distillation: membrane performance and detailed fouling analysis. Journal of Membrane Science 636, 119552.

Gähr, F., Hermanutz, F., Oppermann, W., 1994. Ozonation-an important technique to comply with new German laws for textile wastewater treatment. Water Science and Technology 30 (3), 255.

Goswami, L., et al., 2018. Membrane bioreactor and integrated membrane bioreactor systems for micropollutant removal from wastewater: a review. Journal of Water Process Engineering 26 (October), 314–328. https://doi.org/10.1016/j.jwpe.2018.10.024.

Hakanson, L., 1980. An ecological risk index for aquatic pollution control.a sedimentological approach. Water Research 14 (8), 975–1001. https://doi.org/10.1016/0043-1354(80)90143-8.
Han, G., et al., 2018. Low-pressure nanofiltration hollow fiber membranes for effective fractionation of dyes and inorganic salts in textile wastewater. Environmental Science and Technology 52 (6), 3676–3684. https://doi.org/10.1021/acs.est.7b06518.
Harane, R.S., Adivarekar, R.V., 2017. Sustainable processes for pre-treatment of cotton fabric. Textiles and Clothing Sustainability 2 (1), 1–9.
Hassan, M.M., Carr, C.M., 2018. A critical review on recent advancements of the removal of reactive dyes from dyehouse effluent by ion-exchange adsorbents. Chemosphere 209, 201–219.
Holkar, C.R., et al., 2016. A critical review on textile wastewater treatments: possible approaches. Journal of Environmental Management 182, 351–366.
Hossain, L., Khan, M.S., 2020. Water footprint management for sustainable growth in the Bangladesh apparel sector. Water (Switzerland) 12 (10). https://doi.org/10.3390/w12102760.
Hovers, E., et al., 2003. An early case of color symbolism: ochre use by modern humans in Qafzeh Cave. Current Anthropology 44 (4), 491–522.
Hussain, T., Wahab, A., 2018. A critical review of the current water conservation practices in textile wet processing. Journal of Cleaner Production 198, 806–819.
Johnston, W.T., 2008. The discovery of aniline and the origin of the term "aniline dye". Biotechnic & Histochemistry 83 (2), 83–87.
Kao, C.M., et al., 2001. Regulating colored textile wastewater by 3/31 wavelength ADMI methods in Taiwan. Chemosphere 44 (5), 1055–1063.
Keßler, L., Matlin, S.A., Kümmerer, K., 2021. The contribution of material circularity to sustainability—recycling and reuse of textiles. Current Opinion in Green and Sustainable Chemistry 32, 100535. https://doi.org/10.1016/j.cogsc.2021.100535.
Koszewska, M., 2018. Circular economy—challenges for the textile and clothing industry. Autex Research Journal 18 (4), 337–347.
Kumar, V., et al., 2017. Contribution of traceability towards attaining sustainability in the textile sector. Textiles and Clothing Sustainability 3 (1), 1–10.
Kumar, P.S., et al., 2019. A critical review on recent developments in the low-cost adsorption of dyes from wastewater. Desalination and Water Treatment 172, 395–416.
Lahlou, F.-Z., Mackey, H.R., Al-Ansari, T., 2021. Wastewater reuse for livestock feed irrigation as a sustainable practice: a socio-environmental-economic review. Journal of Cleaner Production 294, 126331. https://doi.org/10.1016/j.jclepro.2021.126331.
Laqbaqbi, M., et al., 2019. Application of direct contact membrane distillation for textile wastewater treatment and fouling study. Separation and Purification Technology 209, 815–825.
Leal Filho, W., et al., 2019. A review of the socio-economic advantages of textile recycling. Journal of Cleaner Production 218, 10–20.
Li, F., et al., 2018a. Direct contact membrane distillation for the treatment of industrial dyeing wastewater and characteristic pollutants. Separation and Purification Technology 195, 83–91.
Li, M., et al., 2018b. A self-standing, support-free membrane for forward osmosis with No internal concentration polarization. Environmental Science and Technology Letters 5 (5), 266–271. https://doi.org/10.1021/acs.estlett.8b00117.
Li, M., et al., 2019. Concentration and recovery of dyes from textile wastewater using a self-standing, support-free forward osmosis membrane. Environmental Science and Technology 53 (6), 3078–3086. https://doi.org/10.1021/acs.est.9b00446.

Li, F., et al., 2021. Metals pollution from textile production wastewater in Chinese southeastern coastal area: occurrence, source identification, and associated risk assessment. Environmental Science and Pollution Research 28 (29), 38689–38697. https://doi.org/10.1007/s11356-021-13488-3.

Lin, S.H., Chen, M.L., 1997. Treatment of textile wastewater by-chemical methods for reuse. Water Research 31 (4), 868–876. https://doi.org/10.1016/S0043-1354(96)00318-1.

Lin, J., et al., 2015. Fractionation of direct dyes and salts in aqueous solution using loose nanofiltration membranes. Journal of Membrane Science 477, 183–193. https://doi.org/10.1016/j.memsci.2014.12.008.

Lin, J., et al., 2019. Sustainable management of textile wastewater: a hybrid tight ultrafiltration/bipolar-membrane electrodialysis process for resource recovery and zero liquid discharge. Industrial & Engineering Chemistry Research 58 (25), 11003–11012. https://doi.org/10.1021/acs.iecr.9b01353.

Lu, X., et al., 2010. Textile wastewater reuse as an alternative water source for dyeing and finishing processes: a case study. Desalination 258 (1–3), 229–232. https://doi.org/10.1016/j.desal.2010.04.002.

Ma, J., et al., 2021. Pilot-scale study on catalytic ozonation of bio-treated dyeing and finishing wastewater using recycled waste iron shavings as a catalyst. Ozonation is an ideal technique due to the high oxidation potential of ozone (2.07 V), environment-friendly product of Scientific Reports 8 (1), 3. https://doi.org/10.1080/15226514.2017.1337076. Edited by R.P. Singh, A.S. Kolok, and S.L. Bartelt-Hunt.

Madhav, S., et al., 2018. A review of textile industry: wet processing, environmental impacts, and effluent treatment methods. Environmental Quality Management 27 (3), 31–41. https://doi.org/10.1002/tqem.21538.

Menon, V., Rao, M., 2012. Trends in bioconversion of lignocellulose: biofuels, platform chemicals & biorefinery concept. Progress in Energy and Combustion Science 38 (4), 522–550.

Mishra, S., Cheng, L., Maiti, A., 2021. The utilization of agro-biomass/byproducts for effective bio-removal of dyes from dyeing wastewater: a comprehensive review. Journal of Environmental Chemical Engineering 9 (1), 104901.

Monika, et al., 2021. Application of wastewater in irrigation and its regulation with special reference to agriculture residues. In: Water Pollution and Management Practices. Springer Singapore, pp. 177–199. https://doi.org/10.1007/978-981-15-8358-2_8.

Okafor, C.C., et al., 2021. Sustainable management of textile and clothing. Clean Technologies and Recycling 1 (1), 70–87. https://doi.org/10.3934/ctr.2021004.

Othman, M.H.D., et al., 2021. Advanced membrane technology for textile wastewater treatment. In: Membrane Technology Enhancement for Environmental Protection and Sustainable Industrial Growth. Springer, pp. 91–108.

O'Neill, C., et al., 1999. Colour in textile effluents - sources, measurement, discharge consents and simulation: a review. Journal of Chemical Technology and Biotechnology 74 (11), 1009–1018. https://doi.org/10.1002/(SICI)1097-4660(199911)74:11<1009::AID-JCTB153>3.0.CO;2-N.

Pang, Y.L., Abdullah, A.Z., 2013. Current status of textile industry wastewater management and research progress in Malaysia: a review. Clean - Soil, Air, Water 41 (8), 751–764. https://doi.org/10.1002/clen.201000318.

Panigrahi, T., Santhoskumar, A.U., 2020. Adsorption process for reducing heavy metals in Textile Industrial Effluent with low cost adsorbents. Progress in Chemical and Biochemical Research 3, 135–139.

Paul, R., Ramesh, K., Ram, K., 1995. Effluent treatment of textile waste waters. Textile Dyer and Printer 28 (24), 18.

Paz, A., et al., 2017. Biological treatment of model dyes and textile wastewaters. Chemosphere 181, 168–177.

Provin, A.P., et al., 2021. Textile industry and environment: can the use of bacterial cellulose in the manufacture of biotextiles contribute to the sector? Clean Technologies and Environmental Policy 23 (10), 2813–2825. https://doi.org/10.1007/s10098-021-02191-z.

Pugh, T.W., Cecil, L.G., 2012. The contact period of central Petén, Guatemala in color. Res: Anthropology and Aesthetics 61 (1), 315–329.

Rajput, H., et al., 2021. Photoelectrocatalysis as a high-efficiency platform for pulping wastewater treatment and energy production. Chemical Engineering Journal 412. https://doi.org/10.1016/j.cej.2021.128612.

Ramlow, H., Machado, R.A.F., Marangoni, C., 2017. Direct contact membrane distillation for textile wastewater treatment: a state of the art review. Water Science and Technology 76 (10), 2565–2579. https://doi.org/10.2166/wst.2017.449.

Ramlow, H., et al., 2020. Dye synthetic solution treatment by direct contact membrane distillation using commercial membranes. Environmental Technology 41 (17), 2253–2265.

Rashid, R., et al., 2021. A state-of-the-art review on wastewater treatment techniques: the effectiveness of adsorption method. Environmental Science and Pollution Research 28 (8), 9050–9066.

Reddy, S., Osborne, W.J., 2020. Heavy metal determination and aquatic toxicity evaluation of textile dyes and effluents using *Artemia salina*. Biocatalysis and Agricultural Biotechnology 25, 101574.

Ren, X., 2000. Development of environmental performance indicators for textile process and product. Journal of Cleaner Production 8 (6), 473–481.

Rioja-Cabanillas, A., et al., 2021. Hydrogen from wastewater by photocatalytic and photoelectrochemical treatment. Journal of Physics: Energy 3 (1), 12006. https://doi.org/10.1088/2515-7655/abceab.

Robinson, T., et al., 2001. Remediation of dyes in textile effluent: a critical review on current treatment technologies with a proposed alternative. Bioresource Technology 77 (3), 247–255.

Samsami, S., et al., 2020. Recent advances in the treatment of dye-containing wastewater from textile industries: overview and perspectives. Process Safety and Environmental Protection 138–163. https://doi.org/10.1016/j.psep.2020.05.034.

Sarayu, K., Sandhya, S., 2012. Current technologies for biological treatment of textile wastewater-A review. Applied Biochemistry and Biotechnology 167 (3), 645–661. https://doi.org/10.1007/s12010-012-9716-6.

Sathya, U., et al., 2021. Development of photochemical integrated submerged membrane bioreactor for textile dyeing wastewater treatment. Environmental Geochemistry and Health 43 (2), 885–896.

Saxena, S., Raja, A.S.M., Arputharaj, A., 2017. Challenges in Sustainable Wet Processing of Textiles, pp. 43–79. https://doi.org/10.1007/978-981-10-2185-5_2.

Schneider, R., 2006. Recycling waste water from textile production. In: Recycling in Textiles, A Volume. Woodhead Publishing Series in Textiles, pp. 73–94. https://doi.org/10.1533/9781845691424.2.73.

Shelley, T.R., 1994. Dye pollution clean-up by synthetic mineral. International Dyer 79, 26–31.

Sivakumar, D., et al., 2013. Constructed wetland treatment of textile industry wastewater using aquatic macrophytes. International Journal of Environmental Sciences 3 (4), 1223.

Tkaczyk, A., Mitrowska, K., Posyniak, A., 2020. Synthetic organic dyes as contaminants of the aquatic environment and their implications for ecosystems: a review. The Science of the Total Environment 717, 137222.

Upadhye, V.B., Joshi, S.S., 2012. Advances in wastewater treatment- A review. International Journal of Chemical Sciences and Applications 3, 264–268.

Vajnhandl, S., Valh, J.V., 2014. The status of water reuse in European textile sector. Journal of Environmental Management 29–35. https://doi.org/10.1016/j.jenvman.2014.03.014. Academic Press.

Valdez-Vazquez, I., et al., 2020. Simultaneous hydrogen production and decolorization of denim textile wastewater: kinetics of decolorizing of indigo dye by bacterial and fungal strains. Brazilian Journal of Microbiology 51 (2), 701–709. https://doi.org/10.1007/S42770-019-00157-4/FIGURES/4.

Vandevivere, P.C., Bianchi, R., Verstraete, W., 1998. Treatment and reuse of wastewater from the textile wet-processing industry: review of emerging technologies. Journal of Chemical Technology & Biotechnology: International Research in Process, Environmental AND Clean Technology 72 (4), 289–302. https://doi.org/10.1002/(sici)1097-4660(199808)72:4<289::aid-jctb905>3.0.co;2-%23.

Velusamy, S., et al., 2021. A review on heavy metal ions and containing dyes removal through graphene oxide-based adsorption strategies for textile wastewater treatment. Chemical Record 21 (7), 1570–1610. https://doi.org/10.1002/tcr.202000153.

Xie, Y., et al., 2022. Enhancing real-time prediction of effluent water quality of wastewater treatment plant based on improved feedforward neural network coupled with optimization algorithm. Water 14 (7), 1053. https://doi.org/10.3390/w14071053.

Yaseen, D.A., Scholz, M., 2019. Textile dye wastewater characteristics and constituents of synthetic effluents: a critical review. International Journal of Environmental Science and Technology 16 (2), 1193–1226. https://doi.org/10.1007/s13762-018-2130-z.

Yasin, S., et al., 2016. Global consumption of flame retardants and related environmental concerns: a study on possible mechanical recycling of flame retardant textiles. Fibers 4 (2), 16.

Younis, S.A., Serp, P., Nassar, H.N., 2020. Photocatalytic and biocidal activities of ZnTiO2 oxynitride heterojunction with MOF-5 and g-C3N4: a case study for textile wastewater treatment under direct sunlight. Journal of Hazardous Materials, 124562. https://doi.org/10.1016/j.jhazmat.2020.124562.

Zengınbal, H., Okcu, G.D., Yalcuk, A., 2018. The impact of textile wastewater irrigation on the growth and development of apple plant. International Journal of Phytoremediation 20 (2), 153–160. https://doi.org/10.1080/15226514.2017.1337076.

Zhao, S., Wang, Z., 2017. A loose nano-filtration membrane prepared by coating HPAN UF membrane with modified PEI for dye reuse and desalination. Journal of Membrane Science 524, 214–224. https://doi.org/10.1016/j.memsci.2016.11.035.

Zhezhova, S., et al., 2021. The development of textile industry in Shtip. Tekstilna industrija 69 (4), 14–19. https://doi.org/10.5937/tekstind2104014k.

# Sustainability aspects, LCA and ecolabels

*Siva Rama Kumar Pariti*[1], *Umesh Sharma*[2], *Laxmikant Jawale*[3] *and Sujata Pariti*[4]
[1]BluWin Ltd, Huddersfield, United Kingdom; [2]Sweet Merchandising Services Ltd, Kent, United Kingdom; [3]Apparel Impact Institute, Kalyan West, Maharashtra, India; [4]Independent Consultant, Navi Mumbai, Maharashtra, India

## 20.1 Introduction

Sustainable living is an important ideology in the modern world. In this chapter, the concept of sustainability and its relevance to the textile industry are explored. Across the world and throughout history, textiles have been the base of some of the most lucrative economies and networks of exchange in the world. Historically too, for many centuries, textiles-related trade was well known, and the silk road and trade through the silk road are one such example. Textiles are important, and it is not just because they are needed to stay warm, but also due to their feel and elegance. Similar to developing responsible habits in the purchase and consumption of food products, it is important to have an awareness of the impact of textile production and consumption on the environment. Textiles provide a good and elegant look, but it is important to feel good about them too. Accordingly, many people around the world are pushing for more sustainable textiles or ethically produced fabrics.

The current state of the textile supply chain is not predominantly based in Western countries. During the 1900s till the 1980s the chemicals and dyestuffs manufacturing and innovations were dominated by few countries in the West but later, manufacturing moved to developing countries like India, China and Brazil, which have historically provided raw materials to Western markets. Outsourcing to factories in the new industrial countries makes it increasingly difficult to manage the proper balance between compliance, ethical practices and best manufacturing practices. During the industrial revolution of the 19th century, many instances of human rights abuse and environmental destruction took place, and they are now resurfacing. During the later decades of the 20th century, a movement of environmental awareness grew in textiles, farming and wider industrial production. This has grown into a 21st-century phenomenon where some of the most creative minds in textiles are working on some of the most pressing issues in textile manufacture.

Chemicals and textiles, if not manufactured using best practices, could cause harm to human health and the environment. Dyestuffs, chemicals and textile wet processing and leather tanneries could pollute the environment if proper chemical management, wastewater treatment and air emission controls are not in place. More often, it is observed that industrial wastewater is discharged with lots of pollutants, entering

Sustainable Innovations in the Textile Industry. https://doi.org/10.1016/B978-0-323-90392-9.00002-1

into water bodies and sludge without being properly disposed of. It is for this reason that sustainability is very important in the textile sector. Sustainability is referred to as the sustainable development of textile companies that will meet current needs without harming the environment and will enable future generations to meet their own needs.

## 20.2 Historical aspects and current scenario

Historically, manufacturing processes have always been focused on productivity, quality, optimisation and profitability aspects, and were not mainly focused on their impact on the environment, how to manage the waste or circularity aspects. The manufacturing processes have also not matured enough to understand the impacts on the environment and climate, and to conduct the life cycle and impact assessments during the earlier years of manufacturing.

In general, for the textile manufacturing sector, the negative impacts associated were well summarised by Draper et al., into five main categories (DEFRA, 2008; Draper et al., 2007): (1) energy consumption in the production of primary materials (especially man-made fibres), in yarn manufacturing, in yarn and fabric finishing, in washing and drying clothes, as well as in marketing and sales processes; (2) water consumption associated with raw materials growth, production stage pretreatment chemicals, dyes and finishes, as well as with laundry; (3) use of chemicals (especially in wet pretreatment, dyeing, finishing and laundry) and their release into water; (4) solid waste arising from yarn, fabric and final product manufacturing, packaging and disposal of products at the end of their life; and (5) direct $CO_2$ emissions, particularly related to transportation processes.

The present-day generation is well informed and highly connected with planet Earth and is looking at the full impact of every process and product being produced. Hence, manufacturing with production, profit and process alone is not good enough but needs also to look into the additional aspects of the impacts on the environment, health, safety and ethical aspects. It is very important that one should look at the bigger picture and overall impact of each product and process in relation with the environment and end-of-life aspects, as well as in order to create a positive impact for future generations. There are many organisations currently studying all aspects of manufacturing in order to evaluate the overall impacts of the products that are being produced.

The impact assessments, life cycle assessments, end-of-life scenarios etc., are looked at before any product or process for manufacturing is considered. Hence, the scope, assessment methodology, system boundaries, assumptions considered, data collection and assessment, analysis processes and final impact need to be evaluated properly for each process in order to not face greenwashing claims. The world today is experiencing a very high level of consumption of textile materials per person. The more that is used, the more will be produced and the more will be the negative impact on the environment. The textiles purchased and used by consumers often contain chemicals, harmful for human health, and after washing, these products end up in the environment and have impacts on the ecological systems.

## 20.3 Pillars of sustainability

Sustainability can be considered to be based on three pillars: environmental, social and economic (Purvis et al., 2018).

***Environmental pillar:*** Companies place a lot of emphasis on the environmental pillar such as reducing carbon footprints, packaging waste, water usage and their overall effect on the environment. Companies have found that it can have a beneficial impact on the planet and can also have a positive financial impact. The five phases of environmental sustainability are the material, manufacturing, retail, consumption and disposal phases.

***Social pillar:*** Social sustainability focuses on the support and approval of its employees, stakeholders and the community it operates in, who can help the business in a number of ways such as raising internal morale and employee engagement, improving risk management, unlocking new markets, etc.

***Economic pillar:*** Economic sustainability is the practice that supports long-term economic growth without negatively affecting social, environmental and cultural aspects of the community. This pillar focuses on certain activities such as compliance, proper governance, risk management, etc. Fig. 20.1 shows the three pillars of sustainability.

**Figure 20.1** The three pillars of sustainability.

### 20.3.1 Sustainable practices in the textile industry

For developing sustainable textiles, the production processes should be made environmentally friendly, which means all materials and processes, inputs and outputs should be healthy and safe for people and the environment at all stages of the life cycle. Production and processing of sustainable textiles which come from renewable or recycled sources not only help to reduce the negative impacts on the environment but also supports millions of workers to earn fair wages and ensure proper working conditions.

#### 20.3.1.1 Need for sustainable practices

- The use of natural resources without impacting the overall carbon emissions has been increasing over the last few years.
- Chemicals used in the textile industry have led to an increase in water pollution which is currently leading to water scarcity.
- Textiles consume a large amount of fossil fuels during wet processing and manufacturing practices. Fossil fuels lead to an increase in the amount of global warming due to the emission of greenhouse gases.
- Apparel industries also need to eliminate environmental hazards and improve process efficiency.

## 20.4 Impacts of textile industry on health and environment

There are over 800 substances that can be used in the manufacture of apparel and footwear that can cause harm to consumers. By conducting chemical analyses of textile articles and production materials against harmful substances for human health, one can understand the impact of these chemicals. Many brands and retailers have adopted restricted substances lists (RSLs) in order to reduce or eliminate the harmful effects on consumer health. Often, it has been observed that workers have been affected by inhaling toxic fumes generated from the use of dangerous chemicals in the garment manufacturing process. These workers experience chest pain, severe headaches, vomiting and bleeding. In another case, a pregnant worker who experienced swelling was told by doctors in intensive care that her baby may not be breathing (Akhter et al., 2017). There are many chemicals which can potentially cause health hazards and the most important ones are listed below.

- Alkylphenol ethoxylates (APEOs): APEOs are now known to be endocrine disruptors as they mimic hormones in the human body, thereby disrupting the endocrine system and hence our hormonal balance. These are used extensively as wetting agents and detergents and as emulsifiers in textile chemical formulations. APEOs are also toxic to aquatic life.
- Phthalates: Often, plastisol prints contain plasticisers to make the prints flexible. These plasticisers contain a harmful chemical group called phthalates which are potential carcinogens, endocrine disruptors and toxic for reproduction. They reduce fertility and cause disorders

such as early puberty in girls. In addition, they cause diseases like asthma and thyroid disorders.

- Formaldehyde: This chemical is most often used in wrinkle-free or wash-n-wear or non-iron fabrics. The fabrics are treated with a resin finish that contains the harmful substance, formaldehyde, which is a skin sensitiser (causes dermatitis and eczema) and a suspected carcinogen.
- Perfluoroalkyl and Polyfluoroalkyl Substances (PFAS): PFAS are a diverse group of chemicals used in a wide range of consumer and industrial products. PFAS do not easily break down and some types have been shown to accumulate in the environment and in our bodies. PFAS are mainly used for oil- and water-repellent garments or stain-repellent jackets that help to protect garments from stains and dirt.
- Isocyanates: Polyurethanes contain a dangerous chemical group called isocyanates that are suspected carcinogens, which affect the lungs and are extremely irritating to the eyes and nose. Fabrics are nowadays coated with polyurethane finishes to make them more durable and abrasion-resistant (all-weatherproof) or to give them a 'leather look'.
- Allergenic disperse dyes: Disperse dyes could also be mutagenic, i.e., they can cause mutations in the gene structure of a cell. Some disperse dyes that are used to colour synthetic fibres such as polyester and nylon have been known to cause allergies and rashes on the skin if there is close and prolonged contact with the garment.
- Banned amines from azo dyes: Certain amines released from the azo dyes are carcinogenic. Some direct dyes, disperse dyes, acid dyes and pigments that are used to colour cotton, silk and wool fabrics are based on an 'azo' chemistry that can release certain compounds called 'amines' under acidic conditions. Such dyes, on intimate and prolonged contact with the skin, can be absorbed through the pores in the skin and can cleave into carcinogenic amines due to the action of enzymes in the stomach.
- Biocides: These are chemical substances used to inhibit or control the growth of harmful bacteria, moulds and germs. Many biocides are used as preservatives for leather, in dyes and pigment formulations and printing thickeners, which can be transferred to a garment or footwear. Sometimes, an antibacterial or antiodour finishing treatment is given to garments (especially sportswear) which also contains biocides. Antimould sachets are used during transportation to prevent fungal growth. Some harmful biocides used in textile chemical formulations are isothiazolinones (which are sensitising), chlorophenols (pentachlorophenol and tetra chlorophenol), dimethyl fumarate, triclosan, organotin and ortho-phenyl phenol.
- Heavy metals: Antimony used as a catalyst in the production of polyester fibre, cadmium from pigments and paints, chromium 6 from tanning agents and metal complex dyes, lead from paints, surface coatings and metallic sundries and arsenic from pesticides and pigments can lead to the presence of heavy metals in garments, leather, sundries and jewellery articles at levels much higher than those permissible or safe for human health. Natural dyeing using metallic mordants also results in heavy metal contamination on textiles and in effluents.
- Chlorinated solvents: Solvents are widely used in textile manufacturing as cleaning agents, spot cleaners, in colourant preparations, dry cleaning, degreasing and in varnishes, adhesives and resins. Some solvents are flammable, volatile and explosive. Perchloroethylene (Perc) and trichloroethylene that are used for dry cleaning have now been identified as carcinogens and are known to cause damage to the central nervous system, liver and kidney. Benzene and carbon tetrachloride – also commonly used solvents – are carcinogens and cause nervous system disorders on short-term exposure.

There are other groups of chemicals such as chlorinated and brominated flame retardants, chlorobenzenes and chlorotoluenes, polycyclic aromatic hydrocarbons, dioxins and furans, N-nitrosamines, unreacted monomers and pesticides which are also used in the production and are considered to impact human health and are restricted by many brands and retailers and considered in RSLs. In an interesting development, the Stockholm Resilience Centre led a group of 28 internationally renowned scientists to identify the nine processes that regulate the stability and resilience of the Earth system. The scientists proposed quantitative planetary boundaries within which humanity can continue to develop and thrive for generations to come. Crossing these boundaries increases the risk of generating large-scale abrupt or irreversible environmental changes. This planetary boundary framework has generated enormous interest within science, policy and practice (Stockholm Resilience Centre, 2012).

## 20.5 Sustainable innovations and circular economy in textiles

In recent years, the circular economy has become a very important part of national and international policies. There is a global awareness of the need to go circular, and the linear model of take—make—dispose is now becoming unpopular. The time to act is now, but the transformation of production and consumption systems will not happen immediately. Many organisations like the Ellen MacArthur Foundation highlight a tendency to think that there is only a simple choice between going linear and circular. The existing models, incentives and structures are primarily designed for linear production and must be rethought for a successful transition to the circular economy.

To open up circular economy opportunities, there is a need to establish the landscape for innovation. To support a circular economy, innovation needs to be pursued with that objective in mind. In order to achieve the full potential of innovation for the circular economy, specific and sustained efforts are needed to create enabling frameworks and incentives for private innovation efforts. It is also important to encourage consumers to rapidly and broadly adopt innovative and more sustainable consumption patterns. There are many innovations which are focused on circular economy principles. A few of the innovations which can be classified into the fibres, colourants/chemicals, materials, processes and others, and their traceability are elaborated below.

### *20.5.1 Fibres*

#### *20.5.1.1 Natural fibres*

Innovators are pushing the limits to search for more sustainable options in every aspect of the textiles field. In material innovations, AlgiKnit is creating eco-conscious, renewable yarns for the circular economy. Innovators are developing durable yet rapidly degradable yarns from kelp, a variety of seaweed. The extrusion process turns the biopolymer mixture into kelp-based thread that can be knitted or 3D printed to minimise waste. The material is built for everyone: a functional and accessible resource without environmental harm. The technology operates in a closed-loop product lifecycle, utilising materials with a significantly lower carbon footprint than conventional textiles (www.keellabs.com). Agricultural crop waste such as stems and leaves are purified into soft fibre bundles ready to spin into cellulosic yarns and AltMat fibres (AltMat) are one such example. The fibre named Agraloop BioFibre and the feedstock include left-overs from various food and medicinal crops including oilseed, hemp, banana and pineapple (Circular Systems). Similarly, Spinnova technology enables cellulose fibre production from wood but also from textile or agricultural waste such as wheat or barley straw (microfibrillated cellulose) without dissolution or other complex chemical processes (spinnova.com).

#### 20.5.1.2 Regenerated fibres

One innovation approach focuses on the preparation of pulp as a feedstock for the manufacture of man-made cellulosic fibre (MMCF) from waste materials. The Hurd Co. is the exclusive producer of Agrilose, a MMC feedstock pulp made from 100% agricultural waste such as plant material that is normally thrown away or burned after food crops are harvested. The material produced such as viscose/rayon, modal or lyocell fabric has a production process that uses less water and significantly less energy than conventional MMCF (The Hurd Co, n.d.). Another inventor, Renewcell, uses post-consumer waste to produce biodegradable raw material Circulose pulp (dissolving pulp product). This involves recycling technology which dissolves used cotton and other cellulose fibres (post-consumer waste) and transforms them into a new raw material, Circulose pulp. This pulp can be used to make biodegradable virgin-quality viscose, lyocell, modal, acetate and other types of regenerated fibres (Circulose).

On a similar concept, the Infinited Fiber Company uses cellulose-rich raw material – worn-out clothes, cardboard or wheat or rice straw – to produce premium-quality fibres for the textile industry. This technology breaks waste down and captures its value at the polymer level, giving it a new life as Infinna – the unique textile fibre that looks and feels like cotton and is known scientifically as cellulose carbamate fibre (Infinited Fiber,n.d.). Similarly, one technology called Circ is working towards apparel recycling and focuses on the manufacturing of products from textile waste to replace virgin materials (Circ).

#### 20.5.1.3 Synthetic alternatives

In the field of synthetic alternatives, Kintra Fibers, a US-based technology company, created a biodegradable and recyclable polyester alternative from poly butylene succinate derived from renewable feedstock. This technology makes compostable biosynthetic resin and yarns for the apparel industry by eliminating petrochemicals and microfibre pollution from the fashion industry (Kintra Fibers – and The Natural Upgrade for Synthetics, n.d.).

### Synthetics from waste/carbon capture

Carbon capture technology has a high potential to produce building blocks for polymers or chemicals used in textile steam. Synthetic fibres have been generated using materials which are obtained from waste as feedstock. Worn Again has developed a chemical recycling technology that can separate and recapture polyester (PET) and cellulose from non-rewearable textiles to produce virgin equivalents – cost-competitive polyester and cellulosic raw materials to be reintroduced into the supply chain as part of a continual process. The main focus of technology is on solving the challenging issue of converting polyester and polycotton blended textiles, and PET plastic, at their end of use, back into circular raw materials (Anon, n.d., a-d). Similarly, PurFi is another combination of the best of chemical and mechanical recycling technologies in which pre-consumer textile wastes are converted back to virgin-quality fibres. This technology can process cotton, PET, polyester/cotton blends as well as separate out elastane (PurFi, n.d.).

One of the polyester chemical recycling technologies uses an enzymatic recycling process for depolymerising the PET (polyethylene terephthalate) contained in various plastics or textiles. The possibility of infinite recycling of all types of PET waste as well as the production of 100% recycled and 100% recyclable PET products, without loss of quality, is currently being evaluated. Plastic and textile waste is now a precious raw material enabling the circular economy to become an industrial reality (Carbios, n.d.).

Another innovative technology developed by Fairbrics helps to accelerate the technology transition towards the circular industry. This technology converts waste $CO_2$ into polyester fabric using molecular chemistry and with the least environmental impact. Here, $CO_2$ is captured from industrial sources, reacted with a catalyst and solvent to generate chemicals that are used for polyester synthesis. These chemicals are polymerised to create polyester pellets which are then spun into yarn and finally into fabric. In the near future, it is expected that this technology will produce carbon-negative 100% sustainable PET (Anon, n.d., a-d). Recycling $CO_2$ from industrial off-gases such as syngas from biomass resources (e.g., municipal waste) to produce fuels, chemical feedstocks and other chemicals additionally using fermentation (Anon, n.d., a-d) techniques and enzyme (Rubi Laboratories, n.d.) systems will become more prominent in the near future.

## *20.5.2 Colourants*

### *20.5.2.1 Microorganism-based colourants*

Synthetic colourants preparation and production is considered to be one of the most polluting industries impacting on overall health and the environment. A notable innovation has been performed to derive the colourants form bacteria. Pili is developing a completely new dye technology, designing enzymatic cascades that turn renewable carbon feedstocks, such as sugar, into textile dyes. These microbial enzymes are re-engineered to produce brilliant and effective dyes from renewable resources (www.pili.bio). One similar innovator, Vienna Textile Lab (Vienna Textile Lab, n.d.), fabricates organic dyes from naturally occurring bacteria to provide a sustainable alternative to conventional synthetic colours. Yet another company, Kbcolos, is engaged in producing colourants from bacteria and focuses on biomass as an inexhaustible source, by feeding waste sources and readily available low-cost substrates to achieve consistent colour production under specific controlled conditions (Anon, n.d., a-d).

In another invention, the innovator Algalife manufactures algae-based pigments to produce an organic fibre and environmentally friendly products without waste and saving water. The additional advantage claimed by this innovator is that the dyes and fibres are created by algae microorganisms, and they release antioxidants, vitamins and minerals that can nourish and protect our body and skin (Berlin, n.d.). Currently, the majority of pigments used within industry are derived from petroleum, such as carbon black. The carbon black pigment made by algae as a colourant, Algae Ink, could replace the fossil fuel-based black pigment which is being used in very high quantities in the textile industry. This black algae-based pigment has a potential negative carbon footprint, is bio-based, resistant to UV-light exposure and safe. Living Ink is a biomaterials company on a mission to use sustainable algae technologies to replace petroleum-derived products, such as ink (Living Ink, n.d.).

Colorifix minimises the environmental impact of industrial dyeing by replacing chemistry with biology, from the creation of dyes to their fixation into fabrics. The company ships a tiny quantity of engineered microorganisms to clients who grow the colour via fermentation. The microorganisms grow on renewable feedstocks such as sugar, yeast and plant by-products. The microorganisms divide every 20 min, resulting in a large quantity of colourful dye liquor within just 1 or 2 days. This is then placed directly into standard dyeing machines, requiring no additional specialist equipment or toxic chemicals (Colorifix).

#### *20.5.2.2 Waste material to colourants*

In another segment, innovators used waste material to produce colourants. Recycrom, through a sophisticated production process, transformed fabric textile waste fibres into an incredibly fine powder that can be used as a pigment dye for fabrics and garments made of any substrate such as cotton, wool, nylon or any natural fibre and blend. Recycrom colourants are applied as a suspension using various methods: exhaustion dyeing, dipping, spray, screen printing and coating (Recycrom.com., 2018). On other hand, high-performing black pigments are created through transforming wood waste which is sourced from certified sources by Nature Coatings. This manufacturing process involves a closed-loop and circular system which emits negligible amounts of $CO_2$ and other greenhouse gases. These pigments are suitable for multiple applications including, but not limited to, screen printing, rotary printing, coatings, paint, resin casting and wood colouring (Nature Coatings, n.d.).

#### *20.5.2.3 Plant-based colourants*

The US-based innovator, Stony Creek Colors produces plant-based indigo with an optimised production in a (non-GMO) closed-loop process which has the potential to be carbon negative (Colors, n.d.). Archroma, a global leader in speciality chemicals and Stony Creek Colors entered into a strategic partnership to produce and bring to the market Stony Creek's IndiGold high-performance plant-based pre-reduced indigo (Mutual, 2022). The innovator BioGlitz produces biodegradable glitter which is based on eucalyptus tree extract. This eco-glitter is fully biodegradable, compostable and allows for the sustainable consumption of glitter without the environmental damage associated with microplastics (BioGlitz, n.d.).

### 20.5.3 Materials

A circular economy drives innovations towards materials as well. Econic's technology involves the use of waste $CO_2$ using a catalyst to produce polyols which can then be used as the starting raw material for making various polymers, epoxides which may contain up to 50% $CO_2$. This process also allows polyol manufacturers to tailor the amount of $CO_2$ incorporated into their polyols according to the performance requirements of a wide range of polyurethane applications (Econic, n.d.). Carbohydrates from plant materials to produce bioresin and engineered bio-composites were reported under the name Pond Cycle. This technology combines shorter chains of molecules into longer, complex chains, configured for specific use cases (Pond.global, n.d.). The elastic fibres are made with a chemical component that consists in part of $CO_2$ instead of oil. This precursor, called cardyon, is already used for foam in mattresses and sports floorings, and now it is being applied to the textile industry. Covestro and the Institute of Textile Technology of RWTH Aachen University, along with various textile manufacturers, have developed elastane fibres based on $CO_2$ on an industrial scale and aim to make the innovative fibres ready for the market. It can be used for stockings and medical textiles, for example, and might replace conventional elastane fibres based on crude oil (Covestro, 2021).

Piñatex (Hijosa et al., 2011) is a pioneer in leather alternatives made from cellulosic fibres extracted from pineapple leaves. The material is a synthetic leather made from pineapple waste, which is much more ethical than real leather made from animal skins. Through repurposing agricultural waste into natural textiles, Ananas Anam created a social impact by introducing new jobs into rural areas, while providing a second and diversified income stream to pineapple farmers. Ananas Anam provides low-impact textile solutions that support the efforts of brands and industries to reduce their emissions and meet their climate and sustainability targets.

Leather material is also not an exception for this change, and innovators such as Amadou Leather produced sustainable leather using a mushroom material which is grown on recycled sawdust and uses existing edible mushroom cultivation techniques. This material is completely compostable at the end of the lifecycle and improves soil quality. The material also gives a feel like suede to the touch with a natural rich brown colour and a malleable surface that is lightweight and flexible. The material is naturally antimicrobial as it stops the proliferation of bacteria (Amadou Leather). Mushroom-based leather is being produced by MycoWorks, made from mycelium, the underground root-like system of fungi; Mylo is a bio-based leather.

Another innovation, Treekind, is a flexible, leather-like material, made from urban plant waste, agricultural waste and forestry waste. Here the innovator combines lignocellulosic feedstock with a natural binder, hence Treekind is local, fully biodegradable and recyclable. It uses less than 1% of the water from leather production and is estimated to have a very low resource footprint (Biophilica, n.d.).

ELeather group from the UK successfully produced sustainable, engineered leather from leather waste. The process uses unique entanglement of leather fibres using only water to produce this material. ELeather re-imagined a waste stream into valuable, high-performance materials that dramatically reduces $CO_2$. The material is made with up to 55% recycled leather, which delivers a luxurious look and feel with enhanced strength and consistent form (ELeather Group, n.d.).

In the advanced innovation by Provenance collagen, the primary component of skin, is used to recreate leather. In this process from a single collagen molecule to distinct surface textures, skin is developed through a multi-level process (provenance.bio, n.d.).

### 20.5.4 Processes

Process innovation is one of the best ways to minimise the environmental impact and produce sustainable materials. There have been numerous innovations reported almost daily, and few among the many innovations are discussed here. Alchemie Technology developed clean-tech dyeing and finishing processes, which are enabled by its unique digital fluid jetting technology. This process is currently focused on polyester colouration. Alchemie's digital manufacturing solutions for dyeing and finishing deliver significant reductions in environmental impacts: reducing wastewater, chemicals and energy consumption (www.alchemietechnology.com).

e$CO_2$Dye developed a waterless textile dyeing process and equipment for dyeing. The process uses $CO_2$ as the solvent for dyeing as opposed to water as the medium in conventional dyeing. e$CO_2$Dye is offering a complete package from process, equipment, dye specification to a colour database for pretreatment and dyeing of natural and synthetic fibres in supercritical $CO_2$. This process significantly reduces water consumption, chemicals and energy use and most of the $CO_2$ is recycled in a closed-loop process (Eco2dye.com, 2023).

An innovation where dyeability, printing enhancements, hydrophilicity, hydrophobicity, fire retardancy and antimicrobial properties were imparted in natural and synthetic textiles uses Multiplexed Laser Surface Enhancement ("MLSE") technology. This technology utilises combined high-power pulsed UV laser and high-frequency electrical discharge plasma, to create a high-energy reaction zone at the substrate interface, promoting rapid synthesis to achieve the required functional treatments. It uses significantly less energy, chemicals and water than conventional pretreatments and is safe (mti-x.com, n.d.).

A spray application for colouration by Imogo technologies uses high-precision spray application of dye and finishing chemistry. This new process for dyeing textile materials can reduce the wasteful use of water, chemicals and energy in the textile dyeing process (imogotech.com, n.d.). NTX Cooltrans worked on waterless textile colouration technology by merging innovations in chemistry and machinery to deliver precise and accurate colouration of nearly any fabric material. This is achieved without heat and could lead to up to a 90% reduction in water use and 40% reduction in dye use (Anon, 2020).

An innovation to reduce salt use for reactive dyeing process called Nano-Dye involves cationic pretreatment for cotton before the colouration process. Nano-Dye's process changes the charge of the cotton molecule to the opposite charge of the dye. This can halve the amount of dye baths needed and can avoid the use of salt and reduce the water use (Nano-Dye.com, n.d.). Similar cationisation chemicals for cotton are available in the market from a few more chemical suppliers such as Dow Chemical Co. (USA), Rishabh Metals and Chemicals Private Ltd (India), to name a few.

Intelligent use of inert gas to produce foam and the use of foam for colouration or transportation of chemicals onto the substrate through a nebulisation process are a few of the technologies which are in the market from few innovators. The indigo dyeing process using a foam process produces zero water discharge and minimal dye waste, while producing deep indigo colours that the fashion industry loves (Indigo Mill Designs). In the case of the garment washing process, nebulisation process, foam or mist are also used to reduce water use and chemical transportation to obtain various finishing processes.

### *20.5.5 Transparency and traceability*

To track the sustainable journey of any material/process/article, transparency and traceability play an integral part. The awareness of transparency and traceability is growing across all the supply chain to make the sustainable claim. In all these processes, several innovators have worked to simplify the communication with various innovations and some such innovations are described below.

Technology innovator Made2Flow uses machine learning-based solutions to gather data directly from production partners and validate them based on multiple sources. The data are then benchmarked and translated into environmental impact indicators such as $CO_2$, water, biodiversity, energy. In this way, this technology helps consumers to get informed of decisions and gives brands an insight over their supply chain impact hotspots enabling them to reach their targets (www.made2flow.com).

The traceability software TrusTrace with the tagline 'product traceability data you can trust' helps to uphold material claims and reduce compliance risks in the textile value chains by automating the chain of custody, from raw materials to final goods, and ensures accurate documentation on products to market (TrusTrace – a leading fashion supply chain traceability software, n.d.).

A blockchain traceability system, Textile Genesis, was specifically created for the apparel sector and focuses on sustainable fibres such as wood-based fibres, organic cotton and organic wool. This system creates radical transparency from fibre-to-retail and ensures authenticity and provenance of sustainable textiles against generics. The pioneering system engages all five to six tiers of the supplier ecosystem to create unprecedented traceability based on five key principles. Consumers can scan the barcode with their mobile device to see the various steps that were taken to create the product (textilegenesis.com).

Another tracer technology, i.e., AWARE traceability technology, can be used for all kinds of sustainable materials. Authentic sustainable materials are verified by unique tracer particles and validated by secure blockchain. These sustainable materials are available directly at nominated spinners in several locations around the world (wearaware.co). Oritain and Haelixa are two of the technologies which work on traceability technology using DNA tracing to manage supply chain traceability.

### 20.5.6 Safe and Sustainable by Design (SSbD)

In pursuit of a more sustainable and healthier world, the concept of Safe and Sustainable by Design (SSbD) in materials and chemicals is fast emerging as an important requirement. Understanding the crucial relevance of incorporating sustainability and circularity into material and chemical design is a perspective acquired through years of industry experience and observation. This strategy, founded on a thorough understanding of environmental and economic realities, is more than a theoretical one, and is a practical requirement for long-term industry viability and environmental stewardship. The SSbD approach advocates for the intrinsic integration of safety, sustainability and functionality right from the design phase of the material and chemical development. The essence of SSbD lies in its proactive stance, and it is not merely about mitigating negative impacts but about creating positive value for both the environment and society. Successful waste management begins at the design stage. By imagining a product's end-of-life during design, use phase and end-of-life greatly reduces their overall carbon footprint. This is not simply an environmental need, but it is also becoming a regulatory requirement.

The SSbD framework has been promoted by the EU since 2022 to guide the innovation process for materials and chemicals. It focuses on iterative redesign and assessment phases to steer innovation towards green and sustainable industrial transition; substitute or minimise the production and use of substances of concern, in line with current and future regulatory frameworks; and minimise the impacts on health, climate and the environment during sourcing, production, use and end-of-life of chemicals, materials and products. Thus, it covers safety and the three pillars of sustainability – environment, society and economy.

#### 20.5.6.1 Holistic approach to safety

SSbD prioritises human health and environmental safety, with the goal of eliminating or significantly reducing the dangers connected with the use of materials and chemicals. This involves understanding and engineering non-regrettable substitutions and reducing the production and use of hazardous materials, thereby protecting ecosystems and lowering health risks. The concept goes beyond safety to include resource efficiency, energy conservation and the use of renewable resources. It promotes the development of materials that are not only less hazardous but also contributes to environmental sustainability, keeping a balance on economic, social interests. Therefore, it aims to create materials and chemicals that are both commercially competitive and socially responsible, as well as in keeping with ethical practices and community well-being.

As global environmental challenges intensify, SSbD stands out as a crucial strategy for minimising ecological footprints and fostering a harmonious relationship between industry and nature. With a growing awareness of the impact of chemicals on health, SSbD offers a safe path to significantly enhance public health through reduced exposure to hazardous substances. Sustainable materials and chemicals can open new market opportunities, reduce dependency on non-renewable resources, and drive economic innovation and growth as it catalyses innovation, pushing the boundaries of material science and chemistry to create solutions that meet safety and sustainability criteria.

In fact, the SSbD concept is critical for maintaining long-term resource availability, and protecting and providing the Earth for future generations. The path to fully realising the potential of SSbD necessitates collaborative efforts in teaching, research, innovation, strategic alliances and policymaking. To create, standardise and promote the implementation of SSbD principles, a multifaceted approach involving stakeholders from multiple sectors is required.

### *20.5.6.2 SSbD for textile sustainability*

The textile industry heavily relies on a myriad of materials and chemicals in its production processes, including polymers, solvents, colourants and finishing chemicals. As the urgency for climate change mitigation grows, there is a greater emphasis on developing sustainable textile products that are part of the solution rather than the issue. The SSbD concept can be used to develop circular-by-design textile products, both technical textiles and fashion garments, by combining inherently recyclable bio-based materials, and innovative functional finishing chemicals and processes. Thus, this concept can integrate safety, circularity and higher efficiency to rejuvenate the textile industry. It can also add value with enhanced multi-functionalities that address different product performances, but also facilitate the production, use and maintenance life cycle stages.

This sustainable transition of the textile industry is not just important for the environment, it is also becoming more of a market expectation. As there is a growing consumer preference for sustainably produced textile products, the companies that are early adopters of sustainable design principles will see a favourable influence on their brand image and consumer loyalty (Caldeira et al., 2022; Safe and Sustainable, 2021; European Commission, n.d.; van Dijk et al., 2022).

## 20.6 Life cycle assessment

Carbon footprint is a simplified form of Product Environmental Footprint (PEF) calculation, and both tools are ultimately based on ISO14040, the global standard for life cycle assessment (LCA). Carbon footprint is simplified in the sense that only one impact category (climate change) is considered, while a PEF or LCA would typically consider other resources, environmental and human health categories, such as energy consumption, impacts on habitat and the emission of carcinogens. It is no wonder that assessing textiles and other products with respect to their contributions to climate change is increasingly popular (Shen and Patel, 2008; Peters et al., 2014). Carbon footprint is the central method for doing this, and it considers the relative importance of different greenhouse gases. Another key feature of carbon footprint is that it typically attempts to take a perspective of emissions that encompasses the entire life cycle of a product or service. Every year, global emissions from textile production are equivalent to 1.2 billion tonnes of $CO_2$, a figure that outweighs the value of the footprint of international flights and shipping combined. The textile industry is increasingly engaged in carbon footprint as a part of policy development and product design (Chinasamy, 2019).

A typical LCA study method has four stages (Escamilla et al., 2014), the first step is defining the goals of the study, describing the product to be assessed. The second step involves creation of the inventory, data from all the processes of the product's life cycle are collected and processed, from the start to the end or cradle to grave. This step evaluates the energy and raw material requirements, environmental emissions and discharges of the product, and other processes are computed and presented for all stages of production. This information is then used to calculate consumption and discharges from all processes of the product's life cycle. The third stage is called the impact assessment, where the inventory data are translated into effects on human health, ecological health and resource depletion. This step evaluates the potential release of all the greenhouse gases identified by the Intergovernmental Panel on Climate Change: carbon dioxide, methane, nitrous oxide, hydrofluorocarbons, perfluorocarbons and sulphur hexafluoride. The final step involves recommendations made based on the results of the inventory and impact stages. Fig. 20.2 shows the areas covered in the life cycle assessment process.

In the atmosphere, a kilogramme of each of these gases causes a different degree of insulation, and each has a different residence time. Carbon footprint typically uses 100 years as a standard time horizon for analysis. The interpretation phase of a carbon footprint study is a time to formally reflect on the meaning of outcomes of the preceding three steps, and probably to do parts of them again. During carbon footprint calculation, global warming potentials are used to calculate an aggregated indicator in units of 'kilogrammes of carbon dioxide equivalent'. These units differentiate the footprint concept in this context from other uses, as 'footprint' otherwise commonly refers to a surface area.

### 20.6.1 Carbon footprints in the textile industry

The manufacturing processes can include fibre production, yarn production, fabric production (nonwoven, knitting, weaving), wet processing, finishing and sewing. A textile manufacturing plant that performs all the processes from yarn production to product is commonly termed a 'vertical plant'. The materials could include cotton, polyester, viscose, lyocell, wool and silk, among others. Calculations of the carbon footprint of textile manufacturing plants are generally made within the context of an EMS, an organisational carbon footprint or for the purpose of a product LCA.

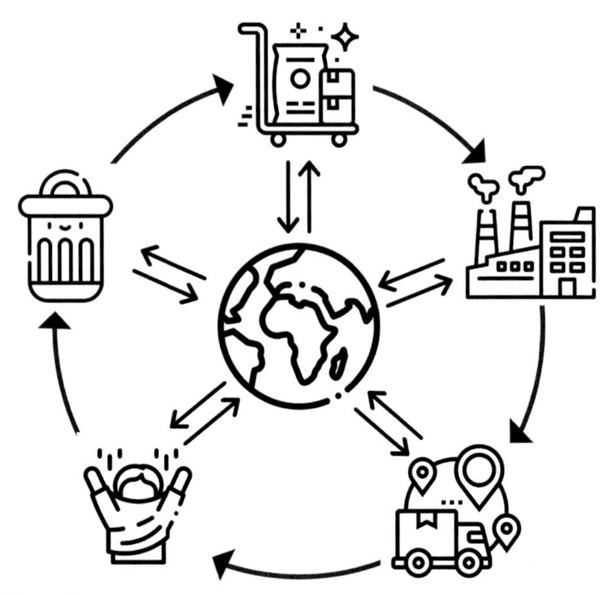

**Figure 20.2** Life cycle assessment.

In order to conduct the LCA process, the scope for 'cradle-to-gate', 'cradle-to-grave' or other scopes needs to be defined and accordingly the assessment to be conducted. Cradle-to-gate is a common expression in LCA indicating that raw materials are included but end use and disposal are excluded. This is similar to 'cradle-to-plate' used in food LCA and is in contrast to 'cradle-to-grave' indicating an all-inclusive system analysis (Table 20.1).

## 20.6.2 Calculating carbon footprints of textile products

In addition to the textile manufacturing processes described previously, the full life cycle of textile products encompasses processes related to raw material extraction (e.g., cultivation, forestry, extraction of oil), distribution, retail, use (e.g., consumer transportation to and from the store, laundry) and end-of-life handling. Typically, manufacturing and use (consumer transportation as well as laundry) are, because of high energy intensity, the most important contributors to the carbon footprint. Depending on the methodology, raw material extraction may also be an important contributor.

Thus, LCA practitioners and decision-makers must be very careful in extracting carbon footprint results from studies, and using them out of context. To interpret results, the specific methodology and assumptions of each study must be considered.

**Table 20.1** Overview of studies on the carbon footprint of textile manufacturing plants (Handbook of Life Cycle Assessment, 2015).

| Publication | Method | Studied item (in kg) (origin) | Results (kg $CO_2$ eq.) |
|---|---|---|---|
| **Peer-reviewed scientific papers** | | | |
| van der Velden et al. (2013) (uses data from ecoinvent [Swiss Centre for Life Cycle Inventories: [Ecoinvent Centre, 2010] and IDEMAT [Vogtlander, 2012]) | GWP100 (IPCC, 2007; 100 years) | Cotton fibres, cultivation and ginning[a] (Ecoinvent) | 3.47 |
| | | Polyethylene terephthalate (Ecoinvent) | 2.7 |
| | | Ring[a] spinning cotton yarn 45 dtex[b] (IDEMAT) | 11.32 |
| | | Ring[a] spinning cotton yarn 300 dtex (IDEMAT) | 1.7 |
| | | Texturing polymer fibres (IDEMAT) | 0.51 |
| | | Circular[a] knitting 83 dtex (IDEMAT) | 0.26 |
| | | Circular[a] knitting 300 dtex (IDEMAT) | 0.07 |
| | | Weaving 45 diex (IDEMAT) | 16.6 |
| | | Weaving 45 dtex (IDEMAT) | 2.49 |
| | | Airjet[a] dyeing of polyester, excluding chemicals (IDEMAT) | 2.25 |
| | | Heat setting and washing synthetic fabrics (IDEMAT) | 0.91 |
| Astudillo et al. (2014) | IPCC (2013); GWP 100a v.1.02 | Silk production, farm practices | 80.9 |
| | | Silk production recommended practices | 52.5 |
| | | Nylon 66 polymer production – (Swiss Centre for Life Cycle Inventories [Ecoinvent Centre, 2010]) | 8 |

*Continued*

**Table 20.1 Continued**

| Publication | Method | Studied item (in kg) (origin) | Results (kg $CO_2$ eq.) |
|---|---|---|---|
| **Other publications** | | | |
| Cherretl et al. (2005) [c] | Wackernagel et al. (2004) | Spun fibres of organic cotton, US | 0.0024 |
| | | Spun fibres of conventional cotton, US | 0.0059 |
| | | Spun fibres of hemp | 0.0045 |
| | | Spun fibres of polyester, Europe | 0.0072 |
| | | Spun fibres of polyester, US | 0.0095 |
| Murugesh and Selvadass (2013) | Not reported | Exhaust dyeing of knitted cotton using softflow dyeing machine, colour 1 | 12.43 |
| | | Exhaust dyeing of knitted cotton using softflow dyeing machine, colour 2 | 16.88 |
| | | Cold pad batch dyeing of knitted cotton, colour 3 | 6.59 |
| | | Squeezer and softener pad | 0.076 |
| | | Relax drying | 0.072 |
| | | Stenter drying | 0.113 |
| | | Compacting | 0.135 |

[a]The marked text is an assumption regarding technology when data are lacking in the source document.
[b]dtex or decitex is the mass in grams per 10,000 m.
[c]The results shown are based on the values per tonne of fibres as stated in that report, but it seems likely that the original report data are actually per kg of fibre.

### 20.6.3 *Carbon footprint labelling of textile products*

Product carbon footprint labelling has been initiated in countries including the United Kingdom, the Netherlands and Japan in order to educate consumers in making informed purchasing decisions (Tan et al., 2014). From a corporate perspective, carbon footprint labelling offers the potential to gain a competitive advantage via green marketing that taps into increasing global concerns about carbon emissions (Vanclay et al., 2010). From a non-governmental organisation perspective, the information provided by carbon footprint labels can assist customers thinking about which product to buy

(Upham et al., 2011), and endorsing companies with carbon footprint labels can encourage them to publicly commit to reducing their carbon emissions (Boardman, 2008).

Consumers' awareness about the environmental impacts of products' life cycles is being increased by the communication of quantitative values on carbon footprint labels, backed by strategies and methods developed for quantifying greenhouse gas emissions of products and services (Tan et al., 2014). There are a number of labels for textiles and other products that include a carbon footprint component. One of the first was the Carbon Reduction Label, which was first introduced in the UK in 2006 by the Carbon Trust (see www.carbontrust.com). The original labels licenced by the Trust for use on products stated the amount of $CO_2$ and other greenhouse gases emitted per product or per use of the product.

Currently, two different types of labels are offered: a 'Reducing $CO_2$' label and a '$CO_2$ Measured' label. The Reducing $CO_2$ label is a communication tool to show the company's commitment for carbon reduction and can also appear with a statement about the size of the carbon footprint of the labelled product or service. Certified applicants are committed to reduce their measured carbon footprint over 2 years or lose the label. The results will be assessed by a third party after 2 years to verify the correct use of the label (Carbon Trust, 2015). The $CO_2$ Measured label indicates that a product's carbon footprint has been calculated but makes no commitment in relation to reducing that footprint. The carbon footprint assessment is based on PAS2050 or the GHG Protocol. This means that the whole life cycle of the product including production, use and disposal is included in measuring the carbon footprint.

Another carbon footprint label is Carbon Care Asia (https://www.ccinnolab.org/AboutCarbonCareLabel), which was developed by a Hong Kong-based company. The Carbon Care labelling system evaluates companies' and organisations' efforts in reducing the carbon footprint in three steps: measuring, reducing and offsetting carbon footprints. An applicant who wishes to use the label needs to provide a verified audit report of its emission inventory according to the Carbon Care label protocol. Six different labelling options are allowed based on the level of emission reduction of the applicant compared with its emission baseline. The awarded labels show the level of achievement in carbon reduction as 5%, 20%, 40%, 60% or 80% compared with the measured emission baselines. A golden label is awarded to applicants who can achieve carbon neutrality.

Carbonfund.org is an American nonprofit organisation that developed the Carbonfree Product Certification to make consumers aware of the companies that are compensating for their greenhouse gas emissions including $CH_4$, $N_2O$, $SF_6$, HFCs, PFCs and biomass $CO_2$ emissions (Carbonfund, 2012). To acquire the Carbonfree certification, an LCA needs to be performed to measure the greenhouse gas emissions throughout the product's life cycle. Carbon footprint analysis of textiles and clothing suggests that many possible improvements to the product life cycle can be achieved because of large differences between possible alternatives.

## 20.7 Ecolabels

Ecolabels are designed to inform consumers whether the purchased clothing is sustainable, and that it does not contain any harmful substances. In the fashion industry, brands are regularly using greenwashing tactics in order to convince consumers that their clothing is sustainable. This happens due to the scope and boundary conditions selected. The term greenwashing is used when brands create false marketing materials or branding that suggest they are environmentally friendly, without actually committing to sustainable practices. In order to combat greenwashing and to understand what the clothes are really made from, it is important to become familiar with clothing labels.

### *Fairtrade certified*

The Fairtrade (https://www.fairtradecertified.org/) logo, often found in apparel is one of the most common ecolabels which ensures that a certain standard has been maintained while manufacturing that piece of clothing. Most Fairtrade standards focus on ensuring that workers and producers get fair terms of trade, good prices and longer lead times. All of these things promote worker security, self-sufficiency and more sustainable practices. Regarding materials, if cotton is made in fair terms, it will be indicated by either the Fairtrade Mark or Fairtrade Sourcing Partnership label.

### *Cradle to Cradle*

The Cradle to Cradle (https://c2ccertified.org/) certification is an ecolabel that shows there has been an effort to create an eco-intelligent product. Fabrics or clothing items can be given Basic, Silver, Gold or Platinum level certification based on efforts across areas including eco-materials, social responsibility, water efficiency, renewable energy and recycling.

### *GOTS certified*

The Global Organic Textile Standard (GOTS) (https://global-standard.org/) has been developed by leading standard organisations with the aim of unifying the existing organic standards in the field of sustainable textile processing. This standard looks at all stages of fabric creation, including harvesting raw materials, the manufacturing process and responsible labelling.

### *Made in Green*

The Made in Green (https://www.oeko-tex.com/en/our-standards/oeko-tex-made-in-green) label certifies that a product has been manufactured in factories that respect the environment and universal rights of workers. This helps consumers understand exactly where their fabrics and clothes are coming from.

## OEKO-TEX

The International Oeko-Tex (https://www.oeko-tex.com/en/our-standards/oeko-tex-standard-100) Association tests textiles for harmful substances and provides certifications for textiles that do not use harmful chemicals in their manufacturing processes. If a textile article carries the Oeko-Tex Standard 100 label, the consumer can be sure that every part of the item is harmless for human health.

## OEKO-TEX STeP

OEKO-TEX STeP (https://www.oeko-tex.com/en/our-standards/oeko-tex-step) stands for Sustainable Textile and Leather Production and is a modular certification system for production facilities in the textile and leather industry. The goal of STeP is to implement environmentally friendly production processes in the long term, to improve health and safety and to promote socially responsible working conditions at production sites. The target groups for STeP certification are textile and leather manufacturers as well as brands and retailers. OEKO-TEX STeP includes a comprehensive analysis and assessment of the production conditions. STeP analyses all important areas of a company using six modules: Chemical management, Environmental performance, Environmental management, Social responsibility, Quality management and Health protection and safety at work.

## Nordic Ecolabel

Nordic Ecolabel (https://www.nordic-ecolabel.org/) works to reduce the environmental impact from production and consumption of goods. The Nordic Swan Ecolabel sets strict environmental requirements in all relevant phases of a product's life cycle. It also sets strict requirements for chemicals used in ecolabelled products. This ecolabel tightens requirements for goods and services continuously to create sustainable development. This makes it easy for consumers and professional buyers to choose the best environmentally friendly goods and services.

## EU Ecolabel

This Ecolabel (https://environment.ec.europa.eu/topics/circular-economy/eu-ecolabel-home_en) established in 1992 is recognised across Europe and worldwide. The EU Ecolabel is a label of environmental excellence that is awarded to products and services meeting high environmental standards throughout their lifecycle: from raw material extraction, to production, distribution and disposal. The EU Ecolabel promotes the circular economy by encouraging producers to generate less waste and $CO_2$ during the manufacturing process. The EU Ecolabel criteria also encourage companies to develop products that are durable, easy to repair and recycle.

## 20.8 Future perspectives

In recent times, many innovators and research-based start-ups have been working on technological innovations to further improve sustainable production. There are many organisations which are constantly scouting for many such innovations (https://fashionforgood.com/; https://www.plugandplaytechcenter.com/; https://hmfoundation.com). Here are the few focus areas to watch out for:

- Creating building blocks from materials which are not possible to reuse or recycle but need to go for incineration and secured landfill using pyrolysis, gasification, valorisation, cracking, carbon capture, carbon sequestering and reacting technologies. The building blocks for chemical industries like ethylene, propylene, butylene, etc. will be produced from the organic waste which is destined to get landfilled.
- Separation of fibre blends and depolymerisation using hydrothermal, hydrolysis, methanolysis, glycolysis or use of ionic liquids or other solvents to get monomers for creating polymers for circular use. Cellulose-rich or polyester-rich blends with various proportions can undergo separation, decolouration, depolymerisation and repolymerisation processes to get back the materials from recycled sources. The choice of process may depend on the blend composition, final product requirement and various other factors.
- Use of microwave, ultrasound or thermal energies for creating new materials to depolymerise and create the monomers in the case of polyester.
- Use of sound waves to separate and collect the microfibres, use of microbubbles to separate the waste in the water stream and also to remove colour from the garments to be able to reuse.
- Use of algae, fungi, seaweeds, enzymes, mushrooms or plant-based derivatives to create new colourants, chemicals or materials which are biodegradable.
- Use of plasma technology for achieving functional properties using various gases.
- Smart transportation of chemicals and colourants onto substrates to minimise resources like water and energy, and minimise water using bubbles, foam, spray forms.
- New chemistries which follow green chemistry principles and atom economics.
- New machinery to reduce use of resources such as water, electricity and steam in the processing.
- Techniques to reduce aerosol loading to the atmosphere.
- New analytical techniques for identifying, sorting and separating post-consumer articles.

## 20.9 Conclusion

Sustainability is a multifaceted concept that involves environmental, social and economic factors. Sustainable practices must be implemented in all stages including production, consumption, energy, transportation and disposal. A more equitable, resilient and sustainable future can be nurtured only by continuously innovating and adopting the technologies which can continuously reduce emissions and use of resources.

Sustainability is not only reducing the carbon footprint of industrial processes and textile products, it is also about people and communities. Sustainable manufacturing also means protecting the planet for future generations, while elevating the living standards of the people. Therefore, it is important that the resources of the planet should be preserved for future generations. It is possible to make this happen only by reducing the reliance on fossil fuels, preventing pollution and promoting sustainable practices.

## References


Akhter, S., Rutherford, S., Chu, C., 2017. What makes pregnant workers sick: why, when, where and how? an exploratory study in the ready-made garment industry in Bangladesh. Reproductive Health 14 (1).

Alchemie Technology, n.d. Endeavour. https://www.alchemietechnology.com/endeavour/.

AltMat, n.d. Home. https://www.altmat.in/ Accessed 13 February 2023.

Amadou Leather, n.d. Amadou Leather. https://www.amadouleather.com/.

Anon, 2020. A Revolutionary Waterless Coloration Technology. NTX Cooltrans. Accessed 14 February 2023. https://www.ntx.global/ntx-cooltrans/.

Anon, n.d.-a. Worn Again Technologies | A Waste-free, Circular Resource World is within Reach. https://www.wornagain.co.uk/.

Anon, n.d.-b. Fairbrics. http://www.fairbrics.co/.

Anon, n.d.-c. Technology and Product – KBCols Sciences Pvt. Ltd. https://www.kbcolssciences.com/technology-and-product-applications/.

Anon, n.d.-d. LanzaTech. https://www.lanzatech.com/.

Astudillo, M.F., Thalwitz, G., Vollrath, F., 2014. Life cycle assessment of Indian silk. Journal of Cleaner Production 81, 158–167. https://doi.org/10.1016/j.jclepro.2014.06.007.

Berlin, I.D.Z., n.d. Algalife – Bio Textile Creation. bundespreis-ecodesign.de. https://www.bundespreis-ecodesign.de/en/projectoverview/algalife-a-force-for-good.

BioGlitz, n.d. Sustainability. https://www.bioglitz.co/sustainability.

Biophilica, n.d. Biophilica. https://www.biophilica.co.uk/.

Boardman, B., 2008. Carbon labelling: too complex or will it transform our buying? Significance 5 (4), 168–171. https://doi.org/10.1111/j.1740-9713.2008.00322.x.

Caldeira, C., Farcal, R., Garmendia Aguirre, I., Mancini, L., Tosches, D., Amelio, A., Rasmussen, K., Rauscher, H., Riego Sintes, J., Sala, S., 2022. Safe and Sustainable by Design Chemicals and Materials - Framework for the Definition of Criteria and Evaluation Procedure for Chemicals and Materials. EUR 31100 EN. Publications Office of the European Union, Luxembourg. https://doi.org/10.2760/404991/JRC128591.

Carbios, n.d. Enzymatic Recycling. https://www.carbios.com/en/enzymatic-recycling/.

Carbon Trust, 2015. Carbon Footprint Labels from the Carbon Trust. Accessed June 2022.

Carbonfund, 2012. CarbonFree Product Certification Carbon Footprint Protocol. Version 4.0. Accessed May 2022.

Cherretl, N., Barrett, J., Clemett, A., Chadwick, M., Chadwick, M.J., 2005. Ecological Footprint and Water Analysis of Cotton, Hemp and Polyester. Stockholm Environment Institute, Sweden.

Chinasamy, J., 2019. London Fashion Week: Fast Facts about Fast Fashion, Unearthed. Accessed April 29, 2023. https://www.unearthed.greenpeace.org/2019/09/12/fast-facts-aboutfast-fashion/.

Circ, n.d. Circ Technology • Circ. https://www.circ.earth/circ-technology/.

Circular Systems, n.d. Agraloop — Circular Systems | Regenerative Impact. https://www.circularsystems.com/agraloop.

Circulose, n.d. Circulose. https://www.circulo.se/faq.

Colorifix, n.d. Colorifix Solutions. https://www.colorifix.com/colorifix-solutions/.

Colors, S.C., n.d. Traceable, Plant-Based Dyes. tony Creek Colors. http://www.stonycreekcolors.com/.
Covestro, 2021. Dress with CO2. Covestro. Accessed April 30, 2023. https://www.covestro.com/press/dress-with-co2/.
DEFRA, 2008. Sustainable Clothing Roadmap Briefing Note December 2007: Sustainability Impacts of Clothing and Current Interventions.
Draper, S., Murray, V., Weissbrod, I., 2007. Fashioning Sustainability – A Review of Sustainability Impacts of the Clothing Industry. Forum for the Future and Mark and Spencer, UK.
Eco2dye.com, 2023. Waterless Textile Dyeing with eCO2Dye. eCO2Dye - Waterless Textile Dyeing. Accessed 14 February 2023. http://www.eco2dye.com/.
Ecoinvent Centre, 2010. Ecoinvent Data v2.2. Final Reports Ecoinvent 2010. Swiss Centre for Life Cycle Inventories (Dubendorf).
Econic, n.d. How it Works. https://www.econic-technologies.com.
ELeather Group, n.d. Technology | Engineered Leather. https://www.eleathergroup.com/technology/.
European Commission, n.d. https://www.research-and-innovation.ec.europa.eu/research-area/industrialresearch-and-innovation/key-enabling-technologies/chemicals-and-advanced-materials/safeandsustainable-design_en.
Escamilla, M., Paul, R., 2014. Chapter 26: Methodology for testing the life cycle sustainability of flame retardant chemicals and nanomaterials. In: Papaspyrides, C.D., Kiliaris, P. (Eds.), Polymer Green Flame Retardants. Elsevier, Netherlands, pp. 891–905.
Handbook of Life Cycle Assessment (LCA) of Textiles and Clothing, 2015, pp. 15–16.
Hijosa, C., Ribé, A., Jiménez, J., Paul, R., Brouta-Agnésa, M., 2011. Natural Nonwoven Materials. Applicant: Ananas Anam Limited, Ireland. International Publication No. WO2011148136 A2.
https://www.c2ccertified.org/ Accessed June 2022).
https://www.environment.ec.europa.eu/topics/circular-economy/eu-ecolabel-home_en (Accessed January 2023).
https://www.fashionforgood.com/ (Accessed January 2023).
https://www.global-standard.org/ (Accessed January 2023).
https://www.hmfoundation.com (Accessed January 2023).
https://www.fairtradecertified.org/ (Accessed June 2022).
https://www.nordic-ecolabel.org/ (Accessed January 2023).
https://www.oeko-tex.com/en/our-standards/oeko-tex-made-in-green (Accessed January 2023).
https://www.oeko-tex.com/en/our-standards/oeko-tex-standard-100 (Accessed January 2023).
https://www.oeko-tex.com/en/our-standards/oeko-tex-step (Accessed January 2023).
https://www.plugandplaytechcenter.com/ (Accessed January 2023).
imogotech, n.d. Products. https://www.imogo.com/product-2/ (Accessed 14 February 2023).
Indigo zero – a talk with Ralph Tharpe, Denim and Jeans. Available at: https://www.denimsandjeans.com/denim-fabric-developments/indigo-zero-a-talk-with-ralph-tharpe/37536 (Accessed: 05 February 2024).
Infinited Fiber, n.d. Our Technology. https://www.infinitedfiber.com/our-technology/.
IPCC, 2007. Climate change 2007: the physical science basis. In: IPCC Fourth Assessment Report: Climate Change 2007. Intergovernmental panel on climate change (IPCC), Geneva, Switzerland. https://www.archive.ipcc.ch/publications_and_data/ar4/wg1/en/ch2s2-10-2.html (Accessed 05 February 2024).
IPCC, 2013. In: Stocker, T.F., Qin, D., Plattner, G.-K., Tignor, M., Allen, S.K., Boschung, J., et al. (Eds.), Climate Change 2013: The Physical Science Basis. Working Group I

Contribution to the Fifth Assessment Report of the Intergovernmental Panel of Climate Change. Cambridge University Press, Cambridge, United Kingdom and New York, NY. USA. Accessed October 2014. http://www.ipcc.ch/report/ar5/wgl/.
Keel Labs, n.d. Keel Labs. https://www.keellabs.com/.
Kintra Fibers — The Natural Upgrade for Synthetics, n.d. Kintra Fibers. https://www.kintrafibers.com/.
Living Ink, n.d. Living Ink — About Living Ink. https://www.livingink.co/about/.
Made2Flow, n.d. Made2Flow. https://www.made2flow.com/.
MTIX, n.d. MTIX LTD - Innovation In Textile Processing Technology. https://www.mti-x.com/.
Murugesh, K., Selvadass, M., 2013. Life cycle assessment for the dyeing and finishing process of organic cotton knitted fabrics. Journal of Textile and Apparel, Technology and Management 8 (2), 7.
Mutual, 2022. Archroma and Stony Creek Colors to enter strategic partnership to produce Indigold™ high-performance plant-based pre-reduced indigo at scale. https://www.archroma.com/press/releases/archroma-and-stony-creek-colors-to-enter-strategic-partnership-to-produce-indigold-high-performance-plant-based-pre-reduced-indigo-at-scale.
Nano-Dye, n.d. Disruptive Sustainable Cotton Dyeing | Nano-Dye. https://www.nano-dye.com.
Nature Coatings, n.d. Nature Coatings. https://www.naturecoatingsinc.com/.
Peters, G., Granberg, H., Sweet, S., 2014. The role of science and technology in sustainable fashion. Chapter 18. In: Fletcher, K., Tham, M. (Eds.), The Handbook of Sustainable Fashion. Routledge, UK.
Pili.bio, n.d. Technology. https://www.pili.bio/9/technology.
Pond.global, n.d. Pond.global. https://www.pond.global/.
provenance.bio, n.d. Provenance - Technology. http://www.provenance.bio/fabrics/index.php?id=13.
PurFi, n.d. PurFi - A Circular Fiber Company. https://www.purfiglobal.com/.
Purvis, B., Mao, Y., Robinson, D., 2018. Three pillars of sustainability: in search of conceptual origins. Sustainability Science 14 (3), 681−695. Available at: https://doi.org/10.1007/s11625-018-0627-5.
Recycrom.com, 2018. Tech Explained − Recycrom − Officina+39. https://www.recycrom.com/tech-explained/.
Rubi Laboratories, n.d. Home. https://www.rubi.earth/.
Safe and Sustainable-By-Design: Boosting Innovation and Growth within the European Chemical Industry, October 2021. https://www.cefic.org/app/uploads/2021/09/Safe-and-Sustainable-by-Design-Report-Boosting-innovation-and-growth-within-the-European-chemical-industry.pdf.
Shen, L., Patel, M.K., 2008. Life cycle assessment of polysaccharide materials: a review. Journal of Polymers and the Environment 16 (2), 154−167. https://doi.org/10.1007/s10924-008-0092-9.
Spinnova, n.d. Spinnova. Cleanest process. Disruptive circularity. https://spinnova.com/.
Stockholm Resilience Centre, 2012. Planetary boundaries - Stockholm Resilience Centre. Stockholmresilience.org. https://www.stockholmresilience.org/research/planetary-boundaries.html.
Tan, M.Q.B., Tan, R.B.H., Khoo, H.H., 2014. Prospects of carbon labelling − a life cycle point of view. Journal of Cleaner Production 72, 76−88. https://doi.org/10.1016/j.jclepro.2012.09.035.
Textilegenesis, n.d. TextileGenesisTM. https://www.textilegenesis.com/.
The Hurd Co, n.d. The Hurd Co - General 2. https://www.thehurdco.com/hurd-technology.
TrusTrace − Leading fashion supply chain traceability software, n.d. TrusTrace − Leading Fashion Supply Chain Traceability Software. https://www.trustrace.com/.

Upham, P., Dendler, L., Bleda, M., 2011. Carbon labelling of grocery products: public perceptions and potential emissions reductions. Journal of Cleaner Production 19 (4), 348–355. https://doi.org/10.1016/j.jclepro.2010.05.014.

van der Velden, N.M., Patel, M.K., Vogtländer, J.G., 2013. LCA benchmarking study on textiles made of cotton, polyester, nylon, acryl, or elastane. International Journal of Life Cycle Assessment 19 (2), 331–356. https://doi.org/10.1007/s11367-013-0626-9.

van Dijk, J., Flerlage, H., Beijer, S., Slootweg, J.C., van Wezel, A.P., 2022. Safe and sustainable by design: a computer-based approach to redesign chemicals for reduced environmental hazards. Chemosphere 296, 134050. https://doi.org/10.1016/j.chemosphere.2022.134050.

Vanclay, J.K., Shortiss, J., Aulsebrook, S., Gillespie, A.M., Howell, B.C., Johanni, R., Maher, M.J., Mitchell, K.M., Stewart, M.D., Yates, J., 2010. Customer response to carbon labelling of groceries. Journal of Consumer Policy 34 (1), 153–160. https://doi.org/10.1007/s10603-010-9140-7.

Vienna Textile Lab, n.d. About Us. https://www.viennatextilelab.at/about-us/.

Vogtlander, J.G., 2012. IDEMAT 2012 Database. https://www.ecocostsvalue.com.

Wackernagel, M., Monfreda, C., Moran, D., Goldfinger, S., Deumling, D., Murray, M., 2004. National Footprint and Biocapacity Accounts 2004: The Underlying Calculation Method. Global Footprint Network, Oakland, CA, USA.

wearaware.co, n.d. Aware. https://www.wearaware.co/.

# Index

'*Note:* page numbers followed by "f" indicate figures and "t" indicate tables.'

Printed in the United States
by Baker & Taylor Publisher Services